CHITIN
and CHITOSAN
DERIVATIVES

Advances in Drug Discovery
and Developments

CHITIN
and CHITOSAN
DERIVATIVES

Advances in Drug Discovery and Developments

Edited by
Se-Kwon Kim

CRC Press
Taylor & Francis Group
Boca Raton London New York

CRC Press is an imprint of the
Taylor & Francis Group, an **informa** business

First published in paperback 2024

First published 2014 by CRC Press
2385 NW Executive Center Drive, Suite 320, Boca Raton FL 33431

and by CRC Press
4 Park Square, Milton Park, Abingdon, Oxon, OX14 4RN

First issued in hardback 2019

CRC Press is an imprint of Taylor & Francis Group, LLC

© 2014, 2024 by Taylor & Francis Group, LLC

Library of Congress Cataloging-in-Publication Data

Chitin and chitosan derivatives : advances in drug discovery and developments / editor, Se-Kwon Kim.
     pages cm
   "A CRC title."
   Includes bibliographical references and index.
   ISBN 978-1-4665-6628-6 (hardcover : alk. paper)
   1. Chitin. 2. Chitosan. 3. Chitin--Derivatives. I. Kim, Se-Kwon, editor of compilation.

QP702.C5C473 2014
573.7'74--dc23   2013038094

ISBN: 978-1-4665-6628-6 (hbk)
ISBN: 978-1-03-291907-2 (pbk)
ISBN: 978-0-429-09943-4 (ebk)

DOI: 10.1201/b15636

Visit the Taylor & Francis Web site at
http://www.taylorandfrancis.com

and the CRC Press Web site at
http://www.crcpress.com

Contents

PART I Synthesis and Characterization of Chitin and Chitosan Derivatives

PART II   Biological Activities of Chitin and Chitosan Derivatives

PART III   Biomedical Applications of Chitin and Chitosan Derivatives

Preface

The utilization of marine resources is commonly considered inexpensive and abundant, with great interest to develop biological and biomedical applications. Chitosan is produced from chitin, which is a natural polysaccharide composed of randomly distributed β-(1–4)-linked D-glucosamine (deacetylated unit) and N-acetyl-D-glucosamine (acetylated unit). Chitin is a long-chain polymer comprising N-acetylglucosamine, a derivative of glucose, and is found ubiquitously. It is the main component of the cell walls of fungi, the exoskeletons of arthropods, such as crustaceans (e.g., crabs, lobsters, and shrimps) and insects, the radulas of mollusks, and the beaks and internal shells of cephalopods, including squid and octopuses. However, marine crustacean shells are widely used as primary sources for the production of chitosan. Crab and shrimp are important marine species of great commercial importance in the tropical and subtropical waters of the Pacific, Atlantic, and Indian oceans. Chitin and chitosan have been widely used for various biological and biomedical applications during the last two decades, owing to their unique properties. However, extensive research on the applicability of chitin and chitosan was limited due to their poor solubility. The current research trends focus mainly on the increase in solubility of chitin and chitosan through the addition of chemical groups at the molecular level. Modified chitin and chitosan derivatives have several applications including biological and biomedical.

In recent times, chitin and chitosan derivatives have been achieving a strong market position attracting various marine researchers and the consumers as well. Chitin and chitosan derivatives are currently receiving a great deal of attention with regard to medical and pharmaceutical applications owing to their interesting properties that make them suitable for use in the biomedical field, such as biocompatibility, biodegradability and nontoxicity, analgesic, antitumor, hemostatic, hypocholesterolemic, antimicrobial, and antioxidant properties.

This book contains various relevant topics such as modification of chitin and chitosan derivatives and their characterization as well as applications toward drug discovery and delivery in preliminary research. These topics can provide a platform to develop various biotechnology industries. The book covers the modification of chitin and chitosan derivatives and the problems associated with it together with probable solutions for the same.

- Chitin and chitosan have a wide range of bioactivity, which may be utilized only once the modification is accomplished in an appropriate manner. Part I of this book provides the basic information about the synthesis and characterization of a variety of chitin and chitosan derivatives.
- Part II deals with the characterization of modified chitin and chitosan derivatives, which forms an important basis for the biological application of these polymers. This part also discusses activities such as biological, specifically antioxidant, anti-inflammatory, anticancerous, antiviral, and anticoagulant and antimicrobial.
- Part III includes the biomedical application of the chemically modified as well as composite materials of chitin and chitosan derivatives in tissue engineering, drug delivery, and wound dressing applications.

The contributors are from various countries (Europe, India, Japan, Korea, Vietnam, Africa, and Malaysia) and have contributed chapters at an excellent standard. This book is a must-read for novices and experts in the fields of biotechnology, chemical sciences, natural products, materials science, pharmaceutical science, nutraceutical, and biomedical engineering sciences. It will be useful for further research in the scientific and industrial arenas.

I would like to acknowledge CRC Press, Taylor & Francis Group, for their encouragement and suggestions to get this wonderful compilation published. I would also like to extend my sincere gratitude to all the contributors for providing help, support, and advice to accomplish this task. I strongly recommend this book for marine biomaterials researchers/students and hope that it helps them to enhance their understanding in this field.

Se-Kwon Kim
Busan, South Korea

Editor

Professor Se-Kwon Kim, PhD, currently serves as a senior professor in the Department of Chemistry and as a director of the Marine Bioprocess Research Center (MBPRC) at Pukyong National University, Republic of Korea. He earned his BSc, MSc, and PhD from the Pukyong National University and joined as a faculty member. He has previously served as a scientist at the University of Illinois, Urbana-Champaign, Illinois (1988–1989), and was a visiting scientist at the Memorial University of Newfoundland in Canada (1999–2000).

Professor Se-Kwon Kim was the president of the Korean Society of Chitin and Chitosan (1986–1990) and the Korean Society of Marine Biotechnology (2006–2007). He was also the chairman of the 7th Asia-Pacific Chitin and Chitosan Symposium that was held in South Korea in 2006. He is one of the board members of the International Society of Marine Biotechnology and the International Society for Nutraceuticals and Functional Foods. Moreover, he was the editor-in-chief of the *Korean Journal of Life Sciences* (1995–1997), the *Korean Journal of Fisheries Science and Technology* (2006–2007), and the *Korean Journal of Marine Bioscience and Biotechnology* (2006–present). His research has been credited with the best paper award from the American Oil Chemist's Society (AOCS) and the Korean Society of Fisheries Science and Technology in 2002.

Professor Se-Kwon Kim's major research interests are investigation and development of bioactive substances derived from marine organisms and their application in oriental medicine, nutraceuticals, and cosmeceuticals via marine bioprocessing and mass-production technologies. He has also conducted research on the development of bioactive materials from marine organisms for applications in oriental medicine, cosmeceuticals, and nutraceuticals. To date, he has authored more than 500 research papers and holds 110 patents. In addition, he has written or edited more than 40 books.

Contributors

Farhan Jalees Ahmad
Department of Pharmaceutics
Hamdard University
New Delhi, India

Mohammad Zaki Ahmad
Department of Pharmaceutics
Najran University
Najran, Saudi Arabia

S. Aisverya
Department of Chemistry
Thiruvalluvar University
Tamilnadu, India

Sohail Akhter
Department of Pharmaceutics
Hamdard University
New Delhi, India

and

Department of Pharmaceutical Sciences
Utrecht University
Utrecht, the Netherlands

Fernanda Andrade
Department of Pharmaceutical Technology
University of Porto
Porto, Portugal

T. V. Anilkumar
Division of Experimental Pathology
Sree Chitra Tirunal Institute for Medical
    Sciences and Technology
Kerala, India

Mohammad Anwar
Department of Pharmaceutics
Hamdard University
New Delhi, India

Inmaculada Aranaz
Departamento de Química Macromolecular
Instituto de Ciencia y Tecnología de
    Polímeros
Madrid, Spain

Francisca Araújo
Department of Pharmaceutical Sciences
Institute of Health Sciences-North
Gandra, Portugal

and

Institute of Biomedical Engineering
University of Porto
Porto, Portugal

Kazuo Azuma
Faculty of Agriculture
Tottori University
Tottori, Japan

Sanat Kumar Basu
Department of Pharmaceutics
Gupta College of Technological Sciences
West Bengal, India

Berglind Eva Benediktsdóttir
School of Health Sciences
University of Iceland
Reykjavík, Iceland

Anumita Chaudhury
Procter & Gamble
Kobe, Japan

Chong-Su Cho
Department of Agricultural
    Biotechnology
and
Research Institute for Agricultural and
    Life Sciences
Seoul National University
Seoul, South Korea

Myung-Haing Cho
College of Veterinary Medicine
Seoul National University
Seoul, South Korea

Yun-Jaie Choi
Department of Agricultural Biotechnology
and
Research Institute for Agricultural and
    Life Sciences
Seoul National University
Seoul, South Korea

Surajit Das
Procter & Gamble
Kobe, Japan

R. Deepa
Division of Experimental Pathology
Sree Chitra Tirunal Institute for Medical
    Sciences and Technology
Kerala, India

Nguyen Anh Dzung
Institute of Biotechnology & Environment
Tay Nguyen University
Dak Lak Province, Vietnam

Maher Z. Elsabee
Department of Chemistry
Cairo University
Cairo, Egypt

Manal Farea
Conservative Department
School of Dental Sciences
Universiti Sains Malaysia
Kelantan, Malaysia

Alberto Gallardo
Departamento de Química
    Macromolecular
Instituto de Ciencia y Tecnología de
    Polímeros
Madrid, Spain

Arijit Gandhi
Department of Pharmaceutics
Gupta College of Technological Sciences
West Bengal, India

Vivek S. Gaware
School of Health Sciences
University of Iceland
Reykjavík, Iceland

T. Gomathi
Department of Chemistry
Thiruvalluvar University
Tamilnadu, India

Ahmad Sukari Halim
School of Medical Sciences
Universiti Sains Malaysia
Kelantan, Malaysia

Shinsuke Ifuku
Department of Chemistry and
    Biotechnology
Tottori University
Tottori, Japan

Nazma N. Inamdar
Government College of Pharmacy
Maharashtra, India

Sougata Jana
Department of Pharmaceutics
Gupta College of Technological
    Sciences
West Bengal, India

Hans E. Junginer (Retired)
Leiden/Amsterdam Center for Drug Research
Leiden University
Leiden, the Netherlands

Sang-Ki Kang
Department of Agricultural
    Biotechnology
and
Research Institute for Agricultural and
    Life Sciences
Seoul National University
Seoul, South Korea

Fatih Karadeniz
Marine Bioprocess Research Center
Pukyong National University
Busan, South Korea

Mustafa Zafer Karagozlu
Department of Chemistry
Pukyong National University
Busan, South Korea

Se-Kwon Kim
Department of Chemistry
and
Marine Bioprocess Research Center
Pukyong National University
Busan, South Korea

Ramona Lieder
Landspitali University Hospital
and
School of Science and Engineering
Reykjavik University
Reykjavik, Iceland

Sushila Maharjan
Department of Agricultural
    Biotechnology
and
Research Institute for Agricultural and
    Life Sciences
Seoul National University
Seoul, South Korea

Panchanathan Manivasagan
Department of Chemistry
Pukyong National University
Busan, South Korea

Már Másson
School of Health Sciences
University of Iceland
Reykjavík, Iceland

Saburo Minami
Graduate School of Engineering
Tottori University
Tottori, Japan

Rania E. Morsi
Egyptian Petroleum Research Institute
Cairo, Egypt

Vishnukant Mourya
Government College of Pharmacy
Andhra Pradesh, India

K. Nasreen
Department of Chemistry
Thiruvalluvar University
Tamilnadu, India

Dai-Nghiep Ngo
Department of Biochemistry
University of Science
Ho Chi Minh City, Vietnam

Nor Shamsuria Omar
Oral Biology Unit
School of Dental Sciences
Universiti Sains Malaysia
Kelantan, Malaysia

Tomohiro Osaki
Faculty of Agriculture
Tottori University
Tottori, Japan

Vandana Patravale
Department of Pharmaceutical Sciences and
    Technology
Institute of Chemical Technology
Maharashtra, India

Willi Paul
Division of Biosurface Technology
Sree Chitra Tirunal Institute for Medical
    Sciences and Technology
Kerala, India

Mónica Perez
Departamento de Química Macromolecular
Instituto de Ciencia y Tecnología de
    Polímeros
Madrid, Spain

M. Prabaharan
Department of Chemistry
Hindustan Institute of Technology and Science
Tamilnadu, India

Ziyaur Rahman
Irma Lerma Rangel College of Pharmacy
and
Texas A&M Health Science Center
Kingsville, Texas

Farshad Ramazani
Department of Pharmaceutical Sciences
Utrecht University
Utrecht, the Netherlands

Juan Alfonso Redondo
Departamento de Química Macromolecular
Instituto de Ciencia y Tecnología de
    Polímeros
Madrid, Spain

Maximas H. Rose
Department of Zoology
Manonmaniam Sundaranar University
Tamilnadu, India

Assal M. M. Sadeghi
Hakim Pharmaceutical Co.
Tehran, Iran

Hiroyuki Saimoto
Graduate School of Engineering
Tottori University
Tottori, Japan

Mohammad Samim
Department of Chemistry
Hamdard University
New Delhi, India

Bruno Sarmento
Department of Pharmaceutical Technology
and
Institute of Biomedical Engineering
University of Porto
Porto, Portugal

and

Department of Pharmaceutical Sciences
Institute of Health Sciences-North
Gandra, Portugal

Hitoshi Sashiwa
Frontier Materials Development Laboratories
Kaneka Corporation
Osaka, Japan

Kalyan Kumar Sen
Department of Pharmaceutics
Gupta College of Technological Sciences
West Bengal, India

Kalimuthu Senthilkumar
Marine Bioprocess Research Center
Pukyong National University
Busan, South Korea

Chandra P. Sharma
Division of Biosurface Technology
Sree Chitra Tirunal Institute for Medical
    Sciences and Technology
Kerala, India

Olafur E. Sigurjonsson
Landspitali University Hospital
and
School of Science and Engineering
Reykjavik University
and
Biomedical Center
University of Iceland
Reykjavik, Iceland

Bijay Singh
Department of Agricultural
    Biotechnology
and
Research Institute for Agricultural and
    Life Sciences
Seoul National University
Seoul, South Korea

P. N. Sudha
Department of Chemistry
Thiruvalluvar University
Tamilnadu, India

Jayachandran Venkatesan
Department of Chemistry
and
Marine Bioprocess Research Center
Pukyong National University
Busan, South Korea

Swati Vyas
Department of Pharmaceutical Sciences and
    Technology
Institute of Chemical Technology
Maharashtra, India

Musarrat Husain Warsi
Department of Pharmaceutics
Hamdard University
New Delhi, India

Part I

Synthesis and Characterization of
Chitin and Chitosan Derivatives

1 Chitin and Chitosan Derivatives

Se-Kwon Kim and Jayachandran Venkatesan

CONTENTS

1.1 INTRODUCTION TO CHITIN AND CHITOSAN

Chitin is a nontoxic, biodegradable polymer of high molecular weight. It is the most abundant second common polysaccharide found in nature after cellulose. Chitin is a fiber, and in addition, it posesses exceptional chemical and biological qualities such as biocompatibility, biodegradability, nontoxicity, and adsorption properties that can be employed in various industrial and medical applications. Chitin is made up of a linear chain of acetylglucosamine groups (Figure 1.1). Chitosan is derived from chitin, abundantly occurring in shrimp's shells. Chitosan is obtained by removing sufficient acetyl groups (CH_3-CO) for the molecule to be soluble in most diluted acids (Figure 1.2). This process, called deacetylation, releases amine groups (NH) and gives the chitosan a cationic characteristic. Chitosan is a linear polysaccharide composed of randomly distributed β-(1–4)-linked D-glucosamine and *N*-acetyl-D-glucosamine. This is especially interesting in an acidic environment in which the majority of polysaccharides are usually neutral or negatively charged (Peter 1995; Kurita 1998; Ravi Kumar 2000; Dutta et al. 2004; Rinaudo 2006; Pillai et al. 2009; Venkatesan and Kim 2010).

1.2 CHITIN AND CHITOSAN DERIVATIVES

The low solubility of chitosan at pH above 6.5 resulted in the synthesis and characterization of many chitosan derivatives. Different kinds of chitin and chitosan derivatives are shown in Table 1.1 (Ravi Kumar 2000). In recent years, significant development has been achieved in the use of chitin and

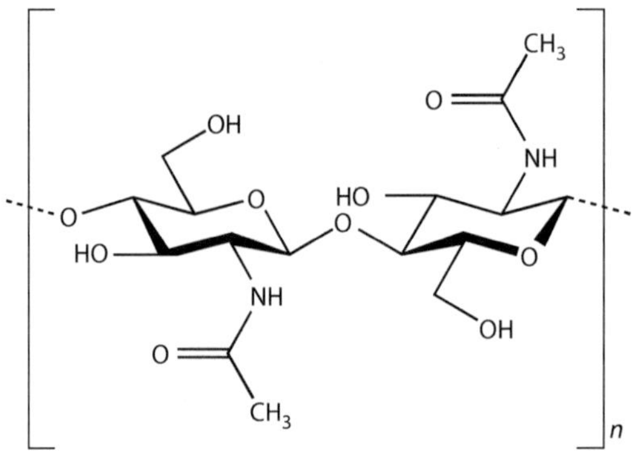

FIGURE 1.1   Structure of chitin.

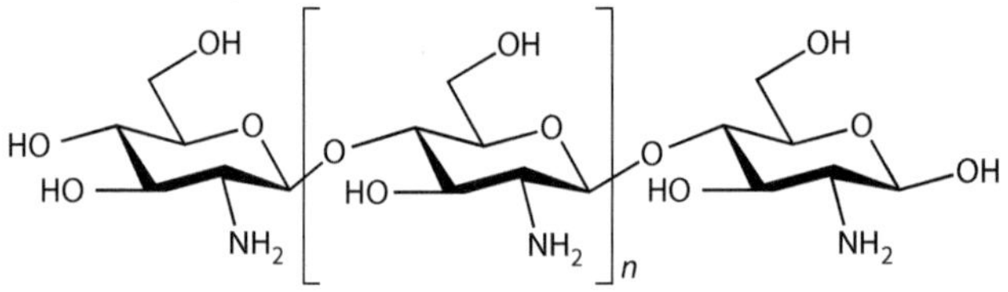

FIGURE 1.2   Structure of chitosan.

TABLE 1.1
Chitin and Chitosan Derivatives

S. No.	Type of Derivatives	Functional Groups
1	N-Acyl chitosans	Formyl, acetyl, propionyl, butyryl, hexanoyl, octanoyl, decanoyl, dodecanoyl, tetradecanoyl, lauroyl, myristoyl, palmitoyl, stearoyl, benzoyl, lauroyl, myristoyl, palmitoyl, stearoyl, benzoyl, monochloroacetoyl, dichloroacetyl, trifluoroacetyl, carbamoyl, succinyl, acetoxybenzoyl
2	N-Carboxyalkyl (aryl) chitosans	N-Carboxybenzyl, glycine-glucan (N-carboxymethyl chitosan), alanine glucan, phenylalanine glucan, tyrosine glucan, serine glucan, glutamic acid glucan, methionine glucan, leucine glucan
3	N-Carboxyacyl chitosans	From anhydrides such as maleic, itaconic, acetyl-thiosuccinic, glutaric, cyclohexane 1,2-dicarboxylic, phthalic, cis-tetrahydrophthalic, 5-norbornene-2,3-dicarboxylic, diphenic, salicylic, tri-mellitic, pyromellitic anhydride
4	o-Carboxyalkyl chitosans	o-Carboxymethyl, crosslinked o-carboxymethyl
5	Sugar derivatives	1-Deoxygalactic-1-yl-, 1-deoxyglucit-1-yl-, 1-deoxymelibiit-1-yl-, 1-deoxylactit-1-yl-, 1-deoxylactit-1-yl-4(2,2,6,6-tetramethylpiperidine-1-oxyl)-, 1-deoxy-6′-aldehydolactit-1-yl-,1-deoxy-6′-aldehydomelibiit-1-yl-, cellobiit-1-yl-chitosans, products obtained form ascorbic acid
6	Metal ion chelates	Palladium, copper, silver, iodine
7	Semisynthetic resins of chitosan	Copolymer of chitosan with methyl methacrylate, polyurea-urethane, poly (amide ester), acrylamide-maleic anhydride

chitosan derivatives (Figures 1.3 through 1.6). The chemical modification of chitosan brings to it new functional properties for various biological and biomedical applications (Cárdenas et al. 2006; Jayakumar et al. 2006, 2008a,b; Li et al. 2006, 2007a,b, 2011, 2013; Chesnutt et al. 2007; Cao et al. 2008; Liu et al. 2009; Datta et al. 2012; Srakaew et al. 2012; Yeh and Lin 2012; Zhao et al. 2012; Muzzarelli et al. 1984; Hirano et al. 1985; Saiki et al. 1990; Kornilaeva et al. 1995; Ryzhenkov et al.

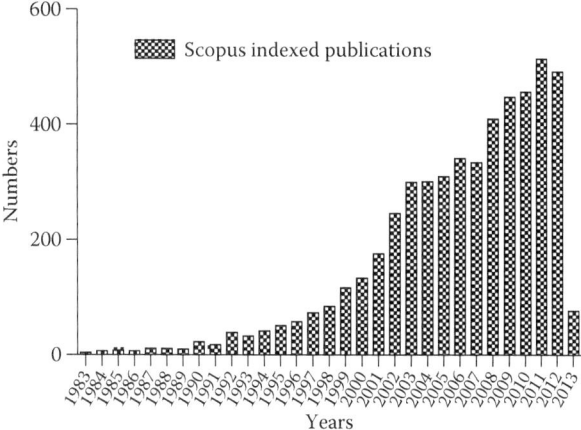

FIGURE 1.3 Scopus indexed articles last three decades related to chitin and chitosan derivatives.

1996; Gamzazade et al. 1997; Mariappan et al. 1999; Vongchan et al. 2002; Zhang et al. 2003; Xing et al. 2004; Xing et al. 2005; Jayakumar et al. 2007; Je et al. 2007; Ma et al. 2007; Yang et al. 2008, 2013; Zhou et al. 2009; Karadeniz et al. 2011; Shelma and Sharma 2011; Wang et al. 2011; Fan et al. 2012; Miao and Chen 2012; Lu et al. 2013; Pires et al. 2013).

1.2.1 CHEMICAL MODIFICATIONS

1.2.2 COMPOSITES

Considerable attention has been given to chitosan composite materials and their applications in the field of biomedical application because of its minimal foreign body reactions, intrinsic antibacterial nature, biocompatibility, biodegradability, and the ability to be molded into various geometries and forms such as porous structures, suitable for cell ingrowth and osteoconduction. The composite of

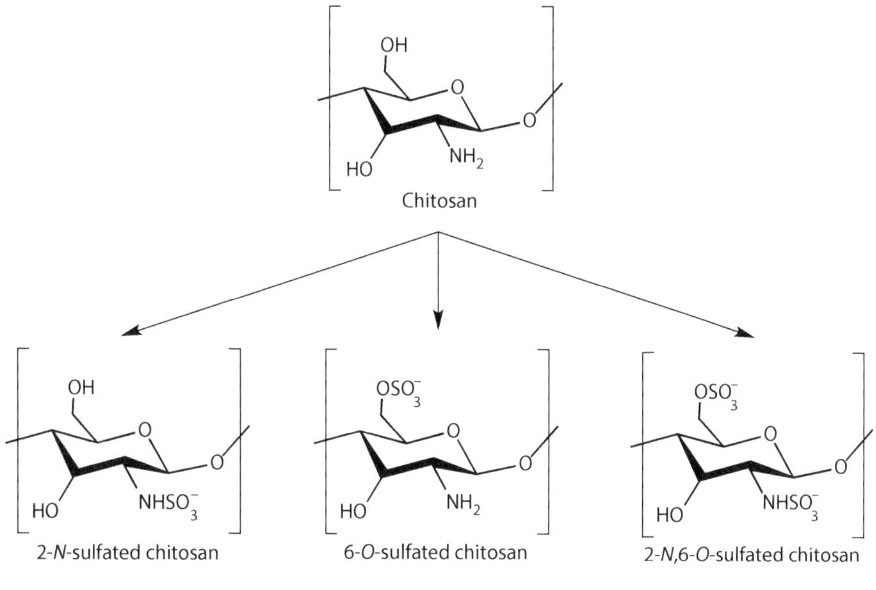

Chitosan

2-N-sulfated chitosan   6-O-sulfated chitosan   2-N,6-O-sulfated chitosan

FIGURE 1.4 Sulfated chitosan.

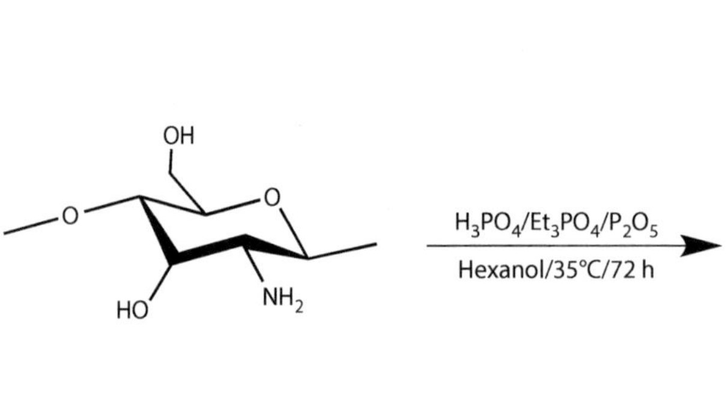

FIGURE 1.5 Phosphorylated chitosan.

chitosan, including hydroxyapatite, is very popular because of its biodegradability and biocompatibility (Venkatesan and Kim 2010). Several composites have been prepared for various biological activities such as chitosan with alginate (Li et al. 2005), collagen (Tan et al. 2001), calcium phosphate (Zhang and Zhang 2002), hydroxyapatite (Kong et al. 2005), and polysulfone (Huang et al. 1999). Most of the chitin and chitosan composite materials have been used in bone tissue engineering. Bone comprises both organic and inorganic portions that are mainly composed of collagen and hydroxyapatite. Several researchers have tried to substitute the function of collagen by chitosan (Di Martino et al. 2005; Thein-Han and Misra 2009; Venkatesan and Kim 2010).

1.2.3 HYDROGELS

Hydrogels form a network of polymer chains that are water insoluble and are superabsorbents, mainly composed of natural or synthetic polymers. They constitute cross-linked polymer networks that in turn have a high number of hydrophilic groups or domains. Hydrogels possess a degree of flexibility similar to natural tissues, due to their high water content. Hydrogels are employed in different applications, including tissue engineering, drug delivery system, biosensors, and contact lenses. The use of the natural polymer, chitosan, as the scaffold material in hydrogels has been highly pursued due to its

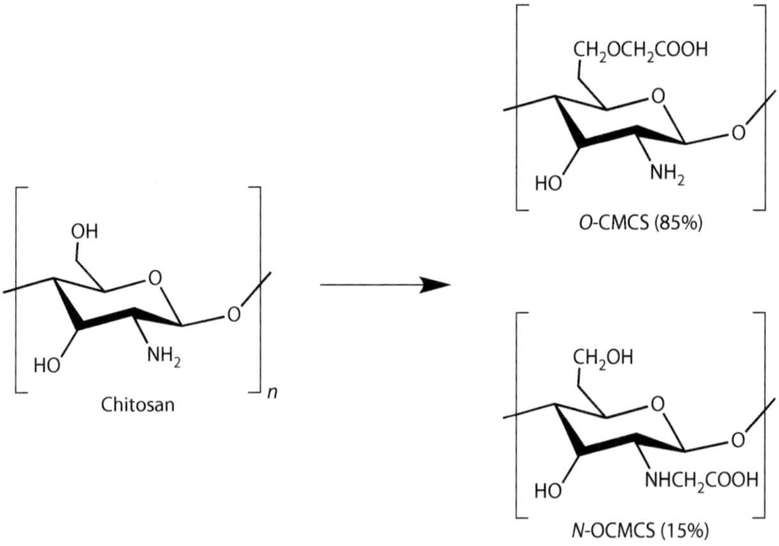

FIGURE 1.6 Carboxymethyl chitosan.

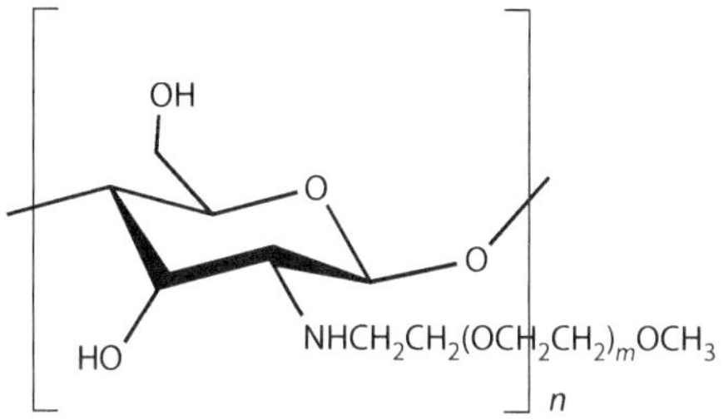

FIGURE 1.7 Chemical structure of chitosan-PEG (thermoreversible hydrogels).

biocompatibility, low toxicity, and biodegradability. Chitosan is distinct from other commonly available polysaccharides because of the presence of nitrogen in its molecular structure, its cationicity, and its capacity to form polyelectrolyte complexes. The cationic nature of the polymer enables it to become water soluble after the formation of carboxylate salts, such as formate, acetate, lactate, malate, citrate, glyoxylate, pyruvate, glycolate, poly(ethylene glycol), and ascorbate (Bhattarai et al. 2010) (Figure 1.7).

1.3 BIOLOGICAL ACTIVITY OF CHITIN AND CHITOSAN DERIVATIVES

1.3.1 ANTIBACTERIAL ACTIVITY

Antibacterial activity is based on molecular weight, degree of deacetylation, type of substituents, which can be cationic or easily form cations, and the type of bacteria. In general, high molecular weight chitosan cannot pass through cell membranes and forms a film that protects the cells against nutrient transport through the microbial cell membrane. The chemical modification of chitosan into chitooligosaccharides are water soluble and can better incorporate the active molecule into the cell. Gram-negative bacteria, often represented by *Escherichia coli*, have an anionic bacterial surface on which cationic chitosan derivatives interact electrostatically. Thus, many chitosan conjugates have cationic components, such as ammonium, pyridinium, or piperazinium substituents introduced into their molecules to increase their positive charge. Gram-positive bacteria such as *Staphylococcus aureus* are inhibited by the binding of low molecular weight chitosan derivatives to DNA or RNA (Vinsova and Vavrikova 2011). Water-soluble quaternary-based *N,N,N*-trimethylchitosan and O-([2-hydroxy-3-trimethylammonium])propyl chitin derivatives were synthesized and characterized and used for antibacterial activities against *Salmonella choleraesuis* and *Bacillus subtilis*. The minimum inhibitory concentration values vary from 0.02 to 20.48 mg/mL, and their minimum bactericidal concentration values vary from 0.08 to 40.96 mg/mL against *S. choleraesuis* and *B. subtilis* (Rúnarsson et al. 2010; Benediktsdóttir et al. 2011; Huang et al. 2013; Geng et al. 2013).

The antibacterial activity of water-soluble chitosan derivatives prepared by Maillard reactions against *Staph. aureus, Listeria monocytogenes, Bacillus cereus, E. coli, Shigella dysenteriae*, and *Salmonella typhimurium* was examined. Relatively high antibacterial activity against various microorganisms was noted for the chitosan–glucosamine derivatives when compared with the acid-soluble chitosan. Water-soluble chitosan produced by Maillard reaction may be a promising commercial substitute for acid-soluble chitosan (Chung et al. 2011).

1.3.2 ANTICANCER ACTIVITY

Traditional chemotherapeutic agents act by killing cells that divide rapidly—one of the main properties of most cancer cells. Cytotoxic drugs continue to play a major role in cancer therapy. However, they produce side effects, especially through the destruction of lymphoid and bone marrow cells. Therefore, strategic improvements in cancer therapy are needed to enhance efficacy and at the same time decrease side effects. Hence, biocompatible anticancer drugs are required to treat various types of cancers. However, chitosan-based drugs that are employed to treat various cancers are still at the laboratory stages of testing. The introduction of several groups into chitosan modifies its structure significantly, thereby increasing the biological activity of chitosan. The introduction of sulfates and phenyl groups in carboxymethyl benzylamide dextrans into chitosan, which is easily modified by different functional groups have good anticancer activity, especially in breast cancer cells. The SCS and the sulfated benzaldehyde chitosan (SBCS) significantly inhibited cell proliferation, induced apoptosis, and blocked the FGF-2-induced phosphorylation of ERK in MCF-7 cells, SBCS had better inhibitory effects and a lower IC_{50} compared with SCS (Farley et al. 2006).

Dialkylaminoalkylation and reductive amination followed by quaternization of chitosan led to the cationic chitosan derivatives eliciting dose-dependent inhibitory effects on the proliferation of tumor cell lines. The cytotoxic activity of the chitosan derivatives increased with the number of carbons in the alkyl group in the order chitosan <NH_2-chitosan < DMAE-chitosan < DMAiP-chitosan < DEAE-chitosan. The *in vitro* tumor-suppressive activity of qDEAE-chitosan was as potent as that of antibiotics, such as adriamycin (IC50 5–100 mg/mL) and cecropin B (IC50 10–300 mg/mL), and much higher than values reported for other polysaccharides (IC50 1000 mg/mL) (Lee et al. 2002) (Figures 1.8 and 1.9, Table 1.2).

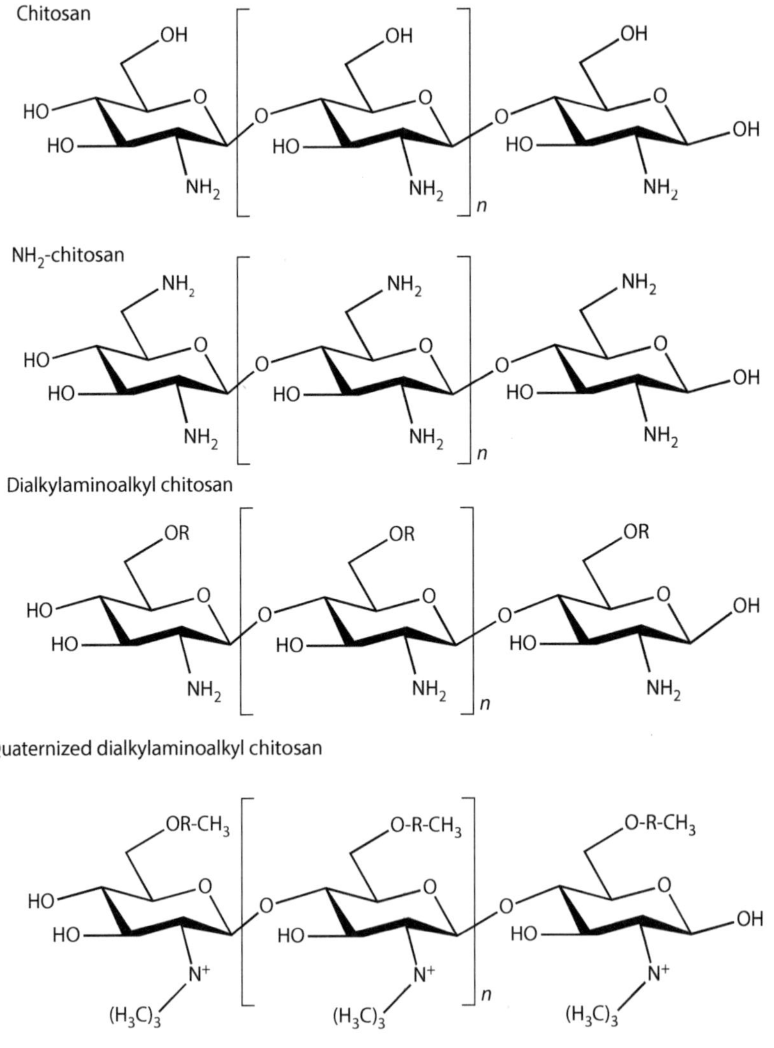

FIGURE 1.8 The cationic chitosan derivatives (DMAE-chitosan; R = DMAE, DMAiP-chitosan; R = DMAiP, DEAE-chitosan; R = DEAE).

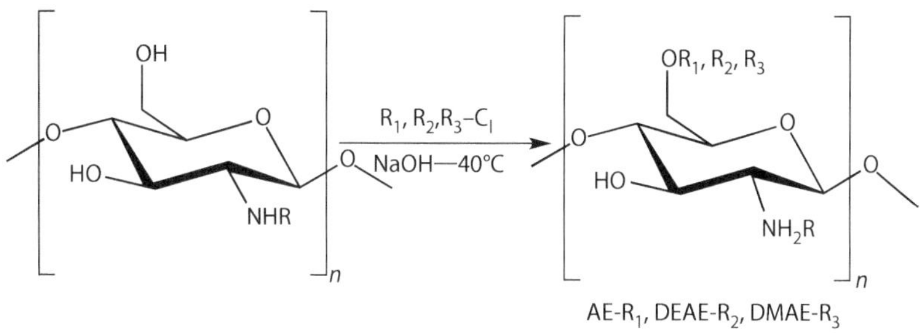

R = H, CONH$_2$

R$_1$ = NH$_2$(CH$_2$)$_2$–, R$_2$ = (CH$_3$CH$_2$)$_2$N(CH$_2$)$_2$–, R$_3$ = (CH$_3$)$_2$N(CH$_2$)$_2$–

AE-R$_1$—Aminoethyl chitooligosaccharide

DEAE-R$_2$—Diethyl aminoethyl chitooligosaccharide

DMAE-R$_3$—Dimethyl aminoethyl chitooligosaccharide

FIGURE 1.9 Synthesis and chemical structure of chitosan derivatives. (Adapted from Je, J.-Y., Y.-S. Cho, and S.-K. Kim. 2006. *Bioorganic & Medicinal Chemistry Letters* 16(8):2122–2126.)

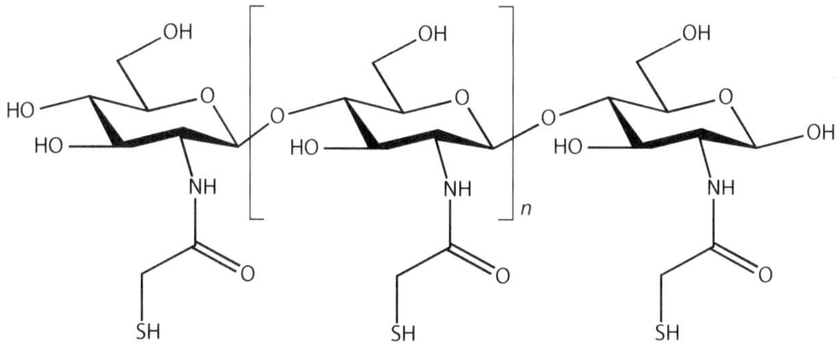

FIGURE 1.10 Chemical structure of chitosan–thioglycolic acid conjugate.

1.3.3 ANTIINFLAMMATORY

Antiinflammatory refers to the property of a substance or treatment that reduces inflammation. Theophylline is a drug that reduces the inflammatory effects of allergic asthma but is difficult to administer at an appropriate dosage without causing adverse side effects. The adsorption of theophylline to chitosan nanoparticles modified by the addition of thiol groups would improve theophylline absorption by the bronchial epithelium and enhance its antiinflammatory effects (Lee et al. 2006) (Figure 1.10).

1.4 BIOMEDICAL APPLICATIONS

1.4.1 TISSUE ENGINEERING

Tissue engineering is the use of a combination of cells, engineering and materials methods, and suitable biochemical and physiochemical factors to improve or replace biological functions. Although it was once categorized as a subfield of biomaterials, having grown in scope and importance it can

TABLE 1.2

Cytotoxic Activities of Chitosan Derivatives against Different Cell Lines

Panel of Cell Lines	Cell Line	Compound/Cytotoxicity (IC$_{50}$, g/mL)	
		DEAE-Chitosan (90%)	DEAE-Chitosan (50%)
Cervix cancer	HeLa	16 (±4)	26 (±4)
Lung cancer	A549	51 (±4)	93 (±5)
Human fibrosarcoma	HT1080	63 (±3)	126 (±5)

Source: Adapted from Je, J.-Y., Y.-S. Cho, and S.-K. Kim. 2006. *Bioorganic & Medicinal Chemistry Letters* 16(8):2122–2126.

be considered as a field in its own right. Different materials have been used in tissue engineering such as natural and synthetic. Both the materials play a role in tissue engineering of artificial organs. In recent years, significant progress has been made in the development of chitosan-based biomaterials for tissue engineering. The main practical use of chitosan has been mainly restricted to the unmodified forms in tissue engineering applications. Recently, there has been a growing interest in the modification of chitosan to improve its solubility, introduce desired properties, and widen the field of its potential applications by choosing various types of side chains; however, certain properties have been altered by these modifications (Kim et al. 2008). Chitosan derivatives have been widely used to prepare several tissue engineering organs, such as skin, bone, liver, nerve, and blood vessels.

1.4.2 DRUG DELIVERY

The chemical modification of chitosan definitely has a great effect on the delivery of different kinds of drugs in a specific place (Sonia and Sharma 2011). The chitosan nanoparticles exhibit an increase in loading capacity and efficacy. Antitumor active compounds, such as doxorubicin, paclitaxel, docetaxel, and norcantharidin, are used as drug carriers. It is evident that chitosan, with its low molecular weight, is a useful carrier for molecular drugs that require targeted delivery. The antioxidant scavenging activity of chitosan has been established by the strong hydrogen-donating ability of chitosan. The low molecular weight and greater degree of quaternization have a positive influence on the antioxidant activity of chitosan. The phenolic and polyphenolic compounds with antioxidant effects are condensed with chitosan to form mutual prodrugs (Mourya and Inamdar 2009; Vinsova and Vavrikova 2011).

The conjugates of certain kinds of anticancer agents with chitin and chitosan derivatives display good anticancer effects with a decrease in the adverse effects of the original drug because of a predominant distribution into the cancer cells and a gradual release of free drug from the conjugates. The conjugates of MMC (mitomycin C) with *N*-succinyl-chitosan showed good antitumor activities against various tumor models because of their predominant distribution into the tumor tissue and sustained-release characteristics, irrespective of water-insoluble and soluble formulations. Chitin and chitosan derivatives discussed in this chapter are good candidates for a polymeric drug carrier in cancer chemotherapy (Kato et al. 2005).

1.5 CONCLUSION

In this chapter, the synthesis, characterization, and potential applications toward biological and biomedical properties of chitin and chitosan derivatives have been discussed. The chemical modification of chitin and chitosan has led to various fields such as tissue engineering and drug delivery.

ACKNOWLEDGMENT

This work was supported by a grant from the Marine Bioprocess Research Centre of the Marine Bio 21 Center funded by the Ministry of Oceans and Fisheries, Transport and Maritime, Republic of Korea.

REFERENCES

Benediktsdóttir, B. E., V. S. Gaware, Ö. V. Rúnarsson, S. Jónsdóttir, K. J. Jensen, and M. Másson. 2011. Synthesis of *N,N,N*-trimethyl chitosan homopolymer and highly substituted *N*-alkyl- *N,N*-dimethyl chitosan derivatives with the aid of di- *tert*-butyldimethylsilyl chitosan. *Carbohydrate Polymers* 86(4):1451–1460.

Bhattarai, N., J. Gunn, and M. Zhang. 2010. Chitosan-based hydrogels for controlled, localized drug delivery. *Advanced Drug Delivery Reviews* 62(1):83–99.

Cao, Z., H. Xu, R. Tang, W. Wei, and W. Xu. 2008. Synthesis, characterization and molluscicidal activity of triphenyltin phosphorylated chitosan. *Chemistry Bulletin/Huaxue Tongbao* 71(7):528–532.

Cárdenas, G., G. Cabrera, E. Taboada, and M. Rinaudo. 2006. Synthesis and characterization of chitosan alkyl phosphate. *Journal of the Chilean Chemical Society* 51(1):815–820.

Chesnutt, B. M., Y. Yuan, N. Brahmandam et al. 2007. Characterization of biomimetic calcium phosphate on phosphorylated chitosan films. *Journal of Biomedical Materials Research—Part A* 82(2):343–353.

Chung, Y.-C., J.-Y. Yeh, and C.-F. Tsai. 2011. Antibacterial characteristics and activity of water-soluble chitosan derivatives prepared by the Maillard reaction. *Molecules* 16(10):8504–8514.

Datta, P., S. Dhara, and J. Chatterjee. 2012. Hydrogels and electrospun nanofibrous scaffolds of *N*-methylene phosphonic chitosan as bioinspired osteoconductive materials for bone grafting. *Carbohydrate Polymers* 87(2):1354–1362.

Di Martino, A., M. Sittinger, and M. V. Risbud. 2005. Chitosan: A versatile biopolymer for orthopaedic tissue-engineering. *Biomaterials* 26(30):5983–5990.

Dutta, P. K., J. Dutta, and V. S. Tripathi. 2004. Chitin and chitosan: Chemistry, properties and applications. *Journal of Scientific and Industrial Research* 63(1):20–31.

Fan, L., P. Wu, J. Zhang et al. 2012. Synthesis and anticoagulant activity of the quaternary ammonium chitosan sulfates. *International Journal of Biological Macromolecules* 50(1):31–37.

Farley, J. H., N. P. Clear, B. Leroy, T. L. Davis, and G. McPherson. 2006. Age, growth and preliminary estimates of maturity of bigeye tuna, *Thunnus obesus*, in the Australian region. *Marine and Freshwater Research* 57(7):713–724.

Gamzazade, A., A. Sklyar, S. Nasibov, I. Sushkov, A. Shashkov, and Yu Knirel. 1997. Structural features of sulfated chitosans. *Carbohydrate Polymers* 34(1–2):113–116.

Geng, X., R. Yang, J. Huang, X. Zhang, and X. Wang. 2013. Evaluation antibacterial activity of quaternary-based chitin/chitosan derivatives in vitro. *Journal of Food Science* 78(1):M90–M97.

Hirano, S., Y. Tanaka, M. Hasegawa, K. Tobetto, and A. Nishioka. 1985. Effect of sulfated derivatives of chitosan on some blood coagulant factors. *Carbohydrate Research* 137(C):205–215.

Huang, J., H. Jiang, M. Qiu et al. 2013. Antibacterial activity evaluation of quaternary chitin against *Escherichia coli* and *Staphylococcus aureus*. *International Journal of Biological Macromolecules* 52:85–91.

Huang, R. Y. M., R. Pal, and G. Y. Moon. 1999. Crosslinked chitosan composite membrane for the pervaporation dehydration of alcohol mixtures and enhancement of structural stability of chitosan/polysulfone composite membranes. *Journal of Membrane Science* 160(1):17–30.

Jayakumar, R., H. Nagahama, T. Furuike, and H. Tamura. 2008a. Synthesis of phosphorylated chitosan by novel method and its characterization. *International Journal of Biological Macromolecules* 42(4):335–339.

Jayakumar, R., N. Nwe, S. Tokura, and H. Tamura. 2007. Sulfated chitin and chitosan as novel biomaterials. *International Journal of Biological Macromolecules* 40(3):175–181.

Jayakumar, R., R. L. Reis, and J. F. Mano. 2006. Chemistry and applications of phosphorylated chitin and chitosan. *E-Polymers* 35:1–16.

Jayakumar, R., N. Selvamurugan, S. V. Nair, S. Tokura, and H. Tamura. 2008b. Preparative methods of phosphorylated chitin and chitosan—An overview. *International Journal of Biological Macromolecules* 43(3):221–225.

Je, J. Y., E. K. Kim, C. B. Ahn et al. 2007. Sulfated chitooligosaccharides as prolyl endopeptidase inhibitor. *International Journal of Biological Macromolecules* 41(5):529–533.

Je, J.-Y., Y.-S. Cho, and S.-K. Kim. 2006. Cytotoxic activities of water-soluble chitosan derivatives with different degree of deacetylation. *Bioorganic & Medicinal Chemistry Letters* 16(8):2122–2126.

Karadeniz, F., M. Z. Karagozlu, S. Y. Pyun, and S. K. Kim. 2011. Sulfation of chitosan oligomers enhances their anti-adipogenic effect in 3T3-L1 adipocytes. *Carbohydrate Polymers* 86(2):666–671.

Kato, Y., H. Onishi, and Y. Machida. 2005. Contribution of chitosan and its derivatives to cancer chemotherapy. *In Vivo* 19(1):301–310.

Kim, I-Y., S-J. Seo, H-S. Moon et al. 2008. Chitosan and its derivatives for tissue engineering applications. *Biotechnology Advances* 26(1):1–21.

Kong, L., Y. Gao, W. Cao, Y. Gong, N. Zhao, and X. Zhang. 2005. Preparation and characterization of nano-hydroxyapatite/chitosan composite scaffolds. *Journal of Biomedical Materials Research Part A* 75(2):275–282.

Kornilaeva, G. V., T. V. Makarova, A. I. Gamzazade, A. M. Sklyar, S. M. Nasibov, and E. V. Karamov. 1995. Sulphated chitosan derivatives as HIV-infection inhibitors. *Immunologiya* 1(1):13–16.

Kurita, K. 1998. Chemistry and application of chitin and chitosan. *Polymer Degradation and Stability* 59(1):117–120.

Lee, D-W., S. A. Shirley, R. F. Lockey, and S. S. Mohapatra. 2006. Thiolated chitosan nanoparticles enhance anti-inflammatory effects of intranasally delivered theophylline. *Respiration Research* 7(1):112.

Lee, J-K., H-S. Lim, and J-H. Kim. 2002. Cytotoxic activity of aminoderivatized cationic chitosan derivatives. *Bioorganic & Medicinal Chemistry Letters* 12(20):2949–2951.

Li, B., L. Huang, X. Wang, J. Ma, and F. Xie. 2011. Biodegradation and compressive strength of phosphorylated chitosan/chitosan/hydroxyapatite bio-composites. *Materials and Design* 32(8–9):4543–4547.

Li, B., L. Huang, X. Wang, J. Ma, F. Xie, and L. Xia. 2013. Effect of micropores and citric acid on the bioactivity of phosphorylated chitosan/chitosan/hydroxyapatite composites. *Ceramics International* 39(3):3423–3427.

Li, Q. L., Z. Q. Chen, B. W. Darvell et al. 2007a. Chitosan-phosphorylated chitosan polyelectrolyte complex hydrogel as an osteoblast carrier. *Journal of Biomedical Materials Research—Part B Applied Biomaterials* 82(2):481–486.

Li, Q. L., Z. Q. Chen, B. W. Darvell et al. 2006. Biomimetic synthesis of the composites of hydroxyapatite and chitosan-phosphorylated chitosan polyelectrolyte complex. *Materials Letters* 60(29–30):3533–3536.

Li, Q. L., N. Huang, Z. Chen, and X. Tang. 2007b. Biomimetic synthesis of the nanocomposite of phosphorylated chitosan and hydroxyapatite and its bioactivity in vitro. *Key Engineering Materials* 330–332(19):721–724

Li, Z., H. R. Ramay, K. D. Hauch, D. Xiao, and M. Zhang. 2005. Chitosan–alginate hybrid scaffolds for bone tissue engineering. *Biomaterials* 26(18):3919–3928.

Liu, J., Y. Zheng, W. Wang, and A. Wang. 2009. Preparation and swelling properties of semi-IPN hydrogels based on chitosan-g-poly(acrylic acid) and phosphorylated polyvinyl alcohol. *Journal of Applied Polymer Science* 114(1):643–652.

Lu, X., H. Guo, L. Sun, L. Zhang, and Y. Zhang. 2013. Protective effects of sulfated chitooligosaccharides with different degrees of substitution in MIN6 cells. *International Journal of Biological Macromolecules* 52(1):92–98.

Ma, B., W. Huang, W. Kang, and J. Yan. 2007. Studies on preparation of sulfated derivatives of chitosan from mucor rouxianus. *Lizi Jiaohuan Yu Xifu/Ion Exchange and Adsorption* 23(5):451–458.

Mariappan, M. R., E. A. Alas, J. G. Williams, and M. D. Prager. 1999. Chitosan and chitosan sulfate have opposing effects on collagen–fibroblast interactions. *Wound Repair and Regeneration* 7(5):400–406.

Miao, J., and G. Chen. 2012. Sulfated chitosan (SCS)/polysulfone (PS) composite nanofiltration membrane surface cross-linked by epichlorohydrin.

Mourya, V. K., and N. N Inamdar. 2009. Trimethyl chitosan and its applications in drug delivery. *Journal of Materials Science: Materials in Medicine* 20(5):1057–1079.

Muzzarelli, R. A. A., F. Tanfani, M. Emanuelli, D. P. Pace, E. Chiurazzi, and M. Piani. 1984. Sulfated N-(carboxymethyl)chitosans: Novel blood anticoagulants. *Carbohydrate Research* 126(2):225–231.

Peter, M. G. 1995. Applications and environmental aspects of chitin and chitosan. *Journal of Macromolecular Science, Part A: Pure and Applied Chemistry* 32(4):629–640.

Pillai, C. K. S., W. Paul, and C. P. Sharma. 2009. Chitin and chitosan polymers: Chemistry, solubility and fiber formation. *Progress in Polymer Science* 34(7):641–678.

Pires, N. R., P. L. R. Cunha, J. S. Maciel et al. 2013. Sulfated chitosan as tear substitute with no antimicrobial activity. *Carbohydrate Polymers* 91(1):92–99.

Ravi Kumar, M. N. V. 2000. A review of chitin and chitosan applications. *Reactive and Functional Polymers* 46(1):1–27.

Rinaudo, M. 2006. Chitin and chitosan: Properties and applications. *Progress in Polymer Science* 31(7):603–632.

Rúnarsson, Ö. V., J. Holappa, C. Malainer et al. 2010. Antibacterial activity of *N*-quaternary chitosan derivatives: Synthesis, characterization and structure activity relationship (SAR) investigations. *European Polymer Journal* 46(6):1251–1267.

Ryzhenkov, V. E., M. A. Solovyeva, O. V. Remesova, and I. V. Okunevich. 1996. Hypolipidemic action of sulfated polysaccharides. *Voprosy Meditsinskoj Khimii* 42(2):118–119.

Saiki, I., J. Murata, M. Nakajima, S. Tokura, and I. Azuma. 1990. Inhibition by sulfated chitin derivatives of invasion through extracellular matrix and enzymatic degradation by metastatic melanoma cells. *Cancer Research* 50(12):3631–3637.

Shelma, R. and C. P. Sharma. 2011. Development of lauroyl sulfated chitosan for enhancing hemocompatibility of chitosan. *Colloids and Surfaces B: Biointerfaces* 84(2):561–570.

Sonia, T. A. and C. P. Sharma. 2011. Chitosan and its derivatives for drug delivery perspective. In *Chitosan for Biomaterials I*, R. Jayakumar, M. Prabaharan, and R. A. A. Muzzarelli, editors. Springer: Berlin, Heidelberg, pp. 23–53.

Srakaew, V., P. Ruangsri, K. Suthin, P. Thunyakitpisal, and W. Tachaboonyakiat. 2012. Sodium-phosphorylated chitosan/zinc oxide complexes and evaluation of their cytocompatibility: An approach for periodontal dressing. *Journal of Biomaterials Applications* 27(4):403–412.

Tan, W., R. Krishnaraj, and T. A Desai. 2001. Evaluation of nanostructured composite collagen–chitosan matrices for tissue engineering. *Tissue Engineering* 7(2):203–210.

Thein-Han, W. W. and R. D. K. Misra. 2009. Biomimetic chitosan–nanohydroxyapatite composite scaffolds for bone tissue engineering. *Acta Biomaterialia* 5(4):1182–1197.

Venkatesan, J. and S-K. Kim. 2010. Chitosan composites for bone tissue engineering—An overview. *Marine Drugs* 8(8):2252–2266.

Vinsova, J. and E. Vavrikova. 2011. Chitosan derivatives with antimicrobial, antitumour and antioxidant activities–A review. *Current Pharmaceutical Design* 17(32):3596–3607.

Vongchan, P., W. Sajomsang, D. Subyen, and P. Kongtawelert. 2002. Anticoagulant activity of a sulfated chitosan. *Carbohydrate Research* 337(13):1239–1242.

Wang, K., H. Li, Q. Ye et al. 2011. Synthesis and anticoagulant action of sulfated chitosan/Eu 2O 3 nano-oxides hybrid materials. *Zhongguo Xitu Xuebao/Journal of the Chinese Rare Earth Society* 29(6):764–768.

Xing, R., S. Liu, H. Yu, Q. Zhang, Z. Li, and P. Li. 2004. Preparation of low-molecular-weight and high-sulfate-content chitosans under microwave radiation and their potential antioxidant activity in vitro. *Carbohydrate Research* 339(15):2515–2519.

Xing, R., H. Yu, S. Liu et al. 2005. Antioxidant activity of differently regioselective chitosan sulfates in vitro. *Bioorganic and Medicinal Chemistry* 13(4):1387–1392.

Yang, J., K. Luo, D. Li et al. 2013. Preparation, characterization and *in vitro* anticoagulant activity of highly sulfated chitosan. *International Journal of Biological Macromolecules* 52(1):25–31.

Yang, Y., Y. Zhou, W. Hou, Y. Zhao, and Y. Ren. 2008. Preparation and anticoagulant activity of sulfated 6-carboxychitosan.

Yeh, H. Y. and J. C. Lin. 2012. Surface phosphorylation for polyelectrolyte complex of chitosan and its sulfonated derivative: Surface analysis, blood compatibility and adipose derived stem cell contact properties. *Journal of Biomaterials Science, Polymer Edition* 23(1–4):233–250.

Zhang, C., Q. Ping, H. Zhang, and J. Shen. 2003. Preparation of N-alkyl-O-sulfate chitosan derivatives and micellar solubilization of taxol. *Carbohydrate Polymers* 54(2):137–141.

Zhang, Y. and M. Zhang. 2002. Calcium phosphate/chitosan composite scaffolds for controlled *in vitro* antibiotic drug release. *Journal of Biomedical Materials Research* 62(3):378–386.

Zhao, D., J. Xu, L. Wang et al. 2012. Study of two chitosan derivatives phosphorylated at hydroxyl or amino groups for application as flocculants. *Journal of Applied Polymer Science* 125(SUPPL. 2):E299–E305.

Zhou, H., J. Qian, J. Wang et al. 2009. Enhanced bioactivity of bone morphogenetic protein-2 with low dose of 2-N, 6-O-sulfated chitosan *in vitro* and in vivo. *Biomaterials* 30(9):1715–1724.

2 Synthesis, Characterization, and Biomedical Applications of Chitosan and Its Derivatives

Hans E. Junginer and Assal M.M. Sadeghi

CONTENTS

2.1   CHITIN AND CHITOSAN

Chitin is a biopolymer consisting of *N*-acetyl glucosamine subunits present in the exoskeleton of crustaceans, such as crab and shrimps, as well as the cell wall of fungi and algae (Jeuniaux 1982). It was first discovered by Henry Braconnot, a French scientist, in 1811 and since then different research groups have been working on chitin and its major derivative, chitosan, independently. However, because of the complexity of the molecule and difficulty in its synthesis, the application of this polymer was limited for many years. In 1977, the 1st International Conference on Chitin and Chitosan (1st ICCC), brought together scientists working on biological and chemical aspects of these polymers. Since then, a major interest was aroused on chitin and chitosan research (Muzzarelli 1978, Ruiz-Herrera 1978).

Chitin is the second most abundant natural polymer after cellulose, but its application in industry is limited because of its insolubility; hence, chitosan was produced from deacetylation of chitin (Roberts 1982) (Figure 2.1). Chitosan is a polysaccharide composed of two common sugars present in mammalian tissues, the deacetylated D-glucosamine linked to the *N*-acetyl D-glucosamine unit using β-1–4 glucosidic bonds (Ulansky and Rosiak 1992). The distribution of these subunits depends on the chitosan preparation method (Dodane and Vilivalam 1998) (Figure 2.2).

Different methods have been used for the deacetylation of chitin and production of chitosan. The first method was by using NaOH (50%w/w) and temperatures of 100–120°C. The application of lower temperatures and reduced quantity of NaOH is also reported in different studies (Alimunir and Zainuddin 1992, Peniston and Johnson 1980). Utilization of water miscible diluents, such as acetone and tributylammonium (TBA), as reaction medium is another approach for uniform distribution of NaOH throughout the reaction medium and deacetylation of chitin particles (Hayes 1992).

The deacetylation of chitin can also be carried out using an enzymatic method with chitin deacetylases extracted from *Colletotrichum lindemuthianum* and *Mucor rouxii*. This method is more favorable because it does not require a chemical reaction or high temperature, is environmentally friendly, and is nondegradative. The disadvantage of this method is the ability of the large enzymes to reach the chitin substrate which limits the deacetylation process (Domard et al. 1986).

More advanced techniques were used for the deacetylation process by presoaking chitin in 15 M NaOH for about 3 days at 30°C so that it swells, followed by deacetylation at 40°C in the alkaline

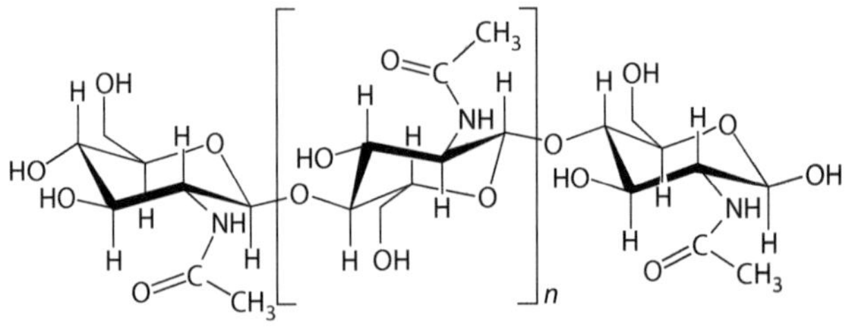

FIGURE 2.1   The chemical structure of chitin. (Adapted from Aranaz, I. et al. 2009. *Current Chemical Biology* 3: 203–230.)

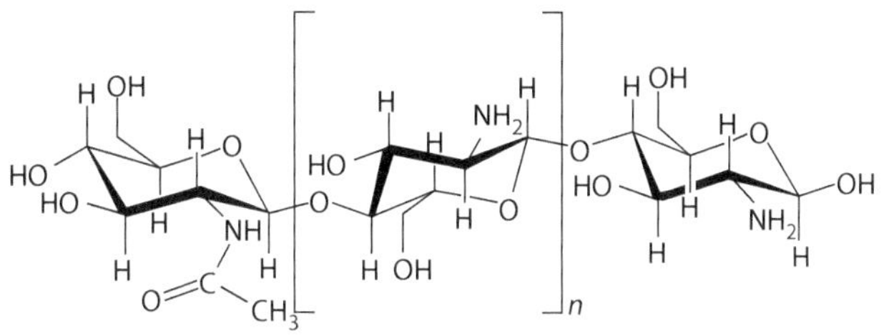

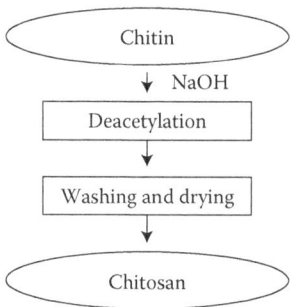

FIGURE 2.2   The chemical structure of chitosan. (Adapted from Aranaz, I. et al. 2009. *Current Chemical Biology* 3: 203–230.)

solution to enhance the swelling process. This product was water soluble at physiological pH. A second, more complicated method involved freeze–pump out–thaw (FPT) cycles; however, this method is not suitable for large-scale production of chitosan (Varum and Smidsrød 1994, Lamarque et al. 2005).

During the deacetylation of chitosan, the polymeric chains are degraded and chitosan's crystalline structure is damaged because of the harsh conditions during the reaction process (Rege and Block 1999). It was shown that the optimum conditions during the deacetylation of chitin were obtained with NaOH 75% (w/v) and a temperature of 110°C (Sukwattanasinitt et al. 2002, Galed et al. 2008). The source of chitin and its isolation process are also factors that affect the quality of chitosan (Galed et al. 2005) (Figure 2.3).

Today, chitosan with different degrees of deacetylation (%DD), varying from 60% to 98%, is commercially available; the degree of deacetylation of chitosan can be determined by different techniques, such as infrared spectroscopy, nuclear magnetic resonance spectroscopy (^{1}H-NMR) and (^{13}CNMR), and conductometric titration (Raymond et al. 1993, Brugnerotto et al. 2001, Duarte et al. 2001, Van de Velde and Kiekens 2004, Kassai 2008) (Figure 2.4). Chitosan is available in different molecular weights depending on the source and preparation method ranging from 50 to 2000 kDa (Chenite et al. 2001). Although the DD is important for chitosan performance in different applications, the average Mw of chitosan is significant in biochemical and biopharmacological fields (Baxter et al. 1992, Tharanathan and Kittur 2003). Chitosan is now produced commercially in countries such as India, Japan, Poland, Norway, Thailand, and Australia (Kumar 2000).

Chitosan is the most abundant amino polysaccharide and because of its high content of nitrogen (~7%) compared to other natural polymers, such as cellulose with only 1% nitrogen content, it is of great commercial interest. Chitosan is used extensively in different fields because of its excellent characteristics, such as biocompatibility, biodegradability, nontoxicity, ability to form films, chelating metal ions, and absorption properties (Kumar 2000).

FIGURE 2.3   A schematic presentation of chitosan preparation from chitin. (Adapted from Aranaz, I. et al. 2009. *Current Chemical Biology* 3: 203–230.)

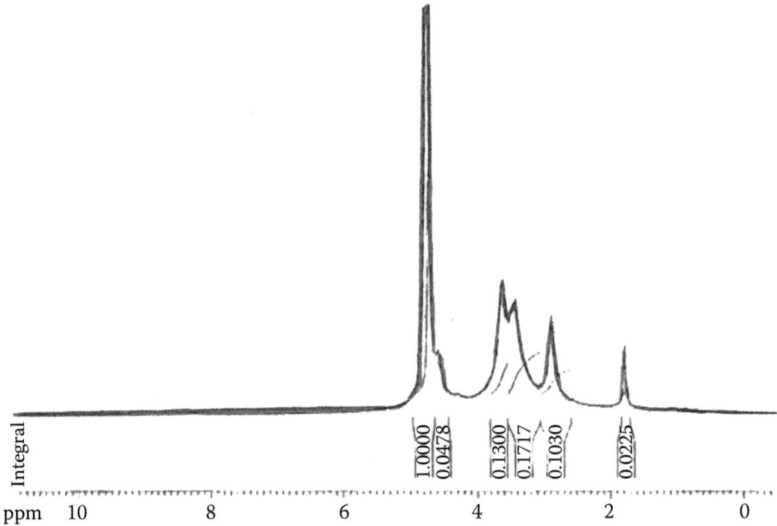

FIGURE 2.4 ¹H-NMR spectrum of chitosan. The peak at 1.9 ppm was attributed to methyl protons from the acetylated chitosan, the peak at 3.2 ppm was attributed to the proton ring connected to the C–OH group, and finally the peak at 4.9 ppm was attributed to the proton on the anomeric carbon. (Adapted from Sadeghi, A.M.M. et al. 2008c. *Journal of Bioactive and Compatible Polymer* 23: 262–275.)

2.2 USAGE OF CHITOSAN

2.2.1 Water Treatment

In the past 30 years, chitosan has been used for water purification; the positive charge of chitosan is employed to remove dyes, oil, grease, metal ions, and fine particulate matter from the water streams that may cause turbidity in water (Roller and Kovill 1999).

2.2.2 Agriculture

In agriculture, chitosan is used as coating agents for fruits and vegetables. It can be used as a hormone stimulant for plant growth and production. Moreover, chitosan can reduce disease severity in some plants by increasing biosynthesis of phenolic compounds or inducing the secondary metabolites to induce plants' immune system. It is also used to protect the plants against microorganisms in the soil (Uthairatanakij et al. 2007).

2.2.3 Cosmetics

Chitosan is used in different cosmetic products, such as skin care, hair products, and lip sticks. In skin products, the hydrating properties of chitosan reduce the water loss from the skin surface and increase skin moisture while its film-forming ability gives a pleasant smoothness to the skin (Dodane and Vilivalam 1998, Muzzarelli and Muzzarelli 1998, Bernkop-Schnurch 2002). Klingels et al. have used chitosan in skin care formulations and found that chitosan can increase water resistance and prevent drying. Also, chitosan can enhance UV filter adherence and prevent their washing out. It is also used in lip sticks to protect lips from drying, moisturizing them and helping with color adherence. Chitosan can be used in deodorants to prevent the enzyme producing bacteria because of its antibacterial effect. It has been shown that chitosan can retain fragrance in perfumes and mask perspiration odor for a long time (Klingels et al. 1999).

Chitosan can also be used in creams and lotions as emulsifying agent. Chitosan with both hydrophilic and hydrophobic segments is used as an emulsion stabilizer by adsorbing at the interfacial surface (Ramos et al. 2003).

2.2.4 BIOLOGICAL APPLICATIONS OF CHITOSAN

Owing to its interesting properties, such as biocompatibility, biodegradability, and nontoxicity, chitosan is used in various biomedical fields. Furthermore, chitosan is reported to have other properties, such as analgesic, antitumor, hemostatic, hypocholesterolemic, antimicrobial, and antioxidant properties (Koide 1998, Kumar 2000, Kumar et al. 2004). Most of these characteristics are because of the cationic nature of chitosan; however, factors such as DD and molecular weight (Mw) are essential for its performance. In their paper, Functional Characterization of Chitin and Chitosan, Aranaz et al., summarized the relationship between the biological properties of chitosan and the factors affecting those properties (Aranaz et al. 2009) (Table 2.1).

Biodegradability of chitosan is because of the presence of numerous proteases, such as pepsin, lysozyme, papain, and others, that are present in the mammalian body and are able to degrade chitosan to nontoxic oligosaccharides that can be incorporated in the metabolic pathways or be further excreted (Pangburn et al. 1982).

Biocompatibility of chitosan has been studied in numerous cell lines *in vitro*, such as fibroblasts, endothelial, epithelial, myocardial, hepatocytes, and chondrocytes, to prove that chitosan is cytocompatible (Chatelet et al. 2001). Schipper et al. conducted studies that show that chitosan with a DD above 35% is less toxic than chitosan with DD less than 35%. These studies further proved that the Mw of chitosan had no effect on its toxicity (Schipper et al. 1996).

The mucoadhesion property of chitosan depends mainly on the DD and the cell type. The interaction between chitosan and the cells is because of the presence of free amino groups on the polymer; hence, a higher DD means more free amino groups and more positive charge available for cell interaction. Moreover, cells with more negative surface charge, such as fibroblasts, show a stronger adhesion with the polymer than cells with less negative charge present on their surface, such as keratinocytes (Aranaz et al. 2009). The mucoadhesive membranes, such as the nasal, vaginal, and gastrointestinal (GI) membranes, contain mucus which is made of negatively charged glycoproteins called mucin. In the acidic environment of the stomach, chitosan becomes more positively charged and is, therefore, able to bind to the negatively charged mucus membrane by electrostatic forces. Chitosan with high Mw can penetrate further in the mucus layer and increase mucoadhesion.

TABLE 2.1

Relationship between Chitosan Biological Properties and Their Characteristics

Property	Characteristic
Biodegradability	DD, distribution of acetyl groups, Mw
Biocompatibility	DD
Mucoadhesion	DD, Mw
Hemostatic	DD, Mw
Analgesic	DD
Adsorption enhancer	DD
Antimicrobial	Mw
Anticholesterolemic	DD, Mw, viscosity
Antioxidant	DD, Mw

Source: Adapted from Aranaz, I. et al. 2009. *Current Chemical Biology* 3: 203–230.

Moreover, higher DD increases the adhesion because of increase in charge density (Lehr et al. 1992, He et al. 1998, Aranaz et al. 2009).

Studies have shown that chitosan has anticoagulant activity because of the presence of the positive charge that can interact with the negative surface charge of the red blood cells (Rao and Sharma 1997, Klokkevoid et al. 1999). Studies conducted on both high and low Mw of chitosan showed that higher Mw chitosan can bind more platelets and was more hemostatic (Jian et al. 2008).

Chitosan is thought to have analgesic effect because of the presence of free amino groups that can be protonated in the presence of proton ions in inflamed areas which will in turn decrease pH and cause the analgesic effect. The absorption of bradykinin that is released during pain may also result in analgesic effect of chitosan (Allan et al. 1984, Ohshima et al. 1987, Okamoto et al. 1993, Aranaz et al. 2009).

Junginger et al. studied the permeation-enhancing effect of chitosan. They concluded that the positive charge of chitosan can bind to the structural protein of the tight junction, disrupt their structure, and enable the opening of the tight junction and the paracellular transport (Junginger and Verhoef 1998, Kotzé et al. 1999, Smith et al. 2004). The permeation-enhancing effect of chitosan depends on its DD and Mw. According to Schipper et al., chitosan with high DD can efficiently transport mannitol molecules across the Caco-2 cell models *in vitro*; however, chitosan with DD lower than 80% can only transport the model drug if the Mw of chitosan was high (Schipper et al. 1997, 1996).

Numerous studies have been conducted on chitosan to show that it has antimicrobial activity against bacteria, yeast, and fungi. Different mechanisms were proposed for the cause of these inhibitions. Helander et al. proposed that the interaction of the positive surface of chitosan with the anionic groups on the cell surface of microorganisms prevents the essential solutes from reaching the organisms (Helander et al. 2001). Labeled chitosan oligomers with Mw 5–8 kDa were shown by Liu et al. to be able to penetrate the cell nucleus of *Escherichia coli* bacteria and prevent the RNA and protein synthesis (Liu et al. 2001). It was also suggested by Liu et al. that an increase in DD increases the electrostatic binding of chitosan to the membrane and the permeabilizing effect, whereas an increase in Mw decreases its nucleus permeation (Liu et al. 2001). Other studies have indicated that an increase in the positivity of chitosan allows it to bind to bacterial cell walls more strongly (Gerasimenko et al. 2004). Avadi et al. also conducted antibacterial studies on chitosan and its derivative Diethyl methyl chitosan (DEMC) against *E. coli*. They concluded that chitosan has a minimum inhibitory concentration (MIC) and minimum bacterial concentration (MBC) of 2500 µg/mL in 0.25% acetic acid, whereas DEMC showed an MIC and MBC of 125 µg/mL. These results suggest that because DEMC has a higher positive charge than chitosan, its antibacterial effect is more pronounced than that of chitosan itself (Avadi et al. 2004b). Different studies were conducted on the relationship of DD and Mw of chitosan and its antibacterial effect; however, there is no conclusive evidence on this relationship yet (Aranaz et al. 2009).

Studies were carried out to understand the mechanisms involved in the reduction of cholesterol by chitosan or the anticholesterolemic effect of chitosan. One hypothesis was that the viscous chitosan solution can trap the cholesterol present in the diet preventing it to be absorbed by the body. Moreover, the presence of the cationic groups on the chitosan can react with the anionic groups on the fatty acids and formation of insoluble chitosan salts (Muzzarelli et al. 2006, Aranaz et al. 2009). However at pH 7 in the small intestine, chitosan is not soluble anymore and the salt formation with cholesterol may be lost.

Thongngam et al. studied the formation of micelle-like clusters made from the interaction between the amino group and the hydroxyl group of chitosan and the anionic surface of the bile salts (Thongngam and McClements 2004, 2005). It was also proposed by Ogawa et al. that chitosan interaction with lipids form a protective layer around the lipid droplet and prevent the lipase to reach the lipid (Ogawa et al. 2003). In a study carried out by Beysseriat et al. it was shown that high Mw chitosan absorbs fat droplet more than low Mw chitosan because of higher cationic loops present in the high Mw activity and its higher surface activity (Beysseriat et al. 2006). Studies have been conducted to show a correlation between chitosan's characteristics, such as DD, Mw, and swelling capacity, and its ability to bind fat; however, no significant correlation could be predicted (Zhou et al. 2006).

Studies were also conducted to establish that chitosan has antioxidant activity against free radicals and super oxides, such as ferrous oxide. Chitosan with higher degrees of DD and lower Mw has shown to possess more antioxidant activity against super oxides. The metal ion chelating property of chitosan is probably the reason for binding the superoxide radicals and chitosan's natural antioxidant activity (Peng 1998, Xing et al. 2005).

2.2.5   BIOMEDICAL APPLICATION OF CHITOSAN

Owing to characteristics such as biocompatibility, biodegradability, and nontoxicity, chitosan has been used for different biomedical purposes, such as pharmaceutical application, wound healing, and tissue regeneration (Berger et al. 2004, Kofuji et al. 2005).

2.2.6   PHARMACEUTICAL APPLICATION

Chitosan has been used extensively in pharmaceutical industry in drug delivery systems, such as tablets, microspheres, micelles, vaccines, nucleic acids (NAs), hydrogels, and nanoparticles. In 2002, the first monograph of chitosan HCl was introduced in the *European Pharmacopeia*; however, this monograph did not distinguish the purity or special grades for biomedical applications.

2.2.7   SOLID DOSAGE FORMULATIONS

In tabletting, chitosan can be used as an excipient to delay the release of the active ingredient from the tablets. Chitosan with high Mw is more viscous and can be used as sustained release factor to prolong the duration of drug activity, improve therapeutic efficiency, and reduce side effects in oral tablets (Kofuji et al. 2005). According to Illum et al. and Jiang et al., the Mw and DD are important factors that can influence the effect of chitosan in therapeutic and intelligent drug delivery systems (Illum et al. 1994, Jiang et al. 2004). Chitosan can also be used as coating materials in drug delivery applications because of its good film-forming and mucoadhesive properties. These characteristics are suitable for controlled release of drugs during a prolonged period of time.

Chitosan-based vaginal metronidazole tablets were prepared by El-Kamel et al. These tablets showed adequate drug release as well as good adhesion properties (Kamel et al. 2002).

2.2.8   CHITOSAN MICROSPHERES

Chitosan microspheres were prepared by complexation between the cationic chitosan and the polyanionic compounds, such as tripolyphosphate (TPP) or alginates. The microspheres were loaded with different drugs to protect them from the hostile environment of the GI tract, improve the drug absorption via the paracellular pathway, and the drug release at a specific site (Bochard et al. 1996, Thanou et al. 2001a,b). The polypeptide calcitonin was used for the treatment of osteoporosis and Paget's disease in animals. These diseases require long-term therapy; therefore, a controlled release system is desirable for drug administration. Chitosan beads containing salmon calcitonin were prepared using ion gelation technique with TPP to protect the drug from the digestive degradation and to provide a sustained release profile. The results suggest that a hydrophobic drug can be incorporated in chitosan beads and the drug release from the beads follows a non-Fickian release, corresponding to a swelling controlled system (Aydin and Akbuga 1996). In a study by Avadi et al. chitosan microspheres were prepared by complexation between chitosan and TPP and loaded with Brilliant Blue as model drug to investigate the release of hydrophilic molecules from the microspheres. In their studies, they have shown that Eudragit S coated capsules containing Brilliant Blue loaded chitosan beads were suitable for colon drug delivery (Avadi et al. 2005a). Lueßen et al. and Kotzé et al. used chitosan microspheres containing different drugs on Caco-2 cell monolayer and showed an increase in transport of buserelin, insulin, and vasopressin derivative across the cells (Lueßen et al. 1997, Kotzé et al. 1997a). Mutara et al. and Shu et al. have shown in their studies that variables such as type and concentration of chitosan, pH of the TPP solution, drug concentration, gelation time, and drying conditions can all determine the fate of drug release from chitosan microspheres (Murata et al. 1999, Shu and Zho 2000).

2.2.9 Chitosan Micelles

Chitosan micelles were also studied extensively in the field of drug delivery and cancer treatment. Polymeric micelles have an internal hydrophobic core and an external hydrophilic shell. Hence, poor water-soluble drugs can be placed in the hydrophobic core to improve their solubility and bioavailability. Electrostatic interactions between oppositely charged polymers can also lead to the formation of micelles. These micelles are termed polyion complex micelles or PIC micelles (Harada and Kataoka 1995, Kataoka et al. 2001). At pH lower than 6.5, chitosan is water soluble and has polycationic properties because of the protonation of its amino groups. Studies were conducted on the formation of PIC micelles prepared from chitosan and PVP-block-poly (styrene-alter-maleic anhydride) PVP-b-PSMA by complexation between negatively charged PSMA and positively charged chitosan to make the core and be surrounded by the hydrophilic PVP shell. Coenzyme A (CoA) has been used as the model drug and the release studies from the micelles were investigated; the results suggest that PIC micelles have great potential in the delivery of CoA (Luo ct al. 2009).

Investigations were carried out by Hu et al. on stearic acid grafted chitosan oligosaccharide micelles as an intracellular delivery carrier of antitumor targeting therapy. It was shown that chitosan oligosaccharide-stearic acid (CSO-SA) complexes with lower amino substitution degree or lower Mw of CSO were more stable than those with higher amino substitution degree or Mw of CSO. Moreover, the acidic pH and ionic strength resulted in more stable complexes (Du et al. 2010). Doxorubicin conjugated stearic acid-grafted chitosan oligosaccharide (CSO-SA) were shown to be able to effectively suppress the tumor growth and reduce the toxicity in comparison to the commercial doxorubicin hydrochloride injection (Hu et al. 2009). These results are mainly because of the small size of the micelles and their hydrophilic surface that protect them from the mononuclear phagocyte systems (MPS) (Li and Huang 2008). Because of the passive enhanced permeability and retention (EPR) of the particles, the polymeric micelles accumulate in the tumor tissue (Omelyaneko et al. 1998). The releases of the drug from the micelles were shown to be more pronounced in the acidic environment which is favorable in the tumor tissue where the pH of the environment is mildly acidic than the healthy tissue (Hu et al. 2009). These CSO-SA micelles were further modified by polyethylene glycol (PEG). The results suggest that the uptake of the PEGylated micelles by the macrophage cell lines RAW 264.4 were reduced substantially compared to the non-PEGylated micelles. However, there was no obvious increase in the uptake of the mitomycin C model drug in rat liver or human liver tumor cell line *in vitro* when the particles were PEGylated (Hu et al. 2008). The PEGylation will improve the stability of micelle circulation in the blood stream because of the steric hindrance of the PEG chains (Shibata et al. 2004). Moreover, PEGylation improves the stability of the delivery system by preventing drug absorption and uptake by reticuloendothelial systems (RES) (Patel 1992, Hong et al. 2004). The modifications of the chitosan molecule were also shown to overcome the anticancer drugs, such as paclitaxel (PTX) from burst releasing and immature leakage in the blood stream (Hou et al. 2010).

2.2.10 Chitosan in Nasal Vaccines

It is now proven that the nasally administered vaccines can stimulate both cell-mediated and humoral immune responses (Illum et al. 2001). The main target site of the nasally administered vaccines is thought to be the nasal-associated lymphoid tissue (NALT) present in the pharynx (Figure 2.5). The epithelium of NALT is made of ciliated epithelial cells, mucous goblet cells, and nonciliated cells. Below the epithelial surface B-cell follicles, inter-follicular T cells as well as macrophages and dendritic cells are present in a loose reticular network (Frieke Kuper et al. 1992, Illum et al. 2001).

Cationic polysaccharides, including chitosan, are used as a delivery system for nasally administered vaccines. Positively charged chitosan can bind strongly to the negatively charged mucous present on the nasal epithelium by electrostatic interaction (Illum et al. 2001). It is believed that the bioadhesive property of chitosan can decrease the clearance of the vaccine from the nasal cavity by enhancing the adhesion of the vaccine formulations to the nasal surface. Furthermore, the positive charges of the chitosan molecule can react with the negatively charged proteins such as F-actin filaments of the tight junctions to disrupt their structure and open the tight junction allowing for the paracellular transport of the antigen molecule across the membrane (Dodane et al. 1999).

Nasal chitosan-based influenza, pertussis, and diphtheria vaccines were studied extensively by Illum et al. According to their studies, the nasal administration of chitosan–antigen complex was able to induce the secretion of IgG to a similar level as the parenteral formulations, whereas the secretion of the IgA was more pronounced in all cases. Moreover, the vaccinated animals showed immunity against the disease after the challenge test (Illum et al. 2001).

Günbeyaz et al. conducted studies on nasal delivery of bovine herpesvirus1 (BHV-1) using chitosan microparticles by spray drying method. BHV-1 is a major cause of conjunctivitis, genital infections, and respiratory infections in cattle (Nandi et al. 2009), and bovine respiratory disease, a major concern in cattle industry worldwide (Ackermann and Engels 2006). They have developed chitosan microparticles with sizes below 10 μm, positive surface charge, and high loading efficiency that could form gel and attach to the nasal mucosa for an extended period of time. The uptake of FITC labeled particles was investigated in Madin Darby bovine kidney (MDBK) cell line and the confocal laser scanning microscopic imaging suggests that particles were seen in the nucleus of the cells. Thus, they have concluded that chitosan-based gel and microparticulate systems maybe promising adjuvant/delivery systems for the delivery of antigens (Günbeyaz et al. 2010).

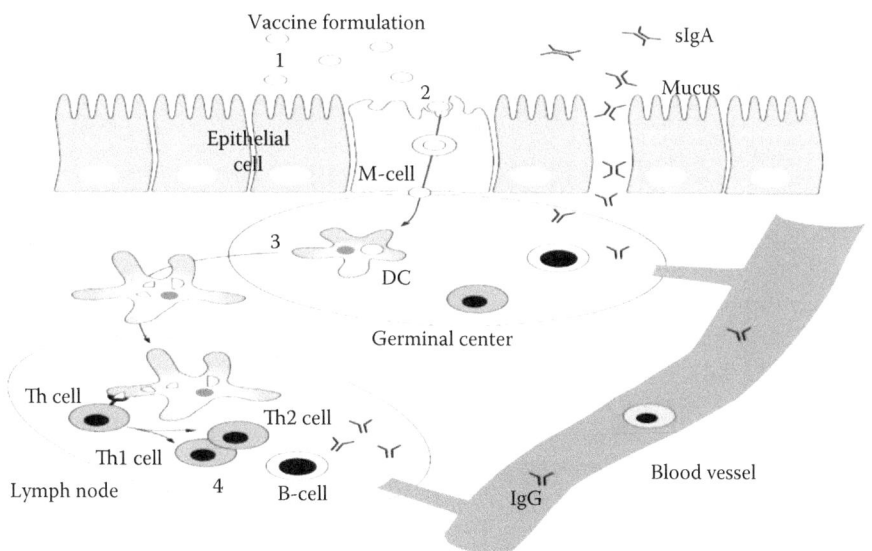

FIGURE 2.5 Schematic overview of the consecutive steps toward successful nasal vaccine delivery. (1) muco-adhesion; (2) antigen uptake, by M-cell transport; (3) delivery to and subsequent activation/maturation of DC; (4) induction of B- and T-cell responses. DC, dendritic cell; M-cell, microfold cell; Th cell, helper T cell. (Adapted from Slütter, B., N. Hagenaars, and W. Jiskoot. 2008. *Journal of Drug Targeting* 16: 1–17.)

Other studies were carried out by Illum and her coworkers on chitosan-based DNA flu vaccines prepared by polyelectrolyte complexation between positive chitosan and negatively charged DNA at a ratio that favors a positive zeta potential for stability of the particles. The result suggested a boost in immune response in comparison to the naked DNA (Illum et al. 2001). The increase in immune response is mainly because of its absorption-enhancing effect and not because of chitosan itself.

2.2.11 Chitosan in Oral Vaccines

Investigations were conducted to demonstrate the effectiveness and safety of chitosan as absorption enhancer to improve oral vaccine delivery. The vaccine–chitosan complex can enhance the antigen uptake by the gut-associated lymphoid tissues (GALT) and induce strong systemic and mucosal immune responses against antigens, similar to the nasal vaccination. Although oral vaccination has several advantages, such as lower cost of production, ease of production, and better patient compliance over the parenteral injection, degradation of vaccine in the gut and low uptake in the lymphoid tissues of the GI tract are hurdles that must be overcome before development of suitable oral vaccines. Studies by van der Lubben et al. using chitosan microparticles loaded with ovalbumin model drug have suggested that particles with size around 1 µm are able to be taken up by the M-cells or the macrophages of the Peyer's patches (PP). More than 90% of the loaded ovalbumin was entrapped in the porous structure of chitosan and was released intracellularly after digestion in the PPs. The mucosal absorption of the particles was thought to be because of the permeation-enhancing effect of chitosan through paracellular transport. These results suggest that chitosan drug delivery systems are promising candidates for oral vaccination (Van der Lubben et al. 2001a,b). The use of chitosan in oral drug delivery is, however, limited owing to its low solubility at alkaline pH corresponding to the intestine. Hence, alkylated derivatives of chitosan were synthesized that are soluble in alkaline pH and used for oral drug delivery and oral vaccination. Baudner et al., evaluated chitosan microparticles as intranasal vaccine delivery for group C meningococcal conjugated vaccine (CRM-MenC) with the mucosal adjuvant LTK63, a nontoxic *E. coli* enterotoxin mutant, in mice. Their studies have indicated that mice receiving chitosan microparticles and the LTK63 mutant showed higher systemic and mucosal antibodies against meningococcal polysaccharide in comparison with mice receiving subcutaneous vaccine. Moreover, higher bactericidal activity was observed in the serum of the mice vaccinated with the microparticle conjugated vaccine along with the adjuvant. Their result demonstrated that the concomitant use of chitosan microparticles and the LTK 63 mutant significantly enhance the immunogenicity of the intranasal vaccine (Baudner et al. 2003, 2005).

2.2.12 Chitosan-Mediated Nucleic Acid Transfer

During last decade, NA therapy has been emerging for the treatment of various genetic disorders. The NA encoding genes include metabolic suicide genes, genes encoding the angiogenesis inhibitors, proapoptotic proteins, pro-drug activators or immune response modulators, small interfering RNA (siRNA), and antisense RNA. Owing of the properties such as bioavailability, biodegradability, nontoxicity, and the ability to bind plasmid DNA (pDNA) through ionic interactions, chitosan has been studied extensively as DNA carrier. The positive surface charge of chitosan allows it to interact with the negatively charged NA. Chitosan particles were prepared using different methods, such as ionic gelation, coacervation, covalent cross-linking, and desolvation (Berthold et al. 1996, Ohya et al. 1999, Mao et al. 2001, Hamidia et al. 2008). Different parameters such as the preparation technique used, pH, the salt concentrations, NA concentrations, and temperature can all determine the size of the particles as well as their transfection efficacy (Lai et al. 2009).

Mumper et al. were one of the first groups of researchers who worked on chitosan as a carrier for pDNA (Mumper et al. 1995). Ever since, many investigations were conducted on chitosan vectors for carrying pDNA; however, to date no chitosan vector was identified that could overcome all the hurdles involved in the process of NA therapy. These problems include enzymatic degradation of NA, inefficient cellular uptake, lysosome encapsulations, failure of NA/chitosan dissociation, and nuclear localization (Koping-Hoggard et al. 2001, Davis 2002, Kaneda 2005, Lai et al. 2009). To overcome some of these problems, different modifications were made on the chitosan vector. Chitosan grafted by PEG was shown to have less aggregation and more transfection efficiency. This is mainly because of reduction of opsonization of the chitosan polyplex as well as prolongation of blood circulation of the vector. Zhang et al. have conducted studies to show that pegylated chitosan vectors could not only avoid the clearance by the Kupffer cells but could also increase the rate of transfection of the vector to the liver tumor cells (Zhang et al. 2007).

Polyethylenimine (PEI) was also used by Wong et al. to modify the chitosan vector. Their studies have shown that the rate of transfection was increased by threefold in comparison to the unmodified chitosan vector (Wong et al. 2006). Moreover, investigations by Lai et al. using PEI modification of chitosan vector have indicated a 1000-fold increase in transfection efficiency of 293A kidney cell line compared with standard chitosan in the presence of 10% fetal bovine serum (Lai et al. 2009). The result is presented in Figure 2.6.

Chitosan conjugation with another polymer usually results in an increase in size of the vector which is not favorable for transfection efficiency; hence, functional group modification of chitosan is another method for vector synthesis. Zheng et al. used the trimethyl chitosan (TMC) derivative vector containing pDNA encoding green fluorescent protein (GFP) to study the efficiency of transfection of the GI tract of the nude mice by GI mucosa administration. They have shown that GFP was expressed mainly in the gastric and upper intestinal mucosa (Zheng et al. 2007). Other derivatives of chitosan, such as PEG-graft-trimethyl chitosan, were shown to be able to increase the transfection in several cell lines *in vitro* compared with the nonpegylated chitosan derivative (Germershaus et al. 2008).

Ligand modification of chitosan vectors were also studied extensively for their effect on transfection efficiency. Several proteins and peptides can be conjugated to chitosan to boost the transfection efficiency in different cell lines. For example, conjugation of KNOB protein, an essential protein to recombinant human adenoviral vector, can increase the transfection efficiency to about 130-folds and 7-folds in Hela and HEK 293 cell lines, respectively. Moreover, the conjugation of transferrin, an iron binding glycoprotein functioning as targeting moiety in some cancer cells, can also increase transfection by 4-folds in both Hela and HEK 293 cell lines (Mao et al. 2001). There are numerous proteins and peptides with functions suitable for vector conjugation but because of the instability of the produced vector, elicitation of immune responses, technical difficulties, and so on they have not been used to conjugate chitosan vectors (Lai et al. 2009).

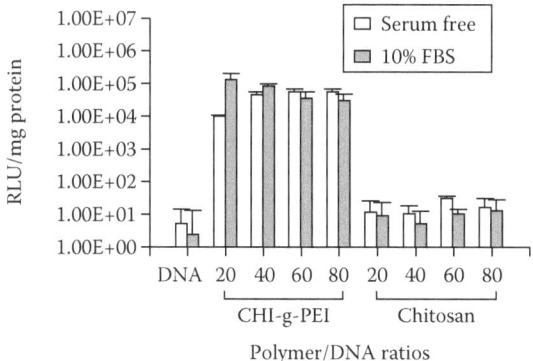

FIGURE 2.6 Luciferase activity assay of CHI-g-PEI and standard chitosan on 293A kidney cells 48 h after transfection in the presence and absence of 10% fetal bovine serum. The MW of PEI used in the fabrication of CHI-g-PEI is 600 Da. (Reproduced from *Journal of Controlled Release* 134, Lai, W-F. et al., Nucleic acid delivery with chitosan and its derivatives, 158–168. Copyright 2009, Elsevier.)

TABLE 2.2

Comparison of Different Strategies in Chitosan Vector Derivation

Strategies	Advantages	Disadvantages
Structural Modification		
Copolymerization	Traditional practice in polymer chemistry Well-supported by accumulated knowledge and experiences of the predecessors	Choice of conventionally used polymeric modifiers could be limited. Significant increase in the size of the NA carrier
Functional group modification	No significant increase in the size of the NA carrier after the modification process in general. Practically unlimited possibility in vector derivation	Require sophisticated consideration of SAR; however, the knowledge of SAR in vector design is momentarily deficient. Involve potentially intricate chemical conversion processes
Ligand Conjugation		
Proteinaceous	No sophisticated forethoughts of complicated chemical conversion processes are generally needed Proteins and peptides can possess diverse functionalities A wide choice of possible ligand candidates	The resultant conjugated vector might be instable Immunogenic Expensive
Nonproteinaceous	No sophisticated forethoughts of complicated chemical conversion processes are generally needed Cheaper than proteinaceous ligands	Choice of conventionally used ligands could be limited

Source: Reproduced from *Journal of Controlled Release* 134, Lai, W-F. et al., Nucleic acid delivery with chitosan and its derivatives, 158–168. Copyright 2009, Elsevier.

Nonprotein ligands, such as hexoses, were also used to conjugate chitosan vectors. Manosylated chitosan (MC) was produced by Kim et al. and shown to be able to induce mannose receptor-mediated endocytosis by enhancing the transfer of pDNA into dendritic cells *in vitro*. Further studies showed that the injection of the MC/pDNA encoding murine IL-12 into BALB/c mice with CT-26 carcinoma suppressed the tumor growth and angiogenesis. The expression of proapoptotic molecules, such as p21 and p27, was elevated, whereas the expression of proliferating cell nuclear antigens, cyclin D1 and cyclin-dependent kinase 4, and antiapoptotic proteins were reduced. The protein expression changes were mainly because of cell cycle arrest and apoptosis after transfection (Kim et al. 2006). More studies were conducted on conjugation of folic acid (FA), deoxycholic acid (DCA), and hyaluronic acid (HA) to chitosan. These investigations indicated that such conjunctions enhanced the transfection efficacy compared with chitosan vectors (Liu and Yao 2002, Mansouri et al. 2004). The nonproteinaceous ligands are more stable and less immunologic than the protein and peptide vectors; however, the choice of these nonproteinaceous ligands is limited. Table 2.2 summarizes the advantages and disadvantages of different strategies used to modify chitosan vectors.

2.2.13 CHITOSAN-MEDIATED RNA TRANSFER

Chitosan can also deliver RNA by increasing its cellular uptake and preventing degradation of naked RNA. Chitosan vector modifications were carried out to enhance RNA transfer. Alpar et al. were among the first groups who modified chitosan through ionic gelation of chitosan salts with TPP to produce nanoparticles. These nanoparticles were more efficient in transferring siRNA compared with the nonmodified chitosan mainly because of more efficient binding capacity and load capacity of RNA (Katas and Alpar 2006). Other studies were conducted on thymine pyrophosphate conjugated chitosan that showed up to 70% transfection of siRNA for silencing the endogenous EGFP gene in HepG2 cells. This high transfection is probably because of an enhanced binding of RNA to chitosan as well as an increase in solubility of the complex in the presence of extra amine groups (Rojanarata et al. 2008). Chitosan has also been coated with PLGA or polyisohexylcyanoacrylate for the delivery of siRNA; these investigations have shown the potential of chitosan to deliver siRNA to the site of the tumor and the ability of siRNA as oncological NA therapy (Pille et al. 2006, Nafee et al. 2007).

2.2.14 CHITOSAN HYDROGELS

In the past decade much attention has been given to the development of biocompatible and biode-gradable hydrogels. These hydrogels are three-dimensional hydrophilic polymers which can absorb and retain up to thousands of times more fluids than their dry weights and are suitable carriers for macromolecules. Due to its biocompatibility, nontoxicity, and biodegradability, chitosan is receiv-ing special attention as a hydrogel. Like most hydrogels, chitosan contains a lot of water and some of this water is bound tightly to the polymer and the rest is found as free water. Chitosan hydrogels can swell and dehydrate depending on the environment and the composition of the polymers (Rohindra et al. 2004). Hydrogels can play a crucial role in malfunctions and injuries of living tissues. The high water content of hydrogels renders them compatible with the host tissues. They can induce the healing process of the damaged tissues and mimic functional and morphological characteristic of the organ tissues. Chitosan hydrogels can be prepared by different methods; in each method chito-san is either physically bound or chemically cross-linked to form hydrogels (Dash et al. 2011).

A stable hydrogel must be sufficiently strong to form a semi-permanent junction in molecular network and also be able to exchange water with the polymer network. Chitosan hydrogels can be prepared physically by ionic, polyelectrolyte, interpolymer, and hydrophobic associations that result in gelation of a chitosan solution (Figure 2.7). Although the preparation of hydrogels with the physical association is easy, their application is limited because of weak mechanical strength and uncontrolled dissolution (Dash et al. 2011).

The physically bonded hydrogels have the advantage of gel formation without the use of cross-linking but they often provide inconsistent *in vivo* results due to a lack of control in gel pore size, dissolution, and degradation properties. To obtain better mechanical results, irreversible networks can be produced by covalently binding the polymer to the cross-linker molecules. These cross-linker molecules include enzyme-sensitive or photo-reactive molecules. The cross-linker molecules

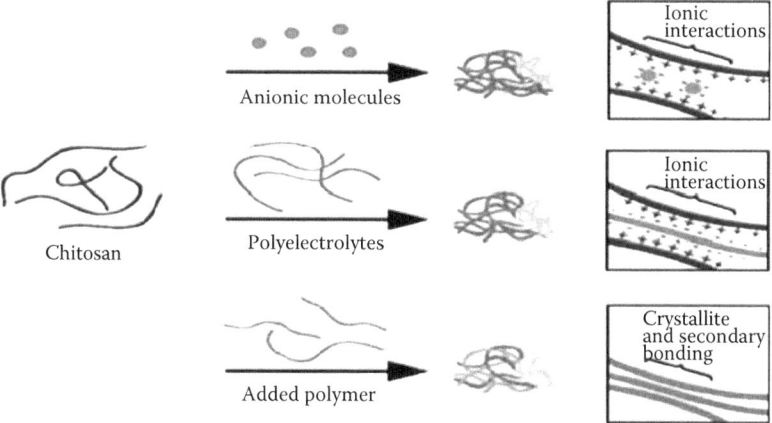

FIGURE 2.7 Schematic representation of chitosan hydrogel networks derived by physical associations: networks formed with ionic molecules, polyelectrolyte polymer, and neutral polymers. (Reproduced from *Progress in Polymer Science* 36, Dash, M. et al., Chitosan—A versatile semi-synthetic polymer in biomedical applications, 981–1014. Copyright 2011, Elsevier.)

can be attached to the available NH$_2$ and –OH groups on chitosan by amide and ester bonding as well as Schiff base formation (Dash et al. 2011).

Investigators have cross-linked chitosan with other polymers to improve the physical and mechanical properties of chitosan hydrogels. Glutaraldehyde cross-linked chitosan hydrogels were prepared by Rohindra et al. to determine the swelling behavior at different temperature and pH. The comparison between the noncross-linked and cross-linked chitosan hydrogels indicated that the noncross-linked chitosans showed a maximum swelling ratio at all temperatures and pH. The amount of swelling was, however, decreased by increasing the degree of cross-linking (Rohindra et al. 2004). Drug loading in hydrogels is achieved by diffusion, entrapment, and tethering. The easiest way to load drugs in hydrogels is by diffusion where hydrogels were placed in a medium saturated with drug. The therapeutic agent diffuses slowly into the gel depending on the porosity of the hydrogels, size, and chemical characteristics, such as hydrophilicity and hydrophobicity of the drug and hydrogel. Once the hydrogel is placed *in vivo* the drug can diffuse into the neighboring tissues. The diffusion method is only to load small drugs. For larger molecules, the drugs can be entrapped during the gelation process. The drug is usually mixed in a polymer solution and a cross-linking or complexation agent is added. The drug release is mainly a burst release *in vivo* because of concentration gradient between the hydrogel and the environment. To overcome the burst release and the possible toxicity, the drugs can also be covalently or physically linked to the polymer before gelation. This is called tethering which can release the drugs only when the hydrogel or the molecular tether is broken (Dash et al. 2011).

Chitosan hydrogels were not only used in controlled delivery systems, but also in tissue engineering. Tissue analogs, such as cartilage, bone, liver, and nerve tissues, were prepared by tissue engineering (Dash et al. 2011).

Fukuda et al. designed a novel photocrossed chitosan to generate spatially controlled 3D co-culture systems. These systems allow 3D cell and are also suitable for co-culture of additional cell types because of the adhesive properties of the polymers. The result of their investigation suggested that cellular attachment properties were changed from cell repulsive to cell adhesive. Because of the properties of chitosan hydrogel, co-cultures of spheroids and support cells were generated. Hepatoblastoma cells and HepG2 cells were initially seeded in the wells to form spheroids that secreted albumin which further induced the adhesion of fibroblastic cells to chitosan hydrogels. Both cell lines were thus co-cultured and proliferated on the surface of hydrogels. These controlled co-cultured systems could be used for biomimicking cellular growth in 3D environment (Fukuda et al. 2006).

The toxicity of chitosan hydrogels for subcutaneous and intraperitoneal implantation in rats was carried out by Azab et al. Tissue damage was not detected in distant organs, such as brain, heart, liver, spleen, and kidney, after subcutaneous implantation of the hydrogels containing a radioisotope. The degradation of the hydrogels was thought to be because of oxidation of chitosan. The result of their study suggests that these hydrogels could be suitable candidates for use in brachytherapy delivery in cancer patients (Azab et al. 2007). It was shown that chitosan can promote cell growth in cell culture. For this reason it has been used extensively in tissue engineering. Luca et al. studied the role of chitosan hydrogels for orthopedic applications in bone defects. They have developed a thermosensitive chitosan-based hydrogel that formed an implant *in situ* after subcutaneous injection in rats. These hydrogels contained recombinant human bone morphogenetic protein (rhBMP-2) and platelet-rich plasma (PRP)-containing growth factors such as platelet-rich plasma growth factor-BB. According to the results, the ectopic bone formation was induced by rhBMP-2 chitosan carrier. These chitosan hydrogel carriers are thought to be promising carriers for injured bone regeneration in the future (Luca et al. 2011).

2.2.15 CHITOSAN NANOPARTICLES

In the past decade, several investigations have focused on nanoparticles and their role as drug delivery vehicles. Nanoparticles were first introduced in the mid-1970s by Birrenbach and Speiser (1976). Polymeric nanoparticles possess unique properties by themselves and additionally have the advantage of protecting the protein and peptide drugs from the chemical and enzymatical degradation in the GI tract, increasing their stability and absorption across the intestinal epithelium as well as controlling the drug release (Couvreur et al. 1979, Takeuchi et al. 1996, Vila et al. 2002, Krauland and Alonso 2007). Chitosan nanoparticles with excellent biodegradable and biocompatible properties have been used extensively as drug delivery vehicles. Chitosan nanoparticles have been found appropriate for noninvasive routes of drug administration: ocular, nasal, oral, and pulmonary routes. These applications are facilitated by the absorption-enhancing effect of chitosan. Additionally, chitosan nanoparticles have been proposed as nonviral vectors in gene therapy and have shown adjuvant effect in vaccines. Nanoparticles can be prepared by different techniques, such as polymerization, nanoprecipitation, and inverse microemulsion; however, most of these methods involve the use of organic solvents, heat, and vigorous agitation that maybe harmful for the protein and peptide drugs (Calvo et al. 1997, Damg et al. 1998). In 1997, the first chitosan nanoparticles were prepared by Alonso et al. using ionotropic gelation of chitosan (polycation) with TPP as polyanion. The protein or peptide drugs were dissolved in either the TPP or chitosan and the ionic interaction of the negative TPP with the positive amine group of the chitosan resulted in a nanosuspension containing the drug (Mao et al. 2006). Mao et al. used the polyelectrolyte complexation (PEC) technique to directly bind positively charged chitosan to the negatively charged protein or peptide drugs via electrostatic interactions (Fattal et al. 2002). The particle size and surface charge are critical factors in nanoparticle absorption and their fate. Studies have shown that particles with a size of 100 nm were taken up 6 times more than particles with 100 μm by the absorptive cells (Delie 1998). Positively charged nanoparticles are more likely to be absorbed by the absorptive cells of the GI tract. This is because of interaction of the positively charged particles with the negatively charged mucous lining of the GI tract. This interaction can disrupt the protein structure of the tight junctions and opening them to enhance the paracellular transport of the nanoparticles. Alonso and coworkers used ionotropic interaction to prepare chitosan nanoparticles loaded with insulin. Their results showed that the loading capacity of insulin was 55% and the nanoparticles were able to enhance the nasal absorption of insulin in rabbits (Urrusuno et al. 1999). Chitosan nanoparticles loaded with tetanus toxoid were also used in nasal vaccine delivery and the result suggested that the antitetanus IgG and IgA levels were increased in comparison with the fluid vaccine (Vila et al. 2004). Chitosan-based cyclosporine A nanoparticles were shown to reach the ocular mucosa of rabbits better than chitosan solution containing the drug. These results suggest that chitosan nanoparticles may be suitable vehicles for the enhancement of the therapeutic index of challenging drugs for the application to the ocular mucosa (de Campos et al. 2001). Avadi et al. prepared insulin nanoparticles by the ionic gelation technique using Arabic gum and chitosan. The particles were about 170–200 nm in size and carried a positive surface charge. They studied the mucoadhesion properties of the nanoparticles in excised rat jejunum. The result suggests that nanoparticles can enhance the permeation of insulin across the intestinal epithelium compared to the free insulin solution (Avadi et al. 2011). Lim et al. conducted studies on chitosan–insulin nanoparticles prepared by ionic gelation. The nanoparticles were administered intragastrically by a gavage needle and blood samples were collected from diabetic rats. The results showed that the nanoparticles are able to lower the blood glucose level; in contrast, rats fed with the control insulin, insulin–chitosan solution, and chitosan solution did not show any significant decrease in serum glucose level. Moreover, an increase in the serum insulin level was not observed which is probably because of the local action of insulin nanoparticles in the intestinal epithelium (Ma et al. 2005).

Intratracheal administration of glycolated chitosan nanoparticles has been shown to have a rapid uptake into the systemic circulation and being excreted via urine peaked at 6 h post-administration. These particles were shown to induce transient neutrophilic pulmonary inflammation up to 3 days post-administration. An increase in pro-inflammatory cytokines IL-1β, IL-6, and TNF-α and chemokine MIP-1α was observed. These findings suggest that chitosan nanoparticles can be used as pulmonary delivery vehicles due to their unique properties such as biocompatibility, transiency, low toxicity, and rapid elimination (Choi et al. 2010).

Although chitosan nanoparticles have been studied extensively as drug carriers for administration in different routes, more investigation is still needed to determine their safety, efficiency as therapeutic agents, drug release, and optimal drug targeting.

2.3 CHITOSAN DERIVATIVES

Although chitosan has been used extensively and successfully in different biomedical and industrial applications, a major drawback of this polymer is that it is insoluble at physiological pH. Chitosan is soluble and active as permeation enhancer agent in acidic environment when it is protonated. Hence, different alkylated derivatives of chitosan have been synthesized. These partially quaternized chitosan derivatives show good water solubility in a wide pH range and have been used in place of chitosan.

2.3.1 SYNTHESIS AND CHARACTERIZATION OF TRIMETHYL CHITOSAN

Perhaps trimethyl chitosan (TMC) is the most studied derivative of chitosan to date. It was initially synthesized by Domard et al. in 1986, to improve the solubility of chitosan by reductive methylation of chitosan with methyl iodide in a strong acidic medium at high temperatures following several steps to obtain chitosan derivative with a final degree of substitution of approximately 60%; this derivative was found to be water soluble at all pH values (Domrad et al. 1996). Almost a decade later, Sieval et al. synthesized TMC with minor modifications to the above method to obtain TMC with different degrees of quaternization between 40% and 80%, depending on the reaction steps and methyl iodide concentration used in the reaction (Sieval et al. 1997). Today, TMC is mainly synthesized according to the following method (Figure 2.8). Briefly, low molecular weight chitosan is dispersed in a basic solution of methylpyrrolidone and sodium hydroxide, methyl iodide, and sodium iodide are, subsequently, added at 60°C for 6 h. The polymer is then precipitated with acetone and separated by centrifugation. For exchanging I⁻ with Cl⁻, the polymer is dissolved in aqueous NaCl (5.0%, w/w) solution and is precipitated with acetone and dried to obtain a water-soluble white powder with quantitative yield of 98% (Sadeghi et al. 2008a) (Table 2.3).

The solubility was improved because of the substitution of the primary amine with methyl group and prevention of the hydrogen bond formation between amine and the carboxylic groups of the chitosan backbone (Figures 2.9 and 2.10).

The absolute molecular weight of TMC decreased with an increase in the degree of quaternization which is mainly because of the degradation of polymer chain caused by an exposure to the harsh reaction conditions during the synthesis (Snyman et al. 2002). The ^{1}H-NMR spectrum of the TMC showed that the degree of quaternization was higher when sodium hydroxide was used as the base in comparison to other bases and also with increase in reaction steps.

Snyman et al. investigated the mucoadhesive properties of TMC (degree of substitution 22–49%) and concluded that TMC was less mucoadhesive in comparison with chitosan and its salts but was more mucoadhesive than pectin polymer. They have reasoned that the decrease in mucoadhesiveness of TMC in comparison to chitosan salts is probably due to the fixed positive charges of the quaternary amine group that may decrease the flexibility of the polymer backbone (Snyman et al. 2002, 2003).

2.3.2 ROLE OF TMC AS PERMEATION ENHANCER

The role of TMC as permeation enhancer was first investigated by Kotzé et al. in Caco-2 cell (human colon carcinoma cell line) as a model for intestinal epithelium at pH 6.7 using [^{14}C] mannitol, fluorescent labeled dextran 4400, and polypeptide drug buserelin. The result suggests an increased transportation rate of 32–60, 167–373, and 28–73 folds, respectively. The confocal laser scanning

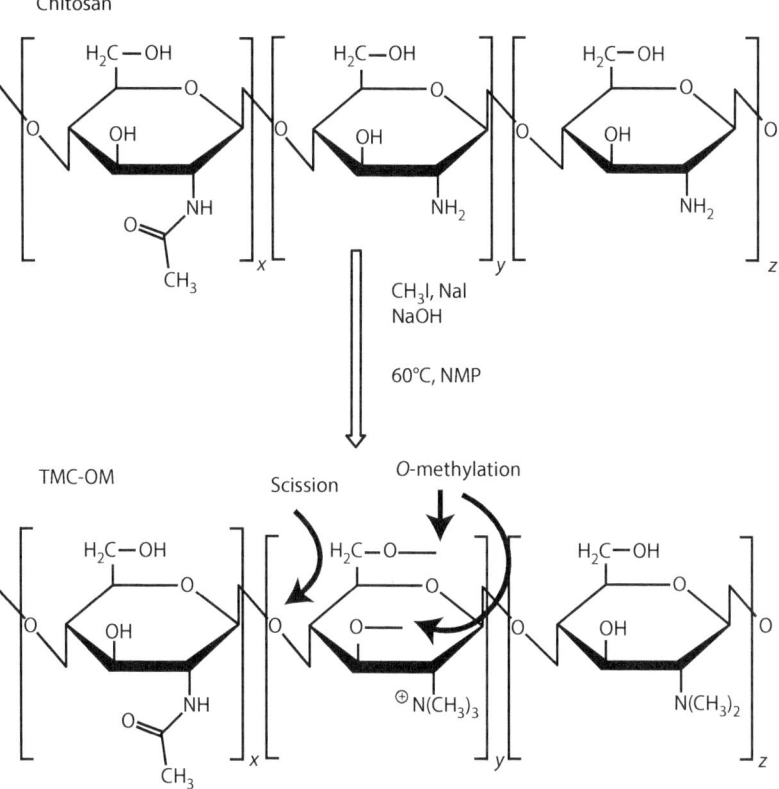

FIGURE 2.8 Synthetic pathway for the preparation of TMC according to the method of Sieval et al. (Reproduced from *Carbohydrate Polymer* 36(2–3), Sieval, A.B. et al., Preparation and NMR characterization of highly substituted *N*-trimethyl chitosan chloride, 157–165. Copyright 1997, Elsevier.)

microscopy (CLSM) revealed that the increase in drug transport was because of the opening of the tight junctions and an increase in paracellular transport and not the intracellular transport (Kotzé et al. 1997b). Furthermore, Kotzé and coworkers conducted studies to compare the permeation-enhancing effect of chitosan chloride and TMC with different degrees of substitution on [^{14}C] mannitol transport across Caco-2 cell layer. They have concluded that while chitosan chloride had no effect on the transport of the labeled mannitol at neutral pH, TMC with higher degree of methylation (60%) was more successful in transporting the mannitol across the membrane and decreasing

TABLE 2.3

Result of TMC Nanoparticles Characterization Using Both the Ionotropic Gelation Method and PEC Method

Polymer	% Insulin Loading	Size (nm)	Polydispersity	Zeta Potential
TMC (ionotropic)	55 ± 5	215	0.1	+22.0
TMC (PEC)	70 ± 5	195	0.32	+29.0

Source: Reproduced from *International Journal of Pharmaceutics* 35, Sadeghi, A.M.M. et al., Preparation, characterization and antibacterial activities of chitosan, N-trimethyl chitosan (TMC) and *N*-diethylmethyl chitosan (DEMC) nanoparticles loaded with insulin using both the ionotropic gelation and polyelectrolyte complexation methods, 299–306. Copyright 2008a, Elsevier.

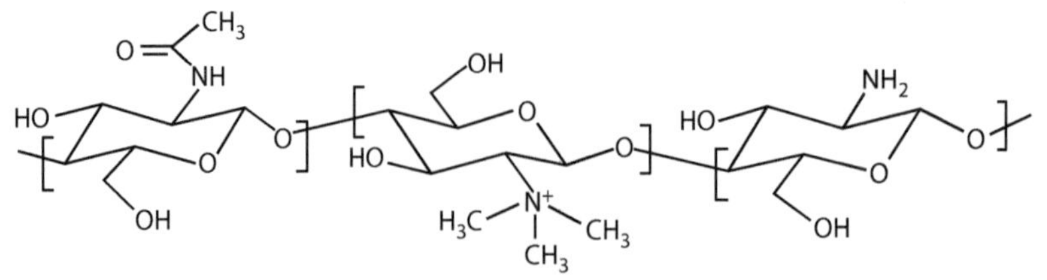

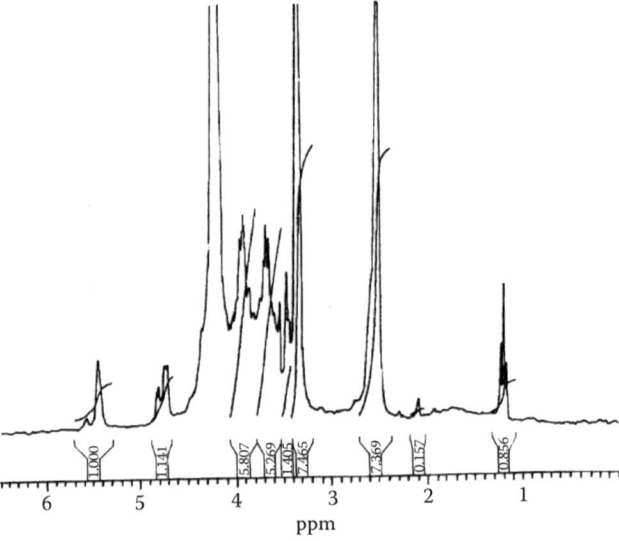

FIGURE 2.9 The chemical structure of TMC. (Reproduced from *Journal of Controlled Release* 64, Thanou, M.M. et al., Effect of degree of quaternization of *N*-trimethyl chitosan chloride for enhanced transport of hydrophilic compounds across intestinal Caco-2 cell monolayers, 15–25. Copyright 2000a, Elsevier.)

the transepithelial electrical resistance (TEER) value than TMC with lower degree of methylation (12%). This suggests that a threshold value of positive charge on the polymer is required to be able to interact with the tight junction and increase the paracellular transport (Kotzé et al. 2000). In further studies by Thanou et al. TMC was synthesized with degrees of substitution of 40% and 60% and the transepithelial transport of [^{14}C] mannitol as the model drug was investigated in Caco-2 cell monolayer. The result suggested that TMC with 60% substitution showed more paracellular transport of the model drug than the 40% TMC at different concentrations of 0.05–1.0% w/v used. The nucleic staining suggested that the increased transport is not because of cytotoxicity but because of the opening of the tight junctions and the paracellular transport. They have concluded that while both TMCs were able to transport the model drug, a higher charge density (TMC 60%) was more favorable for the transport of drugs across the intestinal epithelium (Thanou et al. 2000a).

The intestinal absorption of octreotide, a somatostatin analogue used for the control of endocrine tumors of the GI tract and the treatment of acromegaly, was investigated in pigs by Thanou et al. using 10% TMC with a degree of substitution of 60%. The *in vivo* studies suggested that TMC as permeation enhancer was able to increase the absorption of Octreotide by 14.5-folds and the bioavailability by 24.8 ± 1.8% (Thanou et al. 2001c). The effect of TMC and chitosan as permeation

FIGURE 2.10 The ^{1}H-NMR spectrum of the TMC. The signal at 1.9 ppm is attributed to the acetyl group of the chitin; the peak at 3.6 represents the N(CH$_3$)$_3$ group together with a smaller peak at 3.4 ppm assigned to the N(CH$_3$)$_2$ group. (Reproduced from *Journal of Controlled Release* 64, Thanou, M.M. et al., Effect of degree of quaternization of *N*-trimethyl chitosan chloride for enhanced transport of hydrophilic compounds across intestinal Caco-2 cell monolayers, 15–25. Copyright 2000a, Elsevier.)

enhancers was also investigated in Wistar rats using buserelin. The results suggested that the absolute bioavailability of buserelin applied intraduodenally was 0.8% and increased to 13% with coadministration of 1.0% TMC. Meanwhile, chitosan at pH 7.2 had no significant effect on the intestinal absorption of buserelin (Thanou et al. 2000b). Furthermore, permeation studies using TMC 20% and 60% in reconstituted Calu-3 cell monolayers *in vitro* and intratracheal instillation in rats *in vivo* using octerotide suggested a linear *in vitro/in vivo* correlation between calculated absorption rates, suggesting that the permeation enhancement by polysaccharides, both *in vitro* and *in vivo*, proceeds via an analogous mechanism (Florea et al. 2006).

Sadeghi et al., designed a novel Gas Empowered Drug Delivery (GEDD) system for CO_2 forced transport of insulin, polyethylene oxide (PEO) as a mucoadhesive agent, and TMC as the permeation-enhancing polymer to the surface of the small intestine (Figure 2.11). The *in vivo* studies were done in male rabbits using formulations containing no TMC or PEO, only PEO and both PEO and TMC. The results suggested a higher bioavailability and a 5.5-fold increase in insulin transport across the intestinal membrane with the formulation containing both PEO and TMC. Although, formulations containing only PEO and no TMC also showed increase in insulin transport, but the results were not as pronounced as when TMC was also used in the formulation. They have thus concluded that the increase in insulin permeation was mainly due to the permeation-enhancing effect of TMC that opened the tight junctions and allowed the paracellular transport of insulin across the intestinal membrane (Sadeghi et al. 2009).

Dorkoosh et al. investigated the enhancement of peroral octreotide absorption using delivery systems based on superporous hydrogel (SPH) and SPH composite (SPHC) polymers. Their system consisted of two components; a conveyor system made of SPH and SPHC and a core that contained octreotide. The core was inserted into the conveyor system (core inside), or attached to the surface of the conveyor system (core outside). Four different peroral formulations were investigated: core inside, core outside, core outside including TMC, and only octreotide in the absence of any polymer. They have placed the formulations in enteric-coated capsules and administered them perorally to pigs. The results suggested that while administration of octreotide by itself showed low bioavailability of about 1.0%, the core inside and core outside formulations resulted in 12.7% and 8.7%

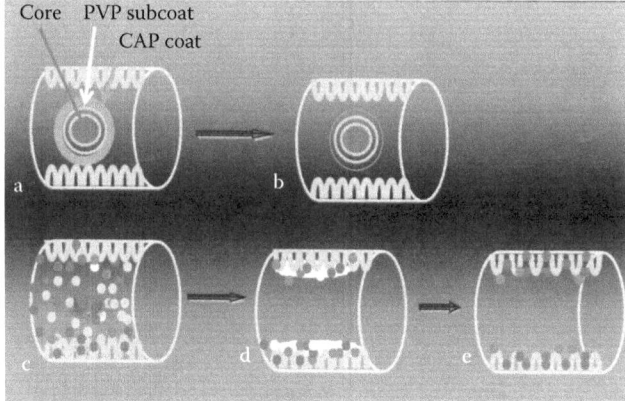

FIGURE 2.11 The GEDD system. (a) Enteric coated and PVP subcoated GEDD system in the stomach pH 1.2. (b) The GEDD system reaching the small intestine at a pH of 6.8. The enteric coating starts to dissolve quickly at this pH. (c) About 5 min later, the PVP subcoat dissolves and the CO_2 gas formation starts to push the insulin drug and the excipients PEO and TMC to the absorbing surface of the intestinal wall (insulin, PEO, and TMC). (d) After about 10–15 min, the tablet has completely dissolved and on the gut surface the mucoadhesive excipients together with insulin are attached. (e) TMC induces the opening of the tight junctions and insulin is able to permeate into the underlying tissue by the paracellular route. (Reproduced from *Journal of Controlled Release* 134(1), Sadeghi, A.M.M. et al., Development of a gas empowered drug delivery in the small intestine, 11–17. Copyright 2009, Elsevier.)

bioavailability, respectively. With the addition of TMC as an extra absorption enhancer the highest bioavailability of 16.1% was achieved (Dorkoosh et al. 2002).

2.3.3 Vaccination Using TMC

Several studies have been carried out on TMC, for mucosal vaccine delivery. TMC was shown to be able to open the tight junctions in between epithelial cells and induce paracellular intestinal absorption (Thanou et al. 2000a, 2001c). Unlike peptide drugs, vaccines must be targeted to M-cells of the Peyer's patches (PP). Mucosal vaccination is simple, noninvasive, and has the potential of inducing both the mucosal immune response as well the systemic one. Van der Lubben et al., prepared TMC microparticles containing diphtheria toxoid (DT) for oral and nasal vaccination. They measured the mucosal and humoral immune responses in both mice and rats using DT loaded in TMC microparticles as well as coadministered with TMC solution. The result suggested that the microparticles had a loading of about 80% with a suitable size for uptake by M-cells of the PP. The *in vivo* studies in mice have demonstrated a strong immune response against the orally administered antigen in the form of microparticle. Moreover, the immune response following the oral and nasal vaccine administration in rats revealed that TMC both in microparticle and solution form is able to induce systemic immune response to DT. They have thus concluded that while TMC microparticles are promising for oral vaccination, TMC in solution and in microparticle form is able to induce systemic and immune response after nasal vaccination (Van der Lubben et al. 2002).

Amidi et al., studied the role of TMC in nasal delivery of FITC–albumin. They prepared TMC nanoparticles using ionic gelation method loaded with labeled albumin and studied the uptake of the nanoparticles across the rat nasal mucosa. They compared the uptake of soluble FITC-albumin with the FITC-albumin nanoparticles and were able to show the presence of fluorescent nanoparticles in nasal-associated lymphoid tissues (NALT) and the underlying cell layers. Only negligible amount of fluorescence was observed with soluble FITC albumin. They thus concluded that TMC is a suitable vehicle for nasal vaccine delivery of proteins (Amidi et al. 2006).

Hagenaars et al. studied the influence of structural properties of TMC on its adjuvanticity. Hence, TMCs with different degrees of quaternization (22–86%), O-methylation (0–76%), and acetylation (9–54%) were formulated with inactivated influenza virus (WIV). The physicochemical characterization of the formulations were studied and evaluated for their immunogenicity in an intranasal vaccination/challenge study in mice. Simple mixing of the TMCs with WIV at a 1:1 (w/w) ratio resulted in comparable positive result as with charged nanoparticles, indicating coating of in activated virus with TMC. The amount of free TMC in solution was comparable for all TMC-virus formulations. After intranasal immunization of mice with free virus and TMC-virus on days 0 and 21, all TMC-virus formulations showed a stronger total IgG, IgG1, and IgG2a/c responses than the virus alone. No significant differences in antibody titers were observed for TMCs that varied in degrees of quaternization or degrees of O-methylation. These results indicate that these structural characteristics play a minor role in their adjuvant properties. TMC with a degree of quaternization of 56% (TMC56) formulated with virus at a ratio of 5:1 (w/w) resulted in significantly lower IgG2a/c:IgG1 ratios compared to TMC56 mixed in ratios of 0.2:1 and 1:1, implying a shift toward a Th2 type immune response. They have concluded that virus-TMC formulations have a more pronounced immunogenicity and may induce protection against viral challenge in mice after intranasal vaccination. Moreover, the adjuvant properties of TMCs as intranasal adjuvant are strongly decreased by reacetylation of TMC; however, the degree of quaternization and O-methylation hardly affected the adjuvanticity of TMC (Hagenaars et al. 2009) (Figure 2.12).

Hagenaars et al. have also studied the effect of TMC in nasal transit time, local distribution, and toxicity of intranasal inactivated influenza vaccine. The inactivated virus was fluorescently labeled and the fate of the vaccine was determined using immunohistochemical (IHC) staining method of nasal cross sections in mice to visualize the antigen both in the presence and absence of TMC as adjuvant in the nasal cavity. The results suggest that no significant difference was observed in

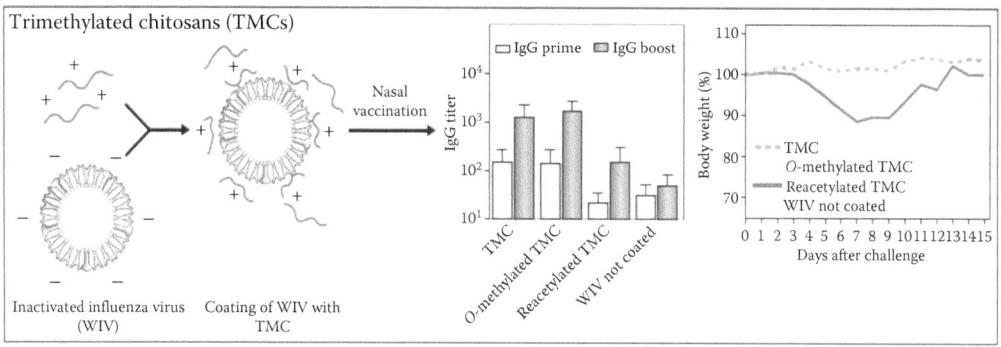

FIGURE 2.12 A schematic summary of influence of structural properties of TMC on its adjuvanticity. (*Journal of Controlled Release* 140(2), Hagenaars, N. et al., Relationship between structure and adjuvanticity of *N,N,N* trimethyl chitosan (TMC) structural variants in a nasal influenza vaccine, 126–133. Copyright 2009, Elsevier.)

residence time of the vaccine in the presence and absence of TMC. The location of the virus without the TMC was mainly in the mucus blobs of the nasal cavity; whereas, the TMC-coated vaccine was found mostly on the epithelial surfaces of the naso- and maxilloturbinates. This suggests a closer interaction of TMC-coated vaccine with the epithelial surface of the nasal cavity, which in turn enhances the uptake and induces the immune response of the vaccine in the presence of the adjuvant. Further toxicity studies revealed that neither vaccine showed local toxicity. They have thus concluded that TMC is a suitable polymer to be used as an adjuvant in nasal vaccine delivery (Hagenaars et al. 2010).

Chitosan derivatives are known to be good vaccine carriers because of their immune stimulating activity, and their bioadhesive properties that can enhance cellular uptake, permeation, and antigen protection. Sayin et al., conducted an investigation to enhance the poor immune response of mucosal vaccination. They prepared nanoparticles with TMC (positively charged) and mono-*N*-carboxymethyl chitosan (MCC, polyampholytic) for mucosal immunization. Mucosal application of a vaccine can effectively induce both systemic and mucosal immune responses. The aqueous dispersions of the derivatives were also prepared for comparison. Nanoparticles were prepared using ionic gelation method and loaded with tetanus toxoid (TT). Nanoparticles had a loading efficacy of about 90% and a size within the range of 40–400 nm. The MCC particles were negatively charged and the TMC nanoparticles were positively charged. The structural integrity of the TT in the formulations was confirmed by SDS-PAGE electrophoresis analysis. The immune responses induced by the nanoparticles loaded with tetanus toxoid were studied *in vivo* in Balb/c mice. Enhanced immune responses were obtained with intranasal application of nanoparticle formulations. TMC nanoparticles with positive surface charge induced higher serum IgG titers in comparison to those prepared with MCC which are negatively charged. They concluded that nanoparticle formulations developed in their investigation can be used as promising adjuvant/delivery systems for mucosal immunization (Sayin et al. 2008). Furthermore, they developed a TMC–MCC nanocomplex for mucosal vaccine delivery of TT. Mice were nasally immunized with TT loaded TMC–MCC complex nanoparticles and compared to that of TMC and MCC nanoparticles. The result of their study suggested that the cell viability was better with TMC–MCC complex nanoparticles, the size of the particles were smaller, and the loading efficiency was higher in comparison to TMC or MCC nanoparticles. Moreover, this complex was able to induce both the mucosal and systemic immune response indicating that this newly developed system has potential for mucosal administration of vaccines (Sayin et al. 2009).

Although several studies have been conducted on the role of TMC as an adjuvant in mucosal vaccine delivery, however, more studies are required for a definitive use of this polymer in humans.

2.3.4 TMC AND GENE DELIVERY

As mentioned earlier in this chapter, chitosan is one of the most commonly studied polymers for nonviral gene delivery. The positive charge of the chitosan can interact with the negative charge of the NAs and condense them into nanoparticles. However, the poor solubility of chitosan at physiological pH and weak buffering capacity makes it less efficient for gene delivery because of partial protonation of the amino groups. This pH-dependence influences NA-binding capacity and therefore the transfection efficiency. Because of these restrictions various investigations have been conducted with TMC for the intracellular delivery of proteins as a gene delivery carrier. The nanoparticles of TMC/DNA were prepared using ionic complexation between the positively charged TMC and negatively charged DNA. The particle size ranges from 150 to 600 nm, depending on the N/P ratio (the ratios of moles of the amine groups of cationic polymers to those of the phosphate ones of DNA). The obtained particles have a spherical morphology and the zeta potential of the particles increases with the N/P ratio and the quaternization degree of TMC (Zhengwei et al. 2007).

Germershaus et al., conducted studies on chitosan, TMC, and TMC grafted with polyethylene glycol that were complexed with DNA; they studied the complex diameter, DNA condensation efficiency, DNA release, cellular uptake of the complex and transfection efficiency from the complexes in NIH/3T3 cell line *in vitro*. The results suggested that with chitosan/DNA complex, at physiological pH, aggregates of approximately 1000 nm in size were formed and the DNA condensation rate was low. Consequently, the cellular uptake of the complex *in vitro* was negligible and the transfection efficiency was minimal. Using TMC/DNA complex the aggregation was reduced substantially, the cellular uptake was increased 8.5-folds compared to chitosan complexes and the transfection efficiency was increased by 678-folds. The PEGylation of the TMC improved the stability of the complexes and increased the cellular uptake in comparison to unmodified TMC and the transfection efficiency was shown to have a 10-fold increase compared to TMC. They have concluded that both TMC and PEGylated TMC are suitable for gene delivery (Germershaus et al. 2008).

A folate conjugated *N*-trimethyl chitosan (folate-TMC) was investigated as a gene delivery carrier using (TMC) as a reference by Zheng et al. The results indicated that both polymers were able to effectively condense pDNA into spherical nanocomplexes. The cellular uptake of the folate-TMC/pDNA complex containing labeled pDNA in KB cells was enhanced compared with that of the TMC/pDNA complex. Moreover, the transfection efficiency of the folate-TMC/pDNA complex in cell lines such as KB cells and SKOV3 cells with overexpressing folate receptor was increased with increasing N/P ratio and were enhanced up to 1.6- and 1.4-fold compared with those of the TMC/pDNA complexes. No significant difference between transfection efficiencies of the two complexes in A549 cells and NIH/3T3 cells (folate receptor deficient cell lines) were observed. They have thus concluded that the increase in transfection efficiencies of the folate-TMC/pDNA complexes is attributed to folate receptor-mediated endocytosis. Furthermore, confocal laser scanning microscope revealed different subcellular localization of each complex at different time points in the process of cellular uptake. This suggested that different intracellular trafficking pathways were employed by the two complexes (Zheng et al. 2009).

The PEGylation of the folate-TMC/pDNA complex was shown to enhance the cellular uptake and transfection efficiency in comparison to the non-PEGylated complex which was mainly because of the stabilizing effect of the PEGylation. It was thus concluded that PEGylation of folate-TMC/pDNA complex has a better potential for improving the transfection efficiency and gene specificity (Zheng et al. 2011).

In another investigation chitosan and TMC-siRNA nanoparticles were prepared by ionic gelation technique using TPP. The complexes were characterized for their size, zeta potential, loading efficiency, stability, cellular uptake, and transfection efficiency *in vitro*. Nanoparticles size obtained were in the range of 109–248 nm; the zeta potential was higher for particles made of TMC compared to those made of chitosan at pH 7.4. In the physiological pH of HEK293 cell line (human embryonic kidney 293) it was observed that the condensation efficiency of siRNA was decreased using chitosan which in turn led to a decrease in transfection efficiency. Although, the use of TPP for preparing nanoparticles improved the stability of the complex it did not have a significant effect on the transfection efficiency. TMC-siRNA were stable at physiological pH, they have shown better condensation efficiency and cellular uptake, however, the transfection efficiency did not improve significantly compared to chitosan nanoparticles. The cytotoxicity studies evaluated by the MTT calorimetric assay revealed that the nanoparticles had low toxicity. It was thus concluded that TMC is a suitable carrier for siRNA transfection *in vivo* (Dehousse et al. 2010).

Although numerous studies were conducted on TMC as a gene delivery carrier, more investigations are needed to be able to use this polymer in humans for gene therapy. There is also some evidence that the binding of the genes to TMC may be too strong to easily get released when the genes have entered the nucleus.

2.3.5 TMC NANOPARTICLES

Similar to chitosan, TMC is also used for the preparation of nanoparticles for delivery of peptides, vaccine, and NAs as described previously. The nanoparticles were mostly prepared by ionic gelation method where the positive charge of the TMC is ionically attached to the negative charges of the peptides or NAs condensing them into nanosized complexes that can protect the drugs from the proteolytic enzymes of the intestine, allow a sustained release of the drugs, and enhance the permeation and uptake of the particles by the cells.

Sadeghi et al. prepared insulin-loaded TMC nanoparticles by both ionotropic gelation method using TPP and polyelectrolyte complexation (PEC). They have characterized the nanoparticles according to their morphology, particle size, zeta potential, loading efficiency, and release studies. Their results suggested that morphology of the particles were round to oval in shape with a smooth surface and very similar using both methods. With the PEC method, the particles were smaller, were more positively charged, and had a higher loading efficiency in comparison to the particles prepared by the ionotropic method using TPP (Table 3, Sadeghi et al. 2008a).

They have reasoned that the higher loading of the insulin by PEC is because of a direct interaction between insulin and TMC; however, in the ionotropic gelation method, both the phosphate group of the TPP and insulin compete with TMC and therefore the loading efficiency of the insulin was less using this method. The difference in size of the particles was not significant. The zeta potential of the particles made by the PEC method was more positive indicating a more pronounced interaction of the positive charged polymer with the negative charge of the tight junction and a more enhanced permeation across the membrane by the paracellular pathway. The increase in zeta potential in PEC method was because of an excess of the positive charge available on the TMC. The release studies conducted in 0.1 N HCl showed that the release from systems made by ionotropic gelation was slower than by the PEC method which was due to the additional cross-linking between TPP and TMC that resulted in a denser matrix structure and lesser availability of insulin in the earlier phase of the release process (Sadeghi et al. 2008a).

Further studies were conducted to compare the role of TMC as a permeation enhancer in nanoparticle form and free soluble form. Insulin nanoparticles were prepared by PEC method and the enhancing effect of TMC was investigated in Caco-2 cells *in vitro*. The TEER studied revealed that TMC was able to reduce TEER and open the tight junctions more in the free soluble form compared to the nanoparticle form. This could be explained by the reduced amount of positive charge present on the surface of the nanoparticles indicated by the zeta potential. Moreover, the apparent permeability (Papp) of insulin across the cell monolayer was shown to be 4-fold higher with the free soluble TMC in comparison to the TMC nanoparticles. The nanoparticles, however, had more enhancing effect in comparison to insulin solution without polymer (Sadeghi et al. 2008b).

Sandri et al. have also studied and compared the role of TMC nanoparticles with different degree of quaternization (TMC 1, 4%; TMC 2, 35%; and TMC 3, 90%) loaded with fluorescein isothiocyanate dextran (FD4, MW 4400 Da), used as the model macromolecule with chitosan nanoparticles.

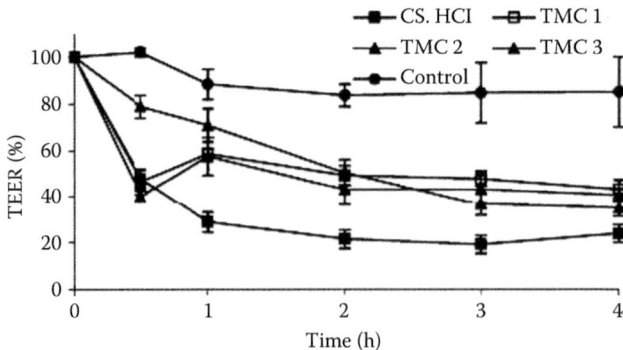

FIGURE 2.13 TEER % versus time profile observed for all the nanoparticulate systems and the control, FD4 solution (mean values ± SE; $n = 6$). (Reproduced from *European Journal of Pharmaceutics and Biopharmaceutics* 65(1), Sandri, G. et al. Nanoparticles based on *N*-trimethylchitosan: Evaluation of absorption properties using *in vitro* (Caco-2 cells) and *ex vivo* (excised rat jejunum) models, 68–77. Copyright 2007, Elsevier.)

The enhancement properties of nanoparticles in Caco-2 cell model and an *ex vivo* rat jejunum was evaluated. The results suggested that all the nanoparticulate systems were able to reduce the TEER value except the one made with TMC 90% (Figure 2.13). The Papp values of Lucifer Yellow, the fluorescent marker of the paracellular pathway, revealed that the Caco-2 monolayer treated with all the nanoparticulate systems were significantly higher than that of the monolayer treated only with the FD4 solution (negative control). This indicates that the contact with the nanoparticulate systems produced an opening of the tight junctions between the cells.

Their *ex vivo* studies revealed that an increase in the degree of TMC quaternization could increase the mucoadhesion properties and prolonged residence time of the nanosystem which in turn could result in more contact between the intestinal epithelium and the nanosystem. So while the absorption through the mucus layer was decreased with an increase in DQ the longer contact allowed for a better internalization. Hence, it was concluded that TMC could be a suitable option for drug delivery in nanoparticulate form (Sandri et al. 2007).

2.3.6 Antibacterial Effect of TMC

The antibacterial effect of TMC was studied by investigators and compared to chitosan. Because of the cationic nature of chitosan, it is able to attach to the anionic structure of the bacterial cell and reduce their activities by different mechanisms (Helander et al. 2001). Because TMC is more positively charged than chitosan it was speculated that TMC may have higher antibacterial activity than chitosan. Sadeghi et al. synthesized TMC with N-substitution degree of approximately 50% and compared its antibacterial activity against *Staphylococcus aureus*. The result suggested that the MIC and minimum bactericidal concentration (MBC) of chitosan and TMC were 1000 and 250 µg/mL, respectively. They concluded that because TMC had a higher zeta potential in comparison to chitosan (+43.2 vs. +24.2 mV) its higher antibacterial activity was mainly because of higher binding of the positive polymer charges with the negative cell wall of the bacteria (Sadeghi et al. 2008a).

In another study conducted by Xu et al., *O*-methyl free TMC was synthesized by initially treating chitosan with formic acid and formaldehyde and then followed by methylation with methyl iodide to obtain trimethylchitosan carboxymethyl (TMCMC) derivative according to Figure 2.10.

Their antibacterial activity was investigated against *Staph. aureus* and *E. coli*. Compared with chitosan, TMC displayed more effective antibacterial activity against both bacteria which were in concordance to previous report (Sadeghi et al. 2008a). However, the antibacterial efficiency decreased as the degree of trimethylation increased at pH 5.5, and increased as the degree of

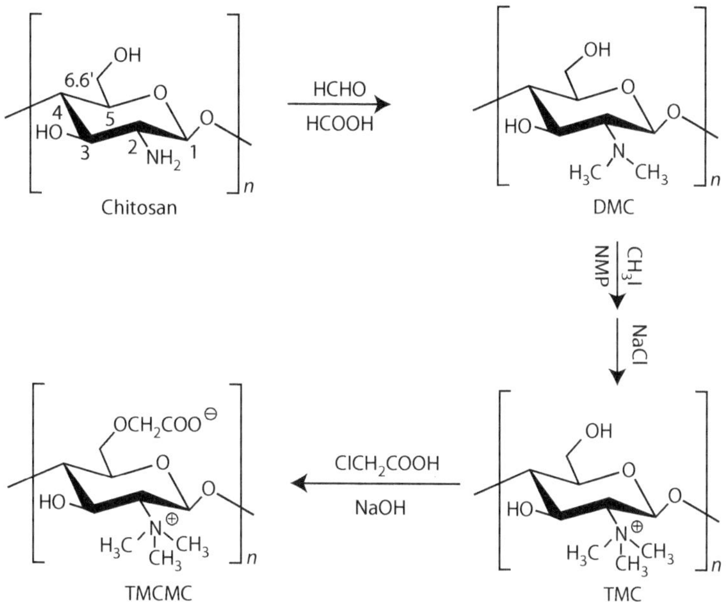

FIGURE 2.14 Synthesis of *N,N,N*-trimethyl *O*-carboxymethyl chitosan. (Reproduced from *Carbohydrate Polymer* 81(4), Xu, T. et al., Synthesis, characteristic and antibacterial activity of *N,N,N*-trimethyl chitosan and its carboxymethyl derivatives, 931–936. Copyright 2010, Elsevier.)

trimethylation increased at pH 7.2. The experimental results showed that the activity of *N,N,N*-trimethyl amino group was weaker than other nonquaternized amino groups, and carboxymethylation did not enhance the antibacterial activity directly (Xu et al. 2010) (Figure 2.14).

Rúnarsson et al. synthesized different methylated derivatives of chitosan with different degrees of methylation and determined their structure–activity relationship (SAR) with their antibacterial effect against *Staph. aureus*. The antibacterial activities of TMC, dimethyl chitosan (DMC), monomethyl chitosan (MMC), O methyl chitosan, and chitosan were compared at two different pH values of 5.5 and 7.2. The result indicated that the antibacterial activity of all the methylated derivatives was higher at pH 5.5 because of the protonation of the N amino group of the polymers resulting in more available positive charge. However, at pH 7.2, the nonquaternized groups are not cationic which resulted in a lower antibacterial activity of the derivatives. There was a negative correlation between the degree of *N*-quaternization and activity at pH 5.5 (Figure 2.15). The SAR at pH 5.5 indicated that the protonated amino groups, which were not trimethylated contribute to the antibacterial activity. When the activity at pH 7.2 was analyzed the N-quaternized group had a positive effect on activity. No correlation was found with degree of mono and dimethylation, and the contribution seemed not to be different from the free amino group (Rúnarsson et al. 2007).

2.3.7 TMC Hydrogels

Rossi et al. conducted studies on thermally sensitive TMC hydrogels for the treatment of oral mucositis induced by chemo-radiotherapy. They mixed TMC with glycerophosphate (GP) at different molar ratios and investigated the gelation time and the mucoadhesive properties of the gel using porcine buccal mucosa. According to their study, the best results were obtained using high molecular weight and low TMC with low DQ mixed with GP at a ratio of 1:2. The mixture was loaded with benzydamine hydrochloride, an antiinflammatory drug with antimicrobial properties and tested for drug release and wash away test. They have concluded that the gelation with TMC

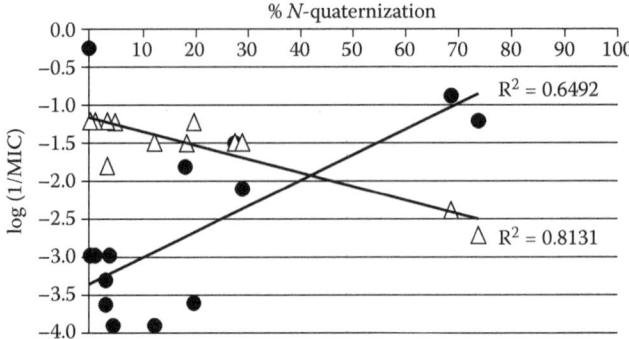

FIGURE 2.15   The relationship between degree of *N*-quaternization and antibacterial activity at pH 7.2 (●) and pH 5.5 (Δ) of the TMC I derivatives. (Reproduced from *European Polymer Journal* 43(6), Rúnarsson, Ö. et al. Antibacterial activity of methylated chitosan and chitooligomer derivatives: Synthesis and structure activity relationships, 2660–2671. Copyright 2007, Elsevier.)

was instantaneous at 37°C, the drug release was prolonged and the system was viscous enough to withstand the saliva washout at the action site for 5 h (Rossi et al. 2010).

In another study by Nazar et al., TMC thermal-sensitive hydrogels were prepared for nasal drug delivery. Three different molecular weight chitosan were used to synthesize TMC and formulate them into hydrogels using polyethylene glycol (PEG) and glycerophosphate (GP). The rheological studies indicate that TMC with low DQ synthesized from medium or high average molecular weight chitosan resulted in a short sol–gel transition time of about 7 min at 32.5°C. Moreover, the resulting hydrogel had good water-holding capacity and strong mucoadhesive property with good affinity for mucosal surfaces. They concluded that TMC hydrogel with these properties maybe suitable for intranasal drug delivery systems (Nazar et al. 2011).

2.4   SYNTHESIS AND CHARACTERIZATION OF OTHER CHITOSAN DERIVATIVES

2.4.1   Synthesis and Characterization of Triethyl Chitosan

N-triethyl chitosan (TEC) is another quaternized derivative of chitosan that was first synthesized and characterized by Avadi et al., in 2003, to overcome the poor solubility of chitosan at physiological pH (Avadi et al. 2003). TEC has three ethyl groups attached to the N-amine group of the chitosan by reductive ethylation (Figure 2.16). Briefly, low molecular weight chitosan was dispersed in *N*-methyl pyrrolidone and stirred with a magnetic stirrer for 4 h at room temperature. Then, aqueous sodium hydroxide (20%, m/V) solution, sodium iodide, and ethyl iodide were added. The mixture was heated at 60°C for 6 h under stirring. The product chitosan–$N^+(CH_2CH_3)_3I^-$ was precipitated with acetone and separated by centrifugation. To exchange I^- with Cl^-, the polymer was dissolved in 10% aqueous sodium chloride solution.

The polymer was precipitated with acetone, centrifuged, and dried to obtain a white water-soluble powder. TEC was prepared by a 2^2 factorial design to optimize the preparative conditions (Table 2.4). They chose ethyl iodide and sodium hydroxide concentrations as the independent variables and the quaternization degree as the response. On the basis of NMR calculations, a high degree of quaternization was achieved through the optimized one-step process (Figure 2.17).

They further studied the permeation-enhancing effect of TEC in inverted rat colon in *ex vivo* study using different molecular weight hydrophilic compounds, sodium fluorescein (M_w332), and Brilliant Blue (M_w772) across the colonic epithelium. The permeation of the drugs across the colonic epithelium was investigated in the presence and absence of permeation enhancers at pH 7.4. The

TABLE 2.4
Result of 2^2 Factorial Design Experiments

Formulations	Exp. 1	Exp. 2	Exp. 3	Mean $\pm$ SD
A1	54.23	58.14	50.32	54.23 $\pm$ 3.91
A2	61.23	58.35	54.69	58.09 $\pm$ 3.29
A3	42.56	48.69	41.32	44.19 $\pm$ 3.95
A4	81.25	78.65	77.10	79.00 $\pm$ 2.10

Source: Reproduced from *Iranian Polymer Journal* 13(5), Avadi, M.R. et al., Synthesis and characterization of *N*-diethyl methyl chitosan, 431–436. Copyright 2004a, Elsevier.

Note: A1, ethyliodide 1.5 mL, NaOH 15%; A2, ethyliodide 1.5 mL, NaOH 30%; A3, ethyliodide 3.0 mL, NaOH 15%; A4, ethyliodide 3.0 mL, NaOH 30%.

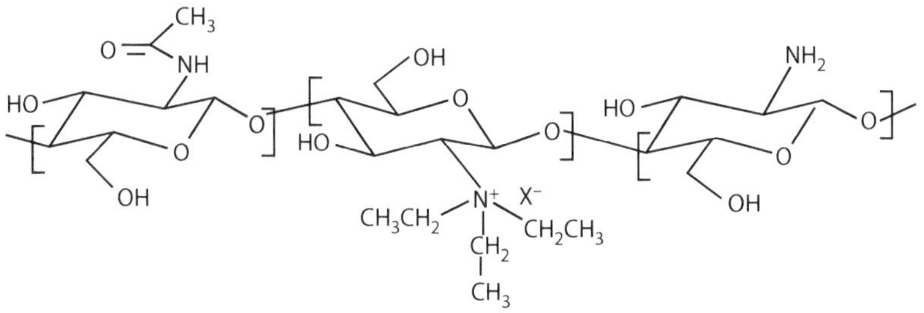

FIGURE 2.16   The chemical structure of TEC. (Adapted from Avadi, M.R. 2003. *Journal of Bioactive and Compatible Polymer* 18: 469–479.)

result suggested a significant increase in the absorption of both model drugs in the presence of TEC in comparison with both chitosan and the free drugs (Figure 2.18). It was speculated that because TEC carries more positive charge than chitosan, it was able to interact with the tight junctions of the colon and disrupt their structure allowing for reversible opening of the tight junction and increase the paracellular transport of the hydrophilic compounds (Younessi et al. 2004).

They thus concluded that TEC maybe a suitable permeation enhancer for colon drug delivery of hydrophilic drugs such as insulin.

2.4.2   TEC NANOPARTICLES

Insulin nanoparticles were prepared by polyelectrolyte complexation between the positive charge of TEC and negative charge of insulin for colon drug delivery by Bayat et al. The nanoparticles were characterized for their size, morphology, zeta potential, insulin loading, and release studies at corresponding pH. Further *ex vivo* studies were conducted on these nanoparticles in inverted rat colon to determine the permeation of insulin across the colon in the presence and absence of 2% TEC and chitosan in free soluble form. Moreover, *in vivo* investigations were carried out to determine the enhanced colon absorption of insulin nanoparticles compared to free insulin in diabetic rats. The insulin absorption from the rat's colon was evaluated by its hypoglycemic effect (Bayat et al. 2008). The results indicated that insulin nanoparticles prepared from TEC were spherical with smooth surface structure of about 170 nm in size, with a positive zeta potential of 25.1 mV and a loading efficiency of 90%. The insulin nanoparticles prepared by the same method exhibited the same shape, with about 270 nm in size and a positive zeta potential of 17.6 mV and loading efficiency of 84%. The release studies from nanoparticles prepared from chitosan and TEC at pH 6.8 suggested a small burst effect followed by a 33.4% and 44.3% insulin release from the TEC and

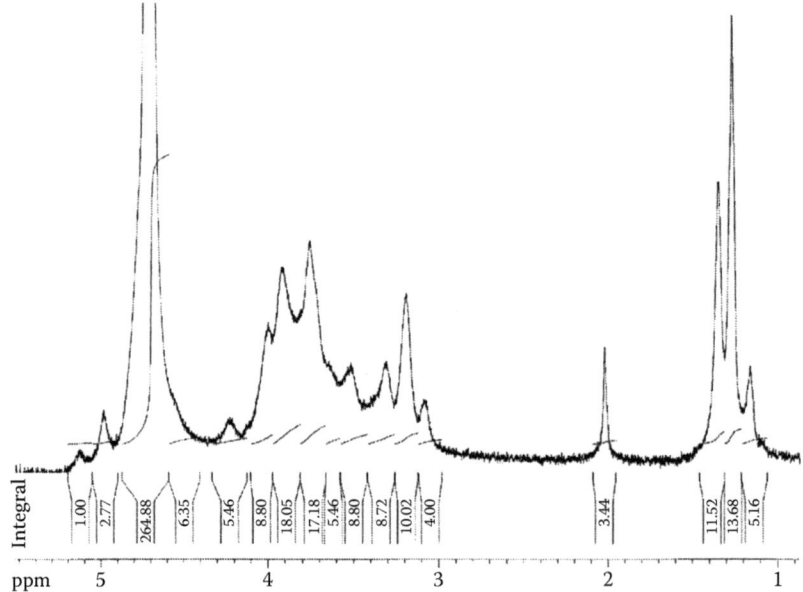

FIGURE 2.17 ¹H NMR spectrum of triethyl chitosan chloride. The signal at 1.2 ppm was attributed to the CH₃ groups of the ethyl substituent, whereas H2–H6 protons of the polysaccharide backbone superimposed the CH₂ groups. The intense band at 4.8 ppm was related to HDO (a proton-exchanged species of D₂O with OH and amine) (Figure 2.17). In this region, as observed more clearly from an extended spectrum, three anomeric protons (H1S) appeared at 4.23, 4.98, and 5.1 ppm. They were attributed to the mono-*N*-acetyl-glucosamine unit, mono and di-*N*-alkyl substituted glucosamine units, respectively. (Adapted from Avadi, M.R. et al. 2003. *Journal of Bioactive and Compatible Polymer* 18: 469–479.)

chitosan nanoparticles, respectively, after 5 h. This is because of better solubility of TEC in neutral and alkaline pH in comparison to chitosan that allows for more insulin release at this pH. The *ex vivo* results suggested that the amount of insulin transported across the colon membrane was more pronounced in the nanoparticulate form in the presence of free TEC polymer in comparison to insulin plus TEC in free soluble form. Also, the transport from the TEC nanoparticles was significantly higher than from chitosan nanoparticles both with and without the respective enhancer (Bayat et al. 2008). The *in vivo* results indicated that the blood glucose level after 60 min did not change significantly when free insulin was used in the presence or absence of the enhancer; however, a significant decrease in the blood glucose level was observed when insulin nanoparticles were injected in the ascending colon. The decrease in blood glucose level was in the following order from the lowest to

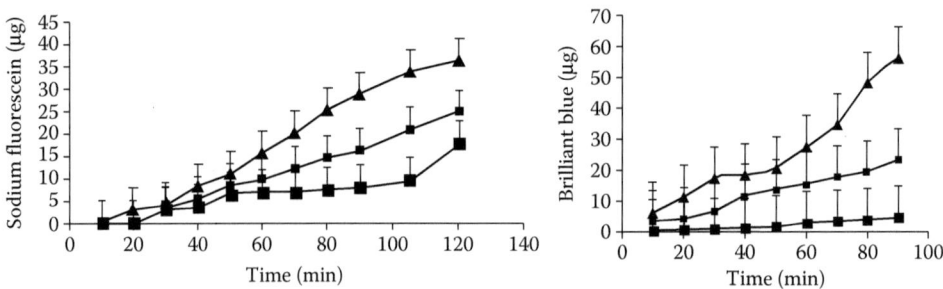

FIGURE 2.18 (a) Sodium fluorescein alone ■, added chitosan ■, added TEC ▲ (*n* = 3); (b) Brilliant Blue alone ■, added chitosan ■, added TEC ▲ (*n* = 3). Bars denote SD values. (Adapted from Younessi, P. et al. 2004. *Acta Pharmaceutica* 54: 339–345.)

the highest: free insulin > free insulin + free chitosan > free insulin + free TEC > insulin–chitosan nanoparticles > insulin–TEC nanoparticles > insulin chitosan nanoparticle + free insulin > insulin–TEC nanoparticles + free TEC. Hence, the best result was obtained using insulin nanoparticles in the presence of TEC. They have, thus, concluded that nanoparticles resulted in better uptake and permeation across the colonic membrane. Although the nanoparticles were mainly permeated across the membrane via endocytosis, presence of the free enhancers resulted in higher paracellular transport by interaction of the permeation enhancers with the tight junctions (Bayat et al. 2008).

In a study carried out by Sadeghi et al., permeation-enhancing effect of chitosan and TEC were compared as soluble form and nanoparticulate systems on insulin absorption in Caco-2 cells. According to this study, chitosan and TEC in free soluble form were able to decrease the TEER value more than in their nanoparticulate form. The insulin transport across the Caco-2 cell membrane was also shown to be higher when insulin and permeation enhancers were used in soluble form. The zeta potential of the polymers in free soluble form was measured to be more positively charged in comparison to their nanoparticles. The reduced available amount of positive charge at the surface of the nanoparticles resulted in weaker opening and less transport of insulin nanoparticles through the tight junctions. In comparison to chitosan, TEC with a more positive surface charge was shown to decrease the TEER value and increase transport of insulin to a higher extent than chitosan. They concluded that the nanoparticles had a low effect on opening of the tight junctions and the paracellular transport of insulin across the Caco-2 cell monolayer (Sadeghi et al. 2008b). Although TEC showed more permeation-enhancing activity compared to chitosan, its permeation-enhancing activity was not as pronounced as TMC. This may be because of less available positive charge of TEC in comparison to TMC indicated by the zeta potential as well as the possible steric hindrance of the TEC as a result of its bulky ethyl groups. In conclusion, this polymer was not studied as extensively as TMC.

2.5 SYNTHESIS AND CHARACTERIZATION OF DIETHYLMETHYL CHITOSAN

Diethyl methyl chitosan (DEMC) was initially synthesized by Avadi et al., to overcome the poor solubility of chitosan at pH values above 5.5. They used a 2^2 factorial design to optimize the synthesis condition. N-alkyl chitosan was initially prepared by introducing a methyl group into the amine group of chitosan via Schiff's base followed by reducing the C–N bond (step one), and finally reacted with ethyliodide to produce diethylmethyl chitosans (step two). The second stage of synthesis was based on a nucleophilic substitution of chitosan amine protons with ethyl groups of an ethyliodide in the presence of sodium iodide and sodium hydroxide in a water/NMP medium. Briefly, chitosan was dissolved in acetic acid, formaldehyde was then added to the chitosan solution, and after 1 h of stirring the pH of solution was adjusted to 4.5 by adding 1 M NaOH solution. Subsequently, 10% $NaBH_4$ solution was added and magnetically stirred for 1.5 h. Finally, methyl chitosan precipitate was obtained by adding 1 M NaOH solution and adjusting the pH of solution at 10. The precipitate was washed with distilled water and then Soxhlet extracted with ethyl alcohol and ether (1:1 v/v) for 3 days. In the second step, methyl chitosan was dispersed in NMP and the NaOH solution, ethyliodide, and sodium iodide were added to the dispersion. The reaction was carried out for 5 h at 60°C. Finally, acetone was added and the precipitate of the chitosan derivative was collected. For exchanging I⁻ with Cl⁻, the polymer was dissolved in 10% sodium chloride aqueous solution. The polymer was precipitated with acetone, centrifuged, and dried to obtain a water-soluble white powder with quantitative yield.

$$\text{Step I: Chit–NH}_2 \xrightarrow[\text{ACOH, 2 h}]{\text{HCOH}} \xrightarrow[\text{1.5 h, pH 4.5}]{\text{NaBH}_4} \text{Chit–NH (CH}_3)$$

$$\text{Step II: Chit–NH (CH}_3) \xrightarrow[\text{NaI, H}_2\text{O, NMP, 60°C, 5 h}]{\text{CH}_3\text{CH}_2\text{I, NaOH}} \text{Chit–N}^+(\text{CH}_3)(\text{C}_2\text{H}_5)_2\text{I}^-$$

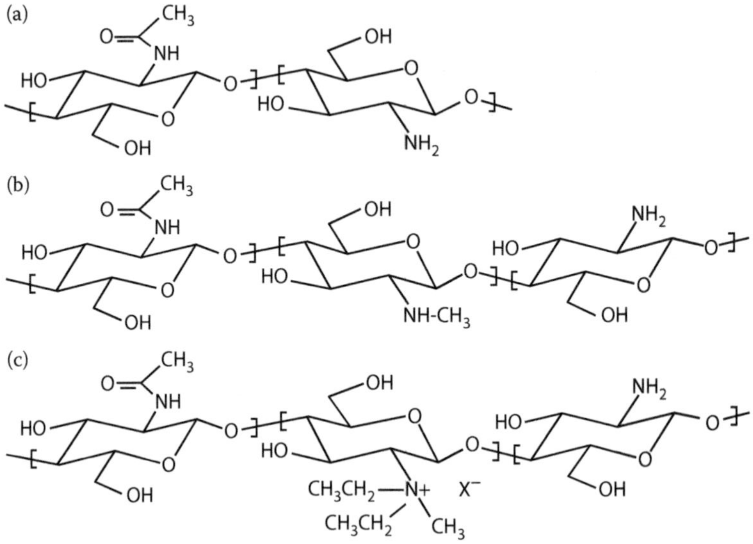

(a)

(b)

(c)

FIGURE 2.19 The structural presentation of (a) chitosan, (b) *N*-methyl chitosan, (c) DEMC. (Reproduced from *Iranian Polymer Journal* 13(5), Avadi, M.R. et al., Synthesis and characterization of *N*-diethyl methyl chitosan, 431–436. Copyright 2004a, Elsevier.)

The second step of the synthesis was optimized by factorial design so the amount of ethyliodide and the concentration of NaOH were chosen as independent variables and the degree of quaternization was used as the response (Avadi et al. 2004a).

The obtained polymer was characterized with ^{1}HNMR to calculate the degree of quaternization of the DEMC (Figures 2.19 and 2.20). They have observed that the best solubility was obtained with the formulation A4 (ethyliodide 3.0 mL, NaOH 30%) resulting in the highest degree of quaternization.

The formulation with the highest DQ was used to study the antibacterial activity of this novel derivative in comparison to chitosan against *E. coli*. The antibacterial activity was determined by

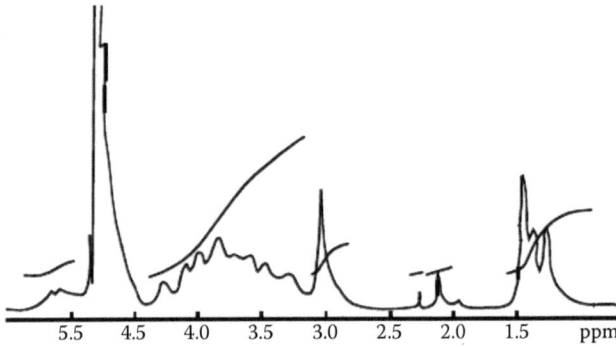

FIGURE 2.20 The HNMR spectra of *N*-diethylmethyl chitosan (DEMC). The signal at 1.3 ppm is attributed to CH$_3$ groups of the ethyl substituted, whereas H2–H6 protons of the polysaccharide backbone superimpose the CH$_2$ groups. The intense band at 4.8 ppm is related to HDO (solvent). In this region, as observed more clearly from an extended spectrum, some different anomeric protons (H1s) appear at 4.05, 4.23, and 5.1 ppm. They can be attributed to mono *N*-acetyl glucosamine unit, mono N-substituted and disubstituted glucosamine units, respectively. (Reproduced from *Iranian Polymer Journal* 13(5), Avadi, M.R. et al., Synthesis and characterization of N-diethyl methyl chitosan, 431–436. Copyright 2004a, Elsevier.)

MIC and MBC measured by turbidometric assay and by plate count, respectively. The lowest concentration of chitosan and DEMC that inhibited the growth of bacteria was considered as MIC. The MBC was determined by assaying the live organisms in those tubes from the MIC test that showed no growth. A loopful from each of those tubes were inoculated on EMB (Eosin–Methylene Blue) agar and looked for signs of growth. Growth of bacteria demonstrates the presence of these germs in the original tube. On the contrary, if no growth was observed, the original tube contained no living bacteria and the chitosan and DEMC considered being bactericidal at that concentration. The MIC and MBC of chitosan and DEMC against *E. coli* were determined in water and different concentrations of acetic acid. Accordingly, the antibacterial activity of chitosan and DEMC was higher in 1.0% acetic acid in comparison with the lower levels of acetic acid concentration. Furthermore, the inhibitory effect of chitosan was slightly lower than that of DEMC in acetic acid media. Moreover, the MIC and MBC values exhibit the lowest antibacterial activity of DEMC in water. Because acetic acid itself has antibacterial activity, a series of control experiments (at the same concentrations of acetic acid) were conducted to verify the results (Table 2.5).

They concluded that the antibacterial activity of both chitosan and DEMC was pH dependent. The MIC of DEMC was decreased from 500 to 62.5 µg/mL when the medium was changed from water to 1.0% acetic acid solution. The results suggested that although acetic acid itself had some antibacterial activity, the activity was significantly increased in the presence of either DEMC or chitosan. DEMC exhibited a higher antibacterial activity than chitosan because of the presence of more positive charge that could bind the negative cell surface membrane of the bacteria and disrupt the membrane making its antibacterial effect more pronounced than chitosan (Avadi et al. 2004b).

The synthesized DEMC was further used for *ex vivo* and *in vivo* studies by Avadi et al., to investigate the permeation-enhancing effect of the polymer (Figure 2.21). They studied the permeation of Brilliant Blue, a hydrophilic model drug, across the intestinal epithelia of everted large intestine of rat using the Barr and Riegelman method (Barr and Riegelman 1970).

They were able to demonstrate that DEMC, as an enhancer, was able to increase permeation of Brilliant Blue into the serosal region much more in comparison with chitosan and the drug with no enhancer (Figure 2.22).

Furthermore, the enhancing effect of DEMC enhancer on insulin absorption from ascending colon section of GIT in rats was investigated. The rats were made diabetic with alloxan and the mixture of insulin (25 units) and DEMC (1%) or insulin alone in phosphate buffer solution (PBS), pH 7.2 was injected into the ascending colon portion. Blood samples were drawn from portal vein at 0, 30, and 60 min and blood glucose concentration was assayed. The results suggested that there was

TABLE 2.5

MIC (µg/mL) and MBC (µg/mL) of Chitosan and N-Diethylmethyl Chitosan (DEMC) against *E. coli* in Water and Acetic Acid (AcOH) Aqueous Solutions

Medium	Control (Blank)		Chitosan		DEMC	
	MIC	MBC	MIC	MBC	MIC	MBC
Distilled water	–	–	–	–	500	500
AcOH 0.25 wt.%	2500	2500	250	250	125	125
AcOH 0.50 wt.%	2500	2500	225	225	110	110
AcOH 0.75 wt.%	1250	1250	158	158	87.5	87.5
AcOH 1.00 wt.%	1250	1250	125	125	62.5	62.5

Source: Reproduced from *European Journal of Polymer* 40, Avadi, M.R. et al., Diethyl methyl chitosan as antimicrobial agent: Synthesis, characterization and antibacterial effects. *European Journal of Polymer* 40: 1355–1361. Copyright 2004b, Elsevier.

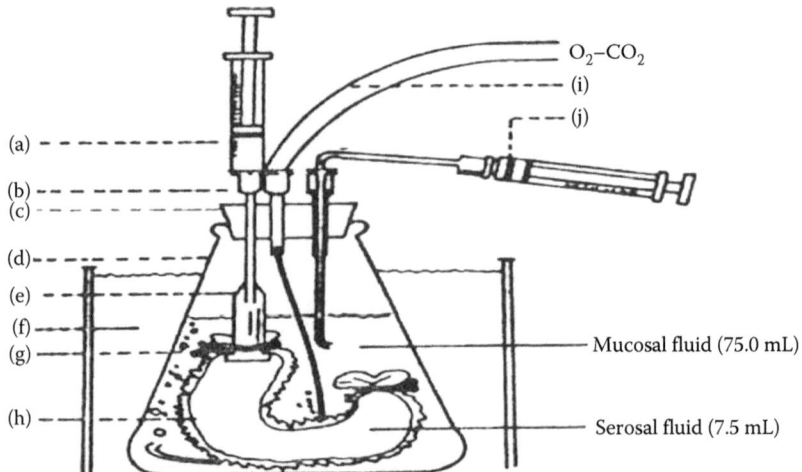

FIGURE 2.21 Apparatus used in the *ex vivo* everted intestine. (a) Disposable syringe for collection of serosal fluid; (b) hypodermic needle; (c) rubber stopper; (d) conical flask; (e) polyethylene centrifuge tube; (f) water bath (37°C); (g) tape used to fasten the intestine to tube; (h) everted intestine; (i) mixture of gas inlet (O_2 95% and CO_2 5%); (j) disposable plastic syringe used to collect mucosal fluid. (Reproduced from *International Journal of Pharmaceutics* 293, Avadi, M.R. et al., Diethyl methyl chitosan as an intestinal paracellular enhancer: *Ex vivo* and *in vivo* studies, 83–89. Copyright 2005b, Elsevier.)

no decrease in blood glucose level after colon injection of insulin. This indicated that insulin was not able to permeate from cell membrane on intestine. In contrast, there was a significant change in blood glucose, when mixture of insulin and DEMC was injected into the ascending colon from 532.62 ± 111.89 mg dL^{-1} (at time 0 min) to 241.19 ± 184.78 mg dL^{-1} (after 60 min). They concluded that DEMC was able to enhance the permeation of drug across the intestinal membrane through interaction of cationic polymer with the anionic structure of the tight junction and allowing for the paracellular transfer (Avadi et al. 2005b).

2.5.1 DEMC NANOPARTICLES

Sadeghi et al., prepared insulin nanoparticles using DEMC and characterized them according to their morphology, size, loading efficiency, zeta potential, and release studies. The results indicated

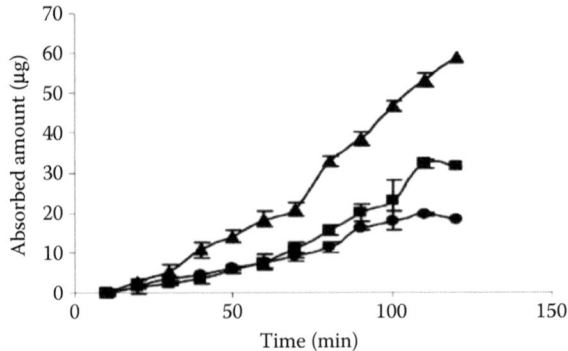

FIGURE 2.22 Profile of the amount of Brilliant Blue absorption as function of time: (●) Brilliant Blue, (■) Brilliant Blue with chitosan, and (▲) Brilliant Blue with DEMC. Experiments were carried out as triplicate. (Reproduced from *International Journal of Pharmaceutics* 293, Avadi, M.R. et al., Diethyl methyl chitosan as an intestinal paracellular enhancer: *Ex vivo* and *in vivo* studies, 83–89. Copyright 2005b, Elsevier.)

that nanoparticles were within the expected size range, had a smooth surface with a positive charge, and loading efficiency of about 75%. The release studies have shown that about 40% of the insulin was released at pH 6.8 after 45 min. The nanoparticles and the free polymer were used in a comparative study in Caco-2 cell monolayer to investigate the transport of insulin across the membrane. The data suggested that while insulin by itself did not pass through the cell membrane, the combination of free insulin and DEMC resulted in higher transport across the membrane in comparison to the insulin nanoparticle prepared by DEMC. The TEER values also followed the same trend; the TEER value did not change with free insulin but was reduced more with free insulin and polymer and less with insulin nanoparticles. This could be because of the presence of more positive charge on the free DEMC (+40.0) compared to DEMC nanoparticles (+19.0). They concluded that the difference in the zeta potential was accountable for the better permeation of insulin across the intestinal epithelium and better paracellular transport (Sadeghi et al. 2008b). The antibacterial activity of DEMC in free form and in nanoparticle form was also investigated. The results suggest that DEMC in free soluble form had more antibacterial activity than in nanoparticle form again because of the presence of higher positive charge of the free polymer. This could account for better complexation of the polymer with the bacterial cell wall and hence better antibacterial effect. According to this study, the antibacterial effect of DEMC was higher than chitosan but lower than TMC which correlated nicely with the zeta potential of the polymers (Sadeghi et al. 2008b).

2.5.2 DEMC in Gene Delivery

Like other chitosan derivatives, DEMC was also used as a nonviral vector for cancer gene therapy. In a study conducted by Safari et al., DEMC was used for gene delivery in human pancreatic cell line (AsPC-1). Enhanced green fluorescent protein plasmid (pEGFP) was used as model plasmid and was complexed with DEMC at different N/P ratios. The complexes were used for transfection of pancreatic cancer cells *in vitro*. The transfection efficiency was dependent on the N/P ratio of the carrier DNA. The data suggested that at charge ratios higher than 10, the number of green fluorescent protein (GFP) positive cells increased and reached a maximum at charge ratio of 40. Fluorescent microscopy of cells transfected with DEMC/pEGFP complexes revealed a 16.7-fold increase in transfection efficiency when the N/P ratio increased from 5 to 40.

The flow cytometry studies had also shown that after transfection the size of the cells increased. They have explained that the excess ratio of cationic polymer to anionic plasmid allowed the nucleopolymer particles to be attached to the negative cell surface and further taken up by the cells through endocytosis. They concluded that DEMC with more positive surface charge than chitosan may be a good candidate as a vector for pancreatic cancer therapy (Safari et al. 2011).

2.6 SYNTHESIS AND CHARACTERIZATION OF DIMETHYL ETHYL CHITOSAN

DMEC is another quaternized derivative of chitosan that was initially synthesized by Bayat et al., using a modified two-step method for oral delivery of peptides (Bayat et al. 2006). In the first step of the synthesis, mono-ethyl chitosan was prepared by introducing an ethyl group onto the amine group of chitosan via a Schiff base and in the next step methyl iodide was added to produce DMEC which was water soluble in a pH range of 4–8 (Figure 2.23).

Briefly, chitosan was dissolved in acetic acid and acetaldehyde was added to the chitosan solution and after 1.5 h of stirring, the pH of the solution was adjusted to 4.5 using 1 M NaOH.

Sodium borohydride solution was then added and magnetically stirred for 2 h. The first step was to obtain mono-ethyl chitosan precipitate by adding 1 M NaOH solution and adjusting the pH of the solution to 10. The precipitate was washed with distilled water and then Soxhlet extracted with ethyl alcohol and ether (1:1 v/v) for 3 days to give ethyl chitosan in high yield (degree of substitution 90%). In the second step, ethyl chitosan was dispersed in NMP for 5 h. Then, NaOH solution, methyl iodide, and sodium iodide were added to the dispersion. The reaction was carried out with

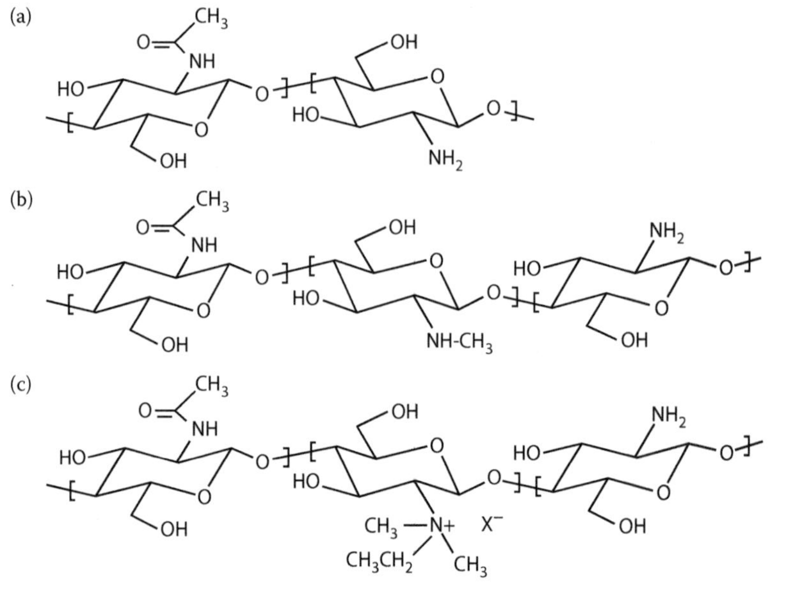

FIGURE 2.23 The structural presentation of (a) chitosan, (b) *N*-methyl chitosan, and (c) dimethylethyl chitosan (DMEC). (Adapted from Bayat, A. et al. 2006. *Journal of Bioactive and Compatible Polymer* 21: 433–444.)

stirring for 5 h at 60°C. They collected DMEC after adding acetone to precipitate the quaternized chitosan derivative (Figure 2.24). To exchange I⁻ with Cl⁻, the polymer was dissolved in aqueous sodium chloride solution, precipitated with acetone, centrifuged, and dried to obtain a water-soluble polymer (Bayat et al. 2006).

The obtained DMEC polymer was used to prepare insulin nanoparticles by the PEC method. The nanoparticles were characterized and shown to have a smooth surface with a positive charge and a loading efficiency of approximately 89%. The size of the obtained nanoparticles was around 170 nm. Insulin release studies from nanoparticles resulted in a small burst release followed by a sustained release up to 5 h and a maximum of 48% at pH 6.8. These nanoparticles were used to investigate the *ex vivo* transport of insulin across the colonic epithelium in rats. The data suggested that insulin transport across the membrane was significantly higher in nanoparticle form plus free DMEC compared to insulin nanoparticles and free soluble form. They hypothesized that the presence of the DMEC enhances the transport of insulin nanoparticles. In their study, they also showed that insulin nanoparticles prepared by DMEC in the presence of DMEC resulted in highest insulin transport compared to nanoparticles prepared by TEC or chitosan in the presence of free TEC and chitosan, respectively (Bayat et al. 2008) (Figure 2.25).

The permeation-enhancing effect of DMEC was studied in diabetic rats using insulin in nanoparticle form in the presence of free soluble enhancer and in free soluble form with and without the enhancer. The results show that insulin nanoparticles prepared by DMEC in the presence of 2% free DMEC showed the highest transport across the colonic membrane. The nanoparticles alone showed less insulin transport but the lowest permeation was seen with the free insulin. Moreover, the permeation effect of TEC and chitosan in free soluble form and nanoparticle form was shown to be less than DMEC. This could be because of the presence of more available positive charge in DMEC in comparison to TEC or chitosan (Bayat et al. 2008).

Sadeghi et al. used insulin nanoparticles prepared by DMEC to investigate the permeation-enhancing effect of the polymer in Caco-2 cell line in a comparative study with insulin nanoparticles prepared by TMC, TEC, DEMC, and chitosan. In their study, they have shown that the insulin transport across the cell line was dependent on the positivity of the polymer and the nanoparticles. Accordingly, the use of insulin in the presence of free soluble polymer was more effective in reducing

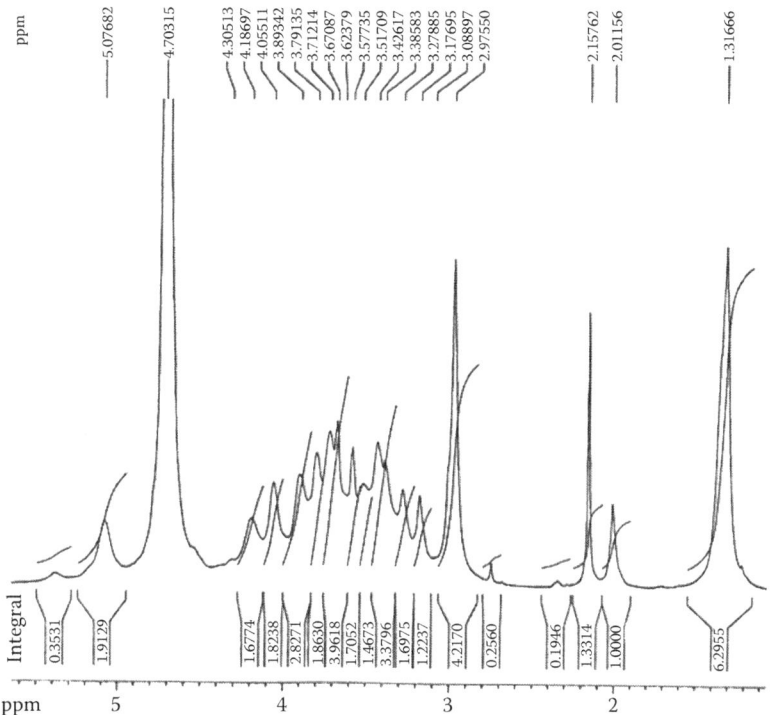

FIGURE 2.24   The ^{1}H-NMR characterization of DMEC revealed a degree of quaternization of 50%. The signal at 1.3 ppm was attributed to CH_3 groups of the ethyl substituent, whereas H2–H6 protons of the polysaccharide backbone superimposed the—CH_2—groups of the ethyl group between 3.15 and 4.2 ppm. The intense band at 4.70 ppm was related to HDO. In this region, as observed more clearly from an extended spectrum, two different anomeric protons (H1) appeared at 5.07 and 5.35 ppm. (Adapted from Bayat, A. et al. 2006. *Journal of Bioactive and Compatible Polymer* 21: 433–444.)

the TEER values and insulin transport in comparison to the respective nanoparticles made from each polymer. The enhancing effect of polymers were according to the following order from lowest to highest, chitosan < TEC < DMEC < DEMC < TMC. They have thus concluded that the enhancing effect of the polymers were because of their cationic interactions with the anionic structure of the tight junction, disrupting their integrity and reversibly opening them for the passage of insulin. In their study they have also shown that although insulin nanoparticles were more effective in insulin permeation than free insulin, their enhancing effect was not too significant. They have reasoned that the positive charge of the polymer was interacted with the negative charge of insulin for the preparation of nanoparticles. Hence, the availability of less positive charge resulted in less interaction with the tight junction and less paracellular transport across the membrane (Sadeghi et al. 2008b).

2.7   SYNTHESIS AND CHARACTERIZATION OF 6-NH$_2$-6-DEOXY CHITOSAN

Sadeghi et al. synthesized and characterized two new derivatives of chitosan, namely trimethylated 6-NH$_2$-6-deoxy chitosan and triethylated 6-NH$_2$-6-deoxy chitosan. The idea was to add two methyl or ethyl groups at the C1 and C6 positions of the chitosan to enhance the solubility of chitosan at pH above 6.0 and also to increase the available positive charge on the chitosan. The *N*-phthaloyl chitosan was synthesized according to Satoh et al. with some modifications (Satoh et al. 2006). Briefly, phthalic anhydride was added to dimethyl formamide (containing 5% w/v) and then chitosan was added to the solution and magnetically stirred for 8 h at 120°C. The phthaloyl chitosan precipitate

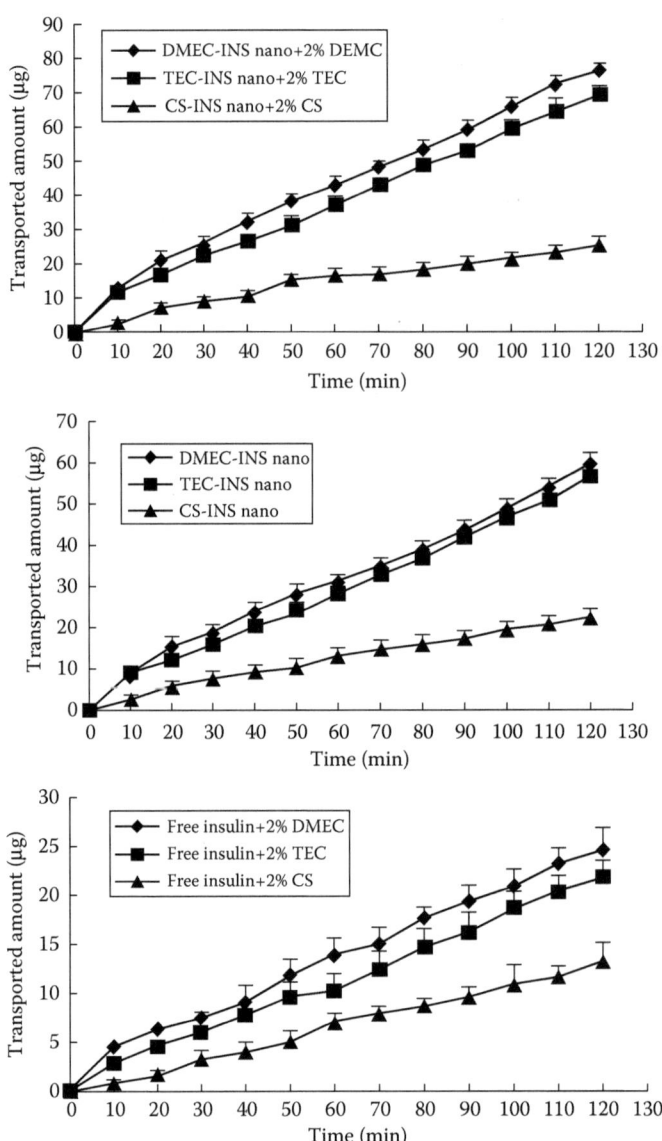

FIGURE 2.25 Amount of insulin transported in *ex vivo* studies using different formulations. (Reproduced from *International Journal of Pharmaceutics* 356(1–2), Bayat, A. et al. Nanoparticles of quaternized chitosan derivatives as a carrier for colon delivery of insulin: *Ex vivo* and *in vivo* studies, 259–266. Copyright 2008, Elsevier.)

was obtained by adding cold water, the precipitate was then washed with methanol for 1 h, filtered, and dried at 40°C. In the second step of the reaction, phthaloyl chitosan was dissolved in pyridine and tosyl chloride was added to the solution. To obtain a viscous solution, the mixture was stirred for 17 h. To precipitate the obtained product, cold water was added and the precipitate was washed with ethanol and diethyl ether and dried at room temperature. In the third step, the product was added to DMSO and ammonia solution (25% w/v) and stirred for 24 h at 80°C. After the mixture was cooled, it was precipitated with acetone, washed with diethyl ether, filtered, and dried at room temperature. In the last step, the product from stage 3 was added to *N*-methyl pyrrolidone (NMP) and stirred at 100°C for half an hour magnetically under nitrogen condition. Consequently, hydrazine monohydrate was added to the mixture and stirred for 12 h at 100°C. Finally, the mixture was

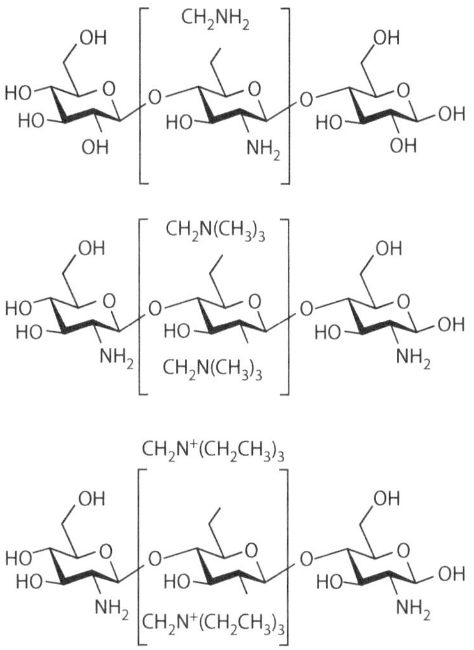

FIGURE 2.26 Chemical structure of 6-NH$_2$-6-deoxy chitosan (a), C2–C6 methylated chitosan (b), and C2–C6 ethylated chitosan (c). (Adapted from Sadeghi, A.M.M. et al. 2008c. *Journal of Bioactive and Compatible Polymer* 23: 262–275.)

cooled and ethyl acetate was added stepwise to obtain precipitate. The obtained precipitate was washed with diethyl ether after filtration and dried at room temperature. The trimethylation and triethylation of the 6-amino-6-deoxy chitosan was carried out according to the one-step method previously described (Sadeghi et al. 2008c, Avadi et al. 2003) (Figure 2.26).

The derivatives were characterized using ^{1}H-NMR and the degrees of quaternization were calculated to be 65% and 51% for C2–C6 trimethylated 6-amino-6-deoxy chitosan and C2–C6 triethylated 6-amino-6-deoxy chitosan, respectively (Figure 2.27).

They measured the zeta potential and antibacterial activity of the new derivatives and compared them to chitosan, TMC, and TEC. The result of the zeta potential suggested the following order: C2–C6 methylated chitosan > C2–C6 ethylated chitosan > TMC > TEC > chitosan. The antibacterial activity of the alkylated 6-amino-6-deoxy chitosan was studied against the Gram-positive bacteria *S. aureus*. The data suggested that C2–C6 methylated chitosan showed the highest antibacterial activity indicated by the low concentration of MIC and MBC followed by C2–C6 ethylated chitosan and TMC, TEC, and chitosan. They concluded that the polymers with higher zeta potential had the higher ability to bind to the peptidoglycans on the bacterial cell wall and may induce severe morphological alterations in Gram-positive bacteria. TEC showed less inhibition than TMC and this was explained by the fact that the ethyl groups are larger in size than the methyl substituted groups and hence the binding of TEC to the bacterial cell wall was sterically hindered. Although these new chitosan alkyl derivatives also showed promising results as permeation enhancer, more *in vitro* and *in vivo* studies are required to confirm their activities (Sadeghi et al. 2008c).

2.8 SYNTHESIS AND CHARACTERIZATION OF MONO-CARBOXYMETHYL CHITOSAN

Mono-carboxymethyl chitosan was synthesized by the substitution of the primary amine group of chitosan with carboxyl groups to yield polymers with polyampholytic properties. Mono-carboxymethylated

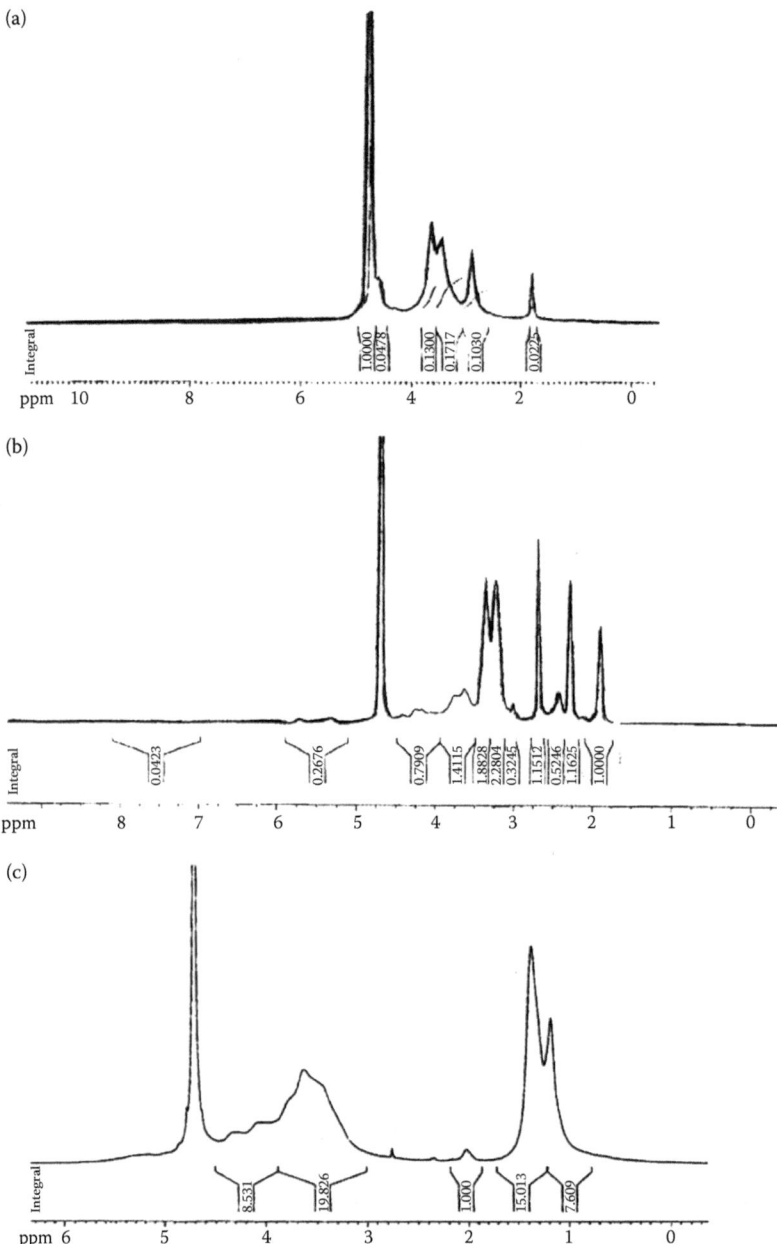

FIGURE 2.27 The ^{1}H-NMR spectrum of (a) chitosan, (b) C2–C6 methylated chitosan (c), and C2–C6 ethylated chitosan. The intense band at 4.6 ppm was due to D$_2$O. In Figure 2.27b, the peaks at 4.4–3.6 ppm were attributed to the glucosamine units and the double peak at 3.6 and 3.2 ppm was attributed to the hydrogen groups of the methyl units substituted at C2–C6; the sharp peak at 2.6 ppm was attributed to dimethylated chitosan and the integral of these signals were used to calculate the degree of quaternization. The signal at 1.97 ppm was because of the methyl protons from the acetylated chitosan. The dual peak at 3.6–3.2 did not shift or change when the solution was acidified with CF$_3$COOD. In Figure 2.27c, the intense band at 4.7 ppm was attributed to the D$_2$O used as the solvent. The peaks at 3.6–4.3 ppm were attributed to the glucosamine units. The double signal at 1.3 was attributed to the CH3 groups of the ethyl substituents, whereas the CH$_2$ groups of the quaternized site were superimposed by the 2-H and the 6-H protons of the polysaccharide backbone. (Adapted from Sadeghi, A.M.M. et al. 2008c. *Journal of Bioactive and Compatible Polymer* 23: 262–275.)

chitosan (MCC) was initially synthesized by Muzzarelli et al. (1982); almost two decades later, Thanou et al., adapted the same method to synthesize MCC and used it as permeation enhancer. To synthesize MCC, chitosan was dissolved in acetic acid, glyoxylic acid was added, and the mixture was stirred at ambient temperature for 1 h. After the formation of the imine (*N*-carboxymethylidene) chitosan, the pH was increased to 4.5 by addition of 1 M NaOH. Subsequently, an aqueous solution of 5% (w/v) sodium borohydride was added drop wise to reduce the formed imine. The mixture was stirred for 1 h, and the product was isolated by precipitation after the addition of ethanol. The product was washed on a glass filter under vacuum with ethanol/water aliquots. This product underwent a second step of carboxymethylation as described above. The obtained MCC was soluble in both acidic and alkaline pH values and had a degree of quaternization of about 58% (Thanou et al. 2001d) (Figure 2.28).

Owing to its unique properties, particularly its biocompatibility, MCC has been extensively used in the biomedical field as moisture-retention agent and bactericide; in wound dressings as artificial bone and skin, and in blood anticoagulants as an element in drug delivery systems. Moreover, MCC is an efficient metal chelater and exhibits high adsorption capacity for dyes. Because it has been shown to possess a variety of unique properties, the compound has attracted worldwide attention (Xue et al. 2009).

MCC has a polyampholytic or zwitterionic character, which allows the formation of clear gels or solutions in the presence of polyanionic compounds like heparins at neutral and alkaline pH values; it, however, aggregates at acidic pH. The permeation-enhancing effect of MCC was studied in transport of LMWH across Caco-2 cell line by Thanou et al. They measured the TEER reduction using different concentrations of MCC (1%, 3%, and 5%) and concluded that 5% MCC was more effective in reversible reduction of TEER than the other concentrations. Furthermore, the apparent permeability of LMWH across the Caco-2 cell monolayer was measured and the results indicated an increase from 2.3 mV ± 0.6 for the control sample to 200 mV ± 19 with 3% MCC and 170 mV ± 10 cm/s with 5% MCC. The difference between 3% MCC and 5% MCC was indicated as not significant. The intestinal absorption of LMWH in rats using 3% MCC was shown to be significantly higher compared to the control sample, that is, AUC increased from 1.0 to 7.0 U/mL min (Thanou et al. 2001d). The authors concluded that MCC with mucoadhesion characteristic significantly increased the intestinal absorption of LMWH by facilitating the paracellular permeation across the intestinal epithelia without being absorbed across the membrane.

2.8.1 Intranasal Vaccination Using MCC

As was previously discussed, Sayin et al. used MCC to prepare tetanus toxoid (TT) nanoparticles for intranasal vaccine delivery in mice and compared them to nanoparticles prepared by TMC and chitosan. In cytotoxicity studies, the 3-(4,5-dimethylthiazol-2-yl)-2,5-diphenyl tetrazolium bromide (MTT) of polymers in Chinese hamster ovary (CHO-K1) cell line showed cell viability in the order of MCC, chitosan, and TMC. Enhanced immune responses were obtained with intranasal application of nanoparticle formulations. TMC and chitosan nanoparticles with positive surface charge

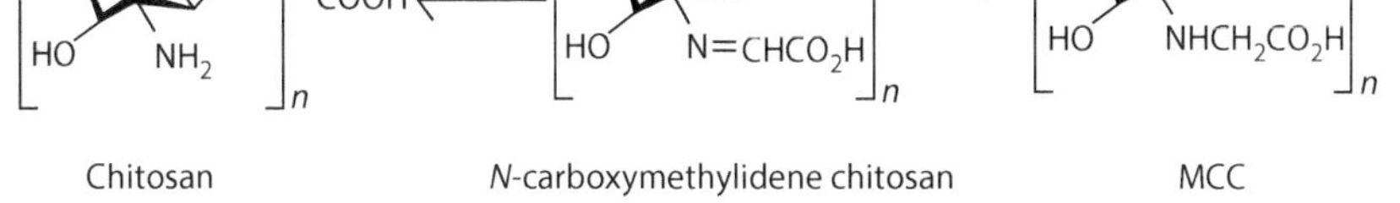

FIGURE 2.28 The schematic presentation of synthesis of the mono-*N*-carboxymethyl chitosan (MCC) chemical structure. (Reproduced from Thanou. M. et al. Mono-*N*-carboxymethyl chitosan (MCC), a polyampholytic chitosan derivative, enhances the intestinal absorption of low molecular weight heparin across intestinal epithelia *in vitro* and *in vivo*. *Journal of Pharmaceutical Science*, 2001d, 90: 38–46. Copyright Wiley-VCH Verlag GmbH & Co. KGaA. Reproduced with permission.)

showed higher IgG serum levels compared to those prepared by MCC with negative surface charge (Sayin et al. 2009). The authors have explained that the negatively charged polymer had repelling electrostatic interactions with the mucin network on the nasal epithelium and hence showed less immune response to the TT. However, because MCC was less cytotoxic than TMC, a TMC–MCC nanocomplex loaded with TT was prepared and was shown to be more effective as an adjuvant than either TMC or MCC alone (Sayin et al. 2009).

2.8.2 MCC Usage in Tissue Engineering

Hydrogels have gained significant attention as candidate materials for cartilage tissue engineering scaffolds. They possess three-dimensional shapes and initial mechanical strength similar to natural extracellular matrix because of their ability to retain great quantities of water. Nanofibrous collagen-coated porous carboxymethyl chitosan microcarriers (CMC-MCs) were successfully prepared by covalent linking of collagen to MCC for tissue engineering by Lu et al. The *in vitro* cell culture revealed that chondrocytes could adhere, proliferate, and remain differentiated on the nanofiber-coated CMC-MCs. Confocal laser microscope has shown that chondrocyte were grown to confluency after 3 days postseeding and after 7 days postseeding CMC-MC were attached to each other to form tissue-like aggregates. It was thus concluded that collagen-treated MCC could be used for application in cartilage tissue engineering as injectable scaffolds for cell delivery (Lu et al. 2008).

In another study conducted by Tang et al., the amino group of MCC was reacted with the aldehyde groups of the gellan gum to form a double network complex hydrogel. The complex was shown to have improved gelation temperature and mechanical properties compared to the gellan gum by itself. The compressive stress/strain measurement showed an increased compressive modulus of the complex hydrogel and an ability to return to the original shape after release of the compressive load. The chondrocyte cell cytotoxicity was shown to be lower with the complex hydrogel than the gellan gum alone and moreover the chondrocyte proliferation was higher using the complex (Tang et al. 2012) (Figure 2.29).

2.8.3 Antibacterial Effect of MCC

Like other derivatives of chitosan, MCC was also studied for its antimicrobial effect. Anitha et al. synthesized *O*-carboxymethytl and *N*–*O* carboxymethyl chitosan, prepared nanoparticles from these derivatives and compared their cytotoxicity and antibacterial effect with chitosan nanoparticles. The nanoparticles were prepared by ionic gelation technique using $CaCl_2$ and TPP for *O*-carboxymethyl and *N*–*O* carboxymethyl chitosan, respectively. The diameters of prepared nanoparticles were found to be 40–50, 90–110, and 90–95 nm for chitosan, *O*-carboxymethyl, and *N*–*O* carboxymethyl chitosan, respectively. The zeta potential values of the prepared nanoparticles were found to be +54.2, −35.12, and +46.74 mV for chitosan, *O*-carboxymethyl, and *N*–*O* carboxymethyl chitosan, respectively. This indicated a positive surface charge for the chitosan and *N*–*O* carboxymethyl chitosan and a negative surface charge for the *O*-carboxymethyl chitosan. The cytotoxicity studies were conducted in breast cancer cells (MCF-7) using MTT assay. The results revealed a 98% cell viability using all three nanoparticles indicating that these nanoparticles were not cytotoxic in MCF-7 cells. Furthermore, the antibacterial effects of the nanoparticles were studied against *S. aureus* using MIC tests. Different concentrations of each nanoparticle were used and the results showed that by increasing the concentrations of the nanoparticles the number of colonies decreased. The antibacterial effect was higher in *N,O*-CMC and lower with chitosan nanoparticles. They concluded that the higher antibacterial effect of *N,O*-CMC was probably because of a higher degree of substitution of *N,O*-CMC compared to chitosan and *O*-carboxymethyl (Anitha et al. 2009).

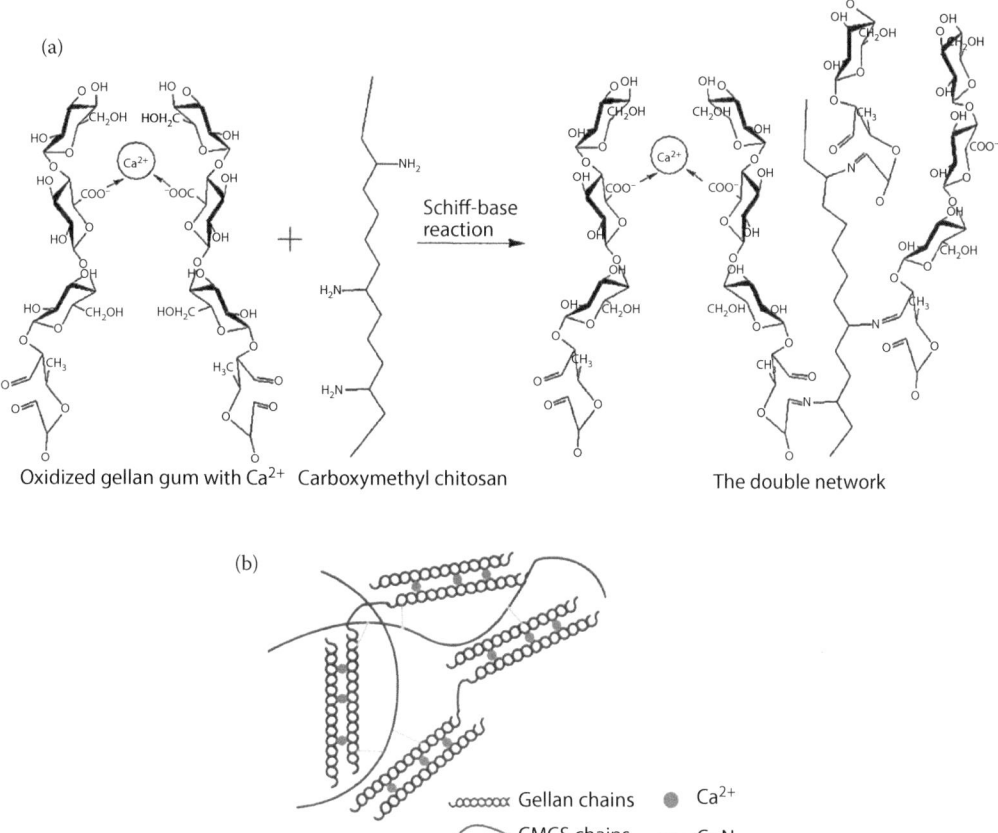

FIGURE 2.29 Schiff-base formation between amino groups of CM–chitosan and aldehyde groups of oxidized gellan gum. (a) In gellan chains, *cis*-dihydroxyl of rhamnose was oxidized to dialdehyde, the addition of Ca^{2+} introduced ionic bonds between the carboxyl groups of gellan via electrostatic interaction, subsequently aldehyde groups and amino groups of CM-chitosan formed the second network via the Schiff-base reaction. The cross linking mechanism of complex hydrogel. (b) Gellan gum chains formed double helix conformations with Ca^{2+}, and then CM-chitosan chains link the aldehyde zones to the formation of a three dimensional network, that created the gel. (Reproduced from *Carbohydrate Polymer* 88, Tang, Y. et al., An improved complex gel of modified gellan gum and carboxymethyl chitosan for chondrocytes encapsulation, 46–53. Copyright 2012, Elsevier.)

2.8.4 MCC NANOPARTICLES

In different studies carried out on MCC it was shown that it can be used as a promising nanocarrier for delivery to tumor cells. *O*-carboxymethyl chitosan was used by Anitha et al. to study its effect as nanocarrier for the delivery of curcumin to cancer cells. Curcumin is a phytochemical with immense biological properties; however, its hydrophobicity and poor oral bioavailability limits its application as a chemotherapeutic agent. Curcumin loaded *O*-carboxymethyl chitosan nanoparticles were prepared by ionic gelation method to increase its bioavailability. The nanoparticles had an optimum size of approximately 150 ± 30 nm, suitable for drug delivery. The *in vitro* release studies showed a slow, controlled, and sustained release of the drug from the nanoparticles and also demonstrated enzyme-triggered degradation and release of the drug in the presence of lysozyme. Fluorescence microscopic imaging and fluorescence activated cell sorting (FACS) analysis indicated that the curcumin was successfully delivered to the cancer cells and resulted in the apoptosis of the cancer cells. The nanoparticles were however nontoxic to normal cells. They have thus concluded that O-carboxymethyl chitosan nanoparticles may be efficient for the controlled delivery of hydrophobic drugs to cancer cells (Anitha et al. 2011).

In another study carried out by Anitha et al., N,O-carboxymethyl chitosan nanoparticles were prepared and loaded with 5-fluorouracil (5-FU) for cancer delivery. The obtained nanoparticles were shown to have a diameter of 80 ± 20 nm, zeta potential: $+52.47 \pm 2$ mV and loading efficiency of 65%. Fluorescent microscopy and flow cytometric analysis were used to show the cellular internalization of nanoparticles. The anticancer activity of the nanoparticles were shown by MTT test and the apoptosis assays showed the toxicity of the drug loaded nanoparticles toward breast cancer cells used in the study. They have thus concluded that the 5-FU loaded N,O-CMC nanoparticles have a potential in breast cancer chemotherapy by which the side effects of conventional chemo treatment could be reduced (Anitha et al. 2012).

O-carboxymethyl chitosan was used to design magneto-fluorescent nanoparticles for cancer-specific targeting, detection, and imaging. The free amine groups of O-carboxymethyl chitosan stabilized magnetite nanoparticles on the surface allow for the covalent attachment of a fluorescent dye such as rhodamine isothiocyanate (RITC) with the aim to develop a magneto-fluorescent nanoprobe for optical imaging. Folic acid and its aminated derivatives were then conjugated on to these magneto-fluorescent nanoparticles using amine, carboxy, or aldehyde groups. The synthesized iron-oxide folate nanoconjugates (FA-RITC-OCMC-SPIONs) showed excellent dispersibility, biocompatibility, and good hydrodynamic sizes under physiological conditions which were extensively studied by a variety of complementary techniques. The cellular internalization efficacy of these folate-targeted and its nontargeted counterparts were studied using a folate-overexpressed (HeLa) and normal (L929 fibroblast) cells by fluorescence microscopy and magnetically activated cell sorting (MACS). Cell-uptake behaviors of nanoparticles clearly demonstrated that cancer cells overexpressing the human folate receptor internalized a higher level of these nanoparticle–folate conjugates than normal cells. These folate-targeted nanoparticles possess specific magnetic properties in the presence of an external magnetic field and the potential of these nanoconjugates as T_2-weighted negative contrast MR imaging agent were evaluated in folate-overexpressed HeLa and normal L929 fibroblast cells (Bhattacharya et al. 2011).

2.9 THIOLATED CHITOSAN

Thiolated polymers are mucoadhesive polymers with thiol side chains. Thiolated polymers could be anionic or cationic; however, most cationic thiomers are based on chitosan. These include, chitosan-cysteine, chitosan-thiogycolic acid, and chitosan-tributylamidine. According to Figure 2.30, sulfhydryl bearing agents can be covalently attached to the primary amino group of chitosan via the formation of amide or amidine bonds. In case of the formation of amide bonds the carboxylic acid

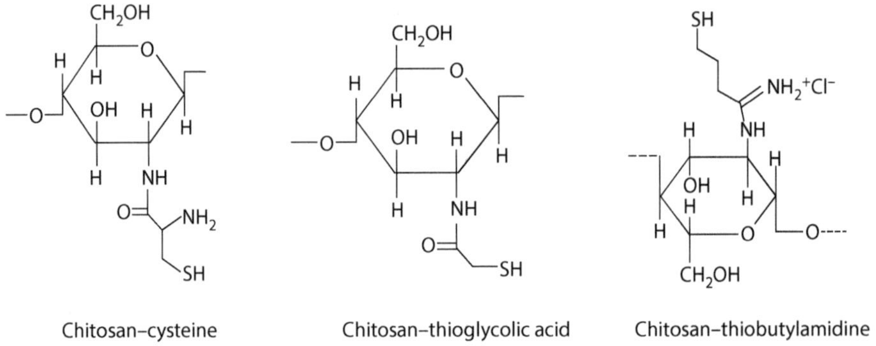

FIGURE 2.30 Thiolated chitosan chemical structures. (Adapted from Bernkop-Schnürch, A., U.M. Brandt, and A.E. Clausen. 1999. *Scientia Pharmaceutica* 67: 196–208.)

group of the ligands cysteine and thioglycolic acid reacts with the primary amino group of chitosan mediated, for instance, by carbodiimides (Bernkop-Schnurp et al. 2005). To avoid oxidation of thiol groups during synthesis the reaction should be performed under inert conditions at pH below 5. At this pH range, the concentration of the reactive form of oxidation of thiol groups, the thiolate anions, was low and the formation of disulfide bonds could almost be excluded. Furthermore, disulfide bonds can be reduced after the synthesis process by the addition of reducing agents, such as dithiotreitol or borohydride (Bernkop-Schnurp et al. 2005). In case of the formation of amidine bonds, 2-iminothiolane is used as coupling reagent. It offers the advantage of a simple one-step coupling reaction. In addition, the thiol group of the reagent is protected toward oxidation because of its chemical structure.

These thiolated chitosans were initially synthesized by Bernkop-Schnürch et al., to improve the mucoadhesive and permeation-enhancing properties of chitosan as excipient in drug delivery systems by covalently attaching thiol moieties to chitosan. The primary amino group at the 2-position of the glucosamine subunits of this polymer is the main target for the immobilization of thiol groups (Bernkop-Schnurp et al. 2001, 2003, 2005, Kast and Bernkop-Schnürch 2001, Roldo et al. 2004).

On the basis of thiol/disulfide exchange reactions and/or a simple oxidation process, disulfide bonds are formed between these polymers and the cysteine-rich glycoproteins of the mucus (Figure 2.31). Hence, thiomers mimic the natural mechanism of secreted mucus glycoproteins, which are also covalently anchored in the mucus layer by the formation of disulfide bonds (Bernkop-Schnurp et al. 2001).

In further studies by Bernkop-Schnürch, chitosan was modified with a 2-iminothiolane to obtain chitosan–4-thio-butyl-amidine conjugates (chitosan–TBA conjugates). The chitosan–TBA conjugates exhibited *in situ* gelling properties due to the formation of disulfide bonds based on an oxidation process of the immobilized thiol groups under physiological conditions (Bernkop-Schnurp et al. 2003). They have used *in vitro* mucoadhesion studies by compressing the thiolated chitosan into 5-mm flat-faced tablets which were then attached to freshly excised intestinal porcine mucosa fixed on a stainless-steel cylinder. The cylinder was then placed in a dissolution vessel containing PBS with pH 6.0 and the detachment of the tablets were observed during one-week period. The results suggested an improved mucoadhesion of chitosan–TBA conjugates in comparison to unmodified chitosan and chitosan–thioglycolic acid conjugates used. The authors have reported a 140-fold increase in the degree of mucoadhesion with the chitosan–TBA conjugates

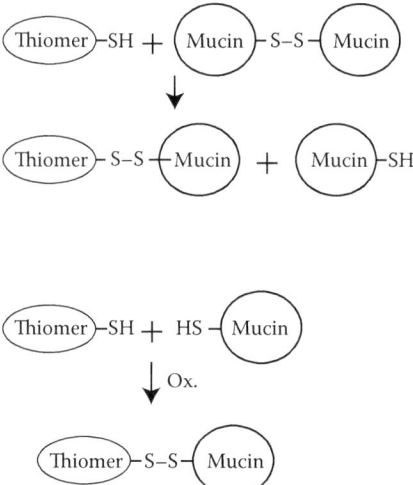

FIGURE 2.31 Mechanism of disulfide bond formation between thiomers and mucus glycoproteins (mucins). (Adapted from Bernkop-Schnürch, A. 2005. *Advanced. Drug Delivery Review* 57(11): 1569–1582.)

which was described by the cationic characteristics and immobilized thiol group on this chitosan derivative (Bernkop-Schnurp et al. 2003). Because the mucoadhesion of the chitosan–TBA polymer was very high, it was suggested that it could be used in vaginal drug delivery. Hence, clotrimazole as model drug was used for drug release studies from the polymer. The data suggested a controlled release of the drug which was suitable for the purpose of vaginal drug delivery (Bernkop-Schnurp et al. 2003).

Chitosan–TBA (chitosan-4-thiobutylamidine) was also used to develop a delivery system for improvement of oral insulin administration. Chitosan–TBA conjugate, enzyme inhibitors, and insulin were mixed and compressed in 10 mg tablets. The tablets were enteric coated with Eudragit L 100-55 and compared *in vivo* with unmodified chitosan–insulin tablet and insulin solution. The tablets were orally administered to the rat and the reduction in glucose level was measured. The result suggested that insulin in solution had no effect in reduction of glucose, the unmodified chitosan–insulin tablet showed a slight glucose reduction; however, the chitosan–TBA insulin tablet showed a significant reduction in glucose level. Due to strong mucoadhesive properties of chitosan–TBA–insulin tablets an intimate contact of the dosage form to the mucosa was provided preventing a presystemic metabolism of insulin on the way to the absorption membrane. Accordingly, the chitosan–TBA conjugate, enzyme inhibitors, and insulin tablet had a bioavailability of 1.69% compared to subcutaneous injection and may be a promising candidate for oral insulin application (Krauland et al. 2004).

2.9.1 THIOLATED CHITOSAN MICROPARTICLES AND NANOPARTICLES

Microparticles and nanoparticles offer the advantage of a prolonged residence time on mucosal membranes, higher drug uptake, and the possibility to reach greater mucosal surface areas. Consequently, thiolated chitosan micro- and nanoparticles have been developed and evaluated *in vitro* and *in vivo*. In a study carried out by Bernkop-Schnürch et al., a novel method was used to develop submicron particles of thiolated chitosan without being ionically cross-linked. They initially used TPP and sulfate by ionic gelation method to form submicron particles and microparticles of thiolated chitosan. The thiol groups were oxidized for stabilizing inter- and intramolecular disulfide bonds. The polyanions were then removed to give the particles a positive charge. The particles had a diameter of approximately 350 nm and a zeta potential of +11.0 mV. The presence of more positive charge and the remaining free thiol group could be responsible for improved mucoadhesion and permeation-enhancing effect of these particles. They have concluded that these nanoparticles may be promising candidates for mucosal drug delivery (Bernkop-Schnurp et al. 2006).

Thiolated chitosan thioglycolic acid (chitosan–TGA) nanoparticles were prepared to enhance the bioavailability of nasal application of leuprolide. The nanoparticles had a diameter of approximately 250 nm with a positive zeta potential of 10 mV. The release studies showed a sustained release of the leuprolide from the thiolated nanoparticles over 6 h which was probably because of molecular disulfide bonds within the nanoparticle networks. Using thiolated chitosan nanoparticles and unmodified chitosan nanoparticles resulted in 5.2- and 2.0-fold increase in nasal transport of leuprolide across the porcine nasal mucosa in comparison to leuprolide solution, respectively. The pharmacokinetic studies in rats showed improved transport of leuprolide from thiolated nanoparticles compared to the free solution which was due to facilitated transport of thiolated nanoparticles across the nasal mucosa (Shahnaz et al. 2012).

The distribution of thiolated mucoadhesive anionic poly (acrylic acid) (PAA) and cationic chitosan (CS) nanoparticles on intestinal mucosa was studied and compared. The nanoparticles were prepared by ionic gelation and the particles were labeled with both hydrophilic and hydrophobic dyes. Chitosan was modified with 2-iminothiolane (CS–TBA) and PAA was modified with cysteine (PAA–Cys) to obtain both cationic and anionic thiolated polymers, respectively. Both unmodified and modified CS nanoparticles and PAA nanoparticles were examined *in vitro* in terms of their mucoadhesive and mucus-penetrating properties on the mucosa of rat small intestine. The mucoadhesion of the particles had the following order CS–TBA > PAA–Cys > CS > PAA; CS-TBA showed twice as much mucoadhesion as PAA–Cys. The result suggested that thiolated polymers can tightly adhere to the intestinal mucus layer for a prolonged time through covalent bonding with mucin glycoproteins via thiol–disulfide exchange reactions. The diffusion rate of the anionic and cationic modified and unmodified polymers were investigated in fresh porcine intestinal mucus secretions. The result showed that the unmodified anionic polymer PAA was able to diffuse deeper in the mucus layer compared to the modified PAA which was less mobile. Moreover, the modified cationic polymer (CS–TBA) showed the least mobilization due to the interaction of thiol groups of the particles with the mucus layer. In general, cationic polymer showed more immobilization than the anionic polymer as the negative charge of the anionic polymers repulses the negative charge of the mucus layer and allowed for more enhanced diffusion of the anionic polymers within the mucus layer. It was thus concluded that mucoadhesion can slow down the particle transit time through the GI tract, resulting in prolonged residence time, localization of the delivery system at a specific target site, and an increase in drug concentration gradient (Dünnhaupt et al. 2011).

Microparticles of chitosan modified with 2-iminothiolane (CS–TBA) were used for nasal insulin delivery in rats. The microparticles were prepared by emulsification solvent evaporation and loaded with insulin. The result suggested that modified chitosan CS–TBA microparticles showed a controlled release of insulin over 6 h and had a bioavailability of 7.24% compared to unmodified chitosan–insulin microparticles with 2.04% bioavailability. It was suggested that these microparticles can be used for nasal peptide delivery (Krauland et al. 2006).

2.9.2 THIOLATED CHITOSAN GEL

Rheological measurements of the chitosan *in vitro* have characterized the *in situ* gelling properties of the polymer. The sol–gel transition of the thiolated chitosan was completed at pH 5.5 after 2 h to form highly cross-linked gels. Also, a significant decrease in the content of thiol groups on the polymers was observed that indicated the formation of disulfide bonds (Bernkop-Schnürp et al. 2003, Hornof et al. 2003). The *in situ* gelling property of the chitosan thiomers was used for the preparation of liquid or semi-solid vaginal, nasal, and ocular formulations. Moreover, thiolated chitosans have been used in cancer therapy. Letrozole (LTZ), a hydrophobic drug, was loaded in acetyl-polyamidoamine (Ac-PMAM)-thiolated chitosan as films for localized chemotherapy based on breast delivery systems. The films were biodegraded *in vivo* by erosion or degradation to release the hydrophobic drug (Saboktakin et al. 2011).

Chitosan hydrogels modified by disulfide cross-linking were prepared for cell viability and protein release. Chitosan was modified with N-acetyl-L-cysteine (NAC). To minimize interference with biological function, the degree of substitution of thiol groups was kept below 50%. The hydrogels were loaded with insulin and bovine serum albumin (BSA); the protein release was also controlled by loading, composition, and disulfide bonds. The compatibility of hydrogels on fibroblast cells NIH 3T3 was evaluated and indicated that the hydrogels were compatible and the cells could migrate into hydrogels. The cells were shown to maintain their 3D structure within the hydrogels indicating the possible application of this disulfide cross-linked chitosan hydrogels for tissue engineering, drug delivery, and cell culture (Wu et al. 2009).

2.10 SUMMARY

In conclusion, a vast amount of research has been conducted on chitosan and its derivatives. Chitosan, this biodegradable and bioavailable polymer has numerous applications in agricultural, medical, and pharmaceutical industries. However, chitosan's low solubility at pH above 6.5 resulted in synthesis and characterization of many chitosan derivatives. These derivatives were mainly obtained by N-quaternization of chitosan and were all soluble at physiological pH values. Similar to chitosan, several research studies have been conducted on these derivatives for the purpose of drug delivery, tissue engineering, and antibacterial effects. While some of these derivatives such as TMC, MCC, and thiolated chitosan were studied in more detail as they gave the most promising results for application in humans in the future, the other chitosan derivatives were not studied so extensively. However, all those studies also contribute to the general knowledge about a very interesting class of polymers. Chitosan derivatives also would continue even stronger research efforts if they would have the GRAS status (generally recognizes as safe) or would be approved by pharmacopeias.

REFERENCES

Ackermann, M. and M. Engles, 2006. Pro and contra IBR-eradication. *Veterinary Microbiology* 113: 293–302.

Alimunir, A. and R. Zainuddin, 1992. An economical technique for producing chitosan. In: *Advances in Chitin and Chitosan* (C.J. Brine, P.A. Sandford and J.P. Zikasis, eds.) pp. 627–633, Elsevier Applied Science, London.

Allan, G.G., C. Altman, R. Besinger, D. Ghosh, Y. Hirabayasi, and S. Neogi. 1984. Biochemical applications of chitin and chitosan. In: *Chitin, Chitosan and Related Enzymes*, Zikakis JP, Ed. Orland, F L: Academic Press, 119–133.

Amidi, M., S.G. Romijn, G. Bochard, H.E., Junginger, W.E., Hennink, and W. Jiskoot. 2006 Preparation and characterization of protein-loaded N-trimethyl chitosan nanoparticles as nasal delivery system. *Journal of Controlled Release* 111: 107–116.

Anitha, A., V.V. Divya Rani, R. Krishna, V. Sreeja, N. Selvamurugan, S.V. Nair, H. Tamura, and R. Jayakumar. 2009. Synthesis, characterization, cytotoxicity and antibacterial studies of chitosan, *O*-carboxymethyl and *N,O*-carboxymethylchitosan nanoparticles. *Carbohydrate Polymer* 78(4): 672–677.

Anitha, A., S. Maya, N. Deepa, K.P. Chennazhi, S.V. Nair, H. Tamura, and R. Jayakumar. 2011. Efficient water soluble *O*-carboxymethylchitosan nanocarrier for the delivery of curcumin to cancer cells. *Carbohydrate Polymer* 83(2): 452–461.

Anitha, A., K.P. Chennazhi, S.V. Nair, and R. Jayakumar. 2012. 5-Flourouracil loaded N,O-carboxymethyl chitosan nanoparticles as an anticancer nanomedicine for breast cancer. *Journal of Biomedical Nanotechnology* 8(1): 29–42.

Aranaz, I., M. Mengíbar, R. Harris, I. Paños, B. Miralles, N. Acosta, G. Galed, and A. Heras. 2009. Functional characterization of chitin and chitosan. *Current Chemical Biology* 3: 203–230.

Artursson, P. 2001. Chitosan as a nonviral gene delivery system. Structure–property relationships and characteristics compared with polyethylenimine *in vitro* and after lung administration in vivo. *Gene Therapy* 8(14): 1108–1121.

Avadi, M.R., A.H. Ghassemi, A.M.M. Sadeghi, D. Beiki, A. Akbarzadeh, P. Ebrahimnejad, Sh. Shahhosseini, Kh. Bayati, and M. Rafiee-Tehrani. 2005a. γ-Scintigraphic evaluation of enteric coated capsules containing chitosan-brilliant blue gel beads on hydrophilic model for colon drug delivery. *Journal of Drug Scientific Technology* 15: 383–387.

Avadi, M.R., A. Jalali, A.M.M. Sadeghi, K. Shamimi, Kh. Bayati, Kh. E. Nahid, A.R. Dehpour, and M. Rafiee-Tehrani. 2005b. Diethyl methyl chitosan as an intestinal paracellular enhancer: *Ex vivo* and *in vivo* studies. *International Journal of Pharmaceutics* 293: 83–89.

Avadi, M.R., Gh. Mahdavinia, A.M.M. Sadeghi, M. Erfan, M. Amini, M. Rafiee-Tehrani, and A. Shafiee. 2004a. Synthesis and characterization of *N*-diethyl methyl chitosan. *Iranian Polymer Journal* 13(5): 431–436.

Avadi, M.R., A. Sadeghi. A. Tahzibi. Kh. Bayati, M. Pouladzadeh, M.J. Zohourian-Mehr, and M. Rafiee-Tehrani. 2004b. Diethyl methyl chitosan as antimicrobial agent: Synthesis, characterization and antibacterial effects. *European Journal of Polymer* 40: 1355–1361.

Avadi, M.R., A.M.M. Sadeghi, N. Mohamadpour, R. Dinarvand, F. Atyabi, and M. Rafiee-Tehrani. 2011. *Ex vivo* evaluation of insulin nanoparticles using chitosan and arabic gum. *ISRN Pharmaceutics* 1: 1–6.

Avadi, M.R., M.J. Zohourian-Mehr, P. Younessi, M. Amini, M. Rafiee Tehrani, and A. Shafiee. 2003. Optimized synthesis and characterization of N-triethyl chitosan. *Journal of Bioactive and Compatible Polymer* 18: 469–479.

Aydin, Z. and J. Akbuga. 1996. Chitosan beads for the delivery of salmon Calcitonin: preparation and release characteristics. *International Journal of Pharmaceutics* 131(1): 101–103.

Azab, A.K., V. Doviner, B. Orkin, J. Kleinstern, M. Srebnik, A. Nissan, and A. Rubinstein. 2007. Biocompatibility of evaluation of crosslinked chitosan hydrogels after subcutaneous and intraperitoneal implantation in the rat. *Journal of Biomedical Material Research* 83(2): 414–422.

Barr, W.H. and S. Riegelman. 1970. Intestinal drug absorption and metabolism. I. Comparison of methods and models to study physiological factors of *in vitro* and *in vivo* intestinal absorption. *Journal of Pharmaceutical Science* 59: 154–163.

Baudner, B., M.M.Giuliani, J.C. Verhoef, R. Rappuoli, and H.E. Junginger. 2003. Del Guidice, G., The concomitant use of the LTK63 mucosal adjuvant and chitosan-based delivery system enhances the immunogenicity and efficacy of intranasally administered vaccines. *Vaccine* 21: 3837–3844.

Baudner, B., J.C. Verhoef, M.M. Giuliani, S. Peppoloni, G. Del Guidice, and H.E. Junginger. 2005. Protective immune responses to meningococcal C conjugate vaccine after intranasal immunization of mice with the LTK36 mutant plus chitosan or trimethyl chitosan chloride as novel delivery platform. *Journal of Drug Target* 13(8–9): 489–498.

Baxter, A., M. Dillon, K.D. Anthony Taylor, and G.A.F. Roberts. 1992. Improved method for i.r. determination of the degree of N-acetylation of chitosan. *International Journal of Biology and Macromolecule* 14(3): 166–169.

Bayat, A., F. Dorkoosh, A.R. Dehpour, L. Moeizi, B. Larijani, H.E. Junginger, and M. Rafiee-Tehrani. 2008. Nanoparticles of quaternized chitosan derivatives as a carrier for colon delivery of insulin: *Ex vivo* and *in vivo* studies. *International Journal of Pharmaceutics* 356(1–2): 259–266.

Bayat, A., A.M.M. Sadeghi, M.R. Avadi, M., Amini, A. Shafiee, M. Rafiee-Tehrani, R. Majlesi, and H.E. Junginger. 2006. Synthesis of N,N-dimethyl N-ethyl chitosan as a carrier for oral delivery of peptide drugs. *Journal of Bioactive and Compatible Polymer* 21: 433–444.

Berger, J., M. Reist, J.M., Mayer, O. Felt, and R. Gurny. 2004. Structure and interactions in chitosan hydrogels formed by complexation or aggregation for biomedical applications. *European Journal of Pharmaceutics and Biopharmaceutics* 57(1): 35–52.

Bernkop-Schnurch, A. 2002. Mucoadhesive polymers. In: *Polymeric Biomaterials*, Second Edition, Revised and Extended, Dimitrius, L.P., Ed. Marcel Dekker, Inc: New York, 147.

Bernkop-Schnürch, A. 2005. Thiomers: A new generation of mucoadhesive polymers. *Advanced. Drug Delivery Review* 57(11): 1569–1582.

Bernkop-Schnürch, A., U.M. Brandt, and A.E. Clausen. 1999. Synthesis and *in vitro* evaluation of chitosan–cysteine conjugates. *Scientia Pharmaceutica* 67: 196–208.

Bernkop-Schnürch, A., A. Heinrich, and A. Greimel. 2006. Development of a novel method for the preparation of submicron particles based on thiolated chitosan. *European Journal of Pharmaceutics and Biopharmaceutics* 63(2): 166–172.

Bernkop-Schnürch, A., and T.E. Hopf. 2001. Synthesis and *in vitro* evaluation of chitosan–thioglycolic acid conjugates. *Scientia Pharmaceutica* 69: 109–118.

Bernkop-Schnürch, A., M. Hornof, and T. Zoidl. 2003. Thiolated polymers–thiomers: Synthesis and *in vitro* evaluation of chitosan-2-iminothiolane conjugates. *International Journal of Pharmaceutics* 260: 229–237.

Berthold, A., K. Cremer, and J. Kreuter. 1996. Preparation and characterization of chitosan microspheres as drug carrier for prednisolone sodium phosphate as model for anti-inflammatory drugs. *Journal of Controlled Release* 39(1): 17–25.

Beysseriat, M., E.A. Decker, and D.J. McClements. 2006. Preliminary study of the influence of dietary fiber on the properties of oil-in-water emulsions passing through an *in vitro* human digestion model. *Food Hydrocolloid* 20(6): 800–809.

Bhattacharya, D., M. Das, D. Mishra, I. Banerjee, S.K. Sahu, T.K. Maiti, and P. Pramanik. 2011. Folate receptor targeted, carboxymethyl chitosan functionalized iron oxide nanoparticles: A novel ultradispersed nanoconjugates for bimodal imaging. *Nanoscale* 3: 1653–1662.

Birrenbach, G. and P. Speiser. 1976. Polymerized micelles and their use as adjuvants in immunology. *Journal of Pharmaceutical Science* 65: 1763–1766.

Bochard, G., H.L. Lueβen, J.C. Verhoef, C.M. Lehr, A.G. Boer, and H.E. Junginger. 1996. The potential of mucoadhesive polymers in enhancing intestinal peptide drug absorption III: Effects of chitosan glutamate and carbomer on epithelial tight junctions in vitro. *Journal of Controlled Release* 39: 131–138.

Brugnerotto, J., J. Lizardi, F.M. Goycoolea, W. Arguelles-Monal, J. Desbrieres, and M. Rinaudo. 2001. An infrared investigation in relation with chitin and chitosan characterization. *Polymer* 42(8): 3569–80.

Calvo, P., C. Remunan-Lopez, J.L. Vila-Jato, and M.J. Alonso. 1997. Novel hydrophilic chitosan-polyethylene oxide nanoparticles as protein carriers. *Journal of Applied Polymer Science* 63: 125–132.

Chatelet, C., O. Damour, and A. Domard. 2001. Influence of the degree of acetylation on some biological properties of chitosan films. *Biomaterials* 22(3): 261–8.

Chenite, A., M. Buschmann, D. Wang, and C. Chaput. 2001. Rheological characterization of thermogelling chitosan/glycerol phosphate solution. *Carbohydrate Polymer* 46: 39–47.

Choi, M., M. Cho, B.S. Han, J. Hong, J. Jeong, S. Park, M.H. Cho, K. Kim, and W.S. Cho. 2010. Chitosan nanoparticles show rapid extrapulmonary tissue distribution and excretion with mild pulmonary inflammation to mice. *Toxicology Letters* 199(2): 144–52.

Couvreur, P., B. Kante, M. Roland, P. Guiot, P. Baudhuin, and P. Speiser. 1979. Polycyanoacrylate nanoparticles as potential lysosomotropic carriers: Preparation, morphological and sorptive properties. *Journal of Pharmacy and Pharmacology* 31: 331–332.

Damgé, A., C. Michel, M. Aprahamian, and P. Couvreur. 1998. New approach for oral administration of insulin with polyalkyl cyanoacrylate nanocapsules drug carriers. *Diabetes* 37: 246–251.

Dash, M., F. Chiellini, R.M. Ottenbrite, and E. Chiellini. 2011. Chitosan—A versatile semi-synthetic polymer in biomedical applications. *Progress in Polymer Science* 36: 981–1014.

Davis, M.E. 2002. Non-viral gene delivery systems. *Current Opinion in Biotechnology* 13(2): 128–131.

de Campos A, A. Sánchez, and M.J. Alonso. 2001. Chitosan nanoparticles: A new vehicle for the improvement of the delivery of drugs to the ocular surface. Application to cyclosporine A. *International Journal of Pharmaceutics* 224: 159–168.

Dehousse, V., N. Garbacki, S. Jaspart, D. Castagne, G., Piel, A. Colige, and B. Evrard. 2010. Comparison of chitosan/siRNA and trimethylchitosan/siRNA complexes behavior in vitro. *International Journal of Biological Macromolecules* 46(3): 342–349.

Delie, F. 1998. Evaluation of nano and microparticle uptake by the gastrointestinal tract. *Advanced Drug Delivery Review* 34: 221–233.

Doamrd, A., M. Rinuado, C. Terrassin. 1986. New method for quaternization of chitosan. *International Journal of Biological Macromolecules* 8(2): 105–107.

Dodane, V., M.A. Khan, and J.R. Mervin. 1999. Effect of chitosan on epithelial permeability and structure. *International Journal of Pharmaceutics* 182: 21–32.

Dodane, V. and V.D. Vilivalam. 1998. Pharmaceutical applications of chitosan. *Pharmaceutical Science and Technology Today* 1: 246–253.

Dorkoosh, F.A., J.C. Verhoef, J.H. Verheijden, M. Rafiee-Tehrani, G. Borchard, and H.E. Junginger. 2002. Peroral absorption of octreotide in pigs in delivery systems on the basis of superporous hydrogel polymers. *Pharmaceutical Research* 19(10): 1532–1536.

Du, Y-Z., P.Lu, J.P. Zhou, H. Yuan, and F-Q. Hu. 2010. Stearic acid grafted chitosan oligosaccharide micelle as a promising vector for gene delivery system: Factors affecting the complexation. *International Journal of Pharmaceutics* 391: 260–266.

Duarte, M.L., M.C. Ferreira, M.R. Marvao, and J. Rocha. 2001. Determination of the degree of acetylation of chitin materials by 13C CP/MAS NMR spectroscopy. *International Journal of Biological Macromolecules* 28(5): 359–363.

Dünnhaupt, S., J. Barthelmes, J. Hombach, D. Sakloetsakun, V. Arkhipova, and A. Bernkop-Schnürch. 2011. Distribution of thiolated mucoadhesive nanoparticles on intestinal mucosa. *International Journal of Pharmaceutics* 408(1–2): 191–199.

Fattal, E. and C. Vauthier. 1864–1882. Nanoparticles as drug delivery systems, *Encyclopedia of Pharmaceutical Technology* Dekker, Inc. Vol. 1: Issue 1.

Florea, B.I., M. Thanou, H.E. Junginger, and G. Bochard. 2006. Enhancement of bronchial octreotide absorption by chitosan and N-trimethyl chitosan shows linear in vitro/in vivo correlation. *Journal of Controlled Release* 110(2): 353–361.

Frieke Kuper, C., P.J.Koornstra, D.M.H. Hamleers, J. Biewenga, B.J. Spit, A.M. Duijvestijn, P.J.C. van Breda Vriesman, and T. Sminia. 1992. The role of nasopharyngeal lymphoid tissue. *Immunology Today* 13: 219–224.

Fukuda, J., A. Khademhosseini, Y. Yeo, X. Yang, J. Yeh, G. Eng. J. Blumbling, C-F. Wang, D.S. Kohane, and R. Langer. 2006. Micromolding of photocrosslinkable chitosan hydrogel for spheroid microarray and co-cultures. *Biomaterials* 27: 5259–5267.

Galed, G., E. Diaz, and A. Heras. 2008. Conditions of N-deacetylation on chitosan production from alpha chitin. *Natural Product Communications* 3(4): 543–550.

Galed, G., B. Miralles, I. Paños, A. Santiago, and A. Heras. 2005. N-Deacetylation and depolymerization reactions of chitin/chitosan: Influence of the source of chitin. *Carbohydrate Polymer* 62(4): 316–20.

Gerasimenko, D.V., I.D. Avdienko, G.E. Bannikova, O.Y. Zueva, and V.P. Varlamov. 2004. Antibacterial effects of water-soluble lowmolecular-weight chitosans on different microorganisms. *Applied Biochemistry and Microbiology* 40(3): 253–257.

Germershaus, O., S. Mao, J. Sitterberg, U. Bakowsky, and T. Kissel. 2008. Gene delivery using chitosan, trimethyl chitosan or polyethylenglycol-graft-trimethyl chitosan block copolymers: Establishment of structure–activity relationships in vitro. *Journal of Controlled Release* 125(2): 145–154.

Günbeyaz, M., A. Faraji, A. Özkul, N. Purali, and S. Senel. 2010. Chitosan based delivery systems for mucosal immunization against bovine herpesvirus 1 (BHV-1). *European Journal of Pharmaceutical Science* 41: 531–545.

Hagenaars, N., M. Mania, P. deJong, I. Que, R. Nieuwland, B. Slütter, H. Glansbeek et al. 2010. Role of trimethylated chitosan (TMC) in nasal residence time, local distribution and toxicity of an intranasal influenza vaccine. *Journal of Controlled Release* 144(1): 17–24.

Hagenaars, N., R.J. Verheul, I. Mooren, P.H. deJong, E. Mastrobattista, H.L. Glansbeek et al. 2009. Relationship between structure and adjuvanticity of N,N,N-trimethyl chitosan (TMC) structural variants in a nasal influenza vaccine. *Journal of Controlled Release* 140(2): 126–133.

Hamidia, M., A. Azadia, and P. Rafieia. 2008. Hydrogel nanoparticles in drug delivery. *Advanced Drug Delivery Reviews* 60(15): 1638–1649.

Harada, A. and K. Kataoka. 1995. Formation of polyion complex micelles in an aqueous milieu from a pair of oppositely-charged block copolymers with poly(ethylene glycol) segments. *Macromolecules* 28: 5294–5299.

Hayes, E.R. 1992. Canadian Patent No. 1274507.

He, P., S.S. Davis, and L. Illum.1998. *In vitro* evaluation of the mucoadhesive properties of chitosan microspheres. *International Journal of Pharmaceutics* 166(1): 75–88.

Helander, I., E. Nurmiaho-Lassila, R. Ahvenainen, J. Rhoades, and S. Roller. 2001. Chitosan disrupts the barrier properties of the outer membrane of Gram-negative bacteria. *International Journal of Food Microbiology* 71: 235–244.

Hong, J.W., J.H. Park, K.M. Huh, H. Chung, I.C. Kwon, and S.Y. Jeong. 2004. PEGylated polyethylenimine for *in vivo* local gene delivery based on lipiodolized emulsion system. *Journal of Controlled Release* 99: 167–176.

Hornof, M.D., C.E. Kast, and A. Bernkop-Schnürch. 2003. *In vitro* evaluation of the viscoelastic behavior of chitosan–thioglycolic acid conjugates. *European Journal of Pharmaceutics and Biopharmaceutics* 55: 185–190.

Hou, M., Y. Zhang, J. Zhou, A. Zou, D. Yu, Y. Wu, J. Li, H. Li. 2010. Synthesis and characterization of low toxic amphiphilic derivatives and their applications as micelle carrier for antitumor drug. *International Journal of Pharmaceutics* 394: 162–173.

Hu, F.-Q., P. Meng, Y-Q. Dai, Y-Z. Du, J. You, X-H. Wei, H. Yuan. 2008. PEGylated chitosan based polymer micelle as an intracellular delivery carrier for anti-tumor targeting therapy. *European Journal of Pharmaceutics and Biopharmaceutics* 70: 749–757.

Hu, F.-Q., L-N.Liu, Y-Z. Du, H. Yuan. 2009. Synthesis and antitumor activity of doxorubicin conjugated stearic acid-g-chitosan oligosaccharide polymeric micelles. *Biomaterials* 30: 6955–6963.

Illum, L., N.F. Farraj, S.S. Davis. 1994. Chitosan as novel nasal delivery system for peptides drugs. *Pharmaceutical Research* 11: 1186–9.

Illum, L., I. Jabbal-Gill, M. Hinchcliffe, A.N. Fisher, and S.S. Davis. 2001. Chitosan as a novel nasal delivery system for vaccines. *Advanced Drug Delivery Review* 51(1–3): 81–96.

Jeuniaux, C. 1982. La Chitine dans la regne animal. *Bulletin de la Société Zoologique de France* 107: 363–386.

Jian, Y., T. Feng, W. Zheng, W. Qing, Z. Yan-Jun, and C. Shi-Qian. 2008. Effect of chitosan molecular weight and deacetylation degree on hemostasis. *Journal of Biomedical Material Research* 84B: 131–137.

Jiang, H, I. Park, and N. Shim. 2004. *In vitro* study of the immune stimulating activity of an athrophic rhinitis; vaccine associated to chitosan microspheres. *European Journal of Pharmaceutics and Biopharmaceutics* 58: 471–476.

Junginger, H. and J. Verhoef. 1998. Macromolecules as safe penetration enhancers for hydrophilic drugs: A fiction? *Pharmaceutical Science and Technology Today* 1(19): 370–376.

Kamel, A., M. Sokar, V. Naggar, and S. Gamal. 2002. Chitosan and sodium alginate based bioadhesive vaginal tablets. *AAPS Pharmaceutical Science* 4(4): 224–230.

Kaneda, Y. 2005. Biological barriers to gene. In: M.M. Amiji (Ed.), *Polymeric Gene Delivery: Principles and Applications*, Chapter 3, Boca Raton, FL: CRC Press LLC, 29–42.

Kassai, M. 2008. A review of several reported procedures to determine the degree of N-acetylation for chitin and chitosan using infrared spectroscopy. *Carbohydrate Polymer* 71: 497–508.

Kast, C.E. and A. Bernkop-Schnürch. 2001. Thiolated polymers–thiomers: Development and *in vitro* evaluation of chitosan–thioglycolic acid conjugates. *Biomaterials* 22(17): 2345–2352.

Kataoka, K., A. Harada, and Y. Nagasaki. 2001. Block copolymer micelles for drug delivery: Design characterization and biological significance. *Advanced Drug Delivery Review* 47: 113–131.

Katas, H. and H.O. Alpar. 2006. Development and characterization of chitosan nanoparticles for siRNA delivery. *Journal of Controlled Release* 115(2): 216–225.

Kim, T.H., H. Jin, H.W. Kim, M.H. Cho, and C.S. Cho. 2006. Mannosylated chitosan nanoparticle-based cytokine gene therapy suppressed cancer growth in BALB/c mice bearing CT-26 carcinoma cells. *Molecular Cancer Therapy* 5(7): 1723–1732.

Klingels, M., U. Griesbach, C. Panzer, and R. Wachter. 1999. Chitosan—A cosmetic agent for modular. Concepts. *Henkel- Referate* 35: 78–83.

Klokkevoid, P.R., H. Fukuyama, E.C. Sung, and C.N. Bertolami. 1999. The effects of chitosan (poly-N-acetyl glucosamine) on lingual hemostasis in heparinized rabbits. *International Journal of Oral Maxillary Surg*ery 57: 49–52.

Kofuji, K., C.J. Qian, M. Nishimura, I. Sugiyama, Y. Murata, and S. Kawashima. 2005. Relationship between physicochemical characteristics and functional properties of chitosan. *European Polymer Journal* 41(11): 2784–2791.

Koide, S.S. 1998. Chitin–chitosan: Properties, benefits and risks. *Nutrition Research* 18: 1091–101.

Koping-Hoggard, M., I. Tubulekas, H. Guan, K. Edwards, M. Nilsson, K.M., Varum, and P. Artursson. 2001. Chitosan as a non viral gene delivery system. Structure-property relationships and characteristics compared with polyethylenimine *in vitro* and after lung administration in vivo. *Gene Therapy* 8(14): 1108–1121.

Kotzé, A.F., B.J. Leeuw, H.L. Lueβen, A.G. Boer, J.C. Verhoef, and H.E. Junginger. 1997a. Chitosans for enhanced delivery of therapeutic peptides across intestinal epithelia: *In vitro* evaluation in Caco-2 cell monolayers. *International Journal of Pharmaceutics* 159: 243–253.

Kotzé, A.F., H.L. Lueβen, A.G. deBoer, B.J. deLeeuw, J.C. Verhoef, and H.E. Junginger. 1997b. N-trimethyl chitosan chloride as a potential absorption enhancer across mucosal surfaces: *In vitro* evaluation in intestinal epithelial cells (Caco-2). *Pharmaceutical Research* 14: 1197–1202.

Kotzé, A., H. Luessen, and M. Thanou et al. 1999. Chitosan and its derivatives as absorption enhancers for peptide drugs across mucosal epithelia. In: E. Mathiowitz, C.M. Lehr, eds. *Bioadhesive Drug Delivery Systems*. New York: Marcel Dekker Inc 1999; 341–385.

Kotzé, A.F., M.M. Thanou, H.L. Lueβen, A.G. deBoer, J.C. Verhoef, and H.E. Junginger. 2000. Enhancement of paracellular drug transport with highly quaternized N-trimethyl chitosan chloride in neutral environments: *In vitro* evaluation in intestinal epithelial cells (Caco-2). *Journal of Pharmaceutical Science* 88(2): 253–257.

Krauland, A.H. and M.J. Alonso. 2007. Chitosan cyclodextrin nanoparticles as macromolecular drug delivery system. *International Journal of Pharmaceutics* 340: 134–142.

Krauland, A.H., D. Guggi, and A. Bernkop-Schnürch. 2004. Oral insulin delivery: The potential of thiolated chitosan–insulin tablets on non-diabetic rats. *Journal of Controlled Release* 95(3): 547–555.

Krauland, A.H., D. Guggi, and A. Bernkop-Schnürch. 2006. Thiolated chitosan microparticles: A vehicle for nasal peptide drug delivery. *International Journal of Pharmaceutics* 307(2): 270–277.

Kumar, M.N.V.R. 2000. A review of chitin and chitosan applications. *Reactive and Functional Polymers* 46(1): 1–27.

Kumar, M.N.V.R., R.A.A. Muzarelli, C. Muzarelli, H. Sashiwa, and A.J. Domb. 2004. Chitosan chemistry and pharmaceutical perspectives. *Chemical Reviews* 104: 6017–6084.

Lai, W-F., M. Lin, Chia-Mi, and M. Lin. 2009. Nucleic acid delivery with chitosan and its derivatives. *Journal of Controlled Release* 134: 158–168.

Lamarque, G., M. Cretnet, C. Viton, and A. Domard. 2005. New route of deacetylation of α and β-chitins by means of freeze-pump out-thaw cycles. *Biomacromolecules* 6: 1380–1388.

Lehr, C.M., J. Bouwstra, E. Schacht, and H. Junginger. 1992. H., *In vitro* evaluation of mucoadhesive properties of chitosan and some other natural polymers. *International Journal of Pharmaceutics* 78: 43–48.

Li, S.D. and L. Huang. 2008. Pharmacokinetics and biodistribution of nanoparticles. *Molecular Pharmacology* 5: 496–504.

Liu, W.G. and K.D. Yao. 2002. Chitosan and its derivatives—A promising non-viral vector for gene transfection. *Journal of Controlled Release* 83(1): 1–11.

Liu, X., L. Yun, Z. Dong, L., Zhi, and D. Kang. 2001. Antibacterial action of chitosan and carboxymethylated chitosan. *Journal of Applied Polymer Science* 79(7): 1324–1335.

Lu, G., B. Sheng, Y., Wei, G. Wang, L. Zhang, Y.Gong, and X. Zhang. 2008. Collagen nanofiber-covered porous biodegradable carboxymethyl chitosan microcarriers for tissue engineering cartilage. *European Polymer Journal* 44(9): 2820–2829.

Luca, L., A-L. Rougemont, B.H. Walpoth, B.H., Boure, L. Tami, J.M. Anderson, O. Jordan, and R. Gurny. 2011. Injectable rhBMP-2-loaded chitosan hydrogel composite: Osteoinduction at ectopic site and in segmental long bone defect. *Journal of Biomedical Material Research Part* A. 96A(1): 66–74.

Lueβen, H.L., C.M.Lehr, C.O. Rentel, A.B.J. Noach, AG. DeBoer, J.C. Verhoef, and H.E. Junginger. 1997. Bioadhesive polymers for peroral delivery of peptide drugs. *Journal of Controlled Release* 29: 329–338.

Luo, Y., A. Wang, J. Yuan, and Q. Gao. 2009. Preparation, characterization and drug release behavior of polyion complex micelles. *International Journal of Pharmaceutics* 374: 139–144.

Ma. Z., T. Lim, and L.-Y. Lim. 2005. Pharmacological activity of peroral chitosan–insulin nanoparticles in diabetic rats. *International Journal of Pharmaceutics* 293: 271–280.

Mansouri, S., P. Lavigne, K. Corsi, M. Benderdour, E. Beaumont, J.C. Fernandes. 2004. Chitosan–DNA nanoparticles as non-viral vectors in gene therapy: Strategies to improve transfection efficacy. *European Journal of Pharmaceutics and Biopharmaceutics* 57(1): 1–8.

Mao, H.Q., K. Roy, V.L. Troung-Le, K.A. Janes, K.Y. Lin, Y. Wang. August, J.T., Leong, K.W. 2001. Chitosan–DNA nanoparticles as gene carriers: Synthesis, characterization and transfection efficiency. *Journal of Controlled Release* 70(3): 399–421.

Mao, S., U. Bakowsky, and T. Kissel. 2006. Self assembled polyelectrolyte nanocomplexes between chitosan derivatives and insulin. *Journal of Pharmaceutical Science* 95: 1035–1048.

Mumper, R.J., J. Wang, J.M. Claspel, and A.P. Rolland. 1995. Novel polymeric condensing carriers for gene delivery. *Proc. Int. Symp, Controlled Release Bioactive Material* 22: 178–179.

Murata, Y., S. Toniwa, E. Miyamoto, and S. Kawashima. 1999. Preparation of alginate gel beads containing chitosan nicotinic acid salt and functions. *European Journal of Pharmaceutics and Biopharmaceutics* 48: 49–52.

Muzzarelli, R., F. Orlandini, and D. Pacetti. 2006. Chitosan taurocholate capacity to bind lipids and to undergo enzymatic hydrolysis: An *in vitro* model. *Carbohydrate Polymer* 66: 363–71.

Muzzarelli, R.A.A. 1978. Chairman's address: Chitin, an important natural polymer, In: *Proceedings of the First International Conference on Chitin/Chitosan*, R.A.A. Muzzarelli and E.R. Pariser, eds., Massachusetts Institute of Technology Press, Cambridge.

Muzzarelli, R.A.A. and B.B. Muzzarelli. 1998. Structural and functional versatility of chitins. In: *Structural Diversity and Functional Versatility of Polysaccharides*. S. Dumitriu ed. New York: Marcel Dekker.

Muzzarelli, R.A.A., F. Tanfani, M. Emmanueli, and S. Mariotti. 1982. N-(carboxymethylidene) chitosans and *N*-(carboxymethyl)-chitosans: Novel chelating polyampholytes obtained from chitosan glyoxylate. *Carbohydrate Research* 107: 199–214.

Nafee, N., S. Taetz, M. Schneider, U.F. Schaefer, and C.M. Lehr. 2007. Chitosan-coated PLGA nanoparticles for DNA/RNA delivery: Effect of the formulation parameters on complexation and transfection of antisense oligonucleotides. *Nanomedicine* 3(3): 173–183.

Nandi, S., M. Kumar, M. Manohar, and R.S. Chauhan. 2009. Bovine herpes virus infections in cattle. *Animal Health Research Review* 10: 85–98.

Nazar, H., D.G.Fatouros, S.M. Van der Merwe, N. Bouropoulos, G. Avgouropoulos, J. Tsibouklis, and M. Roldo. 2011. Thermosensitive hydrogels for nasal drug delivery: The formulation and characterisation of systems based on *N*-trimethyl chitosan chloride. *European Journal of Pharmaceutics and Biopharmaceutis* 77(2): 225–232.

Ogawa, S., E. Decker, and D. McClements. 2003. Influence of environmental conditions on the stability of oil in water emulsions containing droplets stabilized by lecithin–chitosan membranes. *Journal of Agricultural and Food Chemistry* 51: 5522–5527.

Ohshima, Y., H. Nishino, Y. Yonekura, S. Kishimoto, and S. Wakabayasi. 1987. Clinical application of chitin non-woven fabric as wound dressing. *European Journal of Plastic Surgery* 10: 66–69.

Ohya, Y., R. Cai, H. Nishizawa, K. Hara, and T. Ouchi. 1999. Preparation of PEG-grafted chitosan nano-particles for peptide drug carrier. *Proc. International Symposium Controlled Release Bioactive Materials* 26: 655–656.

Okamoto, Y., Y. Minami, and A. Matsuhashi. 1993. Application of polymeric *N*-acetyl-D-glucosamine (chitin) to veterinary practice. *Journal of Veterinary Medicine* 55: 743–747.

Omelyaneko, V., P. Kopeckova, C. Gentry, and J. Kopecek. 1998. Targetable HPMA copolymer adriamycin conjugates. Recognition, internalization and subcellular fate. *Journal of Controlled Release* 53: 25–37.

Pangburn, S.H., P.V. Trescony, and J. Heller. 1982. Lysozyme degradation of partially deacetylated chitin, its films and hydrogels. *Biomaterials* 3(2): 105–108.

Patel, H.M. 1992. Serum opsonins and liposomes: Their interaction and opsonophagocytosis. *Critical Reviews Therapeutic Drug Carrier System* 9: 39–90.

Peng, C. 1998. Synthesis of crosslinked chitosan-crown ethers and evaluation of these products as adsorbents for metal ions. *Journal of Applied Polymer Science* 70: 501–506.

Peniston, Q.P. and Johnson, E.L. 1980. U.S. Patent No. 4195175.

Pille, J.Y., H. Li, E. Blot, J.R. Bertrand, L.L. Pritchard, P. Opolon, A. Maksimenko et al. 2006. Intravenous delivery of anti-RhoA small interfering RNA loaded in nanoparticles of chitosan in mice: Safety and efficacy in xenografted aggressive breast cancer. *Human Gene Therapy* 17(10): 1019–1026.

Ramos, V.M., N.M. Rodriguez, M.S. Rodriguez, A. Heras, and E. Agullo. 2003. Modified chitosan carrying phosphoric and alkyl groups. *Carbohydrate Polymer* 51: 425–429.

Rao, S., and C.P. Sharma. 1997. Use of chitosan as biomaterial: Studies on its safety and hemostatic potential. *Journal of Biomedical Material Research* 34: 21–28.

Raymond, L., F.G. Morin, and R.H. Marchessault. 1993. Degree of deacetylation of chitosan using conducto-metric titration and solid-state NMR. *Carbohydrate Research* 246(1): 331–336.

Rege, P.R. and L.H. Block. 1999. Chitosan processing: Influence of process parameters during acidic and alka-line hydrolysis and effect of the processing sequence on the resultant chitosan's properties. *Carbohydate Research* 321(3–4): 235–245.

Roberts, G. 1982. *Chitin Chemistry*. London: Macmillan.

Rohindra, D.R., A.V. Nand, and J.R. Khurma. 2004. Swelling properties of chitosan hydrogels. *The South Pacific Journal of Natural Science* 22(1): 32–35.

Rojanarata, T., P. Opanasopit, S. Techaarpornkul, T. Ngawhirunpat, U. Ruktanonchai. 2008. Chitosan-thiamine pyrophosphate as a novel carrier for siRNA delivery. *Pharmaceutical Research* 25(12): 2807–2814.

Roldo, M., M. Hornof, P. Caliceti, and A. Bernkop-Schnürch. 2004. Mucoadhesive thiolated chitosans as platforms for oral controlled drug delivery: Synthesis and *in vitro* evaluation. *European Journal of Pharmaceutics and Biopharmaceutics* 57: 115–121.

Roller, S., and N. Covill. 1999. The antifungal properties of chitosan in laboratory media in apple juice. *International Journal of Food Microbiology* 47: 67–77.

Rossi, S., M. Marciello, M.C. Bonferoni, F. Ferrari, G. Sandri, C. Dacarro, P. Grisol, and C. Caramella. 2010. Thermally sensitive gels based on chitosan derivatives for the treatment of oral mucositis. *European Journal of Pharmaceutics and Biopharmaceutics* 74(2): 248–254.

Ruiz-Herrera, J. 1978. *in Proceedings of the International Conference on Chitin Chitosan* (Muzzarelli R. A. A., Pariser E. R., eds) pp. 11–21, MIT Sea Grant Report 78–87, Sea Grant Information Center, MIT Sea Grant Program, Boston.

Rúnarsson, Ö., J. Holappa, T. Nevalainen, M. Hjálmarsdóttir, T. Järvinen, Th. Loftsson, J.M. Einarsson, S. Jónsdóttir, M. Valdimarsdóttir, and M. Másson. 2007. Antibacterial activity of methylated chitosan and chitooligomer derivatives: Synthesis and structure activity relationships. *European Polymer Journal* 43(6): 2660–2671.

Saboktakin, M.R., R. Tabatabaie, A. Maharramov, and M.A. Ramazanov. 2011. Synthesis and *in vitro* stud-ies of biodegradable thiolated chitosan hydrogels for breast cancer therapy. *International Biological Macromolecule* 48(5): 747–752.

Sadeghi, A.M.M., M. Amini, M.R. Avadi, F. Siedi, M. Rafiee-Tehrani, and H.E. Junginger. 2008c. Synthesis, characterization, and antibacterial effects of trimethylated and triethylated 6-NH$_2$-6-deoxy chitosan. *Journal of Bioactive and Compatible Polymer* 23: 262–275.

Sadeghi, A.M.M., M.R. Avadi, SH. Ejtemaimehr, SH. Abashzadeh, A. Partoazar, F. Dorkoosh, M. Faghihi, M., Rafiee-Tehrani, and H.E. Junginger. 2009. Development of a gas empowered drug delivery in the small intestine. *Journal of Controlled Release* 134(1) :11–17.

Sadeghi, A.M.M., F.A. Dorkoosh, M.R. Avadi, P. Saadat, M. Rafiee-Tehrani, and H.E. Junginger. 2008a. Preparation, characterization and antibacterial activities of chitosan, N-trimethyl chitosan (TMC) and N-diethylmethyl chitosan (DEMC) nanoparticles loaded with insulin using both the ionotropic gelation and polyelectrolyte complexation methods. *International Journal of Pharmaceutics* 35: 299–306.

Sadeghi, A.M.M., F.A. Dorkoosh, M.R. Avadi, M. Weinhold, A. Bayat, F. Delie, R. Gurny, M.B. Larijani, M. Rafiee-Tehrani, and H.E. Junginger. 2008b. Permeation enhancer effect of chitosan and chito-san derivatives: Comparison of formulations as soluble polymers and nanoparticulate systems on insulin absorption in Caco-2 cells. *European Journal of Pharmaceutics and Biopharmaceutics* 70: 270–278.

Safari, S., F.A. Dorkoosh, M. Soleimani, M.H. Zarrintan, H. Akbari, B. Larijani, and M. Rafiee-Tehrani. 2011. N-Diethylmethylchitosan for gene delivery to pancreatic cancer cells and the relation between charge ratio and biologic properties of polyplexes via interpolations polynomial. *International Journal of Pharmaceutics* 420(2): 350–357.

Sandri, G., M.C.Bonferonia, S. Rossia, F. Ferraria, S. Gibina, Y. Zambito, G. DiColo, and C. Caramella. 2007. Nanoparticles based on N-trimethylchitosan: Evaluation of absorption properties using *in vitro* (Caco-2 cells) and *ex vivo* (excised rat jejunum) models. *European Journal of Pharmaceutics and Biopharmaceutics* 65(1): 68–77.

Satoh, T., H. Kano, M. Nakatani, N. Sakairi, S. Shinkai, and T. Nagasaki. 2006. 6-Amino-6 deoxy chitosan sequential chemical modifications at the C-6 positions of N-phthaloyl-chitosans and evaluation as a gene carrier. *Carbohydrate Polymer* 341: 2406–2413.

Sayin, B., S. Somavarapu, X.W. Li, D. Sesardic, S. Senel, and O.H. Alpar. 2009. TMC–MCC (*N*-trimethyl chitosan-mono-*N*-carboxymethyl chitosan) nanocomplexes for mucosal delivery of vaccines. *European Journal of Pharmaceutical Science* 38(4): 362–369.

Sayin, B., S. Somavarapu, X.W. Li, M. Thanou, D. Sesardic, H.O. Alpar, and S. Senel. 2008. Mono-N-carboxymethyl chitosan (MCC) and N-trimethyl chitosan (TMC) nanoparticles for non-invasive vaccine delivery. *International Journal of Pharmaceutics* 363(1–2): 139–148.

Schipper, N.G.M., J. Hoogstraate, A. deBoer, K.M. Vårum, and P. Artursson. 1997. Chitosans as absorption enhancers for poorly absorbable drugs 2: mechanism of absorption enhancement. *Pharmaceutical Research* 14(7): 923–929.

Schipper, N.G.M., K. Varum, and P. Artursson. 1996. Chitosans as absorption enhancers for poorly absorbable drugs. 1: Influence of molecular weight and degree of acetylation on drug transport across human intestinal epithelial (Caco-2) cells. *Pharmaceutical Research* 13(11): 1686–1692.

Shahnaz, G., A. Vetter, J. Barthelmes, D. Rahmat, F. Laffleur, J. Iqbal, G. Perera et al. 2012. Thiolated chitosan nanoparticles for the nasal administration of leuprolide: Bioavailability and pharmacokinetic characterization. *International Journal of Pharmaceutics* 428: 164–170.

Shibata, H., Y. Yoshioka, S. Ikemizu, K. Kobayashi, Y. Yamamoto, Y. Mukai, T. Okamoto et al. 2004. Functionalization of tumor necrosis factor-alpha using phage display technique and PEGylation improves its antitumor therapeutic window. *Clinical Cancer Research* 10: 8293–8300.

Shu, X.Z. and K.J. Zhu. 2000. A novel approach to prepare TPP/chitosan complex beads for controlled release drug delivery. *International Journal of Pharmaceutics* 201: 51–58.

Sieval, A.B., M. Thanou, A.F. Kotzé, J.C. Verhoef, J. Brussee, and H.E. Junginger. 1997. Preparation and NMR characterization of highly substituted *N*-trimethyl chitosan chloride. *Carbohydrate Polymer* 36(2–3): 157–165.

Slütter, B., N. Hagenaars, and W. Jiskoot. 2008. Rational design of nasal vaccines. *Journal of Drug Targeting* 16: 1–17.

Smith. J., E. Wood, and M. Dornish. 2004. Effect of chitosan on epithelial cell tight junctions. *Pharmaceutical Research* 21(1): 43–49.

Snyman, D., J.H. Hamman, and A.F.Kotze. 2003. Evaluation of the mucoadhesive properties of N-trimethyl chitosan chloride. *Drug Development and Industrial Pharmacy* 29: 61–69.

Snyman, D., J.H. Hamman, J.E. Rollings, and A.F. Kotze. 2002. The relationship between the absolute molecular weight and the degree of quaternization of *N*-trimethyl chitosan chloride. *Carbohydrate Polymer* 50: 145–150.

Sukwattanasinitt M, H. Zhu, H. Sashiwa, and S. Aiba. 2002. Utilization of commercial non-chitinase enzymes from fungi for preparation of 2-acetamido-2-deoxy—glucose from [beta]-chitin. *Carbohydrate Research* 337(2): 133–1337.

Takeuchi, H., H. Yamamoto, T. Niwa, T. Hino, and Y. Kawashima. 1996. Enteral absorption of insulin in rats from mucoadhesive chitosan coated liposomes. *Pharmaceutical Research* 13: 896–901.

Tang, Y., J. Sun, H. Fan, and X. Zhang. 2012. An improved complex gel of modified gellan gum and carboxymethyl chitosan for chondrocytes encapsulation. *Carbohydrate Polymer* 88: 46–53.

Thanou, M., J.C.Verhoef, and H.E. Junginger. 2001a. Oral drug absorption enhancement by chitosan and its derivatives. *Advanced Drug Delivery Review* 52: 117–126.

Thanou, M., J.C. Verhoef, and H.E. Junginger. 2001b. Chitosan and its derivatives as intestinal absorption enhancers. *Advanced Drug Delivery Review* 50: 91–101.

Thanou, M.M., J.C.Verhoef, Jos H.M. Verheijden, and H.E. Junginger. 2001c. Intestinal absorption of octreotide using trimethyl chitosan chloride: Studies in pigs. *Pharmaceutical Research* 18(6): 823–828.

Thanou. M., M.T. Nihot, M. Jansen, J.C. Verhoef, and H.E. Junginger. 2001d. Mono-N-carboxymethyl chitosan (MCC), a polyampholytic chitosan derivative, enhances the intestinal absorption of low molecular weight heparin across intestinal epithelia *in vitro* and in vivo. *Journal of Pharmaceutical Science* 90: 38–46.

Thanou, M.M., A.F. Kotzé, T. Scharringhausen, H.L. Lueßen, A.G. de Boer, J.C. Verhoef, and H.E. Junginger. 2000a. Effect of degree of quaternization of N-trimethyl chitosan chloride for enhanced transport of hydrophilic compounds across intestinal Caco-2 cell monolayers. *Journal of Controlled Release* 64: 15–25.

Thanou, M.M., B.I. Florea, M.W.E. Langemeÿer, J.C. Verhoef, and H.E. Junginger. 2000b. N-Trimethylated chitosan chloride (TMC) improves the intestinal permeation of the peptide drug buserelin *in vitro* (Caco-2 cells) and *in vivo* (rats). *Pharmaceutical Research* 17(1): 27–31.

Tharanathan, R.N. and F.S. Kittur. 2003. Chitin: The undisputed bimolecular of great potential. *Critical Reviews of Food Science Nutrition* 43(1): 61–87.

Thongngam, M. and D.J. McClements. 2004. Characterization of interactions between chitosan and an anionic surfactant. *Journal of Agricultural Food Chemistry* 52(4): 987–91.

Thongngam, M. and D.J. McClements. 2005. Isothermal titration calorimetry study of the interactions between chitosan and a bile salt (sodium taurocholate). *Food Hydrocolloid* 19(5): 813–819.

Tsigos, I., A. Martinou, K.M. Varum, and V. Bouriotis 1996 Enzymatic deacetylation of chitinous substrates employing chitin deacetylases. In: *Advances in Chitin Science*, Vol. I (A. Domard, C. Jeuniaux, R., Muzzarelli, and G. Roberts, eds) pp. 59–69, Jaques Andre.

Ulansky, P. and J. Rosiak. 1992. Radiation chemistry and physical chemistry of chitosan and other polysaccharides. *Radiation Physical Chemistry* 39(1): 53–57.

Urrusuno, R.F., P. Calvo, C.R. Lopez, J.L. Vila, and M.J. Alonso. 1999. Enhancement of nasal absorption of insulin using chitosan nanoparticles, *Pharmaceutical Research* 16: 1576–1581.

Uthairatanakij, A., J.A. Teixeira da Silva, and K. Obsuwan. 2007. Chitosan for improving orchid production and quality. *Orchid Science and Biotechnology* 1(1): 1–5.

Van de Velde, K., and P. Kiekens. 2004. Structure analysis and degree of substitution of chitin, chitosan and dibutyrylchitin by FTIR spectroscopy and solid state 13C NMR. *Carbohydrate Polymer* 58: 409–416.

Van der Lubben, I.M., J.C.Verhoef, G. Bochard, and H.E. Junginger. 2001a. Chitosan for mucosal vaccination. *Advanced Drug Delivery Review* 52(2): 139–144.

Van der Lubben, I.M., J.C.Verhoef, A.C. van Aelst, G. Bochard, and H.E. Junginger. 2001b. Chitosan microparticles for oral vaccination: Preparation, characterization and preliminary *in vivo* uptake studies in murine Peyer's patches. *Biomaterials* 22(7): 687–694.

Van der Lubben, I.M., J.C.Verhoef, M.M. Fretz, F.A.C. Van Opdorp, I. Mesu, G. Kersten, and H.E. Junginger. 2002. Trimethyl chitosan chloride (TMC) as a novel excipient for oral and nasal immunization against diphtheria. *S.T.P. Pharma Sciences* 12(4): 235–242.

Varum, K.M., O. SmidsrØd. 1994. Chitosan preparation. World patent Application No. WO 03011912.

Verheul, R.J., M. Amidi, S. Van der Wal, E. Van Riet, W. Jiskoot, and W.E. Hennink. 2008. Synthesis, characterization and *in vitro* biological properties of O-methyl free *N,N,N*-trimethylated chitosan. *Biomaterials* 29: 3642–3649.

Vila, A., A. Sanchez, K. Janes, I. Behrens, T. Kissel, J.L. Vila Jato, and M.J. Alonso. 2004. Low molecular weight chitosan nanoparticles as new carriers for nasal vaccine delivery in mice. *European Journal of Pharmaceutics and Biopharmaceutics* 57: 123–131.

Vila, A., A. Sanchez, M. Tobio, P. Calvo, and M.J. Alonso. 2002. Design of biodegradable particles for protein delivery. *Journal of Controlled Release* 78: 15–24.

Wong, K., G. Sun, X. Zhang, H. Dai, Y. Liu, C. He, and K.W. Leong. 2006. PEI-g-chitosan, a novel gene delivery system with transfection efficiency comparable to polyethylenimine *in vitro* and after liver administration in vivo. *Bioconjugate Chemistry* 17(1): 152–158.

Wu, Z.M., X.G. Zhang, C. Zheng, C.X. Li, S.M. Zhang, R.N. Dong, and D.M. Yu. 2009. Disulfide-crosslinked chitosan hydrogel for cell viability and controlled protein release. *European Journal of Pharmaceutical Science* 37(3–4): 198–206.

Xing, R., S. Liu, and Z. Guo. 2005. Relevance of molecular weight of chitosan and its derivatives and their antioxidant activities in vitro. *Bioorganic and Medicinal Chemistry* 13(5): 1573–1577.

Xu, T., M. Xin, M. Li., H. Huang, and S.H. Zhou. 2010. Synthesis, characteristic and antibacterial activity of *N,N,N*-trimethyl chitosan and its carboxymethyl derivatives. *Carbohydrate Polymer* 81(4): 931–936.

Xue, X., L. Li, and J. He. 2009. The performances of carboxymethylchitosan in wash-off reactive dyeing. *Carbohydrate Polymer* 75(2): 203–207.

Younessi, P., M.R. Avadi, K. Shamimi, A.M.M. Sadeghi, L. Moeizi, E. Nahid, KH. Bayati, A.R. Dehpour, and M. Rafiee-Tehrani. 2004. Preparation and *ex vivo* evaluation of TEC as an absorption enhancer for poorly absorbable compounds in colon specific drug delivery. *Acta Pharmaceutica* 54: 339–345.

Zhang, Y., J.Chen, Y. Pan, J. Zhao, L. Ren, M. Liao, Z. Hu, L. Kong, and J. Wang. 2007. A novel PEGylation of chitosan nanoparticles for gene delivery. *Biotechnology and Applied Biochemistry* 46(4): 197–204.

Zheng, F., X.W. Shi, G.F. Yang, L.L. Gong, H.Y. Yuan, Y.J. Cui, Y. Wang, Y.M. Du, and Y. Li. 2007. Chitosan nanoparticle as gene therapy vector via gastrointestinal mucosa administration: Results of an *in vitro* and *in vivo* study. *Life Science* 80(4): 388–396.

Zheng, Y., Z. Cai, X. Song, B. Yu, Q. Chen, D. Zhao, J. Xu, and S. Hou. 2009. Receptor mediated gene delivery by folate conjugated *N*-trimethyl chitosan in vitro. *International Journal of Pharmaceutics* 382(1–2): 262–269.

Zheng, Y., X. Song, G. He, Z. Cai, Y. Zhou, B. Yu, J. Xu, Y. Wei, and S. Hou. 2011. Receptor-mediated gene delivery by folate-poly(ethylene glycol)-grafted-trimethyl chitosan in vitro. *Journal of Drug Targeting* 19(8): 647–656.

Zhengwei, M., M. Lie, J. Yan, and Y. Ming. 2007. Changyou, G., Jiacong, S., *N,N,N*-Trimethylchitosan chloride as a gene vector: Synthesis and application. *Macromolecular Biosci*ence 7(6): 855–863.

Zhou, K., W. Xia, C. Zhang, and L. Yu. 2006. *In vitro* binding of bile acids and triglycerides by selected chitosan preparations and their physicochemical properties. *LWT—Food Science Technology* 39(10): 1087–1092.

3 Utilization of Silyl Ethers and Other Protection Groups in the Synthesis of Chitosan Derivatives

Vivek S. Gaware, Berglind Eva Benediktsdóttir, and Már Másson

CONTENTS

3.1 INTRODUCTION

Strategies based on the use of protection groups have a significant role in the synthesis of organic compounds, especially when it involves selective modification of one group in the presence of several other reactive functional groups. Despite the versatility of protection groups, their utilization in chitosan chemistry has not been extensive and is limited almost exclusively to phthaloyl, trityl, and silyl ether protection groups. Efficient protecting group strategy can have a dual role in chitosan chemistry, that is, to protect a functional group for chemoselective or regioselective modification and to improve the solubility of the polymer in organic solvents, thereby promoting efficient modifications under homogeneous conditions.

Structurally (Figure 3.1), chitin and chitosan differ from cellulose only at the C-2 position of the polymer backbone, where chitin has an N-acetyl group and chitosan an amino group, whereas cellulose has a hydroxyl functional group. Chitin is predominantly composed of linear poly-β-(1 $\rightarrow$ 4)-linked N-acetylglucosamine (GlcNAc) units and occurs as α-, β-, and γ-chitin polymorph, which differ in intermolecular hydrogen-bonding structure (Blackwell and Weih 1980). Chitosan is a partially or fully N-deacetylated form of chitin and is composed of poly-β-(1 $\rightarrow$ 4)-linked D-glucosamine (GlcN) and partially (generally <30%) GlcNAc monomer units (Bough et al. 1978; Badawy and Rabea 2011).

Chitosan can be prepared with different average molecular weights (MW), ranging from chitosan oligomers to high-MW chitosan. Chitooligosaccharides (COS) are degraded products of chitosan/chitin and can be prepared by enzymatic (e.g., cellulase, lipase and protease, and chitosanase) or acidic hydrolysis (Lee et al. 2008; Lin et al. 2009). Generally, chitosan with less than 10 kDa can be considered as COS or low-MW chitosan.

3.1.1 PHYSICOCHEMICAL PROPERTIES OF CHITOSAN AND APPLICATION

Chitosan has three reactive functional groups (Figure 3.2), which are the primary amino (C-2), primary hydroxyl (C-6), and secondary hydroxyl (C-3) groups, all of which are nucleophilic.

The amino group is largely responsible for the specific physiochemical properties as this biopolymer, which is polycationic at pH below 6.5 (pKa = ~6.5) (Anthonsen and Smidsrød 1995). The degree of acetylation (DA) and pH influence the charge density. Thus, chitosan is soluble in dilute aqueous acid due to protonation of the amino group but is insoluble in organic and inorganic solvents owing to high polarity and intermolecular hydrogen bonding. Some salt forms of chitosan are soluble in polar organic solvent and in organic mixture. Water-soluble salts of chitosan

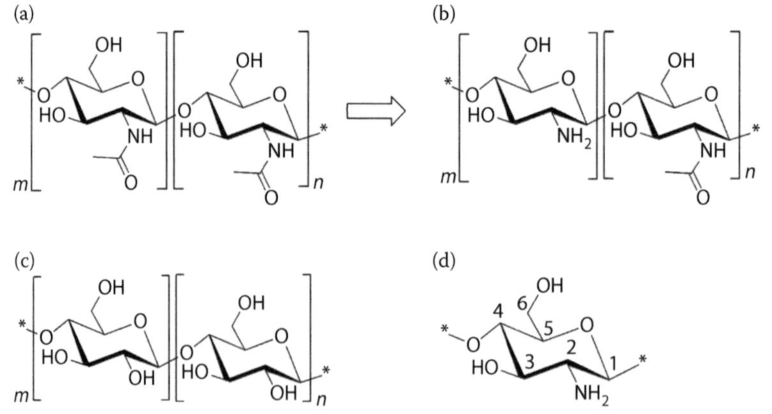

FIGURE 3.1 Structures of the polysaccharides: (a) chitin, (b) chitosan, (c) cellulose, and (d) conventional numbering system to assign carbon atoms in the monomeric unit of the chitosan backbone.

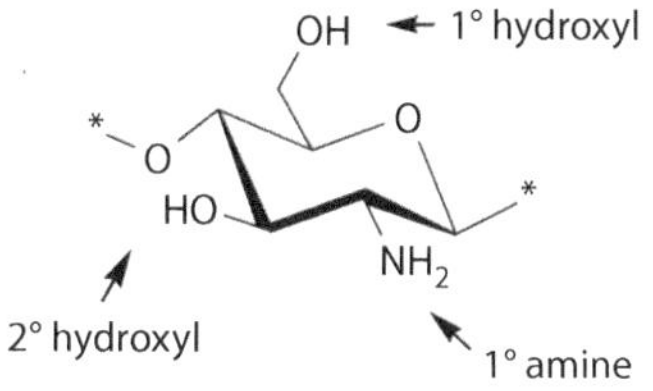

FIGURE 3.2 Multiple functional groups of chitosan.

include acetate, formate, lactate, pyruvate, malonate, citrate, tartarate, glycolate, and ascorbate (Mourya and Inamdar 2008). Some chitosan salts, such as 2-hydroxybenzoate, camphor sulfonate, *p*-toluenesulfonate, and methanesulfonate are soluble in water as well as some organic solvents and can, therefore, be used as precursors for chemical modifications of chitosan (Sashiwa et al. 2000a,b; Runarsson et al. 2008b). Chitosan tends to aggregate even after full protonation under acidic conditions due to hydrogen bonding and hydrophobic interactions exerted by the main backbone and *N*-acetyl groups (Rinaudo 2006b), thereby limiting its chemical reactivity. Because chitosan with high MW is highly viscous and poorly soluble, depolymerization has been done to enable better solubilization (Kubota 1997; Kubota and Eguchi 1997). Therefore, DA, MW, and physical state (salt form, conformation, ionization) of chitosan are important characteristics that influence its chemical and physiological properties.

Pharmaceutical applications of chitosan include its use as mucoadhesive polymer (Lehr et al. 1992), antimicrobial agent (Rabea et al. 2003), absorption enhancer for drug delivery (Thanou et al. 2001), and nonviral vector for gene delivery (Mohammadi et al. 2011). Chitosan is also a promising material for tissue engineering (Di Martino et al. 2005). The antimicrobial properties of cationic chitosan are considered to be caused by its interaction with the anionic cell membrane, disrupting the membrane structure and inducing leakage of electrolytes and proteins from the cytosol (Hernandez-Lauzardo et al. 2011). Interaction between opposite charges may also be a significant factor in the mucoadhesive and absorption-enhancing properties of chitosan (van der Merwe et al. 2004; Bowman and Leong 2006). Again, DA, salt form, and MW can influence the biological effects (Li and Xia 2011). In food and agricultural industry, chitosan has been used as a dietary ingredient, in food preservation, as emulsifying agent, antibacterial agent, seed coating material, to increase flowering and to extend the life of cut flowers (Xia et al. 2011). Chitosan has also been used in cosmetic applications to maintain skin moisture, protect the epidermis, in acne treatment, and to reduce static electricity in hair (Rinaudo 2006a). To reveal novel findings of bioactivities and improve physicochemical properties, different chitosan derivatives with different structures are essential.

3.2 SOME COMMON CHITOSAN DERIVATIVES AND PROCEDURES FOR DIRECT CHEMICAL MODIFICATION

Despite their unique biological and physicochemical properties, chitin and chitosan are underutilized. This can be mainly attributed to the lack of solubility in common solvents and aqueous solutions at neutral or high pH. For example, the low solubility of chitosan at physiological pH of 7.4 limits its applications in nasal and peroral delivery system (Bhattarai et al. 2010). In other applications, the hydrophilicity can lead to less than ideal performance with high swelling and rapid water absorption in aqueous conditions, leading to fast drug release in a sustained release system (Park et al. 2010). Various chemically modified chitosan derivatives have been synthesized to overcome these problems. Although the free amino group has been the main target for modifications, there are also reports of modification aimed at the primary hydroxyl group (Esmaeili et al. 2010). These chemical modification include thiolation (Masuko et al. 2005), derivatization with cyclodextrin (Auzély and Rinaudo 2003; Gonil et al. 2011; Kaftan et al. 2011), glycosylation (Wang et al. 2010), and reducing-end modification (Lohse et al. 2006; Novoa-Carballal and Muller 2012). Many of the common reported chitosan modifications, excluding grafting chitosan nanoparticles and chitosan surfaces, fall into one of the following categories:

Reductive N-alkylation: Schiff reaction of the chitosan amine with aldehydes or ketones yields the corresponding aldimines and ketimines. This can be followed by reduction with sodium borohydride (NaBH$_4$) or sodium cyanoborohydride (NaBH$_3$CN) to obtain the *N*-alkylated derivative. Example of this includes *N*-alkylation and *N,N*-dialkylation by using chitosan and formaldehyde in aqueous acetic acid and hydrogenation by NaBH$_4$ (Muzzarelli and Tanfani 1985; Kim et al. 1997). Verheul and colleagues reported an alternative procedure using formic acid and formaldehyde to give *N,N*-dimethyl chitosan (Verheul et al. 2008).

Quaternization: Quaternization of the amino group of chitosan should improve the aqueous solubility independent of pH and enhance biological properties caused by the polycationic nature of chitosan. The most common approach is to use alkyl iodides in basic media. A number of quaternized trialkyl derivatives have been prepared by this approach, with studies focusing on preparing *N,N,N*-trimethyl chitosan (TMC) (Sieval et al. 1998; Thanou et al. 2000; Runarsson et al. 2008a). Other quaternization procedures include reaction of chitosan with 2,3-epoxypropyl-trimethyl ammonium chloride (Zhao et al. 2010).

N- and O-acylation: These functional groups have been introduced to the chitosan backbone, for example, to enhance its drug carrier properties (Kato et al. 2004). *N*-Acyl chitosan derivatives are obtained by reaction with anhydrides, acyl chlorides, or lactones or by coupling reaction with carboxylic acids using 1-ethyl-3-(3-dimethylaminopropyl carbodiimide) and *N*-hydroxysuccinimide. These reactions are not necessarily *N*-selective but *O*-selective acylation can be achieved by reaction with acid chloride in acidic media (Sashiwa et al. 2002a,b). *N*-Acylated chitosan derivatives, which are soluble in organic solvents, can be obtained by reaction with lauroyl, decanoyl, or hexanoyl chlorides in pyridine/CHCl$_3$ (Zong et al. 2000; Sashiwa et al. 2002a; Kato et al. 2004).

Hydroxyalkylation: Hydroxyalkylation to form chitosan derivatives, such as *O*-hydroxypropyl chitosan (Peng et al. 2005) and *O*-hydroxyethyl chitosan (Ronghua et al. 2003; Xie et al. 2007), is carried out using epoxides that are well-known reactants in carbohydrate chemistry. The main purpose of such a modification is usually to improve aqueous solubility. Depending on the epoxide used and reaction parameters (temperature and pH), the reaction can be controlled to mainly target the primary amino group, the primary hydroxyl group, or both.

Carboxyalkylation: The properties of chitosan can be modified by carboxyalkylation, which introduces an anionic group. Carboxyalkylation can be achieved by reacting chitosan with monohalocarboxylic acids and the reaction selectivity controlled by using different reaction conditions (Kim and Choi 1998). Chemoselective *N*-carboxyalkyl chitosan derivatives can be prepared using carboxyaldehyde in reductive amination (Muzzarelli et al. 1982).

Chitosan–polyethylene glycol (PEG) derivatives: Biocompatibility and pharmacokinetic properties of biomacromolecules can be modified by covalently attaching PEG to the macromolecule. For covalent attachment of PEG to chitosan, modified PEG having a terminal aldehyde or halogen or similar leaving group is reacted with chitosan (Sugimoto et al. 1998). PEGylation of chitosan has decreased the cytotoxicity of the chitosan/drug formulation and prolonged the absorption of the drug (Prego et al. 2006).

Important limitations for direct chemical modification of chitosan are its low solubility under most reaction conditions. Reactions that are compatible with acidic aqueous solutions (such as imination) can be carried out with relative ease but reactions that require basic conditions, such reaction with alkyl halides, are more difficult to perform. Highly polar organic solvents, such as *N*-methyl-2-pyrrolidone (NMP), dimethylformamide (DMF), and dimethylacetamide (DMAc) can be used as solvents. Yet, the outcome is affected by poor solubility of the polymer that results in heterogeneous conditions that can hamper the chemical modification. Although some reactions can be *N*-selective, such as imination, it is often difficult to obtain fully selective modification at the amino group or one of the hydroxyl groups. Polnock et al. has reported that commonly used procedures for *N*-methylation will also lead to partially unwanted *O*-methylation in the synthesis of TMC (Polnok et al. 2004). Alternative procedures for the synthesis of TMC that avoid *O*-methylation are reactions in mixed aqueous/organic medium (Runarsson et al. 2008a) or reductive alkylation followed by reaction with methyl iodide (Verheul et al. 2008), but neither of these studies reported full *N,N,N*-trimethylation. Partial substitution and heteromeric nature of the resulting chitosan derivative complicates the structure characterization and makes it difficult to reach firm conclusions regarding the structure–property and structure–activity relationships. Therefore, alternative approaches are necessary to enable the synthesis of more homomeric chitosan derivatives.

3.3 UTILIZATION OF *N*-PHTHALOYL AND *O*-TRITYL PROTECTION GROUPS FOR THE SYNTHESIS OF CHITOSAN DERIVATIVES

Protection strategies have a significant role in the synthesis of organic compounds, especially when it involves selective modification of one of several reactive functional groups. The ideal characteristics of a protection group are as follows: (1) it can be introduced under mild conditions on the functional group to be protected, (2) it is to be stable under a variety of reaction conditions, and (3) it can be easily and selectively removed without adversely affecting other parts of the molecule. Preferably, both protection and deprotection should be quantitative. Efficient protecting group strategy can have a dual role in chitochemistry, that is, to protect functional groups to enable chemoselective or regioselective modification and to improve solubility of the polymer in organic solvents to allow more efficient modifications under homogeneous conditions. To date, only few protection group strategies have been implemented in chitochemistry. These are phthaloyl protection of the amino group, trityl protection of C-6 hydroxyl group, and silyl ether protection of the C-3 and C-6 hydroxyl groups.

3.3.1 *N*-PHTHALOYL PROTECTION OF CHITOSAN

The rigid crystalline structure of the acetamido or primary NH_2 group plays an important role in the peculiar conformational features of chitosan and in the stabilization of the intra- and/or intermolecular H-bonding (Saito et al. 1981, 1987). A hydrophobic protection group could potentially disrupt the crystalline structure of chitin or chitosan by reducing the hydrogen bonding, thereby improving solubility in organic solvents. This notion was supported by Nishimura et al., who were the first to report *N*-phthaloylation of fully deacetylated chitosan (Nishimura et al. 1991). They reported that the phthaloyl group could be introduced through reaction with phthalic anhydride in DMF at 130°C for 7 h. However, a subsequent investigation found that this reaction was not *N*-selective in pure DMF and gave degree of substitution (DS) 1.54, confirming partial *O*-modification (Kurita et al. 2002). Further improvement of this method, by using the cosolvents DMF/H_2O (95:5, v/v), eliminated the undesired *O*-phthaloylation, and introduced full protection of the amino group (DS 1.0) (Kurita et al. 2002, 2007). The structure of this *N*-phthaloyl-protected chitosan was confirmed with infrared (IR) spectroscopy and solid-state ^{13}C nuclear magnetic resonance (NMR). *N*-phthaloyl chitosan can also be prepared by transesterification, hydrolysis, or alcoholysis from contaminated *N,O*-phthaloylated chitosan (Kurita et al. 2007). *N*-Phthaloyl chitosan is soluble in *m*-cresol, dichloroacetic acid, DMAc/(LiCl), and MeOH/$CaCl_2 \cdot 2H_2O$ (Kurita et al. 2007) and swells in pyridine, DMF, and dimethyl sulfoxide (DMSO) (Kurita et al. 2002, 2007). The relatively low solubility of *N*-phthaloyl chitosan can be attributed to its crystallinity as suggested by x-ray diffraction study (Kurita et al. 2002).

Deprotection of the phthaloyl group can be achieved by hydrazinolysis (with hydrazine hydrate) at 80°C for 16 h. In order to facilitate milder deprotection, phthaloyl aromatic ring with electron withdrawing group (4,5-dichlorophthaloyl group) for chitosan protection was also investigated (Torii et al. 2009). Although the *N*-selective protection of 4,5-dichlorophthaloylated chitosan resulted in

DS 1.0, using similar reaction conditions as for *N*-phthaloyl protection, the deprotection required harsher conditions (hydrazinolysis followed by treatment with 6 M NaOH at 50°C) than for the original phthaloyl group. The 4,5-dichlorophthaloylated chitosan material was soluble in DMAc/ LiCl and partially soluble in DMF and DMSO. The protection and deprotection for both phthaloyl and 4,5-dichlorophthaloyl groups cause MW reduction of chitosan (Holappa et al. 2004; Torii et al. 2009). The *N*-phthaloyl-protected chitosan has been used in the synthesis of a variety of chitosan derivatives as can be seen in Scheme 3.1. In particular, *N*-phthaloyl chitosan has been used for the synthesis of various 6-*O* ester and ether derivatives of chitosan. The *p*-toluenesulfonyl (tosyl) group can also be introduced regioselectively at C-6 position to yield 6-*O*-tosyl-*N*-phthaloyl chitosan. As bulky tosyl ester (OTs) is a very good leaving group, it has been the key reactive intermediate for

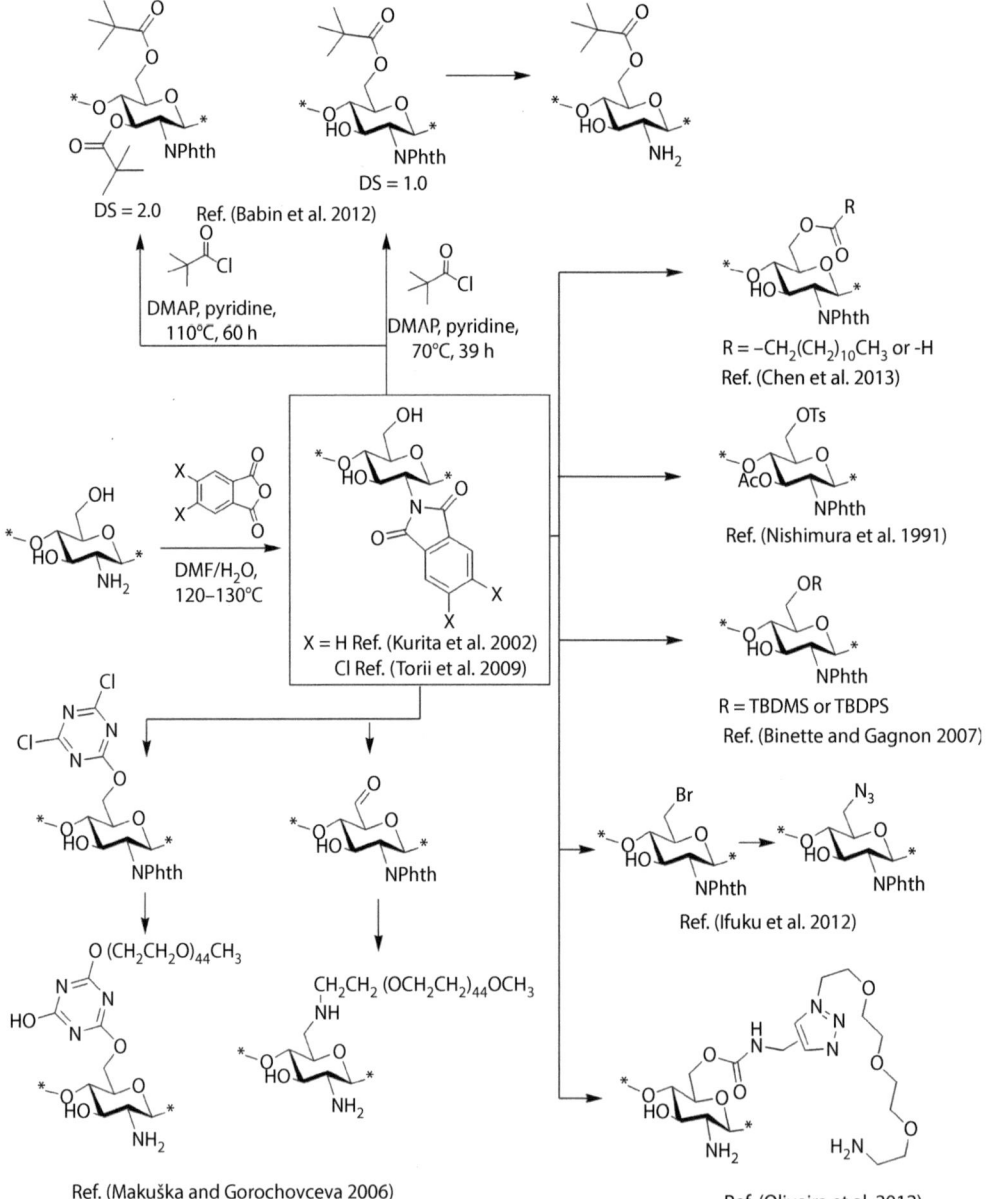

SCHEME 3.1   Synthesis of *N*-phthaloyl chitosan and its derivatives.

TABLE 3.1

Solubility of *N*-Phthaloyl and *O*-Trityl-Protected Chitosan Derivatives

Solvents	Chitosan	*N*-Phthaloyl Chitosan (DS = 1.0)	6-*O*-Trityl-*N*-Phthaloyl Chitosan (DS = 1.0)	6-*O*-Trityl Chitosan (DS = 1.0)
References	a, b	a, b, c	d	d
5% AcOH (aq.)	+	–		
DMAc		+	+	+
DMF	–	± / +	+	–
DMSO	–	± / +	+	–
Pyridine	–	± / +	+	+
CH$_2$Cl$_2$		–	+	–
CHCl$_3$		–	±	–

Note: +, soluble; ±, swelled or partial soluble; –, insoluble. a = (Kurita et al. 2007), b = (Torii et al. 2009), c = (Kurita et al. 2002), d = (Nishimura et al. 1991).

C-6 modification with nucleophiles, such as azides (Yang et al. 2012), potassium thioacetate (Munro et al. 2009), sodium alginate (Jančiauskaitė et al. 2009), halides (Chirachanchai et al. 2001), and hydrides (Zhang et al. 1994).

Although the *N*-phthaloylated chitosan has improved solubility compared to the starting chitosan (Table 3.1), it is still not completely soluble in common solvents, such as pyridine, DMF with the solubility limited to swelling (Kurita et al. 2007). Therefore, reactions using *N*-phthaloyl-protected chitosan generally proceed under heterogeneous conditions with extensive heating and prolonged reaction time is required for significant modification. Although the *N*-phthaloyl chitosan has proven to be useful in the synthesis of chitosan derivatives, the *N*-phthaloyl group is unstable under alkaline conditions in the presence of water and tends to decompose (Kurita et al. 2007). Additionally, the limited solubility of *N*-phthaloyl chitosan can hamper its structure elucidation.

3.3.2 (C-6) *O*-TRITYL PROTECTION OF CHITOSAN AND SYNTHESIS OF TRITYL CHITOSAN DERIVATIVES

The triphenylmethyl (trityl, Tr) group has been used for selective protection of primary (1°) alcohol in the presence of secondary alcohol because of its bulkiness. The trityl group has, therefore, been used in carbohydrate chemistry for the protection of the C-6 hydroxyl group (Scheme 3.2) (Saito et al. 1987).

Trityl ethers are generally prepared by the reaction of primary alcohol with triphenylmethyl chloride (Ph$_3$C-Cl, trityl chloride) in pyridine or DMF with 4-dimethylaminopyridine (DMAP) as catalyst. Alternatively, selective primary alcohol protection can be achieved by reaction with triphenylmethylpyridinium fluoroborate (C$_5$H$_5$N$^+$CPh$_3$BF$_4$-) in CH$_3$CN and pyridine at 60–70°C

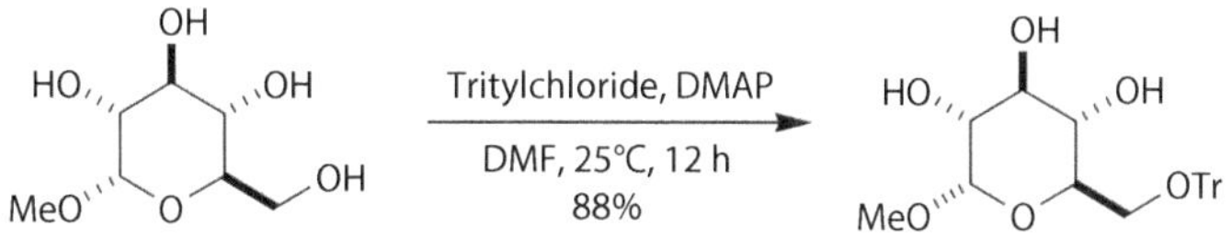

SCHEME 3.2 Selective trityl protection of primary hydroxyl group in the presence of secondary hydroxyl groups. (Adapted from Chaudhary, S.K. and Hernandez, O. 1979. *Tetrahedron Letters* 20: 95–98.)

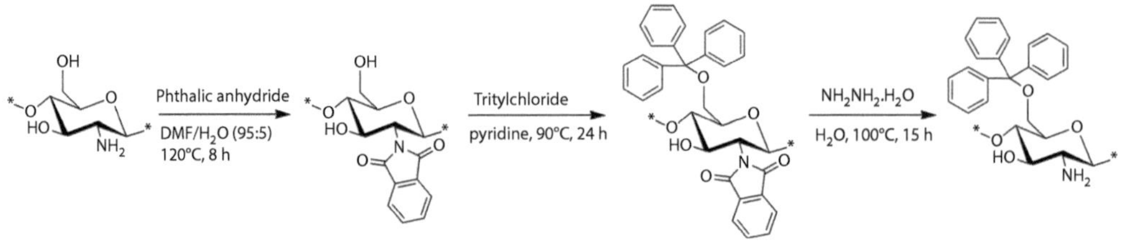

SCHEME 3.3 Synthetic route for regioselective 6-*O*-trityl protection of chitosan.

(Torii et al. 2009). The trityl group is stable under basic conditions and can be easily cleaved under mild acidic conditions or by hydrogenation (Pd/C, H$_2$).

Regioselective protection of primary (C-6) hydroxyl group of chitosan by using trityl chloride was first reported by Nishimura et al. (1991). The *O*-6 trityl protection was introduced in three steps: first, by *N*-phthaloyl protection of the amino group, then reaction with threefold excess of trityl chloride in pyridine at 80°C under inert atmosphere, and finally removal of the phthaloyl group by reaction with hydrazine hydrate (Scheme 3.3).

The 6-*O*-trityl-*N*-phthaloyl chitosan is soluble in pyridine, DMF, DMSO, DMAc, and CH$_2$Cl$_2$ (Table 3.1) with both the primary hydroxyl group and the amino group completely protected (Nishimura et al. 1991). Consequently, selective modification at the C-3 hydroxyl group is possible (Scheme 3.4). Acetylation using acetic anhydride in pyridine afforded 3-*O*-acetyl derivative quantitatively with almost DS 1.0 and enhanced solubility in low-boiling solvents, such as CH$_2$Cl$_2$ and CHCl$_3$. The detritylated 3-*O*-acetyl-2-phthalimido derivative of chitosan was soluble in DMF, DMAc, DMSO, and pyridine (Nishimura et al. 1991), and could be used for subsequent C-6 *O*-regioselective modifications of chitosan.

Holappa and colleagues employed the trityl protection strategy to prepare water-soluble *N*-betainate derivatives of chitosan as is shown in Scheme 3.4. These derivatives were characterized with ^{1}H and ^{13}C NMR as well as 2D ^{1}H-^{1}H COSY and ^{13}C-^{1}H HSQC NMR (Holappa et al. 2004). In a further study, Holappa et al. also synthesized different *N*-chloroacyl-6-*O*-trityl chitosan intermediates under mild conditions and used these for further modification with different tertiary amines to obtain water-soluble quaternary derivatives (Holappa et al. 2005, 2006a, 2006b). Furthermore, 6-*O*-trityl protection has also been effective in reducing *O*-methylation when *N*-trimethylation was carried out (Runarsson et al. 2007).

Although 6-*O*-trityl chitosan is more soluble in organic solvents than chitosan, it is only fully soluble in DMAc and pyridine (Nishimura et al. 1991). Subsequent N-modifications are therefore carried out in polar solvents, such as DMAc, pyridine, or NMP. Characterization of these protected intermediates by liquid-state ^{1}H NMR and subsequent determination of DS can be difficult (Holappa et al. 2004) and often peak broadening of the chitosan backbone can be observed (Kurita et al. 2007; Torii et al. 2009). These derivatives are, therefore, often characterized by solid-state ^{13}C CP/MAS NMR; IR and DS are mostly determined with elemental analysis.

3.4 CHITOSAN DERIVATIVES WITH TRIMETHYLSILYL ETHER PROTECTION

Silyl ethers are an important class of protection groups that are used for the protection of primary, secondary, or tertiary alcohols. Different silyl protecting groups with variations of substituents on silicon are available. These provide a wide spectrum of chemical stability and can fulfill different steric demands. Silylation is possible under a variety of reaction conditions and different reagents may be available for the introduction of a particular silyl group (Greene and Wuts 2007). Thus, in polyhydroxylic structures, selectivity can be achieved easily by choosing the appropriate silylating agent. The most general method for the preparation of silyl ether is by the reaction of an alcohol with the appropriate silyl chloride or silyl triflate reagent. The reaction with silyl chloride is

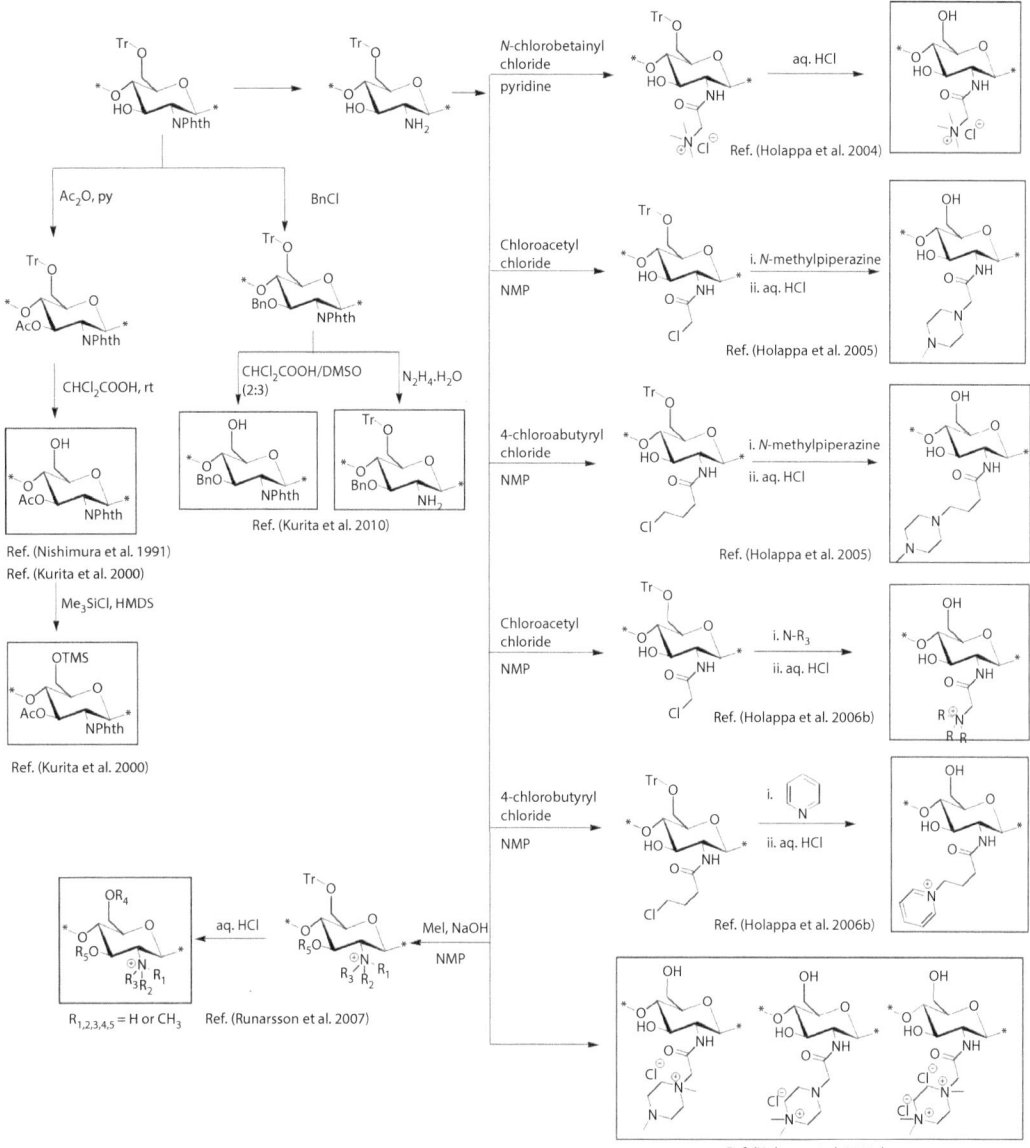

SCHEME 3.4 General synthetic route for *N*-modifications of trityl-protected chitosan derivatives.

commonly done in DMF, with imidazole as a catalyst or in dichloromethane with 2,6-lutidine as base in the case of silyl triflate. Most commonly used silyl ethers and their stability toward acid and base is illustrated in Figure 3.3. Deprotection can be done selectively with acid or base hydrolysis or by reaction with fluoride ion under mild conditions. Because the Si–F bond is about 30 kcal/mol stronger than the Si–O bond, the silyl protection is typically removed with a source of fluoride ion, such as tetra-*n*-butylammonium fluoride (TBAF), HF-pyridine, or *tris*(dimethylamino)sulfonium difluorotrimethylsilicate (TASF) (Nelson and Crouch 1996).

Although silyl protecting groups can also protect amines, they are most frequently used for selective protection of alcohols. In addition to using selective reaction conditions, this *O*-selectivity can be attributed to the stronger Si–O bond compared to the Si–N bond. Silyl ethers are generally less polar and more volatile than the precursor compounds. This enhanced volatility makes them

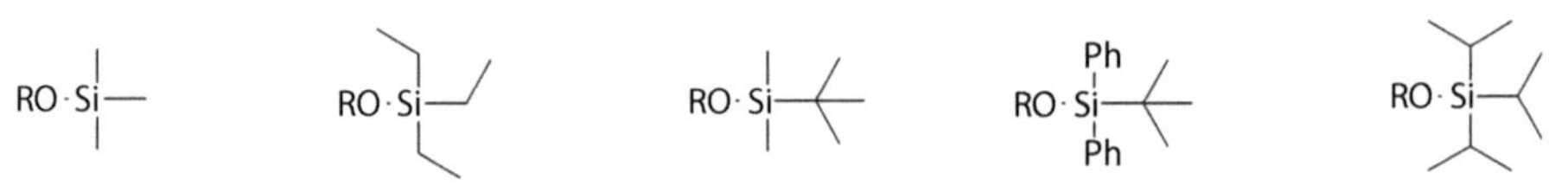

| Trimethylsilyl (TMS) | Triethylsilyl (TMS) | t-Butyldimethylsilyl (TBS/TBDMS) | t-Butyldimethylsilyl (TBDPS) | Triisopropylsilyl (TIPS) |

General stability of silyl ether toward acidic media:

TMS (1) <TES (64) <TBDMS (20,000) <TIPS (700,000) <TBDPS (5,000,000)

General stability of silyl ether toward basic media:

TMS (1) <TES (10–100) <TBDMS~TBDPS (20,000) <TIPS (100,000)

FIGURE 3.3 Examples of commonly used silyl ether protection groups and their general stability under acidic and basic conditions.

suitable for analysis techniques, such as gas chromatography and electron-impact mass spectrometry (EI–MS). Silyl ether protection, mostly trimethylsilyl (TMS), has been used for improvement in thermal stability and organosolubility of polymeric materials, such as cellulose (Keilich et al. 1968; Harmon et al. 1973b; Mormann 2003), amylose (Horton and Lehmann 1978), and dextran (Nouvel et al. 2003) for decades.

3.4.1 TMS CHITIN

TMS protection of both the hydroxyl groups (6-O and 3-O) of chitin has been done to increase its solubility; however, the product obtained is generally highly acid and water labile. Partial protection of the hydroxyl groups (DS $\approx$ 0.6) of chitin was observed when chitin was reacted with hexamethyldisilazane ((Me_3Si)$_2$NH or HMDS) in formamide at 70°C or by HMDS/trimethylsilyl chloride (TMS-Cl) in pyridine (Harmon et al. 1973a). Under similar reaction conditions, cellulose was fully substituted (DS 3.0). This indicates relatively lower reactivity of chitin compared to cellulose (Harmon et al. 1973a). Interestingly, the introduction of TMS to chitin by using HMDS and TMS-Cl in pyridine at 70°C afforded DS 2.0 after 16 h (Kurita et al. 2005). However, lower DS (~0.7) for chitin observed in earlier investigation may be because of partial removal of TMS group when the product was isolated in MeOH. Conversely, extraction with acetone and precipitation with water resulted in fully silylated chitin (Kurita et al. 2005). The fully substituted chitin was soluble in acetone and pyridine and partially soluble in common polar solvents, whereas products with DS below 0.7 only swelled in organic solvent. The fully substituted chitin could thus be characterized with ¹H NMR. The increased solubility of TMS chitin and its derivatives was also observed by Kurita and coauthors who found that trimethylsilylation of chitin and 3-O-acetyl-2-N-phthaloylation of chitosan enhanced the solubility in organic solvents while keeping the reactivity of the hydroxyl group intact (Kurita et al. 1999, 2003, 2004). Recently, triethylsilyl (TES, DS = 1.95) and triphenylsilyl protection (TPS chitin, DS = 1.66) of chitin was reported (Watanabe et al. 2012). The reactivity was in the order of TMS > TES > TPS. Although both TES chitin and TPS chitin had improved solubility compared to the original chitin, TMS chitin had the highest solubility, being soluble in DMAc, DMSO, pyridine, and acetone (Table 3.2). Conversely, desilylation showed that TES was more stable compared to TMS. A summary of the solubility of silylated chitin derivatives can be found in Table 3.2.

Trimethylsilylation of chitin was used to increase its solubility and subsequent reactivity. Full O-acetylation (DS 2.0) was obtained by using DMAP catalyst at 50°C using 50 eq. acetic anhydride with 85% yield (Kurita et al. 2005). The tritylation of TMS chitin at 90°C resulted in 0.95 DS after 48 h reaction. Thus, the preparation of trityl chitin via silylation is superior to direct tritylation or procedure based on N-phthaloyl chitosan (Kurita et al. 2005). Additionally, the tritylated chitin

TABLE 3.2

Solubility of Silyl-Protected Chitin Derivatives

Solvents	Chitin	TMS Chitin (DS = 2.0)	TES Chitin (DS = 1.95)	TPS Chitin (DS = 1.73)
References	a, b	c, d	d	d
DMF	–	±	±	±
DMSO	–	+	±	±
Pyridine	–	+ +	+	+ +
Acetone	–	+ +	±	±
THF		±		
CH₂Cl₂				
CHCl₃		±		

Note: + +, soluble at room temperature; +, soluble on heating; ±, swelled or partial soluble; –, insoluble. a = (Kurita et al. 2007), b = (Torii et al. 2009), c = (Kurita et al. 2005), d = (Watanabe et al. 2012).

product was soluble in polar solvents, such as pyridine and DMSO, and could be characterized with ¹H NMR. However, trimethylsilylation of chitin affects its MW, which was reduced by more than half after the deprotection (Sugita et al. 2008).

3.4.2 TMS CHITOSAN

The impact of trimethylsilylation on solubility and utility for further modification was recently investigated with chitosan (Kurita et al. 2004). Trimethylsilylation was achieved by heating chitosan in pyridine at 100°C and then HMDS and TMS-Cl under heterogeneous conditions (Scheme 3.5), yielding a DS of 2.3. Varying parameters such as temperature, time, and solvents, and using catalysts such as DMAP or other trimethylsilylation reagents such as bromotrimethylsilane did not significantly increase the DS (Kurita et al. 2004). In contrast, prolonged heating at high temperature in sulfolane caused the degradation of the chitosan product. The TMS protection of chitosan markedly enhanced the reactivity relative to chitosan, making it suitable for acetylation at room temperature to give high DS.

The TMS chitosan bond (O-TMS and N-TMS) is very labile in acidic solution. The TMS group can be easily removed in MeOH/H₂O/acetic acid (2:1:1) mixture at room temperature but is stable under neutral and alkaline solutions as well as in saturated aqueous NaHCO₃. TMS chitosan is almost soluble in pyridine but only partially soluble or swelled in acetone, THF, DMSO, and DMAc (Kurita et al. 2004). Owing to the labile nature of TMS, it has therefore mainly been used to increase the solubility and reactivity of chitosan but not as a protection group. However, studies of

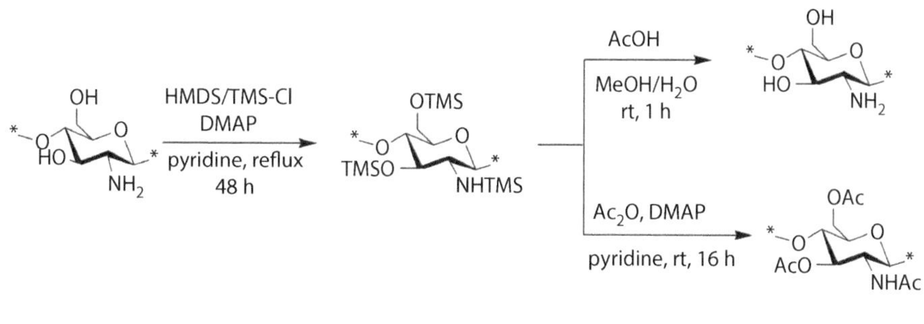

SCHEME 3.5 Synthesis of TMS chitosan, deprotection, and derivatization.

these materials have shown that other silyl protecting groups can tolerate a wider range of reaction conditions that might be useful in chitosan chemistry.

3.5   TBDMS PROTECTION OF CHITOSAN: SYNTHESIS AND APPLICATIONS

E. J. Corey developed a protocol that was highly effective for mild conversion of various alcohols to the corresponding TBDMS ethers (Corey and Venkateswarlu 1972). This involved the use of 2.5 eq. imidazole as catalyst with 1.2 eq. of TBDMSCl reagent and DMF as solvent. The reaction proceeds via *N-tert*-butyldimethylsilylimidazole, a very reactive silylating intermediate (Scheme 3.6). Corey also reported a rapid cleavage of the silyl ethers to alcohols by treatment with 2–3 eq. TBAF in THF at 25°C (Corey and Venkateswarlu 1972). Nucleophilic attack of the small fluoride anion leads to a pentavalent silicon center, which is permitted due to hybridization with the vacant *d*-orbitals of silicon. In addition, the formation of the strong Si–F bond is the driving force for a fast cleavage as stated earlier. TBDMS ethers are stable toward aqueous base but may be converted back to the alcohols under acidic conditions. The general mechanism for TBDMS protection and deprotection is shown in Scheme 3.6.

3.5.1   DiTBDMS Chitosan

Efficient, chemoselective TBDMS protection of chitosan was recently reported (Runarsson et al. 2008b). Initially, silylation of short COS (with 50% *N*-acetylation) with TBDMSCl and imidazole in DMF was investigated. The COS was partly soluble in DMF and the reaction mixture was cloudy in the beginning. The DS after 72 h reaction was 1.64. It was hypothesized that the C-3 hydroxyl group adjacent to the acetamido might be poorly accessible for the reagent. The material obtained after 48 h had good solubility and could thus be modified further with *t*-butyldimethylsilyltriflate (TBDMSOTf) in CH$_2$Cl$_2$ using 2,6-lutidine as base. The TBDMSOTf reagent is more reactive than TBDMSCl and can react with more sterically hindered hydroxyl groups (Mendonca and Laine 2005). The COS was rapidly silylated under these conditions, with DS 2.20 after 24 h. This DS was consistent with full silylation of the C-6 and C-3 hydroxyl groups, also including the reducing-end C-1 hydroxyl and C-4 nonreducing-end hydroxyl groups.

The chitosan polymer is not soluble in DMF and therefore, it could not be silylated even after extensive heating under other similar conditions used for silylation of COS (Runarsson et al. 2008b). Previously, Sashiwa et al. reported that by converting chitosan into the mesylate salt form, a good solubility in DMSO was obtained but not in DMF (Sashiwa et al. 2000b). This salt conversion was used to enable the TBDMS silylation of chitosan in DMSO (Runarsson et al. 2008b). The chitosan mesylate was reacted with TBDMSCl (8 eq.) and imidazole (26 eq.) in DMSO with partially silylated product obtained after 24 h. Subsequently, another portion of an equal amount of TBDMSCl and imidazole was added to the reaction and after a total of 48 h reaction, fully silylated chitosan (3,6-*O*-di-TBDMS chitosan) was isolated. The complete silylation was confirmed with IR

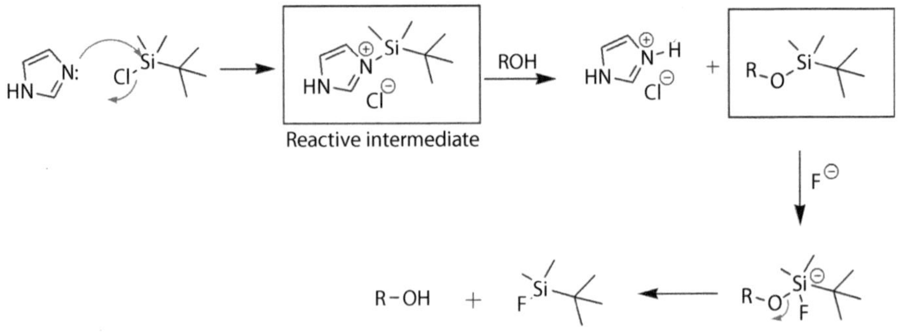

SCHEME 3.6   General reaction mechanism showing TBDMS protection and deprotection of alcohol.

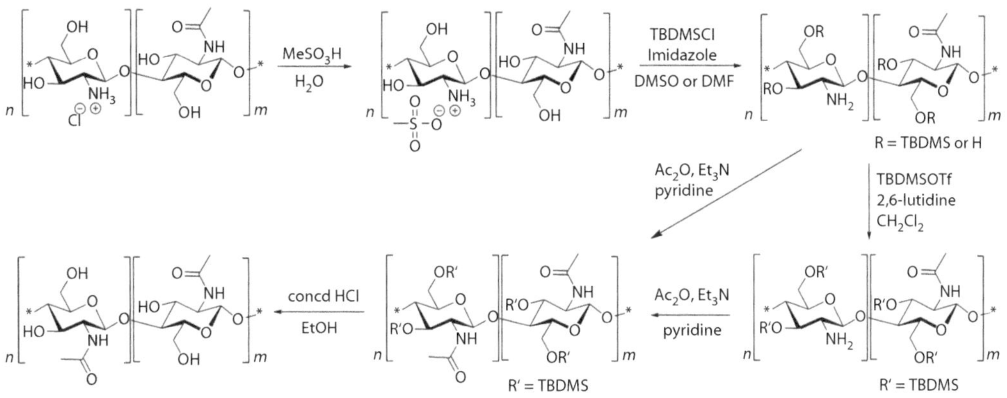

SCHEME 3.7 Synthesis of DiTBDMS-protected chitooligosaccharides. (Adapted from Runarsson, O.V. et al. 2008b. *Carbohydrate Research* 343: 2576–2582.)

and ^{1}H NMR, showing 1.94 DS and a peak for the unsubstituted H-2 at 2.65 ppm corresponding to one proton confirming that the silylation was *O*-selective. The DiTBDMS chitosan was then *N*-acetylated with acetic anhydride and deprotected with HCl/EtOH to confirm full *O*-protection, as there was no *O*-acetyl peak observed in IR after acetylation. Further silylation, using TBDMSOTf to ensure that all sterically hindered hydroxyl groups were silylated, did not affect the DS. Solvent change in COS from DMF to DMSO did not improve earlier results and required TBDMSOTf to obtain complete silylation of the sterically hindered hydroxyl groups (Scheme 3.7).

3.5.2 OPTIMIZATION AND PROPERTIES OF TBDMS CHITOSAN

Optimization of this silylation strategy was recently carried out (Song et al. 2010). The refined procedure, as shown in Scheme 3.8, involved the reprecipitation of the chitosan mesylate salt with dry acetone to obtain a fine powdered material that could be dissolved completely in dry DMSO (Figure 3.4).

Thereafter, a mixture of TBDMSCl (5 eq./10 eq. imidazole) was added under inert atmosphere. As the reaction progressed, the reaction mixture became cloudy and after 30 min, the silylated product separated out from the solvent. The gel-like product was then washed with water and then with CH$_3$CN to give a fine powdered white material with 95% yield (Song et al. 2010) compared to 50–85% yield with the initial method (Runarsson et al. 2008b). The isolation of the silylated chitosan product prepared by optimized protocol is more straightforward as it only requires simple filtration and washing with water and CH$_3$CN, while the original method required time-consuming extraction in hexane. This refined method increased the yield, purity, and solubility of the silylated product, enabling the assignment of individual peaks in the DiTBDMS chitosan backbone by using ^{1}H NMR and ^{1}H-^{1}H COSY NMR (Figure 3.5a and b).

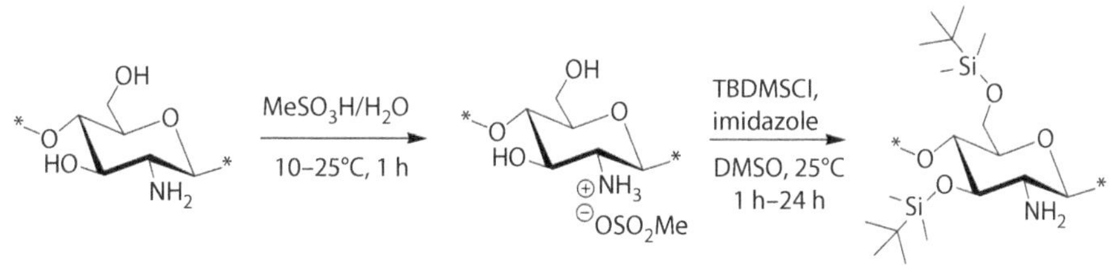

SCHEME 3.8 Synthesis of DiTBDMS chitosan.

Reaction progress   10 min   20 min   30 min

FIGURE 3.4 Progress of the reaction TBDMS protection of chitosan.

^{1}H NMR showed two peaks of equal intensity at 0.89 and 0.90 ppm for *t*-butyl (*t*-Bu-Si) groups with integration area corresponding to 12 protons and four peaks of equal intensity at 0.05, 0.06, 0.10, and 0.13 ppm for methyl (Si-Me) with the integration area corresponding to 12 protons confirmed the two sets of TBDMS attachments corresponding to C-6 and C-3 hydroxyl group. This substitution pattern thereby confirmed 100% substitution of hydroxyl group, which was further assessed with IR that shows distinct peaks for the TBDMS group at 2858–2955 cm^{-1}, 1256 cm^{-1}, 836 and 777 cm^{-1} as well as a diminished O–H band around 3402 cm^{-1} (Figure 3.6).

DiTBDMS chitosan has a better solubility profile compared to TMS chitosan. As can be seen in Table 3.3, DiTBDMS chitosan is highly soluble in common solvents, such as CH_2Cl_2, $CHCl_3$, EtOAc, and pyridine, and gives slightly opaque solutions in NMP, DMF, DMSO, hexane, and THF, showing its potential versatility for reaction under various reaction conditions.

3.5.3 DESILYLATION OF DITBDMS CHITOSAN

DiTBDMS chitosan can be deprotected by using concentrated HCl/EtOH or concentrated HCl/MeOH at room temperature (Runarsson et al. 2008b; Gaware et al. 2013). However, prolonged treatment with acid can cause hydrolysis of the *O*-glycosidic bond causing depolymerization of chitosan (Vårum et al. 2001). Recently, Benediktsdottir and colleagues reported an alternative deprotection using 1 M TBAF in NMP at 50°C (Benediktsdottir et al. 2011). In both deprotection strategies, the deprotection had to be repeated to remove trace amounts of TBDMS (~0.4–7%) (Benediktsdottir et al. 2011; Gaware et al. 2013). Although, the method using TBAF is more efficient (Benediktsdottir et al. 2011), compared to acidic deprotection, it requires time-consuming dialysis to fully remove the NMP solvent.

3.6 DERIVATIVES AND APPLICATIONS OF DITBDMS CHITOSAN

3.6.1 DITBDMS CHITOSAN AS SUPERHYDROPHOBIC FILMS

DiTBDMS chitosan can be used to prepare water-repellent films using a phase separation method. The films can be defined as superhydrophobic as they exhibit topography with a three-level hierarchical roughness organization (Song et al. 2010). These films were extremely water repellent, even at acidic conditions where the chance of protonation on amino group exists and were stable, both over the entire pH range (1–14) as well as over extended periods of time. Moreover, these films contain free amino groups that allow for further chemical modification through surface chemistry and wettability. Owing to their stability and possibilities of surface chemical alterations, these superhydrophobic surfaces may find use in biomedical applications.

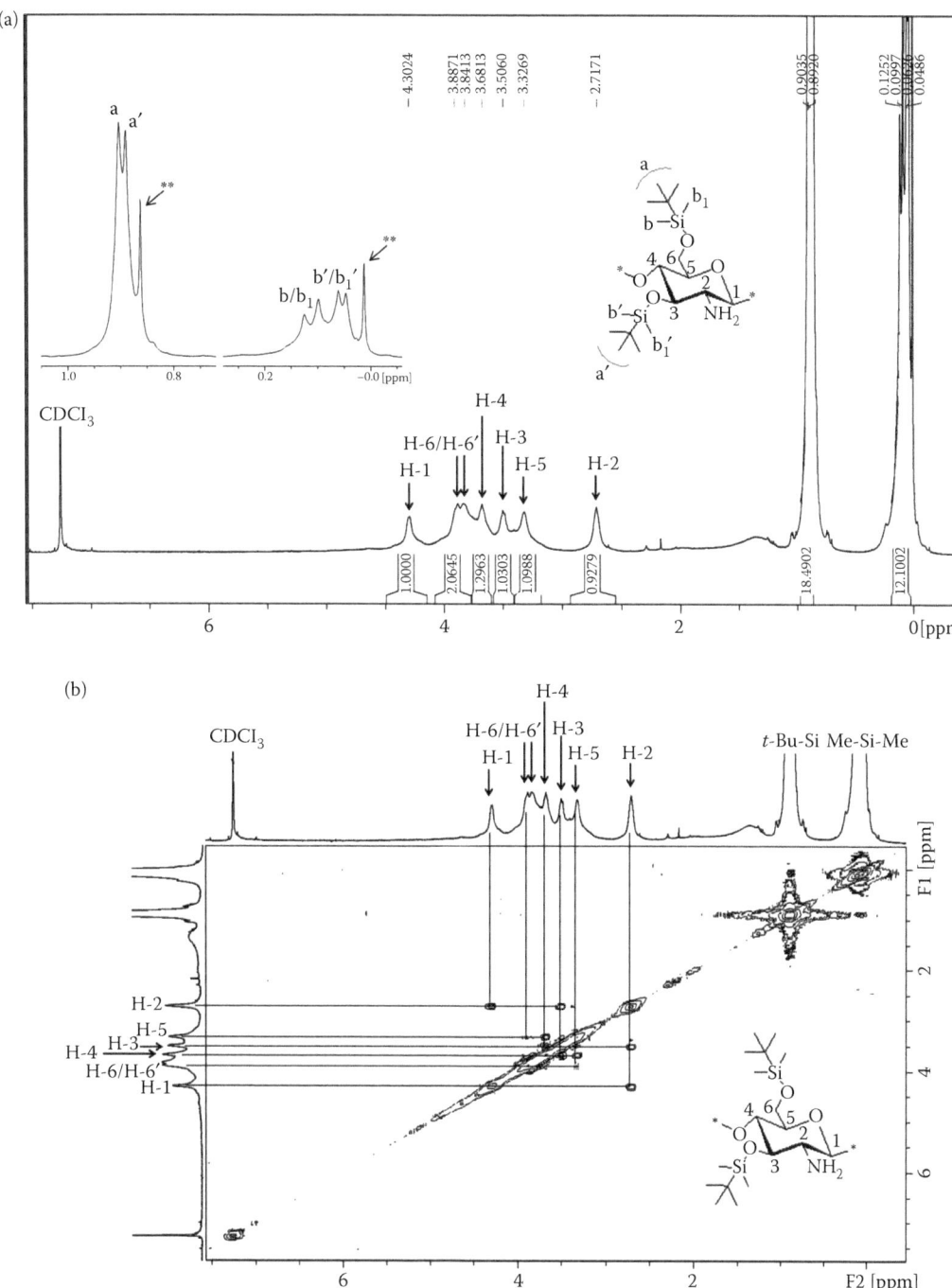

FIGURE 3.5 (a) ^{1}H NMR spectrum of DiTBDMS chitosan; ** shown in (a) can be attributed to undesired by-product di-TBDMS siloxane $(t\text{-Bu})(Me_2)Si\text{-}O\text{-}Si(t\text{-Bu})(Me_2)$. (b) ^{1}H-^{1}H COSY spectrum of DiTBDMS chitosan.

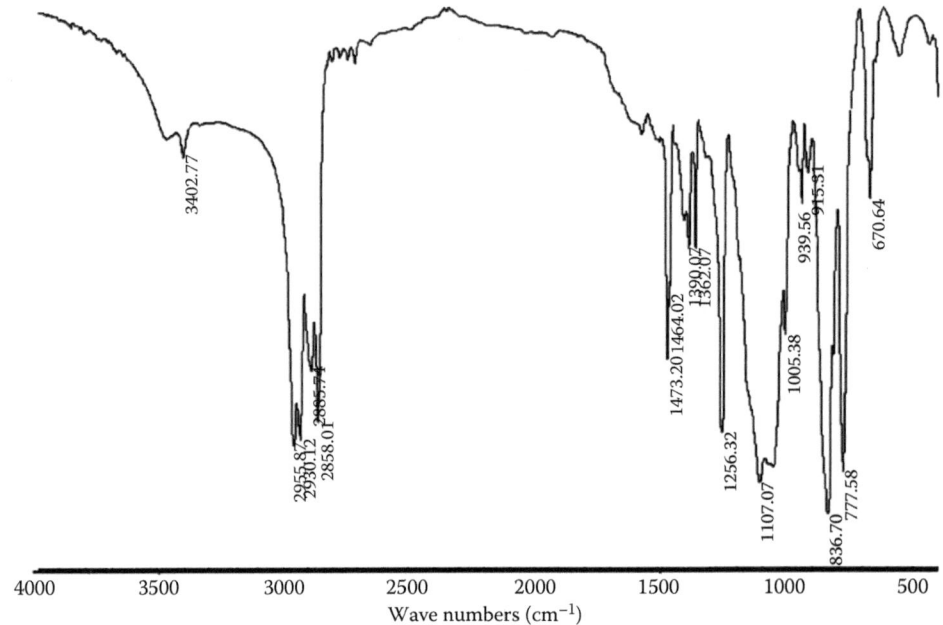

FIGURE 3.6 FT-IR spectrum of DiTBDMS chitosan.

TABLE 3.3
Solubility Table for TBDMS- and TMS-Protected Chitosan Derivatives

Solvent	Boiling Point (°C)	Polarity Index	Chitosan HCl	Chitosan Mesylate	3,6-Di-O-TBDMS Chitosan (DS 1.94)	TMS Chitosan (DS 2.82)
References			a[a]	a[a]	a[a]	b[c]
Water	100	9	++	+	−	
NMP	202	6.7	±	±	+	
DMAc	165	6.5				±
DMF	155	6.4	±	±	+	
DMSO	189	7.2	±	++	+	±
Pyridine	115	5.3	±	±	++	+
THF	65	4	−	−	+	±
Acetone	56	5.1	−	−	−	±
CH₃CN	82	5.8	−	−	−	
EtOAc	77	4.4	−	−	++	
MeOH	65	5.1	−	−	−	
Hexane	69	0	−	−	+	
CH₂Cl₂	41	3.1	−	−	+++[b]	
CHCl₃	61	4.1	−	−	+++[b]	

Note: a (Runarsson et al. 2008b), b (Kurita et al. 2004).

[a]　+++ , completely soluble (in 10% w/v); ++ , completely soluble (in 2.5% w/v); +, slightly opaque; ±, swollen; hardly soluble/insoluble.

[b]　In case of CH₂Cl₂ and CHCl₃, solubility is shown for fully silylated DiTBDMS compound that was obtained after optimization (DS 2.0).

[c]　+, almost soluble; ±, swelled or partial soluble; −, insoluble.

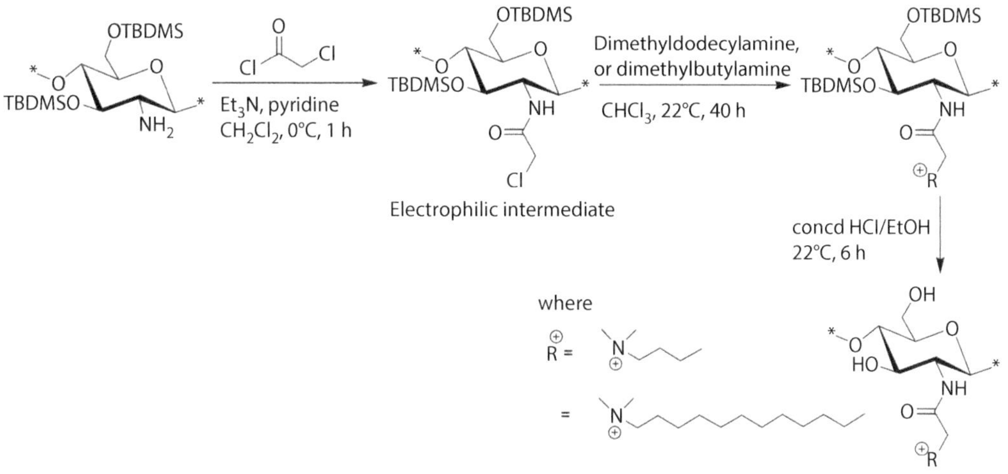

SCHEME 3.9   *N*-Acylation of chitosan using the DiTBDMS protection strategy.

3.6.2   *N*-Acyl- and *N*-Quaternized Chitosan Derivatives and Their Antimicrobial Activity

The DiTBDMS chitosan was recently used as a precursor for the synthesis of a different quaternary *N*-(2-(*N*,*N*,*N*-tri-alkylammoniumyl and 2-pyridiniumyl))-acetyl derivatives of chitosan polymer and chitooligomers (Runarsson et al. 2010). In this study, DiTBDMS chitosan was *N*-acetylated with chloroacetyl chloride (Scheme 3.9). This intermediate was then reacted with tertiary amines and finally, the deprotection to remove the TBDMS groups afforded the desired quaternary products with high DS (>0.90). These derivatives were investigated for antimicrobial activity against *Staphylococcus aureus* (*ATCC 25923*), *Staphylococcus aureus* (*MRSA*) (*ATCC 43300*), *Escherichia coli* (*ATCC 25922*), *Pseudomonas aeruginosa* (*ATCC 27853*), and *Enterococcus facialis* (*ATCC 29212*); and found to have different antibacterial effects that was dependent on the chitosan structure (Runarsson et al. 2010).

3.6.3   Selective Synthesis of Highly Substituted *N*-Alkyl-*N*,*N*-Dimethyl Chitosan Derivatives

Benediktsdottir et al. recently reported highly chemoselective synthesis of TMC and other *N*-alkyl-*N*,*N*-dimethyl chitosan, with high DS, by using DiTBDMS chitosan as a precursor (Scheme 3.10) (Benediktsdottir et al. 2011). The trimethylation was achieved by using CH_3I as methylating agent and Cs_2CO_3 as base, in NMP and the subsequent deprotection was done in 1 M TBAF/NMP. The NMR revealed a fully (100%) *N*-trimethylated chitosan without any *O*-methylation, showing the additional advantages of this protection group.

   In this same study, other *N*-alkyl-*N*,*N*-dimethyl chitosan derivatives were synthesized. First, the *N*-alkyl chains were added by stepwise reductive alkylation to avoid dialkylation (Liberek et al. 2005). Different aldehydes (propyl, butyl, or hexyl aldehyde) were reacted with DiTBDMS chitosan to form *N*-alkylimine-DiTBDMS chitosan and the presence of *n*-alkylimine was confirmed with both IR and NMR. The imines were then reduced with sodium triacetoxyborohydride (STAB-H). *N*,*N*-dimethylation of the *N*-alkyl chitosan derivatives was then done by using dimethylsulfate (DMS) as a methylating agent and Li_2CO_3 as a base in CH_2Cl_2. These derivatives were then deprotected using TBAF/NMP. The derivatives were all completely soluble in water and their structure could be resolved with NMR, with DS of *N*-alkyl-*N*,*N*-dimethylation ranging from 65% to 72%.

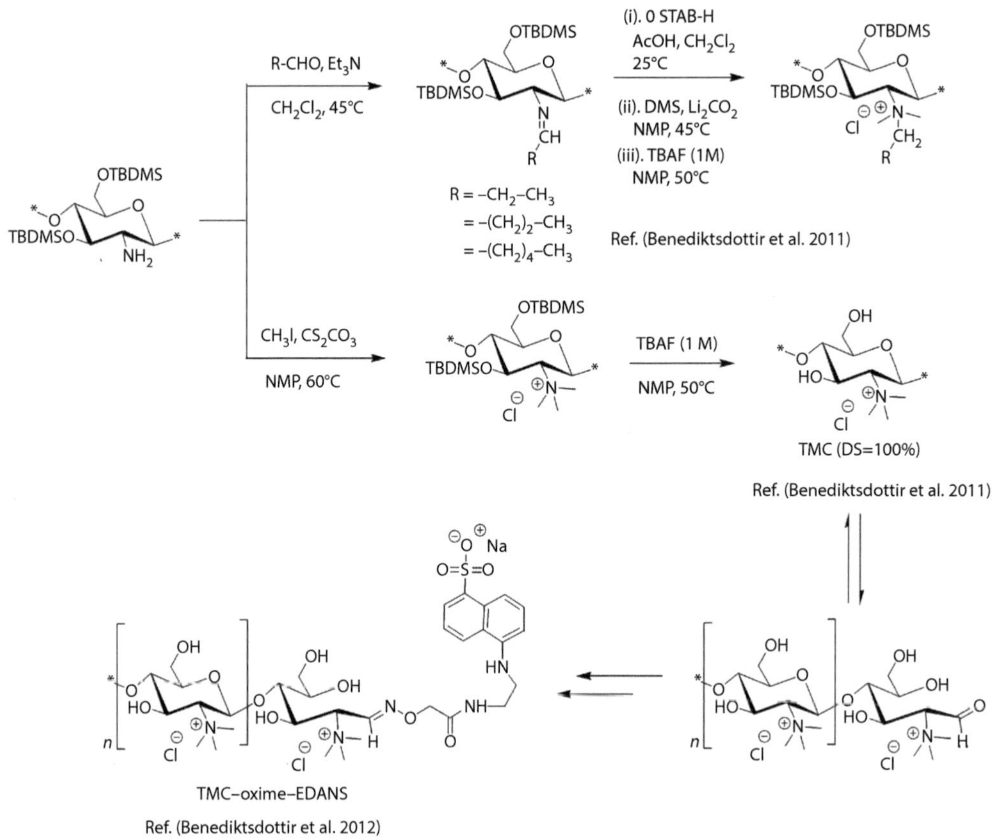

SCHEME 3.10 Synthesis of *N*-alkyl-*N*,*N*-dimethyl- and fluorescently end-labeled chitosan derivatives using DiTBDMS chitosan as precursor.

3.6.4 FLUORESCENT END LABELING OF TMC

One of the advantages of DiTBDMS chitosan is that the C-1 hydroxyl group is protected and the reactivity of the reducing end is maintained. This was demonstrated with fluorescent end label-ing of TMC that was synthesized using the DiTBDMS-protected chitosan (Benediktsdottir et al. 2012). For this purpose, 5-(2-((aminooxyacetyl)amino)ethylamino)naphthalene-1-sulfonic acid (EDANS–O–NH$_2$) fluorophore was reacted to the reducing-end aldehyde of TMC to yield TMC-oxime-EDANS (f-TMC). The reducing-end oxime formation was confirmed by ^{1}H NMR, ^{1}H-^{1}H COSY, and fluorospectrometry. The selective fluorescent end labeling of TMC was used to study the uptake of TMC in human bronchial epithelial cells. TMC was found to both adhere to the cell membrane and be intracellularly localized.

3.6.5 *N*-ACYLATION AND COVALENT LINKING OF HIGHLY LIPOPHILIC PHOTOSENSITIZER MOIETIES

Recently, DiTBDMS chitosan was used in the synthesis of chitosan nanocarriers containing cova-lently linked photosensitizer moieties (Gaware et al. 2013). The DiTBDMS chitosan offers the advantage of good solubility in CH$_2$Cl$_2$, enabling its reaction to the highly lipophilic *meso*-tet-raphenylporphyrin (log P > 9) photosensitizer in a quantitative reaction to give 0.10 or 0.25 DS. Trimethylamine and 1-methylpiperazine were then introduced (0.90 or 0.75 DS) to provide good aqueous solubility of the final deprotected carrier (Scheme 3.11). The synthesis of the carriers was repeated, showing that the procedure was fully reproducible.

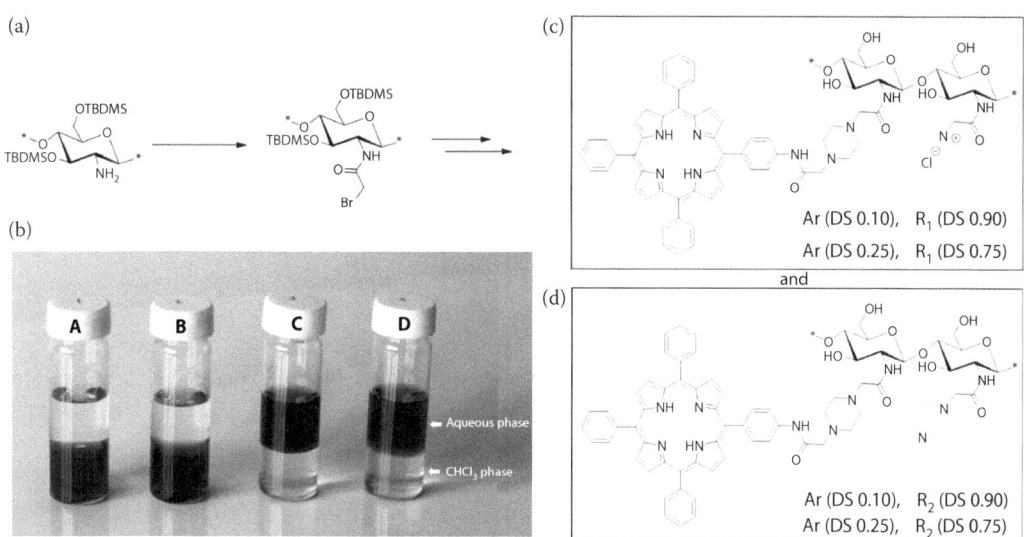

SCHEME 3.11 (a) Synthesis of chitosan–porphyrin nanocarriers through the use of DiTBDMS chitosan. (b) Phase separation of porphyrins (A, B) and chitosan–porphyrin conjugates (C, D) in H_2O:$CHCl_3$. A = 5-(4-aminophenyl)-10,15,20-triphenylporphyrin; B = 5-(4-piperazineacetylamidophenyl)-10,15,20-triphenylporphyrin; C = CS–porphyrin conjugate (c), where Ar (DS = 0.25), R_1 (DS = 0.75); D = CS–porphyrin conjugate (d), where Ar (DS = 0.25), R_2 (DS = 0.75).

Precision in the synthesis was possible only because DiTBDMS-protected intermediates in the synthesis were fully soluble in the reaction medium in all the steps of the synthesis route. The structure of the final compounds was confirmed by ^{1}H NMR, solid-state ^{13}C NMR, UV–VIS, and fluorospectrometry studies. Dynamic light scattering and scanning electron microscopy studies revealed that the carriers can form nanoparticle-like structures in an aqueous solution with size ranging from 140 to 200 nm. The average MW of the starting material was 235 kDa but the conversion to mesylate salt was accompanied by significant degradation to the average MW of 10.5 kDa as determined with gel permeation chromatography (GPC). Limited degradation of the material occurred in multiple reaction steps from conversion to DiTBDMS chitosan to deprotection to produce the final carriers. The average MW for 0.10 DS materials was ~5–9 kDa. There was some interaction with the stationary phase in the case of the more amphiphilic 0.25 DS materials and GPC could, therefore, be used only to assess purity and not the MW. These compounds were evaluated for *in vitro* photochemical gene transfection using HCT116/LUC human colon carcinoma cell line and was found to be highly potent.

3.7 CONCLUSION

The use of protection groups for chemoselective and/or regioselective synthesis of chitosan derivatives is an attractive option, offering some distinct advantages over direct modifications.

These include more selective modification and enhanced solubility in commonly used organic solvents. The increased solubility facilitates synthesis under milder and more homogeneous conditions and leads to better controlled reactions by saving time, cost of reagents, and improved reproducibility of the procedure. Ideally, the protection group introduced should be stable under different reaction conditions and be fully removable without causing much depolymerization. No protection strategy can fulfill these criteria but all currently employed strategies, used for the synthesis of chitosan derivatives, offer some advantage.

The *N*-phthaloyl chitosan has been used as a precursor for *O*-modification, especially modification of the primary C-6 hydroxyl group, and has enhanced solubility in organic solvents compared to the starting chitosan. The 6-*O*-trityl ether protection can be done regioselectively and quantitatively by using *N*-phthaloylated chitosan under homogeneous reaction conditions. The solubility of trityl chitosan enables its further modifications, such as *N*-acylation, under mild and homogeneous conditions. Nevertheless, the introduction of this protection group requires a three-step synthetic procedure. The use of silyl protection groups in the synthesis of chitosan is now emerging. TMS chitosan has been shown to be a useful precursor for further functionalization due to increased solubility; however, the labile nature of TMS diminishes its feasibility as a protection group.

The implementation of silyl protection in chitosan chemistry has been realized with the use of the TBDMS group. This promising protecting group can be introduced under mild reaction conditions to protect both the C-3 and C-6 hydroxyl groups. The high solubility of DiTBDMS chitosan in solvents such as CH_2Cl_2 and $CHCl_3$ has been utilized for a variety of *N*-selective modifications, including synthesis of TMC homopolymer, *N*-alkyl-*N*,*N*-dimethyl chitosan derivatives, *N*-acyl quaternary ammonium derivatives of chitosan, and chitosan nanocarriers with covalently linked highly lipophilic photosensitizers.

The derivatization can, therefore, potentially be better controlled than by the commonly used direct modification and this can justify a multistep synthesis route. Synthesis with protected polymers is demanding, as complete (100%) and selective protection and deprotection must be guaranteed. There are several challenges that remain in chitosan chemistry. Optimal and mild reaction conditions for chemo- and regioselective modification of chitosan with a variety of available protection groups are yet to be developed. Further development of the deprotection procedures is also important to allow full deprotection, in short time and without significant depolymerization. With further advances in the application of protection groups for chemo- and regioselective modification in chitosan chemistry, this may become the preferred approach for the synthesis of chitosan derivatives.

REFERENCES

Anthonsen, M.W. and Smidsrød, O. 1995. Hydrogen ion titration of chitosans with varying degrees of N-acetylation by monitoring induced 1H-NMR chemical shifts. *Carbohydrate Polymers* 26: 303–305.

Auzély, R. and Rinaudo, M. 2003. Controlled chemical modifications of chitosan. Characterization and investigation of original properties. *Macromolecular Bioscience* 3: 562–565.

Babin, M., Ruest, A., Drouin, G. et al. 2012. Regioselective pivaloylation of N-phthaloylchitosan: A promising soluble intermediate for chitosan chemistry. *Carbohydrate Research* 351: 87–92.

Badawy, M.E.I. and Rabea, E.I. 2011. A biopolymer chitosan and its derivatives as promising antimicrobial agents against plant pathogens and their applications in crop protection. *International Journal of Carbohydrate Chemistry* 2011: 1–29.

Benediktsdottir, B.E., Gaware, V.S., Runarsson, O.V. et al. 2011. Synthesis of *N*,*N*,*N*-trimethyl chitosan homopolymer and highly substituted *N*-alkyl-*N*,*N*-dimethyl chitosan derivatives with the aid of di-*tert*-butyldimethylsilyl chitosan. *Carbohydrate Polymers* 86: 1451–1460.

Benediktsdottir, B.E., Sorensen, K.K., Thygesen, M.B. et al. 2012. Regioselective fluorescent labeling of *N*,*N*,*N*-trimethyl chitosan via oxime formation. *Carbohydrate Polymers* 90: 1273–1280.

Bhattarai, N., Gunn, J., and Zhang, M.Q. 2010. Chitosan-based hydrogels for controlled, localized drug delivery. *Advanced Drug Delivery Reviews* 62: 83–99.

Binette, A. and Gagnon, J. 2007. Regioselective silylation of N-phthaloylchitosan with TBDMS and TBDPS groups. *Biomacromolecules* 8: 1812–1815.

Blackwell, J. and Weih, M.A. 1980. Structure of chitin-protein complexes: Ovipositor of the ichneumon fly *Megarhyssa*. *Journal of Molecular Biology* 137: 49–60.

Bough, W.A., Salter, W.L., Wu, A.C.M. and Perkins, B.E. 1978. Influence of manufacturing variables on characteristics and effectiveness of chitosan products.1. Chemical composition, viscosity, and molecular-weight distribution of chitosan products. *Biotechnology and Bioengineering* 20: 1931–1943.

Bowman, K. and Leong, K.W. 2006. Chitosan nanoparticles for oral drug and gene delivery. *International Journal of Nanomedicine* 1: 117–128.

Chaudhary, S.K. and Hernandez, O. 1979. A simplified procedure for the preparation of triphenylmethylethers. *Tetrahedron Letters* 20: 95–98.

Chen, C., Tao, S., Qiu, X., Ren, X., and Hu, S. 2013. Long-alkane-chain modified N-phthaloyl chitosan membranes with controlled permeability. *Carbohydrate Polymers* 91: 269–276.

Chirachanchai, S., Lertworasirikul, A., and Tachaboonyakiat, W. 2001. Carbaryl insecticide conjugation onto chitosan via iodochitosan and chitosan carbonyl imidazolide precursors. *Carbohydrate Polymers* 46: 19–27.

Corey, E.J. and Venkateswarlu, A. 1972. Protection of hydroxyl groups as *tert*-butyldimethylsilyl derivatives. *Journal of the American Chemical Society* 94: 6190–6191.

Di Martino, A., Sittinger, M., and Risbud, M.V. 2005. Chitosan: A versatile biopolymer for orthopaedic tissue-engineering. *Biomaterials* 26: 5983–5990.

Esmaeili, F., Heuking, S., Junginger, H.E., and Borchard, G. 2010. Progress in chitosan-based vaccine delivery systems. *Journal of Drug Delivery Science and Technology* 20: 53–61.

Gaware, V.S., Håkerud, M., Leósson, K. et al. 2013. Tetraphenylporphyrin tethered chitosan based carriers for photochemical transfection. *Journal of Medicinal Chemistry* 56: 807–819.

Gonil, P., Sajomsang, W., Ruktanonchai, U.R. et al. 2011. Novel quaternized chitosan containing β-cyclodextrin moiety: Synthesis, characterization and antimicrobial activity. *Carbohydrate Polymers* 83: 905–913.

Greene, T.W. and Wuts, P.G.M. 2007. *Protective Groups in Organic Synthesis*. New Jersey, John Wiley & Sons, Inc.

Harmon, R.E., De, K.K., and Gupta, S.K. 1973a. New procedure for preparing trimethylsilyl derivatives of polysaccharides. *Carbohydrate Research* 31: 407–409.

Harmon, R.E., De, K.K., and Gupta, S.K. 1973b. Preparation of trimethylsilyl derivatives of polysaccharides. *Starch—Stärke* 25: 429–431.

Hernandez-Lauzardo, A.N., Velazquez-del Valle, M.G., and Guerra-Sanchez, M.G. 2011. Current status of action mode and effect of chitosan against phytopathogens fungi. *African Journal of Microbiology Research* 5: 4243–4247.

Holappa, J., Nevalainen, T., Safin, R. et al. 2006a. Novel water-soluble quaternary piperazine derivatives of chitosan: Synthesis and characterization. *Macromolecular Bioscience* 6: 139–144.

Holappa, J., Nevalainen, T., Savolainen, J. et al. 2004. Synthesis and characterization of chitosan *N*-betainates having various degrees of substitution. *Macromolecules* 37: 2784–2789.

Holappa, J., Nevalainen, T., Soininen, P. et al. 2005. *N*-Chloroacyl-6-*O*-triphenylmethylchitosans: Useful intermediates for synthetic modifications of chitosan. *Biomacromolecules* 6: 858–863.

Holappa, J. Nevalainen, T., Soininen, P., Masson, M., and Jarvinen, T. 2006b. Synthesis of novel quaternary chitosan derivatives via *N*-chloroacyl-6-*O*-triphenylmethylchitosans. *Biomacromolecules* 7: 407–410.

Horton, D. and Lehmann, J. 1978. Selective 6-*O*-acetylation of amylose. *Carbohydrate Research* 61: 553–556.

Ifuku, S., Wada, M., Morimoto, M., and Saimoto, H. 2012. A short synthesis of highly soluble chemoselective chitosan derivatives via "click chemistry". *Carbohydrate Polymers* 90: 1182–1186.

Jančiauskaitė, U., Višnevskij, Č., Radzevičius, K., and Makuška, R. 2009. Polyampholytes from natural building blocks: Synthesis and properties of chitosan-*o*-alginate copolymers. *Chemija* 20: 128–135.

Kaftan, O., Tumbiolo, S., Dubreuil, F. et al. 2011. Probing multivalent host-guest interactions between modified polymer layers by direct force measurement. *Journal of Physical Chemistry B* 115: 7726–7735.

Kato, Y., Onishi, H. and Machida, Y. 2004. N-Succinyl-chitosan as a drug carrier: Water-insoluble and water-soluble conjugates. *Biomaterials* 25: 907–915.

Keilich, V.G., Tihlarik, K., and Husemann, E. 1968. Über die herstellung von tris-O-trimethylsilylpolysaccha-riden. *Die Makromolekulare Chemie* 120: 87–95.

Kim, C., Choi, J., Chun, H., and Choi, K. 1997. Synthesis of chitosan derivatives with quaternary ammonium salt and their antibacterial activity. *Polymer Bulletin* 38: 387–393.

Kim, C.H. and Choi, K.S. 1998. Synthesis and properties of carboxyalkyl chitosan derivatives. *Journal of Industrial and Engineering Chemistry* 4: 19–25.

Kubota, N. 1997. Permeability properties of chitosan-transition metal complex membranes. *Journal of Applied Polymer Science* 64: 819–822.

Kubota, N. and Eguchi, Y. 1997. Facile preparation of water-soluble *N*-acetylated chitosan and molecular weight dependence of its water-solubility. *Polymer Journal* 29: 123–127.

Kurita, K., Hirakawa, M., Aida, K., Yang, J., and Nishiyama, Y. 2003. Trimethylsilylated chitosan: A convenient precursor for chemical modifications. *Chemistry Letters* 32: 1074–1075.

Kurita, K., Hirakawa, M., Kikuchi, S., Yamanaka, H., and Yang, J. 2004. Trimethylsilylation of chitosan and some properties of the product. *Carbohydrate Polymers* 56: 333–337.

Kurita, K., Hirakawa, M., and Nishiyama, Y. 1999. Silylated chitin: A new organosoluble precursor for facile modifications and film casting. *Chemistry Letters*: 771–772.

Kurita, K., Ikeda, H., Shimojoh, M., and Yang, J. 2007. N-Phthaloylated chitosan as an essential precursor for controlled chemical modifications of chitosan: Synthesis and evaluation. *Polymer Journal* 39: 945–952.

Kurita, K., Ikeda, H., Yoshida, Y., Shimojoh, M., and Harata, M. 2002. Chemoselective protection of the amino groups of chitosan by controlled phthaloylation: Facile preparation of a precursor useful for chemical modifications. *Biomacromolecules* 3: 1–4.

Kurita, K., Sugita, K., Kodaira, N., Hirakawa, M., and Yang, J. 2005. Preparation and evaluation of trimethylsilylated chitin as a versatile precursor for facile chemical modifications. *Biomacromolecules* 6: 1414–1418.

Lee, D.-X., Xia, W.-S., and Zhang, J.-L. 2008. Enzymatic preparation of chitooligosaccharides by commercial lipase. *Food Chemistry* 111: 291–295.

Lehr, C.-M., Bouwstra, J.A., Schacht, E.H., and Junginger, H.E. 1992. *In vitro* evaluation of mucoadhesive properties of chitosan and some other natural polymers. *International Journal of Pharmaceutics* 78: 43–48.

Li, X.K. and Xia, W.S. 2011. Effects of concentration, degree of deacetylation and molecular weight on emulsifying properties of chitosan. *International Journal of Biological Macromolecules* 48: 768–772.

Liberek, B., Melcer, A., Osuch, A. et al. 2005. *N*-Alkyl derivatives of 2-amino-2-deoxy-D-glucose. *Carbohydrate Research* 340: 1876–1884.

Lin, S.-B., Lin, Y.-C., and Chen, H.-H. 2009. Low molecular weight chitosan prepared with the aid of cellulase, lysozyme and chitinase: Characterisation and antibacterial activity. *Food Chemistry* 116: 47–53.

Lohse, A., Martins, R., Jorgensen, M.R., and Hindsgaul, O. 2006. Solid-phase oligosaccharide tagging (SPOT): Validation on glycolipid-derived structures. *Angewandte Chemie International Edition English* 45: 4167–4172.

Makuška, R. and Gorochovceva, N. 2006. Regioselective grafting of poly(ethylene glycol) onto chitosan through C-6 position of glucosamine units. *Carbohydrate Polymers* 64(2): 319–327.

Masuko, T., Minami, A., Iwasaki, N. et al. 2005. Thiolation of chitosan. Attachment of proteins via thioether formation. *Biomacromolecules* 6: 880–884.

Mendonca, S. and Laine, R.A. 2005. Synthesis of sterically crowded derivatives of anomeric pairs of D-glucose disaccharides. *Carbohydrate Research* 340: 2055–2059.

Mohammadi, Z., Abolhassani, M., Dorkoosh, F.A. et al. 2011. Preparation and evaluation of chitosan-DNA-FAP-B nanoparticles as a novel non-viral vector for gene delivery to the lung epithelial cells. *International Journal of Pharmaceutics* 409: 307–313.

Mormann, W. 2003. Silylation of cellulose with hexamethyldisilazane in ammonia—activation, catalysis, mechanism, properties. *Cellulose* 10: 271–281.

Mourya, V.K. and Inamdar, N.N. 2008. Chitosan-modifications and applications: Opportunities galore. *Reactive & Functional Polymers* 68: 1013–1051.

Munro, N.H., Hanton, L.R., Moratti, S.C., and Robinson, B.H. 2009. Preparation and graft copolymerisation of thiolated β-chitin and chitosan derivatives. *Carbohydrate Polymers* 78: 137–145.

Muzzarelli, R.A.A. and Tanfani, F. 1985. The N-permethylation of chitosan and the preparation of *N*-trimethyl chitosan iodide. *Carbohydrate Polymers* 5: 297–307.

Muzzarelli, R.A.A., Tanfani, F., Emanuelli, M., and Mariotti, S. 1982. N-(carboxymethylidene)chitosans and n-(carboxymethyl)-chitosans—Novel chelating polyampholytes obtained from chitosan glyoxylate. *Carbohydrate Research* 107: 199–214.

Nelson, T.D. and Crouch, R.D. 1996. Selective deprotection of silyl ethers. *Synthesis* 1996: 1031–1069.

Nishimura, S.I., Kohgo, O., Kurita, K., and Kuzuhara, H. 1991. Chemospecific manipulations of a rigid polysaccharide—Syntheses of novel chitosan derivatives with excellent solubility in common organic-solvents by regioselective chemical modifications. *Macromolecules* 24: 4745–4748.

Nouvel, C., Dubois, P., Dellacherie, E., and Six, J.L. 2003. Silylation reaction of dextran: Effect of experimental conditions on silylation yield, regioselectivity, and chemical stability of silylated dextrans. *Biomacromolecules* 4: 1443–1450.

Novoa-Carballal, R. and Muller, A.H. 2012. Synthesis of polysaccharide-b-PEG block copolymers by oxime click. *Chemical Communications (Camb)* 48: 3781–3783.

Oliveira, J.R., Martins, M.C.L., Mafra, L., and Gomes, P. 2012. Synthesis of an O-alkynyl-chitosan and its chemoselective conjugation with a PEG-like amino-azide through click chemistry. *Carbohydrate Polymers* 87: 240–249.

Park, J.H., Saravanakumar, G., Kim, K., and Kwon, I.C. 2010. Targeted delivery of low molecular drugs using chitosan and its derivatives. *Advanced Drug Delivery Reviews* 62: 28–41.

Peng, Y.F., Han, B.Q., Liu, W.S., and Xu, X.J. 2005. Preparation and antimicrobial activity of hydroxypropyl chitosan. *Carbohydrate Research* 340: 1846–1851.

Polnok, A., Borchard, G., Verhoef, J.C., Sarisuta, N., and Junginger, H.E. 2004. Influence of methylation process on the degree of quaternization of N-trimethyl chitosan chloride. *European Journal of Pharmaceutics and Biopharmaceutics* 57: 77–83.

Prego, C., Torres, D., Fernandez-Megia, E. et al. 2006. Chitosan–PEG nanocapsules as new carriers for oral peptide delivery: Effect of chitosan pegylation degree. *Journal of Controlled Release* 111: 299–308.

Rabea, E.I., Badawy, M.E.T., Stevens, C.V., Smagghe, G., and Steurbaut, W. 2003. Chitosan as antimicrobial agent: Applications and mode of action. *Biomacromolecules* 4: 1457–1465.

Rinaudo, M. 2006a. Chitin and chitosan: Properties and applications. *Progress in Polymer Science* 31: 603–632.

Rinaudo, M. 2006b. Non-covalent interactions in polysaccharide systems. *Macromolecular Bioscience* 6: 590–610.

Ronghua, H., Yumin, D., and Jianhong, Y. 2003. Preparation and anticoagulant activity of carboxybutyrylated hydroxyethyl chitosan sulfates. *Carbohydrate Polymers* 51: 431–438.

Runarsson, O.V., Holappa, J., Jonsdottir, S., Steinsson, H., and Masson, M. 2008a. N-selective 'one pot' synthesis of highly *N*-substituted trimethyl chitosan (TMC). *Carbohydrate Polymers* 74: 740–744.

Runarsson, O.V., Holappa, J., Malainer, C. et al. 2010. Antibacterial activity of *N*-quaternary chitosan derivatives: Synthesis, characterization and structure activity relationship (SAR) investigations. *European Polymer Journal* 46: 1251–1267.

Runarsson, O.V., Holappa, J., Nevalainen, T. et al. 2007. Antibacterial activity of methylated chitosan and chitooligomer derivatives: Synthesis and structure activity relationships. *European Polymer Journal* 43: 2660–2671.

Runarsson, O.V., Malainer, C., Holappa, J., Sigurdsson, S.T., and Masson, M. 2008b. *tert*-Butyldimethylsilyl *O*-protected chitosan and chitooligosaccharides: Useful precursors for *N*-modifications in common organic solvents. *Carbohydrate Research* 343: 2576–2582.

Saito, H., Tabeta, R., and Hirano, S. 1981. Conformation of chitin and N-acyl chitosans in solid-state as revealed by C-13 cross polarization magic angle spinning (Cp-Mas) NMR-spectroscopy. *Chemistry Letters* 10: 1479–1482.

Saito, H., Tabeta, R., and Ogawa, K. 1987. High-resolution solid-state C-13 NMR-study of chitosan and its salts with acids—Conformational characterization of polymorphs and helical structures as viewed from the conformation-dependent C-13 chemical-shifts. *Macromolecules* 20: 2424–2430.

Sashiwa, H., Kawasaki, N., Nakayama, A. et al. 2002a. Chemical modification of chitosan. 14:(1) Synthesis of water-soluble chitosan derivatives by simple acetylation. *Biomacromolecules* 3: 1126–1128.

Sashiwa, H., Kawasaki, N., Nakayama, A. et al. 2002b. Chemical modification of chitosan. 13.(1) Synthesis of organosoluble, palladium adsorbable, and biodegradable chitosan derivatives toward the chemical plating on plastics. *Biomacromolecules* 3: 1120–1125.

Sashiwa, H., Shigemasa, Y., and Roy, R. 2000a. Homogeneous *N,O*-acylation of chitosan in dimethyl sulfoxide with cyclic acid anhydrides. *Chemistry Letters* 29: 1186–1187.

Sashiwa, H., Shigemasa, Y. and Roy, R. 2000b. Dissolution of chitosan in dimethyl sulfoxide by salt formation. *Chemistry Letters* 29: 596–597.

Sieval, A.B., Thanou, M., Kotze, A.F. et al. 1998. Preparation and NMR characterization of highly substituted N-trimethyl chitosan chloride. *Carbohydrate Polymers* 36: 157–165.

Song, W.L., Gaware, V.S., Runarsson, O.V., Masson, M., and Mano, J.F. 2010. Functionalized superhydrophobic biomimetic chitosan-based films. *Carbohydrate Polymers* 81: 140–144.

Sugimoto, M., Morimoto, M., Sashiwa, H., Saimoto, H., and Shigemasa, Y. 1998. Preparation and characterization of water-soluble chitin and chitosan derivatives. *Carbohydrate Polymers* 36: 49–59.

Sugita, K., Yang, J., Shimojoh, M., and Kurita, K. 2008. Influence of trimethylsilylation and detrimethylsilylation on the molecular weight of chitin: Evaluation of viscometry and gel permeation chromatography for molecular weight determination. *Polymer Bulletin* 60: 449–455.

Thanou, M., Florea, B.I., Langemeyer, M.W., Verhoef, J.C., and Junginger, H.E. 2000. *N*-trimethylated chitosan chloride (TMC) improves the intestinal permeation of the peptide drug buserelin *in vitro* (Caco-2 cells) and *in vivo* (rats). *Pharmaceutical Research* 17: 27–31.

Thanou, M., Verhoef, J.C., and Junginger, H.E. 2001. Chitosan and its derivatives as intestinal absorption enhancers. *Advanced Drug Delivery Reviews* 50, Supplement 1: S91–S101.

Torii, Y., Ikeda, H., Shimojoh, M., and Kurita, K. 2009. Chemoselective protection of chitosan by dichlorophthaloylation: Preparation of a key intermediate for chemical modifications. *Polymer Bulletin* 62: 749–759.

van der Merwe, S.M., Verhoef, J.C., Verheijden, J.H.M., Kotze, A.F., and Junginger, H.E. 2004. Trimethylated chitosan as polymeric absorption enhancer for improved peroral delivery of peptide drugs. *European Journal of Pharmaceutics and Biopharmaceutics* 58: 225–235.

Vårum, K.M., Ottøy, M.H., and Smidsrød, O. 2001. Acid hydrolysis of chitosans. *Carbohydrate Polymers* 46: 89–98.

Verheul, R.J., Amidi, M., van der Wal, S. et al. 2008. Synthesis, characterization and *in vitro* biological properties of O-methyl free N,N,N-trimethylated chitosan. *Biomaterials* 29: 3642–3649.

Wang, X., Strand, S.P., Du, Y., and Vårum, K.M. 2010. Chitosan–DNA–rectorite nanocomposites: Effect of chitosan chain length and glycosylation. *Carbohydrate Polymers* 79: 590–596.

Watanabe, T., Kodaira, N., Ikeda, H., and Kurita, K. 2012. Synthesis and some properties of silylated chitins as key intermediates for chemical modifications. *Polymer Bulletin* 68: 1845–1855.

Xia, W.S., Liu, P., Zhang, J.L., and Chen, J. 2011. Biological activities of chitosan and chitooligosaccharides. *Food Hydrocolloids* 25: 170–179.

Xie, Y.J., Liu, X.F., and Chen, Q. 2007. Synthesis and characterization of water-soluble chitosan derivate and its antibacterial activity. *Carbohydrate Polymers* 69: 142–147.

Yang, J., Cai, J., Hu, Y., Li, D., and Du, Y. 2012. Preparation, characterization and antimicrobial activity of 6-amino-6-deoxychitosan. *Carbohydrate Polymers* 87: 202–209.

Zhang, M., Hisamori, H., Yamada, T., and Hirano, S. 1994. C-13 CP/MAS NMR spectral analysis of 6-O-Tosyl, 6-Deoxy-6-iodo, and 6-Deoxy Derivatives of *N*-acetylchitosan in a solid state. *Bioscience, Biotechnology, and Biochemistry* 58: 1906–1908.

Zhao, S.-H., Wu, X.-T., Guo, W.-C. et al. 2010. *N*-(2-hydroxyl) propyl-3-trimethyl ammonium chitosan chloride nanoparticle as a novel delivery system for parathyroid hormone-related protein 1–34. *International Journal of Pharmaceutics* 393: 269–273.

Zong, Z., Kimura, Y., Takahashi, M., and Yamane, H. 2000. Characterization of chemical and solid state structures of acylated chitosans. *Polymer* 41: 899–906.

4 Chemical Aspects of Chitin and Chitosan Derivatives

Hitoshi Sashiwa

CONTENTS

4.1   INTRODUCTION

Chitin exists widely in nature, such as in the shell of crab, shrimp, and insects. Chitosan is an *N*-deacetylated product of chitin. Studies on chitin and chitosan have been enhanced since 1990 because these polysaccharides show excellent biological properties, such as biodegradation in the human body (Sashiwa et al. 1990, Shigemasa et al. 1994), immunological (Nishimura et al. 1984, Mori et al. 1987) and antibacterial properties (Tanigawa et al. 1992, Tokura et al. 1997), and wound-healing activities (Okamoto et al. 1993, Khnor and Lim 2003, Kweon et al. 2003). In recent studies it has been identified that chitosan is a good candidate to be used as a supporting material for gene delivery (Sato et al. 2001), cell culture (Mao et al. 2003) and tissue engineering (Gingras et al. 2003, Wang et al. 2003). Hence, chitin and chitosan are currently in focus and are expected to be used as new functional materials. Nevertheless, commercial or practical use of them is limited at the current stage, so therefore they are still nonutilizing biomass.

Studies on the practical use of chitin and chitosan (include monomer and oligomer) have been carried out in its nonmodified form. To make a breakthrough in their utilization, however, the chemical modification of chitin and chitosan will be a key point by functionalizing them, because they are capable of introducing a variety of functional groups. For this purpose, more existing and fundamental study on the chemical modification of chitin and chitosan is required. For example, the chemistry of cellulose is well studied and still ongoing. Until now, numerous works have been reported on the chemical modification of chitosan. Most of the traditional chemistry of chitosan has been published in reviews or books (Gupta and Kumar 2000, Kumar 2000, Kurita 2001). In this chapter, we highlight various aspects of the chemical modification of chitosan.

4.2   DISSOLUTION OF CHITOSAN IN ORGANIC SOLVENTS

Generally, chitosan dissolves in dilute acidic water, such as HCl and AcOH, owing to the salt formation of the amino group. It has been reported that chitin dissolves in organic solvents such as LiCl/ *N,N*-dimethylacetamide (DMAc) and saturated $CaCl_2 \cdot 2H_2O/MeOH$ (Tokura et al. 1996). Also there have been few reports on the dissolution of chitosan in organic solvents, which is important for the

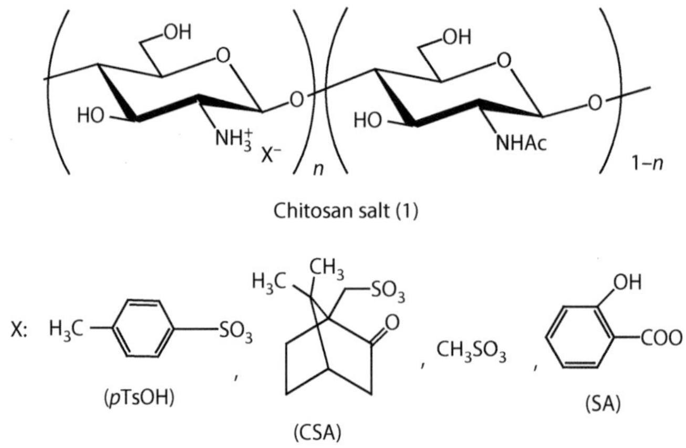

FIGURE 4.1 Chemical structure of chitosan salt that is dissolved in DMSO.

chemical modification. Recently, we first reported the dissolution of chitosan in dimethylsulfoxide (DMSO) by the salt formation with p-toluenesulfonic acid (pTsOH), (1R)-(–)-10-camphorsulfonic acid (CSA), methanesulfonic acid (MeSO$_3$H), and salicylic acid (SA) (Figure 4.1) (Sashiwa et al. 2000b). Chitosan dissolved in water with equimolar quantities of acid as listed in Table 4.1. After the freeze-drying of aqueous chitosan salt solution, a part of chitosan-acid salts, such as pTsOH, CSA, MeSO$_3$H, and SA, dissolved in DMSO. Non-freeze-dried chitosan powder, however, did not perfectly dissolve in DMSO. Furthermore, chitosan-salt film prepared by the drying process from aqueous acidic solution has not shown perfect dissolution. Therefore, freeze-drying is an important process for the dissolution of the chitosan salt in DMSO. These chitosan salts, however, did not dissolve in other organic solvents, such as DMF, MeOH, pyridine, and chloroform. From the ^{1}H NMR spectra in DMSO-d_6, typical signals assigned for chitosan and acids were observed, thereby suggesting that alkyl or allyl sulfonate was effective for the dissolution of chitosan in DMSO. While chitosan SA salt was dissolved in DMSO, other aromatic acids were not. The phenolic hydroxyl group at the *ortho* position should play an important role in the dissolution of chitosan.

TABLE 4.1
Solubility of Chitosan Salts in Organic Solvents[a]

Acid	H$_2$O	DMSO	Others[b]
		Solvent	
pTsOH	O	O	X
CSA	O	O	X
MeSO$_3$H	O	S	X
H$_2$SO$_4$	X	X	X
SA (2-HOC$_6$H$_4$CO$_2$H)	O	O	X
3-HOC$_6$H$_4$CO$_2$H	O	X	X
4-HOC$_6$H$_4$CO$_2$H	O	X	X
C$_6$H$_5$CO$_2$H	O	X	X
HCl	O	X	X

[a] Chitosan was dispersed in solvent and allowed to stand for 1 day.
   O, dissolved; S, partially dissolved; X, undissolved.
[b] Other solvents are DMF, MeOH, pyridine, and chloroform.

The chemical reaction of chitosan with cyclic acid anhydrides is a convenient way to obtain water-soluble chitosan derivatives (Hirano and Moriyasu 1981). In our previous works, *O*-acylation of chitin (degree of deacetylation (DDA) = 20%) in LiCl/DMAc solvent (Shigemasa et al. 1999) or selective *N*-acylation of chitosan with cyclic acid anhydrides (Sashiwa and Shigemasa 1999) was discussed. On the basis of these facts, the reaction of chitosan using the aforementioned DMSO solvent system with cyclic acid anhydrides was tested (Figure 4.2). A special emphasis is that the reaction of chitosan in organic solvent could be achieved first under homogeneous conditions (Sashiwa et al. 2000c). From the ^{1}H NMR analysis, most of the products showed a high degree of substitution (DS) over 2.0 per repeating unit and water-soluble products were obtained. In this case, however, the salt formation is necessary for dissolution in the organic solvent. More recently, complete *N*-acylation of chitosan has been achieved by treating it with cyclic acid anhydrides in aqueous homogeneous media at pH 4–8 (Satoh et al. 2003). Some of the resulting *N*-(carboxy)acyl chitosans were converted into the corresponding imido forms by thermal dehydration.

If nonpretreated (free) chitosan directly dissolves in an organic solvent, it is more useful for not only the chemical reaction but also for the regeneration of chitosan, such as film, gel, and fiber. We discovered the direct dissolution of chitosan in hexafluoro-2-propanol (HFP, Figure 4.3) without any salt formation and pretreatment processes (Sashiwa et al. 2002f). The dissolution behavior of various chitosans in HFP is shown in Table 4.2. Chitosan (SK-10) perfectly dissolved in HFP within 1 day. Flonac C also showed perfect dissolution in 14 days. Other chitosans, however, remained

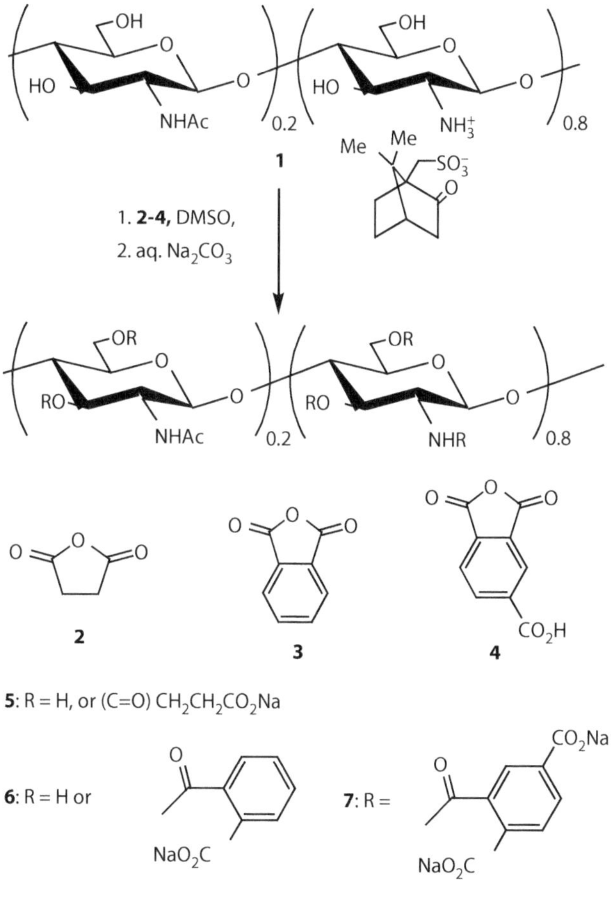

FIGURE 4.2 Homogeneous *N,O*-acylation of chitosan in DMSO.

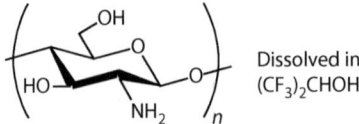

FIGURE 4.3 Chemical structure of chitosan and hexafluoro-2-propanol (HFP).

TABLE 4.2
Dissolution of Chitin and Chitosan in HFP

Chitosan	DDA[a] (%)	Mn	Time (day)	Dissolution in HEP[b]
Flonac C	80	30,000	1	△
Flonac C	80		14	○
SK-10	85	32,000	1	○
10B	96	93,000	14	△
R-10B[c]	96	100,000	14	△
10B-71[d]	71	80,000	14	△
10B-59[d]	59	89,000	14	△
10B-53[d]	53	n.d.	14	△
Chitin-TCL	12	130,000	14	△

[a] DDA, determined by ^{1}H NMR.
[b] ○, perfect dissolution; △, partial dissolution.
[c] R-10B, regenerated sample of 10 B.
[d] Partially *N*-acetylated sample of 10 B.

---

part of the insoluble fractions even after 14 days. From the gravimetric analysis, 86%(w/w) of chitosan 10 B was dissolved. In the case of regenerated chitosan (R-10B), which is low crystallinity owing to its pretreatment, a trace amount of insoluble fraction in R-10 B was also observed. Thus demonstrating that the crystallinity or DDA of chitosan would not influence the solubility. The molecular weight of chitosan (below 32,000) was important on complete dissolution. From the ^{1}H NMR spectrum of chitosan in mixed solvent of HFP and D_2O (1:1), typical signals at δ 2.1 ppm (NHAc), 2.9 ppm (H-2 of GlcN unit), and 3.5–4.0 ppm (sugar unit) were observed, thereby suggesting that chitosan dissolves in HFP with nonprotonated form of amino groups. The interaction of the fluoride moiety with the intra- and intermolecular hydrogen bond in chitosan would be the main factor on dissolution. From these chitosan solutions in volatile HFP (bp = 59°C), the transparent films were formed at room temperature for 1 day, thus demonstrating that chitosan was not decomposed and is stable in HFP. Moreover, HFP is the sole organic solvent to dissolve both chitin and chitosan independent of their DDA. One weak point is the residual amount of insoluble fraction for high MW of chitosan, so that the filtration step is required for further applications. This solvent is useful for membrane, fiber, paper, and other molding technologies owing to its volatile property.

4.3 SUGAR-MODIFIED CHITOSAN

The first report on the modification of chitosan with sugars has been documented by Hall and Yalpani (Figure 4.4) (Hall and Yalpani 1980, Yalpani and Hall 1984). They synthesized sugar-bound chitosan by reductive *N*-alkylation using $NaCNBH_3$ and unmodified sugar (**9**: Method A) or sugar–aldehyde derivative (**10**: Method B). At that time, the sugar-bound chitosans were mainly

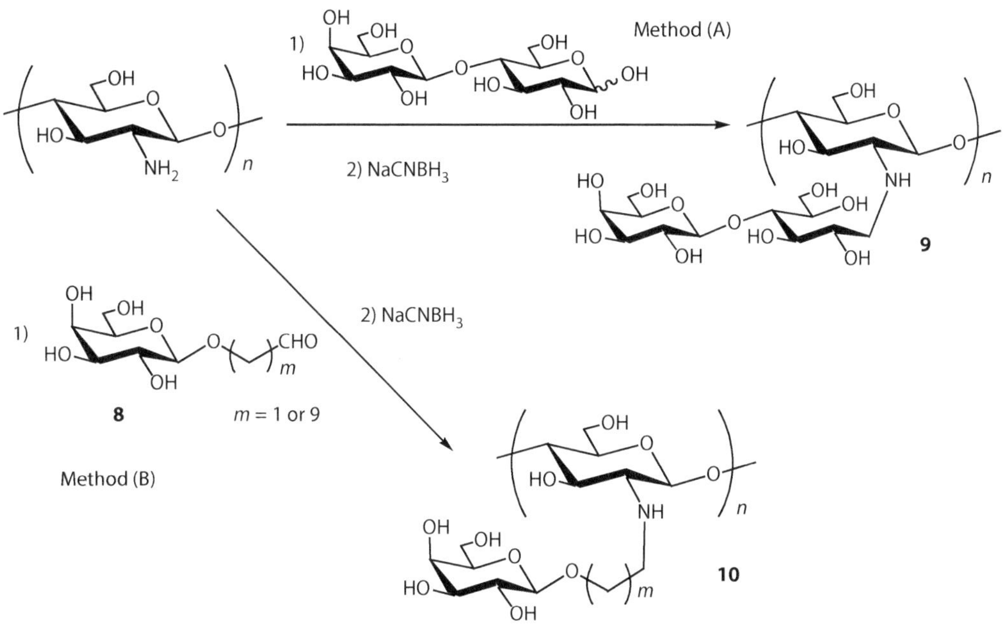

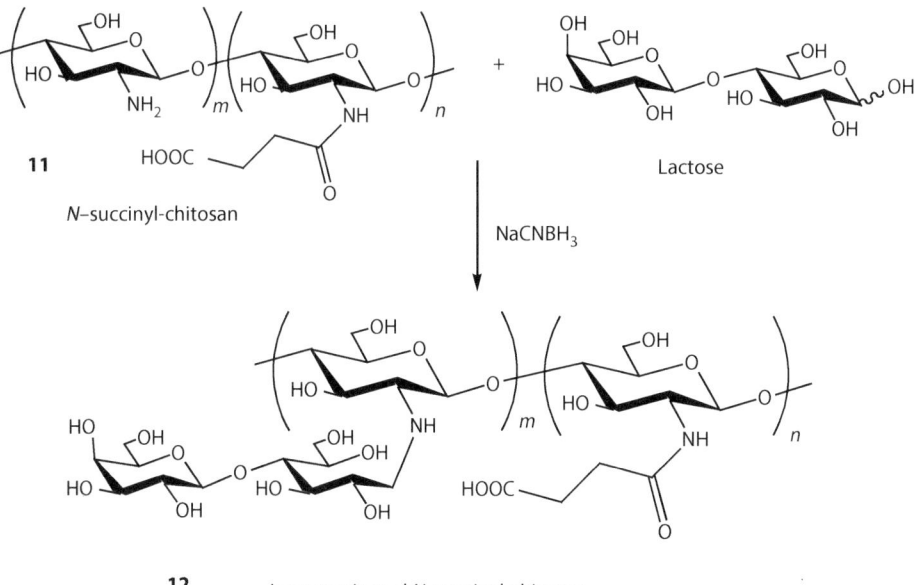

FIGURE 4.4 Strategy for the substitution of sugars to chitosan by reductive *N*-alkylation.

investigated in the rheological study. Since the specific recognition of cell, virus, and bacteria by sugars has been discovered, this modification has generally been used to introduce cell-specific sugars into chitosan. Morimoto reported the synthesis of sugar-bound chitosans, such as D- and L-fucose and their specific interaction with lectin or cells (Morimoto et al. 2001, 2002, Li et al. 1999, 2000). Kato also prepared lactosaminated *N*-succinyl-chitosan (**12**: Figure 4.5) and its fluorescein thiocarbanyl derivative as a liver-specific drug carrier in mice through asialoglycoprotein

FIGURE 4.5 Synthesis of lactosaminated *N*-succinyl-chitosan.

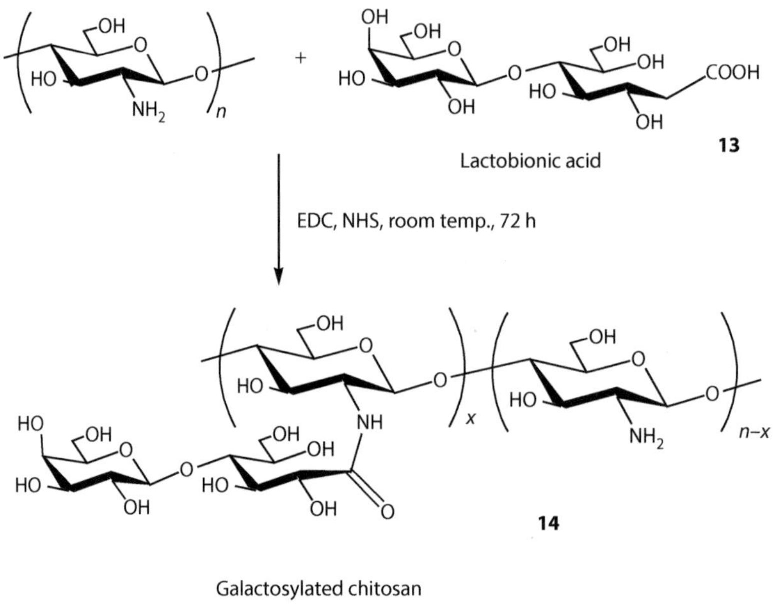

Lactobionic acid **13**

EDC, NHS, room temp., 72 h

14

Galactosylated chitosan

FIGURE 4.6 Synthesis of galactosylated chitosan.

receptor (Kato et al. 2001a). Moreover, derivative **12** was a good drug carrier such as mitomycin C in liver metastasis (Kato et al. 2001b). Galactosylated chitosan (**14**: Figure 4.6) prepared from lactobionic acid and chitosan with 1-ethyl-3-(3-dimethylaminopropyl)carbodiimide (EDC) and *N*-hydroxysuccinimide (NHS) was a good candidate as a synthetic extracellular matrix for hepato-cytes attachment (Park et al. 2003a). Sponge-type complex of cationic **14** and anionic alginate also showed spheroid formation and viability of hepatocytes (Chung et al. 2002). Furthermore, graft copolymers of **14** with poly(ethylene glycol) or poly(vinyl pyrrolidone) were useful for hepatocyte-targeting DNA carrier (Park et al. 2001, 2003b).

Sialic acid is the most ubiquitous sugar present on the mammalian cell surface glycolipids and glycoproteins and is the key epitope recognized as being responsible for various pathogenic infections. Moerover, sialic acid-containing polymers have been shown to be potent inhibitors of hemagglutination of human erythrocytes by *influenza* viruses (Gamian et al. 1991, Roy et al. 1987; Roy and Laferriere, 1988, 1991, 1992; Roy, 1996). We prepared sialic acid-bound chitosan (**16**: Figure 4.7) as a new family of sialic acid-containing polymers using *p*-formylphenyl-α-sialoside (**15**) (Roy et al. 1991) by reductive *N*-alkylation (Sashiwa et al. 2000a). Because deriva-tive **16** was insoluble in water, continuous *N*-succinylation was carried out and water-soluble derivative (**17**) was obtained. The specific binding with lectin of *wheat germ agglutinin* was shown in water-soluble derivative **17**.

Human antibodies against α-galactosyl epitope are responsible for the acute rejection of xeno-transplantated organs from lower animals. Artificial glycopolymers having α-galactosyl epitope are of interest from the viewpoint of medical transplantation of pig liver as they can block immune rejection. This interesting epitope also occurs as a family of bioactive sugar-bound chitosans. Water-soluble α-galactosyl chitosan (**18**: Figure 4.8) prepared by the same strategy as sialic acid showed specific binding against α-galactosyl-specific lectin (*Griffonia simplicifolia*) (Sashiwa et al. 2000f). A different type of spacer has been prepared on sialic acid or α-galactosyl epitope-bound chitosans (Sashiwa et al. 2001b). These epitope-bound chitosans would be useful as a potent inhibitor by *influ-enza* viruses or as a blocking agent for acute rejection.

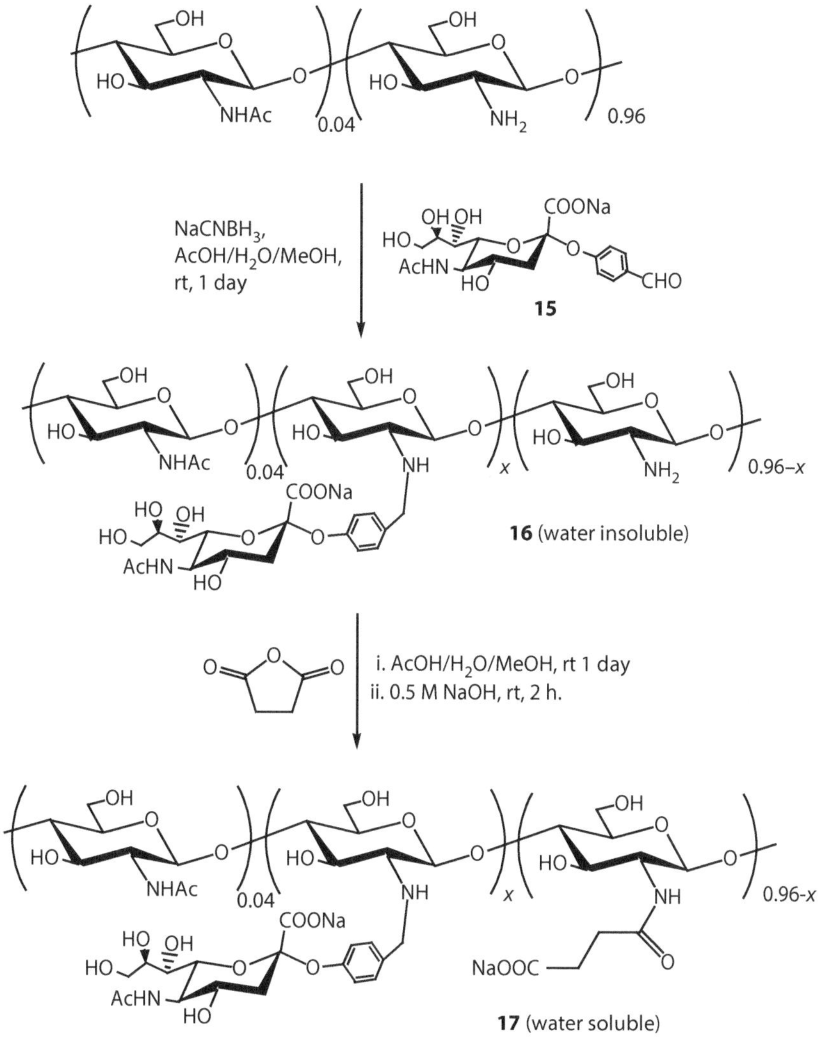

FIGURE 4.7 Synthesis of sialic acid-chitosan and its *N*-succinylation.

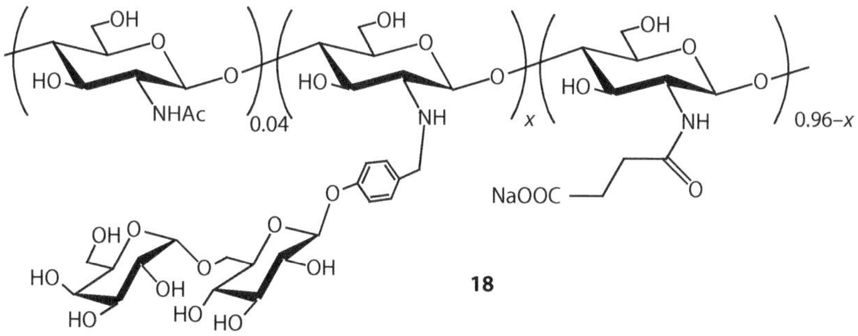

FIGURE 4.8 Structure of water-soluble α-galactosyl chitosan.

4.4 CHITOSAN–DENDRIMER HYBRID

Dendrimers are attractive molecules owing to their multifunctional properties (Tomalia et al. 1985, Astruc and Chardac 2001, Tomalia and Frechet 2002) and have useful applications as viral and pathogenic cell adhesion inhibitors (Reuter et al. 1999, Kitov et al. 2000). Tremendous scientific efforts have gone into the design and synthesis of dendrimers (Zeng and Zimmemen 1997, Fischer and Vogtle 1999, Gorman and Smith 2001). Dendronized polymers, however, are also attractive because of their rod-like conformation and nanostructure (Karakaya et al. 1997, Schluter and Rabe 2000, Vetter et al. 2001, Zhang et al. 2003). Although several investigations have been published toward the synthesis of dendronized polymers (Malenfant and Frechet 2000, Zubarev and Stupp 2002), there is scarce report on dendronized polysaccharide, especially related to the chitin and chitosan backbone. We established the synthesis of a variety of chitosan–dendrimer hybrids mainly by two procedures (Figure 4.9) (Sashiwa et al. 2000d,f, 2001c,d, 2002f,g,h, 2003c,d).

In method A, corresponding denrimers bearing aldehyde and spacer have to be synthesized, and these are then reacted with chitosan by reductive N-alkylation. This procedure is advantageous as it has no cross-linking during the reaction. However, the generation of reactable dendrimer is limited owing to its steric hindrance. However, in method B, the surface of dendrimer-bound type, is beneficial to use in commercial amino-dendrimers, such as poly(amidoamine) (PAMAM) or poly(ethylene imine) dendrimers, and is also possible to bind even in high generations. One weak point in method B is that it has two or more binding points and sometimes may cause cross-linking. The typical example of hybrid by method A is shown in Figure 4.10 (Sashiwa et al. 2000d,f, 2002f). We are imaging this hybrid as "tree-type molecule." Indeed, chitosan is the trunk, the spacer part is the main branch, dendrimer is a subbranch, and flower (or leaf) is functional sugar. In this case, tetraethylene glycol was modified by following the five or seven steps to synthesize the scaffold of dendrimer. PAMAM dendrimers of generation (G) from 1 to 3 bearing tetraethylene glycol spacer were prepared, attached to sialic acid by reductive N-alkylation, and finally attached to the chitosan. The DS of dendrimer per one sugar unit was decreased with increasing the generation as 0.08 ($G = 1$), 0.04 ($G = 2$), and 0.02 ($G = 3$) owing to the steric hindrance of dendrimer. Figure 4.11 shows the different type of chitosan–dendrimer hybrid (Sashiwa et al. 2001c). Sialic acid dendron bearing a focal aldehyde end group was synthesized by a reiterative amide bond strategy. Trivalent ($G = 1$)

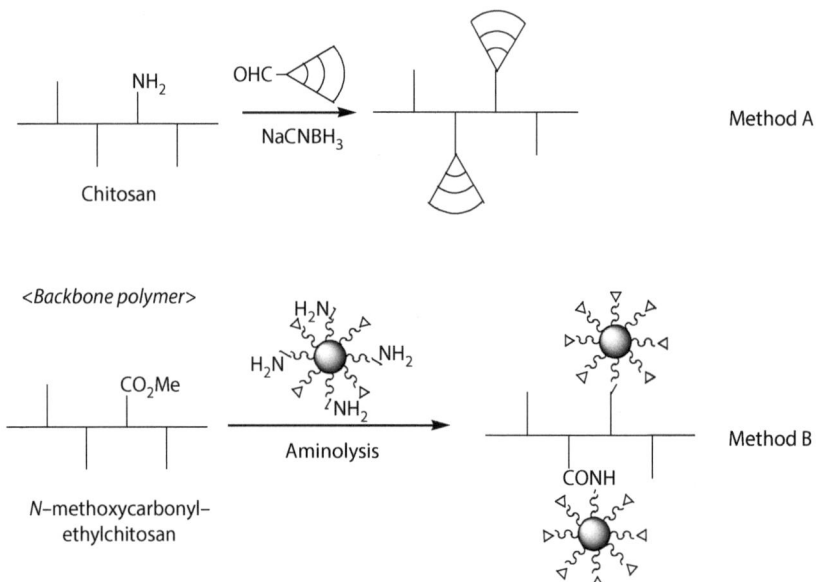

FIGURE 4.9   Synthetic strategy on chitosan–dendrimer hybrid.

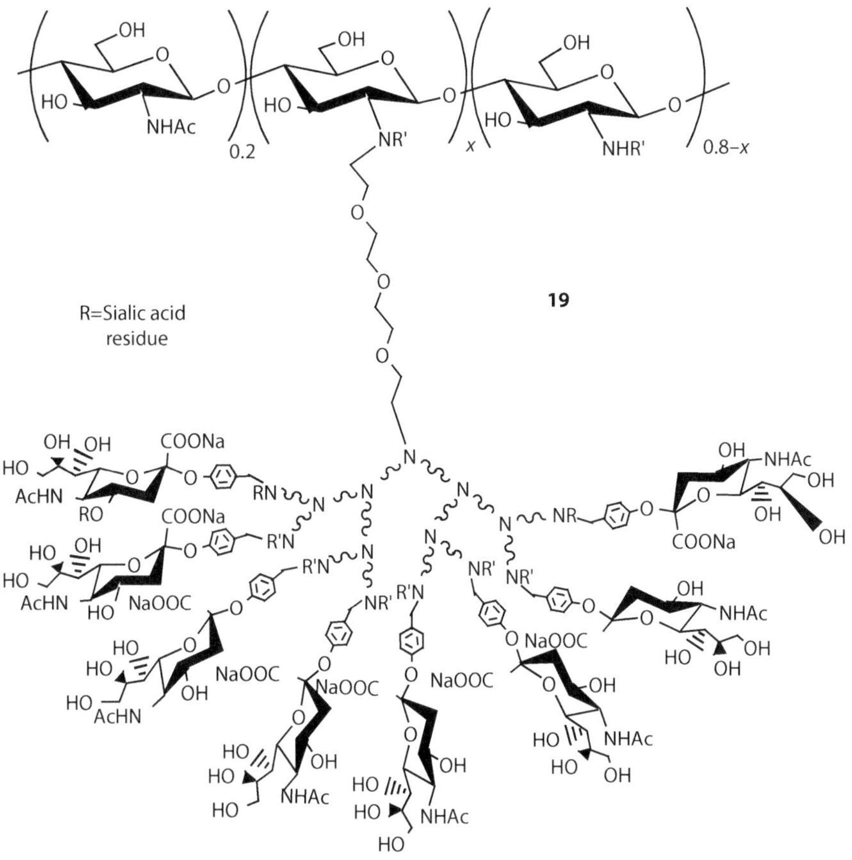

FIGURE 4.10 Chemical structure of chitosan–sialodendrimer hybrid.

and nonavalent ($G = 2$) dendrons having gallic acid as the branching unit and triethylene glycol as the spacer arm were prepared and initially attached to a sialic acid p-phenylisothiocyanate derivative. The focal aldehyde sialodendrons were then convergently attached to the chitosan. The DS of sialodendrimer were 0.13 ($G = 1$) and 0.06 ($G = 2$). Further biological evaluation of these promising hybrids is being investigated toward the inhibition of viral pathogens, including the flu virus.

Chitosan–dendrimer hybrid prepared by following method B is shown in Figure 4.12 (Sashiwa et al. 2001d). As the construction of hybrid was difficult from original chitosan, a derivative, N-methoxycarbonylethylchitosan (**21**), was used as the chitosan backbone. PAMAM dendrimers ($G = 1$–5) having a 1,4-diaminobutane core were attached to **21** by amidation under conditions that prevent cross-linking. The hybrids **22** could be prepared even in high generations ($G = 4$ or 5), although the DS of dendrimer was decreased with increasing the generation of dendrimer from 0.53 ($G = 1$) to 0.17 ($G = 4$) or 0.11 ($G = 5$). Because this hybrid was soluble in acidic water, undesired cross-linking would not occur. However, two or more intermolecular binding points were observed. Anyway, sialic acid was successfully attached to the primary amine of the dendrimer part with DS ranging from 0.7 to 1.4 per glucosamine unit, which means it is a highly convergent synthesis of sialic acid in the chitosan backbone. Given the fact that the flu virus hemagglutinins exist as several clusters of trimers (200–300/virions), it is likely that the novel dendronized chitosan–sialic acid hybrids prepared by following method B would present added beneficial architectures not present in previously reported sialodendrimers (Matrosvich et al. 1990, Msmmen et al. 1998, Kamitakahara et al. 1998). Preliminary biological evaluation of analogous hyperbranched sialodendrimers has already shown increased inhibitory properties (Tomalia and Frechet 2002).

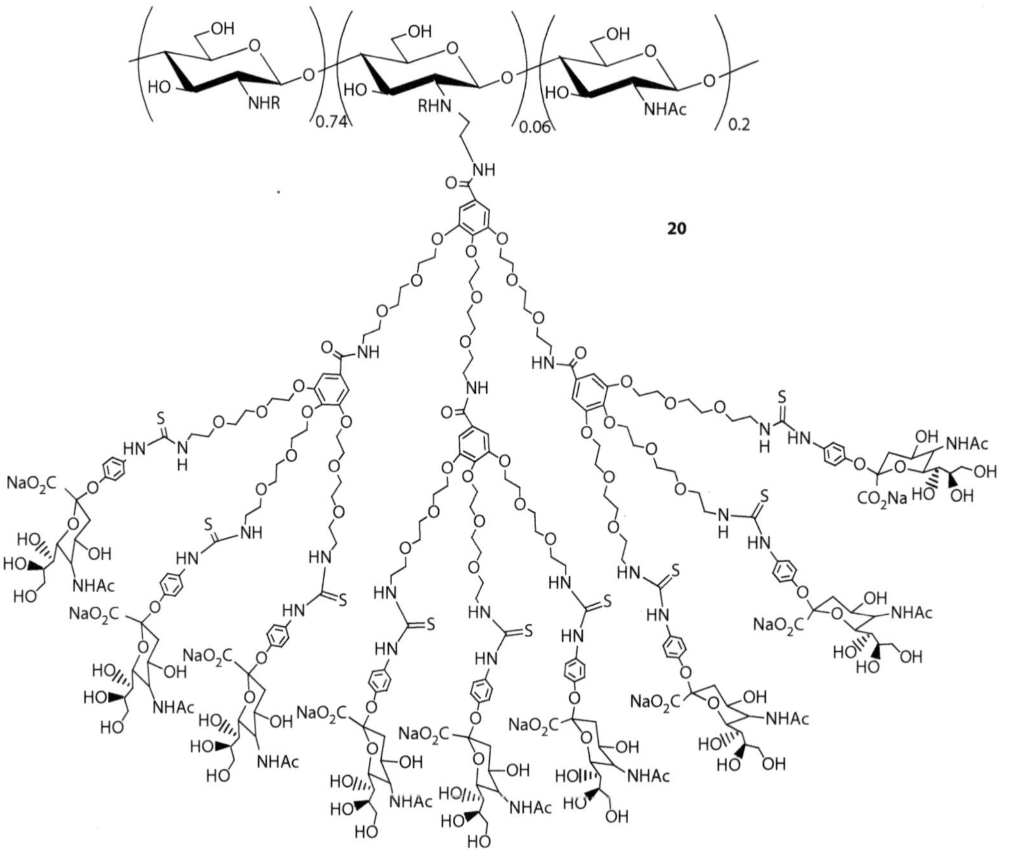

FIGURE 4.11 Hybridization of chitosan with sialodendrimer, composed of gallic acid.

4.5 BIODEGRADATION OF MODIFIED CHITOSANS

In the field of electronic equipments, such as computer and medical instruments, high-speed and high-density information processing is becoming increasingly common, and there is a danger that even low-intensity electromagnetic radiation may cause malfunction. For example, the use of mobile telephones and portable computers is restricted in aircrafts and hospitals. Hence, there is a need for high reliable electromagnetic radiation-shielding materials. Also, the biodegradability of binder between plastics and shielding materials is required for the recycling of plastics. Because chitosan is a biodegradable polymer, it would be a useful candidate for this binder. Chitosan itself, however, is hydrophilic, whereas plastics are hydrophobic. Chemical modification of chitosan to dissolve in organic solvent is necessary for spraying on the surface of the plastics. The hydrophobic ester linkage is advantageous as both the hydrophobic groups contribute organo-solubility and the ester linkage is hydrolyzed by enzymes such as lipase and so on (Figure 4.13). We reported the synthesis of organo-soluble and biodegradable acyl-chitosans toward the electromagnetic shielding materials for electronic equipment (Kamitakahara et al. 1998, Sashiwa et al. 2002d). Acylchitosans bearing long acyl chains (**27**: $n > 4$), hydrophobic pivaloyl (**28**), or benzoyl (**29**) groups showed organo-soluble property. Interestingly, O-acetylchitosan (**26**) first showed water solubility like acetylcellulose (Sashiwa et al. 2002e). Biodegradation of acylchitosans was evaluated by standard activated sludge. The excellent biodegradation was observed in water-soluble **26**. Organo-soluble derivatives (**27** and **29**) also showed good biodegradation. Furthermore, these organo-soluble acylchitosans exhibited good binding properties between plastics and electromagnetic shielding materials, thus suggesting that these acylchitosans are useful as a biodegradable binder.

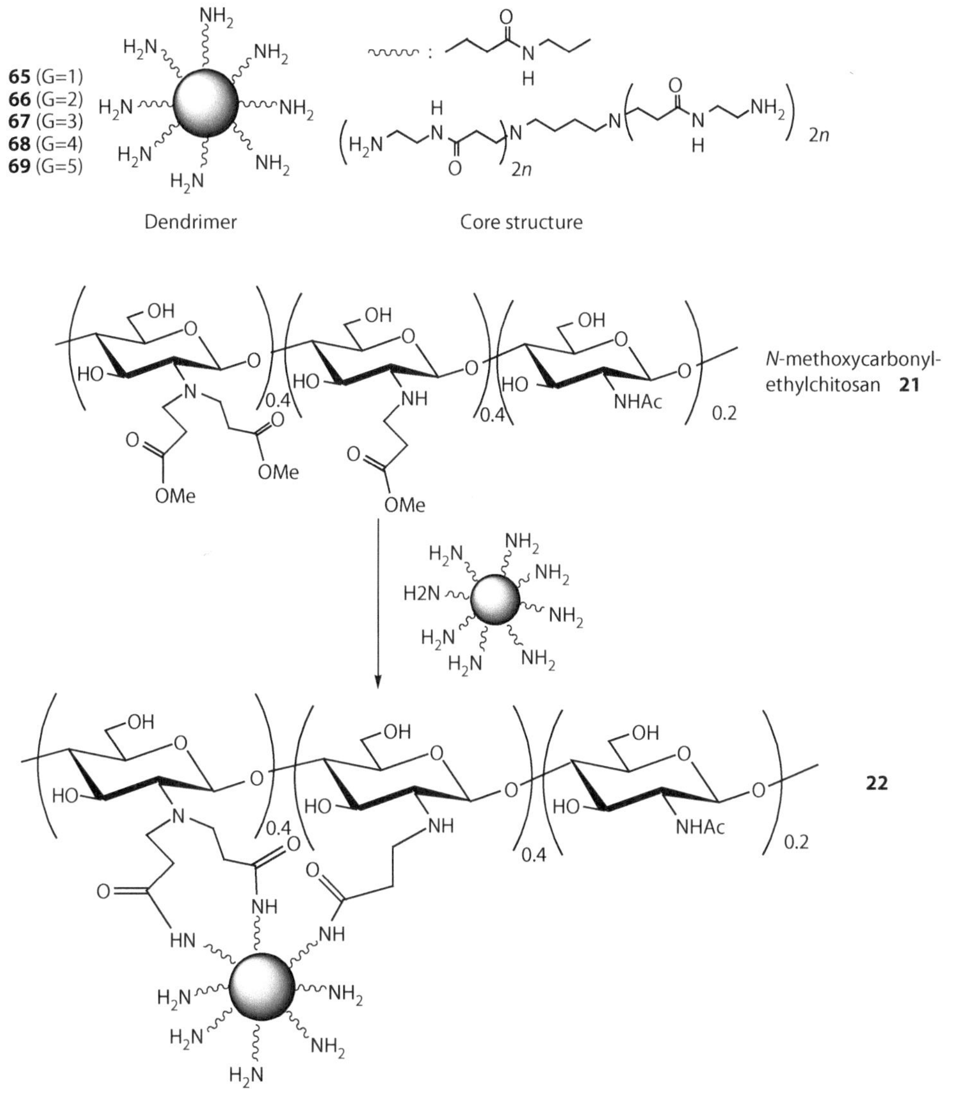

FIGURE 4.12 Reaction of *N*-methoxycarbonylethylchitosan with PAMAM dendrimer.

Alternatively, biodegradation has been investigated for some water-soluble chitosan derivatives prepared by Michael reaction. This reaction has been developed as a new method for the chemical modification of chitosan (Sashiwa et al. 2002g) or partially deacetylated chitin (Aoi et al. 2000). More recently, we reported the Michael reaction of chitosan with acrylic acid using water as a solvent (Sashiwa et al. 2003e). In this case, acrylic acid played as both proton donor for making chitosan dissolve in water and as the reagent for Michael reaction, so that water-soluble *N*-carboxyethylchitosan was successfully obtained. If water-soluble acryl reagents could be applied for this reaction, novel types of functional groups can be introduced by a simple procedure. Various chitosan derivatives were prepared by this reaction in water and AcOH with various acryl reagents (Figure 4.14) (Sashiwa et al. 2003e).

Table 4.3 shows the biodegradation results of these chitosan derivatives by standard activated sludge (Sashiwa et al. 2003f). In any case, the biodegradability was enhanced by chemical modification compared with original chitosan. The excellent biodegradable characteristics were shown in **30** with various DS, although it was gradually decreased with increasing DS. Derivatives **33**, **34**, and **35** modified with poly(ethyleneglycol) (PEG), quaternary ammonium, and amide groups also

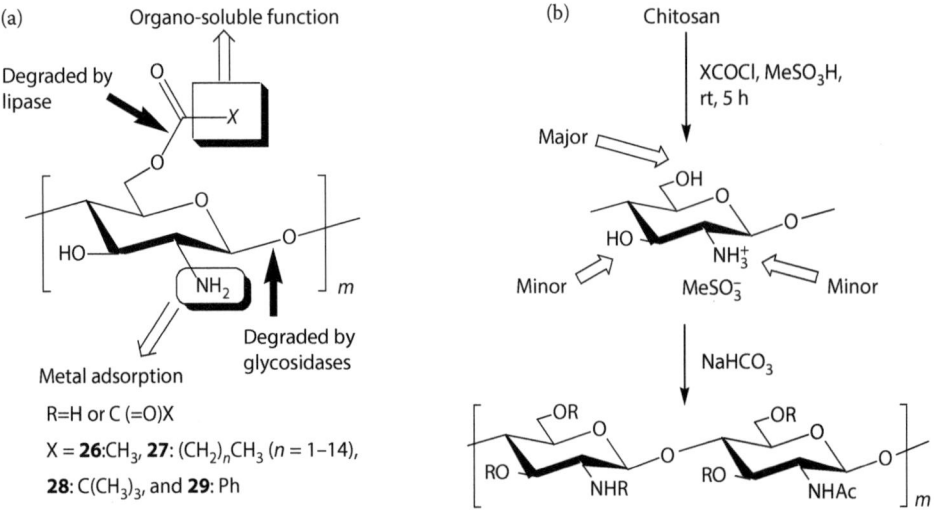

FIGURE 4.13 (a) Concept and (b) synthesis of organo-soluble chitosan derivatives.

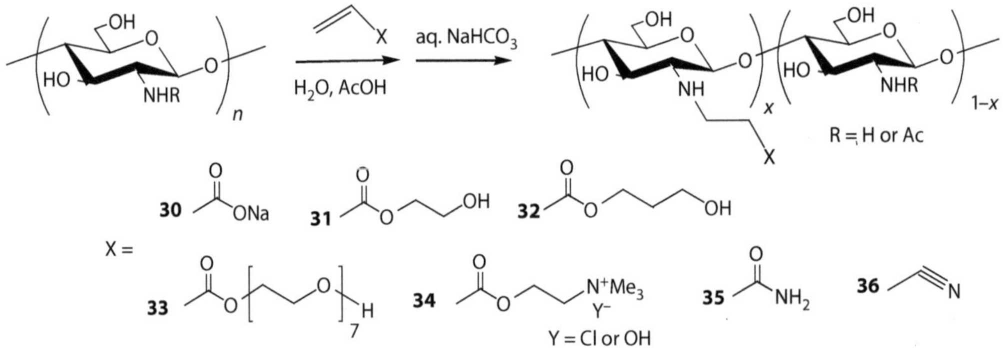

FIGURE 4.14 Michael reaction of chitosan with acrylic acid and its esters in water.

TABLE 4.3

Biodegradation of Chitosan and Its Derivatives[a]

Sample	DS	Solubility in H_2O	Biodegradation (%)
Chitosan	0	No	1.6
30	0.18	Yes	67.3
30	0.27	Yes	62.5
30	0.46	Yes	51.3
31	0.44	Yes	8.6
32	0.26	Yes	5.0
33	0.29	No[b]	24.8
34	0.38	Yes	33.6
35	0.24	No	27.8
36	0.49	No	7.7

[a] Time, 21 days.

[b] Water-insoluble after lyophilization.

showed good biodegradability. Moderate biodegradation was shown in **31** and **36** bearing hydroxy-ethyl and nitrile groups. These results suggest that biodegradation was mostly associated with the chemical structure of chitosan derivatives. Thus, carboxyl, quaternary ammonium, amide, and PEG groups were advantageous for biodegradation, but hydroxyethyl or nitrile group was not. On the contrary, biodegradation was independent of the water solubility of chitosan derivatives. The mechanism for the biodegradation of chitosan derivatives is presumed as follows. First, glycoside or ester linkage in chitosan derivative is hydrolyzed by certain enzymes, such as glycosidase or lipase, and so on. It is then metabolized by certain bacteria and microorganisms in standard activated sludge. The biodegradation study by activated sludge is important from the viewpoint of environ-mental or green chemistry. Chitosan is a so-called biodegradable polymer because it is degradable mainly by chitosanase. However, the biodegradation of chitosan is quite slow by standard activated sludge. Modified chitosan is of more benefit for the biodegradation owing to the destroyed crystal-line structure of chitosan. In our other reports, the ester form of *N*-carboxyethyl chitosan played a good intermediate for further modification with amines (Sashiwa et al. 2003f). Furthermore, the water-soluble derivative, *N*-carboxyethylchitosan, was also beneficial for further modification with the hydrophobic group, laurylaldehyde (Sashiwa et al. 2003g).

4.6 ENZYMATIC PRODUCTION OF *N*-ACETYL-D-GLUCOSAMINE FROM CHITIN

1. In addition to the functions of polymeric chitin and chitosan, monomeric chitin (*N*-acetyl-D-glucosamine: GlcNAc) and chitosan (D-glucosamine: GlcN), which also exist as a component of proteoglycan in cartilage, skin, and connective tissue have interesting bio-logical properties. For example, GlcN has attracted much attention owing to its thera-peutic activity in osteoarthritis and has also been evaluated as a food supplement (Kim and Conrad 1974, Setnikar et al. 1986, 1993). Although the sulfate salt of GlcN is already commercialized for this disease, it is not suitable for oral administration owing to its bitter taste. However, GlcNAc has been a material of focus for the improvement of osteo-arthritis instead of GlcN, because GlcNAc is also a component of proteoglycan and a part of GlcN is transformed to GlcNAc by metabolism. Moreover, GlcNAc will possibly be used as an orally administrated drug, because of its sweet taste. Moreover, GlcNAc is produced by the acid (concentrated HCl) hydrolysis of chitin. This procedure, however, has some problems, such as high cost, low yield (below 65%), and acidic wastes owing to concentrated HCl, and so on. *N*-Acetylation of GlcN also possibly produces GlcNAc. This product, however, is not approved as a natural-type material owing to its chemical reaction process. Although the effective production of GlcNAc is required, there is no report on such works. If enzymatic process could be applied for the hydrolysis of chitin, it would be quite mild conditions and environmentally beneficial process compared with acid hydrolysis.

2. Recently, we have reported on the production of GlcNAc by the use of crude enzymes, including endo- and exo-chitinases (Hiraga et al. 1997, Sashiwa et al. 2001a, 2002a, 2003a). In general, β-chitin derived from squid pen shows more swelling property in water than α-chitin from crab or shrimp shell. Moreover, crude enzymes have some advantage for producing GlcNAc owing to their low cost and including both the *endo*- and *exo*-type of chitinases (Figure 4.15). Therefore, we selected β-chitin and crude enzymes to produce GlcNAc. Table 4.4 shows the production of GlcNAc from β-chitin by various enzymes (Sashiwa et al. 2001a). Among these crude enzymes, cellulase *Tricoderma viride* (T) and *Acremonium* (A) were effective for the production of GlcNAc. Hemicellulase, papain, lipase, and pectinase were not so effective in producing GlcNAc under these conditions. Although the crude preparations of cellulase T and A essentially degrade cellulose, they

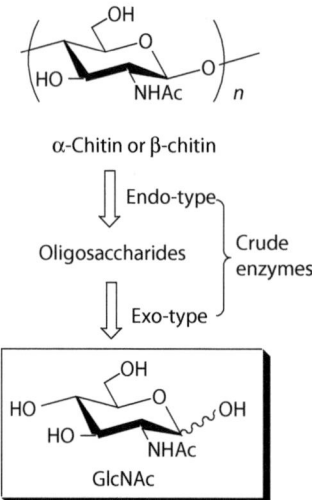

FIGURE 4.15   Production of GlcNAc by crude enzymes.

also degrade chitin owing to the presence of chitinase and produce monomers or oligomers. From the ^1H and ^{13}C NMR (in D_2O) analysis, purified GlcNAc produced by enzymatic hydrolysis showed the same ^1H and ^{13}C signals as commercial GlcNAc, thus suggesting that the chemical structure is GlcNAc.

Table 4.5 shows the enzymatic activity of both endo- and exo-enzymes. Cellulase A showed high activity of endo-enzyme, lipase showed high value of exo-enzyme, and cellulase T showed moderate activity of both enzymes (Sashiwa et al. 2001a). Especially, the hydrolysis by cellulase A was partly stopped at $(GlcNAc)_2$ step, which would be caused by the inefficient exo-enzymatic activity. To enhance the production of GlcNAc, the mixing of two crude enzymes was effective, thus suggesting that the balance of endo- and exo-activity is important to produce GlcNAc. Moreover, the hydrolysis of α-chitin was also enhanced with the mixed enzyme. Toward the maximum production of GlcNAc, we tested the hydrolysis of α- and β-chitin for a long-time reaction (Table 3.6) (Sashiwa et al. 2001a). Almost quantitative production of GlcNAc was observed from β-chitin. The heterogeneous reaction mixture turned to clear solution. Also the yield of GlcNAc from α-chitin was increased to a maximum of 70% by mixing cellulase T and A.

TABLE 4.4

Production of GlcNAc from β-Chitin by Various Enzymes[a]

Run Enzyme	Origin	Yield (%)
Cellulase T	*Tricoderma viride*	26
Cellulase A	*Acremonium*	29
Hemicellulase	*Aspergillus niger*	5
Papain	*Carica papaya* L.	2
Lipase	*Aspergillus niger*	7
Pectinase	*Aspergillus niger*	5

[a] [Chitin] = 10 mg/mL; [enzyme] = 10 mg/mL; pH = 4.8; 37°C; 4 days.

TABLE 4.5
Enzymatic Activity of Crude Enzymes

Crude Enzyme	Endo/mU[a]	Exo/mU
Cellulase T	3.7 (37)	1.4 (14)
Cellulase A	12.3 (18)	0.18 (0.25)
Lipase	2.7 (11)	9.8 (39)
Crude enzyme A	–(27)	–(69)

[a] These values show activity per 1 mg of crude enzymes; parentheses mean activity per 1 mg of protein.

TABLE 4.6
Toward the Maximum Yield of GlcNAc from α- and β-Chitin[a]

Chitin	Cellulase T (mg/mL)	Cellulase A (mg/mL)	Lipase (mg/mL)	Time (day)	Yield (%)
β	20	20	—	7	93
β	—	20	20	7	90
α[b]	20	20	—	18	70

[a] [S] = 10 mg/mL; pH = 4.0 (2 mL); 37
[b] Powdered α-chitin

3. As described earlier, crude enzymes such as cellulase T, cellulase A, and lipase hydrolyzed β-chitin and produced GlcNAc in good yield. These crude enzymes, however, did not efficiently hydrolyze α-chitin, which is more available, has low cost, and is widely distributed in nature compared with β-chitin. Therefore, the effective production of GlcNAc from α-chitin was tested by use of crude enzymes derived from *Aeromonas hydrophila* (crude enzyme A) (Sashiwa et al. 2001a). Because the enzymatic activity of these crude enzymes tends to decrease at higher temperature, the hydrolysis was carried out at 17°C. As for the results, GlcNAc was produced in 66–77% yields from α-chitins after 10 days. The selective production of GlcNAc was observed during continuous hydrolysis and no production of GlcNAc oligomers was found. The effective hydrolysis of α-chitin would be because of the high activity of endo- and exo-enzymes compared with cellulase T, A, lipase, and their mixed enzymes (Sashiwa et al. 2002a, Sukwattanasinitt et al. 2002). On the contrary, excellent production of GlcNAc was also found by using the crude enzyme from *Burkholderia cepacia* TU09 and *Bacillus licheniformis* SK-1 (Pichyangkura et al. 2002). From these studies it was identified that selective production of GlcNAc from both α- and β-chitin could be achieved by these crude enzymes. Needless to say, enzymatic hydrolysis uses quite mild conditions and high selectivity to produce GlcNAc compared with that of the chemical process. This method would be useful for the industrial production of GlcNAc, which helps improve osteoarthritis.

4.7 CONCLUSION

Chitin and chitosan are structurally similar to heparin, chondroitin sulfate, and hyaluronic acid, which are biologically important mucopolysaccharides in the human body. These mucopolysaccharides are anionic polymers bearing carboxyl and sulfyl groups. However, cationic polysaccharide in nature is almost only chitosan. Moreover, this cationic polysaccharide (chitosan) is nontoxic and biodegradable in the human body. This unique property is worthy of remark in biomedical applications. Because chitosan is not dissolved in neutral and basic water, the biomedical use of chitosan is limited. Chemical modification of chitosan is possible by dissolving it in neutral or basic water. Moerover, chemical modification is possible to attach various functional groups and to control the hydrophobic, cationic, and anionic properties. Further studies and development of chitin, chitosan, and their derivatives toward biomedical applications is expected later in the 21st century.

REFERENCES

Aoi, K., T. Seki, M. Okada, H. Sato, S. Mizutani, H. Ohtani, S. Tsuge, and Y. Shiogai. 2000. Synthesis of a novel N-selective ester functionalized chitin derivative and water-soluble carboxyethylchitin. *Macromolecule Chemistry and Physics* 201: 1701–1708.

Astruc, D. and F. Chardac. 2001. Dendritic catalysis and dendrimers in catalysis. *Chemical Reviews* 101: 2991–3023.

Chung, T. W., J. Yang, T. Akaike, Y. Cho, J. W. Nah, S. Kim, and C. S. Cho. 2002. Preparation of alginated/ galactosylated chitosan scaffold for hepatocyte attachment. *Biomaterials* 23: 2827–2834.

Fischer, M. and F. Vogtle. 1999. Dendrimers: From design to application- a progress report. *Angewandte Chemie International Edition* 38: 884–905.

Gamian, A., M. Chomik, C. A. Laferriere, and R. Roy. 1991. Inhibition of influenza A virus hemagglutinin and induction of interferon by synthetic sialylated glycoconjugates. *Canadian Journal of Microbiology* 37: 233–237.

Gingras, M., I. Paradis, and F. Berthod. 2003. Nerve regeneration in a collagen-chitosan tissue-engineered skin transplanted on nude mice. *Biomaterials* 24: 1653–1661.

Gorman, C. B. and J. C. Smith. 2001. Structure-property relationships in dendritic encapsulation. *Account of Chemical Research* 34: 60–71.

Gupta, K. C. and M. N. V. R. Kumar. 2000. An overview on chitin and chitosan applications with an emphasis on controlled drug release formulations. *JMS—Review of Macromolecule Chemistry and Physics* C40: 273–308.

Hirano, S. and T. Moriyasu. 1981. *N*-(Carboxyacyl)chitosans *Carbohydrate Research* 92: 323–327.

Hall, L. D. and M. Yalpani. 1980. Formation of branched-chain, soluble polysaccharide from chitosan. *Journal of Chemical Society, Chemical Communications* 1153–1154.

Hiraga, K., L. Shou, M. Kitazawa, S. Takahashi, M. Shimada, R. Sato, and K. Oda. 1997. Isolation and characterization of chitinase from a flake-chitin degrading marine bacterium, Aeromonas hydrophila H-2330. *Bioscience Biotechnology and Biochemistry* 61: 174–176.

Karakaya, B., W. Claussen, K. Gessler, W. Saenger, and A. D. Schluter. 1997. Toward dendrimers with cylindrical shape in solution. *Journal of American Chemical Society* 119: 3296–3301.

Kamitakahara, H., T. Suzuki, N. Nishigori, Y. Suzuki, O. Kanie, and C.H. Whong. 1998. A lisoganglioside/ poly-L-glutaminic acid conjugate as a picomolar inhibitor of influenza hemagglutinin. *Angewandte Chemie International Edition*. 37: 1524–1527.

Kato, Y., H. Onishi, and Y. Machida. 2001a. Biological characterization of lactosaminated *N*-succinyl-chitosan as a liver-specific drug carrier in mice. *Journal of Controlled Release* 70: 295–307.

Kato, Y., H. Onishi, and Y. Machida. 2001b. Lactosaminated and intact *N*-succinyl-chitosans as a drug carrier in liver metastasis. *International Journal of Pharmacy* 226: 93–106.

Khnor, E. and L. Y. Lim. 2003. Implanted applications of chitin and chitosan. *Biomaterials* 24: 2339–2349.

Kim, J. and E. H. Conrad. 1974. Effect of D-glucosamine concentration on the kinetics of mucopolysaccharide biosynthesis in cultured chick embryo vertebral cartilage. *Journal of Biological Chemistry* 249: 3091–3097.

Kitov, P. I., J. M. Sadowska, G. Mulvey, G. D. Armstrong, H. Ling, N. S. Pannu, R. J. Read, and D. R. Bundle. 2000. Shiga-like toxins are neutralized by tailored multivalent carbohydrate ligands. *Nature* 403: 669–673.

Kumar, M. N. V. R. 2000. A review of chitin and chitosan applications. *Reactive and Functional Polymer* 46: 1–27.

Kurita, K. 2001. Controlled functionalization of the polysaccharide chitin. *Progress in Polymer Science* 26: 1921–1971.

Kweon, D. K., S. B. Song, and Y. Y. Park. 2003. Preparation of water-soluble chitosan/heparin complex and its application as wound healing accelerator. *Biomaterials* 24: 1595–1601.

Li, X., M. Morimoto, H. Sashiwa, H. Saimoto, Y. Okamoto, S. Minami, and Y. Shigemasa. 1999. Synthesis of chitosan-sugar hybrid and evaluation of its bioactivity. *Polymer Advanced Technologies* 10: 455–458.

Li, X., Y. Tsushima, M. Morimoto, H. Saimoto, Y. Okamoto, S. Minami, and Y. Shigemasa. 2000. Biological activity of chitosan-sugar hybrids: Specific interaction with lectin. *Polymer Advanced Technologies* 11: 176–179.

Malenfant, P. R. L. and J. M. J. Frechet. 2000. Dendrimers as solubilizing groups for conducting polymers. Preparation and characterization of polythiophene functionalized exclusively with aliphatic ether convergent dendrons. *Macromolecules* 33: 3634–3640.

Mao, J. S., H. F. Liu, Y. J. Yin, and K. D. Yao. 2003. The properties of chitosan-gelatin membranes and scaffolds modified with hyaluronic acid by different method. *Biomaterials* 24: 1621–1629.

Matrosvich, M. N., L. V. Mochalova, V. P. Marinina, N. E. Byramova, and N. V. Bovin. 1990. Synthetic polymeric sialoside inhibitors of influenza virus receptor-binding activity. *FEBS Letters* 272: 209–215.

Mori, T., M. Okumura, M. Matsuura, K. Ueno, S. Tokura, Y. Okamoto, S. Minami, and T. Fujinaga. 1987. Effects of chitin and its derivatives on the proliferation and cytokine production of fibroblasts in vitro. *Biomaterials* 18: 947–951.

Morimoto, M., H. Saimoto, H. Usui, Y. Okamoto, S. Minami, and Y. Shigemasa. 2001. Biological activities of carbohydrate-branched chitosan derivatives. *Biomacromolecules* 2: 1133–1136.

Morimoto, M., H. Saimoto, and Y. Shigemasa. 2002. Control of functions of chitin and chitosan by chemical modification. *Trends in Glycoscience and Glycotechnologies* 14: 205–222.

Msmmen, M., S. Choi, and G.M. Whiteside. 1998. Polyvalent interactions in biological systems: Implications for design and use of multivalent ligands and inhibitors. *Angewandte Chemie International Edition.* 37: 2754–2757.

Nishimura, K., S. Nishimura, N. Nishi, I. Saiki, S. Tokura, and I. Azuma. 1984. Immunological activity of chitin and its derivatives. *Vaccine* 2: 93–99.

Okamoto, Y., S. Minami, A. Matsuhashi, H. Sashiwa, H. Saimoto, Y. Shigemasa, T. Tanigawa, Y. Tanaka, and S. Tokura. 1993. Polymeric N-acetyl-D-glucosamine (Chitin) includes histrionic activation in dogs. *Journal of Veterinary Medical Science* 55: 739–742.

Park, I. K., T. H. Kim, Y. H. Park, B. A. Shin, E. S. Choi, E. H. Chowdhury, T. Akaike, and C. S. Cho. 2001. Galactosylated chitosan-graft poly(ethylene glycol) as hepatocyte targeting DNA carrier. *Journal of Controlled Release* 76: 349–362.

Park, I. K., J. E. Ihm, Y. H. Park, Y. J. Choi, S. I. Kim, W. J. Kim, T. Akaike, and C. S. Cho. 2003a. Galactosylated chitosan (GC)-graft-poly(vinyl pyrrolidone) (PVP) as hepatocyte targeting DNA carrier. Preparation and physicochemical characterization of GC-graft-PVP/DNA complex (I). *Journal of Controlled Release* 86: 349–359.

Park, I. K., J. Yang, H. J. Jeong, H. S. Bom, I. Harada, T. Akaike, S. Kim, and C.S. Cho. 2003b. Galactosylated chitosan as a synthetic extracellular matrix for hepatocyte attachment. *Biomaterials* 24: 2331–2337.

Pichyangkura, R. S. Kudan, K. Kuttiyangwong, M. Sukwattanasinitt, and S. Aiba. 2002. Quantitative production of 2-acetamido-2-deoxy-D-glucose from crystalline chitin by bacterial chitinase. *Carbohydrate Research* 337: 557–559.

Reuter, J. D., A. Myc, M. M. Hayes, Z. Gan, R. Roy, D. Qin, R. Yin et al. 1999. Inhibition of viral adhesion and infection by sialic acid conjugated dendritic polymers. *Bioconjugate Chemistry* 10: 271–278.

Roy, R., C. A. Laferriere, A. Gamian, M. Chomik, and H. J. Jennings. 1987. *N*-acetylneuraminic acid: Neoglycoproteins and pseudopolysaccharides. *Journal of Carbohydrate Chemistry* 6: 161–165.

Roy, R. and C. A. Laferriere. 1988. Synthesis antigenic copolymers of *N*-acetylneuraminic acid binding wheat germ agglutinin and antibodies. *Carbohydrate Research* 177: C1–C4.

Roy, R., D. F. Tropper, A. Romanowska, M. Letellier, L. Cousineau, S. J. Meunier, and J. Boratynski. 1991. Expedient synthesis of neoglycoproteins using phase transfer catalysis and reductive amination as key reactions. *Glycoconjugate Journal* 8: 75–81.

Roy, R., F. O. Andersson, G. Harm, S. Kelm, and R. Schauer. 1992. Synthesis of esterase-resistant 9-acetylated polysialoside as inhibitor of influenza C virus hemagglutinin. *Angewandte Chemie International Edition* 31: 1478–1481.

Roy, R. 1996. Blue-prints, synthesis and applications of glycopolymers. *Trends in Glycoscience and Glycotechnologies* 8: 79–99.

Sashiwa, H., S. Fujishima, N. Yamano, N. Kawasaki, A. Nakayama, E. Muraki, and S. Aiba. 2001a. Production of *N*-acetyl-D-glucosamine from b-chitin by enzymatic hydrolysis. *Chemistry Letters* 308–309.

Sashiwa, H., S. Fujishima, N. Yamano, N. Kawasaki, A. Nakayama, E. Muraki, K. Hiraga, K. Oda, and S. Aiba. 2002a. Production of *N*-acetyl-D-glucosamine from α-chitin by enzymes from *Aeromonas hydrophila* H-2330. *Carbohydrate Research* 337: 761–763.

Sashiwa, H., S. Fujishima, N. Yamano, N. Kawasaki, A. Nakayama, E. Muraki, M. Sukwattanasinitt, R. Pichyangkura, and S. Aiba. 2003a. Enzymatic production of *N*-acetyl-D-glucosamine from chitin. Degradation study of *N*-acetylchitooligosaccharide and the effect of mixing of crude enzymes-. *Carbohydrate Polymers* 51: 391–395.

Sashiwa, H., N. Kawasaki, A. Nakayama, E. Muraki, and S. Aiba. 2002b. Dissolution of chitosan in hexafluoro-2-propanol. *Chitin and Chitosan Research* 8: 249–251.

Sashiwa, H., N. Kawasaki, A. Nakayama, E. Muraki, N. Yamamoto, I. Aranitoyannis, H. Zhu, and S. Aiba. 2002c. Chemical modification of chitosan 12: Synthesis of organo-soluble chitosan derivatives toward palladium absorbent for chemical plating. *Chemistry Letters* 598–599.

Sashiwa, H., N. Kawasaki, A. Nakayama, E. Muraki, N. Yamamoto, H. Zhu, H. Nagano et al. 2002d. Chemical modification of chitosan 13: Synthesis of organo-soluble, palladium adsorbable, and biodegradable chitosan derivatives toward the chemical plating on plastics. *Biomacromolecules* 3: 1120–1125.

Sashiwa, H., N. Kawasaki, A. Nakayama, E. Muraki, N. Yamamoto, and S. Aiba,. 2002e. Chemical modification of chitosan 14: Synthesis of water-soluble chitosan derivatives by simple acetylation. *Biomacromolecules* 3: 1126–1128.

Sashiwa, H., N. Kawasaki, A. Nakayama, E. Muraki, H. Yajima, N. Yamamori, Y. Ichinose, J. Sunamoto, and S. Aiba. 2003b. Chemical modification of chitosan 15: Synthesis of novel chitosan derivatives by substitution of hydrophilic amine using *N*-carboxyethylchitosan ethyl ester as an intermediate. *Carbohydrate Research* 338: 557–561.

Sashiwa, H., Y. Makimura, Y. Shigemasa, and R. Roy. 2000a. Chemical modification of chitosan 1: Preparation of chitosan-sialic-acid branched polysaccharide. *Chemical Communications* 909–910.

Sashiwa, H., H. Saimoto, Y. Shigemasa, R. Ogawa, and S. Tokura. 1990. Lysozyme susceptibility of partially deacetylated chitin. *International Journal of Biological Macromolecules* 12: 295–296.

Sashiwa, H. and Y. Shigemasa, 1999. Chemical modification of chitin and chitosan 2: Preparation and water soluble property of N-acylated or N-alkylated partially deacetylated chitins. *Carbohydrate Polymers* 39: 127–138.

Sashiwa, H., Y. Shigemasa, and R. Roy. 2000b. Dissolution of chitosan in dimethyl sulfoxide by salt formation. *Chemistry Letters* 596–597.

Sashiwa, H., Y. Shigemasa, and R. Roy. 2000c. Homogeneous *N,O*-acylation of chitosan in dimethyl sulfoxide with cyclic acid anhydrides. *Chemistry Letters* 1186–1187.

Sashiwa, H., Y. Shigemasa, and R. Roy. 2000d. Chemical modification of chitosan 3: Hyperbranched chitosan-sialic acid dendrimer hybrid with tetraethylene glycol spacer. *Macromolecules* 33: 6913–6915.

Sashiwa, H., Y. Shigemasa, and R. Roy. 2000e. Novel *N*-alkylation of chitosan via Michael type reaction. *Chemistry Letters* 862–863.

Sashiwa, H., Y. Shigemasa, and R. Roy. 2001b. Preparation and lectin binding property of chitosan-carbohydrate conjugates. *Bulletin of the Chemical Society of Japan* 74: 937–943.

Sashiwa, H., Y. Shigemasa, and R. Roy. 2001c. Chemical modification of chitosan 10: Synthesis of hyperbranched chitosan-sialodendrimer hybrid using convergent grafting of pre-assembled dendrons built on gallic acid and tri(ethylene glycol) backbone. *Macromolecules* 34: 3905–3909.

Sashiwa, H., Y. Shigemasa, and R. Roy. 2001d. Highly convergent synthesis of hyperbranched chitosan-sialodendrimer hybrid. *Macromolecules* 34: 3211–3214.

Sashiwa, H., J. M. Thompson, S. K. Das, Y. Shigemasa, S. Tripathy, and R. Roy. 2000f. Chemical modification of chitosan: preparation and lectin binding properties of α-galactosyl-chitosan conjugates. Potential inhibition in acute rejection following xenotransplantation. *Biomacromolecules* 1: 303–305.

Sashiwa, H., N. Yamano, S. Fujishima, E. Muraki, N. Kawasaki, A. Nakayama, K. Sakamoto et al. 2001e. Production of *N*-acetyl-D-glucosamine from chitin by enzymatic hydrolysis. *Chitin and Chitosan Research* 7: 257–260.

Sashiwa, H., Y. Shigemasa, and R. Roy, 2002f. Chemical modification of chitosan 8: Preparation of chitosan-dendrimer hybrids via short spacer. *Carbohydrate Polymers* 47: 191–199.

Sashiwa, H., Y. Shigemasa, and R. Roy, 2002g. Chemical modification of chitosan 9: Reaction of *N*-carboxyethylchitosan methyl ester with diamines of acetal ending PAMAM dendrimers. *Carbohydrate Polymers* 47: 201–208.

Sashiwa, H., Y. Shigemasa, and R. Roy, 2002h. Chemical modification of chitosan 11: Chitosan-dendrimer hybrid as a tree like molecule. *Carbohydrate Polymers* 49: 195–205.

Sashiwa, H., H. Yajima, Y. Ichinose, N. Yamamori, J. Sunamoto, and S. Aiba. 2003c. Chemical modification of chitosan 16: Synthesis of polypropyreneimine dendrimer-chitosan hybrid. *Chitin and Chitosan Research* 9: 45–51.

Sashiwa, H., H. Yajima, and S. Aiba. 2003d. Synthesis of chitosan-dendrimer hybrid and its biodegradation. *Biomacromolecules* 4: 1244–1249.

Sashiwa, H., N. Yamamori, Y. Ichinose, J. Sunamoto, and S. Aiba. 2003e. Chemical modification of chitosan 17: Michael reaction of chitosan with acrylic acid in water. *Macromolecular Bioscience* 3: 231–233.

Sashiwa, H., N. Yamamori, Y. Ichinose, J. Sunamoto, and S. Aiba. 2003f. Michael reaction of chitosan with various acryl reagent in water. *Biomacromolecules* 4: 1250–1254.

Sashiwa, H., N. Yamamori, Y. Ichinose, J. Sunamoto, H. Yajima, and S. Aiba. 2003g. Studies on chitin and chitosan 36: Modification of chitosan or *N*-carboxyethylchitosan with laurylaldehyde and their biodegradation. *Chitin and Chitosan Research* 9: 205–210.

Sato, T., T. Ishii, and Y. Okahata. 2001. In vitro gene delivery mediated by chitosan. *Biomaterials* 22: 2075–2080.

Satoh, T., L. Vladimirov, M. Johmen, and N. Sakairi, 2003. Preparation and thermal dehydration of *n*-(carboxy) acyl chitosan derivatives with high stereoregularity. *Chemistry Letters* 318–319.

Schluter, A. D. and J. P. Rabe. 2000. Dendronized polymers: Synthesis, characterization, assembly at interfaces, and manipulation. *Angewandte Chemie International Edition.* 39: 864–883.

Setnikar, I., C. Giacchetti, and G. Zanolo. 1986. Pharmacokinetics of glucosamine in the dog and man. *Arznein-Forsch/Drug Research* 36: 729–735.

Setnikar, I., R. Palumbo, S. Canali, and G. Zanolo. 1993. Pharmacokinetics of glucosamine in man. *Arznein.-Forsch/Drug Research* 43: 1109–1113.

Shigemasa, Y., K. Saito, H. Sashiwa, and H. Saimoto. 1994. Enzymatic degradation of chitins and partially deacetylated chitins. *International Journal of Biological Macromolecules* 16: 43–49.

Shigemasa, Y., H. Usui, M. Morimoto, H. Saimoto, Y. Okamoto, S. Minami, and H. Sashiwa. 1999. Chemical modification of chitin and chitosan 1: Preparation of partially deacetylated chitin derivatives via a ring-opening reaction with cyclic acid anhydrides in lithium chloride/N,N-dimethylacetamide. *Carbohydrate Polymers* 39: 237–243.

Sukwattanasinitt, M., H. Zhu, H. Sashiwa, and S. Aiba. 2002. Utilization of commercial non-chitinase enzymes from fungi for preparation of 2-acetamido-2-deoxy-D-glucose from β-chitin. *Carbohydrate Research* 337: 133–137.

Tanigawa, T., Y. Tanaka, H. Sashiwa, H. Saimoto, and Y. Shigemasa. 1992. Various biological effects of chitin derivatives. In *Advances in Chitin and Chitosan* eds C. J. Brine, P. A. Sandford, and J. P. Zikakis, 206–215. Elsevier, London.

Tokura, S., S. Nishimura, N. Sakairi, and N. Nishi. 1996. Biological activities of biodegradable polysaccharide. *Macromolecular Symposia* 101: 389–396.

Tokura, S., K. Ueno, S. Miyazaki, and N. Nishi. 1997. Molecular weight dependent antimicrobial activity by chitosan. *Macromolecular Symposia* 120: 1–9.

Tomalia, D. A., H. Baker, J. Dewald, M. Hall, G. Kallos, S. Martin, J. Roeck, J. Ryder, and P. Smith. 1985. A new class of polymers: Starburst-dendritic macromolecules. *Polymer Journal* 17: 117–132.

Tomalia, D. A. and J. M. Frechet. 2002. Discovery of dendrimers and dendritic polymers: A brief historical perspective. *Journal of Polymer Science Part A: Polymer Chemistry* 40: 2719–2728.

Vetter, S., S. Koch, and A. D. Schluter. 2001. Synthesis and polymerization of functionalized dendritic macromonomers. *Journal of Polymer Science Part A: Polymer Chemistry.* 39: 1940–1954.

Wang, Y. C., M. C. Lin, D. M. Wang, and H. J. Hsieh. 2003. Fabrication of a novel porous PGA-chitosan hybrid matrix for tissue engineering. *Biomaterials* 24: 1047–1057.

Yalpani, M. and L. D. Hall. 1984. Some chemical and analytical aspects of polysaccharide modification 3. Formation of branched-chain, soluble chitosan derivatives. *Macromolecules* 17: 272–281.

Zeng, F. and S. C. Zimmemen. 1997. Dendrimers in supermolecular chemistry: From molecular reaction to self-assembly. *Chemical Reviews* 97: 1681–1712.

Zhang, A., L. Shu, Z. Bo, and A.D. Schluter. 2003. Dendronized polymers: Recent progress in synthesis. *Macromolecular Chemistry and Physics* 204: 328–339.

Zubarev, E. R. and S. I. Stupp. 2002. Dendron rodcoils: Synthesis of novel organic hybrid structure. *Journal of American Chemical Society* 124: 5762–5773.

5 Preparation of Chitin and Chitosan Derivatives Having Mercapto Groups

Mónica Perez, Juan Alfonso Redondo, Alberto Gallardo, and Inmaculada Aranaz

CONTENTS

5.1 INTRODUCTION

Chitin and its deacetylated derivative chitosan are natural copolymers composed of randomly distributed β-$(1 \rightarrow 4)$-linked 2-acetamido-D-glucose and 2-amino-D-glucose units in variable proportions (Figure 5.1). Chitin occurs naturally in a wide variety of species from fungi to lower animals. However, chitosan only occurs in some fungi and it is normally prepared by the deacetylation of chitin (Muzzarelli 1977; Roberts 1998). Chitin is an abundant polymer and both chitin and chitosan exhibit very interesting properties, such as biocompatibility, biodegradability, and metal-binding ability among others (Aranaz et al. 2009). Therefore, both polymers have potential applications in many fields such as in biomedicine, waste water treatment, or cosmetics (Dodane and Vilivalam 1998; Rinaudo 2006).

As shown in Figure 5.1, the polymers exhibit three functional groups: a free amino group at the C2 position (in low proportion on chitin), a primary alcohol at the C6 position, and a secondary alcohol at the C3 position. Taking advantage of these functionalities, a great deal of work has been carried out in relation to chitin and chitosan modification (Kurita 2001; Alves and Mano 2008; Mourya and Inamdar 2008).

In this chapter, we focus on the chemistry of those chitin and chitosan derivatives containing mercapto groups (–SH). Most of the reactions are carried out through the free amino group but modifications at the C6 position have also been reported.

After chemical modification, some of the properties of chitin and chitosan can be lost or modified. For instance, it has been reported that certain chitosan derivatives that is, carboxymethyl derivatives at the C3 position, showed lower rates of lysozyme degradation than the original polymers (Tokura et al. 1990). Both chitin and chitosan mercapto derivatives seem to have similar degradation

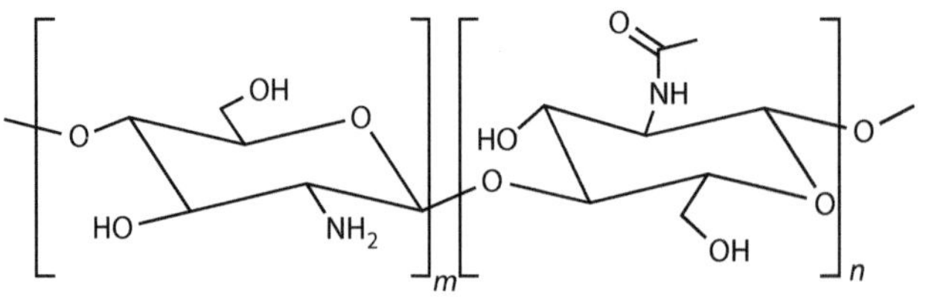

FIGURE 5.1   Chemical structure of chitin and chitosan (m < 0.60 chitin; m > 0.6 chitosan).

rates to the parent polymers and therefore they are very suitable for biomedical applications (Kast et al. 2003; Kurita et al. 1993). Recently, it has been reported that the antimicrobial activity of chitosan is preserved after thiolation at the C2 position although the activity is lower than that reported for chitosan (Han et al. 2012). Moreover, some chitosan properties such as mucoadhesivity, *in situ* gelling properties, permeation-enhancing effect, and transfection-enhancing effect are improved after thiolation (Sarti and Bernkop-Schnürch 2011). Table 5.1 summarizes the effect of thiolation on some chitin and chitosan properties.

Chitin and mainly chitosan mercapto derivatives are very promising molecules to be used in biomedicine (Bernkop-Schnürch et al. 2004; Radhakumary et al. 2011; Talaei, et al. 2011; Zhang et al. 2011), but they have also been proposed in other fields such as waste water treatment (Cárdenas et al. 2001; Sousa et al. 2009), cosmetics (Prinz 2006), or chemistry (Kurita et al. 1996, 1997) as shown in Table 5.2.

TABLE 5.1

Effect of Thiolation on Chitin and Chitosan Properties

Property	Effect of Modification
Biodegradability	Similar to parent polymer
Mucoadhesivity	Better than chitosan
Permeation enhancing effect	Better than chitosan
Efflux pump inhibition	Not described on chitosan
In situ gelling properties	Better than chitosan
Transfection-enhancing properties	Better than chitosan
Antimicrobial activity	Lower than chitosan

TABLE 5.2

Applications of Chitin and Chitosan Mercapto Derivatives

Drug delivery
Gene delivery
Tissue engineering
Wound healing
Metal chelation
Macroinitiators for graft copolymerization
Make up component
Support for enzyme immobilization

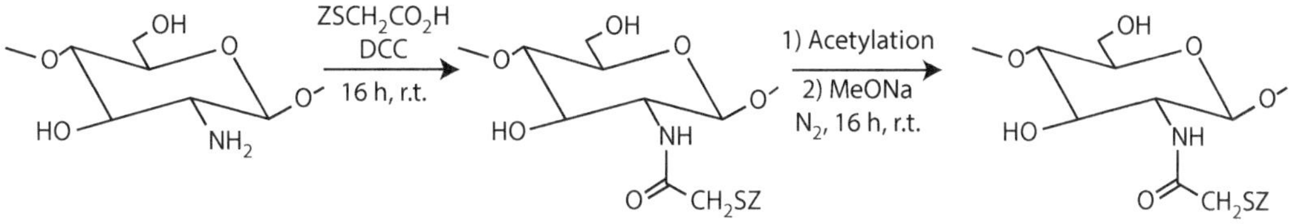

FIGURE 5.2 Introduction of a mercapto group on chitin at C2 position. (Modified from *Carbohydrate Polymers* 20, Kurita, K. et al. Preparation and biodegradability of chitin derivatives having mercapto groups, 239–245. Copyright 1993, with permission from Elsevier.)

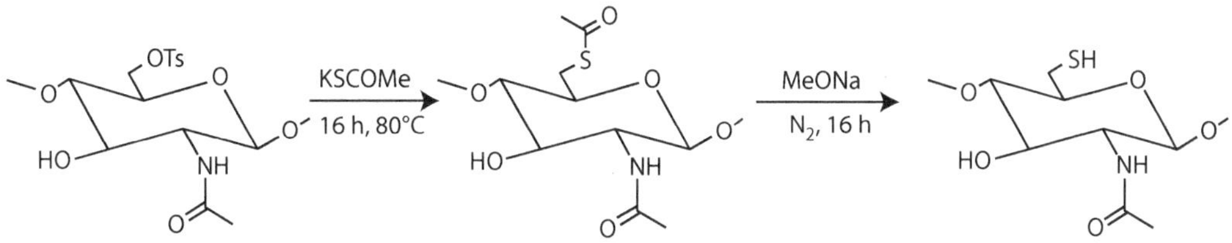

FIGURE 5.3 Introduction of a mercapto group on chitin at C6 position. (Modified from *Carbohydrate Polymers* 20, Kurita, K. et al. Preparation and biodegradability of chitin derivatives having mercapto groups, 239–245. Copyright 1993, with permission from Elsevier.)

5.2   CHEMISTRY OF CHITIN DERIVATIVES

Chitin is insoluble in most solvents and because of its intractable nature the modification of chitin is less common than chitosan because it is quite difficult to find conditions that allow the synthesis of well-defined structures. The introduction of mercapto groups selectively at positions C2 or C6 on chitin has been reported (Kurita et al. 1993). The strategy to introduce mercapto groups at C2 position is based on the mercaptoacetylation of chitosan (Figure 5.2). It has been reported that direct mercaptoacetylation of chitosan using mercaptoacetic acid is unsuccessful, and hence the mecapto group has to be protected with benzyloxy carbonyl group. In the first step, the protected mercaptoacetic acid and dicyclohexylcarbodiimide (DCC) in tetrahydrofuran (THF) were added to a chitosan solution (mixture of acetic acid and methanol) and in the second, the remaining free amino groups were acetylated with acetic anhydride. Finally, the mercapto group was deprotected with sodium methoxide. The degree of substitution ranged from 0.01 to 0.51, depending on the molar ratio of the z-protected mercaptoacetic acid/amino group.

Tosyl chitin has been used as a starting material to introduce mercapto groups on chitin at the C6 position (Figure 5.3). In the first step, tosyl chitin was dissolved in dimethyl sulfoxide (DMSO) and potassium thioacetate was added, and in the second, *S*-deacetylation to generate free mercapto groups was carried out. In this reaction, a derivative with a degree of substitution of 0.87 was produced. Thiolated-chitin derivatives showed a higher swelling and solubility in organic solvents than parent chitin (Kurita et al. 1993).

5.3   CHEMISTRY OF CHITOSAN DERIVATIVES

The derivatization of the primary amino groups of chitosan with coupling agents bearing thiol functions leads to the formation of thiolated chitosan. The attachment of the sulfhydryl-bearing agents to chitosan can be carried out via amide or amidine bonds. Ring opening reactions have also been used to introduce mercapto groups on chitosan but these types of reactions are less frequent.

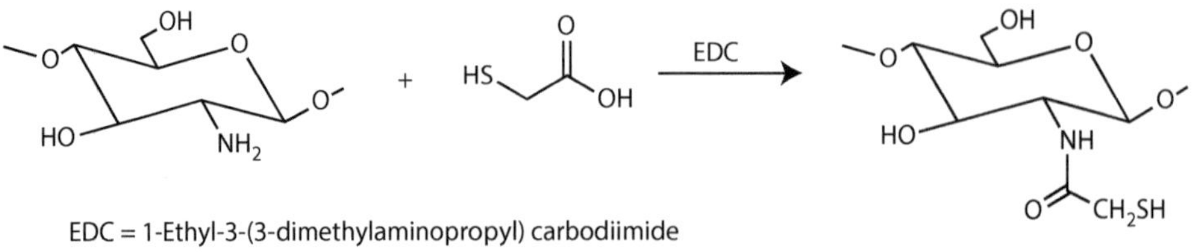

EDC = 1-Ethyl-3-(3-dimethylaminopropyl) carbodiimide

FIGURE 5.4 Thiolation of chitosan via amide bonds.

5.3.1 THIOLATED CHITOSAN VIA AMIDE BONDS

The use of carbodiimines to link to the primary amino groups of chitosan carboxylic acid groups is very common and has been used to prepare different alkyl and aryl thiolated chitosan derivatives (Figure 5.4) (Mourya and Inamdar 2008). In this process, the carbodiimine activates the carboxylic groups forming an *O*-acylurea derivative as an intermediate product that reacts with the primary amino groups of chitosan. To avoid the oxidation of the thiol groups during the synthesis, the reaction should be carried out under inert conditions or can be carried out at pH below 5 because at this pH the presence of thiolated anion (S$^-$) is low and the formation of disulfide bonds is almost excluded.

Initially, this synthetic strategy was used to prepare chitosan cysteine conjugates (CHT–Cys) and thioglycolic acid conjugates (CHT–TGA) (Bernkop-Schnürch et al. 2004). More recently, the process has been extended to thiolactic acid (CHT–TLA) (Sakloetsakun et al. 2009), glutathione (CHT–GSH) (Atyabi et al. 2008), and aromatic molecules such as mercaptonicotinic acid (CHT–MBA) (Millotti et al. 2009) and mercaptobenzoic acid (CHT–MNA) (Millotti et al. 2010). Aryl thiolated chitosan shows a lower pKa (pKa 5–7) than alkyl thiolated (pKa 8–10) and therefore exhibits higher reactivity at physiological pH.

Recently, a so-called *S*-protected thiolated chitosan has been developed (Dünnhaupt et al. 2012). The sulfhydryl ligand thioglycolic acid (TGA) was covalently attached to chitosan and subsequently, the thiol groups were protected by disulfide bond formation with the thiolated aromatic residue 6-mercaptonicotinamide (6-MNA).

Thiolated chitosan was also developed from chitosan by the covalent attachment of homocysteinethiolactone onto the amino groups of chitosan using imidazole as a reactive intermediate (Juntapram et al. 2012).

5.3.2 THIOLATED CHITOSAN VIA AMIDINE BONDS

To prepare thiolated chitosan via amidine bonds, the use of 2-iminothiolane as a coupling agent has been proposed (Figure 5.5). This process is carried out in one step and the chemical structure of the reagent prevents the oxidation of the thiol group (Bernkop-Schnürch et al. 2006).

However, the storage stability of chitosan-4-thiobutyl amidine conjugate (CHT–TBA), even under nitrogen, is low and a decrease of free thiol moieties is observed (Singh et al. 1996). To

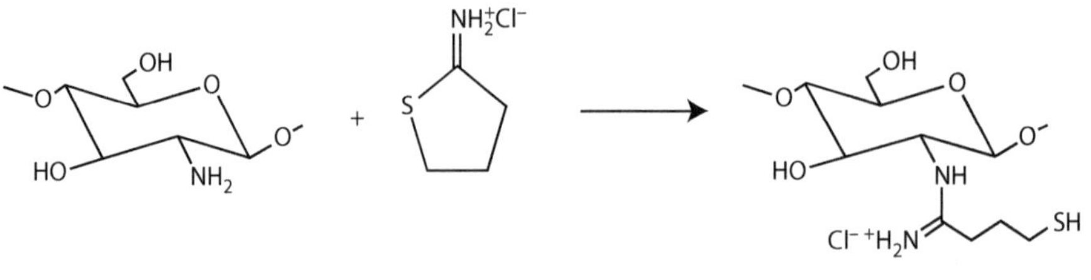

FIGURE 5.5 Thiolation of chitosan via amidine bonds.

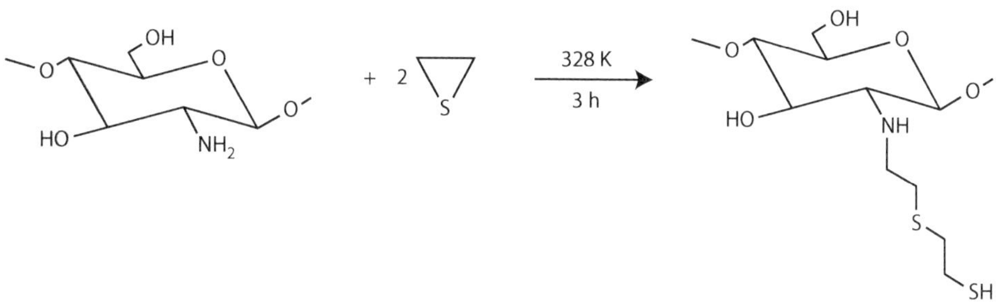

FIGURE 5.6 Thiolation of chitosan via ring opening reaction. (Modified from *Carbohydrate Research* 344, Sousa, K. S., E. C. Silva Filho, and C. Airoldi. Ethylenesulfide as a useful agent for incorporation into the biopolymer chitosan in a solvent-free reaction for use in cation removal, 1716–1723. Copyright 2009, with permission from Elsevier.)

overcome this problem, the use of isopropyl-*S*-acetylthioacetamide HCl as a coupling agent has been proposed (Kafedjiiski et al. 2005).

These reactions are pH dependent because the nucleophilicity of the amino group depends on its protonation state. In general, the reactions are carried out at pH 6.5–7, this value being a compromise between chitosan solubility and thiol oxidation.

5.3.3 RING OPENING REACTIONS

Owing to the nonbonded electron pair on the chitosan amino group, a typical nucleophilic attack on the strained electrophilic carbon atom of the three-membered ring reagent can occur (Figure 5.6).

The excess charge on the sulfur atom attacks an adjacent hydrogen atom, once bonded to nitrogen, to form an S–H bond to give a final neutral product. Again the nonbonded electron pair in the pendant chain has the ability to react with another three-membered ring. Elementary analysis demonstrated the inclusion of only two molecules of ethylene sulfide. This reaction is carried out in the absence of solvents (Sousa et al. 2009).

5.4 EFFECT OF CHITIN AND CHITOSAN PROPERTIES IN CHEMISTRY

As mentioned in the introduction, owing to their natural origin, both chitin and chitosan have differences regarding crystallinity, molecular weight, and deacetylation degree.

It is well known that α-chitin (the most abundant polymorph extracted from crustacean shells) is less reactive than β-chitin (a less abundant polymorph extracted from squid pens). Differences in the synthesis of 6-mercapto chitin because of the crystallinity of the chitin samples have been reported. The amount of tosylation in 6-tosyl chitin prepared from β-chitin varied to a greater extent than previously reported for α-chitin (Kurita et al. 1993; Munro et al. 2009).

The molecular weight of chitosan is another parameter that can affect the modification of the sample. In general, the use of high-molecular-weight samples is avoided because reaction with these samples is more difficult (Masuko et al. 2005). When comparing the amount of modification in chitosan–TBA samples prepared by using different chitosan molecular weights, a direct relationship between the molecular weight and the modification has been reported. The higher the molecular weight, the lower the modification (Bravo-Osuna et al. 2007a; Palmberger et al. 2008).

5.5 CHARACTERIZATION

The sulfur content of the samples can be measured by elementary analysis, but this assay does not give information about the chemical state of sulfur, that is, whether it exists as a reactive sulfhydryl

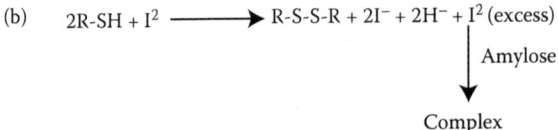

(a)   RS⁻ + DTNB²⁻ ⇌ R-S-TNB⁻ + TNB²⁻

                                          ⇅ RS⁻

                         R-S-S-R + TNB²⁻

    DTNB: 5,5'-dithio-*bis*(2-nitrobenzoic acid)
    TNB: 5-thio-2-nitrobenzoic acid

(b)   2R-SH + I² ⟶ R-S-S-R + 2I⁻ + 2H⁻ + I² (excess)

                                          ↓ Amylose

                         Complex

FIGURE 5.7 Determination of thiol groups. (a) Ellman's method. (b) Iodine tritation. (Modified from *International Journal of Pharmaceutics* 340, Bravo-Osuna, I. et al., Characterization of chitosan thiolation and application to thiol quantification onto nanoparticle surface, 173–181. Copyright 2007, with permission from Elsevier.)

or whether it has been oxidized (unreactive disulfide). To determine the amount of thiol groups, Ellman's method is normally used (Figure 5.7a) (Bernkop-Schnürch et al. 1999). The thiolate anion (R–S⁻) reacts with 5,5'-dithio-*bis*(2-nitrobenzoic acid) forming 5-thio-2-nitrobenzoic acid, which exhibits intense light absorption at a wavelength of 410–420 nm (Ellman et al. 1961). The main limitation of this method is that the intensity of the light absorption of TNB is pH independent only if the pH of the medium is above 7.3 (Riener et al. 2002). Thus, this assay must be only employed above this pH value, with an optimal value of around pH 8–8.5 (Nogueira et al. 2005). Because the solubility of thiolated chitosan in the optimal conditions of Ellman's reaction is poor, Bravo-Osuna et al. proposed a new method for the characterization of thiolated chitosan; this method is based on a classical iodine titration as shown in Figure 5.7b (Bravo-Osuna et al. 2007b).

5.6   CONCLUSIONS

Mercapto groups may possibly be introduced in chitin and chitosan backbones by several methods but only those that are able to produce well-defined structures because of regioselective modification are of interest. Owing to its intractable nature, the amount of mercapto chitin derivatives is low because it is very difficult to find reactions under mild and homogeneous conditions. However, owing to the easy chemistry of the primary amino group and its better solubility, a wider variety of chitosan derivatives have been synthesized.

Although it has not been extensively studied, it seems that the properties of the polymers (crystallinity and molecular weight) have some impact in the final characteristics of the product and therefore it is a factor that should be considered.

REFERENCES

Alves, N. M. and J. F. Mano. 2008. Chitosan derivatives obtained by chemical modifications for biomedical and enviromental applications. *International Journal of Biological Macromolecules* 43: 401–414.

Aranaz, I., M. Mengíbar, and R. Harris et al. 2009. Functional characterization of chitin and chitosan. *Current Chemical Biology* 3: 203–230.

Atyabi, F., F. A. Moghaddam, R. Dinarvand, M. J. Zohuriaan-Mehr, and G. Ponchel. 2008. Thiolated chitosan coated poly hydroxyethyl methacrylate nanoparticles: Synthesis and characterization. *Carbohydrate Polymers* 74: 59–67.

Bernkop-Schnürch, A., M. Hornof, and D. Guggi. 2004. Thiolated chitosan. *European Journal of Pharmaceutics and Biopharmaceutics* 57: 9–17.

Bernkop-Schnürch, A., V. Schwarz, and S. Steininger. 1999. Polymers with thiol groups: A new generation of mucoadhesive polymers? *Pharmaceutical Research* 16: 876–881.

Bernkop-Schnürch, A., A. Weithaler, K. Albrecht, and A. Greimel. 2006. Thiomers: Preparation and *in vitro* evaluation of a mucoadhesive nanoparticulate drug delivery system. *International Journal of Pharmaceutics* 317: 76–81.

Bravo-Osuna, I., G. Ponchel, and C. Vauthier. 2007a. Tuning of shell and core characteristics of chitosan-decorated acrylic nanoparticles. *European Journal of Pharmaceutical Sciences* 30: 143–154.

Bravo-Osuna, I,. D. Teutonico, S. Arpicco, C. Vauthier, and G. Ponchel. 2007b. Characterization of chitosan thiolation and application to thiol quantification onto nanoparticle surface. *International Journal of Pharmaceutics* 340: 173–181.

Cárdenas, G., P. Orlando, and T. Edelio. 2001. Synthesis and applications of chitosan mercaptanes as heavy metal retention agent. *International Journal of Biological Macromolecules* 28: 167–174.

Dodane, V. and V. D. Vilivalam. 1998. Pharmaceutical applications of chitosan. *Pharmaceutical Science Technology Today* 1: 246–253.

Dünnhaupt, S., J. Barthelmes, J. Iqbal et al. 2012. *In vivo* evaluation of an oral drug delivery system for peptides based on S-protected thiolated chitosan. *Journal of Controlled Release* 160: 477–485.

Ellman, G. L., K. D. Courtney, V. Andres, and R.M. Featherstone. 1961. A new and rapid colorimetric determination of acetylcholinesterase activity. *Biochemical Pharmacology* 7: 88–95.

Han, B., Wei, Y., Jia, X., Xu, J. and Li, G. 2012. Correlation of the structure, properties, and antimicrobial activity of a soluble thiolated chitosan derivative. *Journal of Applied Polymer Science* 125: E143–E148.

Juntapram, K., N. Praphairaksit, K. Siraleartmukul, and. Muangsin. 2012. Synthesis and characterization of chitosan-homocysteine thiolactone as a mucoadhesive polymer. *Carbohydrate Polymers* 87: 2399–2408.

Kafedjiiski, K., A. H. Krauland, M. H. Hoffer, and A. Bernkop-Schnürch. 2005. Synthesis and *in vitro* evaluation of a novel thiolated chitosan. *Biomaterials* 25: 819–826.

Kast, C. E., F. W. Losert, and A. Bernkop-Schnürch. 2003. Chitosan thiogycolic acid conjugate: a new scaffold material for tissue engineering?. *International Journal of Pharmaceutics* 256: 183–189.

Kurita, K. 2001. Controlled functionalization of the polysaccharide chitin. *Progress in Polymer Science* 26: 1921–1971.

Kurita, K., S. Hashimoto, H. Yoshino, S. Ishii, and S. I. Nishimura. 1996. Preparation of chitin/polystyrene hybrid materials by efficient graft copolymerization based on mercaptochitin. *Macromolecules* 29: 1939–1942.

Kurita, K., H. Yoshino, S. I. Nishimura, and S. Ishii. 1993. Preparation and biodegradability of chitin derivatives having mercapto groups. *Carbohydrate Polymers* 20: 239–245.

Kurita, K., H. Yoshino, S. I. Nishimura, S. Ishii, T. Mori, and Y. Nishiyama. 1997. Mercapto-chitins: A new type of supports for effective immobilization of acid phosphatase. *Carbohydrate Polymers* 32: 111–175.

Masuko, T., A. Minami, N. Iwasaki, T. Majima, S. I. Nishimura, and Y. C. Lee. 2005. Thiolation of chitosan. Attachment of proteins via thioether formation. *Biomacromolecules* 6: 880–884.

Millotti, G., C. Samberger, E. Fröhlich, and A. Bernkop-Schnürch. 2009. Chitosan-graft-6-mercaptonicotinic acid: Synthesis, characterization, and biocompatibility. *Biomacromolecules* 10: 3023–3027.

Millotti, G., C. Samberger, E. Fröhlich, D. Sakloetsakun, and A. Bernkop-Schnürch. 2010. Chitosan-4-mercaptobenzoic acid: Synthesis and characterization of a novel thiolated chitosan. *Journal of Materials Chemistry* 20: 2432–2440.

Mourya, V. K. and N. N. Inamdar 2008. Chitosan-modifications and applications: Opportunities galore. *Reactive and Functional Polymers* 68: 1013–1051.

Munro, N. H., L. R. Hanton, S. C. Moratti, and B. H. Robinson. 2009. Preparation and graft copolymerisation of thiolated β-chitin and chitosan derivatives. *Carbohydrate Polymers* 78: 137–145.

Muzzarelli, R. A. A. 1977. *Chitin*. Oxford, Pergamon Press.

Nogueira, R., M. Lammerhofel, N. M. Maier, and M. W. Lindner. 2005. Spectrophotometric determination of sulphydryl concentration on the surface of thiol-modified chromatographic silica particles using 2,2-dipyridyl disulfide reagent. *Analytica Chimica Acta* 533: 179–183.

Palmberger, T. F., J. Hombach, and A. Bernkop-Schnürch. 2008. Thiolated chitosan: Development and *in vitro* evaluation of an oral delivery system for acyclovir. *International Journal of Pharmaceutics* 348: 54–60.

Prinz, M. 2006. Chitosan thio-amidine conjugates and their cosmetic as well as pharmaceutic use. Patent number US 7053068.

Radhakumary, C., M. Antonty, and K. Sreenivasan. 2011. Drug loaded thermoresponsive and cytocompatible chitosan based hydrogel as a potential wound dressing. *Carbohydrate Polymers* 83: 705–713.

Riener, C. K., G. Kada, and H. J. Gruber. 2002. Quick measurement of protein sulfhydryls with ellman's reagent and with 4,4-dithiodipyridine. *Analytical and Bioanalytical Chemistry* 373: 266–276.

Rinaudo, M. 2006. Chitin and chitosan: Properties and applications. *Progress Polymer Science* 31: 603–632.

Roberts, G. 1998. *Chitin Chemistry*. London, Macmillan.

Sakloetsakun, D., J. M. R. Hombach, and A. Bernkop-Schnürch. 2009. *In situ* gelling properties of chitosan-thioglycolic acid conjugate in the presence of oxidizing agents. *Biomaterials* 30: 6151–6157.

Sarti, F. and A. Bernkop-Schnürch 2011. Chitosan and thiolated chitosan. *Advances in Polymer Science* 243: 93–110.

Singh, R., L. Kats, W.A. Blättler, and J. M. Lambert. 1996. Formation of N-substituted 2-iminothiolanes when amino groups in proteins and peptides are modified by 2-iminothiolane. *Analytical Biochemistry* 236: 114–125.

Sousa, K. S., E. C. Silva Filho, and C. Airoldi. 2009. Ethylenesulfide as a useful agent for incorporation into the biopolymer chitosan in a solvent-free reaction for use in cation removal. *Carbohydrate Research* 344: 1716–1723.

Talaei, F., E. Azizi, R. Dinarvand, and F. Atyabi. 2011. Thiolated chitosan nanoparticles as a delivery system for antisense therapy: evaluation against EGFR in T47D breast cancer cells. *International Journal of Nanomedicine* 6: 1963–1975.

Tokura, S., Y. Miura, Y. Uraki, K. Watanabe, S. Ikuo, and A. Ichiro. 1990. Biodegradable chitin derivative as various types of drug carriers. *American Chemical Society. Polymer Preprints* 31: 627.

Zhang, H., A. Qadeer, and W. Chen. 2011. *In situ* gelable interpenetrating double network hydrogel formulated from binary components: Thiolated chitosan and oxidized dextran. *Biomacromolecules* 12: 1428–1437.

6 Thiolated Chitosan
Preparation, Properties, and Applications

Nazma N. Inamdar and Vishnukant Mourya

CONTENTS

6.1 INTRODUCTION

The most preferred route for delivering therapeutic agents is oral drug delivery. However, this route is not feasible for delivery of hydrophilic macromolecules, such as therapeutic peptides, proteins, nucleotides, and efflux pump substrates, because of negligible oral bioavailability. This is because of the barriers confronted by these molecules along the oral route, such as diffusion barrier of mucus, enzymatic barrier of peptidases, and permeability barrier of membranes. The strategies to surmount these barriers are by utilization of permeation enhancers, enzyme inhibitors, and multifunctional polymers, ideally guaranteeing both permeation enhancement and enzyme inhibition. In case of multifunctional polymers, these effects can take place only if polymers offer strong mucoadhesive features for providing a tight contact with the mucosa during the whole period of peptide drug release and absorption. Among the group of multifunctional polymers exhibiting all of the aforementioned properties, thiolated polymers (designated thiomers) are the most promising. The thiomers are hydrophilic macromolecules exhibiting free thiol groups on the polymeric backbone as poly(acrylates), carboxymethyl cellulose, alginate, pectin, deacetylated gellan gum, or chitosans.

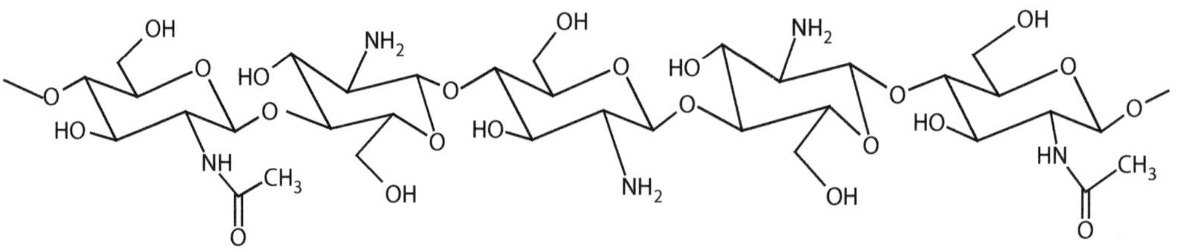

FIGURE 6.1 Structure of chitosan. The monomer units appear randomly along the polymer chain.

All these polymers carry COOH groups on their backbones, except thiolated chitosan. The thiolated chitosans are the only cationic thiomers and hence unique.

Chitosan, a linear polymer consisting of β(1 → 4)-D-glucosamine and β(1 → 4)-N-acetyl-D-glucosamine units, is under investigation for a variety of applications, including biomedical and pharmaceutical (Figure 6.1). However, it is soluble only in acidic conditions of pH below 6 and the solubility is attributed to protonation of amino groups by acids. The limited solubility of chitosan is a major drawback for it being used in physiological conditions. The chemical reactivity of chitosan is harnessed to overcome this drawback. The primary amine and primary as well as a secondary hydroxyl functions present in the monomers of chitosan can be modified chemically to modulate its solubility. The important derivatives of chitosan include acylated, carboxyalkylated, quaternized, sulfated, phosphorylated, and thiolated ones along with certain others (Mourya and Inamdar 2008).

6.2 SYNTHESIS OF THIOLATED CHITOSAN

The primary amino group at the 2-position of the glucosamine subunits of this polymer is the main site for the immobilization of thiol groups. The sulfhydryl-bearing agents can be covalently attached to this primary amino group via the formation of amide or amidine bonds. The thiolated chitosans generated so far include chitosan–cysteine (Ch–Cys), chitosan–N-acetylcysteine (Ch–NAC), chitosan–homocystenine (Ch–HomoCys), chitosan–thioglycolic acid (Ch–TGA), chitosan–thioethylamidine (Ch–TEA), chitosan–4-thiobutylamidine (Ch–TBA), chitosan–glutathione (Ch–GSH), chitosan–thiolactic acid (Ch–TLA), and chitosan–6-mercaptonicotinic acid (Ch–MNA) conjugates. The synthetic schemes for these thiolated chitosans are shown in Figure 6.2. The conjugates, such as Ch–Cys, Ch–TGA, and others, have –SH group immobilized by an uncharged amide bond, whereas Ch–TEA and Ch–TBA conjugates have a raised cationic character because of the covalently attached thiol groups by a cationic amidine substructure.

The sulfhydryl-bearing agents, such as cysteine, thioglycolic acid, and 6-mercaptonicotinic acid, can be covalently attached via the amide bond formation between the primary amino group of chitosan and activated carboxylic acid group of the agent mediated by a water-soluble condensing agent (Bernkop-Schnürch et al. 1999, Bernkop-Schnürch and Hopf 2001, Kast and Bernkop-Schnürch 2001, Millotti et al. 2009).

The carboxylic acid moieties of thiol-containing reagent were activated by 1-ethyl-3-(3-dimethyl-aminopropyl)carbodiimide hydrochloride (EDC), forming an O-acylurea derivative as an intermediate product, which reacts with the primary amino groups of chitosan. For thiolation, EDC was added to 0.1% w/v acidic solution of chitosan in a final concentration of 50–125 mM, followed by the addition of acid (thioglycolic or cysteine) in the ratio 1:1 of polymer to acid. The reaction mixture was incubated at pH 5 for 3 h at room temperature with continuous stirring. The thiolated product was purified by dialysis: once against 5 mM HCl; twice against the same medium, but containing 1% NaCl, to quench ionic interactions between the cationic polymer and the anionic sulfhydryl compound; and twice against 1 mM HCl to adjust the pH of the polymer to 4. The lyophilization of aqueous polymer solutions at −30°C and 0.01 mbar provided the thiomer in powder form. The extent of thiolation was pH dependent for thioglycolic acid between pH 3 and 5 and was optimum at pH 4. At this pH range, the concentration

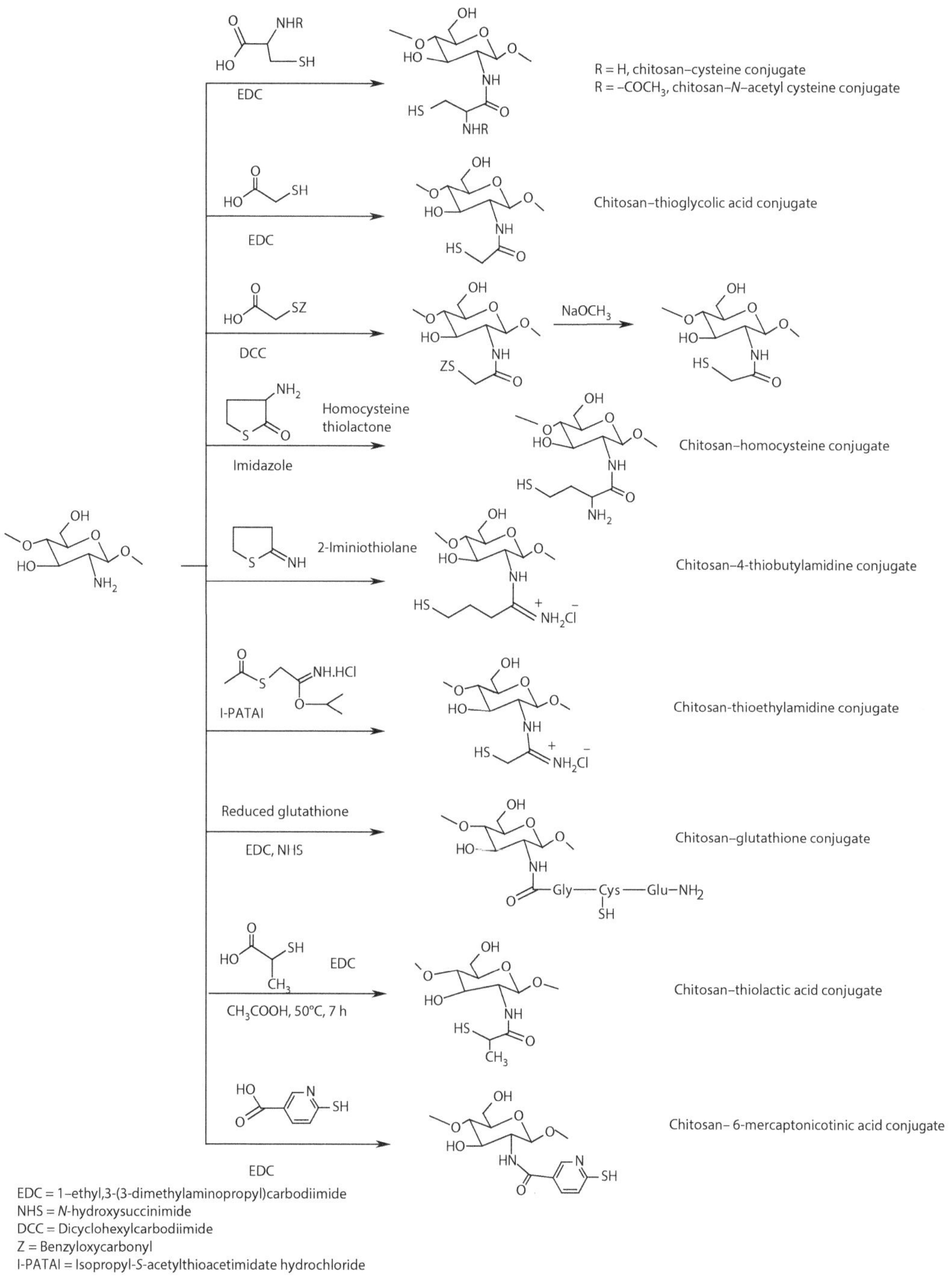

FIGURE 6.2 Synthesis of thiolated chitosans.

of thiolate anions, representing the reactive form for the oxidation of thiol groups, was low, and the formation of disulfide bonds could be almost excluded. Alternately, an unintended oxidation of thiol groups during synthesis could be avoided by performing the reaction under inert conditions.

The reaction conditions were adopted for modification with acetyl cysteine and the reduced form of glutathione employed in large quantities as 4–5 g for 0.5–1 g of chitosan (Kafedjiiski et al.

2005, Schmitz et al. 2008). For the synthesis of chitosan–glutathione conjugate, improvement in the formation of amide bond between glycine carboxylic acid groups of glutathione and amine groups of chitosan was achieved by combining coupling reagents as EDC and N-hydroxysuccinimide. The fast hydrolysis of O-urea derivative formed by activation of –COOH group of glycine residue by EDC limits the –SH immobilization. The N-hydroxysuccinimide esters hydrolyze very slowly and improve the coupling yield. The optimum ratio of EDC:N-hydroxysuccinimde was 1:1 to 1:1.25. Unintended oxidation was curtailed by the treatment of the obtained product with tris (2-carboxyethyl) phosphine hydrochloride (TCEP) in a final concentration of 5 mM at pH 4–5.

The synthesis of Ch–HomoCys was carried out by the covalent attachment of homocysteine thiolactone to the amino groups of chitosan via a ring-opening reaction (Juntapram et al. 2012). Homocysteine thiolactone and imidazole react to give reactive intermediate HS–CH$_2$–CH$_2$–CH(NH$_2$)–C(O)–imidazole, which covalently reacts at the amino group of chitosan.

For Ch–TGA synthesis, dicyclohexylcarbodiimide (DCC) was employed as a condensing agent in the reaction of chitosan with thioglycolic acid protected by benzyloxycarbonyl group at S. The methanolic acidic solution of chitosan was treated with (S-benzyloxycarbonyl)mercaptoacetic acid and DCC in the approximate molar ratio of 1:3:0.1 in tetrahydrofuran at room temperature for 16 h. The protected –SH group was demasked by reacting the product with equal volume of 5% aqueous NaHCO$_3$ to obtain the precipitate of the product which was then purified by extraction with ethanol, followed by acetone, after deionizing with aqueous washings (Kurita et al. 1993).

The modifying agent for Ch–TBA conjugate is 2-iminothiolane or Traut's reagent (a cyclic thio-imidoester or thioimidate), which reacts with amino groups and introduces a sulfhydryl residue via a positively charged amidine substructure (Bernkop-Schnürch et al. 2003, Kast et al. 2003). Here too, ~0.1% chitosan solution in 1% acetic acid with pH adjusted to 7 by 5 M NaOH was reacted with 2-imi-nothiolane (0.1–0.4 g) for 24 h at room temperature under continuous stirring. To avoid an oxidation process during the coupling reaction, 2-mercaptoethanol was added in a final concentration of 3% (v/v). The coupling reactions performed at pH 7 led to the highest yield in polymer-immobilized thiol groups compared to pH 5. At this pH, chitosan is almost completely dissolved in an acetate buffer providing a good access to its amino groups; as well as, the coupling reagent exhibits a high reactiv-ity. The products appeared as white, odorless powder of fibrous structure. They were easily soluble in aqueous solutions at a pH below 6.0 and formed transparent gels of high viscosity.

However, storage stability studies using nitrogen showed an insufficient stability of thiomer, which resulted in a decrease of free thiol moieties. This might be because of the formation of N-chitosanyl-substituted 2-iminothiolane structures. This undesired side reaction occurs after the derivatization of different amines with 2-iminothiolane. It involves the loss of ammonia and yields recyclized N-substituted 2-iminothiolane (Singh et al. 1996) (Figure 6.3).

The chemical modification of chitosan can be done with isopropyl-S-acetylthioacetimidate hydrochloride (i-PATAI) resulting in chitosan–thioethylamidine conjugate (Kafedjiiski et al. 2005a). The nucleophilicity of amino groups is dictated by the protonation state, making the reac-tion pH dependent. The reactions were carried out at pH 6–7 by stirring 1% chitosan solution in acetic acid and an equal amount of i-PATAI for about 3 h at room temperature. Deprotection of S-acetyl group can be done with NH$_2$OH (Kafedjiiski et al. 2006). Deacetylation of the mercapto group, however, may occur spontaneously and quantitatively because the additional deprotecting agent NH$_2$OH led to no further increase of thiol content (Delprino et al. 1993). The reaction was

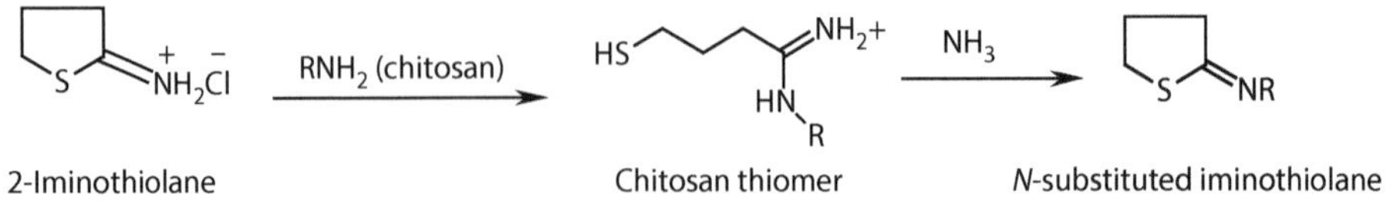

| 2-Iminothiolane | Chitosan thiomer | N-substituted iminothiolane |

FIGURE 6.3   Unstability of the chitosan–4-thiobutylamidine conjugate.

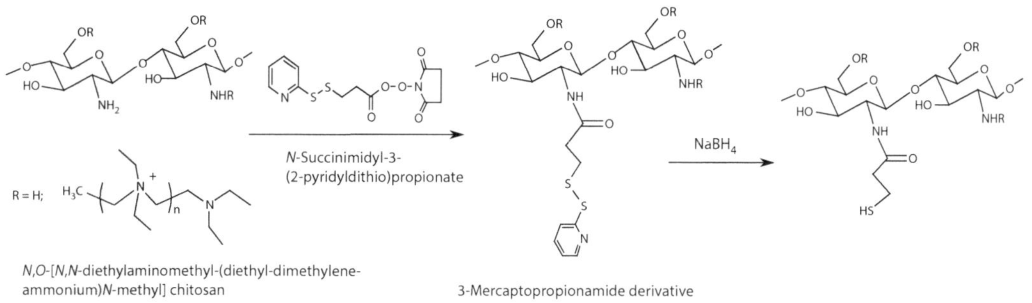

FIGURE 6.4   3-Mercaptopropionamide derivative of *N,O*-[*N,N*-diethylaminomethyl(diethyl-dimethylene-ammonium)*N*-methyl] chitosans.

faster than 2-iminothiolane and the short chain of *i*-PATAI excluded theoretically the possibility of yielding cyclic nonthiol products.

For Ch–TLA, Jayakumar et al. made use of the reaction of chitosan in acidic solution with thio-lactic acid in equal quantities mediated by EDC and precipitated the product in acetone as done for the majority of chitosan derivatives (Jayakumar et al. 2007).

A new type of thiolating reagent and thiolated group has been disclosed by Zambito et al. recently, for chitosans with short pendant chains of diethyl-dimethylene-ammonium groups substituted onto the primary amino group of the chitosan (Zambito et al. 2009). *N,O*-[*N,N*-diethylaminomethyl(diethyl-dimethyleneammonium)$_n$-methyl] chitosans (DS 40% and 60%) were used to transform residual free amino groups into 3-mercaptopropionamide moieties by reacting aqueous solution of modified polymer with *N*-succinimidyl-3-(2-pyridyldithio)propionate (polymer repeating unit-reagent molar ratio 1:20 or 1:10) at ambient temperature for half an hour. This was followed by the reduction of disulfide bond to free thiol group by treatment with reducing agent NaBH$_4$ (disulfide-NaBH$_4$ 1:3, molar ratio) for an hour (Figure 6.4).

The thiolated derivative of trimethyl chitosan (TMC) has been reported by Verheul et al. who made use of partially *N*-carboxymethylated TMC and cystamine with a condensing agent so that the –COOH group of used chitosan derivative reacted with the –SH group (Verheul et al. 2010) (Figure 6.5). The disulfide linkages formed were reduced by dithiothreitol.

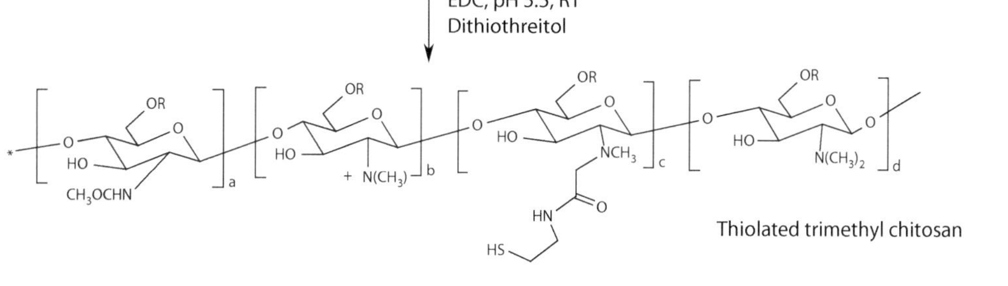

FIGURE 6.5   Synthesis of thiolated trimethyl chitosan from *N*-carboxymethylated trimethyl chitosan and cystamine.

During the thiolation, about 529 µmol/g thiol moieties (as free thiols or disulfides) were introduced to *N*-carboxymethylated TMC –COOH degree of quaternization (DQ) 54%, corresponding to approximately 12% of glucosamine units and indicated quantitative substitution of the carboxylic acids with cystamine. Treatment with dithiothreitol resulted in 283 µmol/g free SH groups in thiolated TMC DQ 54%, corresponding to a degree of thiolation of ~6%. For *N*-carboxymethylated TMC with a DQ 25%, the conversion was slightly less efficient (478 µmol/g thiol moieties as free thiols or disulfide groups, approximately 10% of glucosamine units) and reduction with dithiothreitol resulted in 236 µmol/g free SH groups (approximately degree of thiolation 5%). Degree of thiolation increased up to 341 µmol/g polymer (about 7% degree of thiolation) after a second reduction cycle with dithiothreitol.

6.3 DETERMINATION OF THIOL GROUP

The amount of immobilized thiol groups in reduced and oxidized form can be determined via Ellman's reagent [5,5′-dithio*bis*(2-nitrobenzoic acid) $DTNB^{2-}$] with and without previous quantitative reduction of disulfide bonds with borohydride (Hornof et al. 2003, Leitner et al. 2003). The 250 µL of conjugate solution (2 mg/mL) diluted with 250 µL of 5 M phosphate buffer pH 8.0 is incubated for 2 h at RT with 500 µL of Ellman's reagent [0.3 mg/mL in 0.5 M phosphate buffer pH 8]. The absorbance of supernatant obtained after centrifugation ($24,000 \times g$; 5 min) is measured at 450 nm in microtitration plate reader. The amount of thiol moieties was calculated from a standard curve obtained from solutions with increasing concentrations of L-cysteine hydrochloride hydrate (Bravo-Osuna et al. 2007).

Ellman's reaction is based on the reaction of the thiolate anion (RS^-) $DTNB^{2-}$ according to the reaction scheme given below:

$$RS^- + DTNB^{2-} \rightleftharpoons TNB^- + R\text{–}S\text{–}TNB^- \xrightarrow{RS^-} R\text{–}S\text{–}S\text{–}R + TNB^{2-}$$

The 5-thio-2-nitrobenzoic acid (TNB^{2-}) formed exhibits intense light absorption at a wavelength of 410–420 nm. The main limitation of this method is that the intensity of the light absorption of TNB^{2-} is pH independent only if the pH of the medium is above 7.3. Thus, this assay must be employed only above this pH value, with an optimal value of around pH 8–8.5. The experimental conditions as viscous solution and limited solubility of polymers at pH above 8 may pose problems.

An alternative and very simple thiol quantification method was developed on the basis of the classical iodine titration. Iodine oxidizes –SH groups to disulfuric bonds and forms iodine/amylase (starch) complex with a brilliant blue color. It quantifies the excess of iodine in the reaction medium remaining after reaction with free available thiol groups (Kast and Bernkop-Schnürch 2001). However, the limitation of this technique is the subjective evaluation of the end point of the reaction and also the larger amounts of products generally needed for quantification for example, 3 mg sample dissolved in 1 mL water and pH adjusted to 2–3 with hydrochloric acid.

In an attempt to make the technique independent of subjective considerations, a modified spectrophotometric version of this method that involved standarization was introduced (Bravo-Osuna et al. 2007). In this, 2 mL solution of thiolated polymer (0.5–0.05 mg/mL) in 0.5 mol/L CH_3COOH/CH_3COONa buffer pH 2.7 were oxidized with 0.5 mL of iodine (1 mmol/L) in the presence of 1 mL of starch aqueous solution (1%, w/v) for 24 h at RT and the spectrophotometric measurements were done at 560 nm. Thiol content could be determined by spectrophotometric measurements taken for the use of unmodified chitosan and cysteine hydrochloride treated in the same way as control and standard, respectively. The new method enabled the determination of thiol groups in a small amount of samples at acidic pH, and the monitoring of the thiol determination kinetic with time.

6.4 PROPERTIES

The immobilization of thiol groups improves the various properties of chitosan, such that they are allocated to a promising new category of thiomers.

6.4.1 MUCOADHESIVE PROPERTIES

Mucoadhesion is a result of fine interplay among the interdependent properties of polymer as its mechanism of binding, swelling behavior, cohesiveness, and characteristics of mucosa (Figure 6.6).

The epithelial lining of mucosal tissue is protected by an adherent gel layer of mucus. The mucus is composed of a heterogeneous material with mucin as the primary component. Mucins are a group of large glycoproteins built up in subunits of 500,000 Da or larger. It consists of a linear protein backbone, approximately 800 amino acids long, rich in hydroxylated amino acids—serine and threonine. Most of the hydroxyl residues are linked to oligosaccharide side chains that serve to stiffen the backbone and carry an extensive layer of water for hydration. This generates alternating naked and densely glycosylated regions in the protein. (For gastrointestinal mucins, the contents are 70–80% carbohydrate, 12–25% protein, and upto ~5% ester sulfates.) The oligosaccharide chains are generally upto 19 residues in length. The typical sugars present in mucins are D-galactose, L-fucose, N-acetylglucosamine, N-acetylgalactosamine, N-acetylneuramic acid, sialic acid, and N-acetylglucosamine-6-sulfate. The molecule, although linear, is randomly coiled into a spheroidal in solution, with highly swollen domain on imbibing water. The naked or low glycosylation region of mucin is labile to digestion by trypsin (the digestion products are rich in cysteine commonly referred to as T-domains). Every third or fourth T domain is linked by a disulfide bridge; itself susceptible to reductive disruption by thiols. (The thiol reduction products are commonly referred to as subunits.) The key groups in terms of possible interaction sites for mucoadhesion are carboxylate group (of sialic acid), sulfate group (N-acetylglucosamine-6-sulfate), which contribute to the net negative charge and electrostatic interaction with mucoadhesives. Other potential residues for mucoadhesive interactions are the carbonyl (hydrogen bonding) and methyl (hydrophobic bonding) groups on the N-acetyl residues (N-acetylgalactose, N-acetylglucosamine, and sialic acid) and another methyl group in fucose.

The native chitosan offers mucoadhesive properties due to ionic interactions between the positively charged primary amino groups on the polymer and negatively charged sialic acid and sulfonic acid substructures of the mucus (Hassan and Gallo 1990). But the mucoadhesion is weak and short-lasting (Lehr et al. 1992). The mucoadhesion in thiolated chitosans gets fortified by the presence of thiol groups on the line of strong sulfhydryls, such as mercaptoethanol or dithiothreitol, which bind covalently to mucin glycoproteins. Leitner et al., on the basis of thiol/disulfide exchange reactions and/or a simple oxidation process, suggested that disulfide bonds are formed between thiomer and

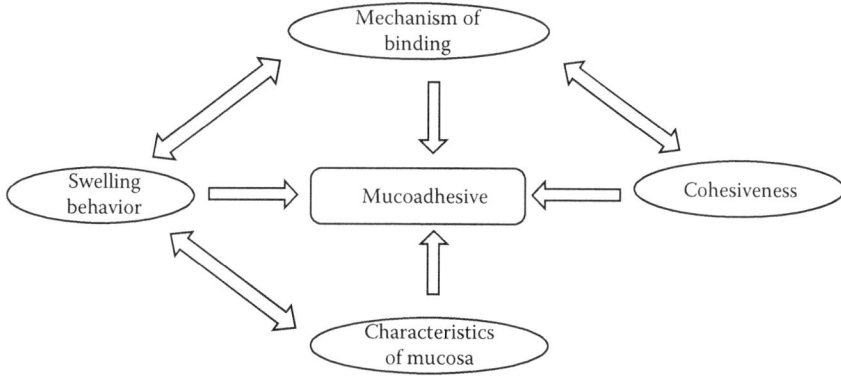

FIGURE 6.6 Factors influencing mucoadhesion.

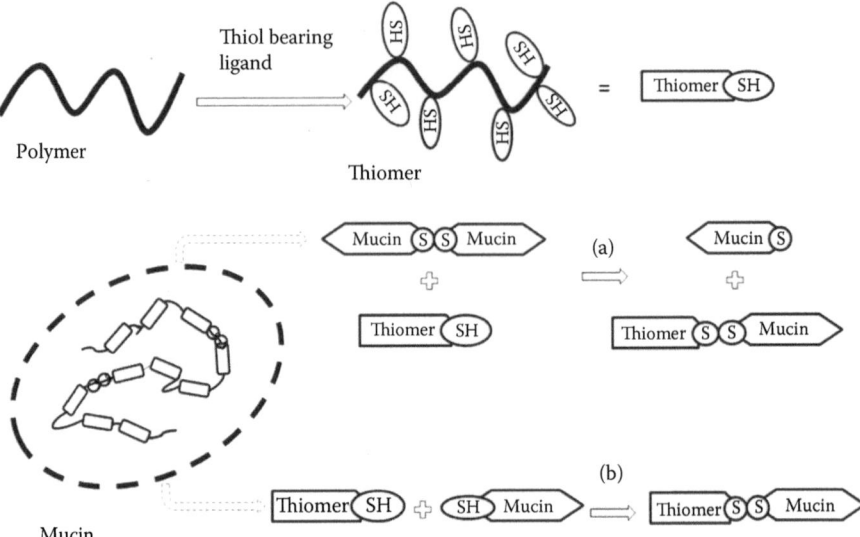

FIGURE 6.7 Structural hierarchy of a mucin from the gastrointestinal tract and hypothetical scheme representing the formation of covalent bonds between thiolated polymers and mucin glycoproteins (a) via thiol/disulfide exchange reaction and (b) via an oxidation process.

cysteine-rich subunits of mucus glycoproteins (Leitner et al. 2003a) (Figure 6.7). Such bonding is affirmed by the addition of the disulfide bond breaker dithiothreitol to already immobilized mucin–thiomer complex, in which the thiolated polymer could be completely removed from the complex.

Thiolated polymers display *in situ* gelling properties owing to the oxidation of thiol groups at physiological pH values, which results in the formation of inter- and intramolecular disulfide bonds. Such disulfide bond formation within the thiomer itself leads to additional anchors, via chaining up with the mucus gel layer (Figure 6.8).

Rheological investigations of Ch–TGA (120, 209, and 439 groups per gram polymer) showed that the *in situ* sol–gel transition, with highly cross-linked gel formation, was complete at pH 5.5

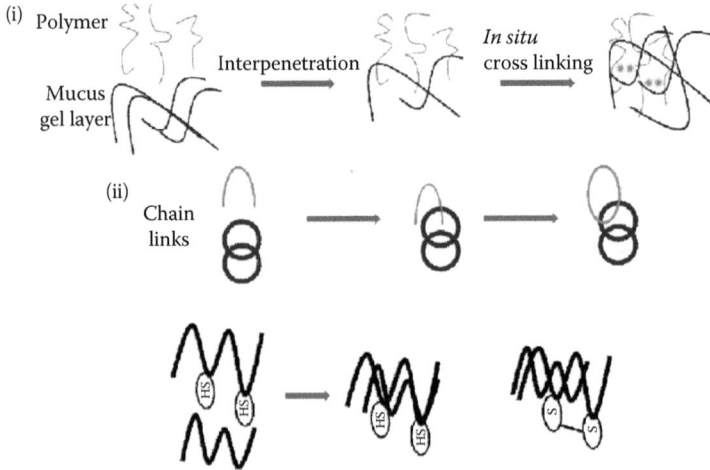

FIGURE 6.8 Schematic presentation of improved mucoadhesion by an *in situ* cross-linking with interpenetration and chain links. (i) Interpenetration. (ii) Formation of intra- and intermolecular disulfide bonds between thiol groups.

after 6 h. In parallel, a significant decrease in the thiol group content of the polymers was observed, indicating the formation of disulfide bonds (Hornof et al. 2003). The rheological properties of unmodified chitosan remained constant during the whole observation period. A clear correlation was demonstrated between the total amount of polymer-linked thiol groups and the increase in elasticity of the formed gel. When more thiol groups were immobilized on chitosan, the higher was the increase in elastic modulus G' in solutions of thiolated chitosan. The increase was 7-, 32-, and 168-fold M thiol groups per gram of polymer, respectively. Swelling properties of Ch–TGA at pH 6.8 were limited, displaying only 5% increment in weight after 2 h of experiment (Schmitz et al. 2008).

It could be anticipated that thiomers with higher amount of immobilized thiol groups should have superior increase in mucoadhesive properties. The anticipation was confirmed by various studies. For instance, the increase in mucoadhesion in comparison to unmodified chitosan obtained was 5–10-fold with Ch–TGA, 3–9-fold with Ch–TEA, and 8-fold with Ch–NAC conjugates (Kast and Bernkop-Schnürch 2001, Hornof et al. 2003, Kafedjiiski et al. 2005a). It has been shown that Ch–TBA conjugates led to a 140- and 42-fold improvement in mucoadhesion compared to the unmodified polymer and Ch–TGA conjugates (Kast and Bernkop-Schnürch 2001, Bernkop-Schnürch et al. 2003).

These promising results of Ch–TBA can be explained by higher cohesive forces and the presence of amidine moieties. After less than 2 h, 1.5% w/v Ch–TBA conjugate solutions of pH 5.5 formed covalently cross-linked gels. The increase in viscosity was more than 100-fold in comparison to unmodified chitosan (Bernkop-Schnürch et al. 2003). Amidine moieties provide additional cationic character to the polymer that very likely leads to the formation of intensified ionic interactions with the anionic moieties of sialic acid and sulfonic acid within the mucus layer. It was observed by Snyder et al. that thiol groups with cationic neighboring moieties react more rapidly with disulfide bonds having anionic substructures as neighbors (Snyder et al. 1983). According to this observation, thiol/disulfide exchange reactions between a thiomer and disulfide bonds within the mucus glycoproteins, often having neighboring asparaginic and glutamic acid substructures, will take place to a greater extent in the case of Ch–TBA conjugates than in the case of Ch–TGA conjugates. This theory was supported by the results of tensile studies (total work of adhesion and maximum detachment force) and mucoadhesive studies (rotating cylinder method) *in vitro* (Table 6.1).

TABLE 6.1
Comparison of Mucoadhesive Properties Using a Rotating Cylinder Method[a]

Conjugate	Degree of Modification Free–SH Groups (μM/g)	Time (h)	Improvement Ratio	Reference
Ch-TGA	9	1	0.88	(Kast and Bernkop-Schnürch
	27.4	4	5	2001)
Ch-TEA	27	4	14	(Hornof et al. 2003)
Ch-TEA	225	31	13	(Hornof et al. 2003)
Ch-NAC	60	5	2.25	(Schmitz et al. 2008)
	320	>100	50	
Ch-GSH	265.5	168	>55	(Kafedjiiski et al. 2005)
Ch-TBA	60	148 ± 25	123	(Bernkop-Schnürch et al.
	95	>168	>140	2003)
Ch-TBA	264	>360	>277	(Roldo et al. 2004)

[a] Test disks consisting of indicated conjugates were attached to excised porcine mucosa, spanned on a cylinder (diameter 4.4 cm, height 5.1 cm) and agitated at 125 rpm in 100 mM phosphate buffered saline pH 6.0 or 6.8 at $37 \pm 0.5°C$. The indicated time of adhesion represents the mean (±S.D.) of at least three experiments. The improvement ratio is calculated by adhesion time of conjugates versus adhesion time of controls.

Mucoadhesion takes place only when thiol groups come into direct contact with the mucus gel layer via interdiffusion and are activated by a pH shift to pH 5–7 on the surface and inside the mucus. The concentration of reactive SH group can be maintained high with proper pH of the thiomer. The pH of the thiolated chitosan, providing the highest mucoadhesive properties, was identified to be pH 3 for Ch–TBA (Bernkop-Schnürch et al. 2004). The lower the pH of the thiomer, the less reactive are the thiol groups; therefore, an oxidation of thiol groups, before getting into contact with the mucus gel layer, could be avoided.

6.4.2 Cohesive Properties (*In Situ* Gelling Properties)

The rheological studies of thiolated chitosans confirm pH-dependent sol–gel transition ability of thiolated chitosan. The inter- and intramolecular disulfide bond formation between thiol functions, on the chitosan backbone, imparts such rheological transition and cohesive properties to the polymer. This can guarantee a limited clearance and prolonged residence on the mucosal layer.

The rheological studies of Ch–TEA ($300.7 \pm 27.4 \ \mu M$ thiol groups per gram polymer) was carried out with 0.5% and 1% polymer solutions adjusted to pH 6.5 to simulate a physiological pH level (Krauland et al. 2005). Both 0.5% and 1% Ch–TEA solutions showed the transition from sol to gel within 30 min. Within 6 h of incubation, the storage modulus of 0.5% and 1% Ch–TEA increased 3354- and 6199-fold, whereas the loss modulus increased 11- and 38-fold, respectively. Frequency sweep measurements demonstrated an increase in cross-linking of the thiolated polymer as a function of time. The formation of inter- and/or intramolecular disulfide bonds was monitored indirectly via determining the decrease of thiol groups. Unmodified chitosan did not exhibit *in situ* gelling properties. The release of a fluorescent marker being incorporated in a 0.5% Ch–TEA solution was significantly slower ($p < 0.001$), when the formulation was preincubated for 1 h and consequently already highly cross-linked. The molecular mass (chain length) of the polymer had a great impact on their cohesive and mucoadhesive properties. The disulfide cross-linking can be catalyzed by the addition of oxidizing agents, such as H_2O_2, sodium periodate, ammonium persulfate, and sodium hypochlorite. One percent (m/v) Ch–TGA ($1053 \pm 44 \ \mu M$ thiol groups per gram of polymer) without any oxidizing agents became gel within 40 min. In contrast, when the oxidizing agents, hydrogen peroxide, sodium periodate, ammonium persulfate, and sodium hypochlorite, were added, respectively, gelation took place within a few minutes. H_2O_2, in a final concentration of 25.2 nmol/L increased the dynamic viscosity of 1% (m/v) Ch–TGA up to 16,500-fold within 20 min (Sakloetsakun et al. 2009).

The tensile strength and mucoadhesive performance of thiolated chitosan (Ch–TBA conjugate) with molecular masses as 150, 400, and 600 kDa displayed the relatively highest mucoadhesive properties for the medium molecular mass thiomer. Use of 400 kDa Ch–TBA conjugate of 264 M of thiol groups per gram of polymer led to a more than 100-fold improvement in mucoadhesion as to unmodified chitosan (Krauland et al. 2005). A reason for this observation might be, on the one hand, the insufficient cohesive properties of low-molecular-mass chitosan leading to a breakup in the adhesive bond within the polymeric network itself rather than between the polymer and the mucus gel layer. On the other hand, if the polymer chains are too long, the extent of interpenetration, which is essential for high mucoadhesive properties, is strongly reduced (Imam et al. 2003).

The observation was explained on the basis of the swelling behavior of the conjugates with the same average amount of immobilized thiol groups but different molecular weights: the maximum swelling with water uptake for the Ch–TBA (MW 400 kDa, 264 mM thiol groups per gram polymer) was reached after 2 h, whereas conjugates of low and high MW attained it after 20 min. The longer the polymer chain length, the higher was the maximum increase in weight on water uptake (9.66-, 10.68-, and 14.17-fold for low, medium, and high MW, respectively) for the corresponding tablets during an observation time of 2 h. Generally, for the higher water uptake, the mucoadhesive properties were weaker. The slow swelling too is a requisite to avoid the formation of an overhydrated form that loses its cohesive and mucoadhesive properties notably in the intestine (Figure 6.9).

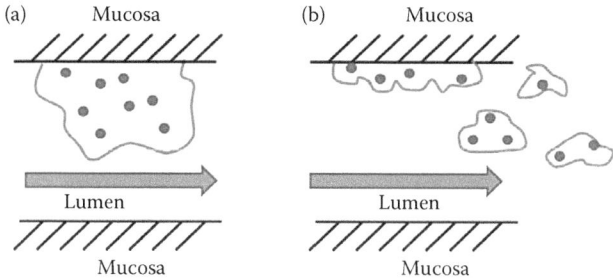

FIGURE 6.9 Importance of cohesiveness in mucoadhesive drug delivery system. (a) Drug delivery system with mucoadhesive and cohesive properties where the drug remains attached to mucosa. (b) Absence of cohesive properties does not allow residence in spite of mucoadhesion.

6.4.3 PERMEATION-ENHANCING PROPERTIES

The chitosan and its salt as glutamate hydrochloride have been reported widely as a permeation enhancer for small and large polar molecules [^{14}C]-mannitol (Artursson et al. 1994, Schipper et al. 1996, Kotze et al. 1998), fluorescently labeled dextran MW 4400 Da (FD4) (Borchard et al. 1996) as well as peptide and protein as insulin (Illum et al. 1994, Aspden et al. 1996), 9-desglycinamide, 8-L-arginine vasopressin (DGAVP) (Lueßen et al. 1997), buserelin (Lueßen et al. 1996), and octreotide acetate (Thanou et al. 2001) across Caco-2 cell monolayer or nasal, duodenal mucosa. In all these studies, absorption enhancement was found only in acidic environments in which the pH was less or of the order of the pK_a value of chitosan (5.5–6.5). Moreover, the absorption-enhancing effects of chitosan were dependent on the concentration administered. But an unlimited increase in concentration did not lead to an unlimited increase in absorption enhancement. This pH-dependent and saturable effect suggests a mechanism of absorption enhancement, namely, paracellular absorption enhancement, via opening of tight junctions (TJs). The opening of TJs by chitosan was attributed to the interaction of a positively charged amino group on the C-2 position of chitosan with negatively charged sites on the cell membranes and TJs of the epithelial cell membrane (Artursson et al. 1994).

To understand the influence of TJs on the paracellular absorption, it is useful to be aware of the molecular architecture of TJs regulating and/or influencing the gate fence function.

The structural components of the TJ and the possible modulating factors are shown in Figure 6.10. The epithelial cells on the apical surface are closely connected by intracellular junctions, the specialized sites and structural components of which are commonly known as the junctional complex. Each junctional complex is composed of three regions: TJs or zonula occludens (ZO) (near the apical surface), zonula adherens (belt desmosomes), and macula adherens (spot desmosomes and hemidesmosomes). The TJ comprises a group of integral membrane proteins: junctional adhesion molecule (JAM), claudin, and occludin. Belt desmosomes form a continuous band around each of the cells in an epithelial sheet typically just below the TJ. The spot desmosomes and hemidesmosomes serve as anchoring sites for actin filaments, which extend from one side of the cell to the other across the cell interior. Furthermore, a dense cytoplasmic network of proteins, referred to as tight junction-associated proteins (TJAPs) and designated ZO-1, ZO-2, and ZO-3, are present at TJ. As can be seen in Figure 6.10, these TJAPs interact among each other and also serve as a link between occludin and the actin filaments of the cytoskeleton. Hemidesmosomes resemble spot desmosomes but instead of joining adjacent epithelial cell membranes together they join the basal surface of epithelial cell to the underlying basal lamina to distribute the shearing forces. The association of TJs with the apical perijunctional actomyosin ring seems to regulate global TJ permeability. These complexes create a semipermeable diffusion barrier that can be regulated between cells.

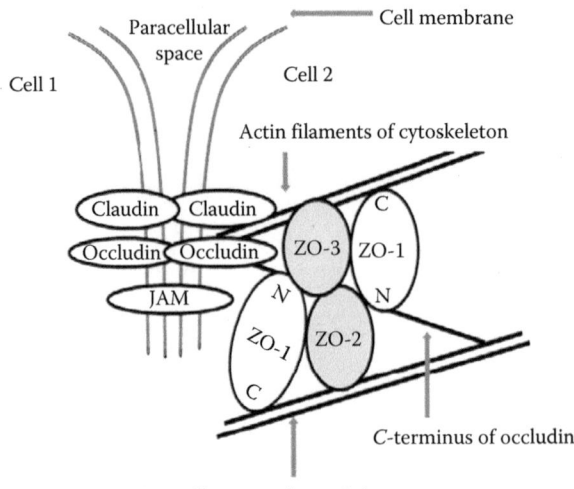

FIGURE 6.10   The molecular architecture of tight junction.

The transmembrane protein, JAM, is immunoglobulin-like in form. The influence of JAM on TJ integrity is not completely understood but there is strong evidence to support the fact that it may play an additional role in cell–cell adhesion (Martin-Padura et al. 1998). The originally identified proteins of TJs are claudin-1 and claudin-2, with a molecular mass of 22–24 kDa, expressing two extracellular loops (Furuse et al. 1998). Currently, 20 claudins, present in different tissues, can be identified in the GenBank database. At least some of the claudins are able to mediate cell adhesion in a Ca^{2+}-independent manner (Kubota et al. 1999). Their function is seen in conferring selectivity to paracellular transport (Simon et al. 1999). The third transmembrane protein, occludin, is a 60–65-kDa protein that was shown to express two extracellular loops from amino acid 81–124 and 184–227. These loops express several tyrosine and glycine residues, as shown in Figure 6.11, and are believed to provide the cohesiveness of the junction barrier. Tyrosine residues are phosphorylated by protein tyrosine kinases, resulting in increased permeability of TJ (Collares-Buzato et al. 1998). Protein tyrosine phosphatases (PTP), however, are able to dephosphorylate these groups, and result in a closing of TJ.

As PTP catalyzing dephosphorylation of occludin bears a cysteine moiety at an active site (Cys 215), GSH was shown to be capable of inhibiting PTP activity by almost 100%, via a disulfide bond formation, with this cysteine substructure, within minutes (Barrett et al. 1999). Being aware of this

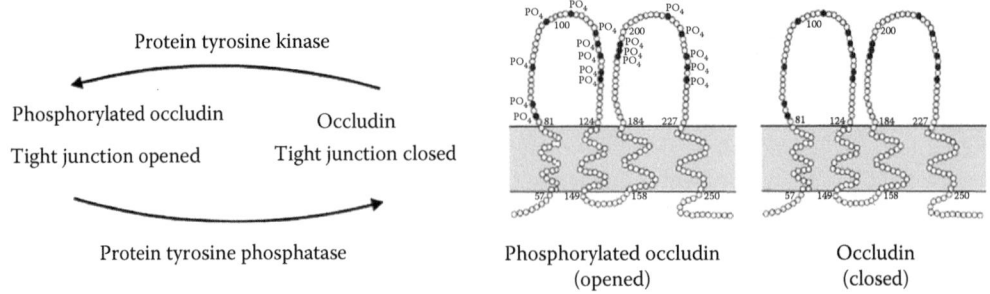

FIGURE 6.11   Schematic presentation of enzymes controlling the degree of phosphorylation of the tyrosine subunits of occludin.

high inhibitory effect of GSH toward PTP, high concentrations of GSH should lead to more open TJ. So far, however, it was only demonstrated that GSH is involved in a H_2O_2-mediated increase in paracellular permeability in Caco-2 cell monolayers (Rao et al. 2000), and recent studies showed that the permeation-enhancing effect of GSH on freshly excised intestinal mucosa is not convincing (Clausen et al. 2002). An explanation for these observations might be given by the rapid oxidation of GSH on the cell surface (Grafstrom et al. 1980), whereas other stable PTP inhibitors, such as phenylarsine oxide or pervanadate, were shown to strongly increase TJ permeability (Staddon et al. 1995). As the enzymes and divalent cations are thought to be required for the regulation of junction permeability, the modulation of enzyme action and agents that chelate divalent cations can be employed for improving permeability.

The understanding of molecular architecture of TJ and confocal laser scanning microscopy confirmed that chitosans are able to bind tightly to the epithelium and induce redistribution of cytoskeletal F-actin and the protein ZO-1, increasing the permeability for enhanced transport, via the paracellular pathway (Schipper et al. 1997). This was further confirmed by Dodane et al. using chitosan hydrochloride (DDA 80%) (Dodane et al. 1999). Along with redistribution of occludin and ZO-1 proteins, they observed the slight perturbation of the plasma membrane, indicating an increased intracellular uptake. Smith et al. using Western blotting of Caco-2 cell fractions observed that ZO-1 protein translocates from the membrane to the cytoskeleton in response to treatment with chitosan DDA 85% (Smith et al. 2004). This tendency was also observed for occludin. Globally, it can be concluded that chitosan is able to enhance the paracellular route of absorption by TJ.

It should be noted that although Caco-2 cell monolayers represent a very good *in vitro* model of intestinal absorption, they lack a mucus barrier presented by gastrointestinal and other epithelial cells *in vivo*. To address this issue, Schipper et al. investigated chitosan-enhancing effects on mannitol permeability using mucus-producing cell line (HT29-H) (Schipper et al. 1999). It was shown that the presence of an intact mucus layer reduces the absorption-enhancing effect of chitosan hydrochloride as the polymer cannot reach the epithelium because of size-limited diffusion and/or competitive charge interactions with mucin. To overcome this important limitation, the authors suggested increasing the effective concentration of the enhancer.

The thiolated chitosan with boosted mucoadhesive activity may prove to be a better permeation enhancer because of extended residence time. The permeation of paracellular markers through intestinal mucosa was enhanced several fold utilizing thiolated chitosan instead of an unmodified one (Table 6.2). The uptake of fluorescence labeled bacitracin, for instance, was improved 1.6-fold utilizing 0.5% of chitosan–cysteine conjugate instead of unmodified chitosan (Bernkop-Schnürch et al. 1999). The uptake of the cationic marker compound rhodamine 123 was 2.6-fold higher in the presence of Ch–TBA versus unmodified chitosans (Föger et al. 2006). The likely mechanism of increase in permeation is opening of TJ by GSH-mediated PTP inhibition. The thiomers are shown to be capable of reducing oxidized glutathione (GSSG) to GSH and oxidizing itself by disulfide bond formation (Clausen et al. 2002). The inhibitory effect of glutathione is limited as it is rapidly oxidized on the cell surface, losing its inhibitory activity. Owing to the combination of GSH with thiolated chitosans, however, the effect of oxidation of the inhibitor on the membrane can be restricted. With added GSH, there would be shifting of balance between GSSG and GSH on the membrane to the side of GSH and a comparatively high availability of GSH on the membrane should, in turn, facilitate TJ opening. The permeation studies with hydrophilic model drugs across intestinal mucosa demonstrated significant improvement in drug uptake with thiomer/GSH system (Bernkop-Schnürch et al. 2003a). The disulfide bond formation in thiomer, while reducing GSSG to GSH, might lead to interpenetration, cross-linking, and promotion of mucoadhesion. As the sojourn of thiomer is extended at the site of drug absorption, the glutathione concentration on the membrane should be raised for that period as well. The postulated glutathione regeneration mechanism is illustrated in Figure 6.12.

This theory is supported by various *in vitro* and *in vivo* studies, where significantly improved permeation was attained. Compared to the buffer, the permeation of tobramycin sulfate, a highly

TABLE 6.2

Permeation-Enhancing Properties of Thiolated Chitosans in Comparison with the Unmodified Polymers Tested on Freshly Excised Intestinal Mucosa

Permeation Enhancer	Test Compound	Enhancement Ratio[a]	Reference
Ch–TBA	Rhodamine-123	2.0	Bernkop-Schnürch et al. (2004)
Ch–TBA/GSH	Rhodamine-123	3.6	Bernkop-Schnürch et al. (2003)
Ch–TEA/GSH	Rhodamine-123	3.1	Kafedjiiski et al. (2006)
Ch–GSH	Rhodamine-123	3.1	Kafedjiiski et al. (2005)
Ch–GSH/GSH	Rhodamine-123	4.9	Kafedjiiski et al. (2005a)
Ch–TBA	Rhodamine-123	2.8	Föger et al. (2006)
Ch–TBA/GSH	Rhodamine-123	3.6	Föger et al. (2006)
Ch–NAC 400 kDa	Rhodamine-123	2.0	Schmitz et al. (2008a)
Ch–NAC 400 kDa	FD4	2.5	Föger et al. (2006)
Ch–6MNA	Rhodamine-123	2.12	Sakloetsakun et al. (2011)
Ch–Cys	FD4	3.42	Sakloetsakun et al. (2011)

[a] Enhancement ratio is the apparent permeation coefficient (P_{app}) of the test compound in the presence of the thiolated chitosan, divided by the P_{app} in the presence of unmodified chitosan.

cationic drug with poor bioavialbilty, using Ch–NAC/GSH was increased 2- and 3.3-fold against rat intestinal mucosa and Caco-2 monolayers, respectively (Hombach et al. 2008).

Permeation of cefadroxil, the drug poorly absorbed from the GI tract, was increased three-fold in tablets comprising 10% cefadroxil, 10% GSH, and 80% Ch–TBA conjugate at pH 3 (Bernkop-Schnürch et al. 2004). Microparticles prepared via the emulsification solvent evaporation technique using Ch–TBA conjugate (304.89 ± 63.45 µM thiol groups per gram polymer)/ GSH system, entrapping fluorescein isothiocyanate (FITC)-labeled insulin showed swelling of 4.39 ± 0.52-fold in size, controlled release of labeled insulin over 6 h, and an absolute bioavailability of 2.04 ± 1.33% (against 1.04 ± 0.27% shown by mannitol–insulin microparticles, respectively) (Krauland et al. 2006).

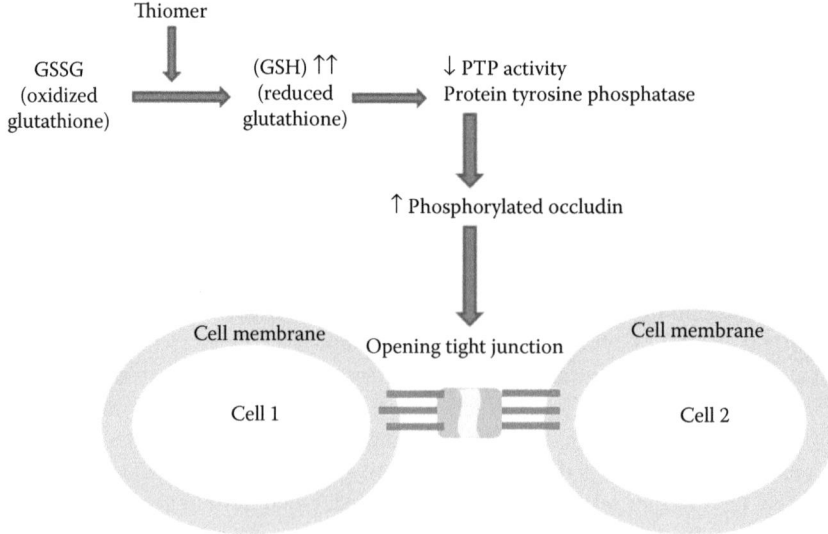

FIGURE 6.12   Postulated effect of thiomer/GSH system on intestinal paracellular permeability.

6.4.4 ENZYME AND EFFLUX PROTEIN INHIBITION PROPERTIES

The other mechanisms proposed for the enhancement of permeation are inhibition of actions of efflux protein as P-glycoprotein (P-gp) and enzymes. P-gp, a 170-kDa transmembrane protein polarized to the apical membrane of mucosal layer, transports the drugs back to the apical side of mucosal cells via an ATP-dependent process. Overall, P-gp plays a major physiological role as a barrier for the entry of xenobiotics, as well as a mechanism to eliminate xenobiotics from systemic circulation. Al-Shawi and coworkers reported that sulfhydryl-substituted purines gave substantial inhibition of P-gp ATPase activity, which was dithiothreitol-reversible, suggesting a covalent reaction with one or more cysteine residues of P-gp. Hence, the effect of Ch–TBA/GSH system on P-gp activity was tested (Al-Shawi et al. 1994). For this, permeation studies were conducted with freshly excised guinea pig ileum mounted in Ussing chambers using the fluorescent dye rhodamine-123 (Rho-123) as specific P-gp substrate. Apparent permeability coefficients (P_{app}) as well as efflux ratios (secretory P_{app}/absorptive P_{app}) were calculated and compared with values gained from experiments with the well-established P-gp inhibitors terfenadine and verapamil. In the presence of terfenadine, verapamil, as well as GSH, the absorptive transportation of Rho-123 across intestinal tissue increases, whereas the secretory transport decreases with efflux ratios around 1.0. Ch–TBA with ~680 µM thiol groups per g polymer and especially Ch–TBA/GSH, not only enhances absorption of Rho-123 but also reduces the basolateral to apical secretion of Rho-123, resulting in efflux ratios of 1.1, 0.8, and 0.5. The study indicates that Ch–TBA/GSH is a potentially valuable tool for inhibiting the ATPase activity of P-gp in the intestine (Werle and Hoffer 2011). Parallel studies with the P-gp substrate saquinavir for transportation across Caco-2 monolayer and rat intestinal mucosa gave the efflux ratios of 6.4 and 2.1, respectively. In the presence of 0.5% Ch–TBA/0.5% GSH, the uptake of saquinavir was 1.6-fold improved in Caco-2 monolayer and 2.1-fold improved in rat intestinal mucosa (Föger et al. 2007).

To provide a proof of principle for a delivery system based on thiolated chitosan *in vivo*, in rats, using Rho-123, was studied. *In vitro*, in comparison to buffer, Rho-123 transport in the presence of 0.5% chitosan, 0.5% Ch–TBA, and 0.5% Ch–TBA/0.5% GSH was 1.8-, 2.6-, and 3.8-fold, respectively. The enteric-coated tablets based on Ch–TBA/GSH, *in vivo* in rats, increased the area under the plasma concentration time curve (AUC_{0-12}) of Rho-123 by 217%, in comparison to buffer control and by 58% in comparison to unmodified chitosan (Föger et al. 2006). *In vivo* in rats, the AUC of saquinavir was increased 1.4-fold and C_{max} 1.6-fold, in comparison with control (Föger et al. 2007).

The comparisons of permeation of Rho-123 was made FD4 (a known trans- and paracellular marker) both in rat intestine and Caco-2 monolayers, using Ch–NAC of MW 150, 400, 600 kDa, and 260–380 µM of thiol group per gram polymer. The highest permeation of Rho-123 was mediated by Ch–NAC of MW 150 kDa going along with the logic of cohesiveness. The highest permeation of FD4 was mediated by Ch–NAC of MW 400, 600 kDa, whereas 150 kDa Ch–NAC seemed to impede the transport of FD4 (Schmitz et al. 2008a). The molecular-weight-dependent permeation enhancement by Ch–TBA/GSH system was seen for acyclovir too (Palmberger et al. 2008).

The improved permeation enhancement, and P-gp-mediated efflux inhibition and effects were seen with Ch–TGA where more than 50% of thiol group was covalently linked by disulfide linkage with 6-mercaptonicotinamide (called as S-protected Ch–TGA; Figure 6.13). The protection of immobilized thiol groups on the polymeric backbone eliminates the risk of an early thiol oxidation and allows easy disulfide exchange reactions through cysteine subunits within the mucus (Dünnhaupt et al. 2012). The use of S-protected Ch–TGA also led to improved absorptive transport of peptide drug, antide, by the oral route (Dünnhaupt et al. 2012a).

Zinc-dependent proteases, such as aminopeptidases and carboxypeptidases, are inhibited by thiomers (Bernkop-Schnürch and Kast 2001). The underlying mechanism is based on the capability of thiomers to bind zinc ions. The enzyme inhibitory effects could be because of high GSH

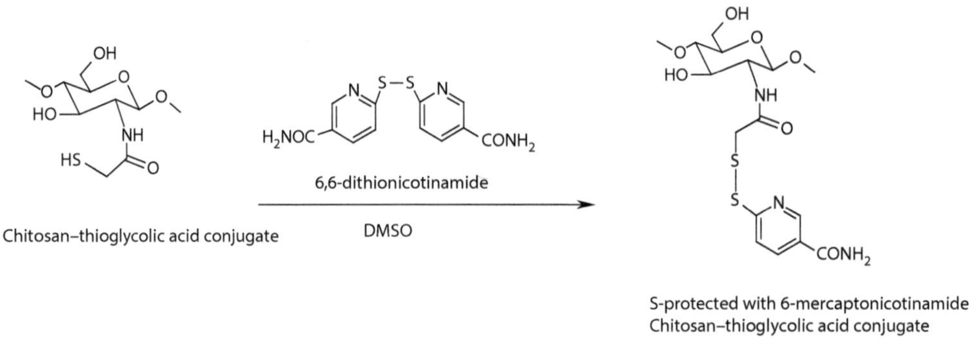

FIGURE 6.13 Synthesis of S-protected Ch–TGA with 6-mercaptonicotinamide.

concentration maintained by thiomers and deprivation of zinc ions by its complex formation with GSH (Dominey and Kustin 1983). This inhibitory effect seems to be highly beneficial for the oral administration of peptide and protein drugs.

6.5 APPLICATIONS

The usefulness of thiolated chitosans as carrier matrices for controlled drug release has been demonstrated for drugs and model drugs.

6.5.1 Noninvasive Peptide Delivery

Noninvasive peptide delivery demands nonionic or cationic polymeric matrix. Anionic thiomers could not be used for such cationic peptides because of the interaction of opposite charges lessening the mucoadhesive properties of polymer and inhibition of drug release from the matrix. Nonionic polymers do not possess sufficient mucoadhesive properties. The thiolated chitosans, with residual amino groups, get a cationic nature and accrued –SH groups impart mucoadhesive, cohesive properties to them. Therefore, thiolated chitosans were chosen for evaluation of peptide delivery by oral, nasal, or buccal route.

A stomach-targeted oral delivery system of Ch–TBA for the peptide drug antide in pigs, displayed absolute and relative bioavailability of 1.1% and 3.2%, respectively (Bernkop-Schnürch et al. 2005). Nasal insulin microparticulate delivery system with Ch–TBA/GSH was evaluated *in vivo* for insulin delivery. The microparticles were prepared by novel precipitation–micronization method with Ch–TBA (304.9 ± 63.5 µM thiol groups per gram polymer) and loaded with insulin. Owing to a hydration process, the size of Ch–TBA/insulin and chitosan/insulin microparticles increased in pH 6.8 phosphate buffer by 2.6- and 2.2-fold, respectively. Fluorescent-labeled insulin-loaded Ch–TBA microparticles showed a controlled release over 4 h. Ch–TBA/insulin administered nasally to rats led to an absolute bioavailability of $6.9 \pm 1.5\%$. The blood glucose level decreased for more than 2 h and the calculated absolute pharmacological efficacy was $4.9 \pm 1.4\%$. Chitosan/insulin, in comparison, displayed a bioavailability of $4.2 \pm 1.8\%$ and a pharmacological efficacy of $0.7 \pm 0.6\%$. Mannitol/insulins showed a bioavailability of $1.6 \pm 0.4\%$ and no reduction of the blood glucose level at all (Krauland et al. 2006a). The additional protection to peptide drug was provided by the coadministration of enzyme inhibitors for their passage by the oral route. Ch–TBA (400 kDa, 118.5 ± 14.3 µM thiol groups per gram polymer) was homogenized with salmon calcitonin, mannitol, chitosan–Bowman–Birk inhibitor conjugate, chitosan–elastatinal conjugate (in mg $6.75 + 0.25 + 1 + 1 + 1$) with and without 0.5% GSH and compressed to 2 mg microtablets enteric coated with a polymethacrylate (Guggi et al. 2003). Biofeedback studies were performed in rats by oral administration of the delivery system and determination of the decrease in plasma calcium level as a function of time. Test formulations led to a significant ($p < 0.005$) decrease in the plasma calcium level of the dosed animals in comparison to control tablets, being based on unmodified chitosan. The addition of glutathione in the tablets led to a further improvement in the oral bioavailability of salmon calcitonin with an earlier onset of action and a decrease in the calcium level of about 10% for at least 10 h. The *in vivo* effect, in part, was attributed to the presence of chitosan–enzyme-inhibitor conjugates to assure protective effect for salmon calcitonin against the most abundant intestinal proteases. The chitosan–pepstatin A conjugate can be employed as enzyme-inhibitor conjugate (Guggi et al. 2003a). The similar testing with tablets of Ch–TBA conjugate ($453.5 \pm 64.1 \, \mu M$ thiol groups per gram polymer), insulin, GSH, and chitosan–Bowman–Birk inhibitor conjugate, chitosan–elastatinal conjugates (in mg $5 + 2.75 + 0.75 + 0.75 + 0.75$), showed a controlled release of insulin over 8 h (Krauland et al. 2004). *In vitro* mucoadhesion studies showed that the mucoadhesive/cohesive properties of chitosan were at least 60-fold improved by the immobilization of thiol groups on the polymer. After oral administration of Ch–TBA–insulin tablets to nondiabetic conscious rats, the blood glucose level decreased significantly for 24 h corresponding to a pharmacological efficacy of $1.69 \pm 0.42\%$ v/s subcutaneous injection. In contrast, neither control tablets nor insulin given in solution showed a comparable effect.

The Ch–GSH conjugate provides a chemical structure, pK_a (8.7) and the rate of solubility, which is favorable for the formation of sufficient concentration of thiolate anions in the physiological medium to provide stronger mucoadhesion. A 7.5-fold higher mucoadhesion could be achieved from ethyl cellulose and Eudragit L100 coated tablets with Ch–GSH as a matrix layer in comparison with control-coated tablets with chitosan as matrix layer. The coated tablet of Ch–GSH conjugate remained attached to the mucosa even after 180 h of incubation. In contrast, the corresponding control was made of chitosan detached from mucosa within 24 h. After water-uptake studies revealed that, the weight of patch systems with Ch–GSH increased approximately 44.5 ± 2.3 mg (127%) after 90 min. This patch system remained even after 180 h on the mucosa and released $49.7 \pm 0.7\%$ of FD4 within 8 h and a 2.5-fold higher transport of FD4 was obtained in contrast to control (Hoyer et al. 2007).

Similar tablet formulations for an antidiabetic peptide, pituitary adenylate cyclase activating polypeptide (PACAP) with matrix of Ch–TBA, GSH, and Brij35 coated by Ch–TBA with or without additional coating of palm wax on all but one side was evaluated for peptide delivery by buccal mucoadhesive path (Langoth et al. 2006). The incorporation of cationic polypeptide PACAP into cationic thiolated chitosan did not affect the mucoadhesive properties of the polymer. The mucoadhesion of the tablets was very strong and could be removed after 6 h from the mucosa with moderate force. The tablet with unmodified chitosan detached from the mucosa after 4 h of application. Bioavailability studies were performed in pigs by buccal administration of the test formulations and compared to formulation with unmodified chitosan. A bioavailability of about 1% was reached with Ch–TBA/GSH, whereas no PACAP was detected in plasma with the use of unmodified chitosan. The buccal route bypassed hepatic first-pass metabolism and degradation in gastrointestinal tract (GIT).

In an attempt to enhance water solubility and assist the mucoadhesive and permeation-enhancing properties of chitosan, its derivative glycol chitosan was thiolated with thioglycolic acid to obtain glycol Ch–TGA (Makhlof et al. 2010). Nanoparticles obtained with such thiomer by the ionic gelation method, demonstrated a particle size in the range of 0.23–0.33 nm with positive surface charge and high calcitonin entrapment efficiency. The increase in mucoadhesion to lung tissue after intratracheal administration to rats was of twofold as compared to nanoparticles of nonthiolated chitosan. The *in vivo* efficacy was also fostered. Calcitonin-loaded glycol chitosan and glycol Ch–TGA nanoparticles resulted in a pronounced hypocalcemic effect for at least 12 and 24 h, and a corresponding pharmacological availability of 27% and 40%, respectively.

Trimethyl chitosan–cysteine conjugate (TMC–Cys) was synthesized in an attempt to combine the permeation-enhancing effects and the mucoadhesion of TMC with thiolated polymers related to different mechanisms for oral absorption. TMC–Cys, with various molecular weights (30, 200, and 500 kDa) and quaternization degrees (15% and 30%), was allowed to form polyelectrolyte nanoparticles with insulin through self-assembly, which demonstrated a particle size of 100–200 nm, a zeta potential of +12 to +18 mV, and a high encapsulation efficiency. TMC–Cys/insulin nanoparticles (TMC–Cys NP) showed a 2.1–4.7-fold increase in mucoadhesion, compared to TMC/insulin nanoparticles (TMC NP), which might be partly attributed to the formation of disulfide between TMC–Cys and mucin as evidenced by DSC measurement. Compared to insulin solution and TMC NP, TMC–Cys NP induced increased insulin transport through rat intestine by 3.3–11.7- and 1.7–2.6-folds, promoted Caco-2 cell internalization by 7.5–12.7- and 1.7–3.0-folds, and augmented uptake in Peyer's patches by 14.7–20.9- and 1.7–5.0-folds, respectively. Such results were further confirmed by *in vivo* experiments with the optimal TMC–Cys NP. Biocompatibility assessment revealed lack of toxicity of TMC–Cys NP. Therefore, self-assembled nanoparticles between TMC–Cys and protein drugs could be an effective and safe oral delivery system (Yin et al. 2009).

The synergism of quaternary ammonium and thiol groups to improve the intestinal drug absorption enhancing properties of the multifunctional chitosan derivatives was demonstrated with N-(3-mercaptopropionoyl, N,O-[N,N-diethylaminomethyl(diethyl-dimethyleneammonium)$_n$-methyl] chitosan derivatives for FD4 and dexamethasone across rat intestinal mucosa and Caco-2 cell layer (Zambito et al. 2009). The *in vivo* experiments displayed enhanced precorneal retention, transcorneal permeation, and intraocular absorption of dexamethasone (Zambito and DiColo 2010). The synergism of quaternary ammonium and thiol groups of the derivative, which was found to be at the basis of the interaction of this polymer with the intestinal epithelium, has been shown to also characterize its interaction with the corneal epithelium. The quaternary ammonium ions were responsible for both permeabilization of corneal epithelium and polymer adhesion to precorneal mucus while the thiols increased the latter.

The thilolated TMC obtained as a result of reaction of cystamine with partially N-carboxymethylated TMC showed a slight reduction in cytotoxicity for thiolated TMCs as compared to the nonthiolated polymers with similar DQs in Calu-3 cells. The fluorescently labeled ovalbumin-loaded nanoparticles prepared with thiolated polymers were positively charged and were stable in 0.8 M NaCl in contrast to particles made from nonthiolated polymers that dissociated under similar conditions. This demonstrated the presence of intermolecular disulfide bonds within particles to hold them together (Verheul et al. 2010).

The thiolated chitosan was employed to protect the basic fibroblast growth factor (bFGF) during its delivery (Ho et al. 2010). The thiomer utilized was 4-thio-butylamidine conjugated 2-N, 3,6-O-sulfated 6-O-carboxymethyl chitosan (45.9 ± 3.7 and $415.6 \pm 12.5\,\mu$mol SH/g sulfated 6-O-carboxymethylchitosan). L929 fibroblast culture tests showed that the polyelectrolyte complex formed between thiol-modified chitosan derivative and bFGF could effectively protect the protein from inactivation over a 120 h period.

6.5.2 CONTROLLED-RELEASE DRUG DELIVERY

As thiomer microparticles and nanoparticles were shown to exhibit the same features as thiolated polymers per se, they might be useful tools for the delivery of various types of challenging drugs. The microparticlulate and nanoparticlulate delivery systems can be generated via different techniques, such as *in situ* gelation and subsequent covalent cross-linking, radical emulsion polymerization, emulsification/solvent evaporation, or air jet milling (Hoyer et al. 2008).

Microparticles based on chitosan disintegrate very rapidly, unless they are combined with multivalent anionic compounds, such as tripolyphosphate (TPP), sodium sulfate, or alginate leading to stabilization by an ionic cross-linking process (Coppi et al. 2001, van der Lubben et al. 2001). The addition of such multivalent anionic compounds, however, decreases the mucoadhesive properties of chitosan. In contrast, microparticles based on thiolated chitosan do not disintegrate because of the formation of disulfide bonds within the polymeric network. Stable microspheres of Ch–TBA

(46.59 ± 15.04 µM thiol groups per gram polymer) for FD4 were prepared by the emulsification/ solvent evaporation method (Maculotti et al. 2005). FD4 was chosen as a model hydrophilic drug. The microparticles of size ~15 µm generated without cross-linking agents were stable with good drug loading efficiency (96.57 ± 1.34%), which retained the spherical morphology on swelling. The degradability by lysozyme appears quite similar for modified and virgin polymers, showing that chemical modification does not influence the biodegradable properties of chitosan. Microspheres were able to control the drug release for at least 1 h, exhibiting comparatively strong mucoadhesive properties. The Ch–TBA conjugate microparticles remain on the mucosa in a 2.5-fold higher concentration with respect to unmodified chitosan microparticles. The microspheres of size 1–59 µM, prepared by the same method with Ch–TBA 100 µM thiol groups per gram polymer, resulted in a controlled release rate for more than 3 h, whereas the unmodified chitosan control reached the maximum peak level of release within an hour (Imam and Bernkop-Schnürch 2005).

The particles of thiolated chitosan, Ch–TBA and Ch–TGA were developed by a simple two-step method (Bernkop-Schnürch et al. 2006, Barthelmes et al. 2011). In the first step, Ch–TBA (10 kDa, 194 ± 26 µM thiol groups per gram polymer) was ionically gelated with TPP or sulfate in aqueous solution, forming submicron particles. In the following step, thiol groups in and on the particles were partially oxidized by the addition of iodine or H_2O_2 solution, or pH change forming stabilizing inter- and intramolecular disulfide bonds. Thereafter, the polyanions were removed to obtain stable particles. (In contrast, particles did not remain stable after removing sulfate as temporary auxiliary ionic cross-linker.) As the degree of oxidation can be controlled during the production process, the share of thiol and disulfide groups can be adjusted on demand. The stable particles of a mean size of 366 ± 30 nm and a zeta potential of around +11.3 ± 1.3 mV can be produced using TPP as ionic cross-linker. On an average, 83% of all thiol groups were oxidized. Even when 91% of all thiol groups on the nanoparticles (size 268 ± 15 nm, zeta potential 9 ± 2 mV) were oxidized, their mucoadhesive properties were still twice as high as the mucoadhesive properties of unmodified nanoparticles (Bernkop-Schnürch et al. 2006a). Cross-linked particles, either ionically or covalently, were degraded by lysozyme under physiological conditions.

The beads of Ch–TLA with TPP cross-linking were prepared at pH 4 for indomethacin delivery. The indomethacin released from these beads of size range 1.5–1.75 mm were pH dependent. The release rate of indomethacin at pH 7.4 was higher than the release rate at pH 1.4 due to the ionization of thiol groups and high solubility of indomethacin in an alkaline medium (Kafedjiiski et al. 2005).

Thiolated chitosans can also help stabilize other drug delivery systems such as liposomes. Grauder et al. reported a thiolated chitosan-coated liposomal delivery system composed of 1,2-dipalmitoyl-*sn*-glycero-3-phosphocholine and a maleimide-functionalized lipid, to which Ch–TGA, S-protected version of Ch–TGA, was covalently coupled (Gradauer et al. 2013). The thiolated chitosan-coated liposomes were nonimmunogenic in mice and was shown to protect encapsulated drugs in the stomach, slowly release them in the small intestine and enhance their absorption through the intestinal tissue by opening TJs and inhibiting efflux pumps.

6.5.3   SITE-SPECIFIC DELIVERY

This pH-dependent sol–gel transition property can be exploited to obtain drug formulations of favorable viscoelastic properties *in situ*, which will stabilize themselves once applied at the site of delivery, and exhibit limited clearance with prolonged residence. The *in situ* gelling or cross-linking process can be observed within a pH range of 5–6, which makes the application of thiolated chitosans on vaginal, nasal, buccal, and ocular mucosa possible (Kast and Bernkop-Schnürch 2001, Hassan and Gallo 1990). The *in situ* gelling property was seen in the pH range of 3–6.8 with Ch–MNA (Millotti et al. 2009).

Mucoadhesive gel formulation of 1% clomiphene citrate in Ch–TGA 41.17 ± 2.34 µM of thiol groups per gram of polymer, by texture analysis, showed enhanced elasticity, cohesiveness, adhesiveness, and mucoadhesion of the gel formulations, but not hardness and compressibility when compared to gel prepared using parent formulation (Cevher et al. 2008). Tablets of Ch–TGA conjugate (280 µM thiol groups per gram polymer) matrix containing 16.67% clotrimazole meant for vaginal use, put forth improved water uptake, cohesive properties, 100-fold prolonged disintegration time, and 26 times longer residence time on vaginal mucosa (Kast et al. 2002).

The nanoparticles of Ch–TGA (17–30 µM thiol groups per gram polymer) were prepared by ionic cross-linking with sodium TPP as carrier of theophylline. The nanoparticles had a diameter of 220 ± 23 nm and a zeta potential of $+15.3 \pm 2$ mV. The antiinflammatory effects of theophylline in BALB/c mice were markedly enhanced with nasal administration of this carrier system compared to unmodified chitosan or theophylline alone apparently because of improved mucoadhesion and theophylline absorption by the bronchial epithelium (Lee et al. 2006). The enhanced bioavailability was also seen for the nasal application of leuprolide with Ch–TGA, whereas improved transnasal transport and nose-to-brain delivery of tizanidine hydrochloride was seen with nanoparticles of drug-loaded Ch–TGA (Patel et al. 2012, Shahnaz et al. 2012). During the ocular application of 0.5%, Ch–NAC in a mouse model of botulinum toxin B induced dry eyes and decreased expression of inflammatory cytokine, which hinted at a protection of the ocular surface (Hongyok et al. 2009).

6.5.4 Polysaccharide Coating of Nanoparticles

Surface modification of colloidal carriers, such as nanoparticles, confers stability and specific functionality to the system. Polysaccharides such as chitosan and derivatives can be deployed for such surface modification. The preformed poly(lactide-*co*-glycolide) (PLGA) nanoparticles were coated with chitosan and the layer was thiolated with 2-iminothiolane in an attempt to protect the encapsulated therapeutic agent from enzymatic degradation and impart mucoadhesion to nanoparticles. Thiolated nanoparticles showed a 3.3-fold prolonged residence time on the mucosa and an unchanged release profile of loaded fluorescein diacetate or curcumin in comparison to unmodified PLGA nanoparticles (Grabovac and Bernkop-Schnürch 2007).

Chauvierre et al. developed a method for the elaboration of nanoparticles composed of a poly(alkylcyanoacrylate) core based on emulsion radical polymerization of alkyl cyanoacrylate monomers from carbohydrate chains (Chauvierre et al. 2003). The method included the formation of a free radical at the end of the polysaccharide chain oxidized by reaction with cerium (IV) ions in aqueous acidic medium. The following polymerization of alkyl cyanoacrylate monomers initiated by these radicals led to linear block copolymers that undergo spontaneous autoassociation, to form nanoparticles with a hydrophobic core, coated by the hydrophilic polysaccharide.

The group of Bravo-Osuna worked with thiolated chitosan for surface coating of poly(isobutyl cyanoacrylate) (PIBCA) nanoparticles. Ch–TBA of low MW was used in the radical emulsion polymerization technique to elaborate PIBCA nanoparticles coated with thiolated chitosan (Bravo-Osuna et al. 2006). The nanoparticles were spherical, with a mean hydrodynamic diameter of approximately 200 nm and positive zeta potential values, indicating the presence of the cationic polysaccharide at the nanoparticle surface. The presence of chitosan and thiolated chitosan on the surface improved the mucoadhesive characteristics of the system in *ex vivo* evaluation under static conditions done by application of nanoparticle suspensions on rat intestinal mucosal surfaces and the determination of the amount of nanoparticles remaining attached to the mucosa after incubation (Bravo-Osuna et al. 2007a). The morphological parameters (hydrodynamic diameter) of nanoparticles governed their diffusion through the intestinal mucus layer, whereas changes in the polymeric shell composition (molecular weight of chitosan, presence of a cross-linked structure, and density of active thiol groups on the surface) influenced the bioadhesive behavior of these colloidal systems. Improved interpenetration ability with the mucus chain during the attachment process was suggested for the chitosan of high molecular weight, enhancing the bioadhesiveness of the system. The presence of thiol groups on the nanoparticle surface at high concentration (200×10^{-6} µM SH/cm^2) increased the mucoadhesion capacity (Bravo-Osuna et al. 2007b). Taking into account the influence that cation's concentration have in the maintenance of both the permeation and the enzymatic barrier of the oral route, the possible calcium-binding capacity of these thiolated colloidal systems was evaluated *in vitro*. In addition, its presentation in the gel layer surrounding the nanoparticles also benefited its binding capacity, obtaining two- to threefolds higher values when the polymer coated the nanoparticles, than when it was in solution. The presence of chitosan on the nanoparticle surface increased the calcium-binding ability, in comparison to noncoated PIBCA nanoparticles. The calcium binding by chitosan or its thio derivative included (i) ion pair formation by exchange mechanism (in the case of protonated forms); (ii) coordination by ligand exchange mechanism, via nitrogen and/or sulfur donor atoms (in the case of unprotonated forms); and (iii) ion pair binding mechanism followed by slow ligand exchange.

The modification of the amino group and cross-linked structure of thiolated chitosan, owing to the formation of inter- and intrachain disulfide bonds, diminished the accessibility of the cation to the active sites of the polymer, decreasing the binding capacity of the cation. However, when the amount of free thiol groups on the nanoparticle surface was sufficiently high, the binding tendency observed was higher than for nanoparticles, elaborated with nonmodified polymer. The antiprotease behavior of this family of core-shell nanoparticles was *in vitro* tested against two model metallopeptidases present in the GIT: carboxypeptidase A (luminal protease) and leucine aminopeptidase M (membrane protease) (Bravo-Osuna et al. 2008). The presence of amino, hydroxyl, and thiol groups on the nanoparticle surface promoted zinc binding and hence the inhibition of the metallopeptidases analyzed. On the contrary, the occurrence of a cross-linked structure in the gel layer surrounding the PIBCA cores of thiolated formulations, partially limited the inhibition of the proteases again, due to low accessibility of cations to the active groups of the cross-linked polymeric shell.

The effect of these nanoparticles on the paracellular permeability of the tracer [^{14}C]-mannitol was assessed using rat intestinal mucosa (Bravo-Osuna et al. 2008a). Results showed that permeation of the tracer and the reduction of transepithelial electrical resistance (TEER) in the presence of nanoparticles was more pronounced in those formulations prepared with intermediate amounts of thiolated polymer. This effect was explained on the basis of high diffusion capacity of these nanoparticles through the mucus layer that allowed them to reach the TJs in higher extent. Results obtained by the work of this group propose the family of surface-modified nanocarriers as promising candidates for the successful administration of peptides and proteins by the oral route. The synthesis and improvement in mucoadhesion and permeation enhancement properties are also reported for chitosan-glutathione-coated poly(2-hydroxyethyl methacrylate) nanoparticles loaded with FD4 (Atyabi et al. 2008, Moghaddam et al. 2009) and chitosan-glutathione-coated poly(methyl methacrylate nanoparticles loaded with paclitaxel (Akhlaghi et al. 2010).

6.5.5 DNA Delivery

Thiolated chitosan, like native chitosan, formed polyelectrolyte complex with DNA and as a carrier exhibited improved stability, enhanced and sustained gene delivery *in vitro* and *in vivo*. The improvement is possibly because of increased mucoadhesion and cell permeation properties. The distinct redox conditions found at the administration site, such as nasal, intestinal mucosa, and in the cytosol, are reasoned for vector activity of thiolated chitosan. The oxidizing conditions favor disulfide bonding in thiolated chitosans, bestowing cohesiveness and tight binding of DNA. However, the disulfide bonds are reversed under apparent reducing conditions, as found in the cytosol, resulting in the dissociation of the DNA.

The Ch–TBA (299.1 ± 11.5 μM thiol groups per gram polymer) formed coacervates with pDNA at a mean size of 125 nm and a zeta potential of +9 mV. Within 10 h, pDNA was completely released from chitosan/DNA particles, while only 12% were released from the thiomer-based particles. At pH 7, the amount of thiol groups significantly ($p < 0.05$) decreased by more than 25% within 6 h. In contrast, in a reducing environment, as found intracellularly, Ch–TBA/DNA nanoparticles dissociated continuously, liberating approximately 50% of pDNA within 3 h. Transfection studies performed in a Caco-2 cell culture evinced the highest efficiency for Ch–TBA/DNA nanoparticles in combination with a glycerol shock solution (Schmitz et al. 2007).

The polyelectrolyte complexes of Ch–TGA (33 kDa, 360 ± 34 μM thiol groups per gram of polymer) and pDNA encoding green fluorescent protein ranged in size from 75–120 nm in diameter and from +2.3 to 19.7 mV in zeta potential, depending on the weight ratio of Ch–TGA to DNA (Lee et al. 2007). These nanocomplexes exhibited effective physical stability and protection against DNase-I digestion at a weight ratio ≥2.5:1. Significantly ($p < 0.01$) higher gene expression was induced in HEK293, MDCK, and Hep-2 cell lines than unmodified chitosan. Nanocomplexes of disulfide-cross-linked Ch–TGA/DNA showed a sustained DNA release and continuous expression in cultured cells lasting up to 60 h posttransfection. Also, intranasal administration of cross-linked Ch–TGA/DNA nanocomplexes to mice yielded gene expression that lasted for at least 14 days. Based on the study with the Ch–NAC/DNA complex, the importance of –SH groups was demonstrated for transfection, whereas the disulfides were important for the stability of complexes (Loretz et al. 2007). The positively charged nanoparticles of thiolated chitosan/pDNA (Ch–NAC, Ch–TGA) evinced cytotoxicity in erythrocyte, lactate dehydrogenase, and 3-[4,5-dimethylthia-zol-2-yl]-2,5-diphenyltetrazolium bromide (MTT) assay (Loretz and Bernkop-Schnürch 2007, Martien et al. 2007).

TMC possesses good solubility over a broad pH range and potent condensation capacity toward pDNA at neutral pH due to fixed positive charges on its backbone (Mourya and Inamdar 2009). Cellular uptake is facilitated through electrostatic affinity between positively charged TMC/pDNA nanocomplexes and negatively charged cell membranes, bringing about a 2.5–35-fold increase in the transfection efficiency compared to chitosan/pDNA complexes (Thanou et al. 2002). However, TMC with higher quaternization degrees (DQ, e.g., 40% or higher) shows appreciable cytotoxicity (Kean et al. 2005) and exerts excessive condensing capacity on pDNA, which deters intracellular gene dissociation and impedes access of RNA polymerase to pDNA, thus limiting the gene expression level (Kawamura et al. 2005). Hence, to combine the advantages of TMC and thiolated chitosan while minimizing their shortcomings, TMC–Cys was evaluated as nonviral gene carriers by the team of Zhao et al. (2010). TMC–Cys (MW 30, 100, and 200 kDa DQ 15% and 30%) were used to form polyelectrolyte nanocomplexes with plasmid encoding enhanced green fluorescence protein (pEGFP). The complexes demonstrated preferable diameters of below 200 nm and zeta potentials of +15 to +20 mV. Cell binding and uptake of TMC–Cys/pEGFP nanocomplexes were enhanced 2.4–3.0- and 1.4–3.0-folds, respectively, compared to TMC/pEGFP nanocomplexes. pEGFP could be easily released from TMC–Cys nanocomplexes at the intracellular glutathione concentration, which promoted its nuclear transport and accumulation. Consequently, TMC–Cys nanocomplexes showed a 1.4- to 3.2-fold increase in the transfection efficiency in HEK293 cells as compared to TMC nanocomplexes and the optimal TMC–Cys (MW100, DQ 30%) nanocomplexes showed a 1.5-fold enhancement than Lipofectamine2000. Such results were further confirmed by *in vivo* transfection with a 2.3- and 4.1-fold higher transfection efficiency of TMC–Cys (MW 100, DQ 30%) nanocomplexes than TMC (MW 100, DQ 30%) nanocomplexes and Lipofectamine2000, respectively.

6.5.6 SCAFFOLD FORMATION AND STENT COATING

The biodegradability of thiolated chitosans had been seen with lysozyme (Kast and Bernkop-Schnürch 2001). The effect of lysozyme on thiolated chitosan and their oxidized form were investigated by viscosity measurements. Thiolated chitosans (Ch–TGA, Ch–GSH, Ch–TBA, Ch–TEA) degraded 12.9–24.7% less than unmodified chitosan. Moreover, the cross-linking process via disulfide bonds additionally reduced the rate of thiomer degradation. A steric hindrance for the binding of the enzyme to chitosan, caused by the covalent attachment of ligands, seems a plausible reason. The range of degradation rates achieved *in vitro* could be modified by alterations of the contents of thiol-disulfide groups, the structure of ligand used, and pH of the reaction medium. The results demonstrate the possibility of manipulation of polymer properties, paving the way for their use as novel tissue engineering material (Kafedjiiski et al. 2007). Owing to the *in situ* gelling properties, it seems possible to provide a certain shape of the scaffold material by pouring a liquid thiolated chitosan cell suspension in a mold. Furthermore, liquid polymer cell suspensions may be applied by injection, forming semisolid scaffolds at the site of tissue damage. Because low concentrated aqueous solutions of thiolated chitosan remain liquid when stored under inert conditions and rapidly gell under access of oxygen, they seem to be promising candidates for such applications.

Further studies in this direction were performed with L-929 mouse fibroblasts seeded onto Ch–TGA sheets. Results of this study showed that thiolated chitosan can provide a porous scaffold structure guaranteeing cell anchorage, proliferation, and tissue formation in three dimensions (Kast et al. 2003). The scaffolds of Ch–TGA/chitosan were prepared by freeze drying and analyzed for tensile strength. Scaffolds obtained from the proportion of Ch–TGA/chitosan 7:3 and a freezing temperature of –20°C had the maximum tensile strength with a pore distribution ranging from a few to several hundred micrometers maintaining growth of fibroblasts (Li et al. 2010).

The stable porous hydrogels with disulfide cross-linking of Ch–NAC (210.6 ± 29.6 and 321.4 ± 32.8 µM/g of polymer 300 kDa) were obtained at pH 7.4. The 3-D hydrogel structure had pores with size ranging from 5 to 30 µm. The NIH 3T3 cells could migrate into the hydrogels, preserving their viability and 3-D cell morphology inside the hydrogels. These hydrogels were loaded with insulin and BSA. During loading, possibly denaturation, aggregation, and subsequent precipitation of protein may have occurred in time, because of the reaction of the reactive groups of protein (ε-amines of the lysine amino acids or the terminal α-amines or disulfide bonds) with the reactive groups of the gel precursors (thiol groups). *In vitro* release showed that insulin and BSA release could be controlled by choosing the composition, loading, and disulfide bond contents (Wu et al. 2009).

The examination of cytotoxicity of thiolated chitosan (Ch–TBA) showed that the thiolated compound had a lower membrane damaging effect, causing a significantly lower hemoglobin release than the unmodified compound. The MTT assay and 5-bromo-2'-deoxyuridine-based enzyme-linked immunosorbent assay revealed comparable toxicity profile for unmodified and thiolated chitosan in a concentration-dependent manner. It could be concluded that such biocompatibility profile will not compromise their pharmaceutical and biomedical uses (Guggi et al. 2004).

Polymer-coated drug-eluting stents are a potential tool to achieve high local tissue concentrations of an effective drug at the precise site and at the time of vessel injury. Thiolated chitosans show promising role in their application as coating material for stents. The first orientating studies demonstrated that by simply dipping the stent in a thiolated chitosan solution and drying it on air, a stable coating with disulfide cross-linking could be achieved. Such coating should allow sustained release of incorporated drugs, such as antiinflammatory agents, or agents avoiding cell proliferation. Recently, it was shown that stents can be successfully coated with thiolated poly(acrylic acid) and that a sustained release of a model peptide drug can be provided out of this thiomeric coating(Shirazi et al. 2003). Similar results can be expected for thiolated chitosans, but have to be verified by ongoing studies. One such study was to assess the ability of thiolated chitosan to conjugate with the peptide and protein. Ch–TBA was covalently conjugated with maleimide-modified bovine serum albumin (obtained by reaction of protein with *N*-(ε-maleimidocaproyl oxy)sulfosuccinimide ester) via thioether (Masuko et al. 2005).

6.5.7 COSMETICS

The change in the rheological properties of thiolated chitosans on oxidation makes them interesting polymers for probable applications in cosmetics. Furthermore, these polymers adhere excellently on skin and hair, by the formation of disulfide links with cysteine subunits of the surface proteins. These properties render thiolated chitosans (Ch–TBA) into promising compounds for hair gels or make-up, which should stabilize themselves after application, avoiding smearing and dissolution (Prinz 2006).

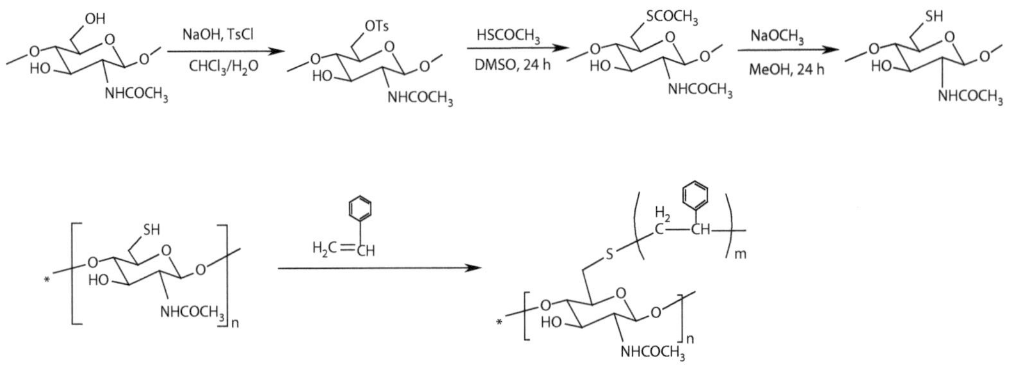

FIGURE 6.14 Synthesis of 6-mercaptochitin or chitosan and grafting with styrene.

6.6 OTHER DERIVATIVES WITH SH GROUP

The introduction of –SH group on the chitosan backbone can be achieved by substituting the 6-OH group. Such derivatives, 6-deoxy-6-mercaptochitin or chitosan, simply called 6-mercaptochitin or chitosan, were synthesized via water-soluble tosyl derivatives of the polymers, obtained by tosyl chloride treatment, followed by sequential reaction with potassium thioacetate and sodium methoxide (Figure 6.14). The derivative showed much enhanced swelling and solubility in organic solvents and was degraded more efficiently by lysozymes than unmodified polymer (Kurita et al. 1993). The thiol groups can also dissociate to give hydrogen and sulfur free radicals, which can initiate polymerization. The work in this direction is provided with the derivatives as 6-mercaptochitin and 6-mercaptochitosan and vinyl monomers, such as styrene and methyl methacrylate, to form comblike copolymers on trunk of chitin and chitosan (Kurita et al. 1996, 1996a, 2002, Munro et al. 2009). The grafts formed may prove to be useful as biomaterials.

The free –SH group of thiolated chitosan can be made to react to obtain novel polymeric preparations too. An injectable *in situ* cross-linked hydrogel has been designed in physiological conditions via Michael-type addition between Ch–NAC and PEG diacrylate (Teng et al. 2010). The gelation time, cross-linking density, and elasticity of hydrogel depended on the content of free thiols in Ch–NAC, temperature, and concentration of Ch–NAC and PEG diacrylate. The hydrogels were thermostable with a porous 3D structure and had a high initial swelling. The swelling was highly temperature dependent and was directly related to the amount of cross-linking. The hydrogels were biodegradable and biocompatibile in HDFs and A549 cells.

6.7 CONCLUSION

The polymer properties of chitosan are spectacularly modified with their conversion to sulfhydryl-bearing derivatives. In the thiol-bearing derivatives, the mucoadhesive, cohesive properties and *in situ* gelling features are strongly improved. These properties further enrich the inherent permeation enhancement capabilities of native chitosan. The permeation-enhancing effect can be raised further by the combination of thiolated chitosans with the permeation mediator glutathione. Owing to these advantageous features, thiolated chitosans have been successfully employed for noninvasive peroral administration of peptide drugs, controlled and site-specific release of smaller drugs, coating of nanoparticles, and DNA delivery. Further applications, where thiolated chitosans might prove successful, include their use as scaffold material and coating material for stents in the field of tissue engineering. The elaboration of thiolated chitosan with derivativatized chitosan as backbone polymer is also gaining attention as with TMC. The graft copolymers of mercaptochitin and mercaptochitosan may serve as the candidates for applications in tissue engineering. The development of hydrogels of thiolated chitosans can also be foreseen.

REFERENCES

Akhlaghi, S. P., S. N. Ostad, R. Dinarvand, and F. Atyabi. 2010. Discriminated effects of thiolated chitosan-coated pMMA paclitaxel-loaded nanoparticles on different normal and cancer cell lines. *Nanomedicine: Nanotechnology, Biology and Medicine* 6:689–697.

Al-Shawi, M. K., I. L. Urbatsch, and A. E. Senior. 1994. Covalent inhibitors of P-glycoprotein ATPase activity. *Journal of Biological Chemistry* 269:8986–8992.

Artursson, P., T. Lindmark, S. S. Davis, and L. Illum. 1994. Effect of chitosan on the permeability of monolayers of intestinal epithelial cells (Caco-2). *Pharmaceutical Research* 11:1358–1361.

Aspden, T. J., L. Illum, and Ø. Skaugrud. 1996. Chitosan as a nasal delivery system: Evaluation of insulin absorption enhancement and effect on nasal membrane integrity using rat models. *European Journal of Pharmaceutical Sciences* 4:23–31.

Atyabi F., F. A. Moghaddam, R. Dinarvand, M. J. Zohuriaan-Mehr, and G. Ponchel. 2008. Thiolated chitosan coated poly hydroxyethyl methacrylate nanoparticles: Synthesis and characterization. *Carbohydrate Polymers* 74:59–67.

Barrett W. C., J. P. DeGnore, S. Konig. et al. 1999. Regulation of PTP1B via glutathionylation of the active site cysteine. *Biochemistry* 38:6699–6705.

Barthelmes, J., S. Dünnhaupt, J. Hombach, and A. Bernkop-Schnürch. 2011. Thiomer nanoparticles: Stabilization via covalent cross-linking. *Drug Delivery* 18:613–619.

Bernkop-Schnürch, A., U. M. Brandt, and A. E. Clausen. 1999. Synthesis and in vitro evaluation of chitosan–cysteine conjugates. *Scientia Pharmaceutica* 67:196–208.

Bernkop-Schnürch, A., A. Heinrich, and A. Greimel. 2006. Development of a novel method for the preparation of submicron particles based on thiolated chitosan. *European Journal of Pharmaceutics and Biopharmaceutics* 63:166–172.

Bernkop-Schnürch, A., D. Guggi, and Y. Pinter. 2004. Thiolated chitosans: Development and in vitro evaluation of a mucoadhesive, permeation enhancing oral drug delivery system. *Journal of Controlled Release* 94:177–186.

Bernkop-Schnürch, A., and T. E. Hopf. 2001. Synthesis and in vitro evaluation of chitosan–thioglycolic acid conjugates. *Scientia Pharmaceutica* 69:109–118.

Bernkop-Schnürch, A., M. Hornof, and T. Zoidl. 2003. Thiolated polymers-thiomers: Synthesis and in vitro evaluation of chitosan-2-iminothiolane conjugates. *International Journal of Pharmaceutics* 260:229–237.

Bernkop-Schnürch, A., and C. E. Kast. 2001. Chemically modified chitosans as enzyme inhibitors. *Advanced Drug Delivery Reviews* 52:127–137.

Bernkop-Schnürch, A., C. E. Kast, and D. Guggi. 2003a. Permeation enhancing polymers in oral delivery of hydrophilic macromolecules: Thiomer/GSH systems. *Journal of Controlled Release* 93:95–103.

Bernkop-Schnürch, A., Y. Pinter, D. Guggi, H. et al. 2005. The use of thiolated polymers as carrier matrix in oral peptide delivery—Proof of concept. *Journal of Controlled Release* 106:26–33.

Bravo-Osuna, I., T. Schmitz, A. Bernkop-Schnürch, C. Vauthier, and G. Ponchel. 2006. Elaboration and characterization of thiolated chitosan-coated acrylic nanoparticles. *International Journal of Pharmaceutics* 316:170–175.

Bernkop-Schnürch, A., A. Weithaler, A. Albrecht, and A. Greimel. 2006a. Thiomers: Preparation and in vitro evaluation of a mucoadhesive nanoparticulate drug delivery system. *International Journal of Pharmaceutics* 317:76–81.

Borchard, G., H. L. Lueßen, A. G. de Boer, J. C. Verhoef, C. M. Lehr, and H. E. Junginger. 1996. The potential of mucoadhesive polymers in enhancing intestinal peptide drug absorption. 3. Effects of chitosan-glutamate and carbomer on epithelial tight junctions in vitro. *Journal of Controlled Release.* 39:131–138.

Bravo-Osuna, I., G. Millotti, C. Vauthier, and G. Ponchel. 2007b. In vitro evaluation of calcium binding capacity of chitosan and thiolated chitosan poly(isobutyl cyanoacrylate) core-shell nanoparticles. *International Journal of Pharmaceutic.* 338:284–290.

Bravo-Osuna, I., C. Vauthier, H. Chacun, and G. Ponchel. 2008a. Specific permeability modulation of intestinal paracellular pathway by chitosan-poly(isobutyl cyanoacrylate) core-shell nanoparticles. *European Journal of Pharmaceutics and Biopharmaceutics* 69:436–444.

Bravo-Osuna, I., C. Vauthier, A. Farabollini, G. Millotti, and G. Ponchel. 2008. Effect of chitosan and thiolated chitosan coating on the inhibition behaviour of PIBCA nanoparticles against intestinal metallopeptidases. *Journal of Nanoparticle Research* 10:1293–1301.

Bravo-Osuna, I., C. Vauthier, A. Farabollini, G.F. Palmieri, and G. Ponchel. 2007a. Mucoadhesion mechanism of chitosan and thiolated chitosan-poly(isobutylcyanoacrylate) core-shell nanoparticles. *Biomaterials* 28:2233–2243.

Bravo-Osuna, I., D. Teutonico, S. Arpicco, C. Vauthier, and G. Ponchel. 2007. Characterization of chitosan thiolation and application to thiol quantification onto nanoparticle surface. *International Journal of Pharmaceutics* 340:173–181.

Cevher E., D. Sensoy, M. A. Taha, and A. Araman. 2008. Effect of thiolated polymers to textural and muco-adhesive properties of vaginal gel formulations prepared with polycarbophil and chitosan. *AAPS PharmSciTech* 9:953–965.

Chauvierre, C., D. Labarre, P. Couvreur, and C. Vauthier. 2003. Novel polysaccharide-decorated poly(isobutyl cianoacrylate) nanoparticles. *Pharmaceutical Research* 20:1786–1793.

Clausen, A. E., C. E. Kast, and A. Bernkop-Schnürch. 2002. The role of glutathione in the permeation enhancing effect of thiolated polymers. *Pharmaceutical Research* 19:602–608.

Collares-Buzato C. B., M. A. Jepson, N. L. Simmons, and B. H. Hirst. 1998. Increased tyrosine phosphorylation causes redistribution of adherens junction and tight junction proteins and perturbs paracellular barrier function in MDCK epithelia, *European Journal of Cell Biology* 76:85–92.

Coppi, G., V. Iannuccelli, E. Leo, M. T. Bernabei, and R. Cameroni. 2001. Chitosan-alginate microparticles as a protein carrier. *Drug Development and Industrial Pharmacy* 27:393–400.

Delprino, L., M. Giacomotti, F. Dosio. et al. 1993. Toxin targeted design for anticancer therapy. I: Synthesis and biological evaluation of new thioimidate heterobifunctional reagents. *Journal of Pharmaceutical Sciences* 82:506–512.

Dodane, V., A. M. Khan, and J. R. Merwin. 1999. Effect of chitosan on epithelial permeability and structure. *International Journal of Pharmaceutics* 182:21–32.

Dominey, L. A. and K. Kustin. 1983. Kinetics and mechanism of Zn(II) complexation with reduced glutathione. *Journal of Inorganic Biochemistry* 18:153–160.

Dünnhaupt, S., J. Barthelmes, D. Rahmat. et al. 2012. S-Protected thiolated chitosan for oral delivery of hydrophilic macromolecules: Evaluation of permeation enhancing and efflux pump inhibitory properties. *Molecular Pharmaceutics* 9:1331–1341.

Dünnhaupt, S., J. Barthelmes, J. Iqbal. et al. 2012a. In vivo evaluation of an oral drug delivery system for peptides based on S-protected thiolated chitosan. *Journal of Controlled Release* 160:477–485.

Föger, F., K. Kafedjiiski, H. Hoyer, B. Loretz, and A. Bernkop-Schnürch. 2007. Enhanced transport of P-glycoprotein substrate saquinavir in presence of thiolated chitosan. *Journal of Drug Targeting* 15:132–139.

Föger, F., T. Schmitz, and A. Bernkop-Schnürch. 2006. In vivo evaluation of an oral delivery system for P-gp substrates based on thiolated chitosan. *Biomaterials* 27:4250–4255.

Furuse, M., K. Fujita, T. Hiiragi, K. Fujimoto, and S. Tsukita. 1998. Claudin-1 and -2: Novel integral membrane proteins localizing at tight junctions with no sequence similarity to occluding. *Journal of Cell Biology* 141:1539–1550.

Grabovac, V. and A. Bernkop-Schnürch. 2007. Development and in vitro evaluation of surface modified poly(lactideco-glycolide) nanoparticles with chitosan-4-thiobutylamidine. *Drug Development and Industrial Pharmacy* 33:767–774.

Gradauer, K., S. Dünnhaupt, C. Vonach. et al. 2013. Thiomer-coated liposomes harbor permeation enhancing and efflux pump inhibitory properties. *Journal of Controlled Release* 165:207–215.

Grafstrom, R., A. H. Stead, S. Orrenius. 1980. Metabolism of extracellular glutathione in rat small-intestinal mucosa. *European Journal of Biochemistry* 106:571–577.

Guggi, D., C. E. Kast, and A. Bernkop-Schnürch. 2003. in vivo evaluation of an oral salmon calcitonin-delivery system based on a thiolated chitosan carrier matrix. *Pharmaceutical Research* 20:1989–1994.

Guggi, D., A. H. Krauland, and A. Bernkop-Schnürch. 2003a. Systemic peptide delivery via the stomach: In vivo evaluation of an oral dosage form for salmon calcitonin. *Journal of Controlled Release* 92:125–135.

Guggi, D., N. Langoth, M.H. Hoffer, M. Wirth, and A. Bernkop-Schnürch. 2004. Comparative evaluation of cytotoxicity of a glucosamine-TBA conjugate and a chitosan-TBA conjugate. *International Journal of Pharmaceutics* 278:353–360.

Hassan, E. E. and J. M. Gallo. 1990. A simple rheological method for the in vitro assessment of mucin–polymer bioadhesive bond strength. *Pharmaceutical Research* 7:491–495.

Ho, Y. C., S. J. Wu, F. L. Mi. et al. 2010. Thiol-modified chitosan sulfate nanoparticles for protection and release of basic fibroblast growth factor. *Bioconjugate Chemistry* 21:28–38.

Hombach, J., H. Hoyer, and A. Bernkop-Schnürch. 2008. Thiolated chitosans: Development and in vitro evaluation of an oral tobramycin sulphate delivery system. *European Journal of Pharmaceutics* 33:1–8.

Hongyok, T., J. J. Chae, Y. J. Shin, D. Na, L. Li, and R. S. Chuck. 2009. Effect of chitosan-N-acetylcysteine conjugate in a mouse model of botulinum toxin B-induced dry eye. *Archives of Ophthalmology* 127:525–532.

Hornof, M. D., C. E. Kast, and A. Bernkop-Schnürch. 2003. in vitro evaluation of the viscoelastic behavior of chitosan-thioglycolic acid conjugates. *European Journal of Pharmaceutics and Biopharmaceutics* 55:185–190.

Hoyer H., W. Schlocker, K. Krum, and A. Bernkop-Schnürch. 2008. Preparation and evaluation of microparticles from thiolated polymers via air jet milling. *European Journal of Pharmaceutics and Biopharmaceutics* 69:476–485.

Hoyer, H., F. Föger, K. Kafedjiiski, B. Loretz, and A. Bernkop-Schnürch. 2007. Design and evaluation of a new gastrointestinal mucoadhesive patch system containing chitosan-glutathione, *Drug Development and Industrial Pharmacy*. 33:1289–1296.

Illum L., N. F. Farraj, and S. S. Davis. 1994. Chitosan as a novel nasal delivery system for peptide drugs. *Pharmaceutical Research* 11:1186–1189.

Imam, M. E. and A. Bernkop-Schnürch. 2005. Controlled drug delivery systems based on thiolated chitosan microspheres, *Drug Development and Industrial Pharmacy* 31:557–565.

Imam, M. E., M. Hornof, C. Valenta, G. Reznicek, and A. Bernkop-Schnürch. 2003. Evidence for the interpenetration of mucoadhesive polymers into the mucus gel layer. *STP Pharma Sciences* 13:171–176.

Jayakumar R., R. L. Reis, and J. F. Mano. 2007. Synthesis and characterization of ph-sensitive thiol-containing chitosan beads for controlled drug delivery applications. *Drug Delivery* 14:9–17.

Juntapram, K., N. Praphairaksit, K. Siraleartmukul, and N. Muangsin. 2012. Synthesis and characterization of chitosan-homocysteine thiolactone as a mucoadhesive polymer. *Carbohydrate Polymers* 87:2399–2408.

Kafedjiiski, K., F. Föger, H. Hoyer, A. Bernkop-Schnürch, and M. Werle. 2007. Evaluation of in vitro enzymatic degradation of various thiomers and crosslinked thiomers. *Drug Development and Industrial Pharmacy* 33:199–208.

Kafedjiiski, K., F. Föger, M. Werle, and A. Bernkop-Schnürch. 2005. Synthesis and in vitro evaluation of a novel chitosan-glutathione conjugate. *Pharmaceutical Research* 22:1480–1488.

Kafedjiiski, K., M. Hoffer, M. Werle, and A. Bernkop-Schnürch. 2006. Improved synthesis and in vitro characterization of chitosan thioethylamidine conjugate. *Biomaterials* 27:127–135.

Kafedjiiski, K., A. H. Krauland, M. H. Hoffer, and A. Bernkop-Schnürch. 2005a. Synthesis and in vitro evaluation of a novel thiolated chitosan. *Biomaterials* 26:819–826.

Kast, C. E., and A. Bernkop-Schnürch. 2001. Thiolated polymers-thiomers: Development and in vitro evaluation of chitosan-thioglycolic acid conjugates. *Biomaterials* 22:2345–2352.

Kast, C.E., W. Frick, U. Losert, and A. Bernkop-Schnürch. 2003. Chitosan thioglycolic acid conjugate: A new scaffold material for engineering? *International Journal of Pharmaceutics* 256:183–189.

Kast C. E., C. Valenta, M. Leopold, and A. Bernkop-Schnürch. 2002. Design and in vitro evaluation of a novel bioadhesive vaginal drug delivery system for clotrimazole. *Journal of Controlled Release* 81:347–354.

Kawamura, K., J. Oishi, J. H. Kang. et al. 2005. Intracellular signal responsive gene carrier for cell-specific gene expression. *Biomacromolecules* 6:908–913.

Kean, T., S. Roth, and M. Thanou. 2005. Trimethylated chitosans as non-viral gene delivery vectors: Cytotoxicity and transfection efficiency. *Journal of Controlled Release* 103:143–153.

Kotze A. F., H. L. Lueßen, B. J. de Leeuw, A. G. de Boer, J. C. Verhoef, and H. E. Junginger. 1998. Comparison of the effect of different chitosan salts and N-trimethyl chitosan chloride on the permeability of intestinal epithelial cells (Caco-2). *Journal of Controlled Release* 51:35–46.

Krauland, A. H., D. Guggi, and A. Bernkop-Schnürch. 2004. Oral insulin delivery: The potential of thiolated chitosan-insulin tablets on non-diabetic rats. *Journal of Controlled Release* 95:547–555.

Krauland, A. H., D. Guggi, and A. Bernkop-Schnürch. 2006. Thiolated chitosan microparticles: A vehicle for nasal peptide drug delivery. *International Journal of Pharmaceutics* 307:270–277.

Krauland A. H., Hoffer M. H., and A. Bernkop-Schnürch. 2005. Viscoelastic properties of a new in situ gelling thiolated chitosan conjugate. *Drug Development and Industrial Pharmacy* 31:885–893.

Krauland, A. H., V. M. Leitner, V. Grabovac, and A. Bernkop-Schnürch. 2006a. In vivo evaluation of a nasal insulin delivery system based on thiolated chitosan. *Journal of Pharmaceutical Sciences* 95:2463–2472.

Kubota, K., M. Furuse, H. Sasaki, N. Sonoda, K. Fujita, A. Nagafuchi, and S. Tsukita. 1999. Ca^{2+}-independent cell-adhesion activity of claudins, a family of integral membrane proteins localized at tight junctions. *Current Biology* 9:1035–1038.

Kurita, K., H. Yoshino, S. I. Nishimura, and S. Ishi. 1993. Preparation and biodegradability of chitin derivatives having mercapto groups. *Carbohydrate Polymers* 20:239–245.

Kurita, K., M. Inoue, and M. Harata. 2002. Graft copolymerization of methyl methacrylate onto mercaptochitin and some properties of the resulting hybrid materials. *Biomacromolecules* 3:147–152.

Kurita, K., S. Hashimoto, H. Yoshino, S. Ishii, and S. I. Nishimura, 1996a. Preparation of chitin/polystyrene hybrid materials by efficient graft copolymerization based on mercaptochitin. *Macromolecules* 29:1939–1942.

Kurita, K., S. Hashimoto, S. Ishii, and T. Mori. 1996. Chitin/poly(methyl methacrylate) hybrid materials: Efficient graft copolymerization of methyl methacrylate onto mercapto-chitin. *Polymer Bulletin* 36:681–686.

Langoth, N., H. Kahlbacher, G. Schoffmann. et al. 2006. Thiolated chitosans: Design and in vivo evaluation of a mucoadhesive buccal peptide drug delivery system. *Pharmaceutical Research* 23:573–579.

Lee, D. W., S. A. Shirley, R. F. Lockey, and S. S. Mohapatra. 2006. Thiolated chitosan nanoparticles enhance anti-inflammatory effects of intranasally delivered theophylline. *Respiratory Research* 7:112–122.

Lee, D., W. Zhang, S. A. Shirley. et al. 2007. Thiolated chitosan/DNA nanocomplexes exhibit enhanced and sustained gene delivery. *Pharmaceutical Research* 24:157–167.

Lehr, C. M., J. A. Bouwstra, E. H. Schacht, and H. E. Junginger. 1992. In vitro evaluation of mucoadhesive properties of chitosan and some other natural polymers. *International Journal of Pharmaceutics* 78:43–48.

Leitner, V. M., M. K. Marschutz, and A. Bernkop-Schnürch. 2003. Mucoadhesive and cohesive properties of poly(acrylic acid)-cysteine conjugates with regard to their molecular mass. *European Journal of Pharmaceutical Sciences* 18:89–96.

Leitner, V. M., G. F. Walker, and A. Bernkop-Schnürch. 2003a. Thiolated polymers: Evidence for the formation of disulphide bonds with mucus glycoproteins. *European Journal of Pharmaceutics and Biopharmaceutics* 56:207–214.

Li, Z., L. Cen, L. Zhao, L. Cui, W. Liu, and Y. Cao. 2010. Preparation and evaluation of thiolated chitosan scaffolds for tissue engineering. *Journal of Biomedical Materials Research Part A* 92:973–978.

Loretz, B., and A. Bernkop-Schnürch. 2007. in vitro cytotoxicity testing of non-thiolated and thiolated chitosan nanoparticles for oral gene delivery. *Nanotoxicology* 1:139–148.

Loretz, B., M. Thaler, and A. Bernkop-Schnürch. 2007. Role of sulfhydryl groups in transfection? A case study with chitosan-NAC nanoparticles. *Bioconjugate Chemistry* 18:1028–1035.

Lueßen H. L., B. J. de Leeuw, M. W. E. Langmeyer, A. G. de Boer, J. C. Verhoef and H. E. Junginger. 1996. Mucoadhesive polymers in peroral peptide drug delivery. VI. Carbomer and chitosan improve the intestinal absorption of the peptide drug buserelin in vivo. *Pharmaceutical Research* 13:1668–1672.

Lueßen, H. L., C. O. Rentel, A. F. Kotze. et al. 1997. Mucoadhesive polymers in peroral peptide drug delivery. IV. Polycarbophil and chitosan are potent enhancers of peptide transport across intestinal mucosae in vitro. *Journal of Controlled Release* 45:15–23.

Maculotti, K., I. Genta, P. Perugini, M. Imam, A. Bernkop-Schnürch, and F. Pavanetto. 2005. Preparation and in vitro evaluation of thiolated chitosan microparticles. *Journal of Microencapsulation* 225:459–470.

Makhlof, A., M. Werle, Y. Tozuka, and H. Takeuchi. 2010. Nanoparticles of glycol chitosan and its thiolated derivative significantly improved the pulmonary delivery of calcitonin. *International Journal of Pharmaceutics* 397:92–95.

Martien, R., B. Loretz, M. Thaler, S. Majzoob, and A. Bernkop-Schnürch. 2007. Chitosan-thioglycolic acid conjugate: An alternative carrier for oral nonviral gene delivery? *Journal of Biomedical Materials Research Part A*. 82:1–9.

Martin-Padura, I., S. Lostaglio, M. Schneemann. et al. 1998. Junctional adhesion molecule, a novel member of the immunoglobulin superfamily that distributes at intercellular junctions and modulates monocyte transmigration. *Journal of Cell Biology* 142:117–127.

Masuko, T., A. Minami, N. Iwasaki, T. Majima, S. I. Nishimura, and Y. C. Lee. 2005. Thiolation of chitosan: Attachment of proteins via thioether formation. *Biomacromolecules* 6:880–884.

Millotti, G., C. Samberger, E. Frohlich, and A. Bernkop-Schnürch. 2009. Chitosan-graft-6-mercaptonicotinic acid: Synthesis, characterization, and biocompatibility. *Biomacromolecules* 10:3023–3027.

Moghaddam, F. A., F. Atyabi, and R. Dinarvand. 2009. Preparation and in vitro evaluation of mucoadhesion and permeation enhancement of thiolated chitosan-pHEMA core-shell nanoparticles. *Nanomedicine: Nanotechnology, Biology and Medicine* 5:208–215.

Mourya, V. K. and N. N. Inamdar. 2009. Trimethyl chitosan and its application in drug delivery. *Journal of Materials Science: Materials in Medicine* 20:1057–1079.

Mourya, V. K. and N. N. Inamdar. 2008. Chitosan-modifications and applications: Opportunities galore. *Reactive & Functional Polymers* 68:1013–1051.

Munro, N. H., L. R. Hanton, S. C. Moratti, and B. H. Robinson. 2009. Preparation and graft copolymerisation of thiolated β-chitin and chitosan derivatives. *Carbohydrate Polymers* 78:137–145.

Palmberger, T. F., J. Hombach, A. Bernkop-Schnürch. 2008. Thiolated chitosan: Development and in vitro evaluation of an oral delivery system for acyclovir. *International Journal of Pharmaceutics* 348:54–60.

Patel, D., S. Naik, and A. N. Misra. 2012. Improved transnasal transport and brain uptake of tizanidine HCl-loaded thiolated chitosan nanoparticles for alleviation of pain. *Journal of Pharmaceutical Sciences* 101:690–706.

Prinz, M. Chitosan-thio-amidine conjugates and their cosmetic as well as pharmaceutic use. US Patent 7053068; May 30, 2006.

Rao, R. K., L. Li, R.D. Baker, S.S. Baker, and A. Gupta. 2000. Glutathione oxidation and PTPase inhibition by hydrogen peroxide in Caco-2 cell monolayer. *American Journal of Physiology—Gastrointestinal and Liver Physiology* 279:332–340.

Roldo, M., M. Hornof, P. Caliceti, and A. Bernkop-Schnürch. 2004. Mucoadhesive thiolated chitosans as platforms for oral controlled drug delivery: Synthesis and in vitro evaluation. *European Journal of Pharmaceutics and Biopharmaceutics* 57:115–121.

Sakloetsakun, D., J. Iqbal, G. Millotti, A. Vetter, and A. Bernkop-Schnürch. 2011. Thiolated chitosans: Influence of various sulfhydryl ligands on permeation-enhancing and P-gp inhibitory properties. *Drug Development and Industrial Pharmacy* 37:648–655.

Sakloetsakun, D., J. M. Hombach, and A. Bernkop-Schnürch. 2009. In situ gelling properties of chitosan-thioglycolic acid conjugate in the presence of oxidizing agent. *Biomaterials* 30:6151–6157.

Schipper, N. G. M., S. Olsson, J. A. Hoogstraate, A. G. de Boer, K. M. Varum, and P. Artursson. 1997. Chitosans as absorption enhancers for poorly absorbable drugs. 2. Mechanism of absorption enhancement. *Pharmaceutical Research* 14:923–929.

Schipper, N. G., K. M. Varum, and P. Artursson. 1996. Chitosans as absorption enhancers for poorly absorbable drugs. 1. Influence of molecular weight and degree of acetylation on drug transport across human intestinal epithelial (Caco-2) cells, *Pharmaceutical Research* 13:1686–1692.

Schipper, N. G. M., K. M. Varum, P. Stenberg, G. Ocklind, H. Lennernas, and P. Artursson. 1999. Chitosans as absorption enhancers for poorly absorbable drugs. 3: Influence of mucus on absorption enhancement. *European Journal of Pharmaceutical Sciences.* 8:335–343.

Schmitz, T., I. Bravo-Osuna, C. Vauthier, G. Ponchel, B. Loretz, and A. Bernkop-Schnürch. 2007. Development and in vitro evaluation of a thiomer-based nanoparticulate gene delivery system. *Biomaterials* 28:524–531.

Schmitz, T., J. Hombach, and A. Bernkop-Schnürch. 2008a. Chitosan-N-Acetyl cysteine conjugates: In vitro evaluation of permeation enhancing and P-glycoprotein inhibiting properties. *Drug Delivery* 15:245–252.

Schmitz, T., V. Grabovac, T. F. Palmberger, M. H. Hoffer, and A. Bernkop-Schnürch. 2008. Synthesis and characterization of a chitosan-N-acetyl cysteine conjugate. *International Journal of Pharmaceutics* 347:79–85.

Shahnaz, G., A Vetter, J. Barthelmes. et al. 2012. Thiolated chitosan nanoparticles for the nasal administration of leuprolide: Bioavailability and pharmacokinetic characterization. *International Journal of Pharmaceutics* 428:164–170.

Shirazi, M., M. Gyongyosi, C. Strehblow, D. Glogar, A. Krauland, and A. Bernkop-Schnürch. 2003. Design and in vitro evaluation of polymercoated drug-eluting intracoronary stents. 30th Annual Meeting & Exposition of the Controlled Release Society. Glasgow, UK, pp. 476.

Simon, D. B., Y. Lu, K. A. Choate. et al. 1999. Paracellin-1, a renal tight junction protein required for paracellular Mg^{2+}resorption. *Science* 285:103–106.

Singh, R., L. Kats, W. A. Blattler, and J. M. Lambert. 1996. Formation of N-substituted 2-iminothiolanes when amino groups in proteins and peptides are modified by 2-iminothiolane. *Analytical Biochemistry* 236:114–125.

Smith, J., E. Wood, and M. Dornish. 2004. Effect of chitosan on epithelial cell tight junctions. *Pharmaceutical Research* 21:43–49.

Snyder, G. H., M. K. Reddy, M. J. Cennerazzo, and D. Field. 1983. Use of local electrostatic environments of cysteines to enhance formation of a desired species in a reversible disulfide exchange reaction. *Biochimica et Biophysica Acta* 749:219–226.

Staddon, J. M., K. Herrenknecht, C. Smales, and L. L. Rubin. 1995. Evidence that tyrosine phosphorylation may increase tight junction permeability. *Journal of Cell Science* 108:609–619.

Teng, D. Y., Z. M. Wu, X. G. Zhang, Y. X. Wang, C. Zheng, Z. Wang, and C. X. Li. 2010. Synthesis and characterization of in situ cross-linked hydrogel based on self-assembly of thiol-modified chitosan with PEG diacrylate using Michael type addition. *Polymer* 51:639–646.

Thanou, M., B. I. Florea, M. Geldof, H. E. Junginger, and G. Borchard. 2002. Quaternized chitosan oligomers as novel gene delivery vectors in epithelial cell lines. *Biomaterials* 23:153–159.

Thanou, M. M., J. C. Verhoef, J. H. M. Verheijden, and H. E. Junginger. 2001. Intestinal absorption of octreotide using trimethyl chitosan chloride: Studies in pigs. *Pharmaceutical Research* 18:823–828.

van der Lubben, I. M., J. C. Verhoef, A. C. van Aelst, G. Borchard, and H. E. Junginger. 2001. Chitosan microparticles for oral vaccination: Preparation, characterization and preliminary in vivo uptake studies in murine Peyer's patches. *Biomaterials* 22:687–694.

Verheul, R. J., S. van der Wal, and W. E. Hennink. 2010. Tailorable thiolated trimethyl chitosans for covalently stabilized nanoparticles. *Biomacromolecules* 11:1965–1971.

Werle, M. and M. Hoffer. 2011. Glutathione and thiolated chitosan inhibit multidrug resistance P-glycoprotein activity in excised small intestine. *Journal of Controlled Release* 111:41–46.

Wu, Z. M., X. G. Zhang, C. Zheng. et al. 2009. Disulfide-crosslinked chitosan hydrogel for cell viability and controlled protein release. *European Journal of Pharmaceutical Sciences* 37:198–206.

Yin, L., J. Ding, C. He, L. Cui, C. Tang, and C. Yin. 2009. Drug permeability and mucoadhesion properties of thiolated trimethyl chitosan nanoparticles in oral insulin delivery. *Biomaterials* 30: 5691–5700.

Zambito, Y. and G. DiColo. 2010. Thiolated quaternary ammonium-chitosan conjugates for enhanced precorneal retention, transcorneal permeation and intraocular absorption of dexamethasone. *European Journal of Pharmaceutics and Biopharmaceutics* 75:194–199.

Zambito, Y., S. Fogli, C. Zaino, F. Stefanelli, M. C. Breschi, and G. DiColo. 2009. Synthesis, characterization and evaluation of thiolated quaternary ammonium chitosan conjugates for enhanced intestinal drug permeation. *European Journal of Pharmaceutical Sciences* 38:112–120.

Zhao, X., L. Yin, J. Ding, C. Tang, S. Gu, C. Yin, and Y. Mao. 2010. Thiolated trimethyl chitosan nanocomplexes as gene carriers with high in vitro and in vivo transfection efficiency. *Journal of Controlled Release* 144:46–54.

7 Graft Copolymerization of Acrylic Monomers on Chitosan and Its Derivatives

Recent Developments and Applications

M. Prabaharan

CONTENTS

7.1 INTRODUCTION

Chitosan, obtained from chitin (poly-β-(1–4)-N-acetyl-D-glucosamine) through deacetylation using strong aqueous alkali solution, is a versatile form of polysaccharide. It has a specific structure and various biological properties, such as immunological activity and antibacterial or wound-healing property (Nishimura et al., 1991; Kurita et al., 2002; Sashiwa et al., 2002). Moreover, it is a nontoxic and biodegradable polymer (Prabaharan, 2008). Because of its biocompatibility, biodegradability, and avirulence, chitosan has been used in many areas, such as in biomedical and agricultural fields. Chemical modification of chitosan so as to impart advantageous properties has attracted the attention of many researchers in recent years. Chitosan grafted with vinyl monomers using various types of initiators has been developed as promising materials for biomedical and industrial applications (Liu et al., 2007; Zhang et al., 2007). Despite the many graft copolymers of chitosan that have been synthesized, the grafting of acrylic monomers, industrially important monomers, onto chitosan has received much importance for biomedical, industrial, and agricultural applications. The chemical combination of chitosan and acrylic-based polymer is a promising method for the preparation of new materials. Owing to the improved physical and chemical properties, these new materials would have potential and multiple applications in various fields.

In recent years, graft copolymerization of acrylic monomers such as methyl methacrylate (MAA), 2-hydroxyethyl methacrylate (HEMA), 3-(trimethoxysilyl)propyl methacrylate (TMSPM), methyl acrylate (MA), (N,N-dimethylamino)ethyl methacrylate (DMA), butyl acrylate (BA), and sodium acrylate (SA) onto chitosan and its derivatives using free radical initiators, γ-radiation, and transition-metal ions have been reported (Liu et al., 2002, 2003). However, no comprehensive review has been reported on the graft copolymerization of acrylic monomers onto chitosan and its derivatives until now. In this chapter, the preparation and properties of various types of graft copolymerized chitosan and its derivatives with acrylic monomers are discussed in detail. Special emphasis has been given to the application of these materials as drug delivery carriers, wound-healing agents, seed coatings, absorbents, and finishing agents.

7.2 GRAFTING OF ACRYLIC MONOMERS ONTO CHITOSAN

7.2.1 METHYL METHACRYLATE–CHITOSAN GRAFT COPOLYMER

A novel redox system, potassium ditelluratocuprate(III) (DTC)–chitosan, was employed to initiate the graft copolymerization of MMA onto chitosan in alkali medium (Liu et al., 2005). In this study, the effects of reaction variables, such as the initiator concentration, ratio of monomer to chitosan, the pH value, as well as reaction temperature and time were investigated, and the grafting conditions were optimized. Graft copolymers with both high grafting efficiency (>90%) and percentage of grafting were obtained. It was observed that the rate of polymerization is higher, which indicated that the DTC–chitosan redox system is an efficient initiator for this graft copolymerization. The scanning electron microscope (SEM) photographs indicated that the graft copolymer improved the compatibility of the blend.

Recently, Liu et al. (2011) prepared the nanoparticles by grafting MMA onto chitosan in aqueous solution using potassium diperiodatocuprate (Cu(III)) as an initiator (Figure 7.1). It was observed that the particle size was dependent on the Cu(III) concentration, temperature, and weight ratio of MMA/chitosan used in the nanoparticle preparation. Transmission electron microscopy (TEM) micrograph showed that chitosan nanoparticles had a very homogeneous morphology with predominantly spherical and uniform particles size distribution. From forward transmission infrared (FTIR), x-ray diffraction, and thermal-stability analysis, it was deduced that MMA could successfully be grafted onto chitosan and nanoparticles constituted by chitosan-g-poly(MMA). These results indicated that Cu(III) can be used to graft vinyl polymers onto natural polymer to form nanoparticles.

7.2.2 2-HYDROXYETHYL METHACRYLATE–CHITOSAN GRAFT COPOLYMER

In recent years, new types of chitosan-based materials have been developed by grafting HEMA and acrylic acid onto chitosan to obtain membranes for wound treatment. The poly(acrylic acid), which is biocompatible with antibacterial properties, is widely used in adhesives and superabsorbent materials owing to its pendant carboxylic groups (Lee et al., 1999; Athawale and Lele, 2001). Polymers grafted by acrylic acid become highly hydrophilic materials and interesting matrices for drug delivery systems. They make a good wound dressing because of their water retaining capacity and, once membranes can absorb large content of water, they are able to retain more drugs in their matrix when dried membranes are immersed in drug aqueous solution. Some of the previous studies have shown that the presence of HEMA in copolymers improves the biocompatibility of these materials

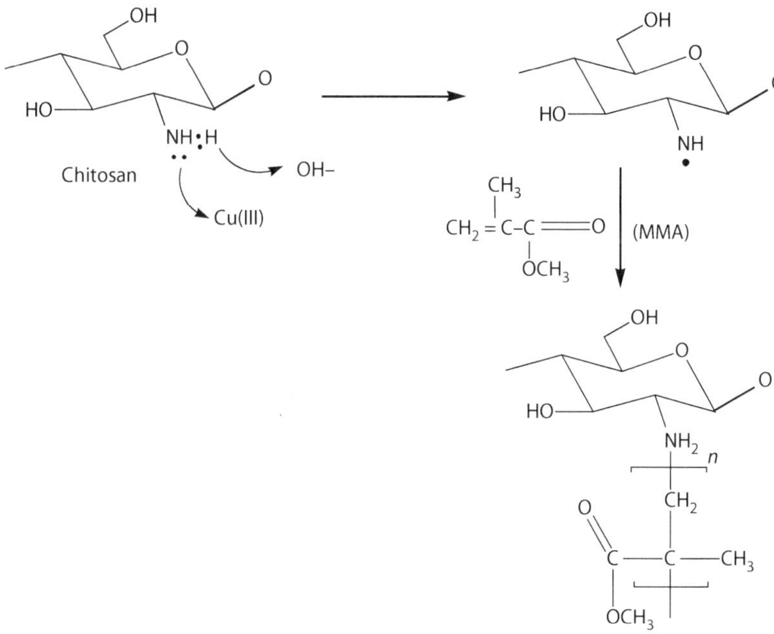

FIGURE 7.1   Synthesis of chitosan-*graft*-poly(MMA).

(Carenza, 1992; Singh and Ray, 1994). Keeping these factors in mind, chitosan-based membranes to be applied on wound healing as topical drug delivery systems were developed by graft copolymerization of acrylic acid and HEMA onto chitosan using cerium ammonium nitrate as chemical initiator as shown in Figure 7.2 (Santos et al., 2006). In this study, swelling degree, cytotoxicity, thrombogenicity, and hemolytic activity of these membranes were evaluated.

In recent years, macromolecules with pronounced amphiphilic character have become a subject of great interest. In fact, amphiphilic synthetic polymers comprise a wide range of graft, block, and star copolymers with a variety of functional groups and an almost unlimited number of their combinations. This opens up a possibility of close imitation of polymer–surfactant interactions in biological tissues. In particular, membrane cells are inter- and intracellularly stabilized by biopolymers, such as proteins and polysaccharides (Darnell et al., 1986). The study of the interaction between analogous polymers and bilayer-forming surfactants is, therefore, highly relevant to a better understanding of the organization of these complex structures. Recently, the grafting of 2-hydroxyethyl acrylate (HEA), HEMA, and *N*-vinylpyrrolidone onto chitosan in aqueous solutions using ^{60}Co γ-radiation technique and the interaction with surfactant were studied (Dergunov et al., 2008). It was observed that the grafting yield increased with the increase in absorbed dose. The degree of grafting increased with increase in the radiation dose. The obtained graft copolymers showed the solubility in water in wide pH interval. The interactions between grafted chitosan copolymers with sodium dodecyl sulfate (SDS) were studied in an aqueous solution. It was found that there is a narrow molar ratio of SDS/cation (~0.40–0.70) depending on different grafted copolymers, at which the turbidity of SDS-grafted chitosan complex has a maximum due to the formation of water-insoluble interpolymer aggregates via the SDS attached to the polymer chain. The turbidity falls sharply with the further addition of excessive SDS, which forms micelle in the solution and causes the deaggregation of the interpolymer aggregates and also because of the precipitation of complexes and returns to the original level. In this study, the morphological properties of thin films of SDS-grafted chitosan complex were also investigated. It was observed that on heating the insoluble complex of SDS and grafted chitosan in water, superstructures were formed.

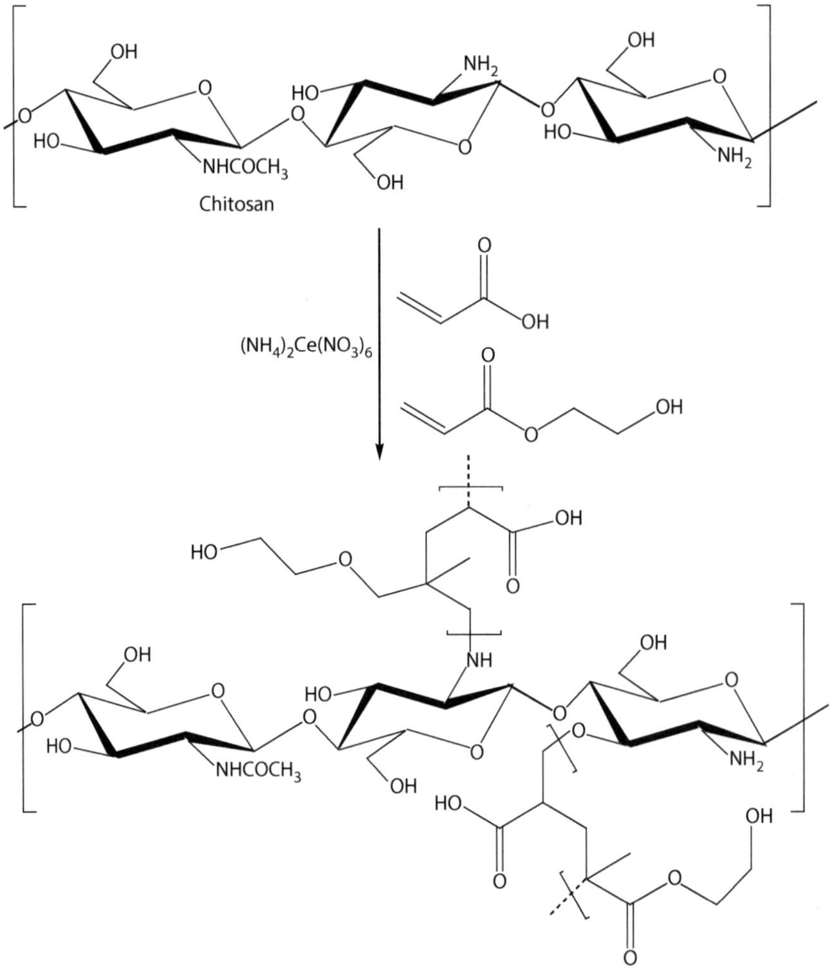

FIGURE 7.2   Graft copolymerization of HEMA and acrylic acid onto chitosan.

Graft copolymerization of HEA onto chitosan using ammonium persulfate (APS) as an initiator was carried out in an aqueous solution (Grigoriy et al., 2008). The effects of APS, HEA concentration, reaction temperature, and duration of graft copolymerization were studied by determining the grafting parameters, such as grafting percentage and grafting efficiency. It was observed that the polymerization rate is much more sensitive to the concentration of the HEA than it is to the concentration of the initiator. The grafted copolymer samples were found to be soluble in water and at alkaline pH, which described an enhanced hydrophilic character as compared with the parent acetylated chitosan. SEM micrographs revealed that native chitosan seemed to have porous morphology, whereas chitosan-g-HEA seemed to have fibrous morphology. These results confirmed that graft polymerization of hydrophilic monomers, such as HEA onto chitosan, is a versatile tool for preparing polysaccharide-based advance multifunctional materials for wide application in medicine and pharmaceutics.

Photoinduced graft copolymerization of vinyl monomers onto polymer backbones has many advantages as compared to other methods of grafting by free radical polymerization. For instance, this grafting technique showed a controlled generation of radical sites on polymer backbones in addition to attaining higher grafting efficiencies (John et al., 1993). Recently, Sherbiny and Smyth (2010) carried out photoinduced graft copolymerization of HEMA onto carboxymethyl chitosan under nitrogen atmosphere in aqueous solution using 2,2-dimethoxy-2-phenyl acetophenone as

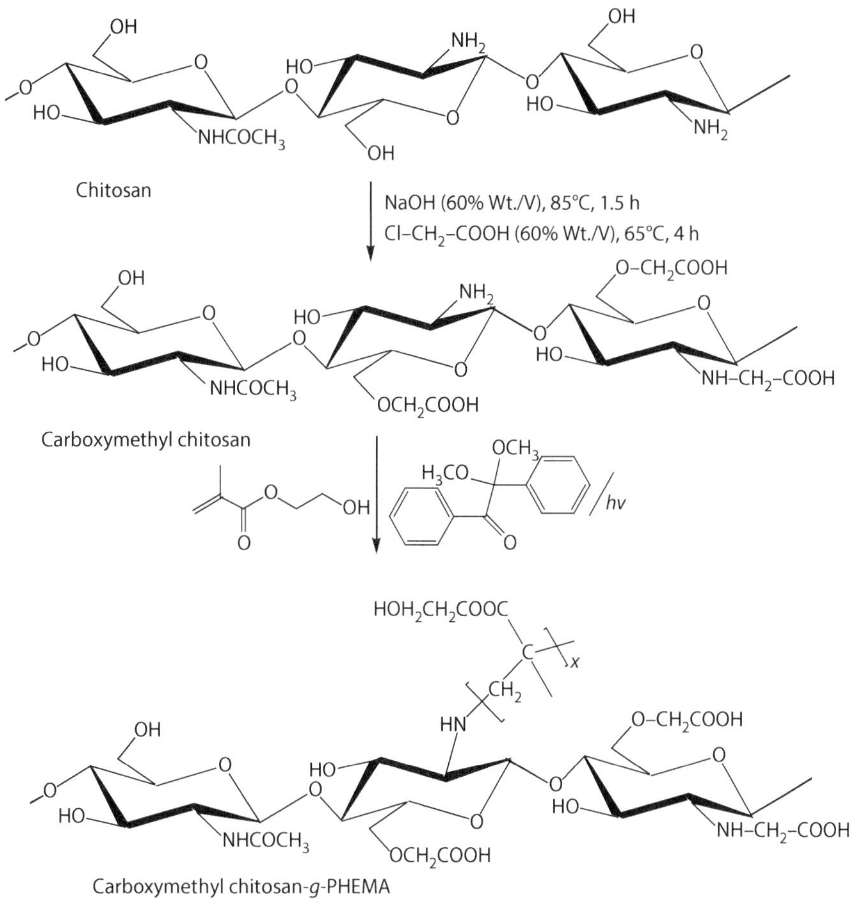

FIGURE 7.3 Synthesis of carboxymethyl chitosan and carboxymethyl chitosan-*g*-HEMA copolymer.

photoinitiator (Figure 7.3). In this study, the effects of HEMA and 2,2-dimethoxy-2-phenyl aceto-phenone concentrations and the reaction time on the grafting yield were investigated by determining the grafting percentage and grafting efficiency. Under the applied experimental conditions, the optimum grafting conditions were obtained at carboxymethyl chitosan = 0.2 g, HEMA = 0.615 mol/L, 2,2-dimethoxy-2-phenyl acetophenone = 0.0078 mol/L, and reaction time = 90 min. The synthesized copolymers revealed a self-ability to form physically cross-linked hydrogels as shown in Figure 7.4. The hydrogel nature of the copolymers was investigated by studying the solubility profiles and the cyclic swelling–deswelling behavior of copolymers with different grafting extents.

7.2.3 3-(TRIMETHOXYSILYL)PROPYL METHACRYLATE–CHITOSAN GRAFT COPOLYMER

Grafting of TMSPM onto chitosan backbone was successfully achieved by ceric ammonium nitrate (CAN), a redox initiator, in 1% acetic acid solution, under homogeneous conditions as shown in Figure 7.5 (Prabaharan and Mano, 2007). The TMSPM-grafted chitosan was characterized by thermogravimetric analysis (TGA), x-ray diffraction (XRD), and solubility studies. TGA results showed that the thermal properties of chitosan are changed by the grafting copolymerization. XRD studies showed that because of the grafting of TMSPM onto chitosan backbone, the original crystallinity of chitosan was destroyed. The incorporation of TMSPM to the chitosan chains increased its solubility in water. The reaction conditions, such as initiator concentration, monomer concentration, reaction temperature, and time, had great influence on grafting copolymerization. The optimum conditions

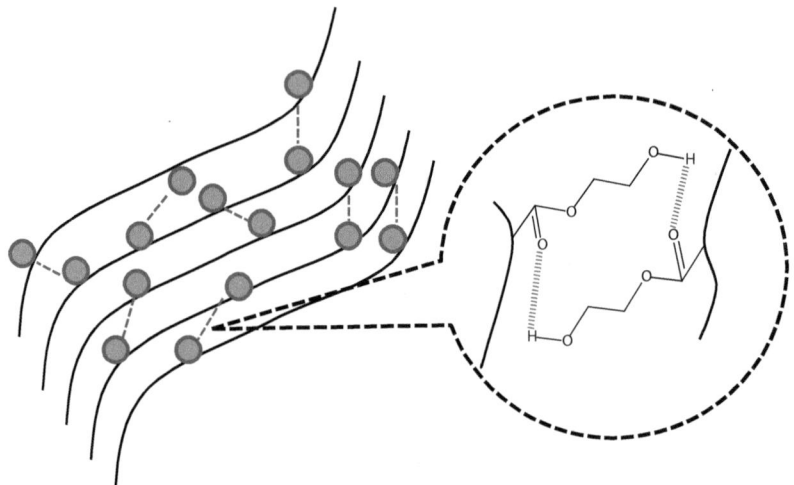

FIGURE 7.4   Cross-linking of carboxymethyl chitosan-*g*-HEMA copolymer through H-bond formation.

for graft copolymerization were determined to be the following: chitosan amount 5 g/L, acetic acid 2% w/w, reaction temperature 70°C, TMSPM 0.3 mol/L, CAN concentration 0.01 mol/L, and time 2 h. The maximum grafting and grafting efficiency obtained under these conditions were 1440% and 97%, respectively. Chitosan-*g*-TMSPM is a novel kind of silicone-containing biomaterial that can be useful for several biomedical applications. Because chitosan-*g*-TMSPM is amphiphilic in nature; it may form the polymeric micelles in the aqueous medium. These micelles can be used as a promising delivery carrier for the entrapment and controlled release of hydrophobic drugs.

7.2.4   (*N*,*N*-DIMETHYLAMINO)ETHYL METHACRYLATE–CHITOSAN GRAFT COPOLYMER

The homopolymer of DMA and some copolymers, including a DMA segment, have exhibited novel thermal sensitivity that is promising for potential applications, such as drug delivery (Singh and Ray, 1997; Sinha et al., 2004). Recently, chitosan-based graft copolymers were synthesized through homogeneous graft copolymerization of DMA onto chitosan derivative, *N*-carboxyethylchitosan, in aqueous solution by using APS as the initiator (Kang et al., 2006). In this study, the effect of polymerization variables, including initiator concentration, monomer concentration, reaction time, and temperature, on grafting percentage was studied. XRD, FTIR, differential scanning calorimetry (DSC), and TGA were used to characterize the graft copolymers. These analyses confirmed the introduction of the poly(DMA) side chain to the *N*-carboxyethylchitosan backbone by graft copolymerization. Surface tension and turbidity measurements, as well as temperature–variable ^{1}H-NMR analysis of the *N*-carboxyethylchitosan-*g*-poly(DMA) aqueous basic solution illustrated the thermal sensitivity of the graft copolymer due to the existence of tertiary amino groups in the poly(DMA) side chain.

7.2.5   METHYL ACRYLATE–CHITOSAN GRAFT COPOLYMER

A highly deacetylated chitosan powder was successfully trichloroacetylated under heterogeneous conditions. The trichloroacetylated chitosan powder was subsequently graft copolymerized heterogeneously with MA using a manganese carbonyl ($Mn_2(CO)_{10}$) coinitiator photoactivated with 436 nm light at room temperature as outlined in Figure 7.6 (Jenkins and Hudson, 2002). Manganese carbonyl can be photoactivated to react with trichlorinated carbons to generate carbon-based radicals capable of initiating vinyl polymerizations (Rosa et al., 1996). This system promotes specific macroradical formation because trichloroacetyl groups were successfully bonded to the chitosan

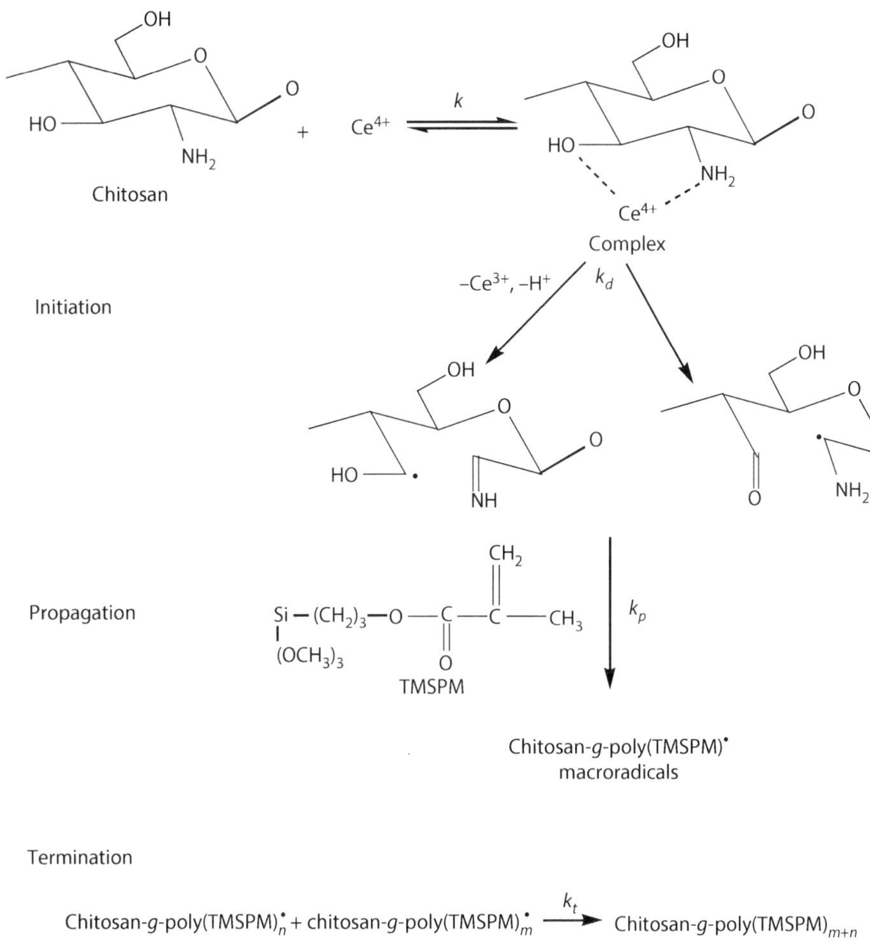

FIGURE 7.5 Mechanism for graft copolymerization of TMSPM onto chitosan.

backbone. Graft and homopolymer yields were obtained by monitoring product weights (of both polymer formed during the reaction and polymer removed with extraction) and the relative amounts of graft and homopolymer in the polymer extract as determined by gel permeation chromatography. Grafting yields achieved levels above 600%, whereas homopolymerization was observed to be on the order of 20–30% based on total weight of polymer formed.

Graft copolymerization of vinyl monomers onto chitosan has been reported using supernormal valence transition metals, such as Mn(VII), Cr(VI), V(V), Ag(III), Cu(III), and so on as initiators (Gao et al., 1998; Liu et al., 2002). Experiments have demonstrated that these transition metals are efficient and cheap initiators in grafting MA or MMA onto macromolecular structures, such as poly(acrylamide), nylon, and starch (Liu et al., 2003). Recently, chitosan-based redox systems such as potassium diperiodatonickelate (Ni (IV))–chitosan, potassium diperiodatocuprate (Cu(III))–chitosan, and potassium diperiodatoargentate (Ag(III))–chitosan were developed to initiate the graft copolymerization of MA onto chitosan in alkali aqueous solution (Liu et al., 2003, 2006a). In these studies, the effects of reaction variables such as monomer concentration, initiator concentration, reaction time, and temperature were investigated and the grafting conditions were optimized. The graft copolymers were shown to be effective compatibilizers in blends of poly(vinyl chloride) and chitosan. The results showed that the grafted products can enhance the thermal stability of pure chitosan and swell in many mixed solvents tested. On

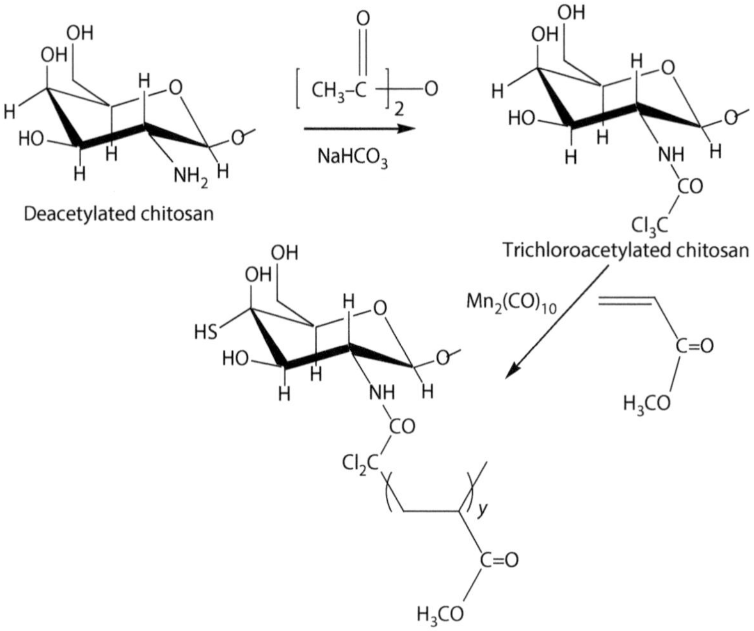

FIGURE 7.6 Synthetic scheme for graft copolymerization of MA onto chitosan.

the basis of the results, chitosan-based redox systems are considered to be efficient redox initiators for the graft copolymerization. The mechanism of initiation of MA grafting onto chitosan
using Ag(III)–chitosan redox system is shown in Figure 7.7. Because the activation energy of the
reaction employing Ag(III)–chitosan as initiator is low, the graft copolymerization is possible at
a mild temperature, compared with other initiators.

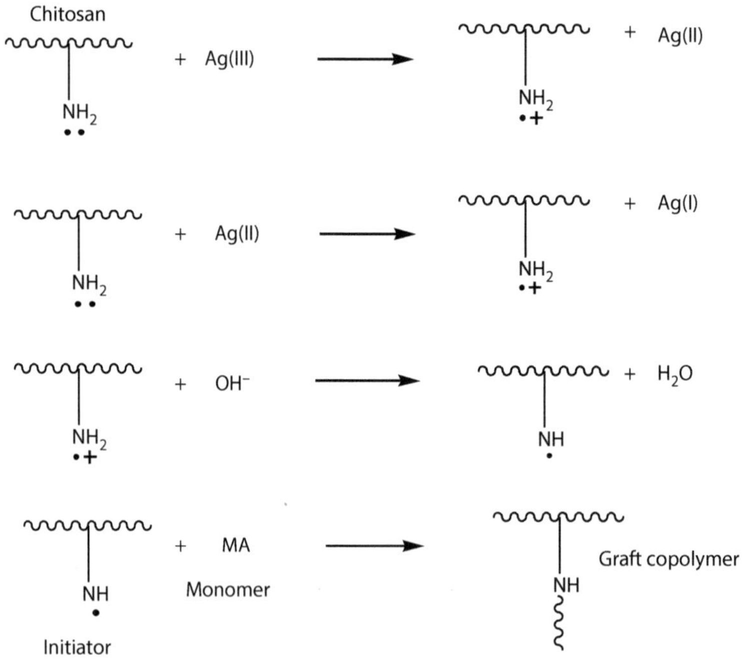

FIGURE 7.7 Graft copolymerization of MA onto chitosan using redox system.

Under the assistance of the efficient transfer reaction between the dithiocarbonate group and radicals in reversible addition–fragmentation chain transfer polymerization (RAFT) polymerization, the graft copolymerization of MA onto chitosan with high grafting efficiency was performed as shown in Figure 7.8 (Kuilin et al., 2007). In this study, the dithiocarbonate groups, the RAFT moiety, was introduced onto chitosan via the reaction of amino group with benzyl chloride and carbon disulfide. Thereafter, the modified chitosan was grafted onto poly(MA) in the presence of the common initiator. The effects of polymerization conditions, such as initiator concentration, monomer concentration, reaction time, and temperature, on grafting parameters were studied. The experimental results showed that the introduction of the dithiocarbonate groups onto chitosan has efficiently increased the graft efficiency, via minimizing the formation of homopolymers in graft copolymerization.

In recent years, graft copolymerization of vinyl monomers onto natural polymers using atom transfer radical polymerization (ATRP) has received much attention. This technique could potentially provide new ways to utilize the abundant natural polymers. It would enable a wide variety of molecular designs to afford novel types of tailored hybrid materials composed of natural polysaccharides and synthetic polymers. Such materials have been used as adhesives, membranes, and to

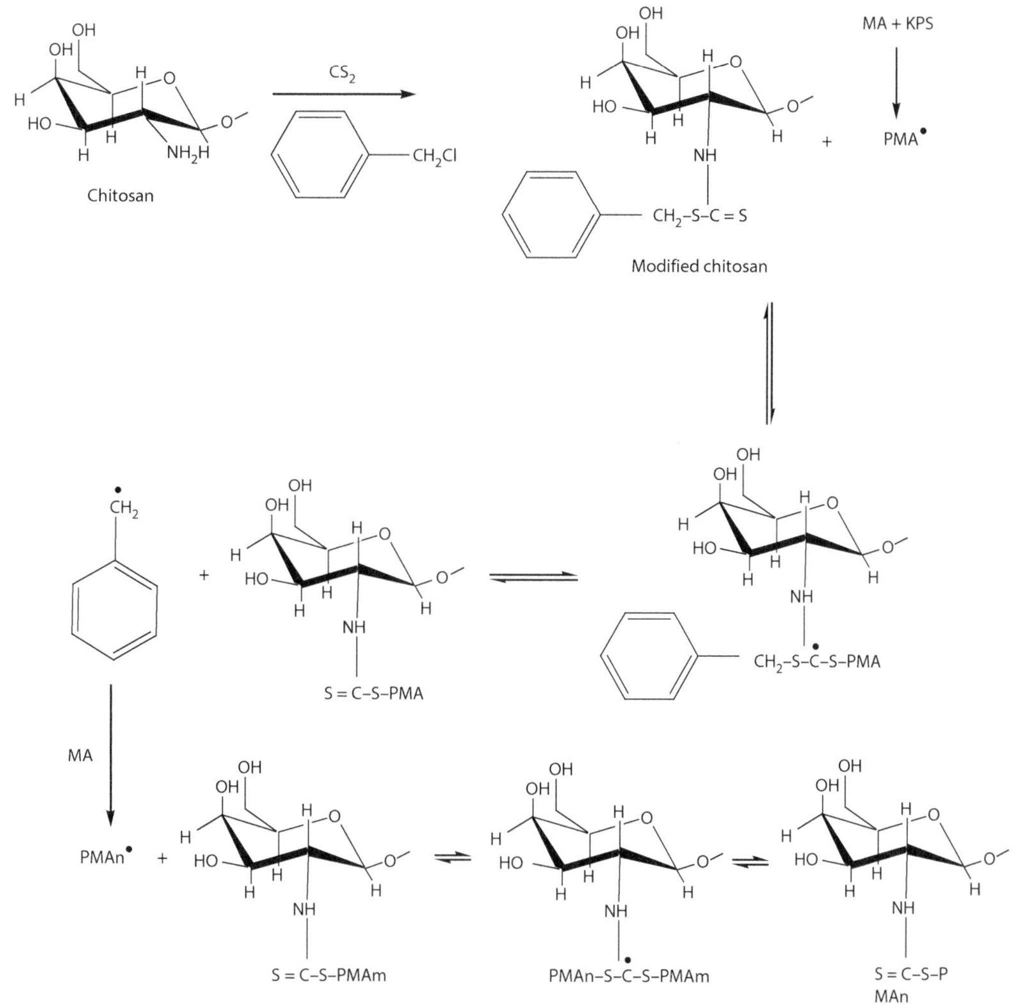

FIGURE 7.8 Mechanism for the graft polymerization of MA onto chitosan.

modify the surfaces of films and fibers. Tahlawy and Hudson (2003) prepared a chitosan macroinitiator by the reaction of chitosan with 2-bromo-isobutyryl bromide, after the chitosan amino group had been protected as the imine. It was demonstrated that chitosan macroinitiators would polymerize a methoxy-poly(ethylene glycol) methacrylate monomer via ATRP. The results indicated that controlled polymerizations occurred with first-order polymerization kinetics.

Recently, Wang et al. (2011) synthesized chitosan-g-MA using CAN as an initiator under nitrogen atmosphere in aqueous solution. The effects of concentration of CAN, the amount of MA, and reaction temperature on graft copolymerization were studied by determining the grafting percentage and grafting efficiency. The maximum grafting percentage and grafting efficiency obtained under the optimum conditions were found to be 640% and 68%, respectively.

7.2.6 BUTYL ACRYLATE–CHITOSAN GRAFT COPOLYMER

The graft copolymerization of BA onto chitosan in acetic acid aqueous solution was investigated, using the γ-irradiation method (Yu et al., 2003). The effect of synthesis variables in the graft copolymerization has been discussed in the light of grafting efficiency, grafting percentage, and homopolymer percentage. Hydrophilicity and impact strength of the films formed from copolymer solution were tested and their feasibility as seed coating was studied. In this work, increasing grafting percentage was observed when the monomer concentration and total dose were increased or when the chitosan concentration and reaction temperature were decreased. Under lower dose rates, the grafting percentage had no significant change, whereas above 35 Gy/min (dose rate), the grafting percentage decreased sharply. Compared with pure chitosan film, the chitosan-g-poly(BA) films had enhanced hydrophobic and impact strength for their practical use.

N-Maleamic acid–chitosan-g-BA was synthesized through the γ-ray irradiation polymerization using N-maleamic acid–chitosan as intermediate (Huang et al., 2005). In this study, the effects of synthesis variables in the graft copolymerization have been discussed in the light of grafting efficiency, grafting percentage, and homopolymer percentage. Increasing grafting percentage was observed when the monomer concentration and total dose were increased or when the reaction temperature was decreased. The DSC results showed that the N-maleamic acid–chitosan-g-poly(BA) has a glass-transition temperature (T_g) of −42°C. On similar lines, El-Shafei et al. (2005) studied the graft copolymerization of BA onto chitosan using potassium persulfate (KPS) as initiator. The results showed that the grafting percentage and the grafting efficiency increased by increasing KPS concentration up to 40 mmol/L and subsequently decreased thereafter.

Graft copolymerization of BA onto chitosan via phthaloylchitosan intermediate has been examined recently (Li et al., 2005). Because the intermediate phthaloylchitosan was soluble in organic solvents, in this study, the graft copolymerization was carried out in homogeneous system. Under appropriate irradiation dose and concentration of BA monomer, the grafting percentage of about 838% was obtained. The graft copolymers exhibited high swelling not only in aqueous acid but also in organic solvents. Owing to the poly(BA) side chains, the resulting graft copolymers exhibited glass-transition phenomena and showed improved thermal stability. Compared with the original chitosan, the graft copolymers exhibited unique chemical properties. Because phthaloylchitosan is easily deprotected to regenerate the free amino groups, the resulting graft copolymers had abundant unreacted amino groups, which made it possible for the graft copolymer to maintain the hydrophilic property and the bioactivity of chitosan.

The graft copolymerization of BA onto chitosan was carried out via 6-O-maleoyl-N-phthaloylchitosan as intermediate using γ-irradiation (Huang et al., 2006). Because the intermediate 6-O-maleoyl-N-phthaloylchitosan is soluble in organic solvents, this method not only enabled the grafting reaction to be carried out in a homogeneous system but also retained the abundant amino groups in the chitosan-g-poly(BA) copolymers. The results showed that the graft extent was dependent on the irradiation dose and the concentration of BA monomer, and copolymers with grafting above 100% were readily prepared.

7.2.7 Sodium Acrylate–Chitosan Graft Copolymer

Natural polymers, such as celluloses or starches, can be prepared as superabsorbent polymers through radical graft polymerization with vinyl monomers and cross-linking (Li et al., 2007; Peng et al., 2008). Carboxymethyl chitosan is a very important chitosan derivative showing very good water solubility and biocompatibility (Pang et al., 2007). Yu et al. (2009) prepared a superabsorbent polymer by graft copolymerization of SA and 1-vinyl-2-pyrrolidone onto the chain of N,O-carboxymethyl chitosan. The surface morphologies before and after the polymerization were examined by SEM. The surface of the N,O-carboxymethyl chitosan was found to be very smooth before modification. However, after the modification, their surface morphology was found to be changed into porous morphology. This change in surface morphology supported the occurrence of graft polymerization. In another study, Yu et al. (2010) grafted acrylic acid and SA onto the chain of hydroxyethyl chitosan to obtain superabsorbent polymer hydroxyethyl chitosan-g-poly(acrylic acid-co-SA) and studied the morphology and swelling properties of the products.

The graft copolymerization of SA onto chitosan in alkali medium using redox system, potassium diperiodacuprate(III)–chitosan, was investigated (Liu et al., 2006b). The effects of reaction variables, such as the initiator concentration, the ratio of monomer to chitosan, and pH, as well as reaction temperature and time were investigated, and the grafting conditions were optimized. The result showed that potassium diperiodacuprate(III)–chitosan redox system is an efficient initiator for the graft copolymerization of SA onto chitosan. In this study, the chitosan-g-SA graft copolymer was used as the compatibilizer in blends of poly(MMA) and chitosan. The SEM studies indicated that the graft copolymer improved the compatibility and biodegradability of the blend.

7.3 APPLICATIONS

7.3.1 Drug Delivery

In recent years, chitosan derivatives have been considered as promising drug delivery systems, which can greatly prolong drug duration in bloodstream and show controlled release properties and improve the utility of drugs and reduce toxic side effects (Prabaharan and Mano, 2005). It is very difficult to control drug release for native chitosan, so that all sorts of chemically modified chitosans have been investigated for controlled drug delivery applications. So far, drug carriers based on chitosan derivatives are mainly prepared by five methods: coacervation (precipitation) method, emulsion-droplet coalescence method, ionic-gelation method, reverse micellar method, and polymeric micelle method (Liu et al., 2011). Among them, polymeric micelle and ionic-gelation methods are commonly used. Most of the approaches in the synthesis of polymeric micelles have been applied to the synthesis of amphiphilic copolymers. There have been many reports of hydrophobic modifications of chitosan and micelle formation by self-aggregation in aqueous solution (Kim et al., 2008). These modifications can be used to synthesize chitosan amphiphilic polymers by introducing hydrophobic groups into chitosan. In the aqueous phase, the hydrophobic segments constitute a hydrophobic core of self-assembled nanoparticles surrounded by external shells. Thus, the internal core can serve as a depot for hydrophobic drugs. In addition, the plasma halftime of intravenously injected nanoparticles can be relatively prolonged because of limited uptake by the liver and spleen.

Because poly(MMA) is a biocompatible polymer and is hydrophobic in nature, it was grafted onto chitosan to form amphiphilic nanocarriers for insulin release (Liu et al., 2011). In this study, insulin was successfully entrapped into the chitosan-g-poly(MMA) nanoparticles because of the ionic interaction between negative insulin and positive hydrophilic chains of nanoparticles. The release studies revealed that insulin was released in a biphasic way, that is, an initial rapid release period followed by a step of slower release. The burst effect was observed in 1 h, in which nearly 61.5%, 53.7%, and 46.9% of the drug was released from nanoparticles with loading content of 14.5%, 8.06%, and 3.42%, respectively. After this initial effect, insulin was released in a continuous way for up to 10 h, reaching the percentage of cumulative release close to 98%. Insulin release rate was influenced by the amount of drug loaded; a higher loading capacity provided a faster release rate. It was found that higher levels of loaded drug lead to a wider concentration gap between the polymeric nanoparticles and the release medium, which caused a higher diffusion rate. The increase in drug content increases the amount of drug close to the surface as well as the drug in the outer hydrophilic area of nanoparticles. The former was responsible for an increased initial burst, whereas the latter caused an increase during the induction period.

Poly(HEMA) has been widely investigated as a biomaterial candidate with various potential applications including soft-tissue replacement, contact lenses, and vascular prostheses (Montheard et al., 1992; Filmon et al., 2002). This significant interest has been paid to poly(HEMA) in the biomedical fields because of its various outstanding characteristics such as biocompatibility and the high hydrophilicity owing to the large number of pendant primary hydroxyl groups. The hydroxyl groups present in poly(HEMA) can allow the formation of self-physically cross-linked hydrogel matrices through the formation of H-bonds in the absence of cross-linking agents. Owing to this beneficial property for drug delivery purposes, Sherbiny and Smyth (2010) grafted poly(HEMA) onto carboxymethyl chitosan to form a cross-linked hydrogel. The self-hydrogel nature of the developed copolymers was investigated by studying their cyclic swelling–deswelling behavior. The graft copolymers with the higher graft percent (857%) attained the lowest swelling values at equilibrium. Decreasing the grafting percent to 19% leads to a marked increase in the equilibrium swelling. The authors suggested that this behavior might be attributed to the association of the hydroxyl groups in the grafted side chains through the formation of H-bonds leading to a reduction in the swelling capacity of the copolymer. These investigations of the carboxymethyl chitosan-g-poly(HEMA) copolymers showed that they can be tailored and exploited as promising carriers for drug delivery purposes.

7.3.2 Wound Healing

Chitosan is currently receiving a great deal of attention for its wound-healing applications. In the area of wound healing, chitosan has the capacity to reduce scar tissue (fibroplasia) by inhibiting the formation of fibrin in wounds (Lloyd et al., 1998). Chitosan can enhance blood coagulation by an independent mechanism of the classical coagulation cascade and appear to be an interaction between negative charges of cell membranes of erythrocytes and positive charges of chitosan filaments (Rao and Sharma, 1997; Okamoto et al., 2003). One hypothesis advanced to explain the ability of chitosan to enhance wound healing is related to its biodegradability (Berger et al., 2004). Lysozyme, normally produced by macrophages, hydrolyzes chitosan and its derivatives to oligomers, which activate macrophages to produce nitric oxide, activated oxygen species, tumor necrosis factor-α, interferon, and interleukin-1. Activated macrophages increase the production of lysozyme, chitinase, and N-acetyl-β-D-glucosaminidase, which catalyze the total depolymerization to monomers. The monomeric aminosugars become available to fibroblasts that proliferate under the action of interleukin-1, for incorporation into chondroitin 4- and 6-sulfate, hyaluronan, and keratin sulfate, thus guiding the ordered deposition of collagen and also accelerating the wound-healing process (Muzzarelli, 1997). All these properties, including its bioadhesivity, make chitosan an excellent biomaterial to treat wounds and scars.

Modifications on the chitosan structure can be carried out to adequate it to the intended application, such as hydrosolubility, adhesivity, and hemocompatibility. The modification of chitosan by graft copolymerization has been studied deeply by several authors (Prabaharan et al., 2007, 2008) because it can provide materials with desired properties through the appropriate choice of the molecular characteristics of the side chain to be grafted (Prabaharan and Jayakumar, 2009). In recent years, the wound-healing materials with controlled drug-release ability have received much attention because of their dual effects. In this context, a novel type of chitosan-based wound-healing material was developed by grafting vinyl monomers, acrylic acid, and HEMA, onto chitosan

(Athawale and Lele, 2001; Lee et al., 2005). Poly(acrylic acid), which is biocompatible and which possesses antibacterial properties, is widely used in adhesives and superabsorbent materials because of its pendant carboxylic groups. Polymers grafted by acrylic acid become highly hydrophilic materials and interesting matrices for drug delivery systems. They make a good wound dressing because of their ability to retain more water and drug in the polymer matrix. It has been found that the presence of HEMA in copolymers improves the biocompatibility of these materials (Carenza, 1992). Recently, Santos et al. (2006) found that chitosan-g-poly(acrylic acid)-g-poly(HEMA) is the best matrix for drug delivery systems than chitosan-g-poly(acrylic acid) because of its good swelling properties and improved cytocompatibility, hemocompatibility, and thrombogenic character.

7.3.3 SEED COATINGS

Butyl acrylate is a hydrophobic and soft monomer; on grafting with chitosan, it is expected to increase the hydrophobicity and flexibility of the macromolecule. Owing to these properties, BA-grafted chitosan would have potential to be used as seed-coating material. Yu et al. (2003) synthesized chitosan-g-poly(BA) as seed-coating materials by γ-ray irradiation-induced graft copolymerization of chitosan with BA. The hydrophilicity and impact strength of the films formed from graft copolymer solution were tested and their feasibility as seed coating was studied. The results of this study showed that the chitosan-g-poly(BA) films have enhanced hydrophobic and impact strength for their practical use compared with pure chitosan. Avirulence is one of the most important properties that are required for seed coating. The material used as seed coating cannot inhibit the germination of seed. In this work, the effect of the BA to chitosan ratio on the germination percentage of coating seeds was investigated. In the range 0.5–1 (BA/chitosan), no significant differences in germination percentage between the treated seeds and the control was observed. When the ratio of BA to chitosan was up to 2, the germination percentage of coating seeds was found to be decreased. This may be because of the increasing hydrophobicity of seed coating, which inhibits the supply of water for seeds. On the basis of these results, the chitosan-g-poly(BA) films can be expected to have broad application for seed coating. In another study, Huang et al. (2005) synthesized N-maleamic acid–chitosan-g-poly(BA) for seed-coating application. The films formed by this copolymer have found to be enhanced hydrophobically for their practical use, which can be expected to have broad application for seed coating, antistaling agent of vegetable and fruit.

7.3.4 ABSORBENTS

Superabsorbent polymers are polymers with a network structure and an appropriate degree of crosslinking, which can absorb a large amount of water (Omidian et al., 2005). These polymers have been extensively used as absorbents in personal care products, matrices for enzyme immobilization, materials for agricultural mulches, and matrices for controlled release devices (Hany, 2007). For superabsorbent polymers used as medical and sanitation materials, they must have high absorption rate and low residual toxic chemicals. Usually, the porous superabsorbent polymers are prepared by adding chemical reagents to produce bubbles in the polymer (Chen et al., 1999). Although the water absorption rate was high in those superabsorbent polymers, toxic chemicals are introduced during this process. To overcome this problem, recently, the porous superabsorbent polymer was fabricated using natural-based polymers through the solvent precipitation and freeze-drying method.

Yu et al. (2009) reported the preparation of a superabsorbent polymer by graft copolymerization of SA and 1-vinyl-2-pyrrolidone along the chains of N,O-carboxymethyl chitosan. Because 1-vinyl-2-pyrrolidone is a nonionic monomer, poly(1-vinyl-2-pyrrolidone) (PVPD) segments were introduced to increase the water affinity of the polymers and, therefore, control the haloduric property of the superabsorbent polymers. In addition, because the PVPD segments can be easily modified, the introduction of PVPD segments provides the possibilities to further functionalize superabsorbent polymers. By studying the water absorption of the polymer synthesized under different conditions, the optimal conditions for synthesizing the polymer with the highest swelling ratio were defined. The results of this study showed that the water absorption rates of the prepared polymers were high, the swelling processes of the polymers showed first-order kinetics, and the swelling ratio of the polymer was pH dependent.

Hydroxyethyl chitosan-*g*-poly(acrylic acid-*co*-SA) superabsorbent polymer was prepared through graft copolymerization of acrylic acid and SA onto the chain of hydroxyethyl chitosan (Yu et al., 2010). This superabsorbent polymer was further treated by the solvent precipitation method and by the freeze-drying method. The water-uptake studies revealed that the water absorption rate of the treated polymer was greatly increased and the microstructure of the treated polymer was changed from small pores to macropores. The swelling processes of the polymers before and after modification fit first-order dynamic processes. In this study, the amount of the residual acrylic acid was found to be greatly decreased after treatments. These studies revealed that chitosan derivatives grafted with acrylic monomers have promising applications in various fields as superabsorbents.

7.3.5 Finishing Agents

Chitosan has a great potential for a wide range of applications because of its biodegradability, biocompatibility, antimicrobial activity, nontoxicity, and versatile chemical and physical properties. Recently, an attempt has been made to use chitosan-*g*-poly(BA) copolymers as ecofriendly textile finishing agents (El-Shafei et al., 2005). In this study, chitosan-*g*-poly(BA) copolymers were applied to cotton fabric in the presence and absence of low formaldehyde cross-linking agent. Fabric performance was assessed through monitoring, nitrogen content, crease recovery angle, tensile strength, and elongation at break. It was observed that the introduction of the copolymer and the cross-linking agent to cotton fabric enhances the performance of the cotton fabric to a great extent. In another study, chitosan-*g*-poly(BA) copolymer was applied to cotton fabrics in the presence and absence of an easy-care finishing agent (Knittex FLC) (El-Shafei et al., 2005a). The consecutive treatment of the fabric with copolymer followed by finishing agent noticeably improved the performance of the fabric assessed from the nitrogen content, crease angle recovery, tensile strength, and breaking strength.

7.4 SUMMARY

During the last few years, an impressive number of chitosan derivatives has been developed for biomedical, industrial, and agricultural applications. The approach of chemical modification of chitosan or grafting chitosan with acrylic monomers, such as MAA, HEMA, TMSPM, MA, DMA, BA, and SA, has received potential importance in drug delivery, wound healing, and seed coatings because the resulting materials exhibit the favored properties of both materials. Chitosan grafted with acrylic monomers has shown an improved drug-loading capacity and controlled release behavior with other unique properties such as hydrophilicity, mechanical strength, biocompatibility, and biodegradability. Owing to the enhanced hydrosolubility, adhesivity, hemocompatibility, and antibacterial properties, graft copolymers of chitosan/acrylic monomers have the potential to be used as wound-healing materials. Because the grafting of chitosan with BA results in hydrophobicity and flexibility of the resulting macromolecules, it would have the potential to be used as seed-coating material. Recent research works have shown that chemically modified chitosan with acrylic monomers could be used as absorbents in personal care products, matrices for enzyme immobilization, and controlled release as the modified chitosans exhibit good swelling behavior. As a result of the versatile physical, chemical, and biological properties, chitosan derivatives have the potential to be used as finishing agents for textile materials. From this chapter, it is clear that a number of studies have been conducted on chitosan grafted with acrylic monomers in the form of drug delivery carriers, wound-healing materials, seed coatings, absorbents, and finishing agents. The results of these works indicate that chitosan grafted with acrylic monomers are promising materials, namely, for biomedical, agricultural, and industrial applications.

REFERENCES

Athawale, V. D. and V. Lele. 2001. Recent trends in hydrogels based on starch-*graft*-acrylic acid: A review. *Starch/Staerke* 53:7–13.

Berger, J., M. Reist, J. M. Mayer, O. Felt, and R. Gurny. 2004. Structure and interactions in chitosan hydrogels formed by complexation or aggregation for biomedical applications. *European Journal of Pharmacy and Biopharmaceutics* 57:35–52.

Carenza, M. 1992. Recent achievements in the use of radiation polymerization and grafting for biomedical applications. *Radiation Physics and Chemistry* 39:485–493.

Chen, J., H. Park, and K. Park. 1999. Synthesis of superporous hydrogels: Hydrogels with fast swelling and superabsorbent properties. *Journal of Biomedical Materials Research* 44:53–62.

Darnell, J., H. Lodish, and D. Baltimore. 1986. *Molecular Cell Biology.* New York: Scientific American Books.

Dergunov, S. A., I. K. Nam, T. P. Maimakov, Z. S. Nurkeeva, E. M. Shaikhutdinov, and G. A. Mun. 2008. Study on radiation-induced grafting of hydrophilic monomers onto chitosan. *Journal of Applied Polymer Science* 110:558–563.

El-Shafei, A., S. Shaarawy, and A. Hebeish. 2005. Grafting copolymerization of chitosan with butyl acrylate: Application to cotton. *Tinctoria* 102:23–30.

El-Shafei, A., S. Shaarawy, and A. Hebeish. 2005a. Graft copolymerization of chitosan with butyl acrylate and application of the copolymers to cotton fabric. *Polymer-Plastics Technology and Engineering* 44:1535–1547.

El-Tahlawy, K. and S. M. Hudson. 2003. Synthesis of a well-defined chitosan graft poly(methoxy polyethylene glycol methacrylate) by atom transfer radical polymerization. *Journal of Applied Polymer Science* 89:901–912.

Filmon, R., F. Grizon, M. F. Basle, and D. Chappard. 2002. Effects of negatively charged groups (carboxymethyl) on the calcification of poly(2-hydroxyethyl methacrylate). *Biomaterials* 23:3053–3059.

Gao, J., J. Yu, W. Wang, L. Chang, and R. Tian. 1998. Comparison of transition metals in the graft copolymerization of vinyl monomers onto starch. *Journal of Macromolecular Science, Part A—Pure and Applied Chemistry* 35:483–494.

Grigoriy, A. M., S. N. Zauresh, A. D. Sergey, K. N. Irina, P. M. Tauzhan, M. S. Erengaip, S. C. Lee, and K. Park. 2008. Studies on graft copolymerization of 2-hydroxyethyl acrylate onto chitosan. *Reactive and Functional Polymers* 68:389–395.

Hany, E. H. 2007. Synthesis and water sorption studies of pH sensitive poly(acrylamide-*co*-itaconic acid) hydrogels. *European Polymer Journal* 43:4830–4838.

Huang, M., X. Shen, Y. Sheng, and Y. Fang. 2005. Study of graft copolymerization of *N*-maleamic acid–chitosan and butyl acrylate by γ-ray irradiation. *International Journal of Biological Macromolecules* 36:98–102.

Huang, M., X. Xia, Z. Zhang, L. Liu, and Y. Fang. 2006. Homogeneous graft copolymerization of chitosan with butyl acrylate by γ-irradiation via a 6-*O*-maleoyl-*N*-phthaloyl-chitosan intermediate. *Journal of Applied Polymer Science* 102:489–493.

Jenkins, D. W. and S. M. Hudson. 2002. Heterogeneous graft copolymerization of chitosan powder with methyl acrylate using trichloroacetyl-manganese carbonyl co-initiation. *Macromolecules* 35:3413–3419.

John, G., C. K. S. Pillai, and A. Ajayaghosh. 1993. Photo-induced graft copolymerization of methyl methacrylate onto cellulose containing benzoyl xanthate chromophore. *Polymer Bulletin* 30:415–420.

Kang, H. M., Y. L. Cai, and P. S. Liu. 2006. Synthesis, characterization and thermal sensitivity of chitosan-based graft copolymers. *Carbohydrate Research*, 341:2851–2857.

Kim, J. H., Y. S. Kim, K. Park, E. Kang, S. Lee, H. Y. Nam, K. Kim et al. 2008. Self-assembled glycol chitosan nanoparticles for the sustained and prolonged delivery of antiangiogenic small peptide drugs in cancer therapy. *Biomaterials* 29:1920–1930.

Kuilin, D., Z. Yaqin, J. Na, L. Jing, Z. Xiangyang, and T. Hua. 2007. A dithiocarbonate group-assisted graft copolymerization of methyl acrylate onto chitosan. *Chemical Journal on Internet* 9:19–30.

Kurita, K., H. Lkeda, Y. Yoshida, M. Shimojoh, and M. Harata. 2002. Chemoselective protection of the amino groups of chitosan by controlled phthaloylation: Facile preparation of a precursor useful for chemical modifications. *Biomacromolecules* 3:1–4.

Lee, J. S., R. N. Kumar, H. D. Rozman, and B. M. N. Azemi. 2005. Pasting, swelling and solubility properties of UV initiated starch-*graft*-poly (AA). *Food Chemistry* 91:203–211.

Lee, J. W., S. Y. Kim, S. S. Kim, Y. M. Lee, K. H. Lee, and S. J. Kim. 1999. Synthesis and characteristics of interpenetrating polymer network hydrogel composed of chitosan and poly(acrylic acid). *Journal of Applied Polymer Science* 73:113–120.

Li, A., J. P. Zhang, and A. Q. Wang. 2007. Utilization of starch and clay for the preparation of superabsorbent composite. *Bioresource Technology* 98:327–332.

Li, Y., L. Liu, X. Shen, and Y. Fang. 2005. Preparation of chitosan/poly(butyl acrylate) hybrid materials by radiation-induced graft copolymerization based on phthaloylchitosan. *Radiation Physics and Chemistry* 74:297–301.

Liu, Y., Y. Li, J. Lv, G. Wu, and J. Li. 2005. Graft copolymerization of methyl methacrylate onto chitosan initiated by potassium ditelluratocuprate (III). *Journal of Macromolecular Science, Part A—Pure and Applied Chemistry* 42:1169–1180.

Liu, Y., Z. Liu, Y. Zhang, and K. Deng. 2002. Graft copolymerization of methyl acrylate onto chitosan initiated by potassium diperiodatonickelate (IV). *Journal of Macromolecular Science, Part A—Pure and Applied Chemistry* 39:129–143.

Liu, Y., Z. Liu, Y. Zhang, and K. Deng. 2003. Graft copolymerization of methyl acrylate onto chitosan initiated by potassium diperiodatocuprate (III). *Journal of Applied Polymer Science* 89:2283–2289.

Liu, J. H., Q. Wang, and A. Q. Wang. 2007. Synthesis and characterization of chitosan-*g*-poly(acrylic acid)/sodium humate superabsorbent. *Carbohydrate Polymers* 70:166–173.

Liu, Z., G. Wu, and Y. Liu. 2006a. Graft copolymerization of methyl acrylate onto chitosan initiated by potassium diperiodatoargentate (III). *Journal of Applied Polymer Science* 101:799–804.

Liu, Y., R. Zhang, J. Zhang, W. Zhou, and S. Li. 2006b. Graft copolymerization of sodium acrylate onto chitosan via redox polymerization. *Iranian Polymer Journal* 15:935–942.

Liu, Z., G. Zhao, J. Yu, J. Zhang, X. Ma, and G. Han. 2011. Preparation and properties of chitosan-*graft*-poly(methyl methacrylate) nanoparticles using potassium diperiodatocuprate (III) as an initiator. *Journal of Applied Polymer Science* 120:2707–2715.

Lloyd, L. L., J. F. Kennedy, P. Methacanon, M. Paterson, and C. J. Knill. 1998. Carbohydrate polymers as wound management aids. *Carbohydrate Polymers* 37:315–322.

Montheard, J. P., M. Chatzopoulos, and D. Chappard. 1992. 2-Hydroxyethyl methacrylate HEMA: Chemical properties and applications in biomedical fields. *Polymer Reviews* 32:1–34.

Muzzarelli, R. A. A. 1997. Human enzymatic activities related to the therapeutic administration of chitin derivatives. *Cellular and Molecular Life Sciences* 53:131–140.

Nishimura, S., O. Kohgo, K. Kurita, and H. Kuzuhara. 1991. Chemospecific manipulations of a rigid polysaccharide: Syntheses of novel chitosan derivatives with excellent solubility in common organic solvents by regioselective chemical modifications. *Macromolecules* 24:4745–4748.

Okamoto, Y., R. Yano, K. Miyatake, I. Tomohiro, Y. Shigemasa, and S. Minami. 2003. Effects of chitin and chitosan on blood coagulation. *Carbohydrate Polymers* 53:337–342.

Omidian, H., J. G. Rocca, and K. Park. 2005. Advances in superporous hydrogels. *Journal of Controlled Release* 102:3–12.

Pang, H. T., X. G. Cheng, H. J. Park, D. S. Cha, and J. F. Kennedy. 2007. Preparation and rheological properties of deoxycholate–chitosan and carboxymethylchitosan in aqueous systems. *Carbohydrate Polymers* 69:419–425.

Peng, G., S. M. Xu, Y. Peng, J. D. Wang, and L. C. Zheng. 2008. A new amphoteric superabsorbent hydrogel based on sodium starch sulfate. *Bioresource Technology* 99:444–447.

Prabaharan, M. 2008. Chitosan derivatives as promising materials for controlled drug delivery. *Journal of Biomaterials Applications* 23:5–36.

Prabaharan, M. and R. Jayakumar. 2009. Chitosan-*graft*-β-cyclodextrin scaffolds with controlled drug release capability for tissue engineering applications. *International Journal of Biological Macromolecules* 44:320–325.

Prabaharan, M. and J. F. Mano. 2005. Chitosan-based particles as controlled drug delivery systems. *Drug Delivery* 12:41–57.

Prabaharan, M. and J. F. Mano. 2007. Synthesis and characterization of chitosan-*graft*-poly(3-(trimethoxysilyl) propyl methacrylate) initiated by ceric (IV) ion. *Journal of Macromolecular Science, Part A—Pure and Applied Chemistry* 44:489–494.

Prabaharan, M., J. J. Grailer, D. A. Steeber, and S. Gong. 2008. Stimuli-responsive chitosan-*graft*-poly(N-vinylcaprolactam) as a promising material for controlled hydrophobic drug delivery. *Macromolecular Bioscience* 8:843–851.

Prabaharan, M., R. L. Reis, and J. F. Mano. 2007. Carboxymethyl chitosan-*graft*-phosphotidylethanolamine: Amphiphilic matrices for controlled drug delivery. *Reactive and Functional Polymers* 67:43–52.

Rao, S. B. and C. P. Sharma. 1997. Use of chitosan as biomaterial: Studies on its safety and haemostatic potential. *Journal of Biomedical Materials Research* A34:21–28.

Rosa, A., G. Ricciardi, E. J. Baerends, and D. J. Stufkens. 1996. Density functional study of the photodissociation of $Mn_2(CO)_{10}$. *Inorganic Chemistry* 35:2886–2897.

Santos, K. S. C. R., J. F. J. Coelho, P. Ferreira, I. Pinto, S. G. Lorenzetti, E. I. Ferreira, O. Z. Higa and M. H. Gil. 2006. Synthesis and characterization of membranes obtained by graft copolymerization of 2-hydroxyethyl methacrylate and acrylic acid onto chitosan. *International Journal of Pharmaceutics* 310:37–45.

Sashiwa, H., H. Kawasaki, A. Nakayama, E. Muraki, N. Yamamoto, and S. Aiba. 2002. Chemical modification of chitosan derivatives by simple acetylation. *Biomacromolecules* 3:1126–1128.

Sherbiny, I. M. E. and H. D. C. Smyth. 2010. Photo-induced synthesis, characterization and swelling behavior of poly(2-hydroxyethyl methacrylate) grafted carboxymethyl chitosan. *Carbohydrate Polymers* 81:652–659.

Singh, D. K. and A. R. Ray. 1994. Graft copolymerization of 2-hydroxyethylmethacrylate onto chitosan films and their blood compatibility. *Journal of Applied Polymer Science* 53:1115–1121.

Singh, D. K. and A. R. Ray. 1997. Radiation-induced grafting of *N,N′*-dimethylaminoethylmethacrylate onto chitosan films. *Journal of Applied Polymer Science* 66:869–877.

Sinha, V. R., A. K. Singla, S. Wadhawan, R. Kaushik, R. Kumria, K. Bansal, and S. Dhawan. 2004. Chitosan microspheres as a potential carrier for drugs. *International Journal of Pharmacy* 274:1–33.

Wang, P., L. Liu, Z. Y. Wei, and M. Qi. 2011. Facile graft copolymerization of chitosan powder with methylacrylate using ceric ammonium nitrate. *Advanced Materials Research* 197/198:563–566.

Yu, C., L. Y. Fei, and T. H. Min. 2010. Hydroxyethyl chitosan-*g*-poly(acrylic acid-*co*-sodium acrylate) superabsorbent polymers. *Journal of Applied Polymer Science* 117:2233–2240.

Yu, C., L. Y. Fei, T. H. Min, and J. J. Xin. 2009. Synthesis and characterization of a novel superabsorbent polymer of *N,O*-carboxymethyl chitosan graft copolymerized with vinyl monomers. *Carbohydrate Polymers* 75:287–292.

Yu, L., Y. He, L. Bin, and Y. Fang. 2003. Study of radiation-induced graft copolymerization of butyl acrylate onto chitosan in acetic acid aqueous solution. *Journal of Applied Polymer Science* 90:2855–2860.

Zhang, J. P., Q. Wang, and A. Q. Wang. 2007. Synthesis and characterization of chitosan-*g*-poly(acrylic acid)/attapulgite superabsorbent composites. *Carbohydrate Polymers* 68:367–374.

8 Preparation of Chitin Nanofibers for Biomedical Application

Shinsuke Ifuku, Hiroyuki Saimoto, Kazuo Azuma,
Tomohiro Osaki, and Saburo Minami

CONTENTS

8.1 INTRODUCTION

Nanofibers are generally defined as fibers with diameter of <100 nm and an aspect ratio of >100 nm. Because nanofibers have extremely high surface-to-volume ratio, their properties are different from those of microsized fibers. Thus, decreasing the width and increasing the aspect ratio of nanofibers compared to microfibers add unique dimensions to optical, mechanical, medical, electrical, and other characteristics for the development of new promising advanced materials from the application viewpoint. Owing to the environmentally benign, biodegradable, biocompatible, renewable, and sustainable biomass, the nanofibers from biopolymers are gaining importance. A variety of naturally occurring nanofibers, such as collagen triple-helix fibers, fibroin fibrils, and keratin fibrils are known. As these nanofibers consist of complex hierarchical organization, they are suggestive of the possibility that nanofibers are extracted from biomass-based organized structural units. Among the variety of biomass-based products, cellulose is the most abundant biopolymer found mainly in wood cell walls. The cellulose nanofibers are highly crystalline structures. The bundles of nanofibers are embedded in matrix substances such as hemicellulose and lignin matrix that form the wood cell wall. Because cellulose nanofibers have extremely tough physical properties, they have potential as high-performance material. Various chemical processes followed by mechanical treatments have been employed mainly for cellulose nanofibers preparation. Abe et al. (2007) isolated cellulose nanofiber bundles of 15-nm-sized nanofibers from wood by a simple method. Apart from wood, the cellulose nanofibers were isolated from rice straw, potato tuber pulp, and parenchymal cells of bamboo and fruits (Abe and Yano 2009, 2010; Ifuku et al. 2011a). Chitin is known to be cellulose analogs with a (1,4)-β-N-acetyl glycosaminoglycan-repeating structure. After cellulose, chitin is the second most abundant biopolymer, occurring mainly in the exoskeletons of shellfish and insects and the cell walls of mushrooms. Although chitin is a semicrystalline biopolymer with nanosized fibrillar morphology and excellent material properties, most chitin is thrown away as industrial waste. Therefore, it is important to make effective use of chitin as an environmentally friendly green material. Because of its linear structure with two hydroxyl groups and an acetamide group, chitin is highly crystalline with strong hydrogen bonding having high binding energy and is arranged as nanosized chitin nanofibers in an antiparallel fashion. Because crab and prawn shells have a hierarchical structure made up of chitin nanofibers, proteins, and minerals (Raabe et al. 2006), we consider that the preparation method of cellulose nanofiber is applicable to several species consisting of chitin nanofibers. We have developed a simple method for the preparation of chitin nanofibers. In this chapter, we describe the available procedures for preparing chitin nanofibers and the biomedical applications of the nanofibers.

8.2 PREPARATION OF CHITIN NANOFIBERS FROM CRAB SHELL

Crab shell has a hierarchical organization with various structural levels as shown in Figure 8.1 (Ifuku et al. 2011b, Shams et al. 2011). Chitin nanofibers have been extracted from crab shells having such complicated hierarchical structures (Raabe et al. 2006). Dried crab shell powder of *Paralithodes camtschaticus* (red king crab) that was commercially available as a fertilizer at low cost was used for the preparation of chitin nanofiber. To extract chitin from crab shell, proteins and minerals were removed by using 2 N NaOH and 2 N HCl according to the conventional method. Purified chitin was kept wet after the removal of proteins and minerals to avoid strong coagulation between chitin fibers. Figure 8.2 shows scanning electron microscopy (SEM) images of thus obtained chitin. Although crab shell foam was still maintained, chitin nanofibers were observed. Thicker chitin nanofibers with approximately 100 nm seemed to be bundles of thinner chitin nanofibers. The wet chitin was dispersed in water with a concentration of 1 wt.% and was passed through a grinder for nanofibrillation. A specially designed pair of grinding stones was used to break down chitin. The pH value of the suspension was adjusted to 3 by adding acetic acid for grinder treatment. After the single-grinder treatment, the chitin slurry became gel foam. This suggests that fibrillation was accomplished because of its high dispersion property of nanofiber with high surface-to-volume ratio in water. Figure 8.3 shows SEM images of the dried sample. Chitin was observed as highly

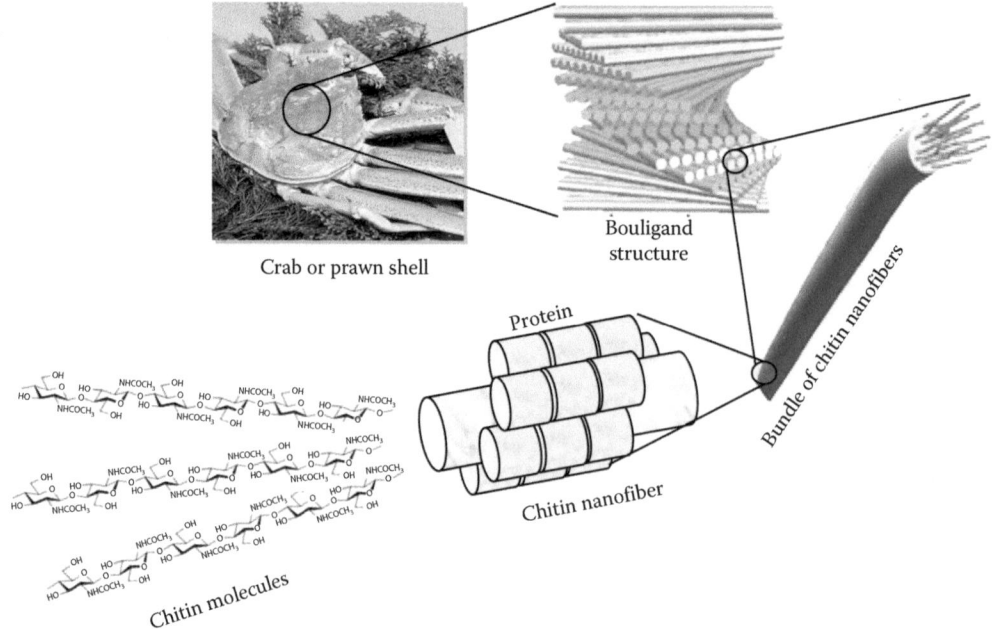

FIGURE 8.1   Schematic presentation of the exoskeleton structure of crab shell.

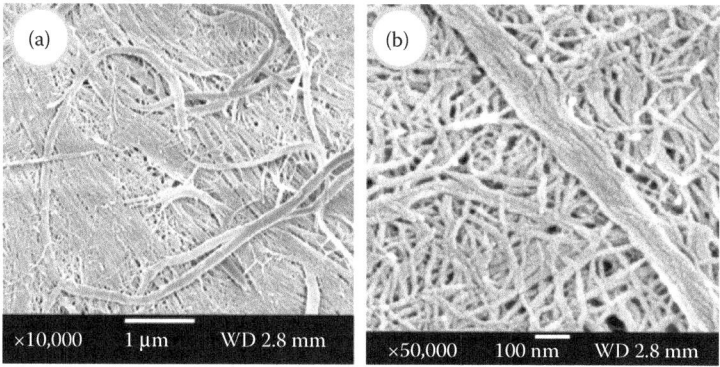

FIGURE 8.2 FE-SEM micrographs of crab shell surface after the removal of matrix without grinder treatment. The length of the scale bars is (a) 1000 nm and (b) 100 nm, respectively. (Reprinted with permission from Ifuku, S. et al., 2011b. Preparation of chitin nanofibers with a uniform width as α-chitin from crab shells. *Biomacromolecules*, 10:1584–1588. Copyright 2009, American Chemical Society.)

thin and uniform nanofibers with a thickness of 10–20 nm. Because chitin nanofibers were successfully isolated from crab shells without altering their natural shape, the aspect ratios are very high. Grinder treatment under acidic condition is the key for nanofibrillation. Cationization of the C2 amino groups further assisted fibrillation of chitin by electrostatic repulsions (Fan et al. 2009). The degree of *N*-acetylation of the nanofibers estimated by comparing the *C* and *N* content was 0.95. Therefore, although the ratio of the amino group was only 5%, electrostatic repulsive force caused by cationization of amino groups facilitates nanofibrillation of chitin fiber and stable dispersion. Domestic high-speed blender was also available for nanofibrillation of chitin as the other option. After blending for 10 min at a rotating speed of 37,000 rpm, chitin nanofibers with a width of 20–30 nm were obtained by the assistance of electrostatic repulsion. Fourier transform-infrared (FT-IR) spectrum of prepared chitin nanofibers was in excellent agreement with the spectrum of commercial pure α-chitin. Moreover, x-ray diffraction profile of chitin nanofibers showed typical crystal patterns of α-chitin and was closely coincident with commercial α-chitin. Thus, chitin nanofibers were extracted from crab shell, and the original molecular structure and α-chitin crystalline structure was maintained even after the chemical treatments and the grinder mechanical treatment.

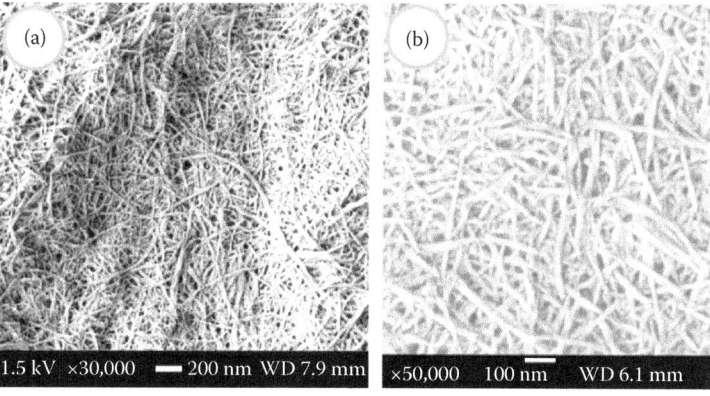

FIGURE 8.3 FE-SEM micrographs of chitin nanofibers from crab shell after one pass through the grinder. The length of the scale bar is (a) 200 nm and (b) 100 nm, respectively. (Reprinted with permission from Ifuku, S. et al., 2011b. Preparation of chitin nanofibers with a uniform width as α-chitin from crab shells. *Biomacromolecules*, 10:1584–1588. Copyright 2009, American Chemical Society.)

8.3 PREPARATION OF CHITIN NANOFIBERS FROM PRAWN SHELL

Chitin nanofibers can be prepared from prawn shell too (Ifuku et al. 2011c). Fresh prawn shells of *Penaeus monodon* (black tiger prawn), *Marsupenaeus japonicus* (Japanese tiger prawn), and *Pandaluseous makarov* (Alaskan pink shrimp) were used as starting materials. They are widely cultured prawn species. Some of the shells are thrown away as industrial waste. These shells were treated using NaOH and HCl solutions to remove proteins and minerals, respectively. Figure 8.4 shows the SEM image of the prawn shell surface after removal of the proteins and minerals. It was observed that crab shell consists of chitin nanofibers with a uniform width and elaborate design. The purified chitin thus obtained was diluted with water with a concentration of 1 wt.% and passed through a grinder for nanofibrillation. We could see the uniform shape of the chitin nanofibers (Figure 8.5). They were highly uniform over an extensive area and the width of the nanofibers was 10–20 nm. The characteristic morphology was similar to nanofibers from crab shell. Thus, chitin nanofibers were successfully prepared from prawn shell. We emphasized that chitin from prawn shell was converted to nanofibers without using an acidic chemical. The explanation of successful fibrillation under neutral pH was as follows. The exoskeleton of crustaceans is made up of the exocuticle and the endocuticle. The exocuticle has a very fine twisted plywood structure. However, the endocuticle has a much coarser structure with a thicker fiber diameter. Although, approximately 90% of the crab shell is made up of the endocuticle, the exoskeleton of prawn having a translucent soft shell is primarily made up of a fine exocuticle. As a result, because of the differences in the cuticle structure and fiber thickness, nanofibrillation of prawn shell is easier than that of crab shell. Chitin nanofibers in acidic water may cause significant problems for application in biomedical materials, nanocomposites, electronic devices, and so on because these materials are sensitive to acid. Therefore, preparation of chitin nanofibers without an acid will expand their application.

8.4 PREPARATION OF CHITIN NANOFIBERS FROM MUSHROOMS

The isolation method of chitin nanofibers from crab and prawn shells was also applicable for fruiting bodies of mushrooms (Ifuku et al. 2011d). However, the morphology and composition of a mushroom cell wall are very different from that of crab and prawn shells. Considering the composition of the mushroom cell wall, isolation of chitin nanofibers must be arranged. Five different species of mushrooms, *Pleurotus eryngii*, *Agaricus bisporus*, *Lentinula edodes*, *Grifola frondosa*, and *Hypsizygus marmoreus*, were used for this study. They are widely used for human consumption. To isolate chitin from these mushrooms, proteins, pigments, glucans, and minerals were removed by a series of treatments with 2% NaOH, 2 M HCl, mixture of 0.5% NaClO$_2$ and 1% AcOH, and

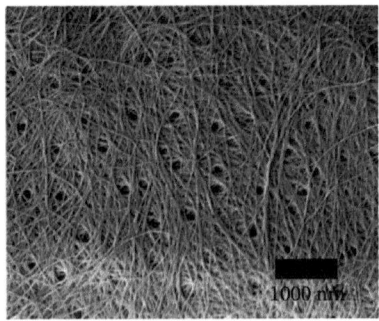

FIGURE 8.4 FE-SEM micrograph of the surface of the black tiger prawn after removing matrix substances. The length of the scale bar is 1000 nm. (Reproduced from *Carbohydrate Polymers*, 84, Ifuku, S. et al., Simple preparation method of chitin nanofibers with a uniform width of 10–20 nm from prawn shell under neutral conditions, 762–764. Copyright 2011, with permission from Elsevier.)

FIGURE 8.5 FE-SEM micrograph of chitin nanofibers from black tiger prawn shell. The length of the scale bar is 1000 nm. (Reproduced from *Carbohydrate Polymers*, 84, Ifuku, S. et al., Simple preparation method of chitin nanofibers with a uniform width of 10–20 nm from prawn shell under neutral conditions, 762–764. Copyright 2011, with permission from Elsevier.)

2% NaOH. The isolated chitin was passed through a grinder along with acetic acid for facilitating nanofibrillation. Figure 8.6 shows SEM images of chitin from five mushrooms after grinder treatment. The isolated chitins were well fibrillated uniform nanofibers. The appearance was similar to that from crab and prawn shells. The nitrogen atom contents of chitin from mushroom samples were smaller than that from crab shell and varied widely with the type of mushrooms. The reason is that because glucan forms complex with chitin, the complete removal of glucan did not occur by chemical treatment. The width also varied depending on the type of mushroom. This indicates that chitin nanofibers with low N content had a considerable amount of glucans on the surface and thus the thickness of the chitin nanofibers was increased. The degree of crystalline indices of chitin nanofibers also decreased with a reasonable correlation to N content ratios, indicating that the crystallinity also decreased with the increase of the amount of amorphous glucan on the surface of the nanofibers. Chitin nanofibers from mushrooms have been added to the list of useful dietary nanofibers. The dietary nanosized fibers obtained from cultivable and edible mushrooms will have a wide range of applications, from the use as novel functional food ingredients to medical applications.

8.5 PREPARATION OF CHITIN NANOFIBERS FROM DRY CHITIN POWDER

As mentioned above, the drying process of chitin generates strong hydrogen bonding between these fibers after removal of the protein and minerals (Ifuku et al. 2010, 2012). The hydrogen bonding makes it difficult to fibrillate chitin to nanofibers. Therefore, the sample must be kept wet for nanofiber preparation. However, this requirement presents a disadvantage in the commercial application of nanofibers. We have succeeded in preparing chitin nanofibers from commercially available dry chitin powder. As can be seen from Figure 8.7a, dry chitin powder from crab shell was also made up of nanofibers. The dry chitin was passed through a grinder with a concentration of 1 wt.% with acetic acid to fibrillate the bundles of chitin nanofibers. Figure 8.7b and c shows SEM images of chitin fibers fibrillated without and with acetic acid, respectively. In Figure 8.7b, the chitin powder was not fibrillated at all due to the strong interfibrillar hydrogen bonding. However, in Figure 8.7c, the aggregates were clearly fibrillated into homogeneous nanofibers with a width of 10–20 nm. This is obviously because of the electrostatic repulsion caused by cationization of amino groups on the

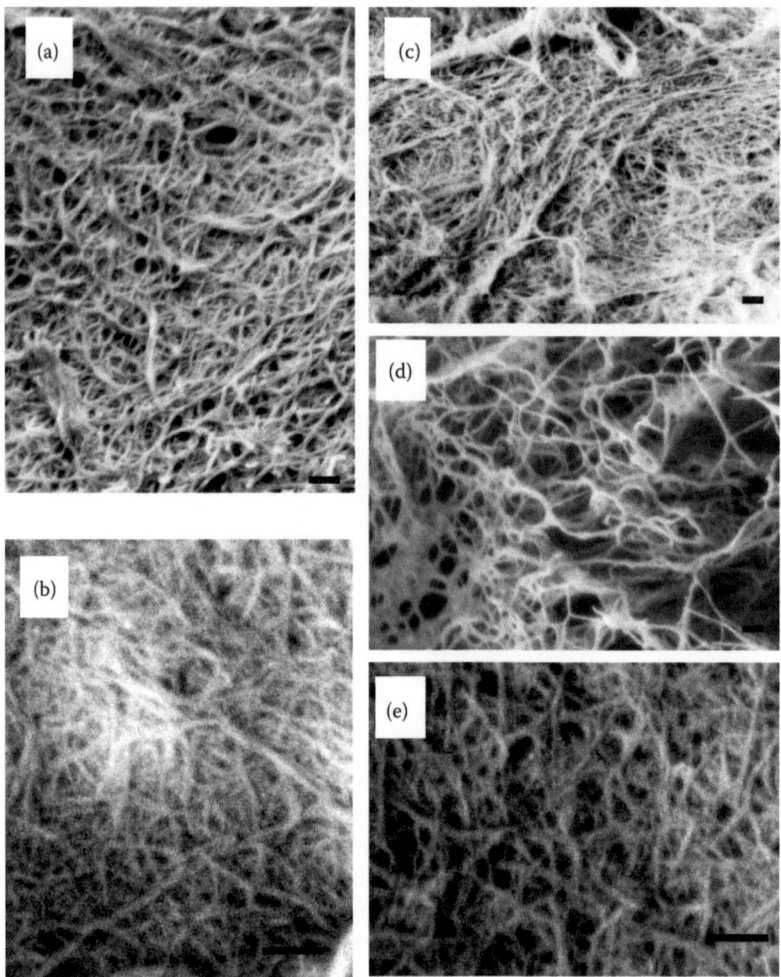

FIGURE 8.6 FE-SEM micrographs of chitin nanofibers prepared from (a) *Pleurotus eryngii*, (b) *Agaricus bisporus*, (c) *Lentinula edodes*, (d) *Grifola frondosa*, and (e) *Hypsizygus marmoreus*. The scale bars are 200 nm in length.

chitin nanofibers although the degree of substitution of amino groups was only 3.9%. In this way, chitin nanofibers could be easily prepared from the dry chitin powder by assistance of the repulsion force. Preparation of chitin nanofiber from commercial prepurified dry chitin is advantageous because a large amount of chitin could be immediately and easily obtained by a simple fibrillation under acidic conditions without any purification process. Several other organic acids were available for facile nanofibrillation, including ascorbic acid, citric acid, and lactic acid. The use of acid-induced electrostatic repulsion yields chitin that is dry, light, low in volume, and nonperishable, and thus constitutes a significant advantage for commercial application in terms of providing a stable supply of chitin that is easy to store and transport. This advantage cannot be utilized for cellulose, because it does not have ionic functional groups to cause electrostatic repulsions.

Recently, a new fibrillation system developed by Sugino Machine Co. Ltd., has attracted much attention for the production of biopolymer-based nanofibers. The novel system, called Star Burst, applies high-pressure water-jet technology for the wet disintegration of several samples. Compared to a grinder, the advantages of the Star Burst system are as follows: (1) The chitin slurry throughput amount can be easily arranged to a large extent from 2 to 840 L/h by changing the scale of the Star

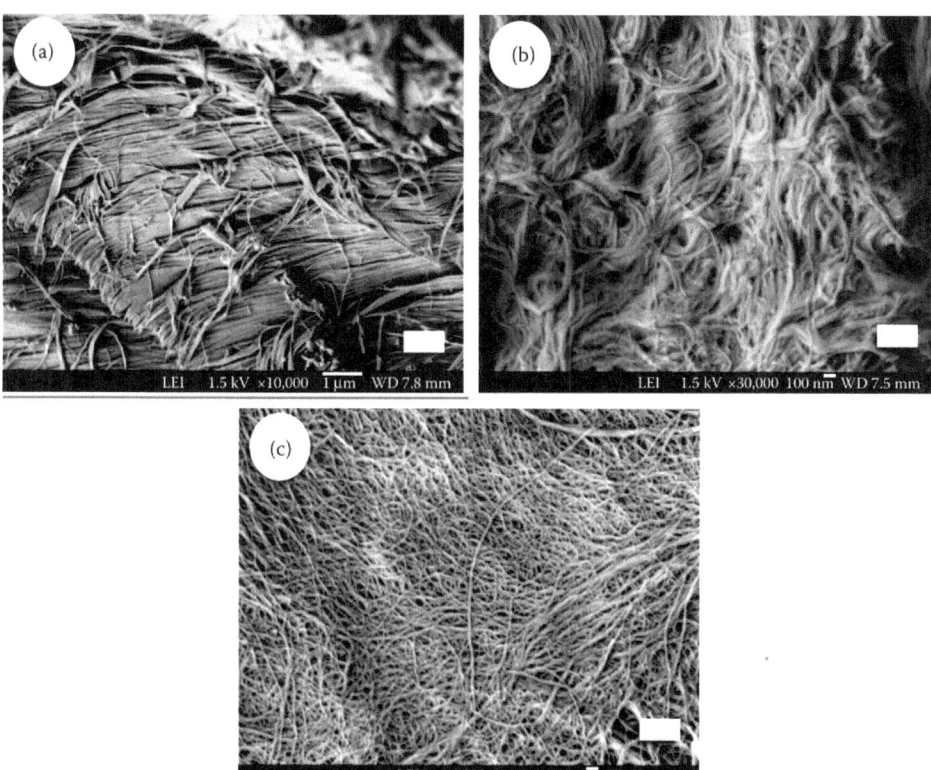

FIGURE 8.7 FE-SEM micrographs of (a) commercially available dry α-chitin powder and chitin fibers after one pass through the grinder (b) without and (c) with acetic acid. The length of the scale bar is (a) 1000 nm and (b and c) 300 nm, respectively. (Reproduced from *Carbohydrate Polymers*, 81, Ifuku, S. et al., 2010. Fibrillation of dried chitin into 10–20 nm nanofibers by a simple grinding method under acidic conditions, 134–139. Copyright 2011, with permission from Elsevier.)

Burst equipment from small to large. (2) In principle, the system can produce the same nanofibers independent of the scale of the equipment, as long as the pressure with which a sample is ejected from the nozzle is constant. (3) This process introduces less contamination than the grinding process does. We have studied the fibrillation of chitin nanofibers into nanofibers by the Star Burst system. Dry chitin powder from crab shell was used as received. The chitin dispersion in water at 1 wt.% with acetic acid was passed through the Star Burst system, equipped with a ball-collision chamber. The slurry was ejected from a small nozzle at 245 MPa of high pressure and collided with ceramic ball. Even after only one pass, chitin seems considerably fibrillated in comparison to the one-pass treatment under the neutral condition. The difference between the results is obviously due to the repulsive force between nanofibers. After five passes, the chitin fibers were further fibrillated. After 10 passes, the chitin was completely fibrillated and had a very fine nanofiber network. The morphology was highly uniform, with a high-aspect ratio. Over a wide area, thicker fibers of about 100 nm diameter were not observed at all. As the number of passes increased, the average thickness of chitin nanofibers became thinner from 19.0 to 16.5 nm. After the Star Burst process under acidic condition, there were no significant differences in the relative degree of crystallinity. This result indicates that at least 10 mechanical treatments did not damage the chitin fibrils, although the system used a super-high-pressure water jet. The system has advantages in quality stability, high-volume production, and low contamination. We expect that this unique system can play a strong role in the commercial use of chitin nanofibers.

8.6   ORAL ADMINISTRATION EFFECT OF CHITIN NANOFIBERS ON INFLAMMATORY BOWEL DISEASE

It is well known that chitin and its derivatives have nonspecific antiviral and antitumor activities (Azuma et al. 2012a,b). It was suggested that the size of chitin influences its effects on immune cells. Therefore, chitin nanofibers are considered to have potential for application in tissue engineering scaffolds, drug delivery, and wound dressing. However, there has been no study on the *in vivo* effects of chitin nanofibers after oral administration. Inflammatory bowel disease is a commonly occurring group of conditions characterized by inflammation in the intestinal tract. Crohn's disease and ulcerative colitis (UC) account for the majority of the cases of these conditions. A model of dextran sulfate sodium (DSS)-induced colitis is one common model of inflammatory bowel disease. We evaluated the preventive effects of chitin nanofibers in a mouse model of DSS-induced acute UC. The effects of chitin nanofibers on the disease activity index (weight loss, loose stools, and bleeding) in DSS-induced acute UC mice were evaluated. The chitin nanofiber-administered group showed a significantly reduced disease activity index. Although the administration of DSS shortened the colon length, colon lengths in the chitin nanofiber groups were significantly greater than those in the control group. Damage in the intestinal mucosa was microscopically evaluated (Figure 8.8). On the sixth day in the control group, severe erosions, crypt destruction, edema, and some ulcers were observed. However, in the chitin nanofiber group, erosions, crypt destruction, and edema were markedly suppressed compared with those in the control group. The numbers of myeloperoxidase (MPO)-positive cells were counted (Figure 8.9). Although the numbers of MPO-positive cells gradually increased with time, in the chitin nanofiber group, the numbers of MPO-positive cells were significantly lower than those in the control group. Moreover, the serum interleukin (IL)-6 concentration was significantly lower in the chitin nanofiber group than in the control group. Thus, chitin nanofibers improved clinical symptoms, colon inflammation, and histological tissue injury in the DSS-induced acute UC mouse model. As MPO is a marker of oxidative stress, high MPO activities were observed in a DSS-induced UC model. IL-6 is a central cytokine in inflammatory bowel disease that contributes to enhanced T-cell survival and apoptosis resistance in the lamina propria at sites of inflammation. Therefore, the chitin nanofibers suppressed the inflammation caused by acute UC by suppressing the MPO-mediated activation of inflammatory cells, such as leukocytes and decreasing serum IL-6 concentrations.

We found that chitin nanofibers improved clinical symptoms and suppressed UC. Then, anti-inflammatory and antifibrosis effects in DSS-induced acute UC mice model were evaluated to know the protective mechanisms of chitin nanofibers. Chitin nanofibers decreased positive areas of nuclear factor (NF)-κB staining in the colon tissue (Figure 8.10). Chitin nanofibers also decreased serum

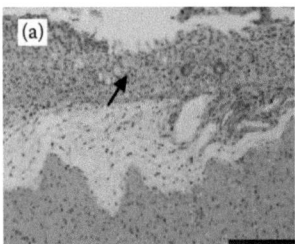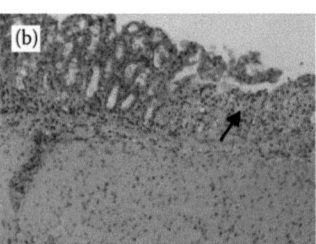

FIGURE 8.8   Effect of chitin nanofibers administration on histopathological changes in DSS-induced acute UC mice. The colon was fixed and tissue sections were stained with hematoxylin and eosin. Data are presented for one mouse each from the control group. (a) Chitin nanofibers and (b) chitin powder (c) groups on day 6. Allows indicate erosions. Bar = 100 mm. (Reproduced from *Carbohydrate Polymers*, 87, Azuma, K. et al., Beneficial and preventive effect of chitin nanofibrils in a dextran sulfate sodium-induced acute ulcerative colitis model, 1399–1403. Copyright 2012, with permission from Elsevier.)

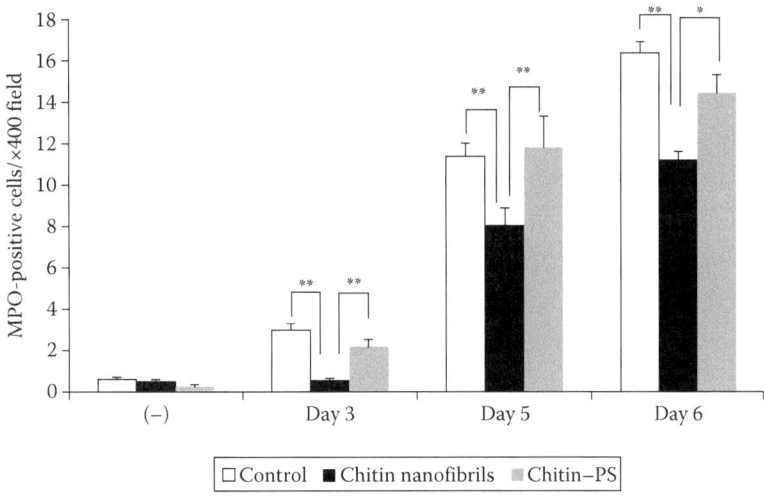

FIGURE 8.9 Effect of chitin nanofibers administration on the MPO-positive cell counts/400× field in the colons of DSS-induced acute UC mice. The data represent the means ±S.E. of 60 fields/400× field in each group. Values are compared among control, chitin nanofibrils, and chitin powder (PS) groups. $**p < 0.01$. (Reproduced from *Carbohydrate Polymers*, 87, Azuma, K. et al., Beneficial and preventive effect of chitin nanofibrils in a dextran sulfate sodium-induced acute ulcerative colitis model, 1399–1403. Copyright 2012, with permission from Elsevier.)

monocyte chemotactic protein-1 (MCP-1) concentration in DSS-induced acute UC (Figure 8.11). Moreover, chitin nanofibers suppressed the increased positive areas of Masson's trichrome staining in colon tissue. However, chitin powder suspension did not show these effects in DSS-induced acute UC mice model. Our results indicated that chitin nanofibers have the anti-inflammatory effect via suppressing NF-κB activation and the antifibrosis effects in DSS-induced acute UC mice model. In the DSS-induced UC mice model, fibrosis in the colon was observed not only in the chronic phase but also in the acute phase. It is known that MCP-1 induces fibrogenic response of the gut in inflammatory bowel disease (IBD) model. Chitin nanofibers suppressed the fibrosis and decreased

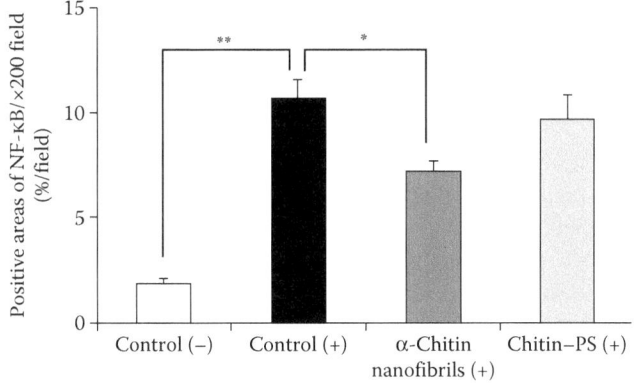

FIGURE 8.10 Effects of chitin nanofibers on colon NF-κB activation in a DSS-induced acute UC mouse model. Data are presented for one mouse each and represent the means ±S.E. of 30 fields/ × 100 field in each group. The statistical analyses were performed with a Steel–Dwass test. $*p < 0.05$, $**p < 0.01$. (Reproduced from *Carbohydrate Polymers*, 90, Azuma, K. et al., α-Chitin nanofibrils improve inflammatory and fibrosis responses in inflammatory bowel disease mice model, 197–200. Copyright 2012, with permission from Elsevier.)

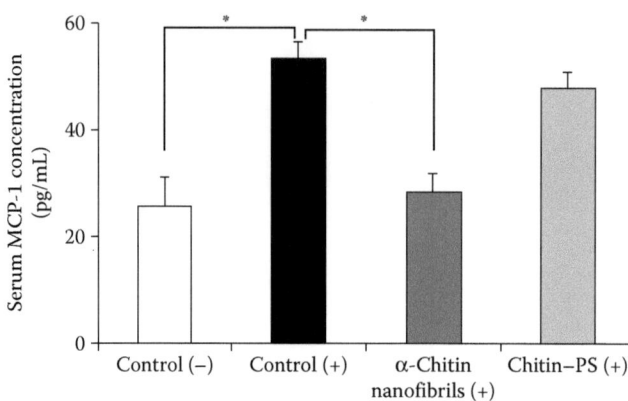

FIGURE 8.11 Effects of chitin nanofibers on serum MCP-1 concentrations in a DSS-induced acute UC mouse model. Data represent the means ±S.E. in each group ($n = 5$) from the control (−), control (+), chitin nanofibers (+), and chitin powder (PS). The statistical analyses were performed with a Tukey–Kramer test. *$p < 0.05$, **$p < 0.01$. (Reproduced from *Carbohydrate Polymers*, 90, Azuma, K. et al., α-Chitin nanofibrils improve inflammatory and fibrosis responses in inflammatory bowel disease mice model, 197–200. Copyright 2012, with permission from Elsevier.)

serum MCP-1 concentration in DSS-induced acute UC mouse model. These results indicated that chitin nanofibers have the suppressive effects of fibrosis in DSS-induced acute UC mouse model. It was indicated that one mechanism of suppressive effects on fibrosis by chitin nanofibers came from suppressing the action of MCP-1. NF-κB occupies a pivotal position in several innate immune-signaling pathways. So far, it has been shown that NF-κB is the critical transcription factor needed to express genes associated with a proinflammatory response. NF-κB activity is increased in the colon during active episodes of IBD. Chitin nanofibers suppressed the activation of NF-κB in colon epithelium in DSS-induced acute colitis model. MCP-1 plays an important role in the pathogenesis of experimental colitis model for the recruitment of immune and enterochromaffin cells. The absence of MCP-1 is associated with a significant reduction in inflammation in experimental colitis model. Proinflammatory cytokine induced the expression of MCP-1 via p38 mytogen-activated protein kinase (MAPK) and NF-κB signaling. Chitin nanofibers decreased serum MCP-1 concentration compared with control (+) group. These results indicated that chitin nanofibers suppressed the increase of MCP-1 in serum via suppressing NF-κB activation.

8.7 CONCLUSIONS

Chitin nanofibers were prepared from the exoskeletons of crabs and prawns and the cell walls of mushrooms (Ifuku and Saimoto 2012). The prepared chitin nanofibers had a fine network structure with 10–20 nm uniform width and a high aspect ratio. Neat chitin nanofibers have application as drug for inflammatory bowel disease. Because chitin is obtained from crab and prawn shells, it is more expensive than cellulose from wood. However, chitin nanofibers have the following several advantages over cellulose nanofibers: (1) Chitin and chitosan are known to have a variety of bio-activities. (2) Chitin nanofibers can be easily prepared from dry chitin samples assisted by electrostatic repulsion. (3) The filtering speed to make a nanofiber sheet is higher than that of cellulose nanofibers. (4) Most of the matrix can be removed from crab and prawn shells by a conventional method. (5) Naturally rare cationic-charged nanofibers are prepared by partial deacetylation of chitin nanofibers. Utilizing these advantages is important to define the separate roles of chitin and cellulose for expanding the application of these nanofibers. This simple and efficient process allowed us to obtain homogeneous chitin nanofibers in their original state. We believe that nanofibers with a characteristic morphology, very high surface area, and excellent mechanical properties have great potential for novel green nanomaterials. In general, chitin precipitates in water. However, because chitin nanofibers can be dispersed homogeneously in water, they are easy to handle and shape into the desired forms. This characteristic led to the above-described applications of chitin nanofibers. We expect that the novel applications of chitin nanofibers will be discovered.

REFERENCES

Abe, K., S. Iwamoto, and H. Yano, 2007. Obtaining cellulose nanofibers with a uniform width of 15 nm from wood. *Biomacromolecules,* 8:3276–3278.

Abe, K., and H. Yano, 2009. Comparison of the characteristics of cellulose microfibril aggregates of wood, rice straw and potato tuber. *Cellulose,* 16:1017–1023.

Abe, K., and H. Yano, 2010. Comparison of the characteristics of cellulose microfibril aggregates isolated from fiber and parenchyma cells of Moso bamboo (*Phyllostachys pubescens*). *Cellulose,* 17:271–277.

Azuma, K., T. Osaki, T. Wakuda, S. Ifuku, H. Saimoto, T. Tsuka, T. Imagawa, Y. Okamoto, and S. Minami, 2012a. Beneficial and preventive effect of chitin nanofibrils in a dextran sulfate sodium-induced acute ulcerative colitis model. *Carbohydrate Polymers,* 87:1399–1403.

Azuma, K., T. Osaki, T. Wakuda, S. Ifuku, H. Saimoto, T. Tsuka, T. Imagawa, Y. Okamoto, and S. Minami, 2012b. α-Chitin nanofibrils improve inflammatory and fibrosis responses in inflammatory bowel disease mice model. *Carbohydrate Polymers,* 90:197–200.

Fan, Y., T. Saito, and A. Isogai, 2009. Preparation of chitin nanofibers from squid pen β-chitin by simple mechanical treatment under acid conditions. *Carbohydrate Polymers,* 9:1919–1923.

Ifuku, S., M. Adachi, M. Morimoto, and H. Saimoto, 2011a. Fabrication of uniform cellulose nanofibers from parenchyma cells of pears and apples. *Sei-i Gakkaishi* 67:86–90.

Ifuku, S., M. Nogi, K. Abe, M. Yoshioka, M. Morimoto, H. Saimoto, and H. Yano, 2011b. Preparation of chitin nanofibers with a uniform width as α-chitin from crab shells. *Biomacromolecules,* 10:1584–1588.

Ifuku, S., M. Nogi, K. Abe, M. Yoshioka, M. Morimoto, H. Saimoto, and H. Yano, 2011c. Simple preparation method of chitin nanofibers with a uniform width of 10–20 nm from prawn shell under neutral conditions. *Carbohydrate Polymers,* 84:762–764.

Ifuku, S., R. Nomura, M. Morimoto, and H. Saimoto, 2011d. Preparation of chitin nanofibers from mushrooms. *Materials,* 4:1417–1425.

Ifuku, S., M. Nogi, M. Yoshioka, M. Moromoto, H. Yano, and H. Saimoto, 2010. Fibrillation of dried chitin into 10–20 nm nanofibers by a simple grinding method under acidic conditions. *Carbohydrate Polymers,* 81:134–139.

Ifuku, S., and H. Saimoto, 2012. Chitin nanofibers: Preparations, modifications, and applications. *Nanoscale,* 4:3308–3318.

Ifuku, S., K. Yamada, M. Morimoto, and H. Saimoto, 2012. Nanofibrillation of dry chitin powder by Star Burst system. *Journal of Nanomaterials,* 2012:1–7.

Raabe, D., P. Romano, C. Sachs, H. Fabritius, A. Al-Sawalmih, S.B. Yi, G. Servos, and H.G. Hartwig, 2006. Microtexture and chitin/calcite orientation relationship in the mineralized exoskeleton of the American lobster. *Materials Science and Engineering A,* 421:143–153.

Shams, M.I., S. Ifuku, M. Nogi, T. Oku, and H. Yano, 2011. Fabrication of optically transparent chitin nanocomposites. *Applied Physics A,* 102:325–331.

Part II

Biological Activities of Chitin and Chitosan Derivatives

9 Anticancer Effects of Chitin and Chitosan Derivatives

Mustafa Zafer Karagozlu, Fatih Karadeniz, and Se-Kwon Kim

CONTENTS

9.1 INTRODUCTION

Chitin is a natural polysaccharide that was first identified in 1821. Henri Braconnot who is the director of the botanical garden in France observed a material in mushrooms that did not dissolve in sulfuric acid. Braconnot named it as fungine. In the late 1830s, it was isolated from insects, and in 1859, chitosan, a derivative of chitin, was produced (Nicol, 1991).

Chitin is synthesized by an enormous number of living organisms and it is the second most abundant polymer after cellulose. Moreover, it occurs in nature as an ordered crystalline microfibril forming structural components in the exoskeleton of arthropods or in the cell walls of fungi and yeast. It is also produced by a number of other living organisms in the lower plant and animal kingdoms, serving many functions where reinforcement and strength are required.

Studies on chitin and chitin derivatives have been intensified since 1990 because these polysaccharides show excellent biological properties, such as biodegradation in the human body (Sashiwa et al., 1990; Shigemasa et al., 1994), and immunological (Nishimura et al., 1984; Mori et al., 1997), antibacterial (Tanigawa et al., 1992; Tokura et al., 1997), and wound-healing properties (Okamoto et al., 1993; Khnor and Lim, 2003; Kweon et al., 2003).

However, chitosan is a natural nontoxic heteropolysaccharide composed of β-1,4-linked-D-glucosamine (GlcN) and N-acetyl-D-glucosamine (GlcNAc) in varying proportions. These polysaccharides have been widely studied and applied in different fields. Although chitosan is a derivative of chitin it has its own unique functions, such as support material for gene delivery (Sato et al., 2001), cell culture (Mao et al., 2003), and tissue engineering (Gingras et al., 2003; Wang et al., 2003). Moreover, in biomedical and pharmaceutical industries, chitosan and chitosan derivatives have been widely used because of their various biological functions, such as antimicrobial (Kong et al., 2010), antibacterial (Jeon et al., 2001), antioxidant (Park et al., 2004), and immunostimulating (Huang et al., 2006) effects. Besides, they have antitumor (Seo et al., 2000; Karagozlu et al., 2010), antidiabetic (Karadeniz et al., 2008), and antiviral (Artan et al., 2010) activities. Unfortunately, poor solubility of chitosan is the principal limiting factor for its wide application. Thereupon, recent studies on chitosan have attracted interest in converting it into more soluble form, such as chitooligosaccharides (COSs). COSs are the degraded oligomers of chitosan, which can be obtained by either chemical (Horowitz et al., 1957; Tsukada and Inoue, 1981; Defaye and Guillot, 1994) or enzymatic (Izume and Ohtakara, 1987) hydrolysis of chitosan. It has been reported that lower oligomers of chitosan are not only water soluble but also exhibit versatile biological activities similar to chitosan

(Qin et al., 2002). The biological activity of COS is known to depend on their structure and molecular weight (Hahn, 1996). Besides, investigators mentioned that properties of COS, such as degrees of polymerization (DP), degrees of acetylation (DA), charge distribution, and nature of chemical modification to the molecule strongly influences its observed biological activities (Muzzarelli, 1997).

9.2   ANTICANCER ACTIVITY AS A THERAPEUTIC AGENT

The main goal of cancer research is to completely prevent recurrence after surgery and to increase the lifetime of the patient. But the major problem of the cure without surgeries is side effects. Cytotoxic anticancer chemotherapeutic agents generally produce severe side effects, while reducing host resistance to cancer and infections. Therefore, it is important to find new, powerful anticancer agents that are highly effective, biodegradable, and biocompatible. Chitosan and chitosan derivative anticancer agents are known to be favorable pharmaceutical material because of their biocompatible and biodegradable properties (Felt et al., 1998).

Several chitin derivatives were investigated in their antitumor activity (Murata et al., 1990). Murata et al. (1990) reported that 6-O-sulfated chitin significantly inhibited the lung tumor colonization in proportion to the degree of sulfation. Furthermore, 6-O-sulfated carboxymethylated chitin (SCM-chitin) with a high degree of sulfation caused a marked decrease of the number of lung tumor colonies in the spontaneous lung metastasis model. SCM-chitin also significantly inhibited the arrest of B16–Bl6 cells in lungs after coinjection with radiolabeled tumor cells.

A large amount of literature exists regarding the effects of antitumor activities of chitosan and its derivatives (Suzuki et al., 1986; Tokoro et al., 1988). Suzuki et al. (1986) found that N-acetyl chitosan oligomer, particularly the hexane and heptamer, display notable antitumor activity against sarcoma 180 solid tumors in BALB/c mice as well as in MM-46 solid tumor implanted in C$_3$H/HC mice. These results indicated that the effect was not by direct cytodial action on the tumor cells and was indeed host mediated. Tokoro et al. (1988) showed that hexameric chitosan oligomer had growth inhibitory effect against Meth-A solid tumor transplanted into BALB/c mice. The antitumor mechanism was assumed to be involved in the increased production of lymphokines, including interleukins 1 and 2, sequentially, leading to manifestation of antitumor effect through the proliferation of cytolytic T-lymphocytes. In addition, the antitumor activity of low-molecular-weight chitosan (LMWC) that was higher than hexamer was investigated. Qin et al. (2002) reported that LMWC was prepared by enzymatic hydrolysis using cellulose and hemicellulose, and investigated the inhibition of the growth of sarcoma 180 tumor cells in mice. Maeda and Kimura (2004) prepared various molecular weight chitosans (such as 21, 46, and 120 kDa) by enzymatic hydrolysis and examined their antitumor activity in sarcoma 180-bearing mice. The antitumor activity of various molecular weight chitosans showed that 21 kDa chitosan significantly reduced tumor growth and final tumor weight. Moreover, 21 and 46 kDa chitosans enhanced the natural killer (NK) activity in intestinal intraphelial lymphocytes or splenic lymphocytes. Harish et al. (2005) generated LMWC by depolymerization induced by potassium persulfate under nitrogen atmosphere. Moreover, Jeon et al. (2001) also carried out a study to identify the correlation between molecular weight of COSs and their antitumor activity. In their research, different molecular weight COSs were prepared by ultrafiltration (UF) membrane reactor system. The researchers suggested that medium-molecular-weight molecular COS ranging from 1.5 to 5.5 kDa could effectively inhibit the growth of sarcoma 180 solid (S180) or uterine cervix carcinoma no. 14 (U14) tumor in BALB/c mice. Hasegawa et al. (2001) reported the growth inhibitory effect of chitosan on bladder tumor cells. They observed deoxyribonucleic acid (DNA) fragmentation, which is a characteristic of apoptosis, and elevated caspase-3-like activity in chitosan-treated cancer cells. In addition, modified chitosans were reported to display the growth inhibitory effect on tumor cells (Sirica and Woodman, 1971), and this property was employed by Ouchi et al. (1992) by conjugating chitosan or chitosanaminooligosaccharide to 5-fluorouracil (5FU) to provide a macromolecular system with strong antitumor activity and reduced side effects. Indeed, the strong antitumor activity exhibited by 5FU is accompanied by undesirable side effects. *In vivo* studies demonstrated that chitosan–5FU conjugate exhibited a strong survival effect against lymphocytic leukemia in mice. Furthermore, chitosan–5FU and COS–5FU conjugates showed remarkable growth inhibitory effects on Met-A fibrosarcoma and MH-134Y hepatoma. Both conjugates displayed no acute toxicity, even in high dose ranges. Therefore, they reported that chitosan–5FU and COS–5FU are expected to act clinically as macromolecular prodrugs of 5FU.

Furthermore, studies on antitumor activity of chitosan and COSs revealed that partially deacetylated chitin and carboxymethyl (CM) chitin with an adequate degree of substitution was effective toward controlling various tumor cells (Nishimura et al., 1984). Unlike many other biological molecules, COSs could exert their biological activities following oral administration and effects that are more or less similar to those of intraperitoneal injection. Moreover, Qin et al. (2002) demonstrated that water-soluble COSs prepared with a mixture of tetramer and pentamer could inhibit the growth of S180 tumor cells in mice after oral and intraperitoneal administration. Therefore, COSs and their *N*-acetylated analogs that are soluble in basic physiological environments could be considered good candidates to develop potential nutraceuticals.

The antitumor mechanism of these COSs was probably related to their induction of T-cell proliferation to produce the tumor inhibitory effects. Through analysis of the splenic cell changes in cancerous mice, Suzuki et al. (1986) proved that the antitumor mechanism of COSs is to enhance acquired immunity by accelerating T-cell differentiation to increase cytotoxicity and maintain T-cell activity. Besides, *in vitro* research demonstrated that charge properties of the chitosan are also important for anticancer activity. Karagozlu et al. (2012) and Huang et al. (2006) studied the anticancer activities of differently charged COS derivatives using four cancer cell lines: HeLa, Hep3B, SW480, and AGS. Neutral red and 3-(4,5-dimethylthiazol-2-Yl)-2,5-diphenyltetrazolium bromide cell-viability studies suggested that highly charged COS derivatives could significantly reduce cancer cell viability, regardless of their positive or negative charge. Furthermore, fluorescence microscopic observations and Western blotting studies confirmed that the anticancer effect of these highly charged COS derivatives were triggering the intrinsic apoptotic pathway.

Laminin is a basal protein in basal lamina and is known to correlate with metastasis of tumor cells. A peptide containing the Tyr–Ile–Gly–Ser–Arg (YIGSR) sequence, corresponding to a partial sequence of laminin, inhibited angiogenesis and thus depressed tumor growth. Nishiyama et al. (2000) prepared YIGSR–chitosan conjugate and assayed antimetastatic activity. The conjugate proved to have higher inhibitory activity against experimental lung metastasis of B16BL6 melanoma cells in mice than did the parent peptide.

Kong et al. (2010) also investigated the matrix metalloproteinase (MMP) inhibition of chitin, water-soluble chitosan, and their carboxymethylated derivatives. In the research, chitosan and chitin, CM–chitosan, and CM–chitin were synthesized by means of carboxymethylation reaction. Their antioxidative and MMP-2 and MMP-9 inhibitory effects were investigated in HT1080 human fibrosarcoma cells. The research suggests that CM–chitosan and CM–chitin is a potent antioxidant and MMP inhibitor via alleviations of radical-induced oxidative damage.

9.3 ANTICANCER ACTIVITY AS A CARRIER

Chitosan and chitin are also used as a drug carrier to provide anticancer and antitumor chemotherapy that can improve drug absorption, stabilize drug components to increase drug targeting, and enhance drug release. As a gene carrier, chitosan can be used for DNA protection and affects the expression period of genes. It has been reported that the conjugates of some kinds of anticancer agents with chitin and chitosan derivatives display good anticancer effects with a decrease in side effects over the original form due to a predominant distribution in the cancer tissue and a gradual release of free drug from the conjugates. For instance, doxorubicin (DOX) is one of the most used anticancer agents that can load in various polymeric or natural hydrogels (Han et al., 2008, Obara et al., 2005). Cho et al. (2009) prepared DOX hydrogel containing COS–DOX to obtain sustained-release profiles of DOX from thermoresponsive and photo-cross-linkable hydrogels and examined its anticancer activity on human lung cancer adenocarcinoma cell line *in vitro* and *in vivo*. The research demonstrated that the released fraction composed of DOX and chitosan–DOX oligomers showed comparable *in vitro* cytotoxicity to free DOX. Besides, DOX hydrogels containing chitosan–DOX conjugates showed superior *in vivo* anticancer effects in human solid tumors compared to free DOX or hydrogel containing free DOX after 3 weeks.

Moreover, doxifluridine and 1-β-D-arabinofuranosylcytosine (Ara-C) is a typical time-dependent antitumor agent. But the major problem of Ara-C is the large dose required because of its resistance in the body. It can be quickly eliminated or inactivated (Aoshima et al., 1976). Thus, various derivatives of Ara-C have been developed in attempts to improve efficacy. One of the derivatives of Ara-C is modified with cytidinedeaminase. This modification catalyzes the transformation from an amino group to a hydroxyl group (Onishi et al., 1990). But the major problem of the usage of the glu-Ara-C in cancer treatment is the release time of the drug. The prolonged release and inhibition of cytidinedeaminase play an essential role in the enhancement of the antitumor effect of Ara-C (Aoshima et al., 1977; Kato et al., 1984; Onishi et al., 1991). Chitin can conjugate with glu-Ara-C (Chi-glu-Ara-C) to extend the release time of the drug. The antitumor effect of Chi-glu-Ara-C was investigated by intraperitoneal administration to mice intraperitoneally inoculated with P388 leukemia (Ichikawa et al., 1993).

Carboxymethyl chitin (CMC) is also used for drug-delivery application. The hydrophobic anticancer drug 5FU was loaded into CMC nanoparticles via emulsion cross-linking method. Drug-release studies showed that the CMC nanoparticles provided a controlled and sustained drug release at pH 6.8 (Jayakumar et al., 2010). Moreover, CMC is a promising biopolymer for cancer diagnosis application. Manjusha et al. developed novel folic acid (FA)-conjugated CMC coordinated to manganese-doped zinc sulfide (ZnS:Mn) quantum dot (FA–CMCS–ZnS:Mn) nanoparticles. The system can be used for targeting, controlled drug delivery, and also imaging of cancer cells. The biocompatible FA–CMCS–ZnS:Mn was used on breast cancer cell line MCF-7 to study the imaging, specific targeting, and cytotoxicity of the drug-loaded nanoparticles. The results showed that the bright and stable luminescence of quantum dots can be used to image the drug carrier in cancer cells without affecting their metabolic activity and morphology (Manjusha et al., 2010).

For gene delivery to cancer cells, several polymers have been used as nonviral vectors (Aoki et al., 2001; Vernejoul et al., 2002; You et al., 2007; Miyata et al., 2008). Even if low solubility and transfection efficiency is limiting the usage of chitosan in gene therapy applications, chitosan is a promising candidate as a vector for gene delivery to cancer cells. Therefore, researchers have modified this polymer to get an effective transfection. For instance, Germershaus et al. (2008), Kean et al. (2005), and Thanou et al. (2002) quaternized chitosan. According to their research on quaternized chitosan derivatives, properties such as their charge, solubility, plasmid interactions, and transfections were increased. In 2011, Safari et al. prepared *N,N*-diethyl *N*-methyl chitosan (DEMC) for gene delivery to human pancreatic cancer cells. According to their biological research and the mathematical modeling results, both showed that after DEMC transfection, cancer cell fluorescence, intensity, and size were changed.

9.4 CONCLUSION

Although the surgical methods are still promising and widely accepted treatments against defined cancer, nonsurgical treatments against cancer have also received much attention with an aim to reduce and eliminate complications after surgical treatments. Therefore, overcoming the side-effect complication of the anticancer agent is the main scope of cancer researchers. Recent studies on the chemical modification of chitin and chitosan are discussed from the viewpoint of biomedical applications because of their excellent biological properties such as biodegradation and biocompatibility in the human body. Such properties can be considered as valuable extensions of the use of chitin and its derivatives. These natural biological properties allow them to be a valuable biomaterial for both anticancer therapy of human solid tumors and cancer diagnosis applications in various ways.

REFERENCES

Aoki, K., S. Furuhata, K. Hatanaka, M. Maeda, J. S. Remy, J. P. Behr, M. Terada, and T. Yoshida. 2001. Polyethylenimine-mediated gene transfer into pancreatic tumor dissemination in the murine peritoneal cavity. *Gene Therapy* 8:508–514.

Aoshima, M., S. Tsukagoshi, Y. Sakurai, J. Oh-ishi, and T. Ishida. 1976. Antitumor activities of newly synthesized N^4-acyl-1-beta-D-arabinofuranosylcytosine. *Cancer Research* 36:2726–2732.

Aoshima, M., S. Tsukagoshi, Y. Sakurai, J. Oh-ishi, and T. Ishida. 1977. N^4-Behenoyl-1-beta-D-arabino furanosylcytosine as a potential new antitumor agent. *Cancer Research* 37:2481–2486.

Artan, M., F. Karadeniz, M. Z. Karagozlu, M. M. Kim, and S. K. Kim. 2010. Anti-HIV-1 activity of low molecular weight sulfated chitooligosaccharides. *Carbohydrate Research* 345:656–662.

Cho, I. Y., S. Park, S. Y. Jeo, and H. S. Yoo. 2009. *In vivo* and *in vitro* anti-cancer activity of thermo-sensitive and photo-crosslinkabledoxorubicin hydrogels composed of chitosan–doxorubicin conjugates. *European Journal of Pharmaceutics and Biopharmaceutics* 73:59–65.

Defaye, J. and J. M. Guillot. 1994. A convenient synthesis for anomeric 2-thioglucobioses, 2-thiokojibiose and 2-thiosophorose. *Carbohydrate Research* 253:185–194.

Felt, C., P. Buri, and R. Gurny. 1998. Chitosan: A unique polysaccharides for drug delivery. *Drug Development and Industrial Pharmacy* 24:979–993.

Germershaus, O., S. Mao, J. Sitterberg, U. Bakowsky, and T. Kissel. 2008. Gene delivery using chitosan, trimethyl chitosan or polyethylenglycol–graft–trimethyl chitosan block copolymers: Establishment of structure–activity relationships *in vitro*. *Journal of Controlled Release* 125:145–154.

Gingras, M., I. Paradis, and F. Berthod. 2003. Nerve regeneration in acollagen–chitosan tissue-engineered skin transplanted on nude mice. *Biomaterials* 24:1653–1661.

Hahn, M. G. 1996. Microbial elicitors and their receptors in plants. *Annual Review of Phytopathology* 34:387–412.

Han, H. D., C. K. Song, Y. S. Park, K. H. Noh, J. H. Kim, T. W. Hwang, T. W. Kim, and H. C. Shim. 2008. A chitosan hydrogel-based cancer drug delivery system exhibits synergistic antitumor effects by combining with a vaccinia viral vaccine. *International Journal of Pharmaceutics* 350:27–34.

Harish Prashanth, K. V. and R. N. Tharanathan. 2005. Depolymerized products of chitosan as potent inhibitors of tumor-induced angiogenesis. *Biochimica et Biophysica Acta* 11:1722–1729.

Hasegawa, M., K. Yagi, S. Iwakawa, and M. Hirai. 2001. Chitosan induces apoptosis via caspase-3 activation in bladder tumor cells. *Japanese Journal of Cancer Research* 92:459–466.

Horowitz, S. T., S. Roseman, and H. J. Blumenthal. 1957. The preparation of glucosamine oligosaccharides I separation. *Journal of American Chemistry Society* 79:5046–5048.

Huang, R., E. Mendis, N. Rajapakse, and S. K. Kim. 2006. Strong electronic charge as an important factor for anticancer activity of chitooligosaccharides (COS). *Life Sciences* 78:2399–2408.

Ichikawa, H., H. Onishi, H. Takahata, Y. Machida, and T. Nagai. 1993. Evaluation of the conjugate between N^4-(4-carboxybutyryl)1-β-D-arabinofuranosylcytosine. *Drug Design and Discovery* 10:343–353.

Izume, M. and A. Ohtakara. 1987. Preparation of D-glucosamine oligosaccharides by the enzymatic hydrolysis of chitosan (biological chemistry). *Bioscience, Biotechnology and Biochemistry* 51:1189–1191.

Jayakumar, R., M. Deepthy, K. Manzoor, S. V. Nair, and H. Tamura. 2010. Biomedical applications of chitin and chitosan based nanomaterials—A short review. *Carbohydrate Polymers* 82:227–232.

Jeon, Y. J., P. J. Park, and S. K. Kim. 2001. Antimicrobial effect of chitooligosaccharides produced by bioreactor. *Carbohydrate Polymers* 44:71–76.

Karadeniz, F., M. Artan, M. M. Kim, and S. K. Kim. 2008. Prevention of cell damage on pancreatic beta cells by chitooligosaccharides. *Journal of Biotechnology* 136:539–540.

Karagozlu, M. Z., F. Karadeniz, C. S. Kong, and S. K. Kim. 2012. Aminoethylated chitooligomers and their apoptotic activity on AGS human cancer cells. *Carbohydrate Polymers* 87:1383–1389.

Karagozlu, M. Z., J. A. Kim, F. Karadeniz, C. S. Kong, and S. K. Kim. 2010. Anti-proliferative effect of aminoderivatized chitooligosaccharides on AGS human gastric cancer cells. *Process Biochemistry* 45:1523–1528.

Kato, Y., M. Saito, H. Fukushima, Y. Takeda, and T. Hara. 1984. Antitumor activity of 1-beta-D-arabinofuranosylcytosine conjugated with polyglutamic acid and its derivative. *Cancer Research* 44:25–30.

Kean, T., S. Roth, and M. Thanou. 2005. Trimethylated chitosans as non-viral gene delivery vectors: Cytotoxicity and transfection efficiency. *Journal of Controlled Release* 103:643–653.

Khnor, E. and L. Lim. 2003. Implantated applications of chitin and chitosan. *Biomaterials* 24:2339–2349.

Kong, C. S., J. A. Kim, B. Ahn, H. G. Byun, and S. K. Kim. 2010. Carboxymethylations of chitosan and chitin inhibit MMP expression and ROS scavenging in human fibrosarcoma cells. *Process Biochemistry* 45:179–186.

Kweon, D. K., S. B. Song, and Y. Y. Park. 2003. Preparation of water-soluble chitosan/heparin complex and its application as wound healing accelerator. *Biomaterials* 24:1595–1601.

Maeda, Y. and Y. Kimura. 2004. Antitumor effects of various low-molecular weight chitosans are due to increased natural killer activity intestinal intraphelial lymphocytes in sarcoma 180-bearing mice. *Nutrition and Cancer* 134:945–950.

Manjusha, E. M., J. C. Mohan, K. Manzoor, S. V. Nair, H. Tamura, and R. Jayakumar. 2010. Folate conjugated carboxymethyl chitosan–manganese doped zinc sulphide nanoparticles for targeted drug delivery and imaging of cancer cells. *Carbohydrate Polymers* 80:414–420.

Mao, J. S., H. F. Liu, Y. J. Yin, and K. D. Yao. 2003. The properties of chitosan–gelatin membranes and scaffolds modified with hyaluronic acid by different method. *Biomaterials* 24:1621–1629.

Miyata, K., M. Oba, M. R. Kano, S. Fukushima, Y. Vachutinsky, M. Han, H. Koyama, K. Miyazono, N. Nishiyama, and K. Kataoka. 2008. Polyplex micelles from triblock copolymers composed of tandemly aligned segments with biocompatible, endosome escaping, and DNA-condensing functions for systemic gene delivery to pancreatic tumor tissue. *Pharmaceutical Research* 25:2924–2936.

Mori, T., M. Okumura, M. Matsuura, K. Ueno, S. Tokura, Y. Okamoto, S. Minami, and T. Fujinaga. 1997. Effects of chitin and its derivatives on the proliferation and cytokine production of fibroblasts *in vitro*. *Biomaterials* 18:947–951.

Murata, J., I. Saiki, K. Matsuno, S. Tokura, and I. Azumo. 1990. Inhibition of tumor cell arrest in lungs by antimetastatic chitin heparimoid. *Japanese Journal of Cancer Research* 80:866–872.

Muzzarelli, R. A. A. 1997. Human enzymatic activities related to the therapeutic administration of chitin derivatives. *Cellular and Molecular Life Sciences* 53:131–140.

Nicol, S. 1991. Life after death for empty shells: Crustacean fisheries create a mountain of waste shells, made of a strong natural polymer, chitin. Now chemists are helping to put this waste to some surprising uses. *New Scientist* 1755:36–38.

Nishimura, K., S. Nishimura, N. Nishi, I. Saiki, S. Tokura, and I. Azuma. 1984. Immunological activity of chitin and its derivatives. *Vaccine* 2:93–99.

Nishiyama, Y., T. Yoshikawa, N. Ohara, K. Kurita, K. Hojo, H. Kamada, Y. Tsutsumi, T. Mayumi, and K. Kawasaki. 2000. A conjugate from a laminin-related peptide, Try-Ile-Gly-Ser-Arg, and chitosan: Efficient and regioselective conjugation and significant inhibitory activity against experimental cancer metastasis. *Journal of Chemical Society Perkin Transactions* 1:1161–1165.

Obara, K., M. Ishihara, Y. Ozeki, T. Ishizuka, T. Hayashi, S. Nakamura, Y. Saito et al. 2005. Controlled release of paclitaxel from photocrosslinked chitosan hydrogels and its subsequent effect on subcutaneous tumor growth in mice. *Journal of Controlled Release* 110:79–89.

Okamoto, Y., S. Minami, A. Matsuhashi, H. Sashiwa, H. Saimoto, Y. Shigemasa, T. Tanigawa, Y. Tanaka, and S. Tokura. 1993. Polymeric *N*-acetyl-D-glucosamine (chitin) induces histionic activation in dogs. *Journal of Veterinary Medical Sciences* 55:739–742.

Onishi, H., P. Pithayanukul, and T. Nagai. 1990. Antitumor characteristics of the conjugate of N^4-(4-carboxybutyryl)-Ara-C with ethylenediamine-introduced dextran and its resistance to cytidinedeaminase. *Drug Design and Delivery* 6:273–280.

Onishi, H., Y. Seno, P. Pithayanukul, and T. Nagai. 1991. Conjugate of ethylenediamine introduced dextran. Drug release profiles and further *in vivo* study of its antitumor effects. *Drug Design and Delivery* 7:139–145.

Ouchi, T., K. Inosaka, T. Banba, and Y. Ohya. 1992. *Design of Chitin or Chitosan/5-Fluorouracil Conjugate Having Antitumor Activity*. United Kingdom: Elsevier, Applied Science.

Park, P. J., J. Y. Je, and S. K. Kim. 2004. Free radical scavenging activities of differently deacetylated chitosans using an ESR spectrometer. *Carbohydrate Polymers* 55:17–22.

Qin, C., Y. Du, L. Xiao, Z. Li, and X. Gao. 2002. Enzymic preparation of water-soluble chitosan and their antitumor activity. *International Journal of Biological Macromolecules* 31:111–117.

Safari, S., F. A. Dorkoosh, M. Soleimani, M. H. Zarrintan, B. Akbari, B. Larijani, and M. R. Tehrani. 2011. *N*-Diethylmethyl chitosan for gene delivery to pancreatic cancer cells and the relation between charge ratio and biologic properties of polyplexes via interpolations polynomial. *International Journal of Pharmaceuticals* 420:350–357.

Sashiwa, H., H. Saimoto, Y. Shigemasa, R. Ogawa, and S. Tokura. 1990. Lysozyme susceptibility of partially deacetylated chitin. *International Journal of Biological Macromolecules* 90:295–296.

Sato, T., T. Ishii, and Y. Okahata. 2001. *In vitro* gene delivery mediated by chitosan. *Biomaterials* 22:2075–2080.

Seo, W. G., H. O. Pae, N. Y. Kim, G. S. Oh, I. S. Park, Y. H. Kim, Y. M. Kim, Y. Lee, C. D. Jun, and H. T. Chung. 2000. Synergistic cooperation between water-soluble chitosan oligomers and interferon-(gamma) for induction of nitric oxide synthesis and tumoricidal activity in murine peritoneal macrophages. *Cancer Letters* 159:189–195.

Shigemasa, Y., K. Saito, H. Sashiwa, and H. Saimoto. 1994. Enzymatic degradation of chitins and partially deacetylated chitins. *International Journal Biological Macromolecules* 16:43–49.

Sirica, A. E and R. J. Woodman. 1971. Selective aggregation of L1210 leukemia cells by the polycation chitosan. *Journal of the National Cancer Institute* 47:377–388.

Suzuki, K., T. Mikami, Y. Okawa, A. Tokoro, S. Suzuki, and M. Suzuki. 1986. Antitumor effect of hexa-*N*-acetylchitohexaose and chitohexaose. *Carbohydrate Research* 151:403–408.

Tanigawa, T., Y. Tanaka, H. Sashiwa, H. Saimoto, and Y. Shigemasa. 1992. *Various Biological Effects of Chitin Derivatives*. Sandford: Elsevier.

Thanou, M., B. I. Florea, M. Geldof, H. E. Junginger, and G. Borchard. 2002. Quaternized chitosan oligomers as novel gene delivery vectors in epithelial cell lines. *Biomaterials* 23:153–159.

Tokoro, A., N. Tatewaki, K. Suzuki, T. Mikami, S. Suzuki, and M. Suzuki. 1988. Growth inhibitory effect of hexa-*N*-acetylchitohexaose and chitohexaose against Meth-A solid tumor. *Chemical and Pharmaceutical Bulletin* 36:784–790.

Tokura, S., K. Ueno, S. Miyazaki, and N. Nishi. 1997. Molecular weight dependent antimicrobial activity by chitosan. *Macromolecular Symposia* 120:1–9.

Tsukada, S. and Y. Inoue. 1981. Conformational properties of chito-oligosaccharides: Titration, optical rotation, and carbon-13 N.M.R. studies of chito-oligosaccharides. *Carbohydrate Research* 88:19–38.

Vernejoul, F., P. Faure, N. Benali, D. Calise, G. Tiraby, L. Pradayrol, C. Susini, and L. Buscail. 2002. Antitumor effect of *in vivo* somatostatin receptor subtype 2 gene transfer in primary and metastatic pancreatic cancer models. *Cancer Research* 62:6124–6131.

Wang, Y. C., M. C. Lin, D. M. Wang, and H. J. Hsieh. 2003. Fabrication of a novel porous PGA–chitosan hybrid matrix for tissue engineering. *Biomaterials* 24:1047–1057.

You, Y. Z., D. S. Manickam, Q. H. Zhou, and D. Oupick`y. 2007. Reducible poly (2 dimethyl aminoethyl methacrylate): Synthesis, cytotoxicity, and gene delivery activity. *Journal of Controlled Release* 122:217–225.

10 Antidiabetic Applications of Chitosan and Its Derivatives

Fatih Karadeniz and Se-Kwon Kim

CONTENTS

10.1 BACKGROUND

Chitosan is a functional and basic linear polysaccharide prepared by *N*-deacetylation of chitin in the presence of alkali sodium hydroxide. Generally, deacetylation cannot be completely achieved even when subjected to harsh treatment. The degree of deacetylation usually ranges from 70% to 95%, depending on the method used. Thus, chitosan is available with various molecular weights and deacetylation degrees. Chitosan is insoluble in water, alkali, and organic solvents but is soluble in most solutions of organic acids when the pH of the solution is below 6. The industrial production and application fields of chitosan have been steadily increasing since the 1970s. Early applications of chitosan were centered on the treatment of wastewater, heavy metal adsorption, food processing, immobilization of cells and enzymes, resin for chromatography, functional membrane in biotechnology, animal feed, and so on. The recent trend is toward producing highly valuable industrial products, such as cosmetics, drug carriers, and pharmaceuticals. Chitin and chitosan are known to exhibit antitumor, antibacterial, hypocholesterolemia, and antihypertensive activity (Kim and Rajapakse 2005). The main motive for the development of new applications for chitosan lies in the fact that it is a very abundant polysaccharide and it is nontoxic and biodegradable. Despite its functions and importance as a biomaterial, the applications of chitosan in food and biomedical industries are narrowed owing to its poor solubility, high molecular weight, and viscosity. There is evidence to validate the non- or indigent absorption of chitin and chitosan in the human intestine owing to lack of enzymes to cleave the β-glucosidic linkage in chitosan. Because chitosan is a water-insoluble large biopolymer, it is difficult to be absorbed by the mammal body. In this respect, enzymatic hydrolysis of chitosan to obtain oligomers has gained considerable interest recently (Jeon and Kim 2000).

Chitosan oligosaccharides (COS) are hydrolyzed derivatives of chitosan composed of β-(1 → 4) D-glucosamine units. They have better properties, such as lower viscosity, relatively smaller molecular size in comparison to chitosan, and short chain length with free amino groups, which makes COS highly soluble in aqueous solutions. COS are effective agents for lowering of blood cholesterol and pressure, controlling arthritis, and enhancing antitumor properties (Kim and Rajapakse 2005). Because COS are biodegradable, water soluble, and nontoxic compounds (Qin et al. 2006), they might be beneficial biomaterials for diseases, such as diabetes and obesity, with increasing morbid-ity and mortality rates.

As a chronic disease, diabetes must be kept under control by improving impaired insulin secre-tion from β-islet cells (pancreas) or elevating insulin efficiency on several tissues. Diabetic disorders, especially hyperglycemia, can lead to serious damage to many parts of the body, in particular, the nerves and blood vessels (Vinik et al. 2003). The cause of diabetes is not fully known, although it is clearly shown that both genetic and environmental factors, notably obesity seems to play important roles. Differentiated adipocytes secrete obesity-related factors called adipokines. Plasma, leptin, tumor necrosis factor (TNF)α, and nonesterified fatty acid levels are all elevated in obesity and play a role in causing insulin resistance (Leong and Wilding 1999). Therefore, suppression and regula-tion of obesity can be achieved by inhibiting adipocyte differentiation and forcing adipocytes to lipolysis to reduce accumulated white adipose tissue (Yamauchi et al. 2001; Langin 2006). Thus, the increased control of the harmful effects of the accumulation of adipose tissue and its metabolism contribute to the search for a better understanding of the prevention of diabetes.

With a long onset and serious complications, which usually result in a high morbidity rate, the treatment of diabetes is a major concern in all countries. Up to now, many kinds of antidiabetic medicines from natural resources have been developed for diabetic patients (Ivorra et al. 1989; Grover et al. 2002; Koski 2004; Li et al. 2004), but most of these biochemical agents are not suited for mass production to be a pharmaceutical agent. The natural compounds demonstrated a signifi-cant practice and show a bright potential in the treatment of diabetes and its complications with their naturally occurring structure and relatively less side effects. In this respect, chitin, chitosan, and its derivatives with available large numbers of different chemical structures and bioactivities offer a great potential to recover and/or prevent obesity and diabetes.

10.2 CHITOSAN DERIVATIVES

Chitosan and its monomer glucosamine have been derived recently as a result of an attempt to find new natural compounds with higher bioactivity than their predecessors (Fenton et al. 2000; Jiang et al. 2007; Prabaharan 2008) (Figure 10.1). The main derivation of chitosan is forming sol-uble forms of chitin, which makes it more biofriendly and easily absorbed by the body after oral administration (Kuroiwa et al. 2002; Hai et al. 2003; Il'ina and Varlamov 2004; Mao et al. 2004).

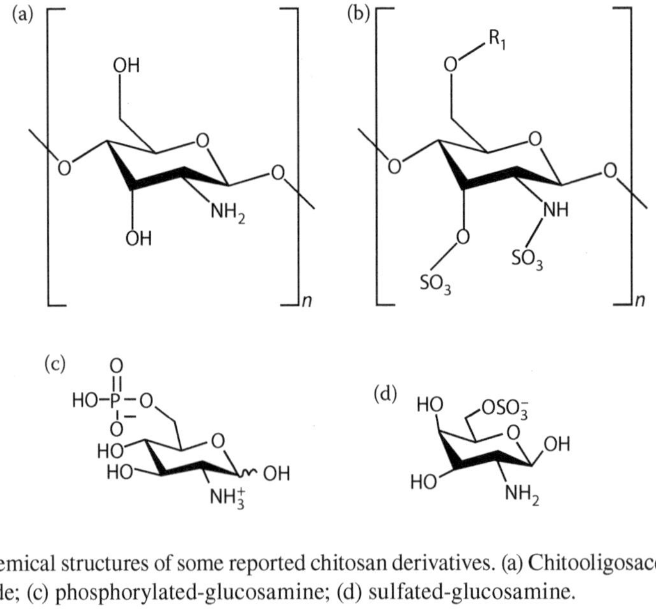

FIGURE 10.1   Chemical structures of some reported chitosan derivatives. (a) Chitooligosaccharide; (b) sulfated-chitooligosaccharide; (c) phosphorylated-glucosamine; (d) sulfated-glucosamine.

Because, effectiveness of the compound, based on its absorption rate by the body is the criterion for evaluation this kind of derivation opened up new angles for chitin derivation toward novel bioactive compounds. In this respect, chitosan is the main derivative of chitin. Rather than the chitosan, COS are also highly bioactive derivatives of chitosan with significantly higher absorption rates and water solubility (Qin et al. 2002).

Besides oligomerization, another main derivation for chitin and its monomer glucosamine is adding negative and/or positively charged side chains. In this manner, glucosamine, chitin, chitosan, and COS are reformed under chemical conditions to give sulfated, phosphorylated, carboxymethyl, deoxymethyl derivatives, and so on (Fei Liu et al. 2000; Kochkina and Chirkov 2000; Huang et al. 2005; Je and Kim 2006; Kim et al. 2010, 2005; Cho et al. 2011). This diversity of derivatives comes with a large variety of bioactivities, including improvement in the effectiveness of the compound in case of already-reported activities.

10.3 APPLICATIONS AGAINST DIABETES

Chitosan-based products are known to have many biological activities, such as antitumor, anti-HIV, antifungal, and antibiotic activities, and activity against oxidative stress (Kendra and Hadwiger 1984; Nishimura et al. 1998; Xie et al. 2001; Kim et al. 2008; Artan et al. 2010). Activities can be grouped into two according to the use of chitin-based products. These products are predominantly used as indirect helping agents to enhance the effectiveness of other active compounds through the chemical modification or nonchemical linkage against diabetes and obesity. However, the main role of chitin-based products is known as therapeutic nutraceutical agents that act directly against diabetes and obesity. In both cases, derivatives of these natural products express a high and significant potential in the light of searching bioactive pharmaceuticals against obesity and obesity-related diabetes.

10.3.1 SUPPORTIVE APPLICATIONS

The preferred route of drug administration for patients is mostly the oral route on chronic therapy of diseases and complications. However, the delivery of many therapeutic peptides and proteins through the digestion system is still an unsolved problem basically because of the size, hydrophilicity, and unstable conditions of these molecules. Thus, several chitosan derivatives have been developed over the years with improved properties for enhanced applicability (Fernández-Urrusuno et al. 1999; Thanou et al. 2001). Therefore, recent studies focused on carrier products for the administration of insulin efficiently in pre- or postdiabetic patients and lately, one of these products is chitosan derivatives (Figure 10.2). It has been reported by Portero et al. (2007) that chitosan sponges are quite successful in buccal administration of insulin. Moreover, up-to-date studies presumed that chitosan-derived particles are intensely usable for insulin administration orally with their high protective effect and harmless structure (Hari et al. 1998; Krauland et al. 2004, 2006). Results of some related studies have suggested that the observed drug delivery activity of chitosan is highly promising in the case of insulin. For example, studies showed that chitosan–insulin nanoparticles have a strong affinity to rat intestinal epithelium after 3 h postoral administration (Ma et al. 2005). This suggests that chitosan as a cofactor for drug delivery makes insulin absorption safe and rapid. Carboxymethyl–hexanoyl chitosan is an amphiphilic chitosan derivative with important swelling ability and water solubility under natural conditions and studies showed that these hydrogels can be used for encapsulating the poorly water-soluble drugs for effective drug delivery (Liu and Lin 2010) that lightens up the way for efficient insulin delivery by chitosan derivatives. Furthermore, Mao et al. (2005) showed that polyethylene glycol (PEG)–trimethyl chitosan complexes are efficiently coupled with insulin and easily taken up by Caco-2 cells.

Besides drug delivery activity for insulin, studies have shown that chitosan complexes can be efficiently used for gene delivery for gene therapy (Köping-Höggård et al. 2001). Therefore, it can be

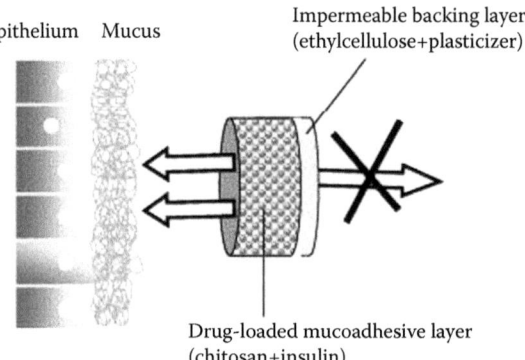

FIGURE 10.2 Structure of the chitosan/ethylcellulose bilayered devices. (From Portero A et al. 2007. *Carbohydrate Polymers* 68(4):617–625. With permission.)

easily adduced that chitosan complex derivatives are potent gene delivery targets for highly prevalent diseases, such as diabetes. Furthermore, it has been reported that these chitosan complexes possess relatively higher uptake and transfection efficiency than that of other polysaccharide complexes used for both drug and gene delivery (Huang et al. 2004). Several researches were conducted to prove chitosan as a nontoxic alternative to other cationic polymers and results demonstrated a prominent potential for further studies of chitosan-based gene delivery systems (Sato et al. 2001). All these results suggest that chitosan and chitosan-based derivatives are the end result of the search for a harmless agent for drug and gene delivery, which is extremely crucial for improved life standards of the diabetic patient.

Moreover, studies on streptozotocin (STZ)-induced diabetic rats expressed that chitosan-based sponges are highly effective for healing diabetic wounds in addition to treatment of diabetic patients (Figure 10.3). Wang et al. (2008) suggest that the application of chitosan–collagen complex is an ideal wound-healing cover to enhance the recovery of healing of wounds, such as diabetic skin wound, which provides a great potential for chitosan and its derivatives to be used clinically for diabetic patients.

To conclude, chitosan-based polymers show great potential for treatment of diabetes therapeutically with their highly efficient drug and gene delivery properties as well as effectiveness on diabetic wound healing.

10.3.2 ANTIDIABETIC APPLICATIONS

Overweight and obesity, two common health-threatening conditions, are considered to result in diabetes worldwide and there are no effective treatments. Therefore, studies of chitosan are focused on its fat-lowering and fat-preventing activities. Several researchers have demonstrated that chitosan tends to bond with the ingested dietary fat and carry it out in the stool while preventing its absorption through the gut (Kanauchi et al. 1995). Relevant research about the fat-lowering activity of chitosan has also shown that chitosan is capable of absorbing fat up to 5 times of its weight. In respect to these results, there are several studies showing chitosan derivatives to be lower than the level of low-density lipoprotein (LDL) while increasing the high-density lipoprotein (HDL) levels. Studies of chitosan and its fat-lowering activity have stated that chitosan and its derivatives are highly effective hypocholesterolemic agents with the ability of decreasing blood cholesterol level up to as much as 50% (Maezaki et al. 1993; Jameela et al. 1995). Moreover, diabetic patient-based studies clearly showed that the daily administration of chitosan could drop the blood cholesterol levels by 6% with an increased level of HDL (Maezaki et al. 1993). In addition to chitosan, COS, an oligomerized derivative of chitosan, shows high activity in regulating blood cholesterol levels. Especially, studies

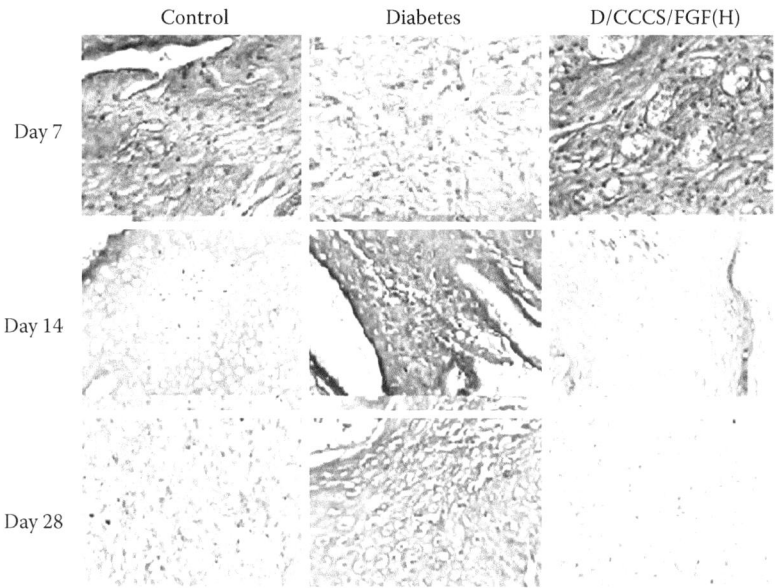

FIGURE 10.3 Immunohistochemical staining of STZ-induced diabetic mouse tissues for TGF-β1 expression for three different test groups after mentioned posttreatment time. Diabetic wounds of STZ-induced diabetic mouses were healed through fibrosis mechanism by D/CCCS/FGF(H) treatment. TGF-B1 amount was increased to heal the wounds by fibrosis. Sample-treated wounds show the same patterns as healthy control groups express. (D/CCCS/FGF(H): diabetes treated with chitosan cross-linked collagen sponge containing recombinant human aFGF). (From Wang W et al. 2008. *Life Sciences* 82(3):190–204. With permission.)

reported that COS are capable of regulating cholesterol levels even in liver. COS prevent the development of fatty liver caused by the action of hepatotropic poisons. Despite few studies that were carried out for the action mechanism of COS in regulating the serum cholesterol level, several of them suggested the possible mechanism of COS lowering the LDL levels. As Remunan-Lopez et al. (1998) suggested, the ionic structure of COS binds bile salts and acid, which inhibit lipid digestion through micelle formation. However, Tanaka et al. (1997) suggest a different mechanism of chitosan and COS where lipids and fatty acids are directly bonded by chitosan.

In addition to fat-lowering mechanisms of chitosan and its derivatives, studies have also proven that chitosan administration can lead to increase the insulin sensitivity of animal models (Neyrinck et al. 2009). It has been shown that 3-month administration of chitosan significantly increased insulin sensitivity in obese patients and expressed a highly notable decrease in body weight and triglyceride levels (Hernández-González et al. 2010).

However, glucosamine and its derivatives are reported to be highly effective at inhibiting adipogenesis *in vitro*. Recent studies showed that a phosphorylated derivative of glucosamine inhibited the adipogenesis of 3T3-L1 cells as well as fat accumulation (Kim et al. 2010) (Figure 10.4). Several researches suggested that acetylated chitin treatment causes adipocytes to break down fats and lower their triglyceride accumulation as much as half of control cells (Kong et al. 2011). Kong et al. (2009) demonstrated clearly that sulfated derivative of glucosamine inhibited the proliferation and adipogenesis mechanism through 5′-adenosine monophosphate-activated protein kinase (AMPK) pathways in 3T3-L1 cells. Glucosamine, acetylated, sulfated, and phosphorylated glucosamine derivatives are reported as successful adipogenic inhibitors with intense potential to prevent weight gain by adipogenesis in patients who are at risk for diabetes. It has also been reported that COS inhibit the fat accumulation and adipogenesis in 3T3-L1 cell line (Cho et al. 2008). In addition, studies have shown that treatment with glucosamines reduced the triglyceride content of adipocytes and enhanced glycerol secretion as a lipid-lowering effect. Most of these studies have expressed the better activity

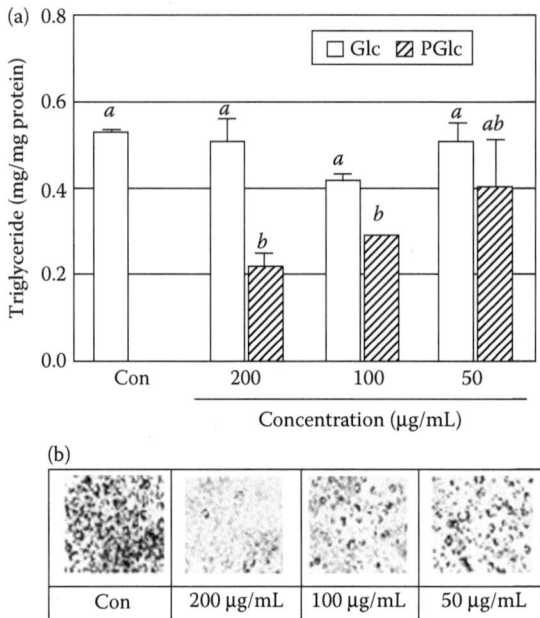

FIGURE 10.4 Effects of glucosamine (Glc) and phosphorylated glucosamine (PGlc) on adipocyte differentiation in 3T3-L1 cells. Lipid accumulation was measured by (a) triglyceride (TG) content and (b) Oil Red O staining for intracellular lipid amount. (From Kong C-S et al. 2010. *The Journal of Nutritional Biochemistry* 21(5):438. With permission.)

of chitosan-based compounds such as COS and glucosamines, after derivation by adding a charged side chain by phosphorylation and sulfation. Therefore, it can be suggested that the cationic power of glucosamine and COS plays the main role in their antiobesity effect. Further, a selective synthesis of phosphorylated or sulfated derivatives of chitosan and glucosamine will open up the way to a better understanding behind the structure–mechanism relation. However, current researchers have strong proofs that chitosan shows its antiobesity effect through the PPAR-γ pathway of adipogenic differentiation that results in fewer adipocytes and lipid accumulation (Cho et al. 2008). Collectively, chitosan and its derivatives such as glucosamines and COS successfully inhibit the differentiation of cells into adipocytes as well as enhancing adipocytes to hydrolyze the triglycerides that show a significant effect against lipid accumulation of the body. This effect of chitosan and its derivatives demonstrates an important impact against obesity in the way of diabetes progression. Hence, it shows a great amount of potential to be used as pharmaceutical agents.

Furthermore, chitosan and its oligosaccharides act as antidiabetic agents for treatment of diabetes in a manner of protecting pancreatic β-cells. In type 2 diabetes, although patients can retain healthy pancreatic β-cells for many years after the disease onset, chronic exposure to high glucose will impair β-cell function in later stages. Impaired β-cell functionality leads to cellular damage in type 2 diabetic patients (Ihara et al. 1999). Therefore, the protection of β-cells is of great importance for elevated insulin secretion as a part of diabetes treatment. Recent studies reported that COS act as a protective agent for pancreatic β-cells against high glucose-dependent cell deterioration (Karadeniz et al. 2011) (Figure 10.5). It is suggested that at the same time, COS could effectively accelerate the proliferation of pancreatic islet cells with elevated insulin secretion in the aid of lowering blood glucose levels. Liu et al. (2007) reported that COS treatment could improve the general situation and diabetic symptoms of rats, decrease the blood glucose levels, and normalize the impaired insulin sensitivity. Moreover, COS were reported as a preventive agent in nonobese diabetic mice from developing type 1 diabetes, which might be related to several bioactivities of COS (Cao et al. 2004). These results supported the hypothesis that COS can prevent pancreatic β-cells of diabetic patients

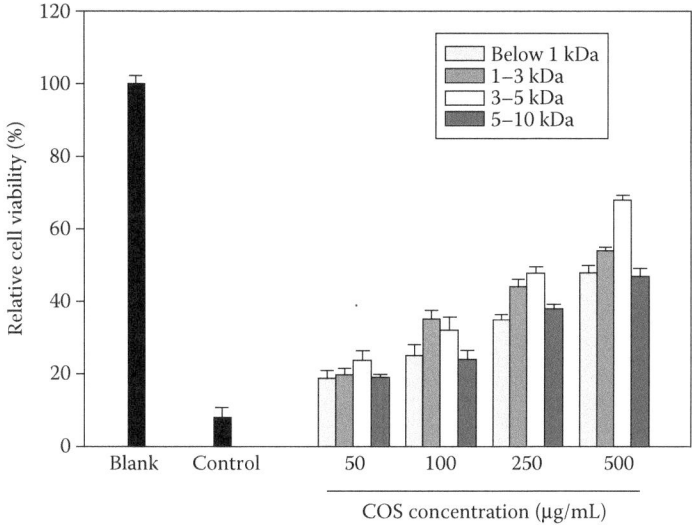

FIGURE 10.5   Effect of COS on cell viability of HIT-T15 pancreatic β-cells after exposure to hydrogen peroxide. (From Karadeniz F et al. 2011. *Carbohydrate Polymers* 86(2):666–671. With permission.)

and normalize the crucial insulin secretion. The mechanism behind this protection is studied and suggested as related to immunopotentiation and antioxidation activity of COS.

Renal failure is one of the most common diseases caused by diabetes mellitus. The metal cross-linked complex of chitosan, chitosan–iron (III), has been recently reported to be highly active in reducing phosphorus serum levels to treat chronic renal failure (Schöninger et al. 2010). This relatively new derivative of chitosan is significantly capable of adsorbing serum phosphorus in alloxan diabetes-induced rats with symptoms of renal failure progression.

Moreover, recent studies indicate that diabetics may be at higher risk for blood coagulation than nondiabetics. This life-threatening condition urges to be treated for diabetic patients. Therefore, a sulfated derivative of chitosan has been shown to possess anticoagulant potency (Vongchan et al. 2002). Furthermore, studies have reported that sulfated chitosan does not show antiplatelet activity unlike heparin, which is an effective anticoagulant agent. Collectively, results proved that sulfated chitosan is a more efficient agent than that of heparin, although heparin has been used for a long time for blood coagulation treatment (Bourin and Lindahl 1993).

In addition to COS, chitosan is also reported to prevent the development and symptoms of non-insulin-dependent diabetes in rats as well as the complications of STZ inducement (Kondo et al. 2000). Briefly, reports suggest that chitosan products protect pancreatic cells and insulin secretion mechanism in diabetic conditions. Furthermore, these compounds can decrease the progression and complication rate of diabetes onset in animal models, demonstrating a great potential for chitosan products to be used as a nutraceutical for the treatment of diabetes.

10.4   CONCLUSION

High mortality and morbidity of diabetes make the diagnosis, prevention, and treatment more important as increasing number of patients are diagnosed by diabetes in the world in recent years. Besides diabetes, factors relating to diabetes, such as obesity and damaged pancreatic cells, must be kept under control to prevent the onset of diabetes. In this manner, chitosan and its derivatives possess various biological activities and have a remarkable potential to be used in several therapeutic applications. Thus, many of the studies carried out to search antidiabetic activities of chitosan-based compounds provide detailed acting mechanisms and activity for prevention and/or treatment of diabetes-based complications. Chitosan and its derivatives such as COS and glucosamines as monomers express high activity in a manner of lowering lipid accumulation and cholesterol as well as pancreatic β-cell prevention. In addition, studies proved that chemical modification of these compounds could express better activity and understanding of the mechanism lying behind antidiabetic effects. Therefore, future research should be directed to enhance the effectiveness of chitosan-based compounds to gain more active and fewer harmful agents. Collectively, in conclusion, this evidence suggests that chitosan-based agents are highly potent nutraceuticals for treatment and prevention of diabetes and diabetes-related complications.

REFERENCES

Artan, M, F Karadeniz, MZ Karagozlu, M-M Kim, and S-K Kim. 2010. Anti-HIV-1 activity of low molecular weight sulfated chitooligosaccharides. *Carbohydrate Research* 345(5):656–662.

Bourin, M-C and U Lindahl. 1993. Glycosaminoglycans and the regulation of blood coagulation. *Biochemical Journal* 289(2):313.

Cao, Z, B Li, and X Qiao. 2004. The effect of chitooligosaccharides on preventing the onset of diabetes in nod mice. *The Journal of Medical Theory and Practice* 12:25–31.

Cho, EJ, SW Kim, HJ Hwang, HS Hwang, and JW Yun. 2008. Chitosan oligosaccharides inhibit adipogenesis in 3T3-L1 adipocytes. *Journal of Microbiology and Biotechnology* 18(1):80–87.

Cho, Y-S, S-H Lee, S-K Kim, C-B Ahn, and J-Y Je. 2011. Aminoethyl-chitosan inhibits LPS-induced inflammatory mediators, inos and cox-2 expression in raw 264.7 mouse macrophages. *Process Biochemistry* 46(2):465–470.

Fei Liu, X, YL Guan, DZ Yang, Z Li, and K De Yao. 2000. Antibacterial action of chitosan and carboxymethylated chitosan. *Journal of Applied Polymer Science* 79(7):1324–1335.

Fenton, JI, KA Chlebek-Brown, TL Peters, JP Caron, and MW Orth. 2000. Glucosamine hcl reduces equine articular cartilage degradation in explant culture. *Osteoarthritis and Cartilage* 8(4):258–265.

Fernández-Urrusuno, R, P Calvo, C Remuñán-López, JL Vila-Jato, and MJ Alonso. 1999. Enhancement of nasal absorption of insulin using chitosan nanoparticles. *Pharmaceutical Research* 16(10):1576–1581.

Grover, JK, S Yadav, and V Vats. 2002. Medicinal plants of India with anti-diabetic potential. *Journal of Ethnopharmacology* 81(1):81.

Hai, L, TB Diep, N Nagasawa, F Yoshii, and T Kume. 2003. Radiation depolymerization of chitosan to prepare oligomers. *Nuclear Instruments and Methods in Physics Research Section B: Beam Interactions with Materials and Atoms* 208:466–470.

Hari, PR, T Chandy, and CP Sharma. 1998. Chitosan/calcium–alginate beads for oral delivery of insulin. *Journal of Applied Polymer Science* 59(11):1795–1801.

Hernández-González, SO, M González-Ortiz, E Martínez-Abundis, and JA Robles-Cervantes. 2010. Chitosan improves insulin sensitivity as determined by the euglycemic–hyperinsulinemic clamp technique in obese subjects. *Nutrition Research* 30(6):392–395.

Huang, M, E Khor, and L-Y Lim. 2004. Uptake and cytotoxicity of chitosan molecules and nanoparticles: Effects of molecular weight and degree of deacetylation. *Pharmaceutical Research* 21(2):344–353.

Huang, R, E Mendis, and S-K Kim. 2005. Factors affecting the free radical scavenging behavior of chitosan sulfate. *International Journal of Biological Macromolecules* 36(1):120–127.

Ihara, Y, S Toyokuni, K Uchida et al. 1999. Hyperglycemia causes oxidative stress in pancreatic beta-cells of gk rats, a model of type 2 diabetes. *Diabetes* 48(4):927–932.

Il'ina, AV, and VP Varlamov. 2004. Hydrolysis of chitosan in lactic acid. *Applied Biochemistry and Microbiology* 40(3):300–303.

Ivorra, MD, M Paya, and A Villar. 1989. A review of natural products and plants as potential antidiabetic drugs. *Journal of Ethnopharmacology* 27(3):243–275.

Jameela, SR, A Misra, and A Jayakrishnan. 1995. Cross-linked chitosan microspheres as carriers for prolonged delivery of macromolecular drugs. *Journal of Biomaterials Science, Polymer Edition* 6(7):621–632.

Je, J-Y and S-K Kim. 2006. Chitosan derivatives killed bacteria by disrupting the outer and inner membrane. *Journal of Agricultural and Food Chemistry* 54(18):6629–6633.

Jeon, Y-J and S-K Kim. 2000. Production of chitooligosaccharides using an ultrafiltration membrane reactor and their antibacterial activity. *Carbohydrate Polymers* 41(2):133–141.

Jiang, L, F Qian, X He et al. 2007. Novel chitosan derivative nanoparticles enhance the immunogenicity of a DNA vaccine encoding hepatitis B virus core antigen in mice. *The Journal of Gene Medicine* 9(4):253–264.

Köping-Höggård, M, I Tubulekas, H Guan et al. 2001. Chitosan as a nonviral gene delivery system. Structure–property relationships and characteristics compared with polyethylenimine *in vitro* and after lung administration *in vivo*. *Gene Therapy* 8:1108–1121.

Kanauchi, O, K Deuchi, Y Imasato, M Shizukuishi, and E Kobayashi. 1995. Mechanism for the inhibition of fat digestion by chitosan and for the synergistic effect of ascorbate. *Bioscience, Biotechnology, and Biochemistry* 59(5):786.

Karadeniz, F, MZ Karagozlu, S-Y Pyun, and S-K Kim. 2011. Sulfation of chitosan oligomers enhances their anti-adipogenic effect in 3T3-L1 adipocytes. *Carbohydrate Polymers* 86(2):666–671.

Kendra, DF and LA Hadwiger. 1984. Characterization of the smallest chitosan oligomer that is maximally antifungal to *Fusarium solani* and elicits pisatin formation in *Pisum sativum*. *Experimental Mycology* 8(3):276–281.

Kim, J, C-S Kong, SY Pyun, and S-K Kim. 2010. Phosphorylated glucosamine inhibits the inflammatory response in lps-stimulated pma-differentiated thp-1 cells. *Carbohydrate Research* 345(13):1851–1855.

Kim, J-H, Y-S Kim, K Park et al. 2008. Antitumor efficacy of cisplatin-loaded glycol chitosan nanoparticles in tumor-bearing mice. *Journal of Controlled Release* 127(1):41–49.

Kim, S-K, and N Rajapakse. 2005. Enzymatic production and biological activities of chitosan oligosaccharides (COS): A review. *Carbohydrate Polymers* 62(4):357–368.

Kim, S-K, P-J Park, W-K Jung, H-G Byun, E Mendis, and Y-I Cho. 2005. Inhibitory activity of phosphorylated chitooligosaccharides on the formation of calcium phosphate. *Carbohydrate Polymers* 60(4):483–487.

Kochkina, ZM and SN Chirkov. 2000. Influence of chitosan derivatives on the development of phage infection in the *Bacillus thuringiensis* culture. *Microbiology* 69(2):217–219.

Kondo, Y, A Nakatani, K Hayashi, and M Ito. 2000. Low molecular weight chitosan prevents the progression of low dose streptozotocin-induced slowly progressive diabetes mellitus in mice. *Biological and Pharmaceutical Bulletin* 23(12):1458.

Kong, C-S, J-A Kim, TK Eom, and S-K Kim. 2010. Phosphorylated glucosamine inhibits adipogenesis in 3T3-L1 adipocytes. *The Journal of Nutritional Biochemistry* 21(5):438.

Kong, C-S, J-A Kim, and S-K Kim. 2009. Anti-obesity effect of sulfated glucosamine by AMPK signal pathway in 3T3-L1 adipocytes. *Food and Chemical Toxicology* 47(10):2401–2406.

Kong, C-S, J Kim, S-S Bak, H-G Byun, and S-K Kim. 2011. Anti-obesity effect of carboxymethyl chitin by AMPK and aquaporin-7 pathways in 3T3-L1 adipocytes. *The Journal of Nutritional Biochemistry* 22(3):276–281.

Koski, RR. 2004. Oral antidiabetic agents: A comparative review. *Journal of Pharmacy Practice* 17(1):39–48.

Krauland, AH, D Guggi, and A Bernkop-Schnürch. 2004. Oral insulin delivery: The potential of thiolated chitosan–insulin tablets on non-diabetic rats. *Journal of Controlled Release* 95(3):547–555.

Krauland, AH, D Guggi, and A Bernkop-Schnürch. 2006. Thiolated chitosan microparticles: A vehicle for nasal peptide drug delivery. *International Journal of Pharmaceutics* 307(2):270–277.

Kuroiwa, T, S Ichikawa, O Hiruta, S Sato, and S Mukataka. 2002. Factors affecting the composition of oligosaccharides produced in chitosan hydrolysis using immobilized chitosanases. *Biotechnology Progress* 18(5):969–974.

Langin, D. 2006. Adipose tissue lipolysis as a metabolic pathway to define pharmacological strategies against obesity and the metabolic syndrome. *Pharmacological Research: The Official Journal of the Italian Pharmacological Society* 53(6):482.

Leong, KS and JP Wilding. 1999. Obesity and diabetes. *Best Practice and Research Clinical Endocrinology and Metabolism* 13(2):221–237.

Li, WL, HC Zheng, J Bukuru, and N De Kimpe. 2004. Natural medicines used in the traditional Chinese medical system for therapy of diabetes mellitus. *Journal of Ethnopharmacology* 92(1):1–22.

Liu, B, W-S Liu, B-Q Han, and Y-Y Sun. 2007. Antidiabetic effects of chitooligosaccharides on pancreatic islet cells in streptozotocin-induced diabetic rats. *World Journal of Gastroenterology* 13(5):725.

Liu, T-Y and Y-L Lin. 2010. Novel pH-sensitive chitosan-based hydrogel for encapsulating poorly water-soluble drugs. *Acta Biomaterialia* 6(4):1423–1429.

Ma, Z, TM Lim, and L-Y Lim. 2005. Pharmacological activity of peroral chitosan–insulin nanoparticles in diabetic rats. *International Journal of Pharmaceutics* 293(1):271–280.

Maezaki, Y, K Tsuji, Y Nakagawa et al. 1993. Hypocholesterolemic effect of chitosan in adult males. *Bioscience, Biotechnology, and Biochemistry* 57(9):1439–1444.

Mao, S, O Germershaus, D Fischer, T Linn, R Schnepf, and T Kissel. 2005. Uptake and transport of PEG-*graft*-trimethyl-chitosan copolymer–insulin nanocomplexes by epithelial cells. *Pharmaceutical Research* 22(12):2058–2068.

Mao, S, X Shuai, F Unger, M Simon, D Bi, and T Kissel. 2004. The depolymerization of chitosan: Effects on physicochemical and biological properties. *International Journal of Pharmaceutics* 281(1):45–54.

Neyrinck, AM, LB Bindels, FDe Backer, BD Pachikian, PD Cani, and NM Delzenne. 2009. Dietary supplementation with chitosan derived from mushrooms changes adipocytokine profile in diet-induced obese mice, a phenomenon linked to its lipid-lowering action. *International Immunopharmacology* 9(6):767–773.

Nishimura, S-I, H Kai, K Shinada et al. 1998. Regioselective syntheses of sulfated polysaccharides: Specific anti-HIV-1 activity of novel chitin sulfates. *Carbohydrate Research* 306(3):427–433.

Portero, A, D Teijeiro-Osorio, MJ Alonso, and C Remuñán-López. 2007. Development of chitosan sponges for buccal administration of insulin. *Carbohydrate Polymers* 68(4):617–625.

Prabaharan, M. 2008. Review paper: Chitosan derivatives as promising materials for controlled drug delivery. *Journal of Biomaterials Applications* 23(1):5–36.

Qin, C, Y Du, L Xiao, Z Li, and X Gao. 2002. Enzymic preparation of water-soluble chitosan and their antitumor activity. *International Journal of Biological Macromolecules* 31(1):111–117.

Qin, C, H Li, Q Xiao, Y Liu, J Zhu, and Y Du. 2006. Water-solubility of chitosan and its antimicrobial activity. *Carbohydrate Polymers* 63(3):367–374.

Remuñán-López, C, A Portero, JL Vila-Jato, and MJ Alonso. 1998. Design and evaluation of chitosan/ethylcellulose mucoadhesive bilayered devices for buccal drug delivery. *Journal of Controlled Release* 55(2):143–152.

Sato, T, T Ishii, and Y Okahata. 2001. *In vitro* gene delivery mediated by chitosan. Effect of pH, serum, and molecular mass of chitosan on the transfection efficiency. *Biomaterials* 22(15):2075–2080.

Schöninger, LMR, RC Dall'Oglio, S Sandri, CA Rodrigues, and C Bürger. 2010. Chitosan iron (III) reduces phosphorus levels in alloxan diabetes-induced rats with signs of renal failure development. *Basic and Clinical Pharmacology and Toxicology* 106(6):467–471.

Tanaka, Y, S-I Tanioka, M Tanaka et al. 1997. Effects of chitin and chitosan particles on balb/c mice by oral and parenteral administration. *Biomaterials* 18(8):591–595.

Thanou, M, JC Verhoef, and HE Junginger. 2001. Oral drug absorption enhancement by chitosan and its derivatives. *Advanced Drug Delivery Reviews* 52(2):117–126.

Vinik, AI, RE Maser, BD Mitchell, and R Freeman. 2003. Diabetic autonomic neuropathy. *Diabetes Care* 26(5):1553–1579.

Vongchan, P, W Sajomsang, D Subyen, and P Kongtawelert. 2002. Anticoagulant activity of a sulfated chitosan. *Carbohydrate Research* 337(13):1239–1242.

Wang, W, S Lin, Y Xiao et al. 2008. Acceleration of diabetic wound healing with chitosan-crosslinked collagen sponge containing recombinant human acidic fibroblast growth factor in healing–impaired STZ diabetic rats. *Life Sciences* 82(3):190–204.

Xie, W, P Xu, and Q Liu. 2001. Antioxidant activity of water-soluble chitosan derivatives. *Bioorganic and Medicinal Chemistry Letters* 11(13):1699–1701.

Yamauchi, T, H Waki, J Kamon et al. 2001. Inhibition of rxr and PPAR gamma ameliorates diet-induced obesity and type 2 diabetes. *Journal of Clinical Investigation* 108(7):1001–1013.

11 Antioxidant, Antimicrobial Properties of Chitin, Chitosan, and Their Derivatives

Dai-Nghiep Ngo and Se-Kwon Kim

CONTENTS

11.1 INTRODUCTION

Chitin, $(1 \rightarrow 4)$-linked 2-acetamido-2-deoxy-β-D-glucan, is found in invertebrates and crustaceans as a structural material in their exoskeletons (Jeon and Kim 2000a,b; Muzzareli 2002). Chitosan, a polymer of glucosamine (GlcN), is prepared by alkaline deacetylation of chitin. Chitin and chitosan have exhibited several biological functions to serve as safe bioactive substances useful in various applications. Chitin and chitosan, naturally occurring biopolymers, have attracted much attention during the past decades owing to their unique properties as safe biomaterials ensuring human health. These outstanding properties of biocompatibility, biodegradability, polyelectrolyte properties, presence of reactive functional groups, and ability to modify the structure via chemical modifications has attracted their applicability in biomedical and pharmaceutical industries (Kim and Park 2001; Huang et al. 2003; Sashiwa and Aiba 2004). Currently, the trend among scientists across the world is to research on naturally occurring bioactive compounds to modify them to make novel compound that enhance their properties. Among them is chitin, a polysaccharide, abundantly found in nature, and chitosan formed by deacetylation of chitin. In addition, chitin oligosaccharides (NA-COSs) and chitooligosaccharides (COSs), partially hydrolyzed products of chitin and chitosan, are of great interest in various areas because of their noncytotoxic and high water-soluble properties (Kim et al. 2006); they and their derivatives have numerous biological properties such as immunoenhancing activity, antitumor activity, antihypertensive effects, anticoagulant (Kim et al. 2006), antibacterial activity, antifungal activity (Yang et al. 2005; Vong and Ngo 2010), and antioxidant activity (Ngo et al. 2008a). In this chapter, the antioxidant activity and antimicrobial effects of chitin, chitosan, and their derivatives have been summarized. Moreover, the chapter also discusses some synthetic and productive methods of certain derivatives of chitin and chitosan, and the potential applications of their antimicrobial and antioxidant properties.

11.2 PREPARATION OF CHITIN AND CHITOSAN DERIVATIVES

11.2.1 PRODUCTION OF NA-COSs AND COSs

In the production of oligomers of chitin and chitosan (NA-COSs and COSs, respectively), like other polysaccharides they can also be broken by hydrolyzing agents owing to the presence of rather unstable glycosidic bonds. Degradation of their O-glycosidic linkages by different methods leads to the production of NA-COSs and COSs varying in the degree of polymerization as well as number and sequence of GlcN and N-acetylglucosamine (GlcNAc) units.

Some of these methods include acid hydrolysis (Chen and Tsaih 1998; Somjit et al. 2005; Einbu et al. 2007), enzymatic hydrolysis (Jeon and Kim 2000a,b), ultrasonic degradation (Liu et al. 2009), oxidative degradation (Shao et al. 2003), chemoenzymatic (Akiyama et al. 1995) and recombinant approaches (Samain et al. 1997), and combination of γ-irradiation and enzymatic methods (Dzung et al. 2007). Moreover, electromagnetic radiation, sonication, and mechanical energy generated by microfiltration can also be used for the production of their oligomers without employing chemical agents. Absorption of energy by chitosan molecules in any of the above methods results in scission of chemical bonds and if the broken bond belongs to the backbone of the polymer (O-glycosidic bond), its molecular weight (MW) will be decreased (Einbu and Vårum 2007; Xing et al. 2005a).

Chitin was ground to powder and inserted into the reactor, adding 12 N HCl (100 g chitin and 1 L HCl) and stirring several times at 40°C, depending on the desired MW of NA-COSs. After reaction is stopped by adding distilled water, hydrolytic chitin solution was neutralized by 25% NaOH solution, centrifuged at 10,000g to remove insoluble residues. The hydrolysates were then separated using ultrafiltration (UF) membranes, the molecular weight cut-off (MWCO) of which were at 3 and 1 kDa, respectively. Two kinds of NA-COSs were prepared by using UF membrane, NA-COS 1–3 kDa (NA-COS passed out through MWCO 3 kDa membrane but not passed out at 1 kDa) and NA-COS <1 kDa (NA-COS passed out through MWCO but not passed out at 1 kDa membrane). The supernatant was desalted and purified using a Micro Acilyzer G3 (Asashi Kasei Corp., Japan) or dialysis, and then the solution was subjected to a spray dryer. Finally, the light-yellow powder of NA-COSs with different MWs was obtained (Ngo et al. 2009). In addition, supernatant was decolored to isolate the white chitin oligosaccharides powder as shown in Figure 11.1.

Jeon and Kim (2000a) prepared COSs of various MWs using enzymatic method combined with an UF membrane reactor system and improved the procedure continuously by connecting to an immobilized enzyme column reactor in which chitosanase from *Bacillus* sp. was absorbed on chitin as a carrier for immobilization. A 1% chitosan solution was prepared by dispersing chitosan in a ratio of 1:1 of water, dissolving it and stirring by adding 400 mL of 1 M lactic acid, and making up to 15:1 with water; the pH was adjusted to be 5.5 with a saturated sodium bicarbonate (NaHCO₃) solution. The chitosan was hydrolyzed by enzymatic reaction in the reactor system and fractioned by passing through UF membrane of MWCO 10, 5, 3, and 1 kDa (Jeon and Kim 2000b).

11.2.2 SYNTHESIS OF CHITIN AND CHITOSAN DERIVATIVES

To improve bioactivity and solubility of chitin and chitosan, chemical modification is one of the most effective approaches, which has been of interest in the recent decades (Alves and Manoa 2008; Mourya and Inamdar 2008). Generally, modified reactions are carried out on two types of reactive groups of chitin backbone; primary and secondary hydroxyl (OH) groups at the C2, C3, and C6 positions, respectively, and adding one type of chitosan graft on the free amino groups. For example, N-trimethyl chitosan derivative strongly affects the dissolution of poorly water-soluble drug cyclosporin A (Zhou et al. 2009); O-6-carboxyl groups enhanced the anticoagulant activity of chitosan derivatives (Huang et al. 2003). COS grafting O-6-aminoethyl groups inhibited the activity of angiotensin-converting enzyme, which played an important physiological role in regulating blood pressure (Ngo et al. 2008b). It has been proposed that quaternary ammonium radicals possess excellent bioactivities and improve solubility of chitosan (Guo et al. 2007; Sajomsang et al. 2008, 2009b).

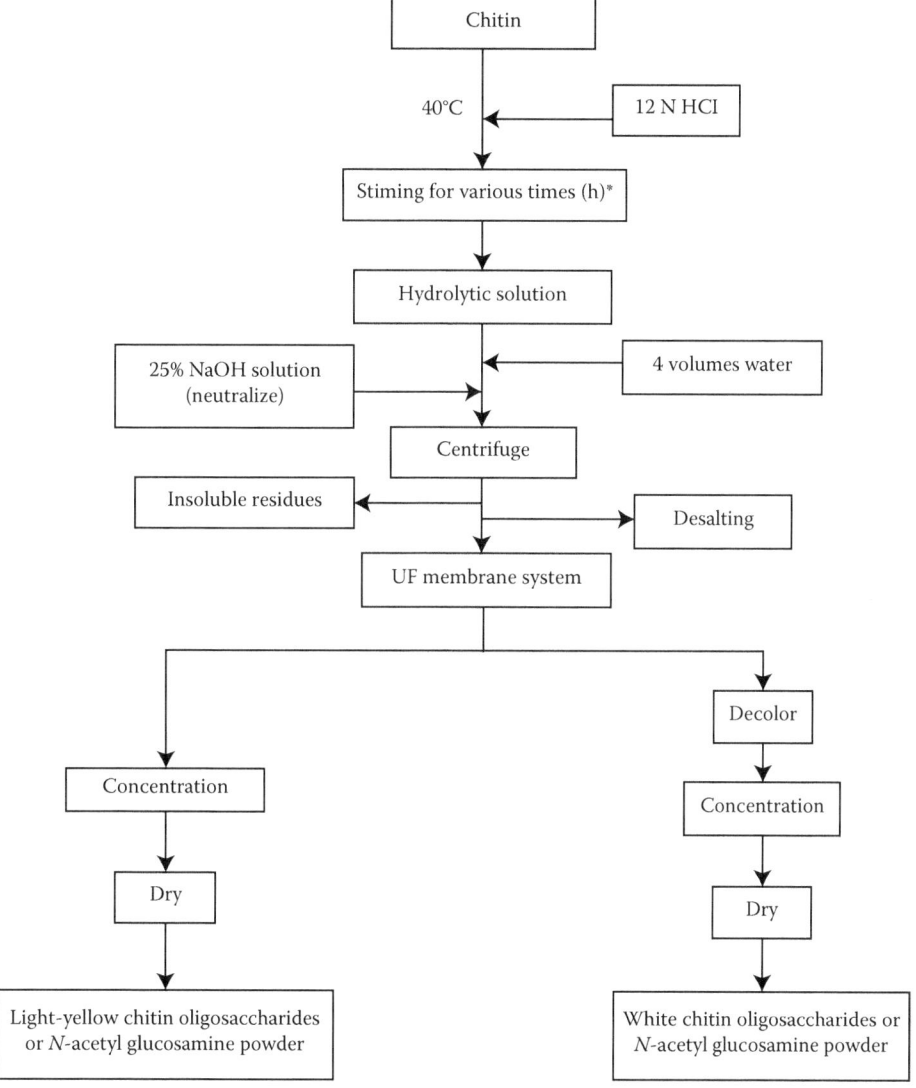

FIGURE 11.1 Process for the production of NA-COSs with different MWs using UF membrane system. (*Depends on hydrolytic time to get different MWs of NA-COSs.)

Recently, *N*-(3-chloro-2-hydroxypropyl) trimethylammonium chloride (Quat-188) has been used as a quaternizing agent and several quaternary ammonium chitosan derivatives have been synthesized (Sajomsang et al. 2009b). The *N*-aryl chitosan derivatives were prepared via Schiff bases formed by the reaction between 2-amino group of chitosan and an aromatic aldehyde followed by reduction of Schiff bases with sodium cyanoborohydride (Sajomsang et al. 2008). Briefly, 1 g of chitosan was dissolved in 1% (v/v) of acetic acid. Subsequently, 0.05 mmol of designated aldehyde was poured into the mixture, followed by stirring for 6 h at 25°C. Thereafter, 0.1 g of sodium borohydride was added and the reaction was maintained by stirring for 12 h, followed by adjusting pH to 7 with 15% of NaOH. *N*-Aryl chitosan derivatives were isolated and purified by distillation to remove unreacted chemicals as shown in Figure 11.2. Then, 20 mL of Quat-188 was added to the reaction flask and the pH of the solution was adjusted to 8 by 15% of NaOH. The solution was stirred for 24 h, followed by adding distilled water and was kept by stirring for 24 h at 50°C. The product was obtained by

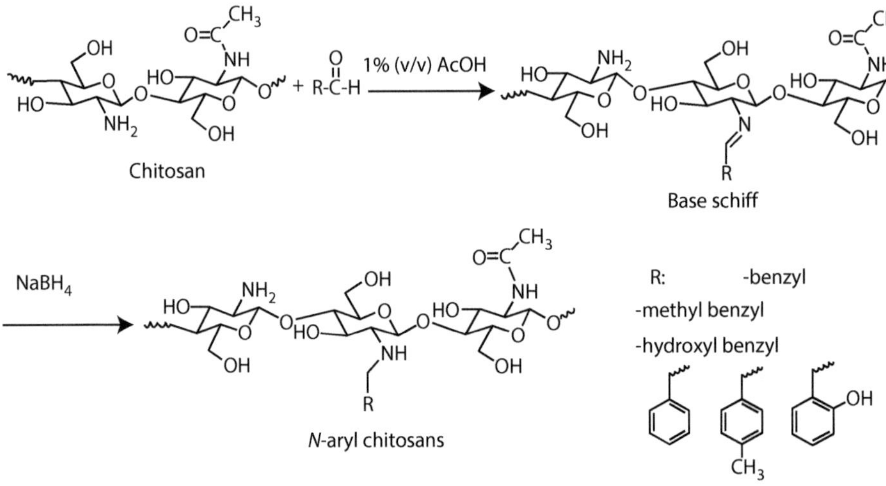

FIGURE 11.2 Synthetic scheme of *N*-aryl chitosan.

precipitation with acetone. The *N*-aryl Ch-Quat derivatives were synthesized via quaternization of *N*-aryl chitosan derivatives by Quat-188 (Sajomsang et al. 2009b).

COS derivatives were also synthesized by grafting at C2 or C6. Here, there are two examples: First, functional groups were grafted to the amino group of COS at C2 as gallate–COS. Briefly, 2.5 g COS was dissolved in 60 mL distilled water, methanol with a ratio of 1:2, and then adjusted to pH 6.8 with triethylamine to obtain a solution A. Gallic acid (~1.0 g) was dissolved in methanol and dicyclohexylcarbodiimide (DCC, ~1.0 g dissolved in 10 mL methanol) reacted with gallic acid to obtain a solution B. The solution B was gradually added to the solution A and stirred in a water bath at 30°C for 5 h and then filtered. The solution obtained was kept at 2°C overnight and was thereafter added with diethyl ether and filtered to obtain a precipitate. The precipitate was dissolved in distilled water and then freeze dried to obtain gallate–COS (Ngo et al. 2011b). Second, COS derivatives were synthesized in accordance with the method developed in the previous study of Je and Kim (2006a) (Ngo et al. 2008b; Yoon et al. 2009) and the method is described in Figure 11.3; functional groups were grafted at C6 position of pyranose. COS was added to each 3 M aqueous substitution solution (AE-COS; 2-chloroethylamino hydrochloride, DMAE-COS; 2-dimethylamino-ethylchloride hydrochloride, and DEAE-COS; 2-diethylamino-ethylchloride hydrochloride) with stirring at 40°C. 3 M of sodium hydroxide was added to the reaction mixture dropwise and was stirred continuously for 48 h. After the reaction, the solution was filtered with filter paper. The reaction solution was acidified with 1 N HCl and dialyzed in water for 2 days. The COS derivatives were freeze dried and obtained as AE-COS, DMAE-COS, and DEAE-COS, respectively.

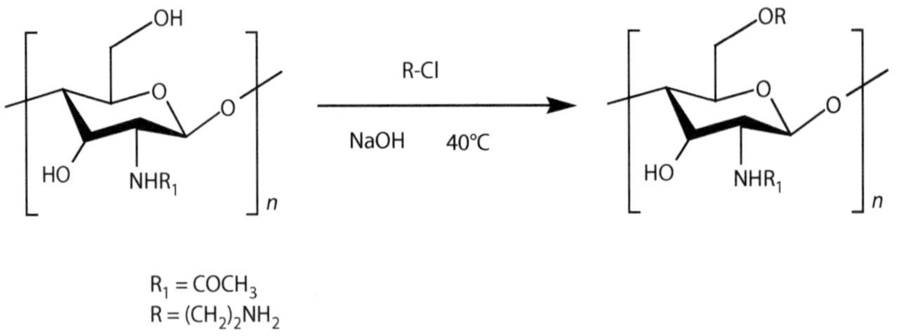

$R_1 = COCH_3$
$R = (CH_2)_2NH_2$

FIGURE 11.3 Synthetic scheme of amino COS derivative.

11.3 ANTIOXIDANT PROPERTIES

In biological systems, an equilibrium between oxidants formation and endogenous antioxidant defense mechanisms exists to protect cellular biomolecules against oxidation and if that balance is disturbed, oxidative stress happens (Kang et al. 2005). Oxidative stress causes injury to important cellular components; thus, the reactive oxygen species (ROS) such as superoxide anion (O^{-2}) and OH radicals generated excessively in tissues can lead to death of cells. Furthermore, ROS play an important role in many chronic diseases, such as cancer, diabetes, neurodegeneration, hypertension, inflammation, and aging (Ngo et al. 2007; 2008a). In addition, synthetic antioxidants, such as butylated hydroxytoluene (BHT), butylated hydroxyanisole (BHA), *tert*-butylhydroquinone (TBHQ), and propyl gallate have been widely used to prevent oxidation. However, the adding of synthetic antioxidants in food is under strict regulation or is even prohibited in some countries because these compounds have potential health hazards. Therefore, today, there is a growing interest in the use of natural antioxidant agents with little or no known harmful health effects to combat oxidative stress. Currently, a great deal of research happening on the antioxidant activity of chitin and its derivatives has attracted a greater attention. Most of the reports about chitin, chitosan, and their oligomers are that they are nontoxic and their derivatives are non- or less toxic. Therefore, they can be easily applied in living systems and food technology. Antioxidant activities of chitin, chitosan, and their oligomers depend on their MW. Kim and Thomas (2007) studied the antioxidant effect of chitosan with three different MWs of 30, 90, and 120 kDa in salmon and recognized that the 30 kDa chitosan was the highest. In addition, the two kinds of chitin oligosaccharides (NA-COSs) with different (MWs) (NA-COS 1–3 kDa and below 1 kDa, NA-COS < 1 kDa) from crab chitin hydrolysis solution were produced and determined their effect against oxidative stress in live cells. Results showed that NA-COS 1–3 kDa was more effective than NA-COS < 1 kDa in protein, deoxyribonucleic acid (DNA) oxidation that damaged those molecules, and production of intracellular free radicals in live cells. The rise of intracellular glutathione (GSH) level and direct intracellular radical scavenging effect was significantly increased in a time-dependent manner in mouse macrophages (RAW 264.7) and inhibited the effect against cellular oxidative stress (Ngo et al. 2009). Ngo and colleagues (2010) used hydrochloric acid to hydrolyze crab chitin for the isolation of chitin oligosaccharides (MW 229.21–593.12 Da); the results showed that reducing the power and scavenging effect on 1,1-diphenyl-2-picrylhydrazyl (DPPH), OH, and alkyl radicals increases in a dose-dependent manner. Their IC50 values for DPPH, OH, and alkyl radicals were 0.8, 1.75, and 1.14 mg/mL, respectively. Furthermore, it has been observed that in COSs their radical scavenging properties are not only dependent on their MWs but also on the degree of deacetylation (DD). Park's group (2004) carried out using electron-spin trapping technique to study antioxidant activity of COS with different MW and DD; results showed that COS with MW range (1–3 kDa) and highly deacetylated (90%) COSs have been identified to have a higher potential to scavenge different radicals such as DPPH, OH, superoxide, and alkyl radicals (Park et al. 2004b,c). The antioxidant activity of chitobiose and chitotriose was investigated using the inhibition of H_2O_2-induced hydroxylation of benzoate and free radical scavenging effects. Chen et al. (2003) reported that chitobiose is an antioxidant more effective than chitotriose and even GlcN hydrochloride.

In recent years, the derivatives of chitin, chitosan, and their oligomers such as *N,O*-carboxymethyl chitosan, hydroxypropylated chitosan, hexanoyl chitin (HCH), *N*-benzoylhexanoyl chitosan, sulfated chitosan, low-molecular weight carboxylmethyl chitosan (LMWCMC), chitosan gallic acid, aminoderivatized chitin, aminoderivatized chitosan, aminoderivatized oligomer, quaternary chitosan, aryl chitosan, quaternization of *N*-aryl chitosan, carboxylated COSs, and quaternized amino-COS have been assayed for their antioxidant capacity (Xing et al. 2005b; Huang et al. 2006; Guo et al. 2007; Je and Kim 2006b; Sun et al. 2007; Rajapakse et al. 2007; Pasanphan and Chirachanchai 2008; Mendis et al. 2008; Sajomsang et al. 2009a,b; Ngo et al. 2011a,b, 2012a,b,c). The sulfated chitosan with different MWs was synthesized and tested for its antioxidant effect. The results showed that their antioxidant activity depended on MW and low MW had stronger effect than MW (Xing et al. 2005b).

Huang et al. (2006) prepared two derivatives of COS 1–3 by graft carboxyl (–COCH$_2$CH$_2$COO$^-$) and quaternized amino (–CH$_2$CH(OH)CH$_2$N$^+$(CH$_3$)$_3$) groups to the amino position of COS with different substitution degrees (CCOS-1–3 and QCOS-1–3) with the aim of investigating free radical scavenging effects. On the basis of the results obtained from studies, scavenging of carbon-centered and DPPH radicals was directly affected by the amount of hydrogen atoms in COS molecules at C2, C3, and C6 position. They also showed that the high substitution of degree of derivatives were not necessarily synonymous with high antioxidant effect. Substitution of degree and special structure of group affected their scavenging effect against OH and alkyl radicals. Je and Kim (2006b) synthesized novel aminoderivatized chitosan (aminoethyl–chitosan, AEC, dimethylaminoethyl–chitosan, DMAEC, and diethylaminoethyl–chitosan, DEAEC) with different DD (50% and 90%). AEC with 90% DD showed the highest scavenging effects against OH and superoxide anion radical with 91.67% and 65.34% at 0.25 and 5 mg/mL, respectively, and their antioxidant effect depends on the substituted group.

Sun et al. (2007) also synthesized LMWCMC, which had superoxide anion scavenging activity. In addition, a novel synthetic chitosan derivative, chitosan–gallic acid has a wide range of antioxidant activity, including alkyl and OH radicals (Pasanphan and Chirachanchai 2008). The carboxylated COS and sulfated GlcN also inhibited free radical-mediated oxidation of cellular biomolecules in live cells (Rajapakse et al. 2007; Mendis et al. 2008). The latest research of Ngo et al. (2011a, b) shows that they synthesized gallyl-COS and demonstrated its effect against the oxidative damage to lipids, proteins, and DNA in cells. It could decrease the activation and expression of nuclear factor kappa-light-chain-enhancer of activated B cells NF-κB (nuclear factor kappa-light-chain-enhancer of activated B cells) and stimulated the production of the endogenous antioxidant enzymes, namely, superoxide dismutase (SOD) and GSH synthetase to make GSH in oxidative stress induced cells using H$_2$O$_2$. Furthermore, Ngo et al. (2012c) have studied the free radical scavenging effect of aminoethyl-COS to prevent the damage of protein, lipid, and DNA of cells. Moreover, the authors also recognized that it stimulated cells to make more intracellular GSH and inhibited direct intracellular free radicals in a dose- and time-dependent manner. Therefore, aminoethyl-COS had potential antioxidant effects by both indirect and direct pathway to inhibit and prevent biological molecular damage of free radicals in live cells (Ngo et al. 2012c).

11.4 ANTIMICROBIAL PROPERTIES

Until now, several mechanisms have been proposed for the explanation of antimicrobial properties of chitosan and its derivatives. The most commonly acceptable mechanism proposed that the positive charge of chitosan or its derivatives interacted with the negative charge of bacterial cell surface that led to dysfunction of bacterial cell membrane. Furthermore, they prevented the entry of materials or caused leakage of intracellular constituents that finally lead to lysis and death of microorganisms (Jeon and Kim 2001; Chung et al. 2004; Je and Kim 2006a; Li et al. 2010). Chung et al. (2004) demonstrated a close relationship between the antibacterial activity of chitosan and the hydrophilicity of cell wall.

Moreover, one of the acceptable mechanisms of antifungal activity is similar to that of antibacterial activity, which is prevented by nutrients by a layer formed via reaction between quaternary ammonium group and polyanion on the microbial cell surface (Roller and Covill 1999). Generally speaking, the mechanism of antimicrobial activity differs with the MW (Jeon et al. 2001; No et al. 2002; Park et al. 2004a; Xia et al. 2011), DD (Tsai et al. 2002), substitution degree of chitin, chitosan and their derivatives, and the type of microorganism (No et al. 2002; Tikhonov et al. 2006; Feng and Xia 2011). Chitosan in solutions increases its positive charge by protonating amino groups. Therefore, chitosan with high deacetylation degrees can be expected to enhance antimicrobial properties (Chung et al. 2004). Tsai et al. (2002) examined antibacterial and antifungal activities of various chitosan with DD ranging from 47% to 53%, 74% to 76%, and 95% to 98%. The results showed that antimicrobial activity raised with increasing DD and was weaker against fungi than against bacteria such as the minimal lethal concentrations (MLC) of chitosan with a high DD, 95% to 98% against *Candida albicans* and *Fusarium oxysporum* was 200 and 500 ppm, respectively, whereas the MLC bacteria were from 50 to 200 ppm. Moreover, some reveal that chitosan is more effective in killing Gram-negative bacteria than Gram-positive bacteria (Chen et al. 2002). Several factors such as DD, MW, polymerization degree, difference in grafted chains, and type of microorganisms affect the antimicrobial effects of chitosan and its derivatives (Table 11.1). There are different kinds of disaccharides or amino acid linked to chitosan or COSs; they have exhibited antimicrobial activities against different bacterial strains. Although chitin and chitosan possess antimicrobial activity against several bacterial strains, research has reported that they enhance the growth of beneficial bacteria (Tsai et al. 2002; Tsai and Hwang 2004). According to Tsai et al. (2002), the beneficial bacteria are much more resistant to chitosan than pathogens and most of probiotics can be inhibited at high concentrations of chitosan. Therefore, currently, chitosan materials are used as antimicrobial agents in several applications as in medicinal and pharmaceutical industries (Mi et al. 2002; Paul and Sharma 2004). In addition, the other mechanism of antibacterial activity was dependence of the hydrophobic–hydrophobic interactions between the benzyl substituent and the hydrophobic groups on the bacterial cell wall (Sajomsang et al. 2009). The report of Guo's group (2007) showed that quaternary ammonium group remarkably improved the antifungal activity of quaternized chitosan derivatives against *Botrytis cinerea* Per and *Colletotrichum lagenarium* (Guo et al. 2007). Recently, researchers reported that chitosan had higher antibacterial activity than COSs and the effect was stronger against Gram-positive bacteria than Gram-negative bacteria at 0.1% concentration (Xia et al. 2011). Ngo et al. (2012a) studied three kinds of aryl-chitosan (benzyl chitosan, BC, *N*-methylbenzyl chitosan, MBC, and *N*-hydroxylbenzyl chitosan, HBC) against *Staphylococcus aureus* ATCC 25923 (MRSA), *S. aureus* ATCC 43300 (MSSA), *Salmonella typhi*, *Escherichia coli* ATCC 25922, and *Pseudomonas aeruginosa* ATCC 27853. The results showed that MIC of BC is lower than that of chitosan. MIC of BC on *S. aureus* MRSA, *S. aureus* MSSA, and *S. typhi* is 375 µg/mL, and on *E. coli* and *P. aeruginosa* it is 250 µg/mL. MIC of MBC with tested bacterial species is 250 µg/mL, except MIC is 375 µg/mL for *S. typhi*. *N*-MBC has antibacterial activity higher than *N*-BC for *S. aureus* MSSA and *S. aureus* MRSA. MIC of *N*-HBC is lower than MIC of chitosan, but it is against *S. typhi* at 250 µg/mL and *N*-BC is not. The antibacterial activity not only depends on the free amino group but also other functional groups such as benzyl, methyl benzyl, and OH benzyl for various bacterial species with different activity (Ngo et al. 2012a). Feng and Xia (2011) synthesized water-soluble *O*-fumaryl chitosan with degree of substitution (DS) from 0.07 to 0.48 and investigated its antibacterial activity against *E. coli* and *S. aureus*. Their results demonstrated that DS increases with an increase of antibacterial effect. Furthermore, propyltrimethylammonium and pentyltrimethylammonium derivatives of chitosan also have increasing effect against *Aspergillus flavus* with an increase of alkyltrimethylammonium groups grafting on chitosan (Rafael et al. 2013).

11.5　FURTHER DEVELOPMENT AND RESEARCH NEEDS

Chitin, chitosan, and their derivatives are currently being used in various applications, such as medical industry and functional foods, based on human health benefits and their biological properties. To improve the biological effect of chitin, chitosan, and their oligomers, especially antioxidant and antimicrobial properties, tremendous efforts are underway in searching for novel chemical derivatives of these materials that have high antioxidant and antimicrobial properties to be applied in new therapies. Furthermore, chitin, chitosan, and their derivatives have demonstrated numerous bioactivities; a few clinical studies have been carried out to confirm their applicability in medical sciences. Moreover, extensive studies should be carried out to overcome some drawbacks and limitations of the medical usage of chitin, chitosan, and their derivatives. Despite few controversial arguments, chitin, chitosan, and their derivatives would be the potential bioactive molecules for numerous applications in future.

TABLE 11.1
Minimum Inhibitory Concentrations (MICs) of Chitin, Chitosan, and Their Derivatives (CD) against Various Microorganisms

Microorganism	DD (%)	MW (kDa)	MIC (ppm) COS, Chitosan	CD	Reference
Gram-Negative Bacteria					
Escherichia coli KCTC 1682	10	~310		62.5[a]	Je and Kim (2006a)
Escherichia coli O.157	90	5–10	1200		Jeon et al. (2001)
Escherichia coli CCRC 10674	95	51	100		Tsai et al. (2002)
Escherichia coli ATCC 25922	~65	18		250[b]	Ngo et al. (2012a)
Escherichia coli ATCC 25922	~65	18		250[c]	Ngo et al. (2012a)
Escherichia coli ATCC 25922	~65	18		250[d]	Ngo et al. (2012a)
Aeromonas hydrophila CCRC 13881	76	285	1000		Tsai et al. (2002)
Aeromonas hydrophila YM1	74	310	500		Tsai et al. (2002)
Pseudomonas aeruginosa CCRC 10944	98	49.1	150		Tsai et al. (2002)
Pseudomonas aeruginosa KCTC 1637	10	~310		7.813[a]	Je and Kim (2006a)
Salmonella typhimurium CCRC 10749	74	310	1500		Tsai et al. (2002)
Salmonella typhi	~65	18		375[c]	Ngo et al. (2012a)
Salmonella typhi	~65	18		250[d]	Ngo et al. (2012a)
Pseudomonas aeruginosa ATCC 27853	~65	18		250[b]	Ngo et al. (2012a)
Pseudomonas aeruginosa ATCC 27853	~65	18		250[c]	Ngo et al. (2012a)
Pseudomonas aeruginosa ATCC 27853	~65	18		250[d]	Ngo et al. (2012a)
Shigella dysenteriae CCRC 13983	95	51	200		Tsai et al. (2002)
Salmonella typhimurium KCTC 1925	10	~310		62.5[a]	Je and Kim (2006a)
Vibrio parahaemolyticus CCRC 10806	98	49.1	100		Tsai et al. (2002)
Vibrio parahaemolyticus	75	1–10	4000		Park et al. (2004c)
Vibrio cholerae CCRC 13860	95	51	150		Tsai et al. (2002)
Gram-Positive Bacteria					
Bacillus cereus CCRC 10250	95	51	200		Tsai et al. (2002)
Bacillus subtilis	75–90	5–10	1250		Park et al. (2004c)
Bacillus subtilis	ND	43		50[e]	Kurita and Shimojoh (2003)
Enterococcus faecalis KCTC 2011	10	~310		62.5[a]	Je and Kim (2006a)
Clostridium perfringens CCRC 10647	95	ND	250		Tsai and Hwang (2002)
Listeria monocytogenes KCTC 3710	10	~310		62.5[a]	Je and Kim (2006a)
Listeria monocytogenes LM-LM	95	51	100		Tsai et al. (2002)
Staphylococcus aureus CCRC 12652	98	49.1	50		Tsai et al. (2002)
Staphylococcus epidermis	75–90	5–10	630		Park et al. (2004c)
Staphylococcus aureus KCTC 1927	10	~310		125[a]	Je and Kim (2006a)
Staphylococcus aureus ATCC 25923 (MRSA)	~65	18		375[b]	Ngo et al. (2012a)
Staphylococcus aureus ATCC 43300 (MSSA)	~65	18		250[c]	Ngo et al. (2012a)
Fungi					
Candida albicans CCRC 20511	95	51	200		Tsai et al. (2002)
Candida albicans	67	ND		1500[f]	Je and Kim (2006a)

TABLE 11.1 (continued)
Minimum Inhibitory Concentrations (MICs) of Chitin, Chitosan, and Their Derivatives (CD) against Various Microorganisms

Microorganism	DD (%)	MW (kDa)	MIC (ppm) COS, Chitosan	MIC (ppm) CD	Reference
Candida glabrata	67	ND		2000[f]	Je and Kim (2006a)
Candida tropicalis	67	ND		2000[f]	Je and Kim (2006a)
Geotrichum spp.	67	ND		500[f]	Je and Kim (2006a)
Rhodotorula spp.	67	ND		1000[f]	Je and Kim (2006a)

Note: Minimum inhibitory concentration (MIC) is defined as the lowest concentration of chitin, chitosan, and their derivatives required for the complete inhibition of microorganisms. ND, not determined.

[a] Aminoethyl–chitin.
[b] Benzyl–chitosan.
[c] *N*-Methylbenzyl–chitosan.
[d] *N*-Hydroxylbenzyl–chitosan.
[e] Mannoside–chitosan.
[f] Chitosan ascorbate.

ACKNOWLEDGMENT

The author wishes to acknowledge Professor Se-Kwon Kim and Professor Moon-Moo Kim.

REFERENCES

Akiyama, K., K. Kawazu, and A. Kobayashi. 1995. A novel method for chemo-enzymatic synthesis of elicitor-active chitosan oligomers and partially *N*-deacetylated chitin oligomers using *N*-acylated chitotrioses as substrates in a lysozyme-catalyzed transglycosylation reaction system. *Carbohydrate Research* 279: 151–160.

Alves, N.M. and J.F. Manoa. 2008. Chitosan derivatives obtained by chemical modifications for biomedical and environmental applications. *International Journal of Biological Macromolecules* 43: 401–414.

Chen, A.S., T. Taguchi, K. Sakai, K. Kikuchi, M.W. Wang, and I. Miwa. 2003. Antioxidant activities of chitobiose and chitotriose. *Biological and Pharmaceutical Bulletin* 26: 1326–1330.

Chen, R.H. and M.T. Tsaih. 1998. Effect of temperature on the intrinsic viscosity and conformation of chitosans in dilute HCl solution. *International Journal of Biological Macromolecules* 23: 135–141.

Chen, Y.M., Y.C. Chung, L.W. Wang, K.T. Chen, and S.Y. Li. 2002. Antibacterial properties of chitosan in waterborne pathogen. *Journal of Environmental Science and Health* 37: 1979–1990.

Chung, Y.C., Y.P. Su, C.C. Chen, G. Jia, H.L. Wang, J.C. Wu, and J.G. Lin. 2004. Relationship between antibacterial activity of chitosan and surface characteristics of cell wall. *Acta Pharmacologica Sinica* 25: 932–936.

Dzung, N.A., V.T.P. Khanh, P.Q. Anh, and P.T.A. Hong. 2007. Study on hydrolysis of chitosan by cellulase combined with gamma irradiation. *Advances in Chitin Science*. Antalya 10: 96–100.

Einbu, A. and K.M. Vårum. 2007. Depolymerization and de-*N*-acetylation of chitin oligomers in hydrochloric acid. *Biomacromolecules* 8: 309–314.

Einbu, A., H. Grasdalen, and K.M. Vårum. 2007. Kinetics of hydrolysis of chitin/chitosan oligomers in concentrated hydrochloric acid. *Carbohydrate Research* 342: 1055–1062.

Feng, Y.W. and W.S. Xia. 2011. Preparation, characterization and antibacterial activity of water-soluble *O*-fumaryl–chitosan. *Carbohydrate Polymers* 83: 1169–1173.

Guo, Z., R. Xing, S. Liu, Z. Zhong, X. Ji, L. Wang, and P. Li. 2007. Antifungal properties of Schiff bases of chitosan, *N*-substituted chitosan and quaternized chitosan. *Carbohydrate Research* 342: 1329–1332.

Huang, R., Y. Du, and J. Yang. 2003. Preparation and anticoagulant activity of carboxybutyrylated hydroxyethyl chitosan sulfates. *Carbohydrate Polymers* 51: 431–438.

Huang, R., N. Rajapakse, and S.K. Kim. 2006. Structural factors affecting radial scavenging activity of chitooligosaccharides (COS) and its derivatives. *Carbohydrate Polymers* 63: 122–129.

Je, J.Y. and S.K. Kim. 2006a. Antimicrobial action of novel chitin derivative. *Biochimica et Biophysica Acta* 1760: 104–109.

Je, J.Y. and S.K. Kim. 2006b. Reactive oxygen species scavenging activity of aminoderivatized chitosan with different degree of deacetylation. *Bioorganic and Medicinal Chemistry* 14: 5989–5994.

Jeon, Y.J. and S.K. Kim. 2000a. Production of chitooligosaccharides using an ultrafiltration membrane reactor and their antibacterial activity. *Carbohydrate Polymers* 41: 133–141.

Jeon, Y.J. and S.K. Kim. 2000b. Continuous production of chitooligosaccharides using a dual reactor system. *Process Biochemistry* 35: 623–632.

Jeon, Y.J. and S.K. Kim. 2001. Effect of antimicrobial activity of chitosan oligosaccharides *N*-conjugated with asparagines. *Journal of Microbiology and Biotechnology* 11: 281–286.

Jeon, Y.J., P.J. Park, and S.K. Kim. 2001. Antimicrobial effect of chitooligosaccharides produced by bioreactor. *Carbohydrate Polymers* 44: 71–76.

Kang, K.B., K.H. Lee, S.W. Chae, R. Zhang, M.S. Jung, Y.K. Lee, S.Y. Kim et al. 2005. Eckol isolated from *Ecklonia cava* attenuates oxidative stress induced cell damage in lung fibroblast cells. *FEBS Letters* 579: 6295–6304.

Kim, K.W. and R.L. Thomas. 2007. Antioxidative activity of chitosans with varying molecular weights. *Food Chemistry* 101: 308–313.

Kim, S.K. and P.J. Park 2001. Subacute toxicity of chitosan oligosaccharide in Sprague–Dawley rats. *Arzneimittel Forschung Drug Research* 51: 769–774.

Kim, S.K., D.N. Ngo, and N. Rajapakse. 2006. Therapeutic prospectives of chitin, chitosan and their derivatives. *Journal of Chitin and Chitosan* 11: 1–10.

Kurita, K. and M. Shimojoh. 2003. Nonnatural branched polysaccharides: Synthesis and properties of chitin and chitosan having disaccharide maltose branches. *Biomacromolecules* 4: 1264–1268.

Li, X.F., X.Q. Feng, S. Yang, G.Q. Fu, T.P. Wang, and Z.X. Su. 2010. Chitosan kills *Escherichia coli* through damage to be of cell membrane mechanism. *Carbohydrate Polymers* 79: 493–499.

Liu, J., H. Sun, F. Dong, Q. Xue, G. Wang, S. Qin, and Z. Guo. 2009. The influence of the cation of quaternized chitosans on antioxidant activity. *Carbohydrate Polymers* 78: 439–443.

Mendis, E., M.M. Kim, N. Rajapakse, and S.K. Kim. 2008. Sulfated glucosamine inhibits oxidation of biomolecules in cells via a mechanism involving intracellular free radical scavenging. *European Journal of Pharmacology* 579: 74–85.

Mi, F.L., Y.B. Wu, S.S. Shyu, J.Y. Schoung, Y.B. Huang, Y.H. Tsai, and J.Y. Hao. 2002. Control of wound infections using a bilayer chitosan wound dressing with sustainable antibiotic delivery. *Journal of Biomedical Materials Research* 59: 438–449.

Mourya, V.K. and N.N. Inamdar. 2008. Chitosan modifications and applications: Opportunities galore. *Reactive and Functional Polymers* 68: 1013–1051.

Muzzareli, R.A.A. 2002. The discovery of chitin, a 570 megayear old polymer. *Chitosan in Pharmacy and Chemistry*, Atec, Italy, 1–8.

Ngo, D.N., T.B.H. Nguyen, and B.L. Vong. 2012a. Antibacterial and antioxidant activity of *N*-aryl chitosan. *Proceedings of the 9th Asia Pacific Chitin and Chitosan Symposium 2011* Nha Trang city, Vietnam, Agricultural Publishing House, Ho Chi Minh City, 208–212.

Ngo, D.H., D.N. Ngo, T.S. Vo, B.M. Ryu, Q.V. Ta, and S.K. Kim. 2012b. Protective effects of aminoethyl–chitooligosaccharides against oxidative stress and inflammation in murine microglial BV-2 cells. *Carbohydrate Polymers* 88: 743–747.

Ngo, D.H., Z.J. Qian, D.N. Ngo, T.S. Vo, I. Wijessekara, and S.K. Kim. 2011a. Gallyl chitooligosaccharides inhibit intracellular free radical-mediated oxidation. *Food Chemistry* 128: 974–981.

Ngo, D.H., Z.J. Qian, T.S. Vo, B.M. Ryu, D.N. Ngo, and S.K. Kim. 2011b. Antioxidant activity of gallate–chitooligosaccharides in mouse macrophage RAW 264.7 cells. *Carbohydrate Polymers* 84: 1282–1288.

Ngo, D.N., M.M. Kim, and S.K. Kim. 2007. Effects of chitin oligosaccharides on production of reactive oxygen species and matrix metalloproteinases in live cells. In: Dans: Senel, S., Varum, K. M., Murat, S. M., Atilla, H. A., (eds.) *Advances in Chitin Science*. Ankara. Vol. X, 355–359.

Ngo, D.N., M.M. Kim, and S.K. Kim. 2008a. Chitin oligosaccharides inhibit oxidative stress in live cells. *Carbohydrate Polymers* 74: 228–234.

Ngo, D.N., M.M. Kim, and S.K. Kim. 2012c. Protective effects of aminoethyl–chitooligosaccharides against oxidative stress in RAW 264.7 cells. *International Journal of Biological Macromolecules* 50: 624–631.

Ngo, D.N., M.M. Kim, Z.J. Qian, W.K. Jung, S.H. Lee, and S.K. Kim. 2010. Free radical scavenging activities of low molecular weight chitin oligosaccharides lead to antioxidant effect in live cells. *Journal of Food Biochemistry* 34, S1: 161–177.

Ngo, D.N., S.H. Lee, M.M. Kim, and S.K. Kim. 2009. Production of chitin oligosaccharides with different molecular weights and their antioxidant effect in RAW 264.7 cells. *Journal of Functional Foods* 1: 188–198.

Ngo, D.N., Z.J. Qian, J.Y. Je, M.M. Kim, and S.K. Kim. 2008b. Aminoethyl chitooligosaccharides inhibit the activity of angiotensin converting enzyme. *Process Biochemistry* 43: 119–123.

No, H.K., N.Y. Park, S.H. Lee, and S.P. Mayers. 2002. Antibacterial activity of chitosans and chitosan oligomers with different molecular weights. *International Journal of Food Microbiology* 74: 65–72.

Park, P.J., J.Y. Je, and S.K. Kim. 2004a. Antimicrobial activity of hetero-chitosans and their oligosaccharide with different molecular weights. *Journal of Microbiology and Biotechnology* 14: 317–323.

Park, P.J., J.Y. Je, and S.K. Kim. 2004b. Free radical scavenging activities of differently deacetylated chitosans using an ESR spectrometer. *Carbohydrate Polymers* 55: 17–22.

Park, P.J., H.K. Lee, and S.K. Kim. 2004c. Preparation of heterochitooligosaccharides and their antimicrobial activity on *Vibrio parahaemolyticus*. *Journal of Microbiology and Biotechnology* 14: 41–47.

Pasanphan, W. and S. Chirachanchai. 2008. Conjugation of gallic acid onto chitosan: An approach for green and water-based antioxidant. *Carbohydrate Polymers* 72: 169–177.

Paul, W. and C.P. Sharma. 2004. Chitosan and alginate wound dressings: A short review. *Trends in Biomaterials and Artificial Organs* 18: 18–23.

Rafael, O.P., T. Mirelle, C.C.G. Teresa, L.D.B. Vanildo, C.T. João, J.T. Marcio, and A.O.T. Vera. 2013. Synthesis, characterization and antifungal activity of quaternary derivatives of chitosan on *Aspergillus flavus*. *Microbiological Research* 168: 50–55.

Rajapakse, N., M.M. Kim, E. Mendis, and S.K. Kim. 2007. Inhibition of free radical-mediated oxidation of cellular biomolecules by carboxylated chitooligosaccharides. *Bioorganic and Medicinal Chemistry* 15: 997–1003.

Roller, S. and N. Covill. 1999. The antifungal properties of chitosan in laboratory media and apple juice. *International Journal of Food Microbiology* 47: 67–77.

Sajomsang, W., P. Gonil, and S. Tantayanon. 2009. Antibacterial activity of quaternary ammonium chitosan containing mono or disaccharide moieties: Preparation and characterization. *International Journal of Biological Macromolecules* 44: 419–427.

Sajomsang, W., U. Ruktanonchai, P. Gonil, V. Mayen, and P. Opanasopit. 2009a. Methylated *N*-aryl chitosan derivative/DNA complex nanoparticles for gene delivery: Synthesis and structure–activity relationships. *Carbohydrate Polymers* 78: 743–752.

Sajomsang, W., S. Tantayanon, V. Tangpasuthadol, and W.H. Daly. 2009b. Quaternization of *N*-aryl chitosan derivatives: Synthesis, characterization and antibacterial activity. *Carbohydrate Research* 344: 2502–2511.

Sajomsang, W., S. Tantayanon, V. Tangpasuthadol, M. Thatteb, and W.H. Daly. 2008. Synthesis and characterization of *N*-aryl chitosan derivatives. *International Journal of Biological Macromolecules* 43: 79–87.

Samain, E., S. Drouillard, A. Heyraud, H. Driguez, and R.A. Geremia. 1997. Gram-scale synthesis of recombinant chitooligosaccharides in *Escherichia coli*. *Carbohydrate Research* 302: 35–42.

Sashiwa, H. and S.I. Aiba. 2004. Chemically modified chitin and chitosan as biomaterials. *Progress in Polymer Science* 29: 887–908.

Shao, J., Y. Yang, and Q. Zhong. 2003. Studies on preparation of oligoglucosamine by oxidative degradation under microwave irradiation. *Polymer Degradation and Stability* 82: 395–398.

Somjit, K., Y. Ruttanapornwareesakul, K. Hara, and Y. Nozaki. 2005. The cryoprotectant effect of shrimp chitin and shrimp chitin hydrolysate on denaturation and unfrozen water of lizardfish *Surimi* during frozen storage. *Food Research International* 38: 345–355.

Sun, T., D. Zhou, F. Mao, and Y. Zhu. 2007. Preparation of low-molecular-weight carboxymethyl chitosan and their superoxide anion scavenging activity. *European Polymer Journal* 43: 652–656.

Tikhonov, V. E., E.A. Stepnova, V.G. Babak, I.A. Yamskov, P.G. Javier, H.B. Jansson, V.L. Luis et al. 2006. Bactericidal and antifungal activities of a low molecular weight chitosan and its *N*-/2(3)-(dodec-2-enyl) succinoyl/-derivatives. *Carbohydrate Polymers* 64: 66–72.

Tsai, G.J. and S.P. Hwang. 2004. *In vitro* and *in vivo* antibacterial activity of shrimp chitosan against some intestinal bacteria. *Fisheries Science* 70: 675–681.

Tsai, G.J., W.H. Su, H.C. Chen, and C.L. Pan. 2002. Antimicrobial activity of shrimp chitin and chitosan from different treatments and applications of fish preservation. *Fisheries Science* 68: 170–177.

Vong, B.L. and D.N. Ngo. 2010. Synthesis and antifungal activity of *N*-aryl chitosan derivatives. *The Science and Technology*, Vietnam. 48: 108–113.

Xia, W., P. Liu, J. Zhang, and J. Chen. 2011. Biological activities of chitosan and chitooligosaccharides. *Food Hydrocolloids* 25: 170–179.

Xing, R., S. Liu, H. Yu, Z. Guo, P. Wang, C. Li, Z. Li, and P. Li. 2005a. Salt-assisted acid hydrolysis of chitosan to oligomers under microwave irradiation. *Carbohydrate Research* 340: 2150–2153.

Xing, R., S. Liu, Z. Guo, H. Yu, P. Wang, C. Li, Z. Li, and P. Li. 2005b. Relevance of molecular weight of chitosan and its derivatives and their antioxidant activities *in vitro*. *Bioorganic and Medicinal Chemistry* 13: 1573–1577.

Yang, T.C., C.C. Chou, and C.F. Li. 2005. Antibacterial activity of *N*-alkylated disaccharide chitosan derivatives. *International Journal of Food Microbiology* 97: 237–245.

Yoon, N.Y., D.N. Ngo, and S.K. Kim. 2009. Acetylcholinesterase inhibitory activity of novel chitooligosaccharide derivatives. *Carbohydrate Polymers* 78: 869–872.

Zhou, X., Y. Hu, Y. Tian, and X. Hu. 2009. Effect of *N*-trimethyl chitosan enhancing the dissolution properties of the lipophilic drug cyclosporin A. *Carbohydrate Polymers* 76: 285–290.

12 Role of Chitosan and Its Derivatives in Cardiovascular Health

Dai-Hung Ngo and Se-Kwon Kim

CONTENTS

12.1 INTRODUCTION

Chitosan is a natural nontoxic biopolymer produced by the deacetylation of chitin, a major component of the shells of crustaceans, such as crab, shrimp, and crawfish. Chitin and chitosan are insoluble in water as well as most organic solvents, which is the major limiting factor for their utilization in living systems. Hence, it is important to produce soluble chitin or chitosan by various methods, such as acidic and enzymatic hydrolysis. Chitooligosaccharides (COS), partially hydrolyzed products of chitosan, or chitin, are of great interest in pharmaceutical and medicinal applications because of their noncytotoxic and high water-soluble properties. Various activities of COS are affected by degree of deacetylation (DD) and molecular weight (MW), or chain length (Razdan and Pettersson, 1994; Muzzarelli et al., 1999; Xia et al., 2011).

Currently, chitosan and its derivatives have attracted considerable attention for their commercial applications in biomedical, food, and chemical industries owing to their numerous biological activities, including antioxidant (Aytekin et al., 2011; Ying et al., 2011), antihypertension (Qian et al., 2010), anticancer (Shen et al., 2009; Cho et al., 2009; Toshkova et al., 2010), anticoagulant (Yang et al., 2012), antibacterial (Yang et al., 2005; Sajomsang et al., 2009; Zhong et al., 2009; Xu et al., 2010b; Yang et al., 2010), anti-inflammatory (Lee et al., 2009; Cho et al., 2011), calcium and ferrous binding agent (Bravo-Osuna et al., 2007; Liao et al., 2007), and hypocholesterolemic agent (Zhou et al., 2006; Zhang et al., 2010).

Cardiovascular diseases (CVD), principally atherosclerosis, stroke, or myocardial infarction, are a significant public health concern worldwide. Attempts to prevent CVD often imply modifications and improvement of causative risk factors, such as high blood pressure, obesity, an unfavorable profile of blood lipids, or insulin resistance (Erdmann et al., 2008). By modulating and improving physiological functions, chitosan and its derivatives may provide novel therapeutic applications for the prevention or treatment of chronic diseases. This chapter focuses on chitosan and its derivatives with properties relevant to cardiovascular health, including antihypertensive, antioxidant, anticoagulant, hypocholesterolemic, and antidiabetic effects.

12.2   ANTIHYPERTENSIVE ACTIVITY

Elevated blood pressure is increasingly prevalent in developed countries and one of the major independent risk factors for CVD (Harris et al., 1985; Kannel and Higgins, 1990). Angiotensin-I converting enzyme (ACE) plays a vital physiological role in the regulation of blood pressure by converting angiotensin-I to angiotensin-II, a potent vasoconstrictor. Therefore, the inhibition of ACE activity is a major target in the prevention of hypertension (Shahidi and Zhong, 2008).

Researches on chitosan and its derivatives have identified their potential to inhibit ACE activity. COS derivatives, such as hetero-COS, aminoethyl COS, carboxylated COS, chitosan trimer oligomers, and chitin derivatives, have been reported as potent ACE inhibitors. A high-salt diet can raise blood pressure because Cl^- activates ACE, whereas chitosan can bind Cl^- and lower the blood pressure (Xia, 2003). Hong et al. (1998) studied ACE inhibitory activity of different COS and identified that chitosan trimer is more effective in lowering blood pressure compared to other oligomers. Specifically, the trimer has a lower IC_{50} value (0.9 μM) than most of the other MW COS. Moreover, COS have remarkable ACE inhibitory activity (Park et al., 2003). The ACE inhibitory activity of hetero-COS was dependent on the DD, and COS with relatively lower DD and medium MW (1–5 kDa) exhibited the highest ACE inhibitory activity with the IC_{50} value of 1.22 mg/mL and the inhibition pattern is competitive according to the Lineweaver–Burk plots. These findings suggested that MW and DD of COS are important factors for the ACE inhibitory activity.

Chitins with different DD have been chemically modified by grafting 2-chloroethylamino hydrochloride onto chitin at the C-6 position to develop ACE inhibitory chitin derivatives (Je et al., 2006). Three kinds of chitin derivatives, including aminoethyl-chitin (AEC) with 10%, 50% (AEC50), and 90% (AEC90) DD were prepared having potential ACE activity. IC_{50} values of ACE were 0.064 μM (AEC), 0.038 μM (AEC50), and 0.103 μM (AEC90). In addition, AEC50 effectively decreased systolic blood pressure in spontaneously hypertensive rats (SHR) in a dose-dependent manner.

Carboxylated COS is a strong antihypertensive compound that shows equal activity to captopril (Huang et al., 2005). Carboxylated COS enhances the activity significantly with increased degree of substitution. Furthermore, Lineweaver–Burk plot studies showed that the inhibition was competitive via the obligatory binding sites of enzymes. Substitution of hydrogen atom at the C-6 position of the pyranose residue with the aminoethyl group promoted the ACE inhibitory effect of COS (800–3000 Da and 90% DD) (Ngo et al., 2008b). Hence, these results exhibited that substitution of the hydrogen atom at C-6 position of pyranose residue by aminoethyl group promotes ACE inhibitory effect of COS.

In addition to the ACE, renin also plays an important role in the renin–angiotensin system. Renin (or angiotensinogenase), is a rate-limiting enzyme in the renin–angiotensin system. It cleaves plasma angiotensinogen to angiotensin-I, which is further converted by ACE to angiotensin-II. Therefore, the inhibition of renin effect is also an attractive target in hypertension therapy (Wijesekara and Kim, 2010). Park et al. (2008) reported that six kinds of COS with potential renin inhibitory activity were prepared using ultrafiltration membrane reactor. According to them, 90% deacetylated and medium MW (1–5 kDa) COS exhibits the highest renin inhibitory activity with IC_{50} value of 0.51 mg/mL, and acts as competitive inhibitor with K_i value of 0.28 mg/mL by Lineweaver–Burk and Dixon plots. Collectively, chitosan and its derivatives are novel therapeutic drug candidates for treating hypertension in the pharmaceutical industry.

12.3   ANTIOXIDANT ACTIVITY

Humans are impacted by many free radicals both from inside our body and surrounding environment, particularly reactive oxygen species (ROS) generated in living organisms during metabolism. It is produced in the forms of H_2O_2, superoxide anion ($O_2^{•-}$), and hydroxyl radicals ($^•OH$). In addition, oxidative stress may cause inadvertent enzyme activation and oxidative damage to cellular systems. Free radicals attack macromolecules, such as DNA, proteins, and lipids, leading to many health disorders, including hypertensive, cardiovascular, inflammatory, aging, diabetes mellitus, neurodegenerative, and cancer diseases. Antioxidants may have a positive effect on human health because they can protect the human body against deterioration by free radicals (Butterfield et al., 2006; Ngo et al., 2011c). Recently, the antioxidant activity of chitosan and its derivatives attracted considerable attention because of their multiple potential activities and availability.

The antioxidant effect of chitin oligosaccharides (NA-COS) produced by acidic hydrolysis from crab shell chitin were evaluated. NA-COS can inhibit myeloperoxidase activity in human myeloid cells (HL-60) and decrease free-radical oxidation of DNA and membrane proteins. In addition, direct intracellular radical scavenging effect and intracellular glutathione level were significantly increased in the presence of NA-COS (Ngo et al., 2008a, 2009).

Gallic acid-grafted COS inhibited intracellular free-radical-mediated oxidation. Gallic acid-grafted COS can be used as a potential natural compound-based antioxidant in functional food and pharmaceutical industries (Ngo et al., 2011a,b). Aminoethyl-COS possesses potential antioxidant activity, and can be used as a scavenger in controlling free radicals that lead to damage to the cellular system (Ngo et al., 2012a,b).

The antioxidant activity of chitosan was studied *in vitro* and *in vivo* (Liu, 2008). Chitosan at an addition of 0.02% had antioxidant effects in lard and crude rapeseed oil but the activity was less than ascorbic acid. In the food industry, chitosan (edible chitosan, more than 83% DD) and COS have been used as dietary food additives and functional factors for their health beneficial effects, as well as drug carriers.

Park et al. (2004a) prepared three kinds of partially deacetylated heterochitosans, such as 90%, 75%, and 50%, deacetylated chitosan from crab chitin, and investigated their scavenging activities against 1,1-diphenyl-2-picrylhydrazyl (DPPH), alkyl, $^{\bullet}OH$, and $O_2^{\bullet-}$ radicals using electron spin resonance (ESR) spectrometer. The scavenging activities of heterochitosans increased from 3% to 69.39% with increasing concentration from 1.25 to 5 mg/mL of alkyl radical. In addition, 90% chitosan with relatively high DD showed the highest radical scavenging effects on the $^{\bullet}OH$ and $O_2^{\bullet-}$ radicals, and the radical scavenging activities of these heterochitosans depend on their DD and concentration.

Yen et al. (2008) reported that chitosan was prepared by alkaline N-deacetylation of crab chitin for 60 (C60), 90 (C90), and 120 (C120) min and its antioxidant effects was determined. Chitosans exhibited antioxidant effects of 58.3–70.2% at 1.0 mg/mL, and reducing powers of 0.32–0.44 at 10 mg/mL. At 10 mg/mL, the scavenging ability of chitosan C60 on DPPH radical was 28.4%, whereas those of other chitosans were 46.4–52.3%. At 0.1 mg/mL, scavenging abilities on $^{\bullet}OH$ radicals were 62.3–77.6%, whereas at 1 mg/mL, chelating abilities on ferrous ions were 82.9–96.5%. All EC50 values (the effective concentration at which the antioxidant activity was 50%) of antioxidant activity were below 1.5 mg/mL. Generally, the effectiveness of chitosans was correlated with N-acetylation times. Overall, crab chitosan was good in antioxidant activity and may be used as a source of antioxidants, as a possible food supplement or ingredient in the pharmaceutical industry.

Kim and Thomas (2007) investigated the antioxidant effect of chitosans with different MWs (30, 90, and 120 kDa) in salmon where the chitosan with lowest MW (30 kDa) exhibited highest antioxidant effect. Anraku et al. (2008) determined that chitosan (2800 Da) may inhibit neutrophil activation and oxidation of serum albumin commonly observed in patients undergoing hemodialysis, resulting in reduction of oxidative stress associated with uremia.

Various chitosan derivatives with high and low MW and a variety of reactive groups have showed significant activities against $^{\bullet}OH$, $O_2^{\bullet-}$ scavenging, and reducing power (Zhong et al., 2007). According to ESR studies, 90% deacetylated medium MW hetero-COS have the highest free-radical scavenging activity of all free radicals tested, such as DPPH, $^{\bullet}OH$, and $O_2^{\bullet-}$, and carbon-centered radicals (Je et al., 2004).

COS can effectively protect human umbilical vein endothelial cells against H_2O_2-induced oxidative stress and apoptosis, which might be of importance in the treatment of CVD (Liu et al., 2009, 2010). COS exerted inhibitory effects on the formation of intracellular ROS and lipid peroxidation, such as malondialdehyde, restoring activities of endogenous antioxidants, including superoxide dismutase and GSH peroxidase, along with the capacity of increasing levels of NO and NO synthase.

Xu et al. (2010a) investigated the protective effect of COS against H_2O_2-induced oxidative stress on human embryonic hepatocytes (L02 cells) and its scavenging activity against the DPPH radical *in vitro*, suggesting that COS might be useful in a clinical setting during the treatment of oxidative stress-related liver damages. Inhibition of free-radical-mediated oxidation of cellular biomolecules, such as lipids, proteins, and direct scavenging of ROS by carboxylated chitooligosaccharides (CCOS) has been reported (Rajapakse et al., 2007). Therefore, the application of chitosan and its derivatives as antioxidants in pharmaceutical industry is promising.

12.4 ANTICOAGULANT ACTIVITY

Blood coagulation is processed by coagulation factors to stop the flow of blood though the injured vessel wall whenever an abnormal vascular condition and exposure to nonendothelial surfaces at sites of vascular injury occur. As endogenous or exogenous anticoagulants interfere with the coagulation factors, the blood coagulation can be prolonged or stopped. These anticoagulants have been used as convenient tools for the exploration of complex mechanisms of coagulation cascade. Coincidentally, the importance of research for anticoagulants also arose with therapeutic purposes; for example, a cure for hemophilia. Heparin has been identified and used for more than 50 years as a commercial anticoagulant and it is widely used for the prevention of venous thromboembolic disorders. However, several side effects of heparin have been identified, such as development of thrombocytopenia, hemorrhagic effect, and ineffectiveness in congenital or acquired antithrombin deficiencies, and inability to inhibit thrombin bound to fibrin. Moreover, heparin is available in very low concentrations in pig intestine or bovine lungs from where it is primarily extracted. Therefore, the necessity of discovering alternative sources of anticoagulants has arisen with interesting demand for safer anticoagulant therapy (Hanson and Sakariassen, 1998; Kim and Wijesekara, 2010).

Some studies have reported the anticoagulant activity of chitosan and chitosan derivatives. Sulfated derivatives of hetero-COS with different MWs were prepared using an ultrafiltration membrane reactor system with significant anticoagulant activity (Park et al., 2004b). Clotting times in thrombin-time assay were prolonged in the presence of various concentrations of the heterochitosans and their COS sulfates using normal human plasma. The 90% deacetylated chitosan sulfate showed the highest anticoagulant effect among all the heterochitosans and their COS sulfates. The study of anticoagulant activity showed that chitosan sulfates with lowered MW (9–35 kDa) demonstrated a regular increase of anti-Xa activity like heparins (Vikhoreva et al., 2005).

Chitosan sulfates have shown obvious anticoagulant activity but the introduction of carboxyl groups could further increase the activity because of better structural similarity to heparin. Carboxymethyl chitosan sulfate showed greater inhibition on the transformation of fibrinogen to fibrin than chitosan sulfate (Nishimura et al., 1986). Generally, C-6 sulfate is a prerequisite for anticoagulant activity (Nishimura et al., 1998) but some works have reported anticoagulant activity with 6-*O* carboxylated and *N*-sulfated chitosan derivatives (Horton and Just, 1973).

Carboxybutyrylated hydroxyethyl chitosan sulfates have shown anticoagulant activity (Ronghua et al., 2003). The introducing of carboxyl groups to amino groups greatly prolonged the activated partial thromboplastin time (APTT) and thrombin time (TT). The best result occurred when the degree of substitution of the carboxyl groups was about 0.4/unit that prolonged APTT and TT with about 5 and 1.5 times compared to that of the uncarboxylated hydroxyethyl chitosan sulfates. Another potential conclusion is that introducing carboxyl groups to *N,O*-position gave better results. Low S% chitosan sulfate and 6-*O*-desulfated chitosan sulfate showed little anticoagulant activity but their *N,O*-carboxybutyrylated derivatives (0.6/unit ds) showed increased APTT or TT, whereas their *N*-carboxybutyrylated derivatives (0.6/unit ds) showed no improvement.

N-Propanoyl-, *N*-hexanoyl-, and *N,O*-quaternary substituted chitosan sulfate have been studied in relation to their anticoagulant activity (Huang et al., 2003). The influences of acyl or quaternary groups on the anticoagulant activity of polysaccharides were studied with respect to APTT, TT, and prothrombin time (PT). The propanoyl and hexanoyl groups increased APTT activity, and the propanoyl group increased TT anticoagulant activity slightly, whereas the *N,O*-quaternary chitosan sulfate showed only a slight TT coagulant activity.

12.5   CHOLESTEROL-LOWERING AND ANTIDIABETIC EFFECTS

An unfavorable profile of blood lipids is an important risk factor for CVD. Many studies have found a positive correlation between hypercholesterolemia and/or hypertriglyceridemia and the likelihood for developing CVD. Not surprisingly, the treatment for hyperlipidemia-accelerated diseases often includes the improvement of serum lipid distribution through diet modifications. The hypocholesterolemic action of chitosan and its derivatives can be explained to be because of the decrease in cholesterol absorption and interference with bile acid absorption, a mechanism similar to those of dietary fiber constituents. Diabetes mellitus is a highly prevalent metabolic disease, in which the pancreas does not produce sufficient insulin to meet its need or the body does not effectively use the insulin it produces. This leads to the elevated levels of blood glucose (hyperglycemia), which can induce the spillage of glucose into the urine (Hayashi et al., 2006; Erdmann et al., 2008).

Chitosan and its derivatives have gained much attention in recent past because of their potential activity against cholesterol. The consumption of chitosan tablets was found to be safe (Tapola et al., 2008). The effect of COS on the level of serum lipids, antioxidant enzyme activities, and lipid peroxidation was investigated in rats fed with high-cholesterol diet for 4 weeks (Kim et al., 2005). Serum total cholesterol, low-density lipoprotein (LDL) cholesterol, and triglyceride levels were significantly decreased and relative high-density lipoprotein (HDL) cholesterol level in total cholesterol significantly increased in COS-supplemented groups. Liver thiobarbituric acid reactive substance level and activities of SOD and catalase of COS-supplemented groups were also significantly reduced. Hence, the results indicated that the supplement of COS reduce levels of serum cholesterol and reduce oxidative damage by activating the hepatic antioxidative defense system in rats fed with high-cholesterol diets.

Zhang et al. (2012) determined the effects of chitosan (CTS) and water-soluble chitosan (WSC) microparticles (MPs) and nanoparticles (NPs) in rats with high-fat diet-induced obesity. The results indicated that CTS and WSC MPs and NPs have greater effects than commercially available CTS and WSC, and can be used as potential antiobesity agents. COS of four different MW ranges (below 1, 1–3, 3–5, and 5–10 kDa) protect pancreatic β-cells from oxidative stress-induced cellular deterioration (Karadeniz et al., 2010).

Cross-linked amphiphilic chitosan significantly absorbs the cholesterol in both polar and nonpolar solvents (Tong et al., 2005). *O*-Carboxymethyl chitosan and *N*-[(2-hydroxy-3-*N,N*-dimethylhexadecyl ammonium) propyl]chitosan chloride are able to alleviate the hepatic fat accumulation (Liu et al., 2011). The chitosan derivatives having 50 kDa increased 3-hydroxy-3-methylglutaryl-coenzyme A (HMG-CoA), hepatic lipase, lecithin cholesterol acyltransferase, and LDL receptor by 543%, 162%, 122%, and 2%, respectively. Further, *N*-lauryl chitosan and *N*-dimethylaminopropyl chitosan possessed higher hydrophobic and cationic properties to alter the composition of olive oil (Muzzarelli et al., 2000).

COS could effectively accelerate the insulin secretion in pancreatic islet cells in streptozotocin (STZ)-induced diabetic rats (Liu et al., 2007). COS could improve the general clinical symptoms of diabetic rats, decrease the plasma glucose and urine glucose, and normalize the disorder of glucose tolerance. COS can be used in the treatment of diabetes mellitus. Jo et al. (2008) investigated prolonged antidiabetic effect of zinc-crystallized insulin-loaded glycol chitosan NPs in type 1 diabetic rats. The prolonged time action profiles and low variability of insulin hydrophobically modified glycol chitosan formulation resulted in improved blood glucose control in diabetic rats and fulfilled a pattern desirable of a basal insulin.

The antidiabetic effect of chitosan (200,000–300,000 kDa) on STZ-induced type 1 diabetic ICR mice by analyzing food consumption, body weight, drinking water consumption, urine volume, nonfasting serum glucose, urine glucose, total serum cholesterol, and triglyceride levels was studied (Do et al., 2008). Chitosan was effective for improving hyperglycemia, hypertriglyceridemia, polydipsia, and polyuria in STZ-induced type 1 diabetic ICR mice and would be useful for relieving type 1 diabetes mellitus. Furthermore, low-MW chitosan prevents the progression of low-dose STZ-induced slowly progressive non-insulin-dependent diabetes mellitus (Kondo et al., 2000).

The human insulin gene can be transfected and expressed successfully by pCMV.Ins (an expression plasmid of the human insulin gene) wrapped with chitosan NPs in NIH3T3 cells and diabetes rats, which indicates that chitosan is a promising, nonviral vector for gene expression (Niu et al., 2008). The hypoglycemic and hypocholesterolemic effects of high- and low-MW chitosan were evaluated in STZ-induced diabetic rats (Yao et al., 2008). The results demonstrated the potential of high-MW chitosan in reducing hyperglycemia and hypercholesterolemia in STZ-induced diabetic rats. The cholesterol-lowering effect of chitosan may be primarily related to an increase in the fecal excretion of cholesterol and bile acid in hamsters (Yao and Chiang, 2006). Collectively, chitosan and its derivatives can be promising candidates as potential material for protecting diabetes mellitus and lowering the cholesterol absorption.

12.6 CONCLUSION

It is assumed that much attention has been paid recently by researchers toward chitosan and its derivatives as the safe and efficient agents in the prevention or treatment of chronic diseases. Chitosan and its derivatives have been found to have considerable and broad biological activities and health beneficial effects. In conclusion, it can be suggested that chitosan and its derivatives are potential therapeutic candidates for preventing CVD and their involvement in future pharmaceuticals are promising.

REFERENCES

Anraku, M., M. Kabashima, H. Namura, T. Maruyama, M. Otagiri, J.M. Gebicki, N. Furutani, and H. Tomida. 2008. Antioxidant protection of human serum albumin by chitosan. *International Journal of Biological Macromolecules* 43:159–164.

Aytekin, A.O., S. Morimura, and K. Kida. 2011. Synthesis of chitosan–caffeic acid derivatives and evaluation of their antioxidant activities. *Journal of Bioscience and Bioengineering* 111:212–216.

Bravo-Osuna, I., G. Millotti, C. Vauthier, and G. Ponchel. 2007. In vitro evaluation of calcium binding capacity of chitosan and thiolated chitosan poly(isobutyl cyanoacrylate) core-shell nanoparticles. *International Journal of Pharmaceutics* 338:284–290.

Butterfield, D.A., H.M. Abdul, W. Opii, S.F. Newman, G. Joshi, M.A. Ansari, and R. Sultana. 2006. Pin1 in Alzheimer's disease. *Journal of Neurochemistry* 98:1697–1706.

Cho, Y.I., S. Park, S.Y. Seo, and H.S. Yoo. 2009. In vivo and in vitro anti-cancer activity of thermo-sensitive and photo-crosslinkable doxorubicin hydrogels composed of chitosan–doxorubicin conjugates. *European Journal of Pharmaceutics and Biopharmaceutics* 73:59–65.

Cho, Y.S., S.K. Kim, C.B. Ahn, and J.Y. Je. 2011. Inhibition of acetylcholinesterase by gallic acid-grafted-chitosans. *Carbohydrate Polymers* 84:690–693.

Do, J.Y., Kwak, D.M., and O.D. Kwon. 2008. Antidiabetic effects of high molecular weight chitosan in streptozotocin-induced type 1 diabetic ICR mice. *Laboratory Animal Research*, 24:311–317.

Erdmann, K., B.W.Y. Cheung, and H. Schröder. 2008. The possible roles of food-derived bioactive peptides in reducing the risk of cardiovascular disease. *Journal of Nutritional Biochemistry* 19:643–654.

Hanson, S.R. and K.S. Sakariassen. 1998. Blood flow and antithrombotic drug effects. *American Heart Journal* 135 (5 Pt 2 Su):S132–S145.

Harris, T., E.F. Cook, W. Kannel, A. Schatzkin, and L. Goldman. 1985. Blood pressure experience and risk of cardiovascular disease in the elderly. *Hypertension* 7:118–124.

Hayashi, K., R. Kojima, and M. Ito. 2006. Strain differences in the diabetogenic activity of streptozotocin in mice. *Biological and Pharmaceutical Bulletin* 25:188–192.

Hong, S.P., M.H. Kim, S.W. Oh, C.H. Han, and Y.H. Kim. 1998. ACE inhibitory and antihypertensive effect of chitosan oligosaccharides in SHR. *Korean Journal of Food Science and Technology* 30:1476–1479.

Horton, D. and E.K. Just. 1973. Preparation from chitin of (1 → 4)-2-amino-2-deoxy-β-D-glucopyranuronan and its 2-sulfoamino analog having blood-anticoagulant properties. *Carbohydrate Research* 29:173–179.

Huang, R., Y. Du, J. Yang, and L. Fan. 2003. Influence of functional groups on the in vitro anticoagulant activity of chitosan sulfate. *Carbohydrate Research* 338:483–489.

Huang, R., E. Mendis, and S.K. Kim. 2005. Improvement of ACE inhibitory activity of chitooligosaccharides (COS) by carboxyl modification. *Bioorganic and Medicinal Chemistry* 13:3649–3655.

Je, J.Y., P.J. Park, and S.K. Kim. 2004. Free radical scavenging properties of hetero-chitooligosaccharides using an ESR spectroscopy. *Food and Chemical Toxicology* 42:381–387.

Je, J.Y., P.J. Park, B. Kim, and S.K. Kim. 2006. Antihypertensive activity of chitin derivatives. *Biopolymers* 83:250–254.

Jo, H.G., K.H. Min, T.H. Nam, S.J. Na, J.H. Park, and S.Y. Jeong. 2008. Prolong diabetic effect of zinc-crystallized insulin loaded glycol chitosan nanoparticles in type I diabetic rats. *Archives of Pharmacal Research* 31:918–923.

Kannel, W.B. and M. Higgins. 1990. Smoking and hypertension as predictors of cardiovascular risk in population studies. *Journal of Hypertension* 8:S3–S8.

Karadeniz, F., M. Artan, C.S. Kong, and S.K. Kim. 2010. Chitooligosaccharides protect pancreatic β-cells from hydrogen peroxide-induced deterioration. *Carbohydrate Polymers* 82:143–147.

Kim, K.N., E.S. Joo, K.I. Kim, S.K. Kim, H.P. Yang, and Y.J. Jeon. 2005. Effect of chitosan oligosaccharides on cholesterol level and antioxidant enzyme activities in hypercholesterolemic rat. *Journal of the Korean Society of Food Science and Nutrition* 34:26–41.

Kim, K.W. and R.L. Thomas. 2007. Antioxidative activity of chitosans with varying molecular weight. *Food Chemistry* 101:308–313.

Kim, S.K. and I. Wijesekara. 2010. Development and biological activities of marine-derived bioactive peptides: A review. *Journal of Functional Foods* 2:1–9.

Kondo, Y., A. Nakatani, K. Hayashi, and M. Ito. 2000. Low molecular weight chitosan prevents the progression of low dose streptozotocin-induced slowly progressive diabetes mellitus in mice. *Biological & Pharmaceutical Bulletin* 23:1458–1464.

Lee, S.H., M. Senevirathne, C.B. Ahn, S.K. Kim, and J.Y. Je. 2009. Factors affecting anti-inflammatory effect of chitooligosaccharides in lipopolysaccharides-induced RAW264.7 macrophage cells. *Bioorganic and Medicinal Chemistry Letters* 19:6655–6658.

Liao, F.H., M.J. Shieh, N.C. Chang, and Y.W. Chien. 2007. Chitosan supplementation lowers serum lipids and maintains normal calcium, magnesium, and iron status in hyperlipidemic patients. *Nutrition Research* 27:146–151.

Liu, B., W.S. Liu, B.Q. Han, and Y.Y. Sun. 2007. Antidiabetic effects of chitooligosaccharides on pancreatic islet cells in streptozotocin-induced diabetic rats. *World Journal of Gastroenterology* 13:725–731.

Liu, H.T., J.L. He, W.M. Li, Z. Yang, Y.X. Wang, X.F. Bai, C. Yu, and Y.G. Du. 2010. Chitosan oligosaccharides protect human umbilical vein endothelial cells from hydrogen peroxide-induced apoptosis. *Carbohydrate Polymers* 80:1062–1071.

Liu, H.T., W.M. Li, G. Xu, X.Y. Li, X.F. Bai, P. Wei, C. Yu, and Y.G. Du. 2009. Chitosan oligosaccharides attenuate hydrogen peroxide-induced stress injury in human umbilical vein endothelial cells. *Pharmacological Research* 59(3):167–175.

Liu, J.N. 2008. Study on the hypolipidemic mechanism of chitosan. Doctor dissertation. Jiangnan University. Wuxi, China.

Liu, X., F. Yang, T. Song, A. Zeng, Q. Wang, Z. Sun, and J. Shen. 2011. Effects of chitosan, O-carboxymethyl chitosan and N-[(2-hydroxy-3-N,N-dimethylhexadecyl ammonium) propyl] chitosan chloride on lipid metabolism enzymes and low-density-lipoprotein receptor in a murine diet-induced obesity. *Carbohydrate Polymers* 85:334–340.

Muzzarelli, R.A.A., N. Frega, M. Milliani, C. Muzzarelli, and M. Cartolari. 2000. Interaction of chitin, chitosan, N-lauryl chitosan and N-dimethilaminoprophyl chitosan with olive oil. *Carbohydrate Polymers* 43:263–268.

Muzzarelli, R.A.A., V. Stanic, and V. Ramos. 1999. Enzymatic depolymerization of chitins and chitosans. In *Methods in Biotechnology: Carbohydrate Biotechnology Protocols*, C. Bucke, ed., Humana Press, Totowa.

Ngo, D.N., M.M. Kim, and S.K. Kim. 2008a. Chitin oligosaccharides inhibit oxidative stress in live cells. *Carbohydrate Polymers* 74:228–234.

Ngo, D.N., M.M. Kim, and S.K. Kim. 2012b. Protective effects of aminoethyl-chitooligosaccharides against oxidative stress in mouse macrophage RAW 264.7 cells. *International Journal of Biological Macromolecules* 50:624–631.

Ngo, D.N., S.H. Lee, M.M. Kim, and S.K. Kim. 2009. Production of chitin oligosaccharides with different molecular weights and their antioxidant effect in RAW 264.7 cells. *Journal of Functional Foods* 1:188–198.

Ngo, D.H., D.N. Ngo, T.S. Vo, B.M. Ryu, Q.V. Ta, and S.K. Kim. 2012a. Protective effects of aminoethyl-chitooligosaccharides against oxidative stress and inflammation in murine microglial BV-2 cells. *Carbohydrate Polymers* 88:743–747.

Ngo, D.H., Z.J. Qian, D.N. Ngo, T.S. Vo, I. Wijesekara, and S.K. Kim. 2011a. Gallyl chitooligosaccharides inhibit intracellular free radical-mediated oxidation. *Food Chemistry* 128:974–981.

Ngo, D.H., Z.J. Qian, T.S. Vo, B.M. Ryu, D.N. Ngo, and S.K. Kim. 2011b. Antioxidant activity of gallate-chitooligosaccharides in mouse macrophage RAW264.7 cells. *Carbohydrate Polymers* 84:1282–1288.

Ngo, D.N., Z.J. Qian, J.Y. Je, M.M. Kim, and S.K. Kim. 2008b. Aminoethyl chitooligosaccharides inhibit the activity of angiotensin converting enzyme. *Process Biochemistry* 43:119–123.

Ngo, D.H., I. Wijesekara, T.S. Vo, Q.V. Ta, and S.K. Kim. 2011c. Marine food-derived functional ingredients as potential antioxidants in the food industry: An overview. *Food Research International* 44:523–529.

Nishimura, S., N. Nish, and S. Tokura. 1986. Inhibition of the hydrolytic activity of thrombin by chitin heparinoids. *Carbohydrate Research* 156:286–292.

Nishimura, S.I., K. Hideaki, K. Shinada, T. Yoshida, S. Tokura, K. Kurita, H. Nakashima, N. Yamamoto, and T. Uryu. 1998. Regioselective synthesis of sulfated polysaccharides: Specific anti-HIV-1 activity of novel chitin sulfate. *Carbohydrate Research* 306:427–433.

Niu, L., Y. Xui, H. Xie, Z. Dai, and H. Tang. 2008. Expression of human insulin gene wrapped with chitosan nanoparticles in NIH3T3 cells and diabetic rats. *Acta Pharmacologica Sinica* 29:1342–1349.

Park, P.J., C.B. Ahn, Y.J. Jeon, and J.Y. Je. 2008. Renin inhibition activity by chitooligosaccharides. *Bioorganic and Medicinal Chemistry Letters* 18:2471–2474.

Park, P.J., J.Y. Je, and S.K. Kim. 2003. Angiotensin I converting enzyme (ACE) inhibitory activity of heterochitooligosaccharides prepared from partially different deacetylated chitosans. *Journal of Agricultural and Food Chemistry* 51:4930–4934.

Park, P.J., J.Y. Je, and S.K. Kim. 2004a. Free radical scavenging activities of differently deacetylated chitosans using an ESR spectrometer. *Carbohydrate Polymers* 55:17–22.

Park, P.J., J.Y. Je, W.K. Jung, C.B. Ahn, and S.K. Kim. 2004b. Anticoagulant activity of heterochitosans and their oligosaccharide sulfates. *European Food Research and Technology* 219:529–533.

Qian, Z.J., T.K. Eom, B.M. Ryu, and S.K. Kim. 2010. Angiotensin I-converting enzyme inhibitory activity of sulfated chitooligosaccharides with different molecular weights. *Journal of Chitin and Chitosan* 15(2):75–79.

Rajapakse, N., M.M. Kim, E. Mendis, and S.K. Kim. 2007. Inhibition of free radical-mediated oxidation of cellular biomolecules by carboxylated chitooligosaccharides. *Bioorganic and Medicinal Chemistry* 15(2):997–1003.

Razdan, A. and D. Pettersson. 1994. Effect of chitin and chitosan on nutrient digestibility and plasma lipid concentrations in broiler chickens. *The British Journal of Nutrition* 72:277–288.

Ronghua, H., D. Yumin, and Y. Jianhong. 2003. Preparation and anticoagulant activity of carboxybutyrylated hydroxyethyl chitosan sulfates. *Carbohydrate Polymers* 51:431–438.

Sajomsang, W., P. Gonil, and S. Saesoo. 2009. Synthesis and antibacterial activity of methylated N-(4-N,N-dimethylaminocinnamyl) chitosan chloride. *European Polymer Journal* 45:2319–2328.

Shahidi, F., and Y. Zhong. 2008. Bioactive peptides. *Journal of AOAC International* 91:914–931.

Shen, K.T., M.H. Chen, H.Y. Chan, J.H. Jeng, and Y.J. Wang. 2009. Inhibitory effects of chitooligosaccharides on tumor growth and metastasis. *Food and Chemical Toxicology* 47:1864–1871.

Tapola, N.S., M.L. Lyyra, R.M. Kolehmainen, E.S. Sarkkinen, and A.S. Schauss. 2008. Safety aspects and cholesterol-lowering efficacy of chitosan tablets. *Journal of the American College of Nutrition* 27:22–30.

Tong, Y., S. Wang, J. Xu, B. Chua, and C. He. 2005. Synthesis of O,O-dipalmitoyl chitosan and its amphiphilic properties and capability of cholesterol absorption. *Carbohydrate Polymers* 60:229–233.

Toshkova, R., N. Manolova, E. Gardeva, M. Ignatova, L. Yossifova, I. Rashkov, and M. Alexandrov. 2010. Antitumor activity of quaternized chitosan-based electrospun implants against Graffi myeloid tumor. *International Journal of Pharmaceutics* 400:221–233.

Vikhoreva, G., G. Bannikova, P. Stolbushkina, A. Panov, N. Drozd, V. Makarovd, V. Varlamo, and L. Gal'braikh. 2005. Preparation and anticoagulant activity of a low-molecular-weight sulfated chitosan. *Carbohydrate Polymers* 62:327–332.

Wijesekara, I. and S.K. Kim. 2010. Angiotensin-I-converting enzyme (ACE) inhibitors from marine resources: Prospects in the pharmaceutical industry. *Marine Drugs* 8:1080–1093.

Xia, W.S. 2003. Physiological activities of chitosan and its application in functional foods. *Journal of Chinese Institute of Food Science and Technology* 3:77–81.

Xia, W., P. Liu, J. Zhang, and J. Chen. 2011. Biological activities of chitosan and chitooligosaccharides. *Food Hydrocolloids*, 25:170–179.

Xu, Q.S., P. Ma, W. Yu, C.Y. Tan, H.T. Liu, C.N. Xiong, Y. Qiao, and Y.G. Du. 2010a. Chitooligosaccharides protect human embryonic hepatocytes against oxidative stress induced by hydrogen peroxide. *Marine Biotechnology* 12:292–298.

Xu, T., M. Xin, M.M. Li, H. Huang, and S. Zhou. 2010b. Synthesis, characteristic and antibacterial activity of N,N,N-trimethyl chitosan and its carboxymethyl derivatives. *Carbohydrate Polymers* 81:931–936.

Yang, J., J. Cai, K. Wu, D. Li, Y. Hu, G. Li, and Y. Du. 2012. Preparation, characterization and anticoagulant activity in vitro of heparin-like 6-carboxylchitin derivative. *International Journal of Biological Macromolecules* 50:1158–1164.

Yang, L., L. Chen, R. Zeng, C. Li, R. Qiao, L. Hu, and Z. Li. 2010. Synthesis, nanosizing and in vitro drug release of a novel anti-HIV polymeric prodrug: Chitosan-O-isopropyl-5′-O-d4T monophosphate conjugate. *Bioorganic and Medicinal Chemistry* 18:117–123.

Yang, T.C., C.C. Chou, and C.F. Li. 2005. Antibacterial activity of N-alkylated disaccharide chitosan derivatives. *International Journal of Food Microbiology* 97:237–245.

Yao, H.T. and M.T. Chiang. 2006. Effect of chitosan on plasma lipids, hepatic lipid and fecal bile acid in Hamsters. *Journal of Food and Drug Analysis* 14:183–189.

Yao, H.T., S.Y. Huang, and M.T. Chiang. 2008. A comparative study on hypoglycemic and hypocholesterolemic effects of high and low molecular weight chitosan in streptozotocin-induced diabetic rats. *Food and Chemical Toxicology* 46:1525–1534.

Yen, M.T., J.H. Yang, and J.L. Mau. 2008. Antioxidant properties of chitosan from crab shells. *Carbohydrate Polymers* 74:840–844.

Ying, G.Q., W.Y. Xiong, H. Wang, Y. Sun, and H.Z. Liu. 2011. Preparation, water solubility and antioxidant activity of branched-chain chitosan derivatives. *Carbohydrate Polymers* 83:1787–1796.

Zhang, J., W. Xia, P. Liu, Q. Cheng, T. Tahi, W. Gu, and B. Li. 2010. Chitosan modification and pharmaceutical/biomedical applications. *Marine Drugs*, 8:1962–1987.

Zhang, H.L., X.B. Zhong, Y. Tao, S.H. Wu, and Z.Q. Su. 2012. Effects of chitosan and water-soluble chitosan micro- and nanoparticles in obese rats fed a high-fat diet. *International Journal of Nanomedicine* 7:4069–4076.

Zhong, Z., X. Ji, R. Xing, S. Liu, Z. Guo, X. Chen, and P. Li. 2007. The preparation and antioxidant activity of the sulfanilamide derivatives of chitosan and chitosan sulfates. *Bioorganic and Medicinal Chemistry* 15:3775–3782.

Zhong, Z., P. Li, R. Xing, and S. Liu. 2009. Antimicrobial activity of hydroxylbenzenesulfonailides derivatives of chitosan, chitosan sulfates and carboxymethyl chitosan. *International Journal of Biological Macromolecules* 45:163–168.

Zhou, K., W. Xia, C. Zhang, and L. Yu. 2006. in vitro binding of bile acids and triglycerides by selected chitosan preparations and their physicochemical properties. *LWT—Food Science and Technology* 39:1087–1092.

13 Biological Applications of Chitin, Chitosan, Oligosaccharides, and Their Derivatives

Panchanathan Manivasagan, Kalimuthu Senthilkumar, Jayachandran Venkatesan, and Se-Kwon Kim

CONTENTS

13.1 INTRODUCTION

Chitosan is a natural nontoxic biopolymer produced by the deacetylation of chitin, a major component of the shells of crustaceans, such as crab, shrimp, and crawfish. Currently, chitosan has received considerable attention for its commercial applications in the biomedical, food, and chemical industries (Knorr 1984; Kurita 1998; Razdan and Pettersson 1994). Chitosan contains three types of reactive functional groups, an amino/acetamido group as well as both primary and secondary hydroxyl groups at the C-2, C-3, and C-6 positions, respectively. The amino contents are the main reason for the differences between their structures and physicochemical properties as well as are correlated with their chelation, flocculation, and biological functions. Chitooligosaccharides (COS) are the degraded products of chitosan or chitin, which have recently been produced by several methods such as enzymatic and acidic hydrolysis. Enzymatic preparation methods have received great attention because of their safety and ease of control. Many nonspecific enzymes, such as cellulases, lipases, and proteases, as well as chitosanases, have been used to prepare COS (Lee et al. 2008b; Lin et al. 2009). Generally, the molecular weights of COS are 10 kDa or less, and during the preparation of chitosans with different molecular weights viscosity is used as a parameter for determining the molecular weight.

Because chitin $(C_8H_{13}O_5N)n$ was first isolated and characterized from mushrooms, the earliest known polysaccharide, by a French chemist Henri Braconnot in 1811 (Domard and Domard 2002), it has been discovered to be the second most abundant natural biopolymer in the world (Lavall et al. 2007; Ngo et al. 2008a), amounting in marine biomass alone to be approximately 106–107 tons. Chitin is a long-chain homopolymer of N-acetyl-D-glucosamine (GlcNAc), (1–4)-linked 2-acetamido-2-deoxy-β-D-glucan, a derivative of glucose. Strong acids can split the chitin, water-insoluble polymer, into acetic acid and chitosan, and chitin can be processed further into two main derivatives of chitosan and amino glucose through many nonspecific enzymes, such as cellulases, lipases, proteases, and chitosanases (Lee et al. 2008b; Lin et al. 2009). Notably, chitosan is a nontoxic biopolymer produced by the deacetylation of chitin, and currently chitosan and its oligosaccharides have received considerable attention due to their biological activities and properties in commercial applications. During the past few decades, as a source of bioactive material, chitosan and COS, which are degradation products of chitin or chitosan produced by enzymatic or acidic hydrolysis, were introduced into a variety of biomedicals, including wound dressings and drug delivery systems (Agnihotri et al. 2004; Kumar et al. 2004), and food and chemical industries (Knorr 1984; Xia et al. 2011). Chitin and its derivatives have delivered biological potential for a wide range of applications, such as in the food and medical field (Ngo et al. 2009; Razdan and Pettersson 1994; Ribeiro et al. 2009), agriculture (Kulikov et al. 2006), aquaculture (Wang and Chen 2005), dental (Arnaud et al. 2010), cosmetics (Morganti and Morganti 2008), wastewater (Peniche-Covas et al. 1992), and membranes (Chatelet et al. 2001). This chapter aims to analyze the most recent advances in biological applications of chitin and its derivatives, particularly those related to antiinflammatory and antioxidant activities, antimicrobial effects, immunity enhancing, as well as anticancer effects and drug delivery, in the field of biological medicine.

13.2 PHYSICO-CHEMICAL PROPERTIES OF CHITIN AND CHITOSAN DERIVATIVES

Chitin, the starting material of chitosan, is a white, hard, and inelastic structural polysaccharide found in cell walls of fungi and in the exoskeletons of crustaceans. The molecular structure of chitin is identified as a high-molecular weight linear polymer of N-acetyl-D-glucosamine units (GlcNAc) linked by β-1,4 bonds. The hydrophobic nature of chitin has made it insoluble in water as well as in most organic solvents. In contrast, chitosan, the N-deacetylated form of chitin is readily soluble in dilute organic acids at low pH (Peniston and Johnson 1980). The most important parameter that determines the solubility of chitosan is the degree of deacetylation (DD). Conversion of chitin into chitosan increases DD, and thereby alters the charge distribution of chitosan molecules. In general, degree of acetylation (DA) of chitin is about 90% and following partial or full deacetylation with alkaline treatment, it is converted into chitosan. In addition to the DD, degree of polymerization (DP) also contributes to the alteration of physico-chemical properties of chitosan. Moreover, COS (relatively lower DP) are better soluble than low molecular weight chitosans (LMWC) with relatively higher DP. However, there is no specific DP to distinguish COS and LMWC. Generally, molecular weight of COS can be considered up to 10 kDa or less, and during preparation of different molecular weight chitosans, viscosity is used as a parameter to determine the molecular weight.

Unlike most polysaccharides, chitosan and COS have positive charges following the removal of acetyl units from D-glucosamine residues. This chemical feature allows chitosan and COS to bind strongly to negatively charged surfaces and is responsible for many of observed biological activities. In addition, nontoxicity, biodegradability, and biocompatibility of chitosan and COS promote their biological applications compared to other synthetic polymers (Kurita 1998).

13.3   CHITIN AND CHITOSAN OLIGOSACCHARIDES PREPARATION METHODS

Similar to all polysaccharides, chitosan can also be cleaved by hydrolyzing agents due to the presence of rather unstable glycosidic bonds. Degradation of *O*-glycosidic linkages of chitosan by different methods leads to the production of COS varying in the DP as well as number and sequence of glucosamine (GlcN) and GlcNAc units. Some of these methods include acid hydrolysis (Il'ina and Varlamov 2004), enzymatic hydrolysis (Kuroiwa et al. 2002; Zhang et al. 1999), oxidative degradation (Mao et al. 2004), ultra sonic degradation (Chen and Chen 2000), chemoenzymatic (Akiyama et al. 1995), and recombinant approaches (Kim and Rajapakse 2005). Moreover, electromagnetic radiation, sonication, and mechanical energy generated by microfiltration can also be used for the production of COS without employing chemical agents (Hai et al. 2003). Absorption of energy by chitosan molecules in any of the above methods results in scission of chemical bonds, and if the broken bond belongs to the backbone of the polymer (*O*-glycosidic bond), a decreased molecular weight will result.

Although chitin and chitosan can be isolated from different sources (Muzzarelli et al. 1994), crab and shrimp shell wastes are currently utilized as the major industrial source of biomass for the large-scale production of COS. These crustacean shell wastes are composed of protein, inorganic salts, chitin, and lipids as the main structural components. Therefore, extraction of chitin and chitosan (the starting materials of COS) is mainly employed in stepwise chemical methods. In the first step, shrimp or crab shells are treated with 3–5% aqueous NaOH solution to remove proteins attached to the shells and thereby prevent the contamination of chitin products with proteins. Deproteinized shells are then neutralized and calcium is removed by treating with 3–5% aqueous HCl solution to afford a white or slightly pink precipitate of chitin. The chitin is then N-deacetylated with 40–45% NaOH to form chitosan with a cationic nature. The resulting crude sample is dissolved in 2% acetic acid and the supernatant is neutralized with aqueous NaOH solution to afford purified chitosan as a white precipitate (Hirano 1996). After alkaline deacetylation, some of the amino groups may remain acetylated and distribute randomly along the whole polymer chain.

For large-scale production of COS, acid hydrolysis is commonly used to cleave glycosidic linkages of chitosan. However, chemical hydrolysis results in low yields of COS and a larger amount of monomeric D-glucosamine units. Therefore, COS prepared by industrial-scale acid hydrolytic methods are generally not considered to serve as bioactive materials because of the possibility of contamination of toxic chemical compounds. As a result, enzymatic hydrolysis of chitosan has been proposed as a preferred method for the production of bioactive COS during the past few decades.

13.4   BIOLOGICAL ACTIVITY

Chitosan is inexpensive and nontoxic and possesses reactive amino groups. It has been shown to be useful in different areas—as an antimicrobial compound in agriculture, as a potential elicitor of plant defense responses, as a flocculating agent in wastewater treatment, as an additive in the food industry, as a hydrating agent in cosmetics, and more recently as a pharmaceutical agent in biomedicine (Aiedeh and Taha 2001; Baba et al. 2002; Ishii et al. 2001; Li et al. 2002).

In this context, the antimicrobial activity of chitosan and its derivatives against different groups of microorganisms, such as bacteria and fungi, has received considerable attention in recent years.

13.4.1   Antimicrobial Activity

13.4.1.1   Antibacterial Activity

Chitosan inhibits the growth of a wide variety of bacteria (Wu et al. 2011). Chitosan has been studied in terms of bacteriostatic/bactericidal activity to control growth of algae and to inhibit viral multiplication (Cuero et al. 1991; Muzzarelli et al. 1990). Moreover, chitosan has several advantages over other types of disinfectants because it possesses a higher antibacterial activity, a broader spectrum of activity, a higher killing rate, and a lower toxicity toward mammalian cells (Franklin and Snow 1981).

Chitosan derivatives containing quaternary ammonium salts, such as N,N,N-trimethyl chitosan, N-propyl-N,N-dimethyl chitosan, and N-furfuryl-N,N-dimethyl chitosan were prepared and tested for their activity against *Escherichia coli* (Jia and Xu 2001). It was shown that the antibacterial activity of quaternary ammonium chitosan in acetic acid medium is stronger than that in water. Their antibacterial activity increased as the concentration of acetic acid is increased. It was also found that the antibacterial activity of quaternary chitosan against *E. coli* is stronger than that of chitosan itself (Jia and Xu 2001).

Certain antibacterial activities of diethylaminoethyl chitin, diethylaminoethyl chitosan, and triethylaminoethyl chitin were evaluated. The triethylaminoethyl chitin was the most active agent. It had a greater activity against *Staphylococcus aureus* than against *E. coli*. A concentration of 500 ppm was needed to kill all *S. aureus* within 120 min. Different molecular weight hydrolysates of diethylaminoethyl chitin showed a dependence of the antibacterial activity on the molecular weight of the hydrolysate (Kim et al. 1997).

In addition, the carbohydrate-branched derivatives 1-deoxyglucit-1-yl chitosan and 1-deoxylactat-1-yl chitosan had activity against *Bacillus circulans* but not against *E. coli,* whereas chito-oligosaccharides of varying degrees of polymerization (DPs) showed activity against *E. coli* but not against *B. circulans* (Franco and Peter 2011).

A much higher concentration of chitosan (1–1.5%) is required for complete inactivation of *S. aureus* after 2 days of incubation at pH 5.5 or 6.5 (Wang 1992). Furthermore, chitosan concentrations of 0.005% were sufficient to elicit complete inactivation of *S. aureus* (Chang et al. 1989). This was in accordance with the findings on the effect of chitosan in meat preservation. The antimicrobial effect on different cultures of bacteria on raw shrimp with different concentrations of chitosan was studied, and variations in their degree of susceptibility to chitosan were observed (Simpson et al. 1997; Strand et al. 2003). According to these findings, chitosan concentrations of 0.02% were required to display a bactericidal effect against *B. cereus*, whereas *E. coli* and *Proteus vulgaris* showed minimal growth at 0.005% chitosan and complete inhibition at 0.0075%. It was also reported that *B. cereus* was inhibited by chitosan. However, much lower concentrations (0.005%) were required, perhaps due to the low molecular weight (35 kDa) of chitosan used in this experiment (Chang et al. 1989).

Numerous studies have also shown the effect of chitosan on *E. coli* inhibition. Complete inactivation was observed after a 2-day incubation period with chitosan concentrations of 0.5% or 1% at pH 5.5. Complete inactivation could be reached even after the first day if the chitosan concentration was more than 1% in the broth (Wang 1992). Meanwhile, a chitosan concentration of 0.1% was required to inhibit *E. coli* growth (Darmadji and Izumimoto 1994) and only 0.0075% was needed to inhibit the growth of *E. coli* (Simpson et al. 1997). These variations were suggested to be due to the existing differences in the degree of acetylation of chitosan; chitosan with a degree of acetylation of 7.5% was more effective than chitosan with a degree of acetylation of 15%.

The antimicrobial effect of water-soluble chitosans (Sudarshan et al. 1992) such as chitosan lactate, chitosan hydroglutamate, and chitosan derived from *Absidia coerulea* fungi was determined on different bacterial cultures. It was observed that chitosan glutamate and chitosan lactate were bactericidal against both Gram-positive and Gram-negative bacteria in the range of 1–5 log cycle reductions within 1 h. In that same study, the authors reported that chitosan was no longer bactericidal at pH 7 because of two major reasons, namely, the presence of a significant proportion of uncharged amino groups and the poor solubility of chitosan. These results are in agreement with findings of a similar study (Papineau et al. 1991) in which a concentration of 0.2 mg/mL chitosan lactate appeared most effective against *E. coli* with a corresponding population drop of 2 and 4 log cycles within 2 min and 1 h exposure, respectively. These authors observed that chitosan glutamate was also effective against yeast cultures such as *Saccharomyces cerevisiae* and *Rhodotorula glutensis*, and inactivation was rapid and complete within 17 min when cultures were exposed to 1 mg/mL chitosan lactate. This was in contrast to the results (Papineau et al. 1991) in which chitosan hydroglutamate was a more effective antagonist than chitosan lactate.

In another study, the antibacterial effects of 69% deactivated shrimp chitosan, 0.63% sulfonated chitosan (SC1), 13.03% sulfonated chitosan (SC2), and sulfobenzoyl chitosan on oyster preservation were reported. Except in the case of *B. cereus*, bacterial growth was effectively inhibited by at least one of the above four compounds tested at 200 ppm. Although the sulfonation increased the solubility of chitosan, totally different antibacterial activities were observed for SC1 and SC2. For most of the bacterial cultures, SC1 had a very pronounced minimal inhibitory concentration (MIC) effect even at the 200 ppm level, whereas SC2 exhibited no antibacterial effect at concentrations below 2000 ppm. It was suggested that because SC2 has more sulfonyl groups, it carries a higher negative charge than SC1; thus, there would be a greater repulsive force between negatively charged SC2 molecules and bacterial cell walls (Tsai et al. 2000).

Chitosan derivatives were claimed as antimicrobials for fish and shellfish against infection from *Vibrio anguillarum*, *Edwardsiella tarda*, *Pasteurella piscicida*, and several bacteria, in agreement with data obtained on brook trout (Muzzarelli et al. 2001). However, chitin and chitosan are accepted diet supplements for cultured fish (Muzzarelli et al. 2001).

13.4.1.2 Antifungal Activity

The antimicrobial activity of chitosan was observed against a wide variety of microorganisms including fungi, algae, and some bacteria. However, the antimicrobial action is influenced by intrinsic factors such as the type of chitosan, the degree of chitosan polymerization, the host, the natural nutrient constituency, the chemical or nutrient composition of the substrates or both, and the environmental conditions (e.g., substrate water activity or moisture or both). Although both native chitosan and its derivatives are effective as antimicrobial agents, there is a clear difference between them. Their different antimicrobial effect is mainly exhibited in live host plants. The fungicidal effect of *N*-carboxymethyl chitosan (NCMC) is also different in vegetables as compared to graminea hosts. In addition, oligomeric chitosans (pentamer and heptamer) have a better antifungal effect than larger units. The chitosan antimicrobial activity is more immediate on fungi and algae than on bacteria (Savard et al. 2002).

Chitosan has been shown to be fungicidal against several fungi (Chen et al. 2002). The minimum inhibitory concentrations (MICs) reported for specific target organisms range from 0.0018% to 1.0% and are influenced by a multitude of factors such as the pH of the growth medium, the DP of chitosan, and the presence or absence of interfering substances such as lipids and proteins (Chen et al. 2005; Sudarshan et al. 1992; Wang 1992).

The inhibitory effect of chitosan was also demonstrated with soilborne phytopathogenic fungi (Stössel and Leuba 1984). The inhibitory activity of chitosan was higher at pH 6.0 (pKa value of chitosan = 6.2) than at pH 7.5, when most amino groups are in the free base form (Stössel and Leuba 1984). The maximal antifungal and pisatin-inducing activities of chitosan were exhibited by chitosan oligomers of seven or more residues. The soilborne phytopathogenic fungi *Fusarium solani* and *Colletotrichum lindemuthianum* were inhibited by chitosan and *N*-carboxymethyl chitosan (Kendra and Hadwiger 1984; Stössel and Leuba 1984).

Chitosan has been utilized in soil amendment, in seed treatment, and as a foliar treatment to control the fungus *F. oxysporum*. Chitosan concentrations ranging from 0.1 to 1 mg/mL indicated that higher protection occurred when seed coating and soil amendment were performed with concentrations of 0.5 and 1 mg/mL. Although chitosan at a concentration of 0.1 mg/mL induced a delay in disease development (root lesions visible by 4 days after inoculation), emergence of wilting symptoms occurred between 7 and 10 days postinoculation, while death of about 80% of the plants was recorded 1 week later (Benhamou et al. 1994).

F. acuminatum, *Cylindrocladium floridanum*, and other pathogens of interest in forest nurseries were inhibited by chitosan *in vitro* (Laflamme et al. 2000). Similarly, *Aspergillus flavus* was completely inhibited in field-growing corn and peanut (Laflamme et al. 2000).

Five chemically modified chitosans were tested for their antifungal activities against *Saprolegnia parasitica* by the fungal growth assay in chitosan-bearing broth. Results indicated that, as for the chitosan-bearing broth assay, *S. parasitica* did not grow normally; on the first day for methylpyrrolidinone chitosan and *N*-phosphonomethyl chitosan and on the second day for *N*-carboxymethyl chitosan, a tightly packed precipitate was present at the bottom of the test tubes instead of the fluffy fungal material as in the control. In contrast, *N*-dicarboxymethyl chitosan seemed to favor fungal growth, whereas dimethylaminopropyl chitosan did not significantly differ from the control data (Muzzarelli et al. 2001).

The use of bioactive substances such as chitosan to control postharvest fungal disease has attracted much attention due to imminent problems associated with chemical agents, which include development of public resistance to fungicide-treated produce, an increasing number of fungicide-tolerant postharvest pathogens, and several fungicides that are still under observation (El Ghaouth et al. 1992). Chitosan (1 mg/mL) reduces the *in vitro* growth of numerous fungi with the exception of Zygomycetes, that is, the fungi containing chitosan as a major component of their cell walls. Hence, chitosan has potential as an edible antifungal coating material for postharvest produce. Recent investigations on chitosan coating of tomatoes have shown that it delayed ripening by modifying the internal atmosphere, which reduced decay (El Ghaouth et al. 1992).

Chitosan treatment (2–8 mg/mL) of wheat seeds significantly improved seed germination to recommended seed certification standards (>85%) and vigor at concentrations >4 mg/mL in two cultivars of spring wheat (Norseman and Max) by controlling seed-borne *F. graminearum* infection. The germination was <80% in the control and >85% in benomyl- and chitosan-treated seeds. The reduction of seedborne *F. graminearum* was >50% at higher chitosan treatments compared to the control (Bhaskara Reddy et al. 1999).

Chitosan and chitosan-laminated films containing antimicrobial agents provide a type of active package so that the preservatives released from the film deposit on the food surface and inhibit the microbial growth (Rabea et al. 2003). The sorbate-loaded edible barrier for mold inhibition on food surfaces was evaluated (Torres et al. 1985), and the use of glucose oxidase/glucose was advocated as a dip for the extension of the shelf-life of fish (Field et al. 1986). The presence of preservatives in chitosan films reduces the intermolecular electrostatic repulsion in the chitosan molecules and facilitates formation of intramolecular hydrogen bonds. It was observed that the packaging film prepared from methylcellulose, chitosan, and a preservative possesses antimicrobial activity.

13.4.2 ANTIOXIDANT ACTIVITY

Both chitosan and its oligosaccharides showed antioxidant effects. In our recent study, the antioxidant activity of chitosan was studied *in vitro* and *in vivo* (Liu et al. 2008). The results showed that chitosan at an addition of 0.02% had antioxidant effects in lard and crude rapeseed oil, but the activity was less than ascorbic acid. When the addition was increased, chitosan and ascorbic acid had similar activities; chitosan could significantly reduce serum FFA and MDA concentrations and elevate SOD, CAT, and GSH-PX activities, the latter being the major antioxidant enzymes in the body, indicating that chitosan regulated the antioxidant enzyme activities and reduced lipid peroxidation.

The cellular antioxidant effects of COS (NA-COS; M_w 229.21e593.12 Da) produced by acidic hydrolysis of crab chitin were also identified (Ngo et al. 2008a). Their study showed that NA-COS have free-radical scavenging effects in a cellular system. They can inhibit myeloperoxidase activity and decrease free-radical oxidation of DNA and membrane proteins. Furthermore, they also stimulate an increase in intracellular GSH levels. Based on the results, they concluded that NA-COS have free radical scavenging effects, acting in both indirect and direct ways to inhibit and prevent biological molecular damage by free radicals in living cells (Park et al. 2004). Hence, chitosan and COS can be used as a scavenger to control radical-induced damage to cellular systems and promises further applications in the future.

13.4.3 Immuno-Stimulating and Anticancer Effects

Polysaccharides represent a structurally diverse class of macromolecules of relatively widespread occurrence in nature. Unlike proteins and nucleic acids, they contain repetitive structural features which are polymers of monosaccharide residues joined to each other by glycosidic linkages. Among these macromolecules, polysaccharides offer the highest capacity for carrying biological information because they have the greatest potential for structural variability. The nucleotides in nucleic acids and the amino acids in proteins can interconnect in only one way while the monosaccharide units in oligosaccharides and polysaccharides can interconnect at several points to form a wide variety of branched or linear structures (Sharon and Lis 1993). The polysaccharide also forms secondary structures, depending on the conformation of sugar residues, the molecular mass, and the inter- and intra-chain hydrogen bonding. The search for potential polysaccharides as antitumor agents probably stems from dissatisfaction with cancer chemotherapy and radiotherapy. A large number of chemical compounds which have been identified as specific agents for killing cancer cells are also toxic to normal cells. Many of the potential anticancer drugs have considerable side effects and therefore have little clinical use. Hence, the discovery and identification of new safe drugs which are active against tumors becomes an important goal of research in biomedical sciences. The enhancement or potentiation of host defense mechanisms emerges as a possible means of inhibiting tumor growth without harming the host. Starting from this point of view, extensive studies have been conducted on polysaccharides extracted from the marine environment.

Chitosan and its oligosaccharides, which are known to possess multiple functional properties, have attracted considerable interest due to their biological activities and potential applications in the food, pharmaceutical, agricultural, and environmental industries. Many researchers have focused on chitosan as a potential source of bioactive materials in the past few decades. Activating the immune system for therapeutic benefit has long been a goal in immunology, especially in cancer treatment (Mellman et al. 2011; Vanneman and Dranoff 2012). Chitin and chitosan, their biodegradability, biocompatibility, and nontoxicity provide chitin and chitosan with huge potential for future development. Chitin and chitosan are widely applied in chemistry, biotechnology, agriculture, veterinary, dentistry, food processing, environmental protection, and medicine (Synowiecki and Al-Khateeb 2003).

Immunity is the state of having sufficient biological defenses to avoid infection, disease, or other unwanted biological invasion. Chitosan are polysaccharides showing an immunity-enhancing effect by enhancement of antibody response. The effect of chitosan as a novel adjuvant to an inactivated influenza vaccine was studied (Chang et al. 2004). The immunostimulating activity of chitosan and COS has been first reported by Nishimura et al. (1984), that is, that chitosan, especially 70%-DD chitosan, could stimulate rats to produce a nonspecific host repellence when infected with *E. coli* and *Sendai* virus. They concluded that 70%-DD chitosan was an immune regulator that can activate macrophages and natural killer cells (NK) and improve the delayed-type hypersensitive reaction, increase cytotoxicity, and induce mitosis in cells producing interleukins, breeding factors, and interferon. Enhanced immune regulation with the increased water solubility of chitosan, evidenced that COS inhibited tumor growth through an increase in immune effects (Suzuki et al. 1986). The COS showed strong inhibition of cancer in BALB/c mice, whereas (GlcNAC)6 and (GlcN)6 showed very strong inhibiting effects for S-180 and MM156 solid tumor growth also in Lewis lung cancer in mice (Tokoro et al. 1988).

Chitosan showed an immunity-enhancing effect by enhancement of antibody response. The effect of chitosan as a novel adjuvant to an inactivated influenza vaccine was studied (Chang et al. 2004). Immunomodulating effect of chitosan shown in BALB/c mice were abdominally inoculated with vaccine and chitosan together twice every three weeks. Blood serum was prepared and the testing for levels of antibodies IgG, IgG1, and IgG2a, as well as IgA antibody in nasal secretions was done. One week after the immunization regimen, the mice were challenged with the deadly flu virus A/PR/8/34(H1N1) and the weights of the mice and levels of antibody protection were measured. The results indicated that using chitosan as an adjuvant increased the antibody content in serum remarkably and increased the antiviral defense in the mice, enhancing the immune reaction to the vaccine. The antitumor mechanism of these COS was probably related to their induction of lymphocyte factor, increasing T-cell proliferation to produce the tumor inhibitory effects. The antitumor mechanism of COS is to enhance acquired immunity by accelerating T-cell differentiation to increase cytotoxicity and maintain T-cell activity (Suzuki et al. 1986). Although several studies have reported the importance of chitosan derivatives for their anticancer activity, no clear information is available describing the relationship between their charge properties and their observed activities. Anticancer activities of differently charged COS derivatives using three cancer-cell lines: HeLa, Hep3B, and SW480 (Huang et al. 2006). However, the exact molecular mechanism for the anticancer activity of strongly charged COS compared to their poorly charged counterparts is not clear. Additionally, chitosan is also known as a drug carrier which can improve drug absorption, stabilize drug components to increase drug targeting, and enhance drug release. As a gene carrier, chitosan can protect DNA and increase the expression period of genes. Hence, chitosan has broad prospects for applications as drug and gene carriers. For example, chitosan could also be used as a drug carrier to provide anticancer and antitumor chemotherapy.

It has been reported that the conjugates of some kinds of anticancer agents with chitin and chitosan derivatives display good anticancer effects with a decrease in side effects over the original form due to a predominant distribution in the cancer tissue and a gradual release of free drug from the conjugates. For instance, doxifluridine and 1-β-D-arabinofuranosylcytosine (Ara-C) could be conjugated with chitosan via a glutaric acid spacer. The conjugates of Ara-C with chitosan, in particular, showed improved antitumor effects against a P388-bearing leukemia model in mice. Glycol-chitosan (G-Chi) was distributed mainly in the systemic circulation and the kidney after *i.v.* administration in normal mice and retained for a long time in the kidney. The therapeutic effects of the conjugates of mitomycin C (MMC) with G-Chi were not necessarily improved over that of the free drug, but the toxic side effects were significantly decreased. Conjugates of MMC with 6-O-carboxymethyl-chitin showed an almost complete suppression of tumor growth at 10 mg eq. MMC/kg, while a lethal adverse effect was also observed. The conjugates of MMC with *N*-succinylchitosan showed good antitumor activities against various tumor models because of their predominant distribution into the tumor tissue and sustained-release characteristics with both water insoluble and soluble formulations (Kato et al. 2005). From the above, it is believed that chitin and chitosan derivatives are good candidates for polymeric drug carriers in cancer chemotherapy.

13.4.4 Application in Drug Delivery System

Drug discovery and development involve highly challenging, laborious, and expensive processes. Most of the drugs in the clinical phase, however, fail to achieve favorable clinical outcomes because they do not reach the target site of action. A significant amount of the administrated drug is distributed over the normal tissues or organs, often leading to severe side effects. An effective approach to overcome this critical issue is the development of targeted drug delivery systems that release the drugs or bioactive agents at the desired site of action. This could increase patient compliance and therapeutic efficacy of pharmaceutical agents through improved pharmacokinetics and biodistribution (DiMasi et al. 2003; Langer 1998). The targeted drug delivery system comprises three components: a therapeutic agent, a targeting moiety, and a carrier system. The drug can be either incorporated by passive absorption or chemical conjugation into the carrier system. The choice of the carrier molecule is of high importance because it significantly affects the pharmacokinetics and pharmacodynamics of the drugs. A wide range of materials, such as natural or synthetic polymers, lipids, surfactants, and dendrimers, have been employed as drug carriers (Duncan 2003, 2006; Torchilin 2008). Chitosan, a linear aminopolysaccharide composed of randomly distributed (1 $\rightarrow$ 4) linked D-glucosamine and *N*-acetyl-D-glucosamine units, is obtained by the deacetylation of chitin, a widespread natural polysaccharide found in the exoskeleton of crustaceans such as crabs and shrimps (Kumar et al. 2004). This cationic polysaccharide has drawn increasing attention within pharmaceutical and biomedical applications, owing to its abundant availability, unique mucoadhesivity, inherent pharmacological properties, and other beneficial biological properties such as biocompatibility, biodegradability, nontoxicity, and low immunogenicity (Felt et al. 1998; Kumar et al. 2004). The presence of reactive functional groups in chitosan offers great opportunity for chemical modification, which affords a wide range of derivatives including quaternized chitosan (*N,N,N*-trimethyl chitosan; TMC), carboxyalkyl chitosan, thiolated chitosan, sugar-bearing chitosan, bile acid-modified chitosan, and cyclodextrin-linked chitosan (Amidi et al. 2006; Bernkop-Schnürch et al. 2004; Kim et al. 2005; Wang et al. 2008).

The chemical modification of chitosan imparts amphiphilicity, which is an important characteristic for the formation of self-assembled nanoparticles, potentially suited for drug delivery applications. The hydrophobic cores of the nanoparticles could act as reservoirs or micro containers for various bioactive substances. Because of their small size, nanoparticles can be administered via the intravenous injection for targeted drug delivery. Conjugation of the targeting moieties to the surface of drug-loaded nanoparticles may improve therapeutic efficiency of the drug (Dufes et al. 2004). Chitosan has been widely utilized as drug delivery systems for low molecular drugs, peptides, and genes (Amidi et al. 2006; Kim et al. 2008; Sang Yoo et al. 2005). Despite the recent emergence of biomacromolecular drugs, the majority of therapeutic drugs that are being developed and marketed are primarily low molecular weight drugs. The successful delivery of low molecular drugs to their respective targets is still of prime importance in therapeutics. Chitosan has prompted the continuous movement for the development of safe and effective drug delivery systems because of its unique physicochemical and biological characteristics. The primary hydroxyl and amine groups located on the backbone of chitosan allow for chemical modification to control its physical properties. When the hydrophobic moiety is conjugated to a chitosan molecule, the resulting amphiphile may form self-assembled nanoparticles that can encapsulate a quantity of drugs and deliver them to a specific site of action. Chemical attachment of the drug to the chitosan throughout the functional linker may produce useful prodrugs, exhibiting the appropriate biological activity at the target site.

Chitosan on drug delivery function have various properties, such as controlled drug release, mucoadhesive, *in situ* gelling, transfection enhancing, permeation enhancing, and efflux pump inhibitory properties, which provides numerous drug delivery systems for various application sites. Also chitosan-based delivery systems have been widely studied in colonic drug targeting. The glycosidic linkage of chitosan is degraded by colonic microflora. By making use of this colon-specific degradation, chitosan has discovered as useful coating for a site-specific delivery. Chitosan and its derivatives are very useful in various drug delivery functions including oral drug, ocular, nasal, vaginal, buccal, parenteral, intravesical, and also vaccine delivery (Bernkop-Schnürch and Dünnhaupt 2012). Various target-specific carriers, based on chitosan and its derivatives concerned with the organ-specific delivery system was reported (Patel et al. 2010). Colon-specific drug delivery systems have gained increasing attention for the treatment of diseases such as Crohn's disease, ulcerative colitis, and irritable bowel syndrome (Chourasia and Jain 2004; Jain and Jain 2008). Chitosan-based delivery systems have been widely studied for colonic drug targeting. This system can protect therapeutic agents from the antagonistic conditions of the upper gastrointestinal tract and release the entrapped agents specifically at the colon through degradation of the glycosidic linkages of chitosan by colonic microflora (Hejazi and Amiji 2003). Chitosan capsules for colon-specific delivery of 5-aminosalicylic acid (5-ASA) was reported. The release of 5-ASA from the capsule was markedly increased in the presence of rat cecal contents (Tozaki et al. 2002). Hydrogel microspheres of chitosan grafted with vinyl polymers for the controlled and targeted delivery of 5-ASA to the colon, which exhibited better therapeutic effects (Jam et al. 2008). Hyaluronic acid-coupled chitosan nanoparticles bearing 5-fluorouracil (5-FU) were also prepared by an ionotropic gelation method for the effective delivery of the drug to the colon tumors (Jain and Jain 2008). These nanoparticles showed enhanced cellular uptake by HT-29 colon cancer cells compared to the uncoupled nanoparticles. The cytotoxicity of 5-FU incorporated in nanoparticles was higher compared to the free 5-FU solution.

The liver is a critical target tissue for drug delivery because many fatal conditions, including chronic hepatitis, enzyme deficiency, and hepatoma, occur in hepatocytes. In general, liver-targeting systems employ passive trapping of microparticles by reticuloendothelium or active targeting based on recognition between hepatic receptor and ligand-bearing particulates (Ogawara et al. 1999). Lactosaminated N-succinyl-chitosan (Lac-Suc), synthesized by reductive amination between N-succinyl-chitosan and lactose in the presence of sodium cyanoborohydride, as a liver-specific drug carrier (Ogawara et al. 1999). Polyion complex micelles (PIC micelles) based on methoxy poly (ethylene glycol) (PEG)-graft chitosan and lactose-conjugated PEG-graft-chitosan for liver-targeted delivery of diammonium glycyrrhizinate (DG) (Yang et al. 2009a). Conjugated glycyrrhizin (GL) to the surface of chitosan nanoparticles (CS-NPs), prepared by an ionic gelation process (Lin et al. 2008). These nanoparticles were developed for a drug delivery system targeting the liver through a specific interaction between GL and hepatocytes.

Kidney-targeted drug delivery is critical when attempting to reduce extra-renal toxicity of the drug and to improve its therapeutic beneficiary for diseases occurring in the kidney. It may be particularly beneficial for drugs such as nonsteroidal antiinflammatory drugs (NSAIDs) (Vriesendorp et al. 1986). N-acetylated low molecular weight chitosan (LMWC) selectively accumulated in the kidneys, especially in the renal tubes after intravenous injection into mice (Yuan et al. 2007). In an attempt to develop drug delivery system for renal targeting, the authors conjugated prednisolone to LMWC (19 kDa) through a succinic acid spacer. The distribution of the conjugates in the kidney was found to be 13-fold higher than that of prednisolone alone. It was concluded that LMWC with a proper molecular weight could be applied as a promising carrier for renal targeting.

Lung cancer is one of the most prevalent cancers and is the leading cause of cancer mortality in the developed world (Jemal et al. 2009). Chitosan-modified poly (lactic-*co*-glycolic acid) nanoparticles containing paclitaxel (C-NPs paclitaxel) with a mean diameter of 200–300 nm by a solvent evaporation method (Yang et al. 2009b). This study demonstrated that the *in vitro* uptake of the nanoparticles by a lung cancer cell line (A549) was significantly increased by chitosan modification. The polymer drug conjugates are composed of a water-soluble polymer that is chemically conjugated to a drug via a biodegradable spacer. The spacer is usually stable in the bloodstream but cleaved at the target site by hydrolysis or enzymatic degradation. Several polymer–drug conjugates have recently entered into phase I/II clinical trials. The representative example is N-(2-hydroxypropyl) methacrylamide (HPMA) copolymer-based drug conjugates such as HPMA copolymer–doxorubicin conjugate and HPMA copolymer–doxorubicin conjugate containing galactosamine as a targeting moiety, developed for the treatment of primary or secondary liver cancer (Seymour et al. 1991).

Chitosan–anticancer drug conjugates have also been investigated. For example, doxorubicin-conjugated glycol chitosan (DOX–GC) with a *cis*-aconityl spacer was synthesized by chemical attachment of N-*cis*-aconityl DOX to GC using carbodiimide chemistry (Son et al. 2003). Low molecular weight chitosan conjugated with paclitaxel (LMWC-PTX) was also synthesized by chemical conjugation of LMWC and PTX through a succinate linker, which can be cleaved at physiological conditions (Lee et al. 2008a). This conjugate was evaluated as a carrier for the oral delivery of paclitaxel. LMWC (MW < 10 kDa) exhibited more favorable characteristics than high molecular weight chitosan, such as lower toxicity and higher water solubility. The N-succinylchitosan conjugates exhibited good antitumor activities against various tumors such as murine leukemias (L1210 and P388), B16 melanoma, Sarcoma 180 solid tumor, a murine liver metastatic tumor (M5076), and a murine hepatic cell carcinoma (MH134) (Kato et al. 2004).

Chitosan and its derivatives can be covalently cross-linked to prepare nano-sized particles as the drug carriers (Prabaharan and Mano 2004). The cross-linking process involves formation of the covalent bonds between the chitosan chains and functional cross-linking agents. The representative chemical cross-linkers that have been widely used for chitosan include bifunctional agents such as PEG dicarboxylic acid, glutaraldehyde, or monofunctional agents such as epichlorohydrin (Bodnar et al. 2005; Goldberg et al. 2007). For cancer therapy, a hydrophilic 5-fluorouracil was successfully loaded into chitosan nanoparticles (250–300 nm in diameter) using the water-in-oil emulsion method, followed by chemical cross-linking of the chitosan in the presence of glutaraldehydes (Ohya et al. 1994). Chitosan-based polyelectrolyte complex (PEC) nanoparticles prepared by electrostatic interactions between oppositely charged polyions, have received considerable attention as carrier systems for drug and gene delivery (Sun et al. 2008). Since chitosan is a hydrophilic and cationic polysaccharide, Kwon et al. (2003) developed hydrophobically modified glycol chitosans (HGCs) by covalent conjugation of bile acid (5α-cholanic acid or deoxycholic acid) to the backbone of glycol chitosan using carbodiimide chemistry.

PEGylated chitosan nanoparticles has attracted increasing attention because of its great potential in therapeutic applications (Bodnar et al. 2005). PEGylation of chitosan nanoparticles can increase their physical stability and prolong their circulation time in blood by reducing the removal by the reticuloendothelial system (Bodnar et al. 2006). PEGylated chitosan nanoparticles have been investigated as carriers for diverse small molecular drugs such as paclitaxel, camptothecin, methotrexate, and all-*trans* retinoic acid (ATRA) (Opanasopit et al. 2007; Qu et al. 2009; Yang et al. 2008). Active targeting receptor-mediated endocytosis (RME), targeting, and the specific receptors should be expressed exclusively on the cancer cells but not on the normal cells. Several targeting moieties or ligands have been identified and successfully utilized for chitosan-based drug delivery systems.

Folic acid, a low molecular weight (441 Da) vitamin, has a high affinity for folate receptors (FRs), which are frequently overexpressed in many types of human cancerous cells, particularly those found in the epithelial tumors of various organs such as colon, lung, prostate, and ovaries. Therefore, folate-conjugated drugs or carriers can be rapidly internalized into cancer cells via receptor-mediated endocytosis. You et al. (2008) developed folate conjugated stearic acid-grafted chitosan oligosaccharides (Fa-CSOSA) by reacting CSOSA with folic acid in the presence of carbodiimide coupling agents. Galactosylated chitosan-coated BSA nanoparticles containing 5-FU was used for the treatment of liver cancer (Zhang et al. 2008). Chitosan-based stimuli-sensitive formulations of physical targets by increasing efforts have been made to exploit physiological signals such as pH, temperature, ionic strength, and metabolites for targeted drug delivery applications (Rotin et al. 1989; Yahara et al. 2003). *N*-acetyl histidine conjugated glycol chitosan (NAcHis–GC), where histidine (with imidazole group, pKa value of 6.5) acting as pH-responsive fusogen, was developed for the efficient intra-cytoplasmic delivery of paclitaxel (Park et al. 2006). The NAcHis–GC conjugate formed self-assembled nanoparticles, with mean diameters of 150–250 nm, at neutral pH due to the hydrophobic nature of the NAcHis group. However, under slightly acidic conditions (similar to endosomes), the imidazole group of NAcHis gets protonated. This may induce the influx of water and ions into endosomes when the nanoparticles are taken up by the cells, causing disruption of endosomal membranes. As a consequence, the disassembled nanoparticles could release the encapsulated paclitaxel into the cytosol.

Magnetic targeting, an attractive physical targeting technique, is garnering substantial attention for drug delivery applications. Here, the therapeutic agents to be delivered are either immobilized on the surface or encapsulated into the magnetic micro- or nanoparticulate carriers. These magnetic carriers, on intravenous administration, concentrate at the specific site of interest (tumor site) using an external high-gradient magnetic field (Pankhurst et al. 2003). Targeted delivery of therapeutic agents to the brain has enormous potential for the treatment of several neurological disorders such as Alzheimer's disease and brain tumor. However, the blood–brain barrier (BBB) significantly impedes the entry of drug molecules into the brain from the bloodstream. Drug-loaded magnetic particulates represent a promising alternative strategy in overcoming the BBB. Gallo et al. developed magnetic chitosan microspheres containing oxantrazole (MCM-OX), an anticancer drug, for the treatment of brain tumors (Hassan and Gallo 1993). Chitosan bound magnetic nanoparticles loaded with epirubicin, an anthracycline drug is used for cancer chemotherapy (Chang et al. 2005).

13.4.5 Antiinflammatory Activity

Although several studies have widely investigated the effects of chitin, chitosan, and their derivatives, few investigating antiinflammatory activity have recently been published. Inflammation is a physiological body immune response against pathogens, toxic chemicals, or physical injury. Although acute inflammation is a short-term normal response that usually causes tissue repair by recruitment of leukocytes to the damaged region, chronic inflammation is a long-term pathological response involving induction of own tissue damage by matrix metalloproteinases (MMPs) (Drayton et al. 2006; Hu et al. 2007). It is generally well known that chronic inflammation is related to periodontal disease, hepatitis, arthritis, gastritis, and colitis. The most important factor in chronic inflammation has been known to be the nuclear factor-kappa B (NF-κB) transcription factor that plays a critical role in regulating genes involved in immune responses (Zhong et al. 2002). NF-κB is known to regulate inflammatory genes encoding proinflammatory cytokines, adhesion molecules, cyclooxygenase-2 (COX-2), and inducible nitric oxide synthase (iNOS) (Epstein et al. 1997; Baldwin Jr 2001). In particular, current approaches to the treatment of inflammation rely on the selective inhibition of COX-2 activity responsible for producing prostanoids, not COX-1. Nonsteroidal anti inflammatory drugs (NSAIDs) are the most widely prescribed drug for treatment of many inflam-matory diseases. However, they display a high incidence of gastric, renal, and hepatic side effects. In recent years, it has been reported that chronic inflammation is associated with an increased risk of malignant transformation (Macarthur et al. 2004). This is because phagocytic leukocytes in chronic inflammatory processes produce large amounts of reactive metabolites of oxygen and nitrogen that induce oxidative stress and lead to oxidation of fatty acids and proteins in cell membrane, thus impairing their normal function. Although the antiinflammatory effects of chitin and its derivatives have been rarely reported, in recent years data have been accumulating. First of all, it was found that chitin is a size-dependent regulator of inflammation (Da Silva et al. 2009). In this study, while both intermediate-sized chitin and small chitin stimulates TNF production in murine peritoneal macrophages, large chitin fragments are inert. Furthermore, it was found that chitin stimulates the expression of TLR2, dectin-1, the mannose receptor and inflammatory cytokines, differentially activated NF-κB, and spleen tyrosine kinase.

Ngo et al. (2009) demonstrated that chitin oligosaccharides can inhibit myeloperoxidase activity in human myeloid cells and oxidation of DNA and protein in mouse macrophages (Ngo et al. 2009). Chitosan was confirmed to partially inhibit the secretion of both IL-8 and TNF-α from mast cells, demonstrating that water-soluble chitosan has the potential to reduce the allergic inflamma-tory response (Kim et al. 2004). Because mast cells are necessary for allergic reactions and have been implicated in a number of neuroinflammatory diseases, chitosan nutraceuticals may help to prevent or alleviate some of these complications. In another study, it was demonstrated that COS enhanced migration of the mouse peritoneal macrophages into inflammatory areas (Moon et al. 2007). LPS-stimulated TNF-α and IL-6 secretion was found to be inhibited in the presence of chitosan oligosaccharide in RAW 264.7 cells (Yoon et al. 2007), suggesting that chitosan oligosac-charide may possess an antiinflammatory effect via the inhibition of TNF-α in the LPS-stimulated inflammation. These functions of chitosan to exert antiinflammatory effects could be utilized in the nutraceutical industry as well as in functional foods for prevention and alleviation of inflam-matory diseases. In addition, it was reported that chitosan promotes phagocytosis and production of osteopontin and leukotriene B by polymorphonuclear leukocytes, production of interleukin-1, transforming growth factor b1, and platelet-derived growth factor by macrophages, and production of interleukin-8 by fibroblasts, enhancing immune responses (Ueno et al. 2001). The effect of chi-tin, chitosan, and their derivatives on MMPs related to chronic inflammation is an interesting topic presently. MMPs are a family of secreted or transmembrane endopeptidases that degrade extracel-lular matrix components. It was described that COS inhibit activation and expression of MMP-2 in primary human dermal fibroblasts (Kim and Kim 2006). In particular, hydrolyzed chitosans with molecular weights as low as 3–5 kDa displayed the highest inhibitory effect on MMP-2. Moreover, the inhibitory effect might be described by the effective chelating capacity of chitosan for Zn^{2+} as a cofactor of MMP-2. Based on these findings, as a result of elucidation of the relationship between their activities and structures, atomic force microscopy demonstrated a direct molecular interac-tion between MMP-2 and chitosan. Affinity chromatography revealed a high-binding specificity of

MMP-2 to chitosan, and a colorimetric assay suggested a noncompetitive inhibition of MMP-2 by chitosan (Gorzelanny et al. 2007).

13.4.6 ANTIVIRAL ACTIVITY

Chitosan and COS are reported to suppress viral infections in various biological systems. In most cases, the mechanism of antiviral activity is not well understood. Furthermore, different studies carried out up-to-date involving factors that are presumed to be involved in determining the antiviral activity of chitosan and COS, have resulted in conflicting observations with regard to the inhibition mechanism. However, one possible explanation is that, cationic charges of amino groups of chitosan and COS may have additional functions to activate the immune and defense systems in plants and animals.

Researches on viral infections in plants have revealed that the treatment of chitosan on leaf surfaces can decrease the number of local necroses caused by different mosaic viruses (Pospieszny et al. 1991). In addition, they have also confirmed that the treatment employing chitosan could suppress infections regardless of the type of virus as well as plant species. Some plants with genetic resistance to particular viruses can form local lesions and that could prevent spreading of viruses. This defense mechanism is referred to as hypersensitivity response of host plants (Pospieszny and Atabekov 1989). A decreased local lesion after treatment of chitosan suggests that it could improve the plant resistance against viral infections. Further, antiviral activity is affected by molecular structural properties of chitosan derivatives. Generally, anionic derivatives of COS result in a lower antiviral activity and increased DD can improve antiviral activity indicating that positively charged groups of COS are responsible for their antiviral activity.

Chitosan and COS are reported to stimulate immune and defense systems in animal cells. Stimulation of functional activity of macrophages with the treatment of COS helps to increase the generation of active oxygen species in mouse models and these reactive radical species lead to viral destruction. Bacon et al. (2000) showed that co-treatments of chitosan with antigen to mice could strongly increase the local and systemic immune and defense responses to influenza A and B viruses. Another possible mechanism of antiviral activity of COS can be explained in relation to interactions between protein receptors on viral coat and blood leucocytes (Bacon et al. 2000). Sosa et al. (1991) observed that carboxymethyl and sulfated derivatives of chitosan could inhibit the replication of HIV-1 in cultured T-cells and human MT-4 lymphocytes. Furthermore, they have suggested that this activity was due to the prevention of interactions between viral coat glycoprotein receptors and target proteins on lymphocytes (Sosa et al. 1991).

COS are also effective in preventing several phage infections. Antiviral activity on phages is known to be dependent on properties of COS such as molecular weight, molecular structure, DD, and also type of bacteriophage. However, it is hard to predict the level of antiviral activity of COS only based on the above factors. Kochkina and Chirkov (2000) have shown that COS are more effective against replication of 1–97A phage in *Bacillus thuringiensis* than that of LMWC. In contrast, LMWC with higher polymerization degree has been observed to become more beneficial in inhibiting coliphage infection than that of LMWC with lower polymerization degree (Kochkina and Chirkov 2000). Therefore, it can be expected that COS and LMWC can be involved in the inhibition of the replication of bacteriophages by different mechanisms. Interestingly, electron microscopic observations have revealed that COS containing lower DP could change the structure of phage particles. Furthermore, these changes make the phage inactive directly through receptor-recognizing structures on the phage particles. However, the ability of COS to prevent phage infection is not well elucidated.

13.4.7 ANGIOTENSIN-I-CONVERTING ENZYME INHIBITION

Chitosan has also been reported to prevent increases in blood pressure. A high-salt diet can raise blood pressure because Cl⁻ activates angiotensin-converting enzyme (ACE), whereas chitosan can bind Cl⁻ and remove it, preventing the blood pressure from rising (Jeon et al. 2000). Moreover, other researchers found that chitosan oligomers also had ACE-inhibitory activity. They reported that the ACE-inhibitory activity of hetero-COS was dependent on the DD and that COS with the relatively lowest DD exhibited the highest ACE-inhibitory activity (Park et al. 2003); substitution of the hydrogen atom at the C-6 position of the pyranose residue with the aminoethyl group promoted the ACE-inhibitory effects of COS (800–3000 Da and 90%DD) (Ngo et al. 2008b).

13.4.8 OTHER BIOLOGICAL ACTIVITIES

Additionally, chitosan and its oligosaccharides have other biological functions such as excluding toxins from the intestines, reducing heavy-metal poisoning in humans, radio-protective properties, preventing tooth decay and tooth diseases, as a bifidus factor (BF) to regulate microbial metabolism in the intestines (Xia et al. 2011), antimutagenic effects (Nam et al. 2001), and so on.

13.5 CONCLUSION

In recent years, chitin and its derivatives—as a high potential resource as well as multiple functional substrates—have generated great interest in various fields, such as biomedical, pharmaceutical, food, and environmental industries. Although chitin is an insoluble polymer in water, which is the major limiting factor for its utilization in living systems, COS are more suited to draw attention for potentially biological applications due to the biocompatibility and nontoxic nature of chitosan. They exert an excellent antioxidant effect as well as antimicrobial effect. In particular, COS and their derivatives are potential candidates capable of preventing or treating diverse chronic inflam-mation such as colitis, periodontal disease, hepatitis, and gastritis, leading to cancer, and through drug delivery system.

ACKNOWLEDGMENTS

The authors acknowledge the Marine Bioprocess Research Center of the Marine Biotechnology Project, funded by the Ministry of Land, Transport and Maritime Affairs, Republic of Korea, for their support and encouragement.

REFERENCES

Agnihotri, SA, NN Mallikarjuna, and TM Aminabhavi. 2004. Recent advances on chitosan-based micro- and nanoparticles in drug delivery. *Journal of Controlled Release* 100(1):5–28.
Aiedeh, K and MO Taha. 2001. Synthesis of iron-crosslinked chitosan succinate and iron-crosslinked hydrox-amated chitosan succinate and their *in vitro* evaluation as potential matrix materials for oral theophylline sustained-release beads. *European Journal of Pharmaceutical Sciences* 13(2):159–168.
Akiyama, K, K Kawazu, and A Kobayashi. 1995. A novel method for chemo-enzymatic synthesis of elici-tor-active chitosan oligomers and partially *N*-deacetylated chitin oligomers using *N*-acylated chitotrio-ses as substrates in a lysozyme-catalyzed transglycosylation reaction system. *Carbohydrate Research* 279:151–160.
Amidi, M, SG Romeijn, G Borchard, HE Junginger, WE Hennink, and W Jiskoot. 2006. Preparation and char-acterization of protein-loaded *N*-trimethyl chitosan nanoparticles as nasal delivery system. *Journal of Controlled Release* 111(1):107–116.
Arnaud, TM, B de Barros Neto, and FB Diniz. 2010. Chitosan effect on dental enamel de-remineralization: An *in vitro* evaluation. *Journal of Dentistry* 38(11):848–852.
Baba, Y, H Noma, R Nakayama, and Y Matsushita. 2002. Preparation of chitosan derivatives containing methyl-thiocarbamoyl and phenylthiocarbamoyl groups and their selective adsorption of copper (II) over iron (III). *Analytical Sciences* 18(3):359–361.
Bacon, A, J Makin, PJ Sizer et al. 2000. Carbohydrate biopolymers enhance antibody responses to mucosally delivered vaccine antigens. *Infection and Immunity* 68(10):5764–5770.

Baldwin Jr, AS. 2001. Perspective series-NF-kB in defense and disease.The transcription factor NF-kB and human disease. *Journal of Clinical Investigation* 107(1):3–6.

Benhamou, N, PJ Lafontaine, and M Nicole. 1994. Induction of systemic resistance to Fusarium crown and root rot in tomato plants by seed treatment with chitosan. *Phytopathology* 84(12):1432–1444.

Bernkop-Schnürch, A and S Dünnhaupt. 2012. Chitosan based drug delivery systems. *European Journal of Pharmaceutics and Biopharmaceutics* 81(3):463–469.

Bernkop-Schnürch, A, D Guggi, and Y Pinter. 2004. Thiolated chitosans: Development and *in vitro* evaluation of a mucoadhesive, permeation enhancing oral drug delivery system. *Journal of Controlled Release* 94(1):177–186.

Bhaskara Reddy, MV, J Arul, P Angers, and L Couture. 1999. Chitosan treatment of wheat seeds induces resistance to *Fusarium graminearum* and improves seed quality. *Journal of Agricultural and Food Chemistry* 47(3):1208–1216.

Bodnar, M, JF Hartmann, and J Borbely. 2005. Preparation and characterization of chitosan-based nanoparticles. *Biomacromolecules* 6(5):2521–2527.

Bodnar, M, JF Hartmann, and J Borbely. 2006. Synthesis and study of cross-linked chitosan-*N*-poly (ethylene glycol) nanoparticles. *Biomacromolecules* 7(11):3030–3036.

Chang, DS, HR Cho, HY Goo, and WK Choe. 1989. A development of food preservative with the waste of crab processing. *Bulletin of the Korean Fisheries Society* 22(2):70–78.

Chang, HY, JJ Chen, F Fang, and Z Chen. 2004. Enhancement of antibody response by chitosan, a novel adjuvant of inactivated influenza vaccine. *Chinese Journal of Biologicals* 17(6):21–24.

Chang, Y-C, D-B Shieh, C-H Chang, and D-H Chen. 2005. Conjugation of monodisperse chitosan-bound magnetic nanocarrier with epirubicin for targeted cancer therapy. *Journal of Biomedical Nanotechnology* 1(2):196–201.

Chatelet, C, O Damour, and A Domard. 2001. Influence of the degree of acetylation on some biological properties of chitosan films. *Biomaterials* 22(3):261–268.

Chen, RH and JS Chen. 2000. Changes of polydispersity and limiting molecular weight of ultrasound-treated chitosan. *Advances in Chitin Science* 4: 361–366.

Chen, S, G Wu, and H Zeng. 2005. Preparation of high antimicrobial activity thiourea chitosan–Ag$^+$/complex. *Carbohydrate Polymers* 60(1):33–38.

Chen, X-G, Z Wang, W-S Liu, and H-J Park. 2002. The effect of carboxymethyl-chitosan on proliferation and collagen secretion of normal and keloid skin fibroblasts. *Biomaterials* 23(23):4609–4614.

Chourasia, MK and SK Jain. 2004. Polysaccharides for colon targeted drug delivery. *Drug Delivery* 11(2):129–148.

Cuero, RG, G Osuji, and A Washington. 1991. N-carboxymethylchitosan inhibition of aflatoxin production: role of zinc. *Biotechnology Letters* 13(6):441–444.

Da Silva, CA, C Chalouni, A Williams, D Hartl, CG Lee, and JA Elias. 2009. Chitin is a size-dependent regulator of macrophage TNF and IL-10 production. *The Journal of Immunology* 182(6):3573–3582.

Darmadji, P and M Izumimoto. 1994. Effect of chitosan in meat preservation. *Meat Science* 38(2):243–254.

DiMasi, JA, RW Hansen, and HG Grabowski. 2003. The price of innovation: New estimates of drug development costs. *Journal of Health Economics* 22(2):151–186.

Domard, A and M Domard. 2002. Chitosan: Structure–properties relationship and biomedical applications. *Polymeric Biomaterials* 9:187–212.

Drayton, DL, S Liao, RH Mounzer, and NH Ruddle. 2006. Lymphoid organ development: From ontogeny to neogenesis. *Nature Immunology* 7(4):344–353.

Dufes, C, J-M Muller, W Couet, J-C Olivier, IF Uchegbu, and AG Schätzlein. 2004. Anticancer drug delivery with transferrin targeted polymeric chitosan vesicles. *Pharmaceutical Research* 21(1):101–107.

Duncan, R. 2003. The dawning era of polymer therapeutics. *Nature Reviews Drug Discovery* 2(5):347–360.

Duncan, R. 2006. Polymer conjugates for drug targeting. From inspired to inspiration! Preface. *Journal of Drug Targeting* 14(6):333–335.

El Ghaouth, A, J Arul, A Asselin, and N Benhamou. 1992. Antifungal activity of chitosan on post-harvest pathogens: Induction of morphological and cytological alterations in *Rhizopus stolonifer. Mycological Research* 96(9):769–779.

Epstein, FH, PJ Barnes, and M Karin. 1997. Nuclear factor-κB—A pivotal transcription factor in chronic inflammatory diseases. *New England Journal of Medicine* 336(15):1066–1071.

Felt, O, P Buri, and R Gurny. 1998. Chitosan: A unique polysaccharide for drug delivery. *Drug Development and Industrial Pharmacy* 24(11):979–993.

Field, CE, LF Pivarnik, SM Barnett, and AG Rand. 1986. Utilization of glucose oxidase for extending the shelf-life of fish. *Journal of Food Science* 51(1):66–70.

Franco, TT and MG Peter. 2011. Advances in chitin and chitosan research. *Polymer International* 60(6):873–874.

Franklin, TJ and GA Snow. 1981. *Biochemistry of Antimicrobial Action*. Chapman & Hall. London.

Goldberg, M, R Langer, and X Jia. 2007. Nanostructured materials for applications in drug delivery and tissue engineering. *Journal of Biomaterials Science, Polymer Edition* 18(3):241–268.

Gorzelanny, C, B Pöppelmann, E Strozyk, BM Moerschbacher, and SW Schneider. 2007. Specific interaction between chitosan and matrix metalloprotease 2 decreases the invasive activity of human melanoma cells. *Biomacromolecules* 8(10):3035–3040.

Hai, L, TB Diep, N Nagasawa, F Yoshii, and T Kume. 2003. Radiation depolymerization of chitosan to prepare oligomers. *Nuclear Instruments and Methods in Physics Research Section B: Beam Interactions with Materials and Atoms* 208:466–470.

Hassan, EE and JM Gallo. 1993. Targeting anticancer drugs to the brain. I: Enhanced brain delivery of oxantrazole following administration in magnetic cationic microspheres. *Journal of Drug Targeting* 1(1):7–14.

Hejazi, R and M Amiji. 2003. Chitosan-based gastrointestinal delivery systems. *Journal of Controlled Release* 89(2):151–165.

Hirano, S. 1996. Chitin biotechnology applications. *Biotechnology Annual Review* 2:237–258.

Hu, J, PE Van den Steen, Q-XA Sang, and G Opdenakker. 2007. Matrix metalloproteinase inhibitors as therapy for inflammatory and vascular diseases. *Nature Reviews Drug Discovery* 6(6):480–498.

Huang, R, E Mendis, N Rajapakse, and S-K Kim. 2006. Strong electronic charge as an important factor for anticancer activity of chitooligosaccharides (COS). *Life Sciences* 78(20):2399–2408.

Il'ina, AV and VP Varlamov. 2004. Hydrolysis of chitosan in lactic acid. *Applied Biochemistry and Microbiology* 40(3):300–303.

Ishii, T, Y Okahata, and T Sato. 2001. Mechanism of cell transfection with plasmid/chitosan complexes. *Biochimica et Biophysica Acta (BBA)-Biomembranes* 1514(1):51–64.

Jain, SK and A Jain. 2008. Target-specific drug release to the colon. *Expert Opinion on Drug Delivery* 5(5):483–498.

Jam, SK, A Jain, Y Gupta, P Khare, and M Kannandasan. 2008. Targeted delivery of 5-ASA to colon using chitosan hydrogel microspheres. *Journal of Drug Delivery Science and Technology* 18(5):315–321.

Jemal, A, R Siegel, E Ward, Y Hao, J Xu, and MJ Thun. 2009. Cancer statistics, 2009. *CA: A Cancer Journal for Clinicians* 59(4):225–249.

Jeon, Y-J, F Shahidi, and S-K Kim. 2000. Preparation of chitin and chitosan oligomers and their applications in physiological functional foods. *Food Reviews International* 16(2):159–176.

Jia, Z and W Xu. 2001. Synthesis and antibacterial activities of quaternary ammonium salt of chitosan. *Carbohydrate Research* 333(1):1–6.

Kato, Y, H Onishi, and Y Machida. 2004. *N*-succinyl-chitosan as a drug carrier: Water-insoluble and water-soluble conjugates. *Biomaterials* 25(5):907–915.

Kato, Y, H Onishi, and Y Machida. 2005. Contribution of chitosan and its derivatives to cancer chemotherapy. *In Vivo* 19(1):301–310.

Kendra, DF and LA Hadwiger. 1984. Characterization of the smallest chitosan oligomer that is maximally antifungal to *Fusarium solani* and elicits pisatin formation in *Pisum sativum*. *Experimental Mycology* 8(3):276–281.

Kim, C-H, S-Y Kim, and K-S Choi. 1997. Synthesis and antibacterial activity of water-soluble chitin derivatives. *Polymers for Advanced Technologies* 8(5):319–325.

Kim, J-H, Y-S Kim, K Park et al. 2008. Antitumor efficacy of cisplatin-loaded glycol chitosan nanoparticles in tumor-bearing mice. *Journal of Controlled Release* 127(1):41–49.

Kim, K, S Kwon, JH Park et al. 2005. Physicochemical characterizations of self-assembled nanoparticles of glycol chitosan-deoxycholic acid conjugates. *Biomacromolecules* 6(2):1154.

Kim, M-S, HJ You, MK You, N-S Kim, BS Shim, and H-M Kim. 2004. Inhibitory effect of water-soluble chitosan on TNF-α and IL-8 secretion from HMC-1 cells. *Immunopharmacology and Immunotoxicology* 26(3):401–409.

Kim, M-M and S-K Kim. 2006. Chitooligosaccharides inhibit activation and expression of matrix metalloproteinase-2 in human dermal fibroblasts. *FEBS Letters* 580(11):2661–2666.

Kim, S-K and N Rajapakse. 2005. Enzymatic production and biological activities of chitosan oligosaccharides (COS): A review. *Carbohydrate Polymers* 62(4):357–368.

Knorr, D. 1984. Use of chitinous polymers in food: A challenge for food research and development. *Food Technology* 38:85–87.

Kochkina, ZM and SN Chirkov. 2000. Effect of chitosan derivatives on the development of phage infection in cultured *Bacillus thuringiensis*. *Mikrobiologiia* 69(2):266–269.

Kulikov, SN, SN Chirkov, AV Il'ina, SA Lopatin, and VP Varlamov. 2006. Effect of the molecular weight of chitosan on its antiviral activity in plants. *Applied Biochemistry and Microbiology* 42(2):200–203.

Kumar, MNVR, RA_A Muzzarelli, C Muzzarelli, H Sashiwa, and AJ Domb. 2004. Chitosan chemistry and pharmaceutical perspectives. *Chemical Reviews-Columbus* 104(12):6017–6084.

Kurita, K. 1998. Chemistry and application of chitin and chitosan. *Polymer Degradation and Stability* 59(1):117–120.

Kuroiwa, T, S Ichikawa, O Hiruta, S Sato, and S Mukataka. 2002. Factors affecting the composition of oligosaccharides produced in chitosan hydrolysis using immobilized chitosanases. *Biotechnology Progress* 18(5):969–974.

Kwon, S, JH Park, H Chung, IC Kwon, SY Jeong, and I-S Kim. 2003. Physicochemical characteristics of self-assembled nanoparticles based on glycol chitosan bearing 5β-cholanic acid. *Langmuir* 19(24):10188–10193.

Laflamme, P, N Benhamou, G Bussières, and M Dessureault. 2000. Differential effect of chitosan on root rot fungal pathogens in forest nurseries. *Canadian Journal of Botany* 77(10):1460–1468.

Langer, R. 1998. Drug delivery and targeting. *Nature* 392(6679):5–9.

Lavall, RL, OBG Assis, and SP Campana-Filho. 2007. β-Chitin from the pens of *Loligo* sp.: Extraction and characterization. *Bioresource Technology* 98(13):2465–2472.

Lee, W-L, H-T Chao, M-H Cheng, and P-H Wang. 2008a. Rationale for using raloxifene to prevent both osteoporosis and breast cancer in postmenopausal women. *Maturitas* 60(2):92–107.

Lee, D-X, W-S Xia, and J-L Zhang. 2008b. Enzymatic preparation of chitooligosaccharides by commercial lipase. *Food Chemistry* 111(2):291–295.

Li, Z, XP Zhuang, XF Liu, YL Guan, and K De Yao. 2002. Study on antibacterial *O*-carboxymethylated chitosan/cellulose blend film from LiCl/N, N-dimethylacetamide solution. *Polymer* 43(4):1541–1547.

Lin, A, Y Liu, Y Huang et al. 2008. Glycyrrhizin surface-modified chitosan nanoparticles for hepatocyte-targeted delivery. *International Journal of Pharmaceutics* 359(1):247–253.

Lin, S-B, Y-C Lin, and H-H Chen. 2009. Low molecular weight chitosan prepared with the aid of cellulase, lysozyme and chitinase: Characterisation and antibacterial activity. *Food Chemistry* 116(1):47–53.

Liu, J-n, W-s Xia, and J-l Zhang. 2008. Effects of chitosans physico-chemical properties on binding capacities of lipid and bile salts in vitro. *Food Science* 1:45–49.

Macarthur, M, GL Hold, and EM El-Omar. 2004. Inflammation and Cancer II. Role of chronic inflammation and cytokine gene polymorphisms in the pathogenesis of gastrointestinal malignancy. *American Journal of Physiology-Gastrointestinal and Liver Physiology* 286(4):G515–G520.

Mao, S, X Shuai, F Unger, M Simon, D Bi, and T Kissel. 2004. The depolymerization of chitosan: Effects on physicochemical and biological properties. *International Journal of Pharmaceutics* 281(1):45–54.

Mellman, I, G Coukos, and G Dranoff. 2011. Cancer immunotherapy comes of age. *Nature* 480(7378): 480–489.

Moon, J-S, H-K Kim, HC Koo et al. 2007. The antibacterial and immunostimulative effect of chitosan-oligosaccharides against infection by *Staphylococcus aureus* isolated from bovine mastitis. *Applied Microbiology and Biotechnology* 75(5):989–998.

Morganti, P and G Morganti. 2008. Chitin nanofibrils for advanced cosmeceuticals. *Clinics in Dermatology* 26(4):334–340.

Muzzarelli, R, R Tarsi, O Filippini, E Giovanetti, G Biagini, and PE Varaldo. 1990. Antimicrobial properties of *N*-carboxybutyl chitosan. *Antimicrobial Agents and Chemotherapy* 34(10):2019–2023.

Muzzarelli, RAA, P Ilari, R Tarsi, B Dubini, and W Xia. 1994. Chitosan from *Absidia coerulea*. *Carbohydrate Polymers* 25(1):45–50.

Muzzarelli, RAA, C Muzzarelli, R Tarsi, M Miliani, F Gabbanelli, and M Cartolari. 2001. Fungistatic activity of modified chitosans against *Saprolegnia parasitica*. *Biomacromolecules* 2(1):165–169.

Nam, K-S, Y-R C and Y-H Shon. 2001. Evaluation of the antimutagenic potential of chitosan oligosaccharide: Rec, Ames and Umu tests. *Biotechnology Letters* 23(12):971–975.

Ngo, Dai-Nghiep, Moon-Moo Kim, and Se-Kwon Kim. 2008a. Chitin oligosaccharides inhibit oxidative stress in live cells. *Carbohydrate Polymers* 74(2):228–234.

Ngo, D-N, S-H Lee, M-M Kim, and S-K Kim. 2009. Production of chitin oligosaccharides with different molecular weights and their antioxidant effect in RAW 264.7 cells. *Journal of Functional Foods* 1(2):188–198.

Ngo, D-N, Z-J Qian, J-Y Je, M-M Kim, and S-K Kim. 2008b. Aminoethyl chitooligosaccharides inhibit the activity of angiotensin converting enzyme. *Process Biochemistry* 43(1):119–123.

Nishimura, K, S Nishimura, N Nishi, I Saiki, S Tokura, and I Azuma. 1984. Immunological activity of chitin and its derivatives. *Vaccine* 2(1):93–99.

Ogawara, K-i, M Yoshida, K Higaki et al. 1999. Hepatic uptake of polystyrene microspheres in rats: Effect of particle size on intrahepatic distribution. *Journal of Controlled Release* 59(1):15–22.

Ohya, Y, M Shiratani, H Kobayashi, and T Ouchi. 1994. Release behavior of 5-fluorouracil from chitosan-gel nanospheres immobilizing 5-fluorouracil coated with polysaccharides and their cell specific cytotoxicity. *Journal of Macromolecular Science—Pure and Applied Chemistry* 31(5):629–642.

Opanasopit, P, T Ngawhirunpat, T Rojanarata, C Choochottiros, and S Chirachanchai. 2007. N-Phthaloylchitosan-g-mPEG design for all-*trans* retinoic acid-loaded polymeric micelles. *European Journal of Pharmaceutical Sciences* 30(5):424–431.

Pankhurst, QA, J Connolly, SK Jones, and J Dobson. 2003. Applications of magnetic nanoparticles in biomedicine. *Journal of Physics D: Applied Physics* 36(13):R167.

Papineau, AM, DG Hoover, D Knorr, and DF Farkas. 1991. Antimicrobial effect of water-soluble chitosans with high hydrostatic pressure. *Food Biotechnology* 5(1):45–57.

Park, JS, TH Han, K Yong Lee et al. 2006. N-acetyl histidine-conjugated glycol chitosan self-assembled nanoparticles for intracytoplasmic delivery of drugs: Endocytosis, exocytosis and drug release. *Journal of Controlled Release* 115(1):37–45.

Park, P-J, J-Y Je, and S-K Kim. 2003. Angiotensin I converting enzyme (ACE) inhibitory activity of heterochitooligosaccharides prepared from partially different deacetylated chitosans. *Journal of Agricultural and Food Chemistry* 51(17):4930–4934.

Park, P-J, J-Y Je, and S-K Kim. 2004. Free radical scavenging activities of differently deacetylated chitosans using an ESR spectrometer. *Carbohydrate Polymers* 55(1):17–22.

Patel, MP, RR Patel, and JK Patel. 2010. Chitosan mediated targeted drug delivery system: A review. *Journal of Pharmacy & Pharmaceutical Sciences* 13(4):536–557.

Peniche-Covas, C, LW Alvarez, and W Argüelles-Monal. 1992. The adsorption of mercuric ions by chitosan. *Journal of Applied Polymer Science* 46(7):1147–1150.

Peniston, QP and EL Johnson. 1980. *Process for the manufacture of chitosan*. US Patent No. 4,195,175 pp. 5–15.

Pospieszny, H and JG Atabekov. 1989. Effect of chitosan on the hypersensitive reaction of bean to alfalfa mosaic virus. *Plant Science* 62(1):29–31.

Pospieszny, H, S Chirkov, and J Atabekov. 1991. Induction of antiviral resistance in plants by chitosan. *Plant Science* 79(1):63–68.

Prabaharan, M and JF Mano. 2004. Chitosan-based particles as controlled drug delivery systems. *Drug Delivery* 12(1):41–57.

Qu, G, Z Yao, C Zhang, X Wu, and Q Ping. 2009. PEG conjugated N-octyl-O-sulfate chitosan micelles for delivery of paclitaxel: *in vitro* characterization and *in vivo* evaluation. *European Journal of Pharmaceutical Sciences* 37(2):98–105.

Rabea, EI, ME-T Badawy, CV Stevens, G Smagghe, and W Steurbaut. 2003. Chitosan as antimicrobial agent: applications and mode of action. *Biomacromolecules* 4(6):1457–1465.

Razdan, A and D Pettersson. 1994. Effect of chitin and chitosan on nutrient digestibility and plasma lipid concentrations in broiler chickens. *British Journal of Nutrition* 72(2):277–288.

Ribeiro, MP, A Espiga, D Silva et al. 2009. Development of a new chitosan hydrogel for wound dressing. *Wound Repair and Regeneration* 17(6):817–824.

Rotin, D, D Steele-Norwood, S Grinstein, and I Tannock. 1989. Requirement of the Na + /H+ exchanger for tumor growth. *Cancer Research* 49(1):205–211.

Sang Y, Hyuk, JE Lee, H Chung, IC Kwon, and SY Jeong. 2005. Self-assembled nanoparticles containing hydrophobically modified glycol chitosan for gene delivery. *Journal of Controlled Release* 103(1):235–243.

Savard, T, C Beaulieu, I Boucher, and CP Champagne. 2002. Antimicrobial action of hydrolyzed chitosan against spoilage yeasts and lactic acid bacteria of fermented vegetables. *Journal of Food Protection* 65(5):828–833.

Seymour, LW, K Ulbrich, SR Wedge, IC Hume, J Strohalm, and R Duncan. 1991. N-(2-hydroxypropyl) methacrylamide copolymers targeted to the hepatocyte galactose-receptor: pharmacokinetics in DBA2 mice. *British Journal of Cancer* 63(6):859.

Sharon, N and H Lis. 1993. Carbohydrates in cell recognition. *Scientific American* 268(1):82–89.

Simpson, BK, N Gagne, INA Ashie, and E Noroozi. 1997. Utilization of chitosan for preservation of raw shrimp (*Pandalus borealis*). *Food Biotechnology* 11(1):25–44.

Son, YJ, J-S Jang, Y Woo Cho et al. 2003. Biodistribution and anti-tumor efficacy of doxorubicin loaded glycol-chitosan nanoaggregates by EPR effect. *Journal of Controlled Release* 91(1):135–145.

Sosa, MAG, F Fazely, JA Koch, SV Vercellotti, and RM Ruprecht. 1991. N-Carboxymethylchitosan-N, O-sulfate as an anti-HIV-1 agent. *Biochemical and Biophysical Research Communications* 174(2):489–496.

Stössel, P and JL Leuba. 1984. Effect of chitosan, chitin and some aminosugars on growth of various soilborne phytopathogenic fungi. *Journal of Phytopathology* 111(1):82–90.

Strand, SP, KM Vårum, and K Østgaard. 2003. Interactions between chitosans and bacterial suspensions: Adsorption and flocculation. *Colloids and Surfaces B: Biointerfaces* 27(1):71–81.

Sudarshan, NR, DG Hoover, and D Knorr. 1992. Antibacterial action of chitosan. *Food Biotechnology* 6(3):257–272.

Sun, W, S Mao, D Mei, and T Kissel. 2008. Self-assembled polyelectrolyte nanocomplexes between chitosan derivatives and enoxaparin. *European Journal of Pharmaceutics and Biopharmaceutics* 69(2):417–425.

Suzuki, K, T Mikami, Y Okawa, A Tokoro, S Suzuki, and M Suzuki. 1986. Antitumor effect of hexa-*N*-acetyl-chitohexaose and chitohexaose. *Carbohydrate Research* 151:403.

Synowiecki, J and NA Al-Khateeb. 2003. Production, properties, and some new applications of chitin and its derivatives. *Critical Reviews in Food Science and Nutrition* 43(2):145–171.

Tokoro, A., Tatewaki, N., Suzuki, K., Mikami, T., Suzuki, S., Suzuki, M. 1988. Growth-inhibitory effect of hexa-*N*-acetylchitohexanse and chitohexaose against Meth-A solid tumor. *Chemical & Pharmaceutical Bulletin* 36(2):784–790.

Torchilin, V. 2008. Antibody-modified liposomes for cancer chemotherapy 5(9): 1003–1025.

Torres, J, M Motoki, and M Karel. 1985. Microbial stabilization of intermediate moisture food surfaces I. Control of surface preservative concentration. *Journal of Food Processing and Preservation* 9(2):75–92.

Tozaki, H, T Odoriba, N Okada et al. 2002. Chitosan capsules for colon-specific drug delivery: Enhanced localization of 5-aminosalicylic acid in the large intestine accelerates healing of TNBS-induced colitis in rats. *Journal of Controlled Release* 82(1):51–61.

Tsai, G-J, Z-Y Wu, and W-H Su. 2000. Antibacterial activity of a chitooligosaccharide mixture prepared by cellulase digestion of shrimp chitosan and its application to milk preservation. *Journal of Food Protection* 63(6):747–752.

Ueno, H, T Mori, and T Fujinaga. 2001. Topical formulations and wound healing applications of chitosan. *Advanced Drug Delivery Reviews* 52(2):105–115.

Vanneman, M and G Dranoff. 2012. Combining immunotherapy and targeted therapies in cancer treatment. *Nature Reviews Cancer* 12(4):237–251.

Vriesendorp, R, AJM Donker, D de Zeeuw, PE de Jong, GK van der Hem, and JRH Brentjens. 1986. Effects of nonsteroidal anti inflammatory drugs on proteinuria. *The American Journal of Medicine* 81(2):84–94.

Wang, G-H. 1992. Inhibition and inactivation of five species of foodborne pathogens by chitosan. *Journal of Food Protection* 55: 916.

Wang, S-H and J-C Chen. 2005. The protective effect of chitin and chitosan against *Vibrio alginolyticus* in white shrimp *Litopenaeus vannamei*. *Fish & Shellfish Immunology* 19(3):191–204.

Wang, Y-S, Q Jiang, R-S Li et al. 2008. Self-assembled nanoparticles of cholesterol-modified O-carboxymethyl chitosan as a novel carrier for paclitaxel. *Nanotechnology* 19(14):145101.

Wu, S, X Liu, A Yeung et al. 2011. Plasma-modified biomaterials for self-antimicrobial applications. *ACS Applied Materials & Interfaces* 3(8):2851–2860.

Xia, W, P Liu, J Zhang, and J Chen. 2011. Biological activities of chitosan and chitooligosaccharides. *Food Hydrocolloids* 25(2):170–179.

Yahara, T, T Koga, S Yoshida, S Nakagawa, H Deguchi, and K Shirouzu. 2003. Relationship between microvessel density and thermographic hot areas in breast cancer. *Surgery Today* 33(4):243–248.

Yang, KW, XR Li, ZL Yang, PZ Li, F Wang, and Y Liu. 2009a. Novel polyion complex micelles for liver targeted delivery of diammonium glycyrrhizinate: *In vitro* and *in vivo* characterization. *Journal of Biomedical Materials Research Part A* 88(1):140–148.

Yang, R, S-G Yang, W-S Shim et al. 2009b. Lung-specific delivery of paclitaxel by chitosan-modified PLGA nanoparticles via transient formation of microaggregates. *Journal of Pharmaceutical Sciences* 98(3):970–984.

Yang, X, Q Zhang, Y Wang et al. 2008. Self-aggregated nanoparticles from methoxy poly (ethylene glycol)-modified chitosan: Synthesis; characterization; aggregation and methotrexate release *in vitro*. *Colloids and Surfaces B: Biointerfaces* 61(2):125–131.

Yoon, HJ, ME Moon, HS Park, SY Im, and YH Kim. 2007. Chitosan oligosaccharide (COS) inhibits LPS-induced inflammatory effects in RAW 264.7 macrophage cells. *Biochemical and Biophysical Research Communications* 358(3):954–959.

You, J, X Li, F de Cui, Y-Z Du, H Yuan, and F qiang Hu. 2008. Folate-conjugated polymer micelles for active targeting to cancer cells: Preparation, *in vitro* evaluation of targeting ability and cytotoxicity. *Nanotechnology* 19(4):045102.

Yuan, Z-X, X Sun, T Gong, H Ding, Y Fu, and Z-R Zhang. 2007. Randomly 50% N-acetylated low molecular weight chitosan as a novel renal targeting carrier. *Journal of Drug Targeting* 15(4):269–278.

Zhang, C, Yao C, G Qu et al. 2008. Preparation and characterization of galactosylated chitosan coated BSA microspheres containing 5-fluorouracil. *Carbohydrate Polymers* 72(3):390–397.

Zhang, U, Y Du, X Yu, M Mitsutomi, and S-i Aiba. 1999. Preparation of chitooligosaccharides from chitosan by a complex enzyme. *Carbohydrate Research* 320(3):257–260.

Zhong, H, MJ May, E Jimi, and S Ghosh. 2002. The phosphorylation status of nuclear NF-κB determines its association with CBP/p300 or HDAC-1. *Molecular Cell* 9(3):625–636.

14 Chitin and Chitosan Derivatives for Wound-Healing Applications

Willi Paul, R. Deepa, T. V. Anilkumar, and Chandra P. Sharma

CONTENTS

14.1 INTRODUCTION

The skin is considered the largest organ of the body and has many different functions. The epidermis or outer layer is made up of mostly dead cells with a protein called keratin. This makes the layer waterproof and is responsible for protection against the environment. The dermis or middle layer is made up of living cells. It also has blood vessels and nerves that run through it and is primarily responsible for structure and support. The subcutaneous fat layer is responsible for insulation and shock absorbency. Cells on the surface of the skin are constantly being replaced by regeneration from below with the top layers sloughing off. Wound is an injury or a break in the skin due to surgical procedures or by external aggression or an involuntary activity. The repair of an epithelial wound is merely a scaling up of the normal process. Science of wound healing is recorded as "three healing gestures" on a clay tablet, one of the oldest medical texts dated 2200 BC. It describes the three gestures as washing the wound; making plasters; and bandaging the wound. Although there has been a significant advancement in today's science of wound healing, the basic theme seems to be similar. The work of Joseph Lister and Louis Pasteur established a sound basis for the management of infection by identifying the cause and developing methods for preventing it (Cohen 1998). Louis Pasteur proved that bacteria did not spontaneously generate but were introduced into wounds from a foreign source. These findings encouraged Lister's advocacy of frequent washing with soap and water and fueled his search for ways to kill bacteria, or the "antiseptic technique"—a major advance in the field of wound healing. The antiseptic technique was followed shortly by the "aseptic technique," in which a sterile environment was used to prevent the onset of infection.

Wounds are generally classified as wounds without tissue loss (e.g., in surgery) and wounds with tissue loss, such as burn wounds, wounds caused as a result of trauma and abrasions, or as secondary events in chronic ailments, for example, venous stasis, diabetic ulcers or pressure sores, and iatrogenic wounds such as skin graft donor sites and dermabrasions. Wounds are also classified by the layers involved: superficial wounds involve only the epidermis, partial thickness wounds involve only the epidermis and the dermis, and full thickness wounds involve the subcutaneous fat or deeper tissue. Although restoration of tissue continuity after injury is a natural phenomenon, infection, quality of healing, speed of healing, fluid loss, and other complications that enhance the healing time represents a major clinical challenge. The majority of wounds heal without any complication. However, chronic nonhealing wounds involving progressively more tissue loss give rise to the biggest challenge to wound-care product researchers. Unlike surgical incisions where there is very little tissue loss and that are easy to heal, chronic wounds disrupt the normal process of healing and are often not sufficient in themselves to effect repair. Delayed healing is generally a result of compromised wound physiology (Winter 1962) and typically occurs with venous stasis, diabetes, or prolonged local pressure. The second major challenge is the prevention of scarring, keloid formation or contractures, and a cosmetically acceptable healing.

14.2   GLOBAL WOUND MARKET

In the year 2009, the global wound-care market was worth US$5.1 billion with an annual growth rate that remained just below double digit (Maddox 2012). This projection was for 234 million surgical procedures and other chronic wounds, such as 40–50 million leg ulcers, 35–40 million pressure sores, and an equal number of burn wounds. The use of growth factors to accelerate the healing of wounds offers tremendous promise as a therapeutic approach in treating chronic wounds. The anticipated wound-care products market in 2015 is expected to reach approximately US$20.3 billion, driven by an aging population and rising incidence of modern epidemics such as diabetes and obesity. Asia-Pacific, the fastest-growing market for wound management products, is expected to reach US$644.7 million by the year 2015.

Today, clinicians are unable to recommend the use of advanced wound-care products, although they are more conducive to healing, due to their high cost. However, using these products enables fewer dressing changes with less care, and because of they facilitate faster healing it reduces the resources required. The analysis of wound-care market indicates the gradual shift toward the advanced wound-care products.

14.3   BIOCHEMICAL PROCESS IN WOUND HEALING

The wound-healing process may be divided into four continuous phases, namely, *hemostasis, inflammation, proliferation*, and *maturation* or *remodeling* (Johnstone and Farley 2005). The process of healing initiates as soon as a tissue is injured. The platelets in the blood come in contact with the collagen and other elements of the extracellular matrix. The platelets present in the exposed blood aggregates and a temporary plug is formed, reducing bleeding. Following *hemostasis*, the neutrophils then enter the wound site. The phagocytes act to clear debris, bacteria, and damaged tissue, and destroy the ingested material. As part of this *inflammatory* phase, the macrophages appear and continue the process of phagocytosis. New vessels are formed and carry oxygenated blood to the wound bed. Once the wound site is cleaned, fibroblasts migrate to begin the *proliferative* phase and lay down a network of collagen fibers surrounding the neovasculature of the wound. Finally, the process of *remodeling* of the collagen fibers laid down in the proliferation phase occurs, which may take years.

14.4   CHITIN AND CHITOSAN

Chitin and chitosan polymers are natural amino-polysaccharides having unique structures, multi-dimensional properties, highly sophisticated functions, and wide-ranging applications in biomedical and other industrial areas (Chandy and Sharma 1990; Paul and Sharma 2000; Muzzarelli and

Muzzarelli 2005). They are considered to be materials of great futuristic potential with immense possibilities for structural modifications, to impart desired properties and functions. Research and development work on chitin and chitosan have reached a stage of intense activities in many parts of the world (Khor 2002; Rinaudo 2006, 2008; Prashanth and Tharanathan 2007). They have excellent biocompatibility and admirable biodegradability with ecological safety and low toxicity. The versatile biological activities such as antimicrobial activity and low immunogenicity have provided ample opportunities for further development (Hirano 1999; Yi et al. 2005; Kurita 2006; Jayakumar et al. 2007; Mourya and Inamdar 2008; Rinaudo 2008). This underutilized resource has become a new functional biomaterial of high potential in various fields (Kurita 1995; Hirano 1996; Kumar et al. 2004). With data emerging from not less than 20 books, more than 300 reviews, more than 12,000 publications, and innumerable patents, the science and technology of these biopolymers are at a turning point where one needs a very critical look on their potential to deliver the goods (Zohuriaan-Mehr 2005; Chen 2006). There have been various attempts to develop wound-healing devices using chitosan because of its inherent wound healing potential. Both chitin and chitosan possess many properties that are advantageous for wound healing, such as biocompatibility, biodegradability (Tomihata and Ikada 1997), hemostatic activity (Abhay 1998), healing acceleration, nontoxicity, adsorption properties, and anti-infection properties (Sathirakul et al. 1996; Suzuki et al. 2000; Ueno et al. 2001b). Chitosan provides a nonprotein matrix for three-dimensional (3D) tissue growth and activates macrophages for tumoricidal activity. It stimulates cell proliferation and histoarchitectural tissue organization (Jayasree et al. 1995). Chitosan is a hemostat, which helps in natural blood clotting and blocks nerve endings, reducing pain (Figure 14.1). Chitosan will gradually depolymerize to release N-acetyl-β-D-glucosamine, which initiates fibroblast proliferation and helps in ordered collagen deposition and stimulates increased level of natural hyaluronic acid synthesis at the wound site. It helps in faster wound healing and scar prevention.

Various dressings based on chitin and chitosan for wound healing has been developed and are in the market. These are available in the form of nonwovens, nanofibrils, composites, films, and

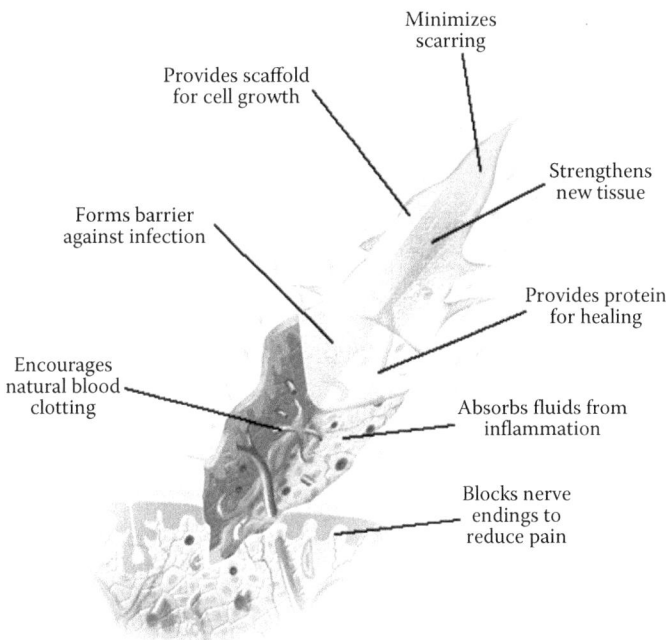

FIGURE 14.1 Schematic representation of the benefits of chitosan in wound healing.

sponges. The most prominent among them is the HemCon® hemostatic bandage (FDA approved in 2002), which was extensively used and claimed to have saved the life of hundreds of U.S. soldiers serving in Iraq or Afghanistan. However, because of mixed results, these hemostatic bandages once hailed as groundbreaking were found largely ineffective and abandoned. The same dressing is now commercially available as a topical antimicrobial dressing having favorable effects on healing of excision wounds. ChitoFlex® hemostatic dressings are also similar to HemCon but available in stuffable strip form. Chitopack C is a cotton-type wound dressing commercialized by Eisai Co. Ltd., Japan in 1993. Chitipack P is a β-chitin-based dressing from the same company. However, it was mainly used for veterinary applications. The Chito-Seal®, a topical hemostasis pad marketed by Abbott Vascular Inc., is intended for use in the management of bleeding wounds, such as vascular access sites and percutaneous catheters or tubes. The pad is coated with Abbott's proprietary chitosan gel. When placed over the puncture site, it becomes a powerful cell-binding agent consisting of positively charged chitosan molecules that attract negatively charged red blood cells and platelets, thus accelerating clot formation and hemostasis. Chitopoly (Fujibo Holdings Inc., Japan) is a fabric consisting of chitosan and polynosic, which is antibacterial and antideodorant, causing no irritation to the skin. Tegasorb™ wound dressing by 3M contains chitosan, which is used with partial and full thickness dermal ulcers, leg ulcers, superficial wounds, abrasions, burns, and donor sites. Chitosan interacts with wound fluid and a soft semitransparent absorbent mass is produced that enhances wound healing. TraumaStat™ hemostatic wound dressing is a unique nonwoven substrate composed of porous polyethylene fibers highly filled with precipitated silica. This substrate is coated with chitosan, manufactured from ChitoClear™, a purest form of chitosan. Clo-Sur® pad is a nonwoven construct sealed by a soluble form of chitosan used as an antimicrobial barrier. This device has received FDA clearance for use in local management of bleeding wounds, such as vascular access sites. ChiGel is a chitosan-based wound dressing that is soluble at neutral environment and forms a gel at pH 7.4 and 37°C. ChiGel is developed by a company in Israel and is indicated for diabetic wound healing, osteoarthritis, and rotator cuff. It is employed as a topical therapy and strongly adheres to the wound and helps stimulate the natural healing process. It also supports the wound's moist environment while allowing critical gas exchange. Studies have shown that it stimulates and supports the intrinsic healing process of slow or nonhealing wounds.

14.5   CHEMICAL MODIFICATIONS

Chitin and chitosan are interesting polysaccharides because of the presence of the amino functionality, which could be suitably modified to impart desired properties and distinctive biological functions, including solubility (Muzzarelli 1977; Roberts 1992; Hudson and Smith 1998; Kurita 2001; Peter 2002; Tharanathan and Kittur 2003; Rinaudo 2006; Campana-Filho et al. 2007). Apart from the amino groups, they have two hydroxyl functionalities for effecting appropriate chemical modifications to enhance solubility (Dumitriu 1996). The possible reaction sites for chitin and chitosan are illustrated in Figure 14.2. As with cellulose (Dumitriu 1996), chitin and chitosan can undergo many

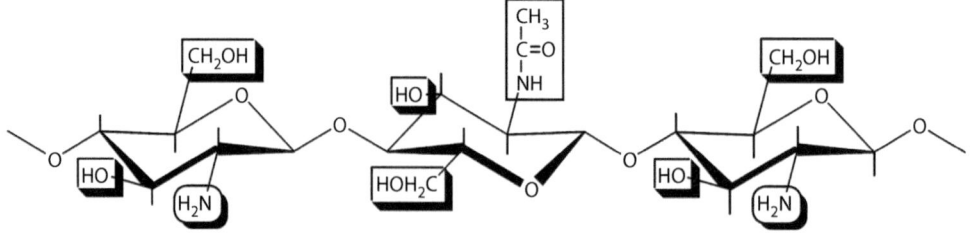

FIGURE 14.2   Illustration of possible reaction sites in chitin and chitosan. (Reprinted from *Progress in Polymer Science*, 34(7), Pillai, C. K. S., W. Paul, and C. P. Sharma, Chitin and chitosan polymers: Chemistry, solubility and fiber formation. 641–678. Copyright 2009, with permission from Elsevier.)

of the reactions such as etherification (Kurita 2001; Gorochovceva and Makusuka 2004; Zhang and Ren 2007), esterification (Kurita 2001; Yoshifuji et al. 2006; Zhang and Ren 2007), cross-linking (Muzzarelli 1990), graft copolymerization (Jenkins and Hudson 2001; Jayakumar et al. 2005), and so on. Muzzarelli (1977) and Hon (1996) have summarized the possible chemical modification reactions. Several authors have reviewed the area, emphasizing various aspects of chemical modification of chitosan (Kurita 1995, 1996, 2001, 2006; Hon 1996; Hirano 1996, 1999; Jenkins and Hudson 2001; Kumar et al. 2004; Sashiwa and Aiba 2004; Jayakumar et al. 2005, 2007; Muzzarelli and Muzzarelli 2005; Yi et al. 2005; Zohuriaan-Mehr 2005; Rinaudo 2006; Prashanth and Tharanathan 2007; Mourya and Inamdar 2008). The amino functionality gives rise to chemical reactions, such as acetylation, quaternization, reactions with aldehydes and ketones (to give Schiff's base), alkylation, grafting, chelation of metals, and so on to provide a variety of products with properties such as antibacterial, antifungal, antiviral, antiacid, antiulcer, nontoxic, nonallergenic, biocompatible, biodegradable, and so on. The hydroxyl functional groups also give various reactions such as o-acetylation, H-bonding with polar atoms, grafting, and so on. Owing to the intractability and insolubility of chitin (Muzzarelli 1977; Muzzarelli et al. 1986; Rinaudo 2006), attention has been given to chitosan with regard to developing derivatives with well-defined molecular architectures having advanced properties and functions.

Recently, chitosan and its derivatives have been reported as attractive candidates for scaffolding materials because they degrade as the new tissues are formed, eventually without inflammatory reactions or toxic degradation (Kubota and Eguchi 1997; Kurita et al. 2002). In tissue engineering applications, the cationic nature of chitosan is primarily responsible for electrostatic interactions with anionic glycosaminoglycans (GAG), proteoglycans, and other negatively charged molecules.

14.6   ACTION OF CHITOSAN

Chitosan is a polymer with several basic amino groups, and hence possesses an overall cationic charge, at acidic pH. This is due to the presence of primary amines on the molecule that bind protons according to the equation

$$\text{Chit} - NH_2 + H_3O^+ \Leftrightarrow \text{Chit} - NH_2^+ + H_2O$$

The charge density of chitosan has been found to depend on the degree of acetylation (DA) and pH. The amino group in chitosan has a pK_a value of ~6.5, which leads to a protonation in acidic solution with a charge density dependent on pH and the %DA value. The amino group gets deprotonated and chitosan tends to precipitate in solution of pH greater than 6.5. The multitude of cationic sites formed due to the protonation of amino groups by acids along the chitosan chain increases its solubility by increasing both the polarity and the degree of electrostatic repulsion. The wound-healing property of chitin and chitosan has been affected by the DA (Howling et al. 2001; Minagawa et al. 2007; Alsarra 2009), its molecular weights (Kubota and Eguchi 1997; Howling et al. 2001; Brandner et al. 2006; Minagawa et al. 2007; Alsarra 2009; Freitas et al. 2011), and the physical state (Azad et al. 2004). Chitin, chitosan, and their derivatives have been reported to accelerate wound healing by enhancing the functions of some inflammatory cells. Infiltration of polymorphonuclear (PMN) cells (Abhay 1998; Ueno et al. 1999, 2001a; Santos et al. 2007), macrophages (Ueno et al. 2001a; Kojima et al. 2004; Wiegand et al. 2010), and fibroblasts (Klokkevold et al. 1996; Howling et al. 2001, 2002; Ueno et al. 2001a, 2001c; Wiegand et al. 2010) or osteoblasts (Klokkevold et al. 1996) were accelerated.

Malette et al. (1983) reported that chitosan solution formed a coagulum in contact with whole blood and concluded that it is a good substitute for graft hemostasis. The hemostatic property of chitosan involved agglutination of red blood cells. The polycationic property of chitosan and its nonspecific binding to cell membranes are possibly related to its hemostatic property. Although various applications of chitosan have been reported, at the physiological pH value of 7.4, chitosan is insoluble and ineffective in many cases. The amino groups get deprotonated and the cationic property decreases significantly. The ineffectiveness and failure of some chitosan sponges as a hemostatic bandage could be because of these deprotonation. Therefore, various studies have been conducted on the development of water-soluble chitosan by chemical modifications and derivatizations. Some of these modified chitosan includes sulfonation, N-acetylation, alkylation, hydroxypropyl chitosan, chitosan-saccharide derivatives, O-succinyl-chitosan, quaternization, and carboxymethylation. The use of functionalized or derivatized chitosan has been reviewed by many authors (Shi et al. 2006; Enescu and Olteanu 2008; Kim et al. 2008; Francesko and Tzanov 2011).

14.7 CHITOSAN DERIVATIVES

A simple introduction of chemical groups such as alkyl or carboxymethyl in chitin and chitosan can significantly increase their solubility in neutral and alkaline pH values without affecting their chemical and biological characteristics. Carboxymethyl chitosan (CMC) is a chitosan-derivative obtained from the carboxymethylation of chitin with chloroacetic acid in alkaline solution. Carboxymethyl derivatives of chitin and chitosan are safe (Rattanatayarom and Wattanasirichaigoon 2007) and have shown promise in wound healing (Jayakumar et al. 2010). It has been shown as efficacious as alginate dressing in the treatment of partial thickness skin graft donor sites (Angspatt et al. 2011). Cell migration study has demonstrated that the migration of fibroblasts was significantly enhanced by the presence of carboxymethyl chitosan in a concentration-dependent manner. Carboxymethyl chitosan also induced increased proliferation and secretion of three kinds of cytokines and enhanced wound healing compared with the control (Chen et al. 2006; Rasad et al. 2010). N,O-carboxymethyl chitosan (NOCC) has been shown to be an effective inhibitor of postsurgical peritoneal adhesion formation (Costain et al. 1997), decrease adhesion in the healing of colonic anastomosis (Xiao et al. 2009), and reduce intra-abdominal adhesions (Falabella et al. 2010). Pectin cross-linked carboxymethyl chitosan hydrogel has been shown to be a promising wound-dressing material that is nontoxic and blood compatible. This hydrogel can maintain a moist environment that is conducive for wound healing (Fan et al. 2012). A composite membrane with chitosan and carboxymethyl chitosan was hemostatic, and had histocompatibility and indicated its usage as dressing of skin repair, and had the potential in promoting wound healing and inhibiting the keloid formation (Pang et al. 2008). A zinc complex of carboxymethyl chitosan has exhibited significantly high antimicrobial activity and can be used as a potential wound-healing device (Patale and Patravale 2011). Carboxymethyl chitosan dermal scaffold has been shown to rapidly induce growth and maturation of blood vessels during wound healing after burn. It is beneficial for wound repair at an early stage with the inhibition of scar proliferation (Teng et al. 2012). CM-chitosan hydrogels promoted cell attachment and rapid growth of fibroblasts and have the potential as skin scaffolds and wound-healing materials (Yang et al. 2010). A photo-curable carboxymethyl chitosan displayed good wound-healing property on burn wound model (Na et al. 2012). Carboxymethyl chitosan promoted wound healing by activating macrophages, accelerating fibroblasts growth, and exerting considerable effects on the secretion of a series of cytokines (Peng et al. 2011). Histological studies demonstrated that NOCC/oxidized alginate hydrogel significantly enhanced the reepithelialization of epidermis and collagen deposition in the wound tissue and the process of wound healing (Li et al. 2012). Amphiphatic carboxymethyl-hexanoyl chitosan hydrogel has also been studied as a rapid stem cell delivery system to efficiently enhance corneal wound healing (Chien et al. 2012).

N-Carboxybutyl chitosan membranes have been developed for the development of topical wound-healing applications (Dias et al. 2010, 2011). N-Carboxybutyl chitosan exhibited the formation of regularly organized cutaneous tissue with reduced anomalous healing (Biagini et al. 1991). Similarly, dibutyryl chitin (DBC) elevated the GAG level in the granulation tissue, documenting its beneficial effect in wound repair (Blasinska and Drobnik 2008). Poly acrylic acid grafted chitin exhibited faster rate and better pattern of epidermal development with higher dermal cell proliferation demonstrating its better efficiency in wound healing than Intrasite® (Pilakasiri et al. 2011). This was similar to alginate dressing in terms of pain score and wound healing (Angspatt et al. 2010). Chitin fibers have shown relatively high keratinocytes and fibroblasts cell attachment and spreading compared to commercial chitin fiber (*Beschitin®*) (Noh et al. 2006). This is helpful in wound healing and regeneration of skin. Partially deacetylated chitin hydrochloride exhibited hemostatic property similar to collagen hydrochloride and has the promotive effect in wound healing (Sugamori et al. 2000). Water-soluble chitin (WSC) prepared by controlling the DA and molecular weight was found to be more efficient than chitin or chitosan as a wound-healing accelerator (Cho et al. 1999). WSC-treated skin had the highest tensile strength and the arrangement of collagen fibers in the skin was similar to normal skin (Han 2005). DBC is obtained by the reaction of chitin with butyric anhydride in the presence of a catalyst. It is easily soluble in organic solvents. Wound dressings made from DBC fibers exhibited ordered accelerated healing of wound and the material biodegraded during the healing process (Schoukens et al. 2009). Topical application of DBC significantly reduced skin wound rank scores and increased the skin remodeling in animal model. It increased the expression of the type 1 collagen and filaggrin. The study demonstrates that DBC efficiently accelerates the proliferation of HaCaT keratinocytes (Jang et al. 2012). DBC efficiently inhibits inflammation and has potential as an effective anti-inflammatory and wound-healing agent (Jeon et al. 2012). DBC has also been studied as a textile dressing (Pielka et al. 2003). Satisfactory wound healing in burn wounds and postoperative/posttraumatic wounds has been achieved clinically (Chilarski et al. 2007).

It has been reported that *N*-sulfosuccinoyl chitosan had significantly higher wound-healing activity compared to other chitosan derivatives (Alekseeva et al. 2010). Polyelectrolyte complexes of chitosan with chondroitin sulfates and hyaluronic acid have been studied for wound-healing application. However, this was not better than pure chitosan, although there was no adverse effect (Denuziere et al. 1998, 2000). Chitosan modified by ionically binding heparin has shown beneficial effect in wound healing. It is hypothesized that the stimulatory effect of this hydrogel is caused by stabilization and activation of growth factors that bind to immobilized heparin and the increased stabilization and concentration of endogenous growth factors in the wound area is caused by heparin–chitosan (Kratz et al. 1997, 1998). Water-soluble chitosan–heparin complex ointment has proved its increased wound-healing effect previously (Kweon et al. 2003). The use of heparin attenuated chitosan's effect in wound healing (Jin et al. 2007). Chitosan, when grafted with poly(ethylene glycol)-modified tyramine enhanced its solubility and had the ability of forming *in situ* curable hydrogel. This hydrogel showed superior healing effects in the skin incision when compared to suture, fibrin glue, and cyanoacrylate (Lih et al. 2012). Cyanoethyl chitosan was prepared by treating activated chitosan with acrylonitrile vapor. The resultant polymer was soluble in trifluoroacetic acid and was used to make fibers by electrospinning. The mats prepared from these fibers exhibited excellent antibacterial property. This was developed as a possible wound-dressing material (Seyam et al. 2012). Chitosan acetate is the simplest derivative of chitosan with high cationic property. *HemCon* bandage is an engineered hemostatic dressing based on chitosan acetate (Burkatovskaya et al. 2008). Phosphorylated chitin and chitosan has also been reported to be having beneficial application in wound healing (Jayakumar et al. 2006).

14.8 SKIN TISSUE ENGINEERING

Tissue engineering is a rapidly expanding area of research in regenerative medicine that aims to create viable substitutes that restore, maintain, or improve the function of human tissues. Skin tissue engineering was the first successfully clinically applied product in regenerative medicine. Cell attachment, proliferation, and tissue formation in a 3D porous scaffold can be engineered for specific application. In this regard, chitosan has emerged as an attractive candidate for scaffolding material as they are capable of degradation as new tissue formation takes place while minimizing inflammatory reactions and toxic degradation products. These cell-based skin substitutes had significant wound-healing and scar-reducing effect on patients. They were prepared by primary cultures of fibroblast and keratinocyte cells on hydrated collagen sponges. However, because of its sheer high cost and complication of utilizing collagen in burn patients, there is still scope for the development of sponge dressing with cocultured fibroblast and keratinocyte cells. This can essentially replace collagen and is highly compatible to human tissue for use in severe burn wounds. The optimization of fibroblast cells and keratinocyte cells plays a pivotal role in controlling collagen production with reduced scar formation. Several reports have been published on the utilization of chitosan and carboxymethyl chitosan in skin tissue engineering. Chitosan has been studied as an additive in poly(lactic acid) for forming fibers that are amorphous in nature (Shalumon et al. 2012). These fibers had fast degradation behavior and the cultured dermal fibroblast cells tend to orient along the direction of fiber and show promise as a material of choice in skin tissue engineering. Macroporous chitosan membrane developed by a selective technology was found suitable for skin tissue engineering (Mei et al. 2012). Initial studies demonstrated that the cells flattened with effective spreading in 24 h. Blend films with silk fibroin were also evaluated as a biomaterial for tissue engineering and has been suggested as a supporting material in skin tissue engineering (Luangbudnark et al. 2012). Glutaraldehyde cross-linked collagen/chitosan scaffold containing fibroblast growth factor was prepared with seeded fibroblast cells. The reepithelialization was significantly faster on seeded scaffold compared to unseeded one (Intaraprasit et al. 2012).

Chitosan scaffolds have also been extensively used for various other tissue engineering applications. The chitosan forms used for tissue engineering include porous scaffolds that help in better cell penetration and attachments, and chitosan in the form of rods, which is usually used in bone tissue engineering and hydrogel that is for soft tissue engineering applications (Liu et al. 2011). They are prepared by lyophilization, heating and drying, electrospinning or gelation method, and so on. Another attractive method is modifying chitosan by combining with extracellular matrix proteins, such as collagen (Chen et al. 2011), glucosaminoglycans (Sampaio et al. 2005; Venkatesan et al. 2012), fibronectin (Custodio et al. 2010), laminin (Ikemoto et al. 2006), hyaluronic acid (Peniche et al. 2007), and so on or biomolecules, such as gelatin (Miranda et al. 2012), alginate (Wang et al. 2010), or peptide sequences such as RGD (Chen et al. 2013) or YIGSR (Itoh et al. 2005). These strategies mimic an extracellular matrix-like environment to enhance cell interaction and proliferation. Chitosan with hydroxyapatite also has various applications in bone tissue engineering (Tripathi et al. 2012) and dental repair (Tian et al. 2012).

The interaction of different types of cell with chitosan scaffold alone or in combination with other biomolecules has been extensively examined and shows very promising results for tissue engineering applications. The neuronal differentiation of muscle derived stem cells was observed in chitosan-based hydrogel scaffolds (Kwon et al. 2012). Nerve regeneration with stem cells was observed along the chitosan conduit (Zheng and Cui 2012). Codelivery of adipose stem cells and growth factors with RGD sequence was attempted with N-methacrylate glycol-modified chitosan gel for chondral repair (Sukarto et al. 2012). Chitosan gel with adipose-derived stem cells showed improvement of myocardial heart (Yang et al. 2011). The modification of appropriate growth factor with chitosan scaffold showed differentiation of neural cells (Fang et al. 2010; Leipzig et al. 2011; Yi et al. 2011). The modification of chitosan with silk fibroin was shown to promote adhesion and migration of stem cell (Altman et al. 2010) and attachment of hepatocytes (She et al. 2009). Chitosan is also known to increase the keratinocytes and transit amplifying cell population in the epidermal portion of skin (Shin et al. 2010). Thus, the properties of chitosan, especially its high affinity to other *in vivo* macromolecules, can be successfully exploited to suit the various applications of tissue repair of different organs.

Biochemistry and histology of chitosan in wound healing has been reviewed in various publications (Muzzarelli et al. 1988; Muzzarelli and Muzzarelli 2005; Muzzarelli 2009). Chitosan provides a nonprotein matrix for 3D tissue growth and activates macrophages for tumoricidal activity. It stimulates cell proliferation and histoarchitectural tissue organization. Chitosan will gradually depolymerize to release N-acetyl-β-D-glucosamine, which initiates fibroblast proliferation and helps in ordered collagen deposition and stimulates increased level of natural hyaluronic acid synthesis at

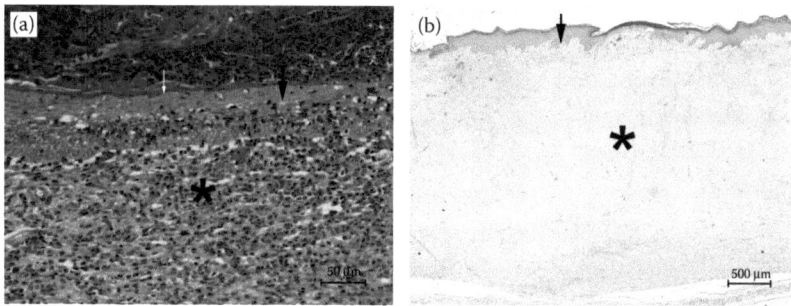

FIGURE 14.3 (a) The 14-day healing reaction of chitosan loaded with cells (40×). The newly formed epidermis (thick arrow) and scaffold (thin arrow) is seen above granulation tissue (black star). Granulation tissue is abundant with proliferating fibroblasts and newly forming blood vessels. (b) The 28-day healing reaction (4×) showing newly formed thicker epidermis (thick arrow) and the neodermis (black star).

the wound site. It helps in faster wound healing and scar prevention. Fibroblasts, keratinocyte cells, epithelial cells, endothelial cells, and growth factors all work together to achieve wound repair. Each of these factors is essential as cell migration, proliferation, and extracellular matrix deposition are accomplished through the activity of these and other growth factors. It has been established that the possible cause of wounds not healing in a timely manner (such as chronic ulcers) is the deficiency in growth factors, or the growth factors cannot perform to their full potential if growth factors are entrapped in wound components such as fibrin cuffs. Thus, supplementing growth factors and utilizing fibroblast and keratinocyte cell-seeded chitosan-based wound healing sponges can help in faster wound healing in cases of burn cases. The quality of healing on a rabbit wound model with PEGylated chitosan sponge cocultured with fibroblast and keratinocyte cells was comparatively better. The histopathological figure (Figure 14.3) shows the 14-day and 28-day healing pattern of PEGylated chitosan.

14.9 CONCLUSION

Various kinds of wound dressings are available for chronic wounds, including venous ulcers, diabetic ulcers, and pressure sores, in wound-care management that help to improve patient mobility, quality of life, and well-being. However, in most cases, these dressings are used for the prevention of infection or to create a moist environment, helping in natural wound healing. Chronic wounds disrupt the normal process of healing and are often not sufficient in themselves to effect repair. A burn injury results in either the loss or disruption of some or all of the functions of the normal skin. Impairment of blood flow in the zone of stasis can occur from shortly after the burn injury up to 48 h postburn (Williams 2002). If blood flow is compromised, this may lead to the eventual necrosis of cells in this zone.

The use of temporary skin substitutes, such as Biobrane, TransCyte, Integra, and Terudermis, has become more popular in the treatment of mild-to-deep dermal burn injury. They are very expensive. Also, they contain collagen that is not recommended for third-degree burns where dermis, epidermis, and hypodermis are completely destroyed (Collagens 2005). Cell-based artificial skin substitute Dermagraft, derived from fibroblast cells, was the first one to be approved by FDA for application in severe burns in 1997. However, because of certain complications noted in burn cases, lately it is used only for diabetic foot ulcers. Apligraft, a bovine collagen matrix cultured with fibroblast cells and keratinocyte cells is approved for venous stasis and diabetic foot ulcers only by the FDA. Epicel is a sheet of autologous keratinocytes (skin cells) used to replace the epidermal or top layer of skin on severely burned patients. This is approved by the FDA; however, it is expensive and some patients are allergic to it. These cell-based skin substitutes had good wound-healing and scar-reducing effect on patients, and are prepared by primary cultures of fibroblast and keratino-cyte cells on hydrated collagen sponges. Because of the very high cost of collagen sponges and its drawback in utilizing for burn cases, there is a need for the development of a new sponge dressing. Chitosan sponge dressing cocultured with fibroblasts and keratinocyte cells, can ultimately replace collagen dressing for applications in severe burn wounds. This also seems to be highly compatible to human tissue.

Wound healing utilizing chitin and chitin derivatives and skin tissue engineering is an expand-ing area of research. With its unique combination of biological, physical, and chemical properties, chitin, chitosan, and their derivatives are widely used in both industrial and medical fields. These properties present a novel, versatile biopolymer that can be tailor-made to suit a particular appli-cation with the required modes of function. The development and commercial applications for chitin, chitosan, and their derivatives have expanded in recent years. Wound healing is a complex process that can be compromised by various factors. Despite proper care, some wounds fail to heal in an appropriate manner and may become chronic. From individual studies reported in the literature, carboxymethyl chitosan seems to be an excellent candidate material for the wound-healing applications in chronic wounds. Carboxymethyl chitosan is a modification of chitosan formed by attaching carboxymethyl groups to the chitosan backbone. Depending on the location of the carboxymethyl group attachment, carboxymethyl chitosan (CMC) can be referred to as "N" when the carboxymethyl group attaches to the amine, "O" when it attaches to the primary hydroxyl group, or NOCC when attached to both (Song et al. 2011). CMC has the advantage of a greater solubility range than native chitosan. The large number of publications in this area sug-gests that carboxymethyl chitosan will be an important derivative in the management of wounds and burns, and skin tissue engineering.

REFERENCES

Abhay, S. P. 1998. Hemostatic wound dressing. US Patent 5836970 A.

Alekseeva, T. P., A. A. Rakhmetova, O. A. Bogoslovskaya et al. 2010. Wound healing potential of chitosan and N-sulfosuccinoyl chitosan derivatives. *Biology Bulletin* 37(4):339–345.

Alsarra, I. A. 2009. Chitosan topical gel formulation in the management of burn wounds. *International Journal of Biological Macromolecules* 45(1):16–21.

Altman, A. M., V. Gupta, C. N. Rios, E. U. Alt, and A. B. Mathur. 2010. Adhesion, migration and mechan-ics of human adipose-tissue-derived stem cells on silk fibroin–chitosan matrix. *Acta Biomaterialia* 6(4):1388–1397.

Angspatt, A., P. Tanvatcharaphan, S. Channasanon, S. Tanodekaew, P. Chokrungvaranont, and W. Sirimaharaj. 2010. Comparative study between chitin/polyacrylic acid (PAA) dressing, lipido-colloid absorbent dress-ing and alginate wound dressing: A pilot study in the treatment of partial-thickness wound. *Journal of the Medical Association of Thailand* 93(6):694–697.

Angspatt, A., B. Taweerattanasil, W. Janvikul, P. Chokrungvaranont, and W. Sirimaharaj. 2011. Carboxymethyl-chitosan, alginate and tulle gauze wound dressings: A comparative study in the treatment of partial-thickness wounds. *Asian Biomedicine* 5(3):413–416.

Azad, A. K., N. Sermsintham, S. Chandrkrachang, and W. F. Stevens. 2004. Chitosan membrane as a wound-healing dressing: Characterization and clinical application. *Journal of Biomedical Materials Research Part B—Applied Biomaterials* 69B(2):216–222.

Biagini, G., A. Bertani, R. Muzzarelli et al. 1991. Wound management with *N*-carboxybutyl chitosan. *Biomaterials* 12(3):281–286.

Blasinska, A. and J. Drobnik. 2008. Effects of nonwoven mats of di-O-butyrylchitin and related polymers on the process of wound healing. *Biomacromolecules* 9(3):776–782.

Brandner, J. M., P. Houdek, C. Scholz, and I. Moll. 2006. Influence of Chitosan on wound healing is dependent on concentration and molecular weight and involves changes in the expression of connexin 43. *Journal of Investigative Dermatology* 126:59–59.

Burkatovskaya, M., A. P. Castano, T. N. Demidova-Rice, G. P. Tegos, and M. R. Hamblin. 2008. Effect of chitosan acetate bandage on wound healing in infected and noninfected wounds in mice. *Wound Repair and Regeneration* 16(3):425–431.

Campana-Filho, S. P., D. de Britto, E. Curti et al. 2007. Extraction, structures and properties of a- and b-chitin. *Quimica Nova* 30:644–650.

Chandy, T. and C. P. Sharma. 1990. Chitosan—As a biomaterial. *Biomaterials, Artificial Cells, and Artificial Organs* 18(1):1–24.

Chen, Po-Yu. *Chitin and Chitosan: From Nature to Technology*, 2006. Available from http://www.meyersgroup. ucsd.edu/literature_reviews/2006/Litreview%20Chitin%20and%20Chitosan%20Po-Yu%20Chen_files/ frame.htm.

Chen, R. N., G. M. Wang, C. H. Chen, H. O. Ho, and M. T. Sheu. 2006. Development of N,O-(carboxymethyl) chitosan/collagen matrixes as a wound dressing. *Biomacromolecules* 7(4):1058–1064.

Chen, W., H. Zhou, M. D. Weir, M. Tang, C. Bao, and H. H. Xu. 2013. Human embryonic stem cell-derived mesenchymal stem cell seeding on calcium phosphate cement-chitosan-RGD scaffold for bone repair. *Tissue Engineering Part A.* 19(7–8):915–927.

Chen, L., C. H. Zhu, D. D. Fan et al. 2011. A human-like collagen/chitosan electrospun nanofibrous scaffold from aqueous solution: Electrospun mechanism and biocompatibility. *Journal of Biomedical Materials Research Part A* 99A(3):395–409.

Chien, Y., Y. W. Liao, D. M. Liu et al. 2012. Corneal repair by human corneal keratocyte-reprogrammed iPSCs and amphiphatic carboxymethyl-hexanoyl chitosan hydrogel. *Biomaterials* 33(32):8003–8016.

Chilarski, A., I. Krucinska, P. Kiekens et al. 2007. Novel dressing materials accelerating wound healing made from dibutyrylchitin. *Fibres & Textiles in Eastern Europe* 15(4):77–81.

Cho, Y. W., Y. N. Cho, S. H. Chung, G. Yoo, and S. W. Ko. 1999. Water-soluble chitin as a wound healing accel- erator. *Biomaterials* 20(22):2139–2145.

Cohen, I. K., ed. 1998. *A Brief History of Wound Healing*. 1st edition. PA: Ortho-McNeil.

Collagens. 2005. In *Wound Care*, ed. C. T. Hess. Philadelphia: Lippincott Williams and Wilkins.

Costain, D. J., R. Kennedy, C. Ciona, V. C. McAlister, and T. D. Lee. 1997. Prevention of postsurgical adhe- sions with N,O-carboxymethyl chitosan: Examination of the most efficacious preparation and the effect of *N,O*-carboxymethyl chitosan on postsurgical healing. *Surgery* 121(3):314–319.

Custodio, C. A., C. M. Alves, R. L. Reis, and J. F. Mano. 2010. Immobilization of fibronectin in chitosan substrates improves cell adhesion and proliferation. *Journal of Tissue Engineering and Regenerative Medicine* 4(4):316–323.

Denuziere, A., D. Ferrier, O. Damour, and A. Domard. 1998. Chitosan-chondroitin sulfate and chitosan- hyaluronate polyelectrolyte complexes: Biological properties. *Biomaterials* 19(14):1275–1285.

Denuziere, A., D. Ferrier, and A. Domard. 2000. Interactions between chitosan and glycosaminoglycans (chon- droitin sulfate and hyaluronic acid): Physicochemical and biological studies. *Annales Pharmaceutiques Françaises* 58(1):47–53.

Dias, A. M., M. E. Braga, I. J. Seabra, P. Ferreira, M. H. Gil, and H. C. de Sousa. 2011. Development of natural- based wound dressings impregnated with bioactive compounds and using supercritical carbon dioxide. *International Journal of Pharmaceutics* 408(1–2):9–19.

Dias, A. M., I. J. Seabra, M. M. Braga, M. H. Gil, and H. C. de Sousa. 2010. Supercritical solvent impregnation of natural bioactive compounds in *N*-carboxybutyl chitosan membranes for the development of topical wound healing applications. *Journal of Controlled Release* 148(1):e33–e35.

Dumitriu, S. 1996. *Polysaccharides in Medicinal Application*. New York: Marcel Dekker.

Enescu, D. and C. E. Olteanu. 2008. Functionalized chitosan and its use in pharmaceutical, biomedical, and biotechnological research. *Chemical Engineering Communications* 195(10):1269–1291.

Falabella, C. A., M. M. Melendez, L. Weng, and W. Chen. 2010. Novel macromolecular crosslinking hydrogel to reduce intra-abdominal adhesions. *Journal of Surgical Research* 159(2):772–778.

Fan, L., Y. Sun, W. Xie, H. Zheng, and S. Liu. 2012. Oxidized pectin cross-linked carboxymethyl chitosan: A new class of hydrogels. *Journal of Biomaterials Science, Polymer Edition* 23(16):2119–2132.

Fang, P., Q. Gao, W. J. Liu et al. 2010. Survival and differentiation of neuroepithelial stem cells on chitosan bicomponent fibers. *Chinese Journal of Physiology* 53(4):208–214.

Francesko, A. and T. Tzanov. 2011. Chitin, chitosan and derivatives for wound healing and tissue engineering. *Advances in Biochemical Engineering/Biotechnology* 125:1–27.

Freitas, R. M., R. Spin-Neto, L. C. Spolidorio, S. P. Campana, R. A. C. Marcantonio, and E. Marcantonio. 2011. Different molecular weight chitosan-based membranes for tissue regeneration. *Materials* 4(2):380–389.

Gorochovceva, N. and R. Makusuka. 2004. Synthesis and study of water-soluble chitosan-*O*-poly(ethylene glycol) graft copolymers. *European Polymer Journal* 40:685–691.

Han, S. S. 2005. Topical formulations of water-soluble chitin as a wound healing assistant—Evaluation on open wounds using a rabbit ear model. *Fibers and Polymers* 6(3):219–223.

Hirano, S. 1996. Chitin biotechnology applications. *Biotechnology Annual Review* 2:237–258.

Hirano, S. 1999. Chitin and chitosan as novel biotechnological materials. *Polymer International* 48:732–734.

Hon, D. N. S. 1996. Chitin and chitosan: Medical applications. In *Polysaccharides in Medicinal Application*, ed. S. Dumitriu. New York: Marcel Dekker.

Howling, G. I., P. W. Dettmar, P. A. Goddard, F. C. Hampson, M. Dornish, and E. J. Wood. 2001. The effect of chitin and chitosan on the proliferation of human skin fibroblasts and keratinocytes *in vitro*. *Biomaterials* 22(22):2959–2966.

Howling, G. I., P. W. Dettmar, P. A. Goddard, F. C. Hampson, M. Dornish, and E. J. Wood. 2002. The effect of chitin and chitosan on fibroblast-populated collagen lattice contraction. *Biotechnology and Applied Biochemistry* 36:247–253.

Hudson, S. M. and C. Smith. 1998. Chitin and chitosan. In *Biopolymers from Renewable Resources*, ed. D. L. Kaplan. Berlin: Springer.

Ikemoto, S., M. Mochizuki, M. Yamada et al. 2006. Laminin peptide-conjugated chitosan membrane: Application for keratinocyte delivery in wounded skin. *Journal of Biomedical Materials Research Part A* 79A(3):716–722.

Intaraprasit, S., A. Faikrua, A. Sittichokechaiwut, and J. Viyoch. 2012. Efficacy evaluation of the fibroblast-seeded collagen/chitosan scaffold on application in skin tissue engineering. *ScienceAsia* 38(3):268–277.

Itoh, S., A. Matsuda, H. Kobayashi, S. Ichinose, K. Shinomiya, and J. Tanaka. 2005. Effects of a laminin peptide (YIGSR) immobilized on crab-tendon chitosan tubes on nerve regeneration. *Journal of Biomedical Materials Research Part B—Applied Biomaterials* 73B(2):375–382.

Jang, S. I., J. Y. Mok, I. H. Jeon et al. 2012. Effect of electrospun non-woven mats of dibutyryl chitin/poly(lactic acid) blends on wound healing in hairless mice. *Molecules* 17(3):2992–3007.

Jayakumar, R., N. Nwe, S. Tokura, and H. Tamura. 2007. Sulfated chitin and chitosan as novel biomaterials. *International Journal of Biological Macromolecules* 40(3):175–181.

Jayakumar, R., M. Prabaharan, S. V. Nair, S. Tokura, H. Tamura, and N. Selvamurugan. 2010. Novel carboxymethyl derivatives of chitin and chitosan materials and their biomedical applications. *Progress in Materials Science* 55(7):675–709.

Jayakumar, R., M. Prabaharan, R. L. Reis, and J. F. Mano. 2005. Graft copolymerized chitosan—Present status and applications. *Carbohydrate Polymers* 62(2):142–158.

Jayakumar, R., R. L. Reis, and J. F. Mano. 2006. Chemistry and applications of phosphorylated chitin and chitosan. *E-Polymers* 35:1–16.

Jayasree, R. S., K. Rathinam, and C. P. Sharma. 1995. Development of artificial skin (template) and influence of different types of sterilization procedures on wound healing pattern in rabbits and guinea pigs. *Journal of Biomaterials Applications* 10(2):144–162.

Jenkins, D. W. and S. M. Hudson. 2001. Review of vinyl graft copolymerization featuring recent advances toward controlled radical-based reactions and illustrated with chitin/chitosan trunk polymers. *Chemical Reviews* 101:3245.

Jeon, I. H., J. Y. Mok, K. H. Park et al. 2012. Inhibitory effect of dibutyryl chitin ester on nitric oxide and prostaglandin E(2) production in LPS-stimulated RAW 264.7 cells. *Archives of Pharmacal Research* 35(7):1287–1292.

Jin, Y., P. X. Ling, Y. L. He, and T. M. Zhang. 2007. Effects of chitosan and heparin on early extension of burns. *Burns* 33(8):1027–1031.

Johnstone, C. C. and A. Farley. 2005. The physiological basics of wound healing. *Nursing Standard* 19(43):7.

Khor, E. 2002. Chitin: A biomaterial in waiting. *Current Opinion in Solid State Material Science* 6:313–317.

Kim, I. Y., S. J. Seo, H. S. Moon et al. 2008. Chitosan and its derivatives for tissue engineering applications. *Biotechnology Advances* 26(1):1–21.

Klokkevold, P. R., L. Vandemark, E. B. Kenney, and G. W. Bernard. 1996. Osteogenesis enhanced by chitosan (poly-N-acetyl glucosaminoglycan) *in vitro*. *Journal of Periodontology* 67(11):1170–1175.

Kojima, K., Y. Okamoto, K. Kojima et al. 2004. Effects of chitin and chitosan on collagen synthesis in wound healing. *Journal of Veterinary Medical Science* 66(12):1595–1598.

Kratz, G., C. Arnander, J. Swedenborg et al. 1997. Heparin-chitosan complexes stimulate wound healing in human skin. *Scandinavian Journal of Plastic and Reconstructive Surgery and Hand Surgery* 31(2): 119–123.

Kratz, G., M. Back, C. Arnander, and O. Larm. 1998. Immobilised heparin accelerates the healing of human wounds *in vivo*. *Scandinavian Journal of Plastic and Reconstructive Surgery and Hand Surgery* 32(4): 381–385.

Kubota, N. and Y. Eguchi. 1997. Facile preparation of water-soluble N-acetylated chitosan and molecular weight dependence of its water-solubility. *Polymer Journal* 29:123–127.

Kumar, M. N., R. A. Muzzarelli, C. Muzzarelli, H. Sashiwa, and A. J. Domb. 2004. Chitosan chemistry and pharmaceutical perspectives. *Chemical Reviews* 104(12):6017–6084.

Kurita, K. 1995. Chemistry and application of chitin and chitosan. *Polymer Degradation and Stability* 59:117–120.

Kurita, K. 1996. Chitin and chitosan graft copolymers. In *Polymeric Materials Encyclopedia*, ed. J. C. Salamone. Boca Raton: CRC Press.

Kurita, K. 2001. Controlled functionalization of polysaccharide chitin. *Progress in Polymer Science* 26: 1921–1971.

Kurita, K. 2006. Chitin and chitosan: Functional biopolymers from marine crustaceans. *Marine Biotechnology (NY)* 8(3):203–226.

Kurita, K., M. Mori, Y. Nishiyama, and Harata M. 2002. N-Alkylation of chitin and some characteristics of the novel derivatives. *Polymer Bulletin* 48(2):159–166.

Kweon, D. K., S. B. Song, and Y. Y. Park. 2003. Preparation of water-soluble chitosan/heparin complex and its application as wound healing accelerator. *Biomaterials* 24(9):1595–1601.

Kwon, J. S., G. H. Kim, D. Y. Kim et al. 2012. Chitosan-based hydrogels to induce neuronal differentiation of rat muscle-derived stem cells. *International Journal of Biological Macromolecules* 51(5):974–979.

Leipzig, N. D., R. G. Wylie, H. Kim, and M. S. Shoichet. 2011. Differentiation of neural stem cells in three-dimensional growth factor-immobilized chitosan hydrogel scaffolds. *Biomaterials* 32(1):57–64.

Li, X., S. Chen, B. Zhang et al. 2012. *In situ* injectable nano-composite hydrogel composed of curcumin, N,O-carboxymethyl chitosan and oxidized alginate for wound healing application. *International Journal of Pharmaceutics* 437(1–2):110–119.

Lih, E., J. S. Lee, K. M. Park, and K. D. Park. 2012. Rapidly curable chitosan-PEG hydrogels as tissue adhesives for hemostasis and wound healing. *Acta Biomaterialia* 8(9):3261–3269.

Liu, X., L. Ma, Z. W. Mao, and C. Y. Gao. 2011. Chitosan-based biomaterials for tissue repair and regeneration. *Chitosan for Biomaterials II* 244:81–127.

Luangbudnark, W., J. Viyoch, W. Laupattarakasem, P. Surakunprapha, and P. Laupattarakasem. 2012. Properties and biocompatibility of chitosan and silk fibroin blend films for application in skin tissue engineering. *Scientific World Journal* 10 pages, Article ID 697201.

Maddox, J. 2012. *The Global Advanced Wound Care Market to 2017*. UK: Espicom Business Intelligence Ltd.

Malette, W. G., H. J. Quigley, R. D. Gaines, N. D. Johnson, and W. G. Rainer. 1983. Chitosan: A new hemostatic. *The Annals of Thoracic Surgery* 36(1):55–58.

Mei, L., D. Hu, J. Ma, X. Wang, Y. Yang, and J. Liu. 2012. Preparation, characterization and evaluation of chitosan macroporous for potential application in skin tissue engineering. *International Journal of Biological Macromolecules* 51(5):992–997.

Minagawa, T., Y. Okamura, Y. Shigemasa, S. Minami, and Y. Okamoto. 2007. Effects of molecular weight and deacetylation degree of chitin/chitosan on wound healing. *Carbohydrate Polymers* 67(4):640–644.

Miranda, S. C., G. A. Silva, R. M. Mendes et al. 2012. Mesenchymal stem cells associated with porous chitosan-gelatin scaffold: A potential strategy for alveolar bone regeneration. *Journal of Biomedical Materials Research Part A* 100(10):2775–2786.

Mourya, V. K. and N. N. Inamdar. 2008. Chitosan-modifications and applications: Opportunities galore. *Reactive Functional Polymers* 68:1013–1051.

Muzzarelli, R. A. A. 2009. Chitins and chitosans for the repair of wounded skin, nerve, cartilage and bone. *Carbohydrate Polymers* 76(2):167–182.

Muzzarelli, R. A. A. 1977. *Chitin*, ed. R. A. A. Muzzarelli. Oxford: Pergamon Press.

Muzzarelli, R. A. A. 1990. Modified chitosans and their chemical behavior. *Polymer Preparation (American Chemical Society Division of Polymer Chemistry)* 31:626.

Muzzarelli, R., V. Baldassarre, F. Conti et al. 1988. Biological activity of chitosan: Ultrastructural study. *Biomaterials* 9 (3):247–252.

Muzzarelli, R. A. A., and C. Muzzarelli. 2005. Chitosan chemistry: Relevance to the biomedical sciences. *Polysaccharides 1: Structure, Characterization and Use* 186:151–209.

Muzzarelli, R. A. A., C. Jeuniaux, and G. W. Gooday. 1986. *Chitin in Nature and Technology*. New York: Plenum.

Muzzarelli, R. A. A. and C. Muzzarelli. 2005. Chitosan chemistry: Relevance to the biomedical sciences. *Advances in Polymer Science* 186:151–209.

Na, H. N., S. H. Park, K. I. Kim, M. K. Kim, and T. I. Son. 2012. Photocurable O-carboxymethyl chitosan derivatives for biomedical applications: Synthesis, *in vitro* biocompatibility, and their wound healing effects (vol 20, pg 1144, 2012). *Macromolecular Research* 20(11):1209–1209.

Noh, H. K., S. W. Lee, J. M. Kim et al. 2006. Electrospinning of chitin nanofibers: Degradation behavior and cellular response to normal human keratinocytes and fibroblasts. *Biomaterials* 27(21):3934–3944.

Pang, H. T., X. G. Chen, Q. X. Ji, and Y. Zhong de. 2008. Preparation and function of composite asymmetric chitosan/CM-chitosan membrane. *Journal of Materials Science: Materials in Medicine* 19(3):1413–1417.

Patale, R. L. and V. B. Patravale. 2011. O,N-Carboxymethyl chitosan-zinc complex: A novel chitosan complex with enhanced antimicrobial activity. *Carbohydrate Polymers* 85(1):105–110.

Paul, W. and C. P. Sharma. 2000. Chitosan, a drug carrier for the 21st century: A review. *STP Pharma Sciences* 10:18.

Peng, S., W. Liu, B. Han, J. Chang, M. Li, and X. Zhi. 2011. Effects of carboxymethyl-chitosan on wound healing *in vivo* and *in vitro*. *Journal of Ocean University of China* 10(4):10.

Peniche, C., M. Fernadez, G. Rodriguez et al. 2007. Cell supports of chitosan/hyaluronic acid and chondroitin sulphate systems. Morphology and biological behaviour. *Journal of Materials Science—Materials in Medicine* 18(9):1719–1726.

Peter, M. G. 2002. Chitin and chitosan from animal sources. In *Biopolymers: Polysaccharides II*, eds. S. De Baets, E. J. Vandamme, and A. Steinbuchel. Weinheim: Wiley-VCH.

Pielka, S., D. Paluch, J. Staniszewska-Kus et al. 2003. Wound healing acceleration by a textile dressing containing dibutyrylchitin and chitin. *Fibres & Textiles in Eastern Europe* 11(2):79–84.

Pilakasiri, K., P. Molee, D. Sringernyuang, N. Sangjun, S. Channasanon, and S. Tanodekaew. 2011. Efficacy of chitin-PAA-GTMAC gel in promoting wound healing: Animal study. *Journal of Materials Science—Materials in Medicine* 22(11):2497–2504.

Pillai, C. K. S., W. Paul, and C. P. Sharma. 2009. Chitin and chitosan polymers: Chemistry, solubility and fiber formation. *Progress in Polymer Science* 34(7):641–678.

Prashanth, K. V. H. and R. N. Tharanathan. 2007. Chitin/chitosan: Modifications and their unlimited application potential—an overview. *Trends in Food Science and Technology* 18:117–131.

Rasad, M. S. B. A., A. S. Halim, K. Hashim, A. H. A. Rashid, N. Yusof, and S. Shamsuddin. 2010. *In vitro* evaluation of novel chitosan derivatives sheet and paste cytocompatibility on human dermal fibroblasts. *Carbohydrate Polymers* 79(4):1094–1100.

Rattanatayarom, W. and S. Wattanasirichaigoon. 2007. Evaluation of dermal irritancy potential of carboxymethyl-chitosan hydrogel and poly-(acrylic acid) chitin hydrogel. *Journal of the Medical Association of Thailand* 90(4):724–729.

Rinaudo, M. 2006. Chitin and chitosan: Properties and applications. *Progress in Polymer Science* 31:603–632.

Rinaudo, M. 2008. Main properties and current applications of some polysaccharides as biomaterials. *Polymer International* 57(3):397–430.

Roberts, G. A. F. 1992. *Chitin Chemistry*, ed. G. A. F. Roberts. London: Macmillan Press.

Sampaio, S., P. Taddei, P. Monti, J. Buchert, and G. Freddi. 2005. Enzymatic grafting of chitosan onto *Bombyx mori* silk fibroin: Kinetic and IR vibrational studies. *Journal of Biotechnology* 116(1):21–33.

Santos, T. C., A. P. Marques, S. S. Silva et al. 2007. *In vitro* evaluation of the behaviour of human polymorphonuclear neutrophils in direct contact with chitosan-based membranes. *Journal of Biotechnology* 132(2): 218–226.

Sashiwa, H. and S. I. Aiba. 2004. Chemically modified chitin and chitosan as biomaterials. *Progress in Polymer Science* 29(9):887–908.

Sathirakul, K., N. C. How, W. F. Stevens, and S. Chandrkrachang. 1996. Application of chitin and chitosan bandages for wound healing. *Advances in Chitin Science* 1:490–492.

Schoukens, G., P. Kiekens, and I. Krucinska. 2009. New bioactive textile dressing materials from dibutyrylchitin. *International Journal of Clothing Science and Technology* 21(2–3):93–101.

Seyam, A. F. M., S. M. Hudson, H. M. Ibrahim, A. I. Waly, and N. Y. Abou-Zeid. 2012. Healing performance of wound dressing from cyanoethyl chitosan electrospun fibres. *Indian Journal of Fibre & Textile Research* 37(3):205–210.

Shalumon, K. T., D. Sathish, S. V. Nair, K. P. Chennazhi, H. Tamura, and R. Jayakumar. 2012. Fabrication of aligned poly (lactic acid)-chitosan nanofibers by novel parallel blade collector method for skin tissue engineering. *Journal of Biomedical Nanotechnology* 8(3):405–416.

She, Z. D., W. Q. Liu, and Q. L. Feng. 2009. Self-assembly model, hepatocytes attachment and inflammatory response for silk fibroin/chitosan scaffolds. *Biomedical Materials* 4(4):045014.

Shi, C., Y. Zhu, X. Ran, M. Wang, Y. Su, and T. Cheng. 2006. Therapeutic potential of chitosan and its derivatives in regenerative medicine. *Journal of Surgical Research* 133(2):185–192.

Shin, D. W., J. H. Shim, Y. Kim et al. 2010. Chitosan increases alpha 6 integrin(high)/CD71(high) human keratinocyte transit-amplifying cell population. *Biomolecules & Therapeutics* 18(3):280–285.

Song, Q., Z. Zhang, J. Gao, and C. Ding. 2011. Synthesis and property studies of N-carboxymethyl chitosan. *Journal of Applied Polymer Science* 119(6):3282–3285.

Sugamori, T., H. Iwase, M. Maeda, Y. Inoue, and H. Kurosawa. 2000. Local hemostatic effects of microcrystalline partially deacetylated chitin hydrochloride. *Journal of Biomedical Materials Research* 49(2):225–232.

Sukarto, A., C. Yu, L. E. Flynn, and B. G. Amsden. 2012. Co-delivery of adipose-derived stem cells and growth factor-loaded microspheres in RGD-grafted N-methacrylate glycol chitosan gels for focal chondral repair. *Biomacromolecules* 13(8):2490–2502.

Suzuki, Y., Y. Okamoto, and M. Morimoto. 2000. Influence of physicochemical properties of chitin and chitosan on compliment activation. *Carbohydrate Polymers* 42:307–310.

Teng, J. Y., R. Guo, J. Xie, D. J. Sun, M. Q. Shen, and S. J. Xu. 2012. Effects of different artificial dermal scaffolds on vascularization and scar formation of wounds in pigs with full-thickness burn. *Zhonghua Shao Shang Za Zhi* 28(1):13–18.

Tharanathan, R. N. and F. S. Kittur. 2003. Chitin—The undisputed biomolecule of great potential. *Critical Reviews in Food Science and Nutrition* 43(1):61–87.

Tian, K., M. Peng, W. Fei, C. H. Liao, and X. H. Ren. 2012. Induced synthesis of hydroxyapatite by chitosan for enamel remineralization. *Smart Technologies for Materials* 530:40–45.

Tomihata, K. and Y. Ikada. 1997. *In vitro* and *in vivo* degradation of films of chitin and its deacetylated derivatives. *Biomaterials* 18(7):567–575.

Tripathi, A., S. Saravanan, S. Pattnaik, A. Moorthi, N. C. Partridge, and N. Selvamurugan. 2012. Bio-composite scaffolds containing chitosan/nano-hydroxyapatite/nano-copper-zinc for bone tissue engineering. *International Journal of Biological Macromolecules* 50(1):294–299.

Ueno, H., T. Mori, and T. Fujinaga. 2001a. Topical formulations and wound healing applications of chitosan. *Advanced Drug Delivery Reviews* 52(2):105–115.

Ueno, H., M. Murakami, M. Okumura, T. Kadosawa, T. Uede, and T. Fujinaga. 2001b. Chitosan accelerates the production of osteopontin from polymorphonuclear leukocytes. *Biomaterials* 22(12):1667–1673.

Ueno, H., F. Nakamura, M. Murakami, M. Okumura, T. Kadosawa, and T. Fujinag. 2001c. Evaluation effects of chitosan for the extracellular matrix production by fibroblasts and the growth factors production by macrophages. *Biomaterials* 22(15):2125–2130.

Ueno, H., H. Yamada, I. Tanaka et al. 1999. Accelerating effects of chitosan for healing at early phase of experimental open wound in dogs. *Biomaterials* 20(15):1407–1414.

Venkatesan, J., R. Pallela, I. Bhatnagar, and S. K. Kim. 2012. Chitosan-amylopectin/hydroxyapatite and chitosan-chondroitin sulphate/hydroxyapatite composite scaffolds for bone tissue engineering. *International Journal of Biological Macromolecules* 51(5):1033–1042.

Wang, J. Z., X. B. Huang, J. Xiao et al. 2010. Spray-spinning: A novel method for making alginate/chitosan fibrous scaffold. *Journal of Materials Science—Materials in Medicine* 21(2):497–506.

Wiegand, C., D. Winter, and U. C. Hipler. 2010. Molecular-weight-dependent toxic effects of chitosans on the human keratinocyte cell line HaCaT. *Skin Pharmacology and Physiology* 23(3):164–170.

Williams, W. 2002. Pathophysiology of the burn wound. In *Total Burn Care*, ed. D. Herndon. London: Saunders.

Winter, G. D. 1962. Formation of the scab and the rate of epithelization of superficial wounds in the skin of the young domestic pig. *Nature* 193:293–294.

Xiao, H., C. Hou, and F. Xue. 2009. Effect of carboxymethylchitosan-carboxymethylcellulose film on colonic anastomosis healing. *Zhongguo Xiu Fu Chong Jian Wai Ke Za Zhi* 23(4):451–455.

Yang, J. J., Z. Q. Liu, Z. L. Yuan, Y. D. Chen, and C. Y. Wang. 2011. Improvement of infarcted heart by the transplantation of adipose-derived stem cells with injectable temperature-responsive chitosan hydrogel. *Heart* 97(21):A12.

Yang, C., L. Xu, Y. Zhou et al. 2010. A green fabrication approach of gelatin/CM-chitosan hybrid hydrogel for wound healing. *Carbohydrate Polymers* 82(4):1297–1305.

Yi, H., L. Q. Wu, W. E. Bentley et al. 2005. Biofabrication with chitosan. *Biomacromolecules* 6(6):2881–2894.

Yi, X., G. H. Jin, M. L. Tian, W. F. Mao, and L. B. Qin. 2011. Porous chitosan scaffold and NGF promote neuronal differentiation of neural stem cells *in vitro*. *Neuroendocrinology Letters* 32(5):705–710.

Yoshifuji, A., Y. Noishiki, M. Wada, L. Heux, and S. Kuga. 2006. Esterification of beta-chitin via intercalation by carboxylic anhydrides. *Biomacromolecules* 7(10):2878–2881.

Zhang, M. and H. X. Ren. 2007. Structural modification and application of chitosan. *Journal of Rehabilitation Tissue Engineering and Research* 11:9817–9820.

Zheng, L. and H. F. Cui. 2012. Enhancement of nerve regeneration along a chitosan conduit combined with bone marrow mesenchymal stem cells. *Journal of Materials Science-Materials in Medicine* 23(9):2291–2302.

Zohuriaan-Mehr, M. J. 2005. Advances in chitin and chitosan modification through graft copolymerization: A comprehensive review. *Iranian Polymer Journal* 14(3):235–265.

Part III

Biomedical Applications of Chitin
and Chitosan Derivatives

15 Chitosan
Amazing Controlled Delivery System

Maher Z. Elsabee and Rania E. Morsi

CONTENTS

15.1 INTRODUCTION

Controlled delivery system (CDS) may be defined as a bioactive carrier system that has the ability to release the incorporated bioactive agent in a controlled manner so as to have a prolonged therapeutic effect. The efficiency of drugs, for example, would increase enormously if they were directed selectively to their cellular targets, a concept first introduced by Paul Ehrich at the beginning of the twentieth century.

An ideal CDS should release the bioactive agent in such a way as to maintain the systemic bioactive concentration within minimum effective concentration level and maximum therapeutic level for a prolonged period of time, which may help in decreasing the dosing frequency.

Various polymers have been used for the development of CDS. Polymeric materials provide the most important avenues for research concerning drug delivery systems, primarily because of their ease of processing and the ability to control their chemical and physical properties. Increasing attention has been given to designing polymeric drug carriers that are biodegradable, nontoxic, and tissue- and body-compatible. Currently, scientists are focusing on the development of CDSs using polymers of natural origin because of their easy availability, low cost, and excellent biocompatibility. These systems further present the versatility of allowing the incorporation of suitable amounts of drugs and improving the bioavailability of degradable drugs and the permeation of hydrophilic substances across epithelial layers (Dhawan 2004). The applicability of natural polysaccharides, such as agar, konjac, and pectin, in the design of dosage forms for sustained release has been reported (Takahashi et al. 1978, 1978, Nakano et al. 1979a,b).

Chitosan (CS) (Felt et al. 1998) is one of the most favorable candidates, which possesses these characteristics (Figure 15.1). CS, one of the widely used polysaccharide-based delivery systems across length scale, that is, nano to macro is widely known in CDS. Before going into the details of the CS-based delivery system, it becomes necessary to understand the chemistry and properties of CS so as to effectively develop CS-based CDSs.

CS is obtained from chitin by heterogeneous alkaline deacetylation using concentrated NaOH solutions (Abdou et al. 2008). Among numerous possible applications, CS seems to be a very promising candidate for the development of drug delivery systems (Muzzarelli 1997, Alishahi et al. 2011). The solubility, conformation, and dimensions of CS chains in aqueous media have been extensively studied as a function of the degree of acetylation (DA) (Shigemasa and Minami 1995). However, the application of CS was limited owing to the insolubility at neutral or high pH region. To improve

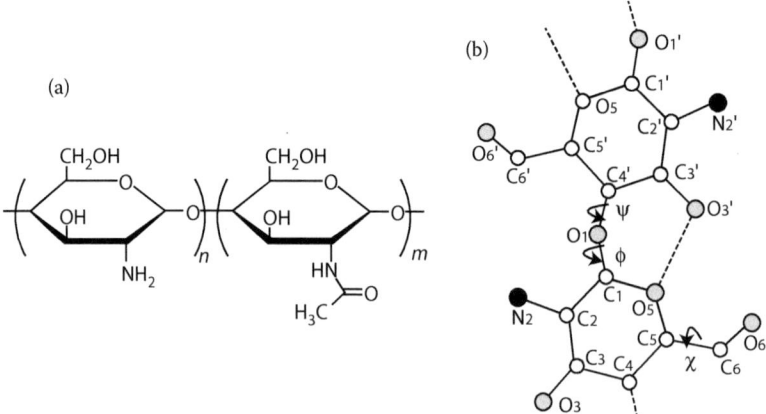

FIGURE 15.1 (a) Chitosan structure. (b) Chemical structure of chitosan. Individual atoms are numbered. Dashed lines denote O3–O5 hydrogen bonds. Two dihedral angles (ϕ, ψ) defining the main chain conformation and one dihedral angle χ defining the O6 orientation are indicated. (From Agrawal, P., G. J. Strijkers, and K. Nicolay. 2009. *Advanced Drug Delivery Reviews* 62:42–58. With permission.)

the soluble property of CS, generally, various CS derivatives with enhanced water solubility are introduced through the chemical modification process. It was reported (Sashiwa and Shigemasa 1999) that, for example, the *N*-acyl and *N*-alkyl derivatives of CS having carboxy or sulfate group are soluble at basic pH region. Moreover, some of CS derivatives substituted with saccharides are soluble at all pH ranges and were stable for at least 2 weeks. These derivatives would be useful for the treatment at neutral or basic pH region to test biological activities *in vitro* and *in vivo*.

The development of radiopharmaceuticals designed to bind specific receptors, including membrane transport systems, is receiving much interest due to their potential to achieve improved *in vivo* monitoring of biochemical and physiological functions (Nunn 1992).

CS has been considered as one of the material of choice for the preparation of NPs in various applications due to its biodegradable and nontoxic properties. Chitosan NPs (CNPs) are formed spontaneously on the incorporation of polyanion such as tripolyphosphate (TPP) in CS solution under continuous stirring condition. These NPs are then harvested and used for gene therapy and drug delivery applications (De Campos et al. 2001, Jayakumar et al. 2010).

15.2 ANTIMICROBIAL ACTIVITY OF CHITOSAN AND ITS DERIVATIVES

CS in its free polymer form has been proved to have antifungal activity against *Aspergillus niger*, *Alternaria alternata*, *Rhizopus oryzae*, *Phomopsis asparagi*, and *Rhizopus stolonifer* (Guo et al. 2006, Seyfarth et al. 2008, Guerra-Sanchez et al. 2009, Ing et al. 2012). It has been shown by many researchers that the antifungal activity of CS was influenced by its molecular weight (MW), degree of substitution, concentration, type of fungus, and type of functional groups in CS derivative chains (Eweis et al. 2006, Tikhonov et al. 2006, Zhong et al. 2007, Ziani et al. 2009, Tayel et al. 2010, Garca-Rincon et al. 2010). In addition, the inhibitory effect was also influenced by particle size and zeta potential of CNPs (Ing et al. 2012). Basically, the antifungal activity is contributed by the polycationic nature of CS. Therefore, CS exhibits natural antifungal activity without the need of any chemical modification (Liu et al. 2004). There are three mechanisms proposed as the inhibition mode of CS. In the first mechanism, the plasma membrane of the fungus is the main target of CS. The positive charge of CS enables it to interact with the negatively charged phospholipid components of fungal membrane. This will increase the permeability of the membrane and cause the leakage of cellular contents, which subsequently leads to cell death. For the second mechanism, CS acts as a chelating agent by binding to trace elements, making the essential nutrients unavailable for normal growth of fungi (Roller and Covill 1999). Lastly, the third mechanism proposed that CS could penetrate the cell wall of fungi and bind to its deoxyribonucleic acid (DNA). This will inhibit the synthesis of mRNA and, thus, affect the production of essential proteins and enzymes (Kong et al. 2010).

15.3 CHITOSAN MUCOADHESIVE PROPERTIES AND ABSORPTION ENHANCEMENT

The mucoadhesive properties are based on CS's cationic character. The mucus gel layer exhibits anionic substructures in the form of sialic acid and sulfonic acid substructures. Based on ionic interactions between the cationic primary amino groups of CS and these anionic substructures of the mucus, mucoadhesion can be achieved. This is beneficial for prolonged residence time at the site of drug absorption owing to increased contact with the absorbing mucosa, resulting in a steep concentration gradient to favor drug absorption, and localization in specified regions to improve the bioavailability of drugs. For optimum mucoadhesion, there has to be an intimate contact between the adhesive and the substrate and interpenetration of the polymer chains with the mucin glycoprotein network.

CS interacts with mucin by multiple modes, namely, due to molecular attractive forces formed by electrostatic interaction between positively charged CS and negatively charged mucosal surfaces. These properties may be attributed to strong hydrogen bonding groups such as –OH, –COOH, strong charges, high MW, sufficient chain flexibility, and to surface energy properties favoring spreading into mucus (Qaqish and Amiji 1999). Factors such as cross-linking status and ionic modification can significantly influence the mucoadhesivity of polymeric microparticles. Dhawan et al. (2004) showed that the microspheres prepared by emulsification followed by ionotropic gelation were found to be more mucoadhesive as compared with those obtained by other methods. This can be explained by the fact that the extent of adsorption of mucin is dependent on the charge of CS microspheres. The CS microspheres cross-linked by thermal and chemical (glutaraldehyde) process showed the lowest charge value, whereas the highest charge values were obtained from ionotropic gelated microparticles (He et al. 1998). They also demonstrated the relationship between the charge of CS microspheres, and negative charge of mucus glycoprotein and mucoadhesion extent. Thus, the electrostatic attraction between the positively charged mucoadhesive CS microspheres and the negatively charged mucus glycoprotein plays an important role in the adsorption of CS microspheres on mucin. The exceptional mucoadhesive properties of CS, in comparison to other polymers such as polycarbophil, used as a reference substance, were found for the first time employing CS films in pig intestinal mucosa (Lehr et al. 1992). Grabovac et al. (2005) investigated the mucoadhesive properties of several polymers by the measurement of time of mucoadhesion on porcine small intestinal mucosa and of total work of adhesion (TWA) by tensile studies. It was shown that the mucoadhesive profile of the precipitated and lyophilized form of CS was pH dependent. At acid medium (pH 3.0), CS was immediately disintegrated. The best results of CS occurred for the lyophilized form, at higher pH values (pH 6.5 and 7.0). Takeuchi et al. (2003) compared the mucoadhesive properties of carbopol and CS-coated liposomes using intestines isolated from male Wistar rats. The mucoadhesive properties of carbopol-coated liposomes were comparable with that of CS-coated liposomes and both formulations improved the enteric absorption of the encapsulated peptide. CS solutions, gels, films, sponges, and micro- and nanoparticles have demonstrated to promote absorption of small polar molecules and peptide/protein drugs through ocular (Majumbar et al. 2008, Sahoo et al. 2008, Yuan et al. 2008), nasal (Gavini et al. 2006, Costantino et al. 2007), pulmonary (Grenha et al. 2007, Ventura et al. 2008), oral (Thanou et al. 2001, Portero et al. 2007), and intestinal mucosa (Ventura et al. 2008, Borchard et al. 1996, Şenel and Hincal 2001) by using animal models and human volunteers. The absorption enhancement of the peptide analog buserelin (gonadotropin-releasing hormone agonist) was studied after transduodenal coadministration with CS gels (pH 6.7) in rats (Borchard et al. 1996). CS substantially increased the bioavailability of the peptide in comparison to control (no polymer) or carbopol 934P-containing formulations. CS was also found to exert marked permeabilization effect on buccal mucosa for transforming growth factor-β (TGF-β) by *in vitro* test in porcine oral mucosa treated with CS gels at 2% concentration entrapping TGF-β (Şenel et al. 2000). As already mentioned, the mechanism of CS absorption enhancement has been suggested to be a combination of mucoadhesion and an effect on the gating properties of tight junctions from epithelium cells (Lehr et al. 1992). On another work, Shapiro et al. (1994) established that CS with high DD is effective as permeation enhancer at low and high MW and also showed clear dose-dependent toxicity, whereas CS of low DD is effective at only high MW and showed low toxicity. Dodane et al. (1999) studied the effects of CS on Caco-2 cells and observed, by confocal microscopy, a decrease of staining of tight junction proteins on cells that received CS treatment. Furthermore, CS induced redistribution of F-actin. Because actin has been shown to be important in regulating paracellular flow across cultured intestinal epithelia, the above effects of CS nonepithelial barrier function might be due to a partial alteration of the cytoskeleton. The increased paracellular permeability was not accompanied by apparent changes in the junctional morphology. All these effects were reversible, indicating that CS had a transient effect on the cellular barrier. These data suggest that CS could be used as a permeability enhancer without causing membrane perturbations such as the effect described after treatment with sodium dodecyl sulfate. CS causes relatively mild and reversible effects on epithelial morphology, which makes it an advantageous absorption-enhancing compound for mucosal delivery of drugs (Dodane et al. 1999).

Furthermore, to achieve high mucoadhesive property, the polymer needs to exhibit high cohesive property; otherwise, the adhesive bond fails within the mucoadhesive polymer rather than between the mucus gel layer and the polymer. In case of CS, however, these cohesive properties are also comparatively weak. Although they can be strongly improved by the formation of complexes with multivalent anionic drugs, multivalent anionic polymeric excipients and multivalent inorganic anions, this strategy is only to a quite limited extent effective as the cationic substructures of CS being responsible for mucoadhesion via ionic interactions with the mucus are in this way blocked. Lueβen et al. (1996), for instance, demonstrated a significantly improved oral bioavailability of buserelin when being administered with mucoadhesive polymers such as CS and carbomer to rats. This effect, however, could not be observed anymore when CS was combined with the polyanionic carbomer in the same formulation. Trimethylation of the primary amino group of CS provides an even more cationic character of the polymer. When trimethylated chitosan (TMC) is additionally PEGylated, its mucoadhesive properties are even up to 3.4-fold improved (Jintapattanakit et al. 2009). Owing to the immobilization of thiol groups on CS, its mucoadhesive properties can also be strongly improved, as the thiolated polymer is capable of forming disulfide bonds with mucus glycoproteins of the mucus gel layer, placing it among the most mucoadhesive polymers known so far (Werle and Bernkop-Schnürch 2008). In addition, as inter- and intrachain disulfide bonds are also formed within CS itself, thiolated CS exhibits substantially improved cohesive properties. Recently, the mucoadhesive properties of thiolated CSs were even significantly further improved by the preactivation of the thiol groups on CS via the formation of disulfide bonds with mercaptonicotinamide.

15.4 BIODEGRADABILITY OF CHITOSAN

Biodegradation plays a major role in the metabolic fate of CS in the body and is important with respect to all polymers used in drug delivery systems, scaffolds in tissue engineering and in case of systemic absorption of hydrophilic polymers, such as CS. For example, a suitable MW is required for renal clearance from 30,000 to 40,000, depending on the polymer used. If the administered polymer's size is larger than this, then the polymer must undergo degradation (Dash et al. 2011). Both chemical and enzymatic biodegradation would provide fragments suitable for renal clearance. Chemical degradation is referred to acid-catalyzed degradation such as in the stomach. Enzymatically, CS can be degraded by enzymes able to hydrolyze glucosamine–glucosamine, glucosamine–N-acetylglucosamine, and N-acetylglucosamine–N-acetylglucosamine linkages (Kean and Thanou 2009). Although depolymerization through oxidation–reduction reaction (Hsu et al. 2002) and free radical degradation (Zoldners et al. 2005) of CS has been reported, these are unlikely to play a significant role in the *in vivo* degradation. CS is known to be degraded in vertebrates predominantly by lysozyme and by certain bacterial enzymes in the colon (Kean and Thanou 2009). Eight human chitinases (in the glycosidehydrolase 18 families) have so far been identified, three of which have shown enzymatic activity (Funkhouser and Aronson 2007). A variety of microorganisms synthesizes and/or degrades chitin, the biological precursor of CS. Both rate and extent of CS biodegradability in living organisms are degree of acetylation (DA) dependent, with a decrease in the degradation rate with increasing DA (Yang et al. 2007, Xu et al. 1996). Basically, given adequate time and appropriate conditions, CSs, in most cases, would degrade sufficiently to be excreted (Kean and Thanou 2009).

15.4.1 CHITOSAN *IN VIVO* BIODEGRADATION

Very little has been reported on the *in vivo* degradation of CS. The mechanism of degradation is currently unclear, especially after intravenous (i.v.) injection. However, studies do indicate that distribution, degradation, and elimination processes are strongly dependent on MW. The liver and kidney were found to be the possible sites of degradation inferred due to the localization of CS. In rabbits, when injected intravenously, CS oligosaccharides enhanced lysozyme activity in the blood

(Hirano et al. 1988). In another study, of subcutaneous implantation of CS, a proposed skin substitute such as glutaraldehyde cross-linked CS/collagen was found to be relatively stable over time compared to collagen alone when implanted subcutaneously in rabbits (Ma et al. 2003).

One of the upcoming areas of study related to CS is its biodistribution, especially using methods other than i.v. administration. The distribution of CS formulates in the body is related to all aspects of the CS formulation from the MW and DD to the size of the delivery vehicle. For instance, in case of nanoparticulate formulations, the kinetics and biodistribution will initially be controlled by the size and charge of the NPs and not by CS structural features. However, after particle decomposition to CS and free drug, inside the cells or target tissues, free CS will distribute in the body and eliminate accordingly. The elimination processes may be preceded by biodegradation. To understand CS biodistribution, the kinetics of its labeled (radio or fluorescent) modifications should be followed, assuming that the label is neither labile nor affecting the physicochemical properties of the CS (Kato et al. 2000).

Banerjee et al. (2002) investigated the distribution of intravenously injected [99 m]Tc-labeled NPs (<100 nm) in Swiss albino mice. Radio-label stability of the NPs was tested and 80% of the radioactivity was associated with the particles after 3 h. NPs were administered in mice and an apparent scavenging played of the reticuloendothelial system (RES) was suggested as radioactivity decreased in organs of this system but remained stable in the blood after 1 h. "Reticuloendothelial system" is an older term for the mononuclear phagocyte system (MPS). The MPS is a part of the immune system that consists of the phagocytic cells located in reticular connective tissue. The cells are primarily monocytes and macrophages and they accumulate in the lymph nodes and the spleen. The Kupffer cells of the liver and tissue histocytes are also part of the MPS. In most of the studies, liver was found to be a significant site of accumulation due to the action of Kepfer cells; this could be due to this organ being a primary site of metabolism as seen with radio-labeled dextran (Line et al. 2000).

15.4.2 Chitosan *In Vitro* Biodegradation

Viscometry and/or gel permeation chromatography are the commonly used assays to determine the decrease in MW (Onishi and Machida 1999). Following an *in vitro* incubation using lysozyme at pH 5.5 in a phosphate buffer at 37°C for 4 h, 50% acetylated CS lost 66% in viscosity (Onishi and Machida 1999). This degradation is DD dependent with the degradation of the higher acetylated CS behaving more like CS (Zhang and Neau 2002, Senel and McClure 2004). In addition, a range of proteases was found to degrade CS films, with leucine aminopeptidase being the most effective with 38% over 30 days (Rao and Sharma 1997). Pectinase isozyme from *Aspergillus niger* has also indicated CS digestion at low pH releasing lower-MW fragments (Kittur et al. 2003, 2005). From the digestion of CS with rat cecal and colonic bacterial enzymes, Zhang and Neau (2002) observed that extracellular enzymes were responsible for degradation that was related to both DD and MW. McConnell et al. (2008) used human fecal preparations and studied the degradation of CS films, cross-linked by glutaraldehyde and TPP, with interesting results. The rate of porcine pancreatic enzymes to degrade CS film was influenced by cross-linker type; for example, glutaraldehyde degraded more readily than TPP.

15.4.3 *In Situ* Gelling Properties

In case of hydrogels, CS offers the advantage of *in situ* gelling properties when its pH-dependent hydratability is addressed properly from the formulation point of view. Gupta and Vyas (2010) developed an *in situ* gelling delivery system by the combination of polyacrylic acid and CS. The resulting formulation was in liquid state at pH 6.0 and underwent rapid transition into the viscous gel phase at physiological pH of 7.4. These *in situ* gelling properties can even be further improved by thiolation. Because of access to oxygen on mucosal surfaces, such as the ocular or nasal mucosa, after having been applied in liquid form out of, for instance, oxygen-free single unit dosage forms, a cross-linking process via disulfide bond formation takes place, resulting in a strong increase in viscosity. Utilizing this cross-linking mechanism, Sakloetsakun et al. (2009) could show an even 16,500-fold increase in viscosity within 20 min of an aqueous 1% (m/v) CS–thioglycolic acid (TGA) conjugate.

15.4.4 TRANSFECTION-ENHANCING PROPERTIES

Transfection is the process of deliberately introducing nucleic acids into cells. The term is used notably for nonviral methods in eukaryotic cells. In contrast to small molecules, where a controlled release of anionic drugs can be achieved, comparatively large polyanionic molecules such as DNA-based drugs and siRNA form stable complexes with CS. In this way, NPs exhibiting a positive zeta potential can be formed, when the ratio of the cationic polymer is sufficiently high. Because of this net positive charge and the small size of these particles, endocytosis can be achieved, in particular when these particles are below 100 nm in size (Mao et al. 2010). As CS is comparatively less toxic than other cationic polymers, such as polyethyleneimine, polylysine, or polyarginine (Yu et al. 2007), it is therefore a promising excipient for nonviral gene delivery systems. In addition, CS–DNA-based drug complexes protect at least to some extent toward degradation by DNAses in this way improving the bioavailability of DNA-based drugs delivered into the body (Lee and Mohapatra 2008). As the transfection efficiency of conventional CS is generally found to be low, the polysaccharide was modified following different strategies to improve its properties. Malmo et al. (2011) investigated the self-branching of CS as a strategy to improve its gene transfer properties without compromising its safety profile. Self-branched and self-branched trisaccharide-substituted CS with molecular mass of 11–71 kDa were synthesized, characterized, and compared with their linear counterparts with respect to transfection efficiency, showing that self-branched CS can yield gene expression levels two and five times higher than those of Lipofectamine and Exgen, respectively. Lipofectamine is a common transfection reagent; it is a cationic liposome formulation used in molecular and cellular biology. It improves the efficiency of transfection by helping the transfected genetic material penetrate the nuclear envelope during mitosis by forming complexes with the negatively charged nucleic acid molecules to allow them to overcome the electrostatic repulsion of the cell membrane (Fallini et al. 2010).

Following another approach, Martien et al. (2007) stabilized CS/plasmid NPs by utilizing thiolated CS forming intrachain disulfide bonds within the complex. In this way, a higher stability toward nucleases was achieved. In addition, the plasmid was mainly released in the target cells, as because of the reducing conditions of the cytoplasm, the disulfide bonds were mainly cleaved there leading to the release of the plasmid at the target site. The transfection rate of thiolated CS/plasmid NPs was in this connection demonstrated to be five times higher than that of unmodified CS/pDNA NPs. This strategy could even be further improved by raising the cationic character of thiolated CS by trimethylation of the remaining primary amino groups (Varkouhi et al. 2010). Furthermore, PEGylated CS and CS/cyclodextrin NPs were identified as promising tools for DNA-based drug delivery (Teijeiro-Osorio et al. 2009, Malhotra et al. 2011).

15.5 BIOLOGICAL PROPERTIES OF CHITOSAN

CS has proved to be a safe excipient in drug formulations during the last few decades. Clinical tests carried out using CS-based biomaterials do not report any inflammatory or allergic reactions following implantation, injection, topical application, or ingestion in the human body (Malhotra et al. 2011). The *in vitro* and *in vivo* cytocompatibility of chitosan films with keratinocytes and fibroblasts has demonstrated that the DD has no significant influence. The CS films with a low DD are very good biomaterials for superficial wound healing (Chatelet et al. 2001). Once placed on the wound, they adhere to fibroblasts and favor the proliferation of keratinocytes and thereby epidermal regeneration. One of the upcoming areas of study related to CS is its biodistribution, especially using methods other than i.v. administration. The distribution of chitosan formulates in the body is related to all aspects of the chitosan formulation from the MW and DD to the size of the delivery vehicle. For instance, in case of nanoparticulate formulations, the kinetics and biodistribution will initially be controlled by the size and charge of the NPs and not by chitosan structural features. However, after particle decomposition to chitosan and free drug, inside the cells or target tissues, free chitosan will distribute in the body and eliminate accordingly. Elimination processes may be preceded by biodegradation. To understand chitosan biodistribution, the kinetics of its labeled (radio or fluorescent) modifications should be followed, assuming that the label is neither labile nor affecting the physicochemical properties of the chitosan (Kato et al. 2000). Banerjee et al. (2005) investigated the distribution of intravenously injected ^{99m}Tc-labeled NPs (<100 nm) in Swiss albino mice. Radio-label stability of the NPs was tested and 80% of the radioactivity was associated with the particles after 3 h. NPs were administered in mice and an apparent scavenging played of the RES was suggested as radioactivity decreased in organs of this system but remained stable in the blood after 1 h (Banerjee et al. 2002). Unfortunately, the NPs were not sufficiently stable to look at long-term distributions. In a later study, by the same author, sodium borohydride was used in the preparation conditions. The modified method prevented aggregation, improved the kinetics of the NPs, and resulted in both less blood clearance and liver accumulation, that is, avoidance of the RES. The radioactivity in blood was present for up to 10% even after 2 h (Banerjee et al. 2005). In most of the studies, liver was found to be a significant site of accumulation due to the action of Kepfer cells; this could be due to this organ being a primary site of metabolism as seen with radio-labeled dextran (Line et al. 2000).

A suggestion was made by Kean and Thanou who proposed that a potential method to study native chitosan without significant modification would be to use ^{14}C as a label, for example, in the food source for the animal/fungi producing the chitin so that the saccharide backbone is labeled, as the detection of native chitosan is appearing to be a challenge. To study the intraperitoneal administration of chitosan, fluorescein isothiocyanate (FITC)-labeled chitosan (50% DD, 100 kDa) was prepared by FITC coupling. It was observed that this labeled chitosan was completely absorbed from the peritoneal cavity (no evidence in abdominal fluid after 14 h). FITC-chitosan was found to be predominantly localized in the kidney at 1 h in a mouse model. There was a rapid renal excretion rate (25% at 1 h, 100% in 14 h) with evidence of degradation due to a decrease in the MW (Kean and Thanou 2010).

Some studies suggest that chitosan binds fat and reduces cholesterol but the mechanism is still questionable (Ravi Kumar 2000, Xu et al. 2007). Apart from the effect that chitosan may have on bile salts and gastrointestinal milieu, the uptake of chitosan into the bloodstream is generally not investigated in oral administration studies. MW has been a parameter that largely influences chitosan systemic absorption and distribution from this route of delivery. It has been seen in some cases that oligomers showed some absorption, whereas larger-MW chitosan were excreted without being absorbed. This effect was investigated using FITC-labeled chitosan with 3.8 kDa (88.4% DD), which was shown to result into the greatest plasma concentration after oral administration compared to 230 kDa (84.9% DD) for which nearly no uptake was detected. In one of the studies aimed at investigating plasma concentration after oral administration, the increasing MW was seen to decrease the plasma concentration (Chae et al. 2005). Although intracellular uptake and distribution of native chitosan have not been investigated, chitosan/DNA complexes have been studied *in vitro*. Chitosan polyplex uptake at 37°C was threefold higher than at 4°C, which could be because of increased interaction and not an ATP-dependent endocytotic mechanism (Ishii et al. 2001). The authors suggested nuclear localization and they also stated little dissociation of DNA from chitosan. In a more extensive study, Leong et al. (1998) stained lysosomes and found some colocalization with chitosan/DNA NPs. However, the majority of the polyplexes were found in the cytosol. A complex of doxorubicin with chitosan was studied, where it was observed that complexes enter cells through an endocytotic mechanisms that was not further elucidated (Janes et al. 2001).

15.5.1 CHITOSAN TOXICITY

Chitosan is considered as being a nontoxic, biologically compatible polymer (Thanou et al. 2001). It is approved for dietary applications in Japan, Italy, and Finland (Illum 1998), and it has been approved by the FDA for use in wound dressings (Wedmore et al. 2006). However, certain modifications implemented on chitosan could make it more or less toxic and any residual reactants should be removed carefully.

15.5.2 IN VITRO TOXICITY

In a series of studies, Schipper et al. (1996) observed the effects of chitosan samples characterized by different MW and DD on Caco-2 cells, HT29-H, and *in situ* rat jejunum. Toxicity was found to be dependent on DD and MW. At high DD, the toxicity is related to the MW and the concentration; at lower DD, toxicity is less pronounced and less related to the MW. Nevertheless, most of the chitosan tested did not increase the dehydrogenase activity significantly in the concentration range tested (1–500 μg/mL) on Caco-2 cells. The *in situ* rat jejunum study showed no increase in lactate dehydrogenase (LDH) activity with any of the chitosan samples tested (50 μg/mL) (Schipper et al. 1999). Red blood cell hemolysis assay is a study that reveals the safety of materials. No hemolysis was observed (<10%) over 1 and 5 h with chitosan of <5, 5–10, and >10 kDa at concentrations of up to 5 mg/mL (Richardson et al. 1999). Furthermore, no red blood cell lysis was observed with paclitaxel chitosan micelles at 0.025 mg/mL (Zhang et al. 2008a,b). Interestingly, chitosan and its derivatives seem to be toxic to several bacteria (Jumaa et al. 2002), fungi (Guo et al. 2006), and parasites (Pujals et al. 2008). This pathogen-related toxicity is an effect that could be beneficial in the control of infectious disease. Bacterial inhibition took place in acidic solutions of pH 5–5.3, and 87 kDa 92% DD chitosan was more effective than a 532 kDa 73% DD chitosan against both *Pseudomonas aeruginosa* and *Staphylococcus aureus*. Antimycotic effect against *Candida albicans* and *Aspergillus niger* was observed in a lipid emulsion of the same chitosan (Jumaa et al. 2002).

15.5.3 IN VIVO TOXICITY

In a study by Hirano et al. (1991), which was relatively long (65 days), no detrimental effect on body weight was found when chitosan oligosaccharides were injected (7.1–8.6 mg/kg over 5 days). An increase in lysozyme activity was apparent on the first day postinjections. Rao and Sharma (1997) stated no "significant toxic effects" of chitosan in acute toxicity tests in mice, and no eye or skin irritation in rabbits and guinea pigs, respectively. He also concluded in the same study that chitosan was not pyrogenic. However, no concentration or DD of the chitosan used was noted. No detrimental effects were also noted by Richardson et al. (1999), although there was a mention of dose effect in his work. The LD50 of paclitaxel chitosan micelles in mice was 72.2 mg/kg, no anaphylaxis was observed in guinea pigs, and no i.v. irritation was observed histopathologically in rabbits at 6 mg/kg. In a study on fat chelation, 4.5 g/day chitosan (MW and DD not specified) in humans was reported not to be toxic, although no significant reduction in fat was observed (Gades and Stern 2003). Arai et al. (1968) found that chitosan has an LD50 comparable to sucrose of >16 g/kg in oral administration to mice. No oral toxicity was found in mice treated with 100 mg/kg CNPs (80 kDa, 80% DD) (Sonaje et al. 2009). Exposure of rat nasal mucosa to chitosan solutions at 0.5% (w/v) over 1 h caused no significant changes in mucosal cell morphology compared to control (Illum et al. 1994). From most studies reported, it appears that chitosan shows minimal toxic effects and this approves its adoption as a safe material in drug delivery.

15.6 DRUG DELIVERY

Researchers have strived to engineer the physical and chemical properties of CDSs to specifically regulate their permeability, environmental response, surface functionality, biodegradability, and biorecognition sites to produce "intelligent" DDSs. Chitosan represents CDS class that has excelled at intelligent drug delivery. Chitosan has been used in the preparation of mucoadhesive formulations improving the dissolution rate of the poorly soluble drug (Genta et al. 1994), drug targeting (Gallo and Hassan 1988, Hassan et al. 1992), and enhancement of peptide absorption (Agnihotri et al. 2004). Chemical attachment of the drug to the chitosan throughout the functional linker may produce useful prodrugs exhibiting the appropriate biological activity at the target site (Park et al. 2010).

Chitosan was used effectively to deliver tetracycline into the stomach to eradicate the microorganism, known as *Helicobacter pylori*, which is the main factor responsible for the formation of gastric ulcer and gastric carcinoma (Hejazi and Amiji 2002) *H. pylori* resides mainly in the gastric mucosa or at the interface between the mucous layer and the epithelial cells of the antral region of the stomach. In acidic medium, the protonated amino groups of chitosan will interact with sialic acid (*N*-acetylneuraminic acid) in the gastric mucus by electrostatic interaction. Thus, chitosan microspheres improve the gastric residence time of a drug that is loaded into the polymer microsphere (Lehr et al. 1992). Chitosan microspheres can also provide pH-responsive release profile by swelling in acidic environment of the gastric fluid (Patel and Amiji 1996). Bhise et al. (2008) designed sustained release systems for the anionic drug naproxen using chitosan as drug carrier matrix. Using polyanionic drugs, the interactions between chitosan and the therapeutic agent are more pronounced and based on an ionic cross-linking. In addition, even stable complexes are formed from which the drug can be released even over a more prolonged time period. Sun et al. (2010), for instance, designed enoxaparin/chitosan nanoparticulate delivery systems, providing very stable complexes that led to a significantly improved drug uptake.

Chitosan can be homogenized with anionic polymeric excipients such as polyacrylates, hyaluronic acid, alginate, pectin, or carrageenan, resulting in comparatively stable complexes of high density. Mainly based on diffusion and erosion processes, incorporated drugs are released in a sustained manner from such complexes (Tapia et al. 2005). Alternative to anionic polymers, multivalent inorganic anions such as TPP or sulfate can be used to achieve the same effect (Shavi et al. 2011).

15.6.1 Colon Targeting

Chitosan-based delivery systems have been widely studied for colonic drug targeting because this system can protect therapeutic agents from the hostile conditions of the upper gastrointestinal tract and release the entrapped agents specifically at the colon through degradation of the glycosidic linkages of chitosan by colonic microflora (Hejazi and Amiji 2003, Park et al. 2010). Strategies for specific drug delivery to the various regions of the GI tract and, in particular, to the colon have been reviewed by Rubinstein (1995). An early strategy for targeting drugs to the colon uses the ecosystem of the specific microflora in the large intestine. There are two main classes of bacterial enzymes: the azoreductases and the polysaccharidases, which are in sufficient quantity so as to be exploited in colonic drug targeting. Based on this idea, different natural and synthetic polymers have been evaluated for their susceptibility of being cleaved by these bacterial enzymes (Friend 1991) and, thus, for their use as major constituents of colon-specific drug delivery systems (Brondsted and Kopecek 1992).

A system consisting of CS microcores entrapped within acrylic microspheres has been proposed and investigated by Lorenzo-Lamosa et al. (1998). Sodium diclofenac (SD), used as a model drug, was efficiently entrapped within CS microcores using spray-drying and then microencapsulated into Eudragit® L-100 and Eudragit S-100 using an oil-in-oil solvent evaporation method. The size of the CS microcores was small (1.8–2.9 mm) and they were efficiently encapsulated within Eudragit microspheres (size between 152 and 223 mm) forming a multireservoir system. Although CS dissolves very fast in acidic media, at pH 7.4, SD release from CS microcores was delayed, the release rate being adjustable (50% dissolved within 30–120 min) by changing the CS MW or the type of CS salt. Furthermore, by coating the CS microcores with Eudragit, perfect pH-dependent release profiles were attained. The probable mechanism of drug release from these CS multicore microspheres at the solubility pH of the coating polymer was given by the authors as follows: once the microspheres reach the small intestine, Eudragit will slowly and continuously dissolve over the time, thus, leaving the CS microcores increasingly exposed to the release medium. The CS microcores, partially cross-linked with Eudragit, will swell upon contact with the basic release medium and form a gel barrier through which the drug will diffuse. After 3–4 h, the CS microcores will reach the colonic region where the CS will undergo degradation process, thereby, triggering the release of the entrapped drug. According to this, there are several factors that may affect the release of the entrapped drug: (a) the pH solubility of the Eudragit coating; (b) the size and swelling behavior of CS microcores; (c) the core/coat ratio; and (d) the microcore coating (Lorenzo-Lamosa et al. 1998).

Tozaki et al. (2002) investigated the use of chitosan capsules for colon-specific delivery of 5-aminosalicylic acid (5-ASA). The surface of the chitosan capsules containing 5-ASA was coated with hydropropyl methylcellulose phthalate as an enteric coating material. The experimental results demonstrated that the capsules were able to reach the large intestine 3.5 h after oral administration into 2,4,6-trinitrobenzenesulfonic acid (TNBS)-induced ulcerative rats. The release of 5-ASA from the capsule was markedly increased in the presence of rat cecal contents. Chitosan capsule-based formulations showed better therapeutic effect than a carboxymethyl cellulose suspension *in vivo*. Chitosan capsules can also be used as carriers for colon-specific delivery of absorption enhancers. Oral delivery of the absorption enhancer along with poorly absorbable drugs using chitosan capsules could improve the absorption characteristics of the drugs (Fetih et al. 2005).

Other chitosan capsules were used for the same target, which is the delivery of the 5-ASA into the colon (Varshosaz et al. 2006). In this case, the capsules were coated with cellulose acetate butyrate. Capsules made from chitosan blend with Ca-alginate labeled with ^{131}I showed that 5-ASA was localized in the colon with systemic bioavailability (Mladenovska et al. 2007). Hydrogel microspheres of chitosan grafted with vinyl polymers were used for the controlled and targeted delivery of 5-ASA to the colon. These microspheres exhibited good therapeutic effects (Jain et al. 2008).

Another chitosan hydrogel beads with a pH-sensitive property and specific biodegradability for colon-targeted delivery of satranidazole has been developed by Jain et al (2007). Chitosan hydrogel beads were prepared by the cross-linking method followed by enteric coating with Eudragit S100. The amount of the drug released after 24 h from the formulation was found to be 97.67% ± 1.25% in the presence of extracellular enzymes as compared with 64.71% ± 1.91% and 96.52% ± 1.81% release of drug after 3 and 6 days of enzyme induction, respectively, in the presence of 4% cecal content (cecum is the large blind pouch forming the beginning of the large intestine; it is also called the *blind gut*). Degradation of the chitosan hydrogel beads in the presence of extracellular enzymes as compared with rat cecal and colonic enzymes indicates the potential of this multiparticulate system to serve as a carrier to deliver macromolecules specifically to the colon. A similar multiparticulate system by coating cross-linked chitosan microspheres with Eudragit L-100 and S-100 as pH-sensitive polymers, for targeted delivery of metronidazole, a broad-spectrum antibacterial agent, to the colon, has been investigated by the same group (Chourasia and Jain 2004). *In vitro* drug-release studies were performed in conditions simulating stomach-to-colon transit in the presence and absence of rat cecal contents. The results showed a pH-dependent release of the drug attributable to the presence of the Eudragit coating. Moreover, the release of drug was found to be higher in the presence of rat cecal contents, indicating the susceptibility of the chitosan matrix to colonic enzymes.

Hyaluronic acid-coupled CNPs bearing 5-flurouracil were prepared by the ionotropic gelation method, for the effective delivery of drug to the colon tumors (Jain and Jain 2008). The NPs were spherical in shape with mean size around 150 nm. The *in vitro* drug release was investigated in different simulated gastrointestinal (GIT) fluids. The biocompatibility of NPs formulations were evaluated for *in vitro* cytotoxicity by MTT assay using HT-29 cell lines and cell uptake was assessed by fluorescent microscopy. The NPs showed significant higher uptake by cancer cells as compared to uncoupled NPs and the uptake of HA coupled NPs by HT-29 colon cancer cells were observed to be 7.9 times more as compared to uncoupled NPs at the end of 4 h.

Microspheres of chitosan and succinyl-prednisolone (Ch-SP-MS) and Eudragit L100-coated Ch-SP-MS (Ch-SP-MS/EuL) were prepared with a mean size of 1.5 and 26.6 μm, respectively, and a drug content of 4.6% and 3% (w/w), respectively (Oosegi et al. 2008). Prednisolone (PD) was released very slowly at pH 1.2 and gradually at pH 6.8. The addition of cecal or colonic content did not accelerate the release. Rats with 2,4,6-trinitrobenzenesulfonic acid (TNBS)-induced colitis were used in animal studies. Gastrointestinal distribution and plasma concentration were investigated by the oral administration of PD alone and Ch-SP-MS/EuL. For PD alone, PD was distributed at the stomach and small intestine, and disappeared from the gastrointestinal tracts within 8 h. When administering Ch-SP-MS/EuL, the drug was distributed mainly in the lower intestine between 3 and 24 h. Plasma concentration was much lower in Ch-SP-MS/EuL than in PD alone, suggesting lower toxic side effects of Ch-SP-MS/EuL. Thus, Ch-SP-MS/EuL delivered PD specifically near the diseased site and PD was released gradually with much less plasma concentration of PD. Ch-SP-MS/EuL are suggested, thus, as a useful delivery system to the site of inflammatory bowel disease.

15.6.2 LIVER TARGETING

Recently, there have been many studies on targeting liver systems using methods such as passive trapping of NPs or active targeting based on hepatic receptor recognition (Kato et al. 2001a,b, Kim et al. 2005, Wang et al. 2010, Zheng et al. 2011). Kato et al. (2001a,b) investigated the biodistribution of fluorescently labeled N-succinyl-chitosan (Suc-FTC) and lactosaminated N-succinyl-chitosan (Lac-Suc-FTC) after i.v. administration to mice intravenously inoculated with M5076 cells at 3 and 12 days postinoculation. At both time points, Lac-Suc-FTC was specifically localized to the liver. The antitumor effects of mitomycin C (MMC), Lac-Suc-MMC conjugate (Lac-Suc-MMC), and highly succinylated Suc (Suc(II))-MMC) conjugate were examined on single i.v. administration for both metastatic stages. Both carriers, Suc showing systemic long circulation and Lac-Suc with an ability of liver-specific localization, are thought to be drug carriers with potentialities for therapeutics at an early stage of metastasis. NPs with moderate size can be delivered to specific sites by size-dependent passive targeting. It has been reported that when the diameter is less than 200 nm, the NPs can be captured easily by Kupffer cells in the liver.

Active targeting systems of receptor recognition can be attained by using the molecules with receptor trapping specific ligands. It is known that hepatocytes can recognize the asialoglycoprotein receptor (ASGP-R) among the liver-associated cell surface receptors. For this reason, the ASGP-R has been exploited as a hepatocyte-specific marker for drug and gene delivery (Park et al. 2003, 2001). A ligand containing galactose moiety can be recognized by, and bound to, the liver-specific galactose receptor. The ligand–receptor complex is taken up rapidly into the cells and the receptor recycles back to the cellular surface (Zhang et al. 2008a,b).

Yang et al. (2009a,b) prepared lactose-conjugated poly(ethylene glycol) (PEG)-graft-chitosan for liver-targeted delivery of diammonium glycyrrhizinate (DG). The lactose conjugated polyion complex (PIC) micelles delivered more DG to the liver than conventional PIC micelles, indicating that LAC-PIC micelles were promising liver-targeted nanocarriers for DG. The limited solubility of chitosan in water prompted the use of water-soluble chitosan derivatives while keeping their favorable biological characteristics. Carboxymethyl chitosan (CMC) fulfills this requirement. CMC is often used as pharmaceutical excipients (Prabaharan and Gong 2008, Prabaharan and Mano 2007, Zhao et al. 2010).

When compared with other modified chitosan materials, thiolated chitosan, as mentioned before, has numerous advantageous features such as significantly improved mucoadhesive and permeation-enhancing properties. Moreover, solutions of thiolated chitosan display *in situ* gelling properties at physiological pH values (Hornof et al. 2003, Kast and Schnurch 2001). The strong cohesive properties of thiolated chitosan make them highly suitable carriers for controlled drug release applications (Kast et al. 2002). Covalently bonded disulfides can be formed spontaneously by autoxidation of sulfhydryls, primarily via oxidation on exposure to air, which can reversibly be cleaved in the presence of reducing agents. The advantage of the NPs with a disulfide cross-linkage in drug delivery is that the disulfide bond is stable in the blood, but cleaved inside the cell. This is because the concentrations of glutathione, the most abundant reducing agent in most cells including mammals, are in a millimolar range inside the cells, whereas those in blood plasma are in a micromolar range (Zheng et al. 2011, Zhang et al. 2008a,b).

A novel thiolated LAC-CMC as hepatic targeting and controlled drug release carriers for glycyrrhizic acid has been synthesized by Zheng et al. (2011). It has been suggested that the dissociation of NPs in the blood is suppressed through the covalent link between thiolated LAC-CMC, since the disulfide bond should be cleaved inside the cell, leading to the release of the entrapped glycyrrhizic acid. The thiolated LAC-CMC/glycyrrhizic acid NPs were prepared by gelification with calcium ions and were characterized by shape, particle size, zeta potential, glycyrrhizic acid encapsulation efficiency (EE), loading capacity (LC), and in vitro release. The pharmacokinetic parameters in rabbits and tissue distribution in mice were evaluated (Zheng et al. 2011). Galactosylated chitosan microspheres were prepared and demonstrated for their utilization for active targeted drug delivery to liver (Zhang et al. 2004). However, not all galactose moieties in the galactosylated chitosan were bound to the galactose receptor and the efficacy of active targeting was reduced because a part of the active moieties may be buried within the microspheres matrix. To increase the efficacy of targeting bovine serum albumin (BSA) with negative charges may strongly adsorb on the soluble galactosyl chitosan with positive charges so that the liver targeting properties on its surface may be significantly changed. The BSA microspheres with galactosyl chitosan was loaded with 5-fluorouracil (5-FU), one of the primary drugs used for the liver cancer treatment to act as both passive and active targeting properties against liver cells and potentially improve the safety and efficiency (Zhang et al. 2008a,b).

Colorectal cancer (CRC) is frequently complicated by the occurrence of metastatic disease, which most commonly presents in the liver and is responsible for approximately 70% of CRC-related deaths. Interleukin-12 (IL-12) is a multifunctional cytokine that enhances T helper cell 1 (Th1) differentiation, cellular immunity, proliferation of natural killer and activated T cells, and has been demonstrated to be one of the most effective inducers of potent antitumor immunity. Unfortunately, excessive toxicity following administration of i.v. IL-12 has proven to be a major obstacle to its clinical application. IL-12 was encapsulated into chitosan using TPP as a coacervated cross-linking agent to form CS-TPP/IL-12 NPs (Xu et al. 2012).

The association efficiency, rate of release, liver-targeting, and toxicity were found to be dependent on the particle size, zeta potential, pH of solution, and whether or not modified with TPP. Systemic delivery of CS-TPP/IL-12 NPs significantly reduced the number and volume of CRC liver metastasis foci compared to the CS-TPP-treated mouse group.

Mechanistically, CS-TPP NPs blocked the toxicity of IL-12 and induced the infiltration of NK cells and some T cells, which are most likely the effector cells that mediate tumor metastasis inhibition during CS-TPP/IL-12 immunotherapy. The results obtained from this study demonstrate the potential benefit of using chitosan modification technology as a cytokine delivery system for the successful prevention of CRC liver metastasis by exploiting liver immunity (Xu et al. 2012).

15.7 CONTROLLED DRUG RELEASE

When sustained drug release cannot be provided by making use of a simple drug dissolution process by diffusion, erosion, membrane control, or osmotic systems, retardation mediated by ionic interactions is often the ultimate ratio. Such a controlled release can be achieved for cationic drugs by using anionic polymeric excipients such as polyacrylates, sodium carboxymethyl cellulose, or alginate. In case of anionic drugs, however, chitosan is the only choice. Bhise et al. (2008), for instance, designed sustained release systems for the anionic drug naproxen using chitosan as drug carrier matrix. Using polyanionic drugs, the interactions between chitosan and the therapeutic agent are more pronounced, and based on an ionic cross-linking in addition, even stable complexes are formed from which the drug can be released even over a more prolonged time period. Sun et al. (2010) has designed enoxaparin/chitosan nanoparticulate delivery systems providing very stable complexes that led to a significantly improved drug uptake. In addition, chitosan can be homogenized with anionic polymeric excipients, such as polyacrylates, hyaluronic acid, alginate, pectin, or carrageenan, resulting in comparatively stable complexes of high density. Mainly based on diffusion and erosion processes, incorporated drugs are released in a sustained manner from such complexes (Tapia et al. 2005). Alternative to anionic polymers, multivalent inorganic anions such as TPP or sulfate can be used to achieve the same effect (Shavi et al. 2011).

15.7.1 Ionic Gelation

TPP is a polyanion that interacts electrostatically with the cationic chitosan (Kawashima et al. 1985a,b). The reversible physical cross-linking by electrostatic interaction, instead of chemical cross-linking, should decrease the potential toxicity impact of reagents and other undesirable effects. For example, Bodmeier and Paeratakul (1989) reported the preparation of TPP–chitosan complex by dropping chitosan droplets into a TPP solution; many researchers have explored its potential pharmaceutical usage (Shu and Zhu 2000).

Various formulations of CNPs produced by the ionic gelation of TPP and chitosan were studied by Xu and Du (2003). The spherical-shaped particles were observed by TEM. The factors that affected the release of BSA as a model protein have been studied, which include MW, DD, and concentrations of chitosan and BSA as well as the presence of PEG in the encapsulation medium.

15.7.2 Chitosan Micro/Nanoparticles: Drug Loading and Release

Drug loading in micro/nanoparticulate systems is achieved using one of the two following methods: (a) incorporating the drug during the preparation of the particles or (b) after the formation of the particles by incubating the drug with them. In both systems, the drug is physically embedded into the matrix as well as adsorbed onto the surface. Both water-soluble and water-insoluble drugs can be loaded by employing these techniques. The water-soluble drugs are generally incorporated by mixing with an aqueous chitosan solution to form a homogeneous mixture, followed by particles production. Water-insoluble drugs and drugs that precipitate in acidic solutions are usually loaded by incubation that involves soaking the preformed particles in a saturated solution of drug (Kumbar et al. 2002).

15.8 CHITOSAN IN BIOIMAGING APPLICATIONS

Chitosan has been shown to improve the dissolution rate of poorly soluble drugs and, thus, can be exploited for bioavailability enhancement of drugs and their delivery. Various therapeutic agents, such as anticancer, anti-inflammatory, antibiotics, antithrombotic, steroids, proteins, amino acids, antidiabetic, and diuretics, have been incorporated in chitosan-based systems to achieve controlled release. A new application related to chitosan is based on the fact that chitosan particles provide an excellent template for bioimaging. Chitosan tissue engineering potential, as a biomaterial to generate structures with predictable pore sizes and degradation rates, makes it particularly suitable for bone and cartilage regeneration.

15.8.1 Preparation Method for Chitosan Nanoparticles Intended as Drug/Protein Carriers

The chitosan for medical applications must be of the highest purity since contaminated protein could impose potential danger in further usage. It is therefore recommended that before the preparation of any nano- or microcapsules, it must be rigorously purified by heating in sodium hydroxide solution for several hours followed by filtration and washing until neutral. CNPs can nowadays be prepared via a number of processes. The emulsion-droplet coalescence method was developed by Tokumitsu et al. (1999), which employs the principles of both emulsion cross-linking and precipitation. CNPs were also prepared by innovative methods such as self-assembly through chemically modifying chitosan molecules (Lee et al. 1998) and reverse micellar through utilization of suitable surfactant for the formation of reversed micelles and drug loading with chitosan attached to the micelles via a cross-linking agent (Mitra et al. 2001). Chitosan–acrylic acid NPs had also been made by polymerization of acrylic acid in chitosan solution (Hu et al. 2002). The resulting NPs have small sizes and carry positive surface charge. Preparation methods by polyionic cross-linking of cationic chitosan molecules with proteins and genes were particularly useful for carrying and delivering the macrosolutes as therapeutic agents. Aside from its strong electrostatic interaction and binding with negatively charged proteins/genes, the chitosan–protein or chitosan–gene system has the ability to gel spontaneously on contact with multivalent polyanions due to the formation of inter- and intramolecular cross-linkage mediated by these polyanions. Among some polyanions investigated, TPP is mostly widely used because of its nontoxic property and quick gelling ability (Kawashima et al. 1985a,b, Gan et al. 2005). Many researchers have explored the capacity of chitosan–TPP nanosystem for association of peptides, proteins, oligonucleotides, and plasmids DNA for potential pharmaceutical usage (Shu and Zhu 2000). Mao et al. (2001) reported CS–DNA NPs prepared by coacervation of CS and DNA in acidic solution. The size of the CS–DNA NPs was in the range of 100–250 nm and such NPs can protect the encapsulated plasmid DNA from nuclease degradation.

In comparison to other NP systems, the chitosan–TPP system forms under mild conditions, is homogeneous with adjustable size, and possesses positive surface charge that can be easily modulated by varying the processing conditions (Quan and Tao 2007).

Protein loading in CNP system can be done by one of two methods of incorporation or incubation. In the incorporation method, BSA was premixed with chitosan solution, adjusted for pH 5.5 and maintained at temperature $20 \pm 2°C$. The formation of BSA loading CNPs started spontaneously when TPP was flush mixed with the BSA–chitosan solution. In the incubation method, CNPs were formed first via TPP coacervation and the particle-containing solution was then mixed with solutions containing BSA at predetermined concentrations. The mixed solutions were gently stirred for 60 min to allow protein adsorption on the NPs to reach isothermal equilibrium. In this method, protein loading is solely via adsorption on the surface of NPs (Quan and Tao 2007).

Protein molecules are large macromolecules able to fold and unfold at different solution conditions; their interactions with long cationic chitosan chain and the consequential encapsulation can be complicated depending on 3-D conformation, and electrostatic and solution conditions. TPP form further hydrogen bonds with free amine groups on both protein and chitosan molecules, resulting in more compact protein–CNPs. Additional adsorption of protein molecules on the surface of the formed particles may occur in sequence, leading to additional protein loading on the particles. The particle size and their zeta potential are affected by the absorption of protein, for example, the particle size without BSA loading is 222.6 nm, which increases to 388.4 nm after incorporation of BSA and to 691.4 nm with BSA by the incubation method, and in the meantime, the zeta potential does not vary by more than 1 or 2 degrees. The authors were not sure about the changes in the zeta potential; however, they speculated that the BSA engagement with long-chain chitosan molecules is not uniform and less spread. Chitosan molecules may likely adopt a spread conformation in solution because of electrostatic repulsion force existing between amine groups along the molecular chain. The carboxyl groups on the surface of a large protein molecule may form hydrogen bond with amine groups at certain sites at the spread chitosan chain, but still maintaining a compact 3-D structure without spreading at the solution pH condition (pH5.5) so as to keep an inner hydrophobic core. Therefore, protein molecule attachment did not sufficiently suppress the positive surface charge of chitosan molecules. There could still be a high proportion of free amine group on the chitosan chain which remains unoccupied.

It was found that BSA EE changes in the range from 61.1% to 69.9% and 78.2% as the MW increases from low, medium, to high. Kim et al. (2003), however, found that higher MW of chitosan showed less albumin release than the lower one. However, in this case, pectin and chitosan formed the encapsulation beads. The release of albumin from the chitosan-coated pectin beads was dependent on pH of coating solution and release medium, which might affect the degree of swelling of pectin beads (Sabnis and Block 2000). Vila et al. (2004) stated that the EE of a protein, tetanus toxoid, was high, irrespective of the chitosan MW and indicated that the entanglement of the protein within the chitosan chains is not affected by their MW. A very interesting study on the sequential TEM observation of the morphological changes, swelling, and degradation of the BSA–CNPs was conducted (Figure 15.2) (Quan and Tao 2007) and the results revealed that the BSA-loaded NPs maintained its shape, size, and integral structure before 6 h during which the majority of loaded protein molecules were burst released. This is followed a significant particle swelling process between 6 and 12 h with reduced density and increased particle size. The particles swelled further and loss of density as the picture at 24 h shows a larger, loose, and translucent particle structure. Finally,

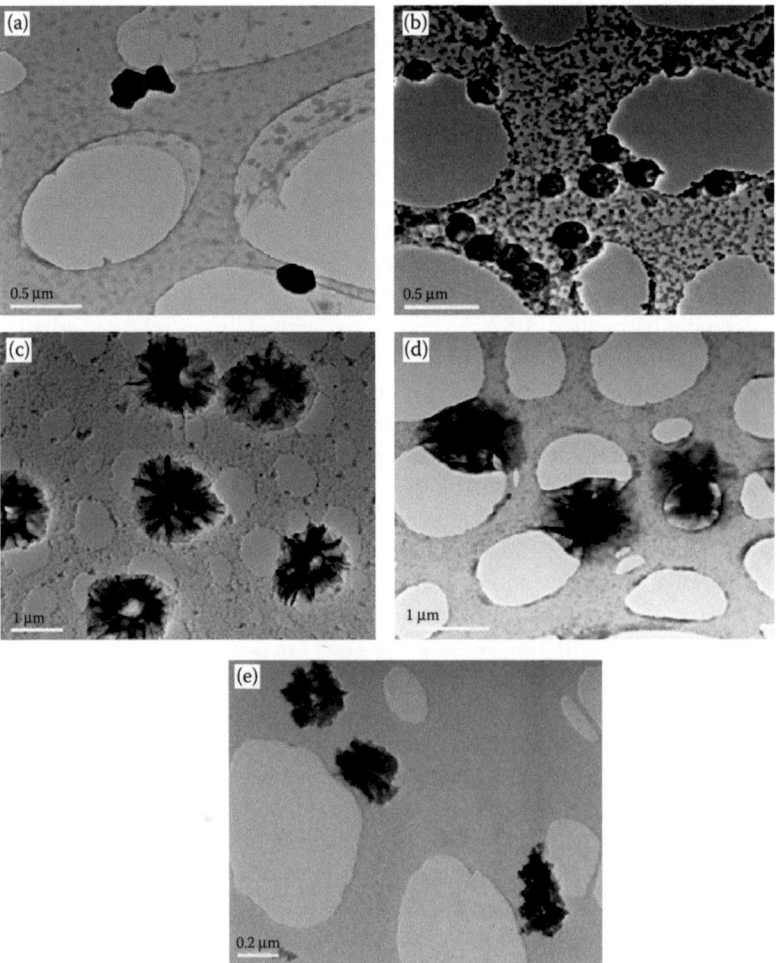

FIGURE 15.2 Time sequential TEM images of chitosan–BSA–TPP nanoparticles taken during protein release studies: (a) 3rd hour; (b) 6th hour; (c) 12th hour; (d) 24th hour; and (e) 48th hour. Particle preparation conditions: chitosan concentration = 3.0 mg/mL, BSA concentration = 1.0 mg/mL, chitosan to TPP mass ratio = 3:1, release medium $T = 37 \pm 1°C$, pH 7.0. (From Quan, G. and W. Tao. 2007. *Colloids and Surfaces B: Biointerfaces* 59:24–34. With permission.)

continued hydrolytic reaction resulted in the particle to lose shape with reduction in size. It is also interesting to observe that the process did not show a distinctive pattern of structural disintegration into fractions of smaller particles, as the degradation largely follows a swelling, loss of compact structure, severance of cross-linkage, and finally surface stripping of materials.

An important observation is that later-stage particle degradation did not offer a second wave of significant release, suggesting that the remaining protein molecules entrapped inside the particle structure are extensively bound and entangled with the chitosan polymer chain.

15.8.2 Liposomes for Drug Delivery

The most common vehicle currently used for targeted drug delivery is the liposome (Torchilin 2006) (Figure 15.3). The word *liposome* is derived from two Greek words: *lipo* ("fat") and *soma* ("body"); it is so named because its composition is primarily of phospholipid. Liposomes are non-toxic, nonhemolytic, and nonimmunogenic even on repeated injections; they are biocompatible and biodegradable and can be designed to avoid clearance mechanisms (RES, renal clearance, chemical or enzymatic inactivation, etc.). Lipid-based, ligand-coated nanocarriers can store their payload in the hydrophobic shell or the hydrophilic interior depending on the nature of the drug/contrast agent being carried.

A liposome is an artificially prepared vesicle composed of a lipid bilayer. The liposome can be used as a vehicle for the administration of nutrients and pharmaceutical drugs. Liposomes can be prepared by disrupting biological membranes (such as by sonication). Liposomes are composed of natural phospholipids, and may also contain mixed lipid chains with surfactant properties (e.g., egg phosphatidylethanolamine). A liposome design may employ surface ligands for attaching to unhealthy tissue. The major types of liposomes are the multilamellar vesicle (MLV), the small unilamellar vesicle (SUV), the large unilamellar vesicle (LUV), and the cochleate vesicle. The only problem in using liposomes *in vivo* is their immediate uptake and clearance by the RES system and their relatively low stability *in vitro*. To combat this, PEG can be added to the surface of the

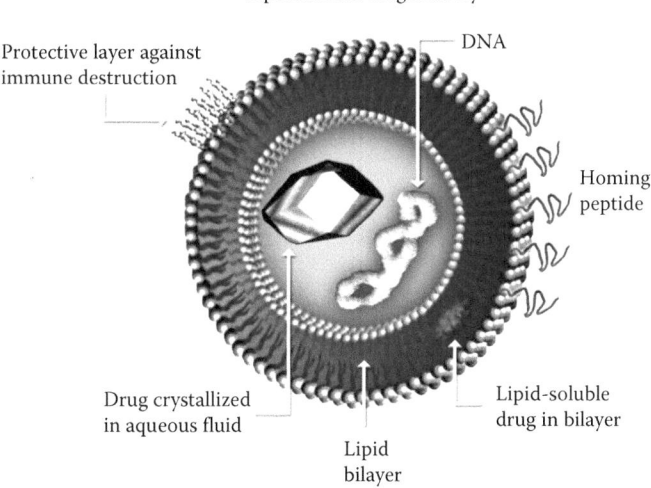

FIGURE 15.3 Liposomes are composite structures made up of phospholipids and may contain small amounts of other molecules. Though liposomes can vary in size from low micrometer range to tens of micrometers, unilamellar liposomes, as pictured here, are typically in the lower size range with various targeting ligands attached to their surface allowing for their surface attachment and accumulation in pathological areas for treatment of diseases. (From Torchilin, V. P. 2006. *Advanced Drug Delivery Reviews* 58:1532–1555. With permission.)

liposomes. Increasing the mole percent of PEG on the surface of the liposomes by 4–10% led to a significant increase of the circulation time *in vivo* from 200 to 1000 min.

A promising tool for the development of mucoadhesive drug delivery systems has been developed by Kast and Schnurch (2001). Improved mucoadhesive property of chitosan was obtained by reacting it with TGA. This was achieved by the formation of amide bonds between the primary amino groups of the chitosan and the carboxylic acid group of TGA. Dependent on the pH value and the weight ratio of polymer to TGA during the coupling reaction, the resulting thiolated polymers, the so-called thiomers, displayed 6.58, 9.88, 27.44, and 38.23 µmol thiol groups per gram polymer. Tensile studies carried out with these chitosan–TGA conjugates on freshly excised porcine intestinal mucosa demonstrated a 6.3-, 8.6-, 8.9-, and 10.3-fold increase in the TWA compared to the unmodified polymer, respectively. This thiomer was shown to be still biodegradable by the glycosidase lysozyme.

This so-called thiomer was further reacted with liposomes consisting of phosphatidylcholine (POPC) and a maleimide-functionalized lipid (Gradauer et al. 2012). The reaction took place between the SH groups of the polymer and the maleimide group of the liposomes as shown in Figure 15.4.

The covalent coupling between the liposomes and the chitosan–TGA polymer was confirmed by the size and zeta potential measurements as well as negative staining transmission electron microscopy images. *In vitro* mucoadhesion studies of thiomer-coated and uncoated liposomes, using an interesting method, the falling liquid film technique, which was developed previously by Begamwar et al. (2009), showed that the residence time on porcine small intestine could be almost doubled by the addition of the polymeric layer around the liposome. These findings show that coating liposomes with chitosan–TGA polymer provides a promising strategy to create a mucoadhesive oral delivery system for sustained drug release (Gradauer et al. 2012).

In recent years, great developments have been made in the field of mucoadhesive polymer systems in formulations that increase the residence time of drugs on mucosal membranes and subsequently, enhance the bioavailability of drugs with poor oral absorption (Aboubakar et al. 2000, Bernkop-Schnurch et al. 2004).

Nanomaterial is matter at dimensions of roughly 1–100 nm, where a unique phenomenon offers novel applications (Sonia and Sharma 2011). The major advantages of NPs as a delivery system are in controlling particle size, surface properties, and release of pharmacologically active agents to achieve the site-specific action of the drug at the therapeutically optimal rate and dose regimen. Although NPs offer many advantages as drug carrier systems, there are still many limitations to be solved such as poor oral bioavailability, instability in circulation, inadequate tissue distribution, and toxicity (Barratt 2003).

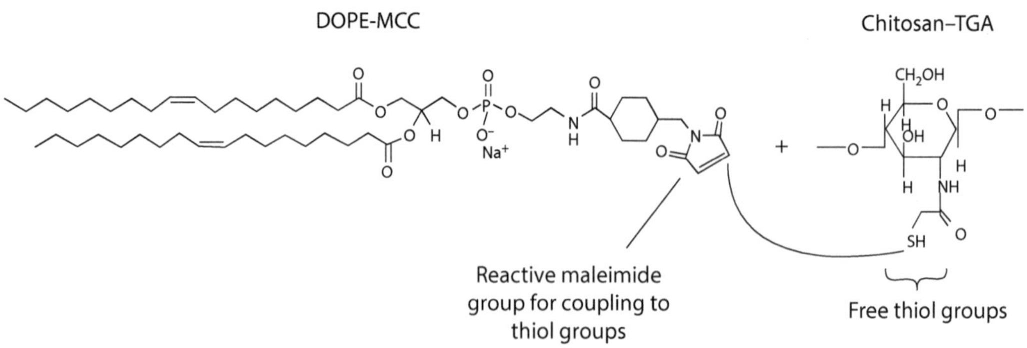

FIGURE 15.4 Reaction scheme for the covalent coupling of chitosan–TGA to a maleimide-functionalized phospholipid to form a stable thioether bond. DOPE-MCC, 1,2-dioleoyl-*sn*-glycero-3-phosphoethanolamine-*N*-[4-(*p*-maleimidomethyl)cyclohexane-carboxamide]-TGA (thioglycolic acid). (From Gradauer, K. et al. 2012. *International Journal of Nanomedicine* 7:2523–2534. With permission.)

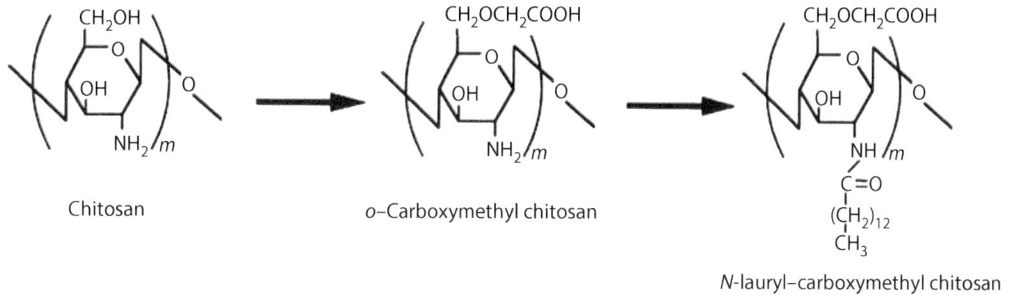

FIGURE 15.5 Chemical structure of chitosan, carboxymethyl chitosan, and *N*-lauryl-carboxymethyl chitosan. (From Hawary, D. L. et al. 2011. *Journal of Radioanalytical and Nuclear Chemistry* 290:557–567. With permission.)

CMC and *N*-lauryl–carboxymethyl chitosan (LCMC) have been prepared as water-soluble derivatives of chitosan (see Figure 15.5). These biodegradable chitosan derivatives were characterized and investigated for nuclear imaging and body distribution. They were labeled with $^{99\,m}$Tc to use them as targeted delivery to some organs *in vivo* for nuclear imaging and to follow their biodistribution within the body. The factors controlling the labeling efficiency have been investigated (Hawary et al. 2011).

Fluorescein thiocarbamyl-Lac-Suc (Lac-Suc-FTC) was prepared by labeling Lac-Suc with fluorescein isothiocyanate. Lac-Suc-FTC was injected intravenously at a dose of either 1 (high dose) or 0.2 (low dose) mg/mouse. At both doses, Lac-Suc-FTC initially underwent fast hepatic clearance, showed maximum liver localization at 8 h, and the amounts localized there were maintained even at 48 h postinjection. Very slow excretion into feces and urine was observed (Kato et al. 2001a,b).

15.9 GENE THERAPY

Gene therapy refers to the transmission of DNA encoding a therapeutic gene of interest into the targeted cells or organs with consequent expression of the transgene. In the past several years, gene therapy has received significant attention due to its potential application in the replacement of dysfunctional gene and treatment of acquired diseases (Gill et al. 1997, Ozbas-Turan et al. 2003). The central problem of gene therapy lies in the development of safe and efficient gene transfection (the process by which a bacterial cell is infected with purified DNA or ribonucleic acid (RNA) isolated from a virus or a viral vector after a specific pretreatment. Transfection is the process of deliberately introducing nucleic acids into cells. The term is used notably for nonviral methods in eukaryotic cell. The word *transfection* is a blend of *trans* and *infection*. Although naked DNA has been found to transfect certain kinds of cells, such as the skeletal muscle cells of the cardiac and the diaphragm region, its electronegativity tends to inhibit itself from entering most negatively charged cell membranes. Moreover, the unprotected DNA is rapidly degraded by nucleases present in plasma. Recently, many techniques have been developed for the introduction of DNA into cells. Systems currently under study for *in vitro* and *in vivo* use include both viral and nonviral vectors. Virus vectors are very effective in terms of transfection efficiency but they have fatal drawbacks such as immune response and oncogenic effects when used *in vivo*. Gelsinger's death from a gene therapy clinical trial in 1999 prompted a hard look at the safety record of the viral vectors and spurred a renewed interest in nonviral methods to ferry genes into tissue (Porteous et al. 1997). Among the nonviral vectors currently investigated, polyelectrolyte complexes (PEC) between DNA and polycations have been extensively investigated. DNA is tightly packed in the PEC complex, so that the entrapped DNA is shielded from contact with DNAses (Wolff 1990, Bordignon et al. 1995, Quong and Neufeld 1998). The cationic polymers include polyethyleneimine (Choate and Khavari 1997), poly(L-lysine) (Oligino et al. 2000), dendrimers (Somia and Verma 2000), polybrene (Ferber 2001), gelatin (Lee et al. 1998), tetra minofullerene (Deshpande et al. 1998), and poly(L-histidine)-graft-poly(L-lysine) (De Smedt et al. 2000). Although PEC systems have some advantages over virus vector, for example, low immunogenicity and easy manufacture (Garnett 1999, Liu and Yao 2002), several problems such as toxicity, lack of biodegradability, low biocompatibility and in particular, low transfection efficiency need to be solved prior to practical use (Zelphati et al. 2001).

Since Mumper et al. (1995) pioneered to apply chitosan to gene delivery system in 1995, an effort has been made to explore the potential of this naturally occurring polysaccharide as a new nonviral vector (Erbacher et al. 1998, Sato et al. 2001, Liu et al. 2003). A variety of chitosan of high MW and low MW or their derivatives have been used to mediate gene delivery into various cell types, including human embryonic kidney cells (HEK293), human lung carcinoma cells (A549), B16 melanoma cells, COS-1, HeLa cells, and mesenchymal stem cells (MSCs) (MacLaughlin et al. 1998, Mao et al. 2001, Liu and Yao 2002, Corsi et al. 2003, Kim et al. 2003a).

Chitosan-based gene delivery systems have been proven to be effective for nonviral gene therapy. The chitosan–DNA complexes are very easy to synthesize and are more effective compared to the commonly used polygalactosamine–DNA complexes; however, their use is limited because of the lower transfection efficiency (Prabaharan and Mano 2005).

Various factors affecting chitosan delivery of nucleic acids were studied by Mao et al. (2010) and Sato et al. (2001). The MW of chitosan, the charge ratio between the luciferase plasmid to chitosan, and the pH of the culture media were found to be the factors related to the *in vitro* transfection efficiency. MacLaughlin et al. (1998) also tested the effect of chitosan MW and charge ratio on particle formation by depolymerizing 102×10^3 mol/g, 89.4% deacetylated chitin into smaller oligomers. Chitosan enhanced adenovirus infectivity to mammalian cells in gene therapy (Kawamata et al. 2002) and low concentration and low-MW chitosans were better in enhancing adenovirus activity (Khor and Lim 2003).

There are numerous reports on the preparation of chitosan–DNA NPs (Borchard 2001, Cui and Mumper 2001, Mao et al. 2001, Chew et al. 2003, Liu et al. 2005) and quaternized chitosan–DNA NPs (Ouchi et al. 1999) using complex coacervation between the positively charged amine groups on chitosan and negatively charged phosphate groups on DNA. NPs are spontaneously formed after addition of DNA solution into chitosan, dissolved in acetic acid solution, under mechanical stirring at room temperature. The size of the complexes can be varied from 50 to 700 nm. The particles have a negative charge for chitosan nitrogen to DNA phosphate ratio (N/P ratio) below 2. At N/P = 2, the particles are neutral and become positively charged for higher N/P values, reaching +35 mV at N/P = 5.49.

To develop CNP for siRNA delivery, Katas and Alpar prepared CNPs by two methods of ionic cross-linking: simple complexation and ionic gelatin using TPP. *In vitro* studies using two types of cells lines, CHO K1 and HEK 293, revealed that the preparation method of siRNA association to the chitosan plays an important role on the silencing effect. Chitosan–TPP NPs with entrapped siRNA had better vectors than chitosan–siRNA complexes; this may be due to their high binding capacity and loading efficiency. The chitosan–TPP NPs seem to have good potential as viable vector candidates with safer and cost-effective siRNA delivery (Katas and Alpar 2006).

The influence of the DA of chitosan on NP formation has also been investigated. For chitosans of the same MW (390×10^3 mol/g), increasing the degree of polymer acetylation required increasing the amount of chitosan to achieve complete DNA complexation. For chitosan with an MW of 390×10^3 mol/g, the N/P ratio to achieve complete DNA complexation for DA of 10%, 30%, and 28% was 3.3:1, 5.0:1, and 9.0:1, respectively. The size and morphology of these NP formulations were not significantly different. The increased DA results in a decrease in overall luciferase expression levels in HEK293, HeLa, and SW756 cells due to particle destabilization in the presence of serum proteins (Kiang et al. 2004a,b).

Liu et al. (2007) reported that the physicochemical properties and *in vitro* gene silencing of chitosan/siRNA NPs are strongly dependent on the MW and DD of chitosan. Chitosan with high MW and high level of DD resulted in the formation of discrete stable NPs of 200 nm in size. Chitosan/

siRNA formulations (N/P: 50) prepared with low-MW chitosan (10 kDa) showed almost no knockdown of endogenous enhanced green fluorescent protein (EGFP) in H1299 human lung carcinoma cells, whereas those prepared from higher MW (65–170 kDa) and DD (80%) showed greater gene silencing ranging between 45% and 65%. The highest gene silencing efficiency (80%) was achieved using chitosan/siRNA NPs at N/P 150 using higher MW (114 and 170 kDa) and DD (84%) that correlated with the formation of stable NPs of 200 nm. From their studies, it can be concluded that there is still scope for improvement and for the optimization of gene silencing using chitosan/siRNA NPs and certain improvements in the polymeric properties would make significant differences.

The protection of encapsulated pDNA from nuclease attack offered by CNPs was confirmed by Bozkir and Saka (2004) by assessing degradation in the presence of DNase I and the transformation of the plasmids with incubated NPs was examined by β-galactosidase assay. Model pDNA existed as a mixture of both supercoiled (84.2%) and open circular (15.8%) forms.

The stability of the DNA–chitosan complexes depends on several factors such as the chitosan chain length and the amount of chitosan. Increasing the chitosan MW and chitosan concentration yielded more stable complexes, indicating that varying the chitosan chain length may provide a tool for controlling the ability of the polyplex to deliver therapeutic gene vectors to cells (Mansouri et al. 2004, Danielsen et al. 2005). Spherical chitosan/DNA NPs with an average 38 nm diameter were prepared by using a simple osmosis-based process (Masotti et al. 2008). About 30% DNA was incorporated with prolonged release time.

Chitosan-alginate core-shell NPs obtained using water-in-oil reverse microemulsion template was used to encapsulate plasmid DNA for gene delivery via the cell endocytosis pathway (You et al. 2006). Carboxymethyl cellulose and alginate have also been complexed with chitosan to prepare NPs. Chitosan–carboxymethyl cellulose NPs were subsequently coated with plasmid DNA (pDNA) [198]. NPs have also been obtained in which the polyion is the active principle itself (heparin, DNA). Vila and coworkers (2007) patented a procedure for obtaining chitosan–heparin NPs cross-linked with TPP.

The *in vitro* and *in vivo* transfection efficiency of chitosan and its derivatives as vectors for gene therapy has also been investigated (Liu and Yao 2002). Zheng et al. (2007) prepared three types of CNPs. These were a 60% trimethylated chitosan oligomer (TMCO-60% with a degree of substitution of 60% and MW = 221×10^3 mol/g) and two chitosan samples differing in MW and DA (chitosan-43: MW = 43×10^3 mol/g, DA = 13%; and chitosan-230: MW = 230×10^3 mol/g, DA = 90%). These authors used them to encapsulate pDNA encoding green fluorescent protein (GFP) using the complex coacervation technique. An *in vivo* study, carried out by oral administration of pDNA–CNPs to mice, showed outstanding GFP expression in the gastric and upper intestinal mucosa. GFP expression in the mucosa of the stomach and duodenum, and jejunum, ileum, and large intestine was found, respectively, in 100%, 88.9%, 77.8%, and 66.7% of the nude mice examined. TMCO-60%–pDNA NPs had a higher *in vitro* and *in vivo* transfection activity with minimal toxicity, which made them an attractive nonviral vector for gene therapy via oral administration.

More recently, PEG-*g*-CNPs prepared by ionotropic gelation with TPP were investigated as potential gene carriers (Csaba et al. 2009). High and long-lasting gene expression levels for these NPs were reported. Moreover, PEG-*g*-chitosan–TPP NPs also mediated high gene expression levels *in vivo*, following nasal administration. These results indicate the potential of PEG-*g*-chitosan–TPP NPs as transmucosal gene delivery systems.

In an attempt to track the efficiency of DNA delivery, Lee et al. (2008) employed fluorescence resonance energy transfer (FRET) to monitor the molecular dissociation of a chitosan/DNA complex with chitosan characterized by different MWs. Plasmid DNA and chitosan were labeled with Quantum Dots and Texas Red, respectively, and confocal microscopy and fluorescence spectroscopy were used to monitor the dissociation of the complexes. As the chitosan MW in the chitosan/DNA complex increased, the Texas Red-labeled chitosan gradually lost FRET-induced fluorescence light. This observation was also observed when HEK293 cells were incubated with chitosan/DNA complex. This indicated that the dissociation of the chitosan/DNA complex was greater with the higher-MW chitosan/DNA complex. Fluorescence spectroscopy analysis confirmed that the molecular dissociation of the chitosan/DNA complex at pH 7.4 and 5.0 and confirmed that the dissociation occurred in acidic environments. Therefore, it appears that the high-MW chitosan/DNA complexes dissociate better in lysosomes than the low-MW complexes. Furthermore, the high-MW chitosan/DNA complex showed superior transfection efficiency in relation to the low-MW complex. It was concluded that the dissociation of the chitosan/DNA complex is a critical event in obtaining the high transfection efficiency of the gene carrier/DNA complex (Lee et al. 2008).

The term "target" carries several connotations in the overall context of drug discovery. Remarkable progress in molecular biology has led to the identification of numerous proteins with key roles in the function of both normal and abnormal cells, which has allowed the formation of specific hypotheses about how modulating the function of defined proteins that are linked to disease could be a route to new drugs. Such disease-linked proteins are commonly referred to as targets. The basis of the hypotheses can range from an attractive scientific theory to information obtained from genetic analysis of tissues obtained from patients with a particular disease, and the process of confirming such hypotheses (to varying degrees of confidence) is usually termed "target validation" (Knowles and Gromo 2003).

15.9.1 Chitosan-Based Systems for the Delivery of Growth Factors

Growth factors are polypeptides that control the growth, differentiation, and metabolism of cells and regulate the process of tissue repair (McCarthy et al. 1996, Brown et al. 1988, Fu et al. 1998, Ueno et al. 1999).

15.9.2 Platelet-Derived Growth Factor-BB

Platelet-derived growth factor-BB (PDGF-BB) is a protein encoded in humans by the *PDGFB* gene. Growth factor such as PDGF-BB is a potent mitogen and chemotactic factor for cells of mesenchymal origin, including periodontal ligament cells and osteoblasts (Lynch et al. 1991). The term *mesenchyme* essentially refers to the morphology of embryonic cells; however, they do persist as stem cells into adulthood. Mesenchymal cells are able to develop into the tissues of the lymphatic and circulatory systems, as well as connective tissues throughout the body, such as bone and cartilage. A sarcoma is a cancer of mesenchymal cells.

PDGF has demonstrated great potential in promoting gingival, alveolar bone, and cementum regeneration in a variety of wound-healing models (Taba et al. 2005). Enhancement of periodontal tissue regeneration using PDGF-BB has been demonstrated in beagle dogs and monkeys (Rutherford et al. 1992). Despite the superior activity of PDGF-BB in tissue regeneration, rapid clearance of PDGF-BB due to short half-life (below 4 h *in vivo*) resulting in a difficulty in maintaining therapeutic concentrations from injection has lead to the administration of extremely high doses (above 10 mg) that can result in adverse side effects (Mehrara et al. 1999). It is essential that a carrier system is developed to localize the delivery of this short half-lived factor to target cells and maintain PDGF-BB at therapeutic concentration levels (1–10 ng/mL) at wound sites for a healing period of up to 4 weeks to obtain enhanced bone regeneration. Such carrier has been developed by Park et al. (2000a) in the form of chitosan sponge for osteoconductive material that induces or stimulates bone formation.

Chitosan solution was freeze dried, cross-linked, and freeze dried again to fabricate the chitosan sponge, which was then soaked into the PDGF-BB solution.

The release rate of PDGF-BB could be controlled by varying the initial loading content of PDGF-BB to obtain optimal therapeutic efficacy. Initial burst release was observed with rapid release during the first day, followed by sustained release up to 6 days and a leveling off of the release rate. The burst effect indicates rapid water uptake of Chitosan sponge and dissolution of the exposed PDGF-BB particles at the surface of the chitosan sponge. As the PDGF-BB-loading content increased, release rate increased correspondingly for the 6 days. The release of PDGF-BB after 6 days maintained therapeutic level (1–2 ng) for 3 weeks. PDGF-BB is reported to increase mitogenesis and chemotaxis of bone cells proportionally to its concentration within 0.1–100 ng/mL range (Matsuda et al. 1992).

Examination of the cellular attachment and proliferation onto chitosan sponge is important as one of the requirements for bone-substituting materials would be an adaptation to a wide variety of bone tissue defects. Bone-substituting materials appear to have been suited in that case where wound filling could be obtained over the wound site. Cellular attachment and migration over the bone-substituting material surface are essential to obtain effective wound filling and bone tissue adaptation (Muzzarelli et al. 1988).

PDGF-BB-loaded chitosan sponge induced significantly high cell attachment and proliferation level, which indicated good cellular adaptability as well as marked increase in new bone formation and rapid calcification. Degradation of the chitosan sponge was preceded at defect site and subsequently replaced with new bone.

Park et al. (2000b), in another work, used porous chondroitin-4-sulfate (CS)–chitosan sponge for controlling growth factor delivery (PDGF-BB) to improve bone formation. Chondroitin-4-sulfate (CS) is proposed considering its ionic interaction with PDGF-BB or chitosan. Since CS is negatively charged, interaction with positively charged chitosan or PDGF-BB is anticipated. Interaction between PDGF-BB and CS is expected to induce prolonged release of PDGF-BB from the sponge. More steady release of PDGF-BB was obtained by using CS–chitosan system than by using chitosan alone.

Release rate of PDGF-BB could be controlled by varying the composition of CS in the sponge or initial loading content of PDGF-BB. CS–chitosan sponge induced increased osteoblast migration and proliferation as compared with chitosan sponge alone. Furthermore, the release of PDGF-BB from CS–chitosan sponge significantly enhanced osteoblast proliferation.

Chitosan–poly(L-lactide) (PLLA) composite matrices as scaffolds to enhance bone formation by releasing growth factors in desirable kinetics has been prepared and investigated by Lee et al. (2002).

The initial burst release was reduced to 8.9 ng and PDGF-BB was released steadily from PLLA–chitosan matrix at a rate of 3–5 ng/day compared to initial release of 90 ng, from chitosan–CS (20%) at the rate of 4–5 ng/day (Figure 15.6). The reduced initial burst release is probably due to the entrapment of PDGF-BB within the hydrophobic PLLA network. Chitosan enhanced the wettability of PLLA and the porous structure serves as channels to release entrapped PDGF-BB. Prolonged

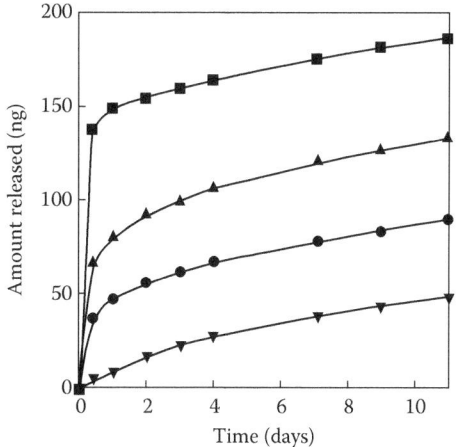

FIGURE 15.6 Release of PDGF-BB from the porous chitosan matrices. Effects of CS content in porous chitosan matrices on PDGF-BB release: ■, 0%; ▲, 20%; and ●, 40% of CS content. Release of PDGF-BB from the PLLA–chitosan composite matrices (▼). PDGF-BB loading content in each matrix was 200 ng. (From Lee, J. Y. et al. 2002. *Journal of Controlled Release* 78:187–197. With permission.)

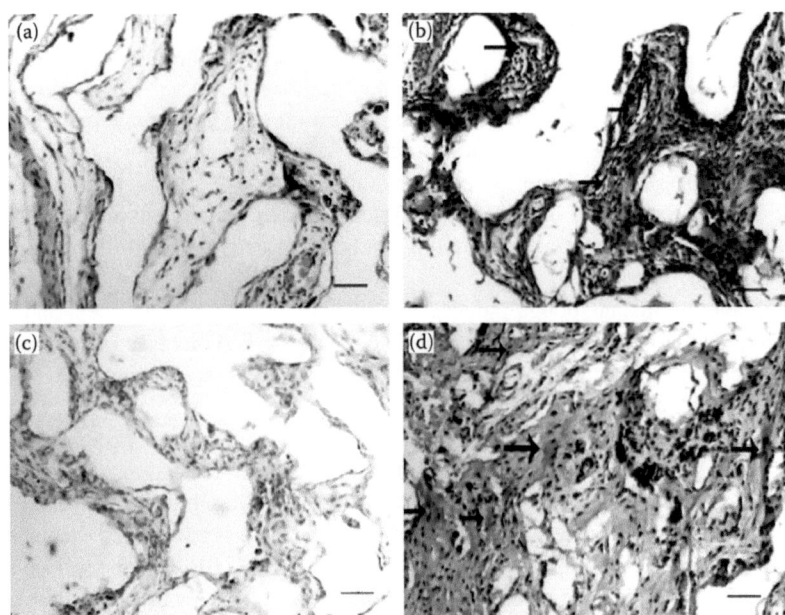

FIGURE 15.7 Light micrographs of implants excised at 4 weeks after implantation. The cross section was stained with hematoxylin and eosin (bar = 150 mm): (a) group 1 scaffold and (b) group 2 scaffolds. The cross section was immunohistochemically stained with PDGFB antibody. (c) Group 1 scaffold was negative. (d) Group 2 scaffold was brightly stained. (From Zhang, Y. et al. 2007. *Biomaterials* 28:1515–1522. With permission.)

release of PDGF-BB can be expected from PLLA–chitosan composite matrices as compared to the chitosan matrices.

Zhang et al. (2007) used porous chitosan/coral composites combined with plasmid-encoding platelet-derived growth factor B (PDGFB) gene as a regenerative material for periodontal regeneration. These scaffolds were evaluated *in vitro* by analysis of microscopic structure and cytocompatibility.

The scaffolds were divided into two groups: pure coral scaffold group and coral scaffolds containing chitosan–DNA composite group. The expression of PDGFB and type-I collagen were detected with RT-PCR after human periodontal ligament cells (HPLCs) were seeded in this scaffold HPLCs incubated in group 2 produced higher level PDGFB during the entire culture period. The maximum concentration of PDGFB in the culture media was detected after 6–9 days incubation and then followed by a marked decrease after 9 days.

After implanted *in vivo* into athymic mice, HPLCs not only proliferate but also increased the expression of PDGFB. The results from the 4 weeks' implantation suggest that two kinds of scaffolds promoted the in-growth of surrounding tissues, with the group 2 scaffolds being the better performer. Vascularized tissue in-growth was also noticed for explanted sites that had two kinds of scaffolds (Figure 15.7a and b). Immunohistological staining with antibodies to PDGFB showed that only group 2 was stained after 4 weeks transplantation (Figure 15.7d).

15.9.3 Tissue Regeneration by Epithelial Growth Factor and Fibroblast Growth Factor

Tissue engineering promises the regeneration of injured organs to avoid organ transplantation and its complications and that merges the fields of cell biology, engineering and materials science, and surgery to fabricate new functional tissue (Chaignaud et al. 1997). Cell-based tissue engineering has been grown to be the most promising alternative therapy for the treatment of many diseases (Pittenger et al. 1999). Epithelial growth factor (EGF) and fibroblast growth factor (FGF), and the combination of these growth factors, can induce multiple cells and tissues because of their pletrophic effect (when a single gene is influenced by many genes).

EGF is a small polypeptide of 53 amino acid residues and has a MW of 6216 Da (Gönül et al. 1995). It has been reported to accelerate cellular proliferation and synthesis of the extracellular matrix (ECM) (Brown et al. 1988, Çelebi et al. 1993, 2002). It acts by binding to the EGF receptor—tyrosine kinase—thereby initiating a series of events that regulate cell proliferation (Carpenter and Cohe 1979, Araki et al. 1989, Wenczak et al. 1992). *In vivo*, several studies have proven that EGF is effective for the acceleration of epithelization in human and animal wounds (Nassif et al. 1998). EGF stimulates the proliferation of keratinocytes in culture, and topical administration of EGF accelerates dermal regeneration of partial-thickness burns or split-thickness incisions *in vivo* (Schultz et al. 1987).

Several formulations of EGF have been studied regarding their ability to accelerate wound healing. The most commonly used form is solution; there are only a few studies reported dealing with bioadhesive gel, microemulsion, and liposome (Brown et al. 1988, Çelebi et al. 1993, 2002).

Chitosan enhances the functions of inflammatory cells such as polymorphonuclear leukocytes, macrophages, and fibroblasts; thus, it promotes granulation and organization (Ishihara et al. 2001, Ueno et al. 2001).

Chitosan membranes have been modified with mouse EGF by a photochemical technique (Karakeçili et al. 2008). The results obtained from cell culture experiments showed that immobilized EGF stimulated fibroblast growth on chitosan membranes, and a considerable difference in cell proliferation was detected on EGF-modified chitosan membranes.

Alemdaroğlu et al. (2006) developed an effective chitosan gel formulation containing EGF, and determined its effect on healing of second-degree burn wounds in rats. A uniform deep second-degree burn of the back skin was performed with water heated to 94°C during a 15-s exposure. The EGF formulations were repeatedly applied on the burned areas with a dose of $0.160 \, mg/cm^2$ for 14 days (one application per day). Healing of the wounds was evaluated immunohistochemically, histochemically, and histologically on the tissue samples. When the results were evaluated immunohistochemically, there were significant increases in cell proliferation observed in the EGF containing gel applied group ($p < 0.001$). The histochemical results showed that the epithelization rate in the EJ group was the highest compared to the ES group results ($p < 0.001$) (Alemdaroğlu et al. 2006).

Basic fibroblast growth factor (bFGF) is a polypeptide growth factor with a high affinity for substances of the ECM, such as heparan sulfate (Dijke and Iwata 1989). Heparan sulfate (HS) is a linear polysaccharide found in all animal tissues. It occurs as a heparan sulfate proteoglycan (HSPG) in which two or three HS chains are attached in close proximity to cell surface or ECM proteins. It is in this form that HS binds to a variety of protein ligands and regulates a wide variety of biological activities, including developmental processes, angiogenesis, blood coagulation, and tumor metastasis. HS has been shown to serve as a cellular receptor for several viruses, including the respiratory syncytial virus. FGF is chemotactic for many cells involved in wound healing, such as endothelial cells, epidermal cells, and fibroblasts. It is also a potent mitogen for these cell types. FGF applied to epidermal wounds of healthy pigs enhances wound healing by 20% (Hebda et al. 1990). FGF is especially effective in wounds showing impaired healing, for example, in diabetic ulcers, venous ulcers, and pressure sores (Broadley et al. 1988, Klingbeil et al. 1991, Hayward et al. 1992). Owing to the antimicrobial effect of chitosan and the proliferative action of bFGF, a combination of chitosan and bFGF may have overproportional effects on wound healing and could be very useful for the therapy of chronic wounds.

Three-dimensional scaffolds from the PEC of chitosan and alginate for the delivery of bFGF have been prepared by Ho et al. (2009). The bFGF-binding efficiency of the chitosan–alginate PEC scaffold, after being conjugated with high concentration of heparin (83.6 μg/mg scaffold), was increased up to 15 times higher than that of the original scaffold (65.6 ng bFGF/mg scaffold vs

4.5 ng bFGF/mg scaffold). The release of bFGF from the original scaffold was quick with the initial burst release. By functionalizing the scaffold with various concentrations of heparin (17.6, 50.3, and 83.6 µg heparin/mg scaffold), the rate of bFGF release from the scaffold decreased in a controlled manner with reduced burst effect.

Tang et al. (2010) prepared heparin-functionalized chitosan (CS)/poly(g-glutamic acid) (g-PGA) NPs (HP-CS/g-PGA NPs) for multifunctional delivery of bFGF and heparin that may be a potential therapeutic method for enhancing ischemic tissue regeneration and preventing blood vessel rethrombosis. The mean particle sizes and bFGF loading efficiency increased with the increase of functionalized heparin contents. The HP-CS/g-PGA NPs were pH sensitive that could sustain bFGF release at pH = 6.7 (simulate the pH of ischemia tissue) and were rapidly disintegrated at pH 7.4 (simulate the pH of repaired tissue). Sustained release of bFGF from the NPs enhanced the proliferation of human foreskin fibroblast (HFF) cells and angiogenic tube formation by human umbilical vein endothelial cells (HUVEC), suggesting the retaining of bFGF mitogenic activity.

Ultraviolet light (UV) irradiation was used to photo cross-link chitosan (Az-CH-LA) aqueous solution containing fibroblast growth factor-2 (FGF-2) to yield an insoluble, flexible hydrogel (Obara et al. 2003). To evaluate its accelerating effect on wound healing, full-thickness skin incisions were made on the back of healing-impaired diabetic (db/db) mice. Application of the chitosan hydrogel significantly induced wound contraction and accelerated wound closure in db/db. However, the addition of FGF-2 in the chitosan hydrogel further accelerated wound closure in db/db mice.

Methylpyrrolidinone chitosan (MPG), a water-soluble derivative of chitosan, has been shown to prevent excessive scar formation (Biagini et al. 1991, Muzzarelli 1990). It was investigated as a carrier material for bFGF, a combination intended for the treatment of wound-healing deficiencies. Soft and flexible fleeces of MPC were prepared by freeze-drying. The spongy structure of the fleeces allows for a high water uptake, which is very useful in the management of highly exudative wounds and also the incorporation of bFGF into a spongy fleece allows a homogeneous distribution of the growth factor over the whole wound area with a minimum of irritation. The MPG fleece gels immediately when it comes into contact with physiological fluids and is well tolerated in full-thickness wounds in mice. This is an important advantage compared to the application of ointments, gels, or solutions containing bFGF. It is especially difficult to distribute an ointment in a wound without causing discomfort or pain to the patient. However, the application of a solution does not cause any irritation to the wound. The fleeces investigated in this study released most of the bFGF in the first 48 h. Therefore, it is conceivable to treat slowly healing wounds with bFGF-loaded MPG fleeces during a period of 48–72 h. This interval matches the intervals for wound dressing change in the conventional therapy of chronic wounds (Berscht et al. 1994).

Kim et al. (2012) developed immobilized chitosan scaffolds with controlled pore architectures for enhanced viability of human mesenchymal stem cells (hMSCs). Chitosan of lower MW was effective in the formation of porous chitosan scaffolds, resulting in an increase of interconnecting micropores (~10 µm) between macropores. Using a layer-by-layer method, heparin-coated chitosan scaffolds were prepared as depots for bFGF. Enzyme-linked immunosorbent assays confirmed that heparin-coated chitosan scaffolds could bind higher amount of bFGF (24.21 ng/mg) compared to 2.53 ng/mg of noncoated scaffold. *In vitro* studies showed that chitosan scaffolds induced the improved viability and proliferation of hMSCs through their synergetic effects of the interconnecting micropores and the sustained released of bFGF.

CNP incorporated with both EGF and FGF, either individually or in combination, have been prepared (Rajam et al. 2011). The *in vitro* study demonstrated the efficiency of CNP to deliver dual growth factors that could be successfully used for tissue engineering and drug delivery applications. Growth factors incorporated NPs did not show any toxicity against fibroblasts up to 4 mg/mL culture medium. Increased proliferation of fibroblasts *in vitro* evidenced the delivery of growth factors from CNP for cellular signaling. The attachment and growth of fibroblast in EGF + FGF–CNP system were greatly improved, whereas the inflammatory induction was not found compared with other groups (Rajam et al. 2011).

15.9.4 PERIPHERAL NERVE REGENERATION BY NERVE GROWTH FACTOR

Peripheral nerve injuries are very common clinical cases. Incomplete or even no functional recovery occurs often so that some patients could be subjected to lifelong disability (Kim et al. 2003b). In the case of short nerve gaps, the nerve stumps are usually connected by end-to-end direct suture and the injured nerves can be repaired by spontaneously regrown axons. With respect to longer nerve gaps, nerve grafts are commonly required.

The capacity of peripheral nerves for self-regeneration is dependent on several factors, including permissive environment and the activation of the intrinsic growth capacity of neurons. Neurotrophic factors are known to play a vital role in controlling the neuronal survival and axonal regeneration at the nerve injury site (Fu and Gordon 1997, Markus et al. 2002). Nerve growth factor (NGF), as an important member of neurotrophin family, not only promotes the survival and neurite outgrowth of sensory neurons both *in vitro* and *in vivo*, but also enhances peripheral nerve regeneration, as shown in many previous studies (Rich et al. 1989, Derby et al. 1993). However, the application of NGF in promoting nerve regeneration was limited by its short biological half-life and its vulnerability to structural disruption or modification, leading to loss of bioactivity. Therefore, protein drug delivery systems are needed not only to improve the biological utilization of NGF by sustained release of bioactive NGF to target site, but also to protect its bioactivity from degradation by direct exposure to harsh environments, for example, light, oxygen, and chemicals. In recent years, microsphere-based drug delivery systems, fabricated by biodegradable synthetic polymers such as poly(lactide-*co*-glycolic acid) (PLGA), have been widely reported (Choi et al. 2002, Han et al. 2010). However, these synthetic materials have several inherent flaws, such as acidic degradation products, retarded clearance rate, and limited biological function (Freiberg and Zhu 2004).

Recent studies have shown that chitosan is a potential candidate biomaterial for nerve tissue engineering (Lu et al. 2007, Li et al. 2009, Huang et al. 2010a,b). In addition, chitooligosaccharides (COSs), the biodegradation products of chitosan, have been demonstrated *in vitro* to promote neuronal differentiation and neurite outgrowth (Yang et al. 2009a,b). All these properties make chitosan an ideal substance for encapsulating NGF and for its sustained release in nerve injury repair.

Ionically, cross-linked chitosan microspheres were used for the controlled release of NGF (Zenga et al. 2011). The microspheres were prepared by the emulsion ionic cross-linking method with sodium tripolyphosphate (STPP) as an ionic cross-linking agent. The NGF EE ranged from 63% to 88% depending on the concentration of STPP as lower STPP concentration showed the highest EE. The *in vitro* release profiles of NGF from NGF–CMSs were influenced by the concentration of STPP. NGF–CMSs that were cross-linked with higher concentration of STPP showed slower but sustained release of NGF. In addition, the released NGF from NGF–CMSs was capable of maintaining the viability of PC12 cells, as well as promoting their differentiation (Zenga et al. 2011).

Genipin (GP), a natural and low toxic agent, was used to cross-link chitosan and concurrently to immobilize NGF onto modified chitosan, followed by fabrication of chitosan (CS)–GP–NGF nerve conduits (Yang et al. 2011). These CS–GP–NGF nerve conduits were noncytotoxic to primary cultured Schwann cells and showed *in vitro* neuroaffinity to PC12 cells in terms of keeping the activity of NGF within nerve conduits, that is, the ability to stimulate neuronal differentiation of PC12 cells. The continuous release profile of NGF from CS–GP–NGF nerve conduits, within a 60-day time span, consisted of an initial burst that was controlled by concentration gradient-driven diffusion, followed by a zero-order release that was controlled by the degradation of chitosan (Zenga et al. 2011).

Liao et al. (2013) and Mottaghitalab et al. (2013) used chitosan microspheres embedded into chitosan–polycaprolactone (CH–PCL) multichannel conduits that can potentially be used for long-gap nerve repair. PCL percentages in the CH–PCLs were optimized into a region changing from around 30 to 42 wt. % while the porosity and average channel diameter of the resulting multichannel conduits were selected as about 80% and 200 μm, respectively. The compressive properties in wet state and *in vitro* degradation rates of CH–PCL multichannel conduits were found to be mainly manipulated by the PCL content in the CH–PCLs, whereas the cumulative amount of released

NGF from the conduits could be independently regulated by altering the initial NGF load inside the embedded microspheres. The optimal microsphere-embedded CH–PCL multichannel conduits with a dimension of around 6 mm in outer diameter and 30 mm in length were able to administrate bioactivity-preserved NGF release in a sustained and controlled manner without significant initial burst release, and the release rates of the conduits could be maintained with approximately linear characteristic over a period of time longer than 6 weeks.

Nanostructured chitosan/poly(vinyl alcohol) (CS/PVA) conjugated (NGF) scaffolds for enhancement of neural cell lines proliferation was prepared using electrospinning technique. To improve the electrospun system, NGF was conjugated to CS/PVA nanocomposites as biochemical cue to promote the survival, growth, and proliferation of both cell lines (Mottaghitalab et al. 2011).

15.9.5 CARTILAGE REGENERATION BY TRANSFORMING GROWTH FACTOR-β1

Current clinical therapies for repairing articular cartilage lesions are usually conducted by drilling or abrading through the matrix of articular cartilage into the underlying subchondral bone to contact the cells and bioactive molecules from the bone marrow. These rigorous techniques, however, are not able to restore the normal physiological structures and functions of articular cartilage and the formed tissue is usually a combination of hyaline and fibrocartilage (Temenoff and Mikos 2000), which is deficient in long-term repair and possibly results in unwanted side effects (Buckwalter 1998).

In search of alternative therapies, the utilization of protein growth factors or other bioactive molecules for the repair of articular cartilage has gained significant importance since many kinds of growth factors in human body can guide and improve the development and regeneration of neonatal tissues (Gillogly et al. 1998, Kumari et al. 2010). Of different growth factors, TGF-β1 has been considered as a key one in guiding and controlling cellular behavior in the regeneration of cartilage tissue since it is capable of instructing chemotaxis and synergism of both resident and transplanted cells (Temenoff and Mikos 2000). In general, TGF-β1 can recruit cells from the synovial membrane and subsynovial space and help them proliferate and differentiate into chondrocytes when some lesions to articular cartilage occur. To date, it is basically clear that the mitosis and chemotaxis of recruited cells are mediated by relatively low concentrations of TGF-β1, and the follow-up chondrogenic differentiation of the cells is generally regulated by higher concentrations of TGF-β1 (Hunziker 2001), meaning that TGF-β1 needs to be effectively administrated if it is used *in vivo*. However, *in vivo* utilization and administration of ectogenic TGF-β1 are frequently limited because TGF-β1 is usually short-lived if it is exposed to an *in vivo* physiological environment (Whitaker et al. 2001). To protect TGF-β1 from proteolysis and antibody neutralization and to maintain its activity for the required period of time, different biodegradable polymers have been employed to deliver TGF-β1, systemically or locally (Temenoff and Mikos 2000). Some biodegradable polyesters have been tested for delivering different protein growth factors (Zhang et al. 2006, Babensee et al. 2000, Puppi et al. 2010). Nevertheless, several of their common drawbacks, such as hydrophobicity, lack of functional groups on the polyester chains, neutral charge distribution, acidic degradation products, and slow degradation rate, have limited their applications in vehicles for carrying hydrophilic and charged protein growth factors (Puppi et al. 2010). In addition, when these polyesters are used for delivering growth factors, the fabrication of polyester-based vehicles usually involves various organic solvents, which could attenuate the bioactivity of the incorporated protein growth factors since most proteins are known to be denatured when exposed to harsh environmental conditions such as heating, sonication, and organic solvents (Tabata and Ikada 1998, Fu et al. 2000, Zhu and Schwendeman 2000). Different chitosan microspheres loaded with hydrophilic ingredients, such as proteins and peptides, have been fabricated using various ionic cross-linkers (Tabata and Ikada 1998, Kim et al. 2003c, Gan and Wang 2007). However, these cross-linked chitosan microspheres usually show severe burst-release characteristics and are hard of maintaining a sustained and prolonged release of TGF-β1 at a therapeutic level over a long period. CH–PCL copolymer microspheres loaded with TGF-β1 were fabricated with an emulsification method using STPP as cross-linker. Their loading efficiency could be regulated by the amount of cross-linker and the composition of CH–PCLs, and some microspheres showed their loading efficiency higher than 80%. It was observed that in neutral PBS media, the composition of CH–PCLs predominantly governed swelling behavior and release profiles of the microspheres while the effect of cross-linker on the swelling and release behavior was limited. The initial fast releases of TGF-β1 from different microspheres could be significantly decreased with increasing polycaprolactone content in CH–PCLs, and some microspheres were able to maintain sustained releases of TGF-β1 by mainly controlling their composition. In addition, in a simulated acidic environment (pH 6.5) for cartilage lesions, release patterns of the microspheres were notably modulated by pH but some selected microspheres could still well administrate the release of TGF-β1 in a sustained way without severe burst features (Wu et al. 2011).

Kim et al. (2003c) prepared porous chitosan scaffold, containing TGF-β1, and microspheres containing BSA had spherical shapes with a size ranging from 0.2 to 1.5 mm. From the release experiments, it was found that both proteins were slowly released from the microspheres during a period of 5 days in a PBS solution (pH 7.4), in which the release rate of TGF-β1 was much lower than that of BSA. Second, MS-TGFs were seeded onto the porous chitosan scaffold, prepared by the freeze-drying method, to observe the effect on the proliferation and differentiation of chondrocytes. It was demonstrated from *in vitro* tests that, compared to the scaffold without MS-TGF, the scaffold containing MS-TGF significantly augments the cell proliferation and production of ECM, indicating the role of TGF-β1 released from the microspheres (Kim et al. 2003c, Gan and Wang 2007).

Recently, gene-activated matrix (GAM) blends these two strategies of tissue engineering and local therapeutic gene delivery, serving as local bioreactor with therapeutic agent's expression and also providing a structural template to fill the lesion defects for cell adhesion, proliferation, and synthesis of ECM (Kim et al. 2003d).

Chitosan–gelatin complex as biomaterials was used to fabricate three-dimensional scaffolds and plasmid DNA were entrapped in the scaffolds encoding TGF-β1, which has been proposed as a promoter of cartilage regeneration for its effect on the synthesis of matrix molecules and cell proliferation (Guoa et al. 2006). The plasmid DNA incorporated in the scaffolds showed a burst release in the first week and a sustained release for the other 2 weeks. The gene transfected into chondrocytes expresses TGF-β1 protein stably in 3 weeks. The histological and immunohistochemical results confirmed that the primary chondrocytes cultured into the chitosan–gelatin scaffold maintained round and owned characters of high secretion of specific ECM. The results show that chondrocytes culture on this GAM maintained their round shape and cell clusters were surrounded with extracellular matrix composed of proteoglycans and type II collagen (Guoa et al. 2006). This technique combined tissue engineering and gene therapy protocols.

15.10   CONCLUSION

Owing of its cationic, biodegradable, and nontoxic character, chitosan exhibits mucoadhesive, permeation-enhancing, *in situ* gelling, and efflux pump inhibitory properties. Because if derivatization, almost all properties of chitosan, however, can be strongly improved, and this trend will certainly hold on resulting in even more potent chitosan derivatives. It is just a question of time when products being based on potent chitosan derivatives will enter the market, and in fact, the initial step in this direction has already been taken. The very first product containing a chitosan derivative (hydroxypropyl–chitosan) has already been registered; being meanwhile commercially available under the brand name Ciclopoli® and various compounds further will certainly follow.

ACKNOWLEDGMENT

The authors express their appreciation to M. Fathy for his great help in the preparation of this manuscript.

REFERENCES

Abdou, E. S., K. Nagy, and M. Z. Elsabee. 2008. Extraction of chitosan from local sources. *Bioresource Technology* 99:1359–1367.

Aboubakar, M., P. Couvreur, H. Pinto-Alphandary, B. Gouriton, B. Lacour, R. Farinotti, F. Puisieux, and C. Vauthier. 2000. Insulin-loaded nanocapsules for oral administration: *In vitro* and *in vivo* investigation. *Drug Development Research* 49:109.

Agnihotri, S. A., N. N. Mallikarjuna, and T. M. Aminabhavi. 2004. Recent advances on chitosan-based micro- and nanoparticles in drug delivery. *Journal of Controlled Release* 100:5–28.

Agrawal, P., G. J. Strijkers, and K. Nicolay. 2009. Chitosan-based systems for molecular imaging. *Advanced Drug Delivery Reviews* 62:42–58.

Alemdaroğlu, C., Z. Değim, N. Çelebi, F. Zor, S. Öztürk, and De. Erdoğan. 2006. An investigation on burn wound healing in rats with chitosan gel formulation containing epidermal growth factor. *Burns* 32:319–327.

Alishahi, A., A. Mirvaghefi, M. R. Tehrani, H. Farahmand, A. Shojao-sadati, F. A. Dorkoosh, and M. Z. Elsabee. 2011. Shelf life and delivery enhancement of vitamin C using chitosan nanoparticles. *Food Chemistry* 126:935–940.

Arai, K., T. Kinumaki, and T. Fujita. 1968. Toxicity of chitosan. *Bulletin of Tokai Regional Fisheries Research Laboratory* 56:89–94.

Araki, F., H. Nakamura, N. Nojima, K. Tsukumo, and S. Sakamoto. 1989. Stability of recombinant human epidermal growth factor in various solutions. *Chemical & Pharmaceutical Bulletin* 37:404–406.

Babensee, J. E., L. V. McIntire, and A. G. Mikos. 2000. Growth factor delivery for tissue engineering. *Pharmaceutical Research* 17:497–504.

Banerjee, T., S. Mitra, K. A. Singh, R. K. Sharma, and A. Maitra. 2002. Preparation, characterization and biodistribution of ultrafine chitosan nanoparticles. *International Journal of Pharmaceutics* 243:93–105.

Banerjee, T., A. K. Singh, R. K. Sharma, and A. N. Maitra. 2005. Labeling efficiency and biodistribution of technetium-99 m labeled nanoparticles: Interference by colloidal tin oxide particles. *International Journal of Pharmaceutics* 289:189–195.

Barratt, G. 2003. Colloidal drug carriers: Achievements and perspectives. *Cellular and Molecular Life Sciences* 60:21–37.

Begamwar, V., V. Shah, and S. J. Surana. 2009. Formulation and evaluation of oral mucoadhesive multiparticu-late system containing metropolol tartarate: An in vitro-ex vivo characterization. *Current Drug Delivery* 6:113–121.

Bernkop-Schnurch, A., D. Guggi, and Y. Pinter. 2004. Thiolated chitosans: Development and *in vitro* evaluation of a mucoadhesive, permeation enhancing oral drug delivery system. *Journal of Controlled Release* 94:177.

Berscht, P. C., B. Nies, A. Liebéndörfer, and J. Kreutert. 1994. Incorporation of basic fibroblast growth factor into methylpyrrolidinone chitosan fleeces and determination of the *in vitro* release characteristics. *Biomaterials* 15:8.

Bhise, K., R. Dhumal, A. Paradkar, and S. Kadam. 2008. Effect of drying methods on swelling, erosion and drug release from chitosan–naproxen sodium complexes. *AAPS Pharm SciTech* 9:1–12.

Biagini, G., A. Bertani, and R. Muzzarelli. 1991. Wound management with N-carboxybutyl chitosan. *Biomaterials* 12:281–291.

Bodmeier, R. and O. Paeratakul. 1989. Spherical agglomerates of water-insoluble drugs. *Journal of Pharmaceutical Sciences* 78:964–967.

Borchard, G. 2001. Chitosans for gene delivery. *Advanced Drug Delivery Reviews* 52:145–150.

Borchard, G., H. L. Lueben, A. G. de Boer, J. C. Verhoef, C. M. Lehr, and H. E. Junginger. 1996. The potential of mucoadhesive polymers in enhancing intestinal peptide drug absorption. III: Effects of chitosan–glutamate and carbomer on epithelial tight junctions in vitro. *Journal of Controlled Release* 39:131–138.

Bordignon, C., L. D. Notarangelo, N. Nobili, G. Ferrari, G. Casorati, P. Panina, E. Mazzolari, D. Maggioni, C. Rossi, and P. Servida. 1995. Gene therapy in peripheral blood lymphocytes and bone marrow for ADA immunodeficient patients. *Science* 270:470–475.

Bozkir, A. and O. M. Saka. 2004. Chitosan-DNA nanoparticles: Effect on DNA integrity, bacterial transforma-tion and transfection efficiency. *Journal of Drug Targeting* 12:281–288.

Broadley, K. N., A. M. Aquino, and B. Hicks. 1988. Growth factors bFGF and TGF-/I accelerate the rate of wound repair in normal and in diabetic rats. *International Journal of Tissue Reactions* 6:345–353.

Brondsted H. and J. Kopecek. 1992. Hydrogels for site-specific drug delivery to the colon: *In vitro* and *in vivo* degradation. *Pharmaceutical Research* 9:1540–1545.

Brown, G. L., L. J. Curtsinger, M. White, R. O. Mitchell, J. Pietsch, and R. Nordquist. 1988. Acceleration of tensile strength of incisions treated with EGF and TGF-β. *Annals of Surgery* 208:788–794.

Buckwalter, J. A. 1998. Articular cartilage repair: Injuries and potential for healing. *Journal of Orthopaedic & Sports Physical Therapy* 28:192–202.

Carpenter, G. and S. Cohe. 1979. Epidermal growth factor. *Annual Review of Biochemistry* 48:193–216.

Çelebi, N., N. Erden, B. Gönü, and M. Koz. 1993. Effects of epidermal growth factor dosage forms on dermal wound strength in mice. *Journal of Pharmacy and Pharmacology* 46:386–387.

Çelebi, N., A. Türkylmaz, B. Gönül, and C. Özoğul. 2002. Effects of epidermal growth factor microemulsion formulation on the healing of stress-induced gastric ulcers in rats. *Journal of Controlled Release* 83:197–210.

Chae, S. Y., M. K. Jang, and J. W. Nah. 2005. Influence of molecular weight on oral absorption of water soluble chitosans. *Journal of Controlled Release* 102:383–394.

Chaignaud, B. E., R. Langer, and J. P. Vacanti. 1997. The history of tissue engineering using synthetic biodegradable polymer scaffolds and cells. In: Atala, A., Mooney, D. L. (Eds.). *Synthetic Biodegradable Polymer Scaffolds*. Birkhauser, Boston: pp. 1–14.

Chatelet, C., O. Damour, and A. Domard. 2001. Influence of the degree of acetylation on some biological properties of chitosan films. *Biomaterials* 22:261–268.

Chew, J. L., C. B. Wolfowicz, H. Q. Mao, K. W. Leong, and K. Y. Chua. 2003. Chitosan nanoparticles containing plasmid DNA encoding house dust mite allergen, Der p 1 for oral vaccination in mice. *Vaccine* 21:2720–2729.

Choate, K. A. and P. A. Khavari. 1997. Direct cutaneous gene delivery in a human genetic skin disease. *Human Gene Therapy* 8:1659–1665.

Choi, H. S., S. A. Seo, G. Khang, J. M. Rhee, and H. B. Lee. 2002. Preparation and characterization of fentanyl-loaded PLGA microspheres: *In vitro* release profiles. *International Journal of Pharmaceutics* 234:195–203.

Chourasia, M. K. and S. K. Jain. 2004. Design and development of multiparticulate system for targeted drug delivery to colon. *Journal of Drug Delivery* 11:201–207.

Corsi, K., F. Chellat, L. Yahia, and J. C. Fernandes. 2003. Mesenchymal stem cells, MG63 and HEK293 transfection using chitosan-DNA nanoparticles. *Biomaterials.* 24:1255–1264.

Costantino, H. R., L. Illum, G. Brandt, P. H. Johnson, and S. C. Quay. 2007. Intranasal delivery: Physicochemical and therapeutic aspects. *The International Journal of Pharmaceutics* 337:1–24.

Csaba, N., M. Köping-Höggård, E. Fernandez-Megia, R. Novoa-Carballal, R. Riguera, and M. J. Alonso. 2009. Ionically crosslinked chitosan nanoparticles as gene delivery systems: Effect of PEGylation degree on *in vitro* and *in vivo* gene transfer. *Journal of Biomedical Nanotechnology* 5:162–171.

Cui, Z. and R. J. Mumper. 2001. Chitosan-based nanoparticles for topical genetic immunization. *Journal of Controlled Release* 75:409–419.

Danielsen, S., S. Strand, C. de Lange Davies, and B. T. Stokke. 2005. Glycosaminoglycan destabilization of DNA-chitosan polyplexes for gene delivery depends on chitosan chain length and GAG properties. *Biochimica et Biophysica Acta* 1721:44–54.

Dash, M., F. Chiellini, R. M. Ottenbrite, and E. Chiellini. 2011. Chitosan— A versatile semi-synthetic polymer in biomedical Applications. *Progress in Polymer Science* 36:981–1014.

De Campos, A. M., A. Sanchez, and M. J. Alonso. 2001. Chitosan nanoparticles: A new vehicle for the improvement of the delivery of drugs to the ocular surface. Application to cyclosporine A. *International Journal of Pharmaceutics* 224:159–168.

De Smedt, S. C., J. Demeester, and W. E. Hennink. 2000. Cationic polymer based gene delivery systems. *Pharmaceutical Research* 17:113–126.

Derby, A., V. W. Engleman, G. E. Frierdich, G. Neises, S. R. Rapp, and D. G. Roufa. 1993. Nerve growth factor facilitates regeneration across nerve gaps: Morphological and behavioral studies in rat sciatic nerve. *Experimental Neurology* 119:176–191.

Deshpande, D., P. Blezinger, R. Pillai, J. Duguid, B. Freimark, and A. Rolland. 1998. Target specific optimization of cationic lipid-based systems for pulmonary gene therapy. *Pharmaceutical Research* 15:1340–1347.

Dhawan, S., A. K. Singla, and V. R. Sinha. 2004. Evaluation of mucoadhesive properties of chitosan microspheres prepared by different methods. *AAPS PharmSciTech* 5:1–7.

Dijke, P. and K. K. Iwata. 1989. Growth factors for wound healing. *Journal of Biotechnology* 7:793–798.

Dodane, V., M. Amin Khan, and J. R. Merwin. 1999. Effect of chitosan on epithelial permeability and structure. *The International Journal of Pharmaceutics* 182:21–32.

Erbacher, P., S. Zou, T. Bettinger, A. M. Steffan, and J. S. Remy. 1998. Chitosan based vector–DNA complexes for gene delivery: Biophysical characteristics and transfection ability. *Pharmaceutical Research* 15:1332–1339.

Eweis, M., S. S. Elkholy, and M. Z. Elsabee. 2006. Antifungal efficacy of chitosan and its thiourea derivatives upon the growth of some sugar-beet pathogens. *International Journal of Biological Macromolecules* 38:1–8.

Fallini, C., Bassell, G. J., and Rossoll, W. 2010. High-efficiency transfection of cultured primary motor neurons to study protein localization, trafficking, and function. *Mol Neurodegener.* 5:17.

Felt, O., P. Buri, and R. Gurny. 1998. Chitosan: A unique polysaccharide for drug delivery. *Drug Development and Industrial Pharmacy* 24:979–993.

Ferber, D. 2001. Gene therapy: Safer and virus-free. *Science* 294:1638–1642.

Fetih, G., S. Lindberg, K. Itoh, N. Okada, T. Fujita, F. Habib, P. Artersson, M. Attia, and A. Yamamoto. 2005. Improvement of absorption enhancing effects of n-dodecyl-beta-D-maltopyranoside by its colon-specific delivery using chitosan capsules. *International Journal of Pharmaceutics* 293:127–135.

Freiberg, S. and X. X. Zhu. 2004. Polymer microspheres for controlled drug release. *International Journal of Pharmaceutics* 282:1–18.

Friend, D. R. 1991. Colon-specific drug delivery. *Advanced Drug Delivery Reviews* 7: 149–199.

Fu, S. Y. and T. Gordon. 1997. The cellular and molecular basis of peripheral nerve regeneration. *Molecular Neurobiology* 14:67–116.

Fu, K., A. M. Klibanov, and R. Langer. 2000. Protein stability in controlled-release systems. *Nature Biotechnology* 18:24–25.

Fu, X., Z. Shen, Y. Chen, J. Xie, Z. Guo, and M. Zhang. 1998. Randomised placebo-controlled trial of use of topical recombinant bovine basic fibroblast growth factor for second-degree burns. *Lancet* 352: 1661–1664.

Funkhouser, J. D. and N. N. Aronson. 2007. Chitinase family GH18: Evolutionary insights from the genomic history of a diverse protein family. *BMC Evolutionary Biology* 7:96–112.

Gades, M. D. and J. S. Stern. 2003. Chitosan supplementation and fecal fat excretion in men. *Obesity Research* 11:683–688.

Gallo, J. M. and E. E. Hassan. 1988. Receptor-mediated magnetic carriers: Basis for targeting. *Pharmaceutical Research* 5:300–304.

Gan, Q. and T. Wang. 2007. Chitosan nanoparticle as protein delivery carrier: Systematic examination of fabrication conditions for efficient loading and release. *Colloids and Surfaces B: Biointerfaces* 59: 24–34.

Gan, Q., T. Wang, and P. McCarron. 2005. Modulation of surface charge, particle size and morphological properties of chitosan–TPP nanoparticles intended for gene delivery. *Colloids and Surfaces B: Biointerfaces* 44:65–73.

Garca-Rincon, J., J. Vega-Perez, M. G. Guerra-Sanchez, A. N. Herńandez-Lauzardo, A. Pena-Daz, and M. G. Velazquez-Del Valle. 2010. Effect of chitosan on growth and plasma membrane properties of *Rhizopus stolonifer* (Ehrenb.:Fr.) Vuill. *Pesticide Biochemistry and Physiology* 97:275–278.

Garnett, M. C. 1999. Gene delivery systems using cationic polymers. *Critical Reviews in Therapeutic Drug Carrier Systems* 16:147–207.

Gavini, E., A. B. Hegge, G. Rassu, V. Sanna, C. Testa, and G. Pirisino. 2006. Nasal administration of carbamazepine using chitosan microspheres: In vitro/in vivo studies. *The International Journal of Pharmaceutics* 307:9–15.

Genta, I., F. Pavanetto, B. Conti, P. Giunchedi, and U. Conte. 1994. Spray drying for the preparation of chitosan microspheres. *Proc. Int. Symp. Controlled Release Bioactive Material* 21:616–617.

Gill, D. R., K. W. Southern, K. A. Mofford, T. Seddon, L. Huang, F. Sorgi, A. Thompson et al. 1997. A placebo-controlled study of liposome-mediated gene transfer to the nasal epithelium of patients with cystic fibrosis. *Gene Therapy* 4:199–209.

Gillogly, S. D., M. Voight, and T. Blackburn. 1998. Treatment of articular cartilage defects of the knee with autologous chondrocyte implantation. *Journal of Orthopaedic & Sports Physical Therapy* 28:241–251.

Gönül, B., D. Erdoğan, C. Ozoğul, M. Koz, A. Babül, and N. Celebi. 1995. Effect of EGF dosage forms on alkali burned corneal wound healing of mice. *Burns* 21:7–10.

Grabovac, V., D. Guggi, and A. Bernkop-Schnürch. 2005. Comparison of the mucoadhesive properties of various polymers. *Advanced Drug Delivery Reviews* 57:1713–1723.

Gradauer, K., C. Vonach, G. Leitinger, D. Kolb, E. Fröhlich, E. Roblegg, A. Bernkop-Schnürch, and R. Prassl. 2012. Chemical coupling of thiolated chitosan to preformed liposomes improves mucoadhesive properties. *International Journal of Nanomedicine* 7:2523–2534.

Grenha, A., C. I. Grainger, L. A. Dailey, B. Seijo, G. P. Martin, and C. Remunan-Lopez. 2007. Chitosan nanoparticles are compatibles with respiratory epithelial cells in vitro. *European Journal of Pharmaceutics and Biopharmaceutics* 31:73–84.

Guerra-Sanchez, M. G., J. Vega-Perez, M. G. Velazquez-delValle, and A. N. Hernandez-Lauzardo. 2009. Antifungal activity and release of compounds on *Rhizopus stolonifer* (Ehrenb.:Fr.) Vuill. by effect of chitosan with different molecular weights. *Pesticide Biochemistry and Physiology* 93:18–22.

Guo, Z., R. Chen, and R. Xing. 2006. Novel derivatives of chitosan and their antifungal activities in vitro. *Carbohydrate Research* 341:351–354.

Guoa, T., J. Zhao, J. Chang, Z. Ding, H. Hong, J. Chen, and J. Zhang. 2006. Porous chitosan-gelatin scaffold containing plasmid DNA encoding transforming growth factor-β1forchondrocytes proliferation. *Biomaterials* 27:1095–1103.

Gupta, S. and S. P. Vyas. 2010. Carbopol/chitosan based pH triggered *in situ* gelling system for ocular delivery of timolol maleate. *Scientia Pharmaceutica* 78:959–976.

Han, B., H. T. Wang, H. Y. Liu, H. Hong, W. Lv, and Z. H. Shang. 2010. Preparation of pingyangmycin PLGA microspheres and related in vitro/in vivo studies. *International Journal of Pharmaceutics* 398: 130–136.

Hassan, E. E., R. C. Parish, and J. M. Gallo. 1992. Optimized formulation of magnetic chitosan microspheres containing the anticancer agent, oxantrazole. *Pharmaceutical Research* 9:390–397.

Hawary, D. L., M. A. Motaleb, H. Farag, O. W. Guirguis, and M. Z. Elsabee. 2011. Water-soluble derivatives of chitosan as a target delivery system of$^{99\,m}$Tc to some organs *in vivo* for nuclear imaging and biodistribution. *Journal of Radioanalytical and Nuclear Chemistry* 290:557–567.

Hayward, P., J. Hokanson, and J. Heggers. 1992. Fibroblast growth factor reverses the bacterial retardation of wound contraction. *The American Journal of Surgery* 163:288–293.

He, P., S. S. Davis, and L. Illum. 1998. *In vitro* evaluation of the mucoadhesive properties of chitosan microspheres. *The International Journal of Pharmaceutics* 166:75–88.

Hebda, P. A., C. K. Klingbeil, J. A. Abraham, and J. C. Fiddes. 1990. Basic fibroblast growth factor stimulation of epidermal wound healing in pigs. *Journal of Investigative Dermatology* 95:626–631.

Hejazi, R. and M. Amiji. 2002. Stomach-specific anti-H. pylori therapy. I: Preparation and characterization of tetracycline-loaded chitosan microspheres. *International Journal of Pharmaceutics* 235:87–94.

Hejazi, R. and M. Amiji. 2003. Chitosan-based gastrointestinal delivery systems. *Journal of Controlled Release* 89:151–165.

Hirano, S., M. Iwata, K. Yamanaka, H. Tanaka, T. Toda, and H. Inui. 1991. Enhancement of serum lysozyme activity by injecting a mixture of chitosan oligosaccharides intravenously in rabbits. *Agricultural Biology and Chemistry* 55:2623–2625.

Hirano, S., H. Seino, Y. Akiyama, and I. Nonaka. 1988. Bio-compatibility of chitosan by oral and intravenous administration. *Polymer Engineering & Science* 59:897–901.

Ho, Y. C., F.-L. Mi, H.-W. Sung, and P.-L. Kuo. 2009. Heparin-functionalized chitosan–alginate scaffolds for controlled release of growth factor. *International Journal of Pharmaceutics* 376:69–75.

Hornof, M., C. E. Kast, and A. B. Schnurch. 2003. *In vitro* evaluation of the viscoelastic behavior chitosan-thioglycolic acid conjugates. *European Journal of Pharmaceutics and Biopharmaceutics* 55:185–190.

Hsu, S. C., T. M. Don, and W. Y. Chiu. 2002. Free radical degradation of chitosan with potassium persulfate. *Polymer Degradation and Stability* 75:73–83.

Hu, Y., X. Jiang, Y. Ding, H. Ge, Y. Yuan, and C. Yang. 2002. Synthesis and characterization of chitosan–poly(acrylic acid) nanoparticles. *Biomaterials* 23:3193–3201.

Huang, J., X. Hu, L. Lu, Z. Ye, Q. Zhang, and Z. Luo. 2010a. Electrical regulation of Schwann cells using conductive polypyrrole/chitosan polymers. *Journal of Biomedical Materials Research Part A* 93:164–174.

Huang, J., L. Lu, X. Hu, Z. Ye, and Z. Luo. 2010b. Electrical stimulation accelerates motor functional recovery in the rat model of 15 mm sciatic nerve gap bridged by scaffolds with longitudinally oriented microchannels. *Neurorehabilitation and Neural Repair* 24:736–745.

Hunziker, E. B. 2001. Growth-factor-induced healing of partial-thickness defects in adult articular cartilage. *Osteoarthritis and Cartilage* 9:22–32.

Illum, L. 1998. Chitosan and its use as a pharmaceutical excipient. *Pharmaceutical Research* 15:1326–1331.

Illum, L., N. F. Farraj, and S. S. Davis. 1994. Chitosan as a novel nasal delivery system for peptide drugs. *Pharmaceutical Research* 11:1186–1189.

Ing, L. Y., N. M. Zin, A. Sarwar, and H. Katas. 2012. Antifungal activity of chitosan nanoparticles and correlation with their physical properties. *International Journal of Biomaterials* 9 pages, Article ID 632698, 2012, doi:10.1155/2012/632698.

Ishihara, M., K. Ono, M. Sato, K. Nakanishi, Y. Saito, and H. Yura. 2001. Acceleration of wound contraction and healing with a photocrosslinkable chitosan hydrogel. *Wound Repair and Regeneration* 9:513–521.

Ishii, T., Y. Okahata, and T. Sato. 2001. Mechanism of cell transfection with plasmid/chitosan complexes. *Biochimica et Biophysica Acta* 1514:51–64.

Jain, A. and S. K. Jain. 2008. *In vitro* and cell uptake studies for targeting of ligand anchored nanoparticles for colon tumors. *European Journal of Pharmaceutical Sciences* 35:404–416.

Jain, S. K., A. Jain, Y. Gupta, and M. Ahirwar. 2007. Design and development of hydrogel beads for targeted drug delivery to the colon. *AAPS PharmSciTech* 8:E56.

Jain, S. K., A. Jain, Y. Gupta, A. Jain, P. Khare, and M. Kannandasan. 2008. Targeted delivery of 5-ASA to colon using chitosan hydrogel microspheres. *Journal of Drug Delivery Science and Technology* 18:315–321.

Janes, K. A, M. P. Fresneau, A. Marazuela, A. Fabra, and M. J. Alonso. 2001. Chitosan nanoparticles as delivery systems for doxorubicin. *Journal of Controlled Release* 73:255–267.

Jayakumar R., K. P. Chennazhi, R. A. A. Muzzarelli, H. Tamura, V. Nair, and N. Selvamurugan. 2010. Chitosan conjugated DNA nanoparticles in gene therapy. *Carbohydrate Polymer* 79:1–8.

Jintapattanakit, A., V. B. Junyaprasert, and T. Kissel. 2009. The role of mucoadhesion of trimethyl chitosan and PEGylated trimethyl chitosan nanocomplexes in insulin uptake. *Journal of Pharmaceutical Sciences* 98:4818–4830.

Jumaa, M., F. H. Furkert, and B. W. Muller. 2002. A new lipid emulsion formulation with high antimicrobial efficacy using chitosan. *European Journal of Pharmaceutics and Biopharmaceutics* 53:115–123.

Karakeçili, A. G., Satriano, C., Gümüşderelioğlu M., and Marletta, G. 2008. Enhancement of fibroblastic proliferation on chitosan surfaces by immobilized epidermal growth factor. *Acta Biomaterialia* 4:989–996.

Kast, C. E. and A. B. Schnurch. 2001. Thiolated polymers-thiomers: Development and *in vitro* evaluation of chitosan-thioglycolic acid conjugates. *Biomaterials* 22:2345–2352.

Kast, C. E., C. Valenta, M. Leopold and A. B. Schnurch. 2002. Design and *in vitro* evaluation of a novel bioadhesive vaginal drug delivery system for clotrimazole. *Journal of Controlled Release* 81:347–354.

Katas, H. and H. O. Alpar. 2006. Development and characterisation of chitosan nanoparticles for siRNA delivery. *Journal of Controlled Release* 115:216–225.

Kato, Y., H. Onishi, and Y. Machida. 2000. Evaluation of N-succinyl-chitosan as a systemic long-circulating polymer. *Biomaterials* 21: 1579–1585.

Kato, Y., H. Onishi, and Y. Machida. 2001a. Biological characteristics of lactosaminated N-succinyl-chitosan as a liver-specific drug carrier in mice. *Journal of Controlled Release* 70:295–307.

Kato, Y., H. Onishi, and Y. Machida. 2001b. Lactosaminated and intact N-succinyl-chitosans as drug carriers in liver metastasis. *International Journal of Pharmaceutics* 226:93–106.

Kawamata, Y., Y. Nagayama, K. Nakao, H. Mizuguchi, T. Hayakawa, T. Sato, and N. Ishii. 2002. Receptor-independent augmentation of adenovirus mediated gene transfer with chitosan in vitro. *Biomaterials* 23:4573–4579.

Kawashima, Y., T. Handa, A. Kasai, H. Takenaka, and S. Y. Lin. 1985a. The effects of thickness and hardness of the coating film on the drug release rate of theophylline granules coated with chitosan-sodium tripolyphosphate complex. *Chemical and Pharmaceutical Bulletin* 33:2469–2474.

Kawashima, Y., Y. T. Handa, H. Takenaka, S. Y. Lin, and Y. Ando. 1985b. Novel method for the preparation of controlled-release theophylline granules coated with a polyelectrolyte complex of sodium polyphosphate-chitosan. *Journal of Pharmaceutical Sciences* 74:264–268.

Kean T. and M. Thanou. 2009. Chitin and chitosan— sources, production and medical applications, In *Desk Reference of Natural Polymers, Their Sources, Chemistry and Applications*. London: Kentus Books pp. 327–361.

Kean, T. and M. Thanou. 2010. Biodegradation, biodistribution and toxicity of chitosan. *Advanced Drug Delivery Reviews* 62:3–11.

Khor, E. and L. Y. Lim. 2003. Implantable applications of chitin and chitosan. *Biomaterials* 24:2339–2349.

Kiang, T., C. Bright, C. Y. Cheung, P. S. Stayton, A. S. Hoffman, and K. W. Leong. 2004a. Formulation of chitosan-DNA nanoparticles with poly(propyl acrylic acid) enhances gene expression. *Journal of Biomaterials Science, Polymer Edition* 15:1405–1421.

Kiang, T., J. Wen, H. W. Lim, and K. W. Leong. 2004b. The effect of the degree of chitosan deacetylation on the efficiency of gene transfection. *Biomaterials* 25:5293–5301.

Kim, T. H., J. E. Ihm, Y. J. Choi, J. W. Nah, and C. S. Cho. 2003a. Efficient gene delivery by urocanic acid-modified chitosan. *Journal of Controlled Release* 93:389–402.

Kim, D. H., K. Han, R. L. Tiel, and J. A. Murovic, D. G. Kline. 2003b. Surgical outcomes of 654 ulnar nerve lesions. *Journal of Neurosurgery* 98:993.

Kim, E. M., H. J. Jeong, I. K. Park, C. S. Cho, C. G. Kim, and H. S. Bom. 2005. Hepatocyte-targeted nuclear imaging using [99mTc]-galactosylated chitosan: Conjugation, targeting and biodistribution. *Journal of Nuclear Medicine* 46:141–145.

Kim, S. E., J. H. Park, Y. W. Cho, H. Chung, S. Y. Jeong, E. B. Lee, and I. C. Kwon. 2003c. Porous chitosan scaffold containing microspheres loaded with transforming growth factor-β1: Implications for cartilage tissue engineering. *Journal of Controlled Release* 91:365–374.

Kim, M. S.,S. J. Park, B. K. Gu, and C.-H. Kim. 2012. Inter-connecting pores of chitosan scaffold with basic fibroblast growth factor modulate biological activity on human mesenchymal stem cells. *Carbohydrate Polymer* 87:2683–2689.

Kim, T. H., Y. H. Park, K. J. Kim, and C. S. Cho. 2003d. Release of albumin from chitosan coated pectin beads in vitro. *International Journal of Pharmaceutics* 250:371–383.

Kittur, F. S., A. B. Vishu Kumar, and R. N. Tharanathan. 2003. Low molecular weight chitosans—preparation by depolymerization with Aspergillus niger pectinase, and characterization. *Carbohydrate Research* 338:1283–1290.

Kittur, F. S., A. B. Vishu Kumar, M. C. Varadaraj, and R. N. Tharanathan. 2005. Chitooligosaccharides—preparation with the aid of pectinase isozyme from Aspergillus niger and their antibacterial activity. *Carbohydrate Research* 340:1239–1245.

Klingbeil, C. K., L. B. Cesar, and J. C. Fiddes. 1991. In: A. Barbul, E. Pines, M. Caldwell, Th. K. Hunt, eds, *Growth Factors and Other Aspects of Wound Healing Biological and Clinical Implications*. A.R. Liss, New York, chapter 6:443–458.

Knowles, J. and G. Gromo. 2003. Target selection in drug discovery. *Nature Reviews Drug Discovery* 2:63.

Kong, M., X. G. Chen, K. Xing, and H. J. Park. 2010. Antimicrobial properties of chitosan and mode of action: A state of the art review. *International Journal of Food Microbiology* 144:51–63.

Kumari, A., S. K. Yadav, and S. C. Yadav. 2010. Biodegradable polymeric nanoparticles based drug delivery systems. *Colloids and Surfaces B: Biointerfaces* 75:1–18.

Kumbar, S. G., A. R. Kulkarni, and T. M. Aminabhavi. 2002. Crosslinked chitosan microspheres for encapsulation of diclofenac sodium: Effect of crosslinking agent. *Journal of Microencapsulation* 19:173–180.

Lee, J. I., K. S. Ha, and H. S. Yoo. 2008. Quantum-dot-assisted fluorescence resonance energy transfer approach for intracellular trafficking of chitosan/DNA complex. *Acta Biomaterialia* 4:791–798.

Lee, K. Y., I. C. Kwon, Y. H. Kim, W. H. Jo, and S. Y. Jeong. 1998. Preparation of chitosan self-aggregates as a gene delivery system. *Journal of Controlled Release* 51:213–220.

Lee, D. and S. S. Mohapatra. 2008. Chitosan nanoparticle-mediated gene transfer. *Methods in Molecular Biology* 433:127–140.

Lee, J. Y., S. H. Nam, S. Y. Im, Y. J. Park, Y. M. Lee, Y. J. Seol, C. P. Chung, and S. J. Lee. 2002. Enhanced bone formation by controlled growth factor delivery from chitosan-based biomaterials. *Journal of Controlled Release* 78:187–197.

Lehr, C. M., J. A. Bouwstra, E. H. Schacht, and H. E. Junginger.1992. *In vitro* evaluation of mucoadhesive properties of chitosan and some other natural polymers. *The International Journal of Pharmaceutics* 78:43–48.

Leong, K. W., H. Q. Mao, V. L. Truong-Le, K. Roy, S. M. Walsh, and J. T. August. 1998. DNA-polycation nanospheres as non-viral gene delivery vehicles. *Journal of Controlled Release* 53:183–93.

Li, X., Z. Yang, A. Zhang, T. Wang, and W. Chen. 2009. Repair of thoracic spinal cord injury by chitosan tube implantation in adult rats. *Biomaterials* 30:1121–1132.

Liao, C., Huang, J., Sun, S., Xiao, B., Zhou, N., Yin, D., and Wan, Y. Multi-channel *chitosanpolycaprolactoneconduits* embedded with microspheres for controlled release of nerve growth factor. *Reactive and Functional Polymers* 73(1): 149–159.

Line, B. R., P. B. Weber, R. Lukasiewicz, and R. N. Dansereau. 2000. Reduction of background activity through radiolabeling of antifibrin Fab' with 99mTc-dextran. *Journal of Nuclear Medicine* 41:1264–1270.

Liu, H., Y. Du, X. Wang, and L. Sun. 2004. Chitosan kills bacteria through cell membrane damage. *International Journal of Food Microbiology* 95:147–155.

Liu, X., K. A. Howard, M. Dong, M. O. Andersen, U. L. Rahbek, M. G. Johnsen, O. C. Hansen, F. Besenbacher, and J. Kjems. 2007. The influence of polymeric properties on chitosan/siRNA nanoparticle formulation and gene silencing. *Biomaterials* 28:1280–1288.

Liu, W., S. Sun, Z. Cao, X. Zhang, K. Yao, and W. W. Lu. 2005. An investigation on the physicochemical properties of chitosan/DNA polyelectrolyte complexes. *Biomaterials* 26:2705–2711.

Liu, W. G. and K. D. Yao. 2002. Chitosan and its derivatives— a promising non vector for gene transfection. *Journal of Controlled Release* 83:1–11.

Liu, W. G., X. Zhang, Sun S. J., Sun G. J., Yao K. D., Liang D. C., Guo G., and Zhang J. Y. 2003. N-Alkylated chitosan as a potential nonviral vector for gene transfection. *Bioconjugate Chemistry* 14:782–789.

Lorenzo-Lamosa, M. L., C. Remunan-Lopez, J. L. Vila-Jato, and M. J. Alonso. 1998. Design of microencapsulated chitosan microspheres for colonic drug delivery. *Journal of Controlled Release* 52:109–118.

Lu, G. Y., L. J. Kong, B. Y. Sheng, G. Wang, Y. D. Gong, and X. F. Zhang. 2007. Degradation of covalently cross-linked carboxymethyl chitosan and its potential application for peripheral nerve regeneration. *European Polymer Journal* 43:3807–3818.

Lueβen, H. L., B. J. de Leeuw, M. W. E. Langemeer, A. G. de Boer, J. C. Verhoef, and H. E. Junginger. 1996. Mucoadhesive polymers in peroral peptide drug delivery. VI. Carbomer and chitosan improve the intestinal absorption of the peptide drug buserelin in vivo. *Pharmaceutical Research*. 13:1668–1672.

Lynch, S. E., G. R. Castilla, R. C. Williams, C. P. Kiritsy, T. H. Howell, M. S. Reddy, and H. N. Antoniades. 1991. The effects of short- term application of a combination of platelet-derived and insulin-like growth factors on periodontal wound healing. *Journal of Periodontology* 62:458–467.

Ma, L., C. Gao, Z. Mao, J. Zhou, J. Shen, X. Hu, and C. Han. 2003. Collagen/chitosan porous scaffolds with improved biostability for skin tissue engineering. *Biomaterials* 24: 4833–4841.

MacLaughlin, F. C., R. J. Mumper, J. J. Wang, J. M. Tagliaferri, I. Gill, M. Hinchcliffe, and A. P. Rolland. 1998. Chitosan and depolymerized chitosan oligomers as condensing carriers for *in vivo* plasmid delivery. *Journal of Controlled Release* 56:259–272.

Majumbar, S., K. Hippalgaonkar, and M. A. Repka. 2008. Effect of chitosan, benzalkonium chloride and ethylenediaminetetraacetic acid on permeation of acyclovir across isolated rabbit cornea. *International Journal of Pharmaceutics* 348:175–178.

Malhotra, M., C. Lane, C. Tomaro-Duchesneau, S. Saha, and S. Prakash. 2011. A novel method for synthesizing PEGylated chitosan nanoparticles: Strategy, preparation, and *in vitro* analysis. *International Journal of Nanomedicine* 6:485–494.

Malmo, J., K. M. Vrum, and S. P. Strand. 2011. Effect of chitosan chain architecture on gene delivery: Comparison of self-branched and linear chitosans. *Biomacromolecules* 12:721–729.

Mansouri, S., P. Lavigne, K. Corsi, M. Benderdour, E. Beaumont, and J. C. Fernandes. 2004. Chitosan-DNA nanoparticles as non-viral vectors in gene therapy: Strategies to improve transfection efficacy. *European Journal of Pharmaceutics and Biopharmaceutics* 57:1–8.

Mao, H. Q., K. Roy, V. L. Troung-Le, K. A. Janes, K. Y. Lin, Y. Wang, A. J Thomas, and K. W. Leong. 2001. Chitosan-DNA nanoparticles as gene carriers: Synthesis, characterization and transfection efficiency. *Journal of Controlled Release* 70:399–421.

Mao, S., W. Sun, and T. Kissel. 2010. Chitosan-based formulations for delivery of DNA and siRNA. *Advanced Drug Delivery Reviews* 62:12–27.

Markus, A., T. D. Patel, and W. D. Snider. 2002. Neurotrophic factors and axonal growth. *Current Opinion in Neurobiology* 12:523–531.

Martien, R., B. Loretz, M. Thaler, S. Majzoob, and A. Bernkop-Schnürch. 2007. Chitosan–thioglycolic acid conjugate: An alternative carrier for oral nonviral gene delivery. *Journal of Biomedical Materials Research Part A* 82:1–9.

Masotti, A., F. Bordi, G. Ortaggi, F. Marino, and C. Palocci. 2008. A novel method to obtain chitosan/DNA nanospheres and a study of their release properties. *Nanotechnology* 19:1–6.

Matsuda, N., W. L. Lin, N. M. Kumar, M. I. Cho, and R. J. Genco. 1992. Mitogenic, chemotactic and synthetic responses of rat periodontal ligament fibroblastic cells to polypeptide growth factors in vitro. *Journal of Periodontology* 63:515–25.

McCarthy, D. W., M. T. Downing, D. R. Brigstock, M. H. Luquette, K. D. Brown, and M. S. Abad. 1996. Production of heparin-binding epidermal growth factor-like growth-factor (Hb-EGF) at sites of thermal injury in pediatric patients. *Journal of Investigative Dermatology* 106:49–56.

McConnell, E. L., S. Murdan, and A. W. Basit. 2008. An investigation into the digestion of chitosan (noncrosslinked and crosslinked) by human colonic bacteria. *Journal of Pharmaceutical Sciences* 97: 3820–3829.

Mehrara, J. B., B. P. Saadeh, S. D. Steinbrech, M. Dudziak, and A. J. Spector. 1999. Adenovirus-mediated gene therapy of osteoblasts *in vitro* and in vivo. *Journal of Bone and Mineral Research* 14:1290–1301.

Mitra, S., U. Gaur, P. C. Ghosh, and A. N. Maitra. 2001. Tumor targeted delivery of encapsulated dextran-doxorubicin conjugate using chitosan nanoparticles as carrier. *Journal of Controlled Release* 74:317–323.

Mladenovska, K., R. S. Raicki, E. I. Janevik, T. Ristoski, M. J. Pavlova, Z. Kavrakovski, M. G. Dodov, and K. Goracinova. 2007. Colon-specific delivery of 5-aminosalicylic acid from chitosan-Ca-alginate microparticles. *International Journal of Pharmaceutics* 342:124–136.

Mottaghitalab, F., M. Farokhi, V. Mottaghitalab, M. Ziabari, A. Divsalar, and M. A. Shokrgozar. 2011. Enhancement of neural cell lines proliferation using nano-structured chitosan/poly(vinyl alcohol) scaffolds conjugated with nerve growth factor. *Carbohydrate Polymer* 86:526–535.

Mottaghitalab, C., Liao, J. Huang, S. Sun, B. Xiao, N. Zhou, D. Yin, and Y. Wan. 2013. Multi-channel chitosan-polycaprolactone conduits embedded with microspheres for controlled release of nerve growth factor. *Reactive and Functional Polymers* 73:149–159.

Mumper, R. J., J. Wang, J. M. Claspell, and A. P. Rolland. 1995. Novel polymeric condensing carriers for gene delivery. *Proc. International Symposium on Control Release Bioactive Material.* 22:178–179.

Muzzarelli, R. A. A. 1990. *Chitin and Chitosan: Unique Cationic Polysaccharides. Towards a Carbohydrate-Based Chemistry*. CEE, Brussels, pp. 199–232.

Muzzarelli, R. A. A. 1997. Human enzymatic activities related to the therapeutical administration of chitin derivatives. *Cellular and Molecular Life Sciences* 53:131–140.

Muzzarelli, R., V. Baldassarre, F. Conti, P. Ferrera, and B. Biagini. 1988. Biological activity of chitosan: Ultrastructural study. *Biomaterials* 9:247–252.

Nakano, M., Y. Nakamura, K. Takikawa, M. Kouketsu, and T. Arita. 1979a. Sustained release of sulphamethizole from agar beads. *Journal of Pharmacy and Pharmacology* 31(1):869–872.

Nakano, M., K. Takikawa, and T. Arita. 1979b. Release characteristics of dibucaine dispersed in konjac gels. *Journal of Biomedical Materials Research* 13(5):811–819.

Nassif, P. S., S. Q. Simpson, A. A Izzo, and P. J. Nicklaus. 1998. Epidermal growth factor and transforming growth factor-a in middle ear effusion. *Otolaryngology—Head and Neck Surgery* 119:564–568.

Nunn, A. D.1992. *Radiopharmaceuticals, Chemistry and Pharmacology*. Marcel Dekker Inc, New York.

Obara, K., M. Ishihara, T. Ishizukab, M. Fujitaa, Y. Ozeki, T. Maehara, Y. Saito, H. Yura, T. Matsui, H. Hattori, M. Kikuchi, and A. Kurit. 2003. Photocrosslinkable chitosan hydrogel containing fibroblast growth factor-2 stimulates wound healing in healing-impaired db/db mice. *Biomaterials* 24:3437–3444.

Oligino, T. J., Q. Yao, S. C. Ghivizzani, and P. Robbins. 2000. Vector systems for gene transfer to joints. *Clinical Orthopaedics* 379:S17–S30.

Onishi H. and Y. Machida. 1999. Biodegradation and distribution of water soluble chitosan in mice. *Biomaterials* 20:175–182.

Oosegi, T., H. Onishi, and Y. Machida. 2008. Novel preparation of enteric-coated chitosan–prednisolone conjugate microspheres and *in vitro* evaluation of their potential as a colonic delivery system. *European Journal of Pharmaceutics and Biopharmaceutics* 68:260–266.

Ouchi, T., J.-I. Murata, and Y. Ohya. 1999. Gene delivery by quaternary chitosan with antennary galactose residues, In: M. A. El-Nokaly and H. A. Soini, eds, *Polysaccharide Applications: Cosmetics and Pharmaceuticals, ACS Symposium Series*, vol. 737, Washington, DC: American Chemical Society, pp. 15–23.

Ozbas-Turan, S., C. Aral, L. Kabasakal, and M. Keyer-Uysal. 2003. Coencapsulation of two plasmids in chitosan microspheres as a nonviral gene delivery vehicle. *Journal of Pharmaceutical Sciences* 1:27–32.

Park, I. K., J. E. Ihm, Y. H. Park, Y. J. Chois, S. I. Kim, and W. J. Kim. 2003. Galactosylated chitosan (GC)-graft-poly(vinyl pyrrolidone) (PVP) as hepatocyte-targeting DNA carrier: Preparation and physicochemical characterization of GC-graft-PVP/DNA complex(1). *Journal of Controlled Release* 86:349–359.

Park, I. K., T. H. Kim, Y. H. Park, B. A. Shin, E. S. Choi, and E. H. Chowdhury. 2001. Galactosylated chitosan-graft-poly (ethylene glycol) as epatocyte-targeting DNA carrier. *Journal of Controlled Release* 76:349–362.

Park, Y. J., Y. M. Lee, J. Y. Lee, Y. J. Seol, C. P. Chung, and S. J. Lee. 2000a. Controlled release of platelet-derived growth factor-BB from chondroitin sulfate–chitosan sponge for guided bone regeneration. *Journal of Controlled Release* 67:385–394.

Park, Y. K., Y. H. Park, B. A. Shin, E. S. Choi, Y. R. Park, and T. Akaike. 2000b. Galactosylated chitosan-graft-dextran as hepatocyte-targeting DNA carrier. *Journal of Controlled Release* 69:97–108.

Park, J. H., G. Saravanakumar, K. Kim, and I. C. Kwon. 2010. Targeted delivery of low molecular drugs using chitosan and its derivatives. *Advanced Drug Delivery Reviews* 62:28–41.

Patel, V. R. and M. M. Amiji. 1996. Preparation and characterization of freeze-dried chitosan-poly (ethylene oxide) hydrogels for site-specific antibiotic delivery in the stomach. *Pharmaceutical Research* 13: 588–593.

Pittenger, M. F., A. M. Mackay, S. C. Beck, R. K. Jaiswal, R. Douglas, J. D. Mosca, M. A. Moorman, D. W. Simonetti, S. Craig, and D. R. Marshak. 1999. Multilineage potential of adult human mesenchymal stem cells. *Science* 284:143–147.

Porteous, D. J., J. R. Dorin, G. Mclachlan, H. Davidson-Smith, H. Davidson, B. J. Stevenson, A. D. Carothers et al. 1997. Evidence for safety and efficacy of DOTAP cationic liposome mediated CFTR gene transfer to the nasal epithelium of patients with cystic fibrosis. *Gene Therapy* 4:210–218.

Portero, A., D. Teijeiro-Osorio, M. J. Alonso, and C. Remunan-Lopez. 2007. Development of chitosan sponges for buccal administration of insulin. *Carbohydrate Polymer* 68:617–625.

Prabaharan, M. and S. Q. Gong. 2008. Novel thiolated carboxymethyl chitosan-g-β-cyclodextrin as mucoadhesive hydrophobic drug delivery carriers. *Carbohydrate Polymer* 73:117–125.

Prabaharan, M. and J. F. Mano. 2005. Chitosan-based particles as controlled drug delivery systems. *Drug Delivery* 12:41–57.

Prabaharan, M. and J. F. Mano. 2007. A novel pH and thermo-sensitive N,O-carboxymethyl chitosan-graft-poly (N-isopropylacrylamide) hydrogel for controlled drug delivery. *E-Polymers* 043:1–14.

Pujals, G., J. M. Sune-Negre, P. Perez, E. Garcia, M. Portus, J. R. Tico, M. Minarro, and J. Carrio. 2008. *In vitro* evaluation of the effectiveness and Cytotoxicity of meglumine antimoniate microspheres produced by spray drying against *Leishmania infantum*. *Parasitology Research* 102:1243–1247.

Puppi, D., F. Chiellini, A. M. Piras, and E. Chiellini. 2010. Polymeric materials for bone and cartilage repair. *Progress in Polymer Science* 35:403–440.

Qaqish, R. B. and M. M. Amiji. 1999. Synthesis of a fluorescent chitosan derivative and its application for the study of chitosan-mucin interactions. *Carbohydrate Polymer* 38:99–107.

Quan, G. and W. Tao. 2007. Chitosan nanoparticle as protein delivery carrier-Systematic examination of fabrication conditions for efficient loading and release. *Colloids and Surfaces B: Biointerfaces* 59:24–34.

Quong, D. and R. J. Neufeld. 1998. DNA protection from extracapsular nucleases, within chitosan or poly-L-lysine-coated alginated beads. *Biotechnology and Bioengineering* 60:124–134.

Rajam, M., S. Pulavendran, C. Rose, and A. B. Mandal. 2011. Chitosan nanoparticles as a dual growth factor delivery system for tissue engineering applications. *International Journal of Pharmaceutics* 410:145–152.

Rao, S. B. and C. P. Sharma. 1997. Use of chitosan as a biomaterial: Studies on its safety and hemostatic potential. *Journal of Biomedical Materials Research* 34:21–28.

Ravi Kumar M. N. V. 2000. A review of chitin and chitosan applications. *Reactive and Functional Polymers* 46:1–27.

Rich, K. M., T. D. Alexander, J. C. Pryor, and J. P. Hollowell. 1989. Nerve growth factor enhances regeneration through silicone chambers. *Experimental Neurology* 105:162–170.

Richardson, S. C., H. V. Kolbe, and R. Duncan. 1999. Potential of low molecular mass chitosan as a DNA delivery system: Biocompatibility, body distribution and ability to complex and protect DNA. *International Journal of Pharmaceutics* 178:231–243.

Roller, S. and N. Covill. 1999. The antifungal properties of chitosan in laboratory media and apple juice. *International Journal of Food Microbiology* 47:67–77.

Rubinstein, A. 1995. Approaches and opportunities in colon-specific drug delivery. *Critical Reviews in Therapeutic Drug Carrier Systems* 12:101–149.

Rutherford, R. B., C. E. Niekrash, J. E. Kennedy, and M. F. Charette. 1992. Platelet-derived and insulin-like growth factors stimulate regeneration of periodontal attachment in monkeys. *Journal of Periodontal Research* 27:285–290.

Sabnis, S. S. and L. H. Block. 2000. Chitosan as an enabling excipient for drug delivery systems I: Molecular modifications. *International Journal of Biological Macromolecules* 27:181–186.

Sahoo, S. K., F. Dilnawaz, and S. Krishnakumar. 2008. Nanothecnology in ocular drug delivery. *Drug Discovery Today* 13:144–151.

Sakloetsakun, D., J. Hombach, and A. Bernkop-Schnürch. 2009. *In situ* gelling properties of chitosan–thioglycolic acid conjugate in the presence of oxidizing agents. *Biomaterials* 30:6151–6157.

Sashiwa, H. and Y. Shigemasa. 1999. Chemical modification of chitin and chitosan 2: Preparation and water soluble property of N-acylated or N-alkylated partially deacetylated chitins. *Carbohydrate Polymer* 39:127–138.

Sato, T., T. Ishii, and Y. Okahata. 2001. *In vitro* gene delivery mediated by chitosan. Effect of pH, serum, and molecular mass of chitosan on the transfection efficiency. *Biomaterials* 22:2075–80.

Schipper, N. G., K. M. Varum, and P. Artursson. 1996. Chitosans as absorption enhancers for poorly absorbable drugs. 1: Influence of molecular weight and degree of acetylation on drug transport across human intestinal epithelial (Caco-2) cells. *Pharmaceutical Research* 13:1686–1692.

Schipper, N. G., K. M. Varum, P. Stenberg, G. Ocklind, H. Lennernas, and P. Artursson. 1999. Chitosans as absorption enhancers of poorly absorbable drugs. 3: Influence of mucus on absorption enhancement. *European Journal of Pharmaceutical Sciences* 8:335–343.

Schultz, G. S., M. White, R. Mitchell, G. Brown, J. Lynch, and D. R. Twardzik. 1987. Epithelial wound healing enhanced by transforming growth factor-a and vaccinia growth factor. *Science* 235:350–352.

Şenel, S. and A. A. Hincal. 2001. Drug permeation enhancement via buccal route: Possibilities and limitations. *Journal of Controlled Release* 72:133–144.

Şenel, S., M. J. Kremer, S. Kas, P. W. Wertz, A. A. Hincal, and C. A. Squier. 2000. Enhancing effect of chitosan on peptide drug delivery across buccal mucosa. *Biomaterials* 21:2067–2071.

Senel, S. and S. J. McClure. 2004. Potential applications of chitosan in veterinary medicine. *Advanced Drug Delivery Reviews* 56: 1467–1480.

Seyfarth, F., S. Schliemann, P. Elsner, and U. C. Hipler. 2008. Antifungal effect of high- and low-molecular-weight chitosan hydrochloride, carboxymethyl chitosan, chitosan oligosaccharide and N-acetyl-d-glucosamine against Candida albicans, Candida krusei and Candida glabrata. *International Journal of Pharmaceutics* 353:139–148.

Shapiro, S., A. Meier, and B. Guggenheim. 1994. The antimicrobial activity of essential oils and essential oil components towards oral bacteria. *Oral Microbiology and Immunology* 9:202–208.

Shavi, G., U. Nayak, M. Reddy, A. Karthik, P. Deshpande, A. Kumar, and N. Udupa. 2011. Sustained release optimized formulation of anastrozole-loaded chitosan microspheres: *In vitro* and *in vivo* evaluation. *Journal of Materials Science: Materials in Medicine* 22:865–878.

Shigemasa,Y. and S. Minami. 1995. Application of chitin and chitosan for biomaterials. *Biotechnology & Genetic Engineering Reviews* 13:383–420.

Shu, X. Z. and K. J. Zhu. 2000. A novel approach to prepare tripolyphosphate/chitosan complex beads for controlled release drug delivery. *International Journal of Pharmaceutics* 201:51–58.

Somia, N. and I. M. Verma. 2000. Gene therapy: Trials and tribulations. *Nature Reviews Genetics* 1:91–99.

Sonaje, K., Y. H. Lin, J. H. Juang, S. P. Wey, C. T. Chen, and H. W. Sung. 2009. *In vivo* evaluation of safety and efficacy of self-assembled nanoparticles for oral insulin delivery. *Biomaterials* 30:2329–2339.

Sonia, T. A. and C. P. Sharma. 2011. Chitosan and its derivatives for drug delivery. *Perspective Advanced Polymer Science* 243:23–54.

Sun, W., S. Mao, Y. Wang, V. B. Junyaprasert, T. Zhang, L. Na, and J. Wang. 2010. Bioadhesion and oral absorption of enoxaparin nanocomplexes. *International Journal of Pharmaceutics* 386: 275–281.

Taba, M., Q. Jin, V. J. Sugai, and V. W. Giannobile. 2005. Current concepts in periodontal bioengineering. *Orthodontics and Craniofacial Research* 8:292–302.

Tabata,Y., Tabata, and Y. Ikada. 1998. Protein release from gelatin matrices. *Advanced Drug Delivery Reviews* 31:287–301.

Takahashi,Y., N. Nambu, and T. Nagai. 1978. Interaction of several nonsteroidal antiinflammatory drugs with pectin in aqueous solution and in solid state. *Chemical & Pharmaceutical Bulletin* 26:3836.

Takeuchi, H., Y. Matsui, H. Yamamoto, and Y. Kawashima. 2003. Mucoadhesive properties of carbopol or chitosan-coated liposomes and their effectiveness in the oral administration of calcitonin to rats. *Journal of Controlled Release* 86:235–242.

Tang, D.-W., S.-H. Yu, Y.-C. Ho, F.-L. Mi, P.-L. Kuo, and H.-W. Sung. 2010. Heparinized chitosan/poly(g-glutamic acid) nanoparticles for multi-functional delivery of fibroblast growth factor and heparin. *Biomaterials* 31:9320–9332.

Tapia, C., V. Corbalán, E. Costa, M. N. Gai, and M. Yazdani-Pedram. 2005. Study of the release mechanism of diltiazem hydrochloride from matrices based on chitosan_alginate and chitosan_carrageenan mixtures. *Biomacromolecules* 6:2389–2395.

Tayel, A. A., S. Moussa, W. F. El-Tras, D. Knittel, K. Opwis, and E. Schollmeyer. 2010. Anticandidal action of fungal chitosan against Candida albicans. *International Journal of Biological Macromolecules* 47:454–457.

Teijeiro-Osorio, D., C. Remuñán-López, and M. J. Alonso. 2009. Chitosan/cyclodextrin nanoparticles can efficiently transfect the airway epithelium in vitro. *European Journal of Pharmaceutics and Biopharmaceutics* 71:257–263.

Temenoff, J. S. and A. G. Mikos. 2000. Review: Tissue engineering for regeneration of articular cartilage. *Biomaterials* 21:431–440.

Thanou, M., J. C. Verhoef, and H. E. Junginger. 2001. Oral drug absorption enhancement by chitosan and its derivatives. *Advanced Drug Delivery Reviews* 52:117–126.

Tikhonov, V. E., E. A. Stepnova, and V. G. Babak. 2006. Bactericidal and antifungal activities of a low molecular weight chitosan and its N-/2(3)-(dodec-2-enyl) succinoyl/derivatives. *Carbohydrate Polymer* 64:66–72.

Tokumitsu, H., H. Ichikawa, and Y. Fukumori. 1999. Chitosan–gadopentetic acid complex nanoparticles for gadolinium neutron capture therapy of cancer: Preparation by novel emulsion-droplet coalescence technique and characterization. *Pharmaceutical Research* 16:1830–1835.

Torchilin, V. P. 2006. Multifunctional nanocarriers. *Advanced Drug Delivery Reviews* 58:1532–1555.

Tozaki, H., T. Odoriba, N. Okada, T. Fujita, A. Terabe, T. Suzuki, S. Okabe, S. Muranishi, and A. Yamamoto. 2002. Chitosan capsules for colon-specific drug delivery: Enhanced localization of 5-aminosalicylic acid in the large intestine accelerates healing of TNBS-induced colitis in rats. *Journal of Controlled Release* 82:51–61.

Ueno, H., T. Mori, and T. Fujinaga. 2001. Topical formulations and wound healing applications of chitosan. *Advanced Drug Delivery Reviews* 52:105–115.

Ueno, H., H. Yamada, I. Tanaka, N. Kaba, M. Matsuura, and M. Okumura. 1999. Accelerating effects of chitosan for healing at early phase of experimental open wound in dogs. *Biomaterials* 20:1407–1414.

Varkouhi, A. K., R. J. Verheul, R. M. Schiffelers, T. Lammers, G. Storm, and W. E. Hennink. 2010. Gene silencing activity of siRNA polyplexes based on thiolated N,N,N-trimethylated chitosan. *Bioconjugate Chemistry* 21:2339–2346.

Varshosaz, J., A. J. Dehkordi, and S. Golafshan. 2006. Colon-specific delivery of mesalazine chitosan microspheres. *Journal of Microencapsulation* 23:329–339.

Ventura, C. A., S. Tommasini, E. Crupi, I. Giannone, V. Cardile, and T. Musumeci. 2008. Chitosan microspheres for intrapulmonary administration of moxifloxacin: Interaction with biomembrane models and *in vitro* permeation studies. *European Journal of Pharmaceutics and Biopharmaceutics* 68:235–244.

Vila, A., A. Sanchez, K. Janes, I. Behrens, T. Kissel, J. L. Vila, and M. J. Alonso. 2004. Low molecular weight chitosan nanoparticles as new carriers for nasal vaccine delivery in mice. *European Journal of Pharmaceutics and Biopharmaceutics* 57:123–131.

Vila, A. I., S. Súarez, and M. J. Alonso. 2007. Chitosan and heparin nanoparticles. *Spanish Patent* WO/2007/042572.

Wang, Q., L. Zhang, W. Hu, Z. H. Hu, Y. Y. Bei, and J. Y. Xu. 2010. Norcantharidin associated galactosylated chitosan nanoparticles for hepatocyte-targeted delivery. *Nanomedicine* 6:371–381.

Wedmore, I., J. G. McManus, A. E. Pusateri, and J. B. Holcomb. 2006. A special report on the chitosan-based hemostatic dressing: Experience in current combat operations. *The Journal of Trauma* 60:655–668.

Wenczak, B. A., J. B. Lynch, and L. B. Nanney. 1992. Epidermal growth factor receptor distribution in burn wounds. *Journal of Clinical Investigation* 90:2392–2401.

Werle, M. and A. Bernkop-Schnürch. 2008. Thiolated chitosans: Useful excipients for oral drug delivery. *Journal of Pharmacy and Pharmacology* 60:273–281.

Whitaker, M. J., R. A. Quirk, S. M. Howdle, and K. M. Shakesheff. 2001. Growth factor release from tissue engineering scaffolds. *Journal of Pharmacy and Pharmacology* 53:1427–1437.

Wolff, J. A., R. W. Malone, P. Williams, W. Chong, G. Acsadi, A. Jani, and P. L. Felgner. 1990. Direct gene transfer into mouse muscle in vivo. *Science* 247:1465–1468.

Wu, H., S. Wang, H. Fang, X. Zan, J. Zhang, and Y. Wan. 2011. Chitosan–polycaprolactone copolymer microspheres for transforming growth factor-β1 delivery. *Colloids and Surfaces B: Biointerfaces* 82:602–608.

Xu, Y. and Y. Du. 2003. Effect of molecular structure of chitosan on protein delivery properties of chitosan nanoparticles. *International Journal of Pharmaceutics* 250:215–226.

Xu, Q., L. Guo, X. Gu, B. Zhang, X. Hu, J. Zhang, J. Chen, Y. Wang, C. Chen, B. Gao, Y. Kuang, and S. Wang. 2012. Prevention of colorectal cancer liver metastasis by exploiting liver immunity via chitosan-TPP/ nanoparticles formulated with IL-12. *Biomaterials* 33:3909–3918.

Xu, G., X. Huang, L. Qiu, J. Wu, and Y. Hu. 2007. Mechanism study of chitosan on lipid metabolism in hyperlipidemic rats. *Asia Pacific Journal of Clinical Nutrition* 16:313–317.

Xu J., S. P. McCarthy, R. A. Gross, and D. L. Kaplan. 1996. Chitosan film acylation and effects on biodegradability. *Macromolecules* 29:3436–3440.

Yang, Y. M., W. Hu, X. D. Wang, and X. S. Gu. 2007. The controlling biodegradation of chitosan fibers by N-acetylation *in vitro* and in vivo. *Journal of Materials Science: Materials in Medicine* 18: 2117–2121.

Yang, K., X. Li, Z. Yang, P. Li, F. Wang, and Y. Liu. 2009a. Novel polyion complex micelles for liver-targeted delivery of diammonium glycyrrhizinate: In vitro and *in vivo* characterization. *Journal of Biomedical Materials Research A* 88:140–148.

Yang, Y., M. Liu, Y. Gu, S. Lin, F. Ding, and X. Gu. 2009b. Effect of chitooligosaccharide on neuronal differentiation of PC-12 cells. *Cell Biology International* 33:352–356.

Yang, Y., W. Zhao, J. He, Y. Zhao, F. Ding, and X. Gu. 2011. Nerve conduits based on immobilization of nerve growth factor onto modified chitosan by using genipin as a crosslinking agent. *European Journal of Pharmaceutics and Biopharmaceutics* 79:519–525.

You, J. O., Y. C. Liu, and C. A. Peng. 2006. Efficient gene transfection using chitosan–alginate core-shell nanoparticles. *International Journal of Nanomedicine* 1:173–180.

Yu, H., X. Chen, T. Lu, J. Sun, H. Tian, J. Hu, Y. Wang, P. Zhang, and X. Jing. 2007. Poly(Llysine)-graft-chitosan copolymers: Synthesis, characterization, and gene transfection effect. *Biomacromolecules* 8:1425–1435.

Yuan, X. B., Y. B. Yuan, W. Jiang, J. Liu, E. J. Tian, and H. M. Shun. 2008. Preparation of rapamycin-loaded chitosan/PLA nanoparticles for immunossupression in corneal transplantation. *The International Journal of Pharmaceutics* 349:241–248.

Zelphati, O., Y. Wang, S. Kitada, J. C. Reed, P. L. Felgner and J. Corbeil. 2001. Intracellular delivery of proteins with a new lipid-mediated delivery system. *The Journal of Biological Chemistry* 14:35103–35110.

Zenga, W., J. Huanga, X. Hu, W. Xiao, M. Rong, Z. Yuan, and Z. Luo. 2011. Ionically cross-linked chitosan microspheres for controlled release of bioactive nerve growth factor. *International Journal of Pharmaceutics* 421:283–290.

Zhang, C., Y. Cheng, G. Qu, X. Wu, Y. Ding, Z. Cheng, L. Yu, and Q. Ping. 2008a. Preparation and characterization of galactosylated chitosan coated BSA microspheres containing 5-fluorouracil. *Carbohydrate
Polymer* 72:390–397.

Zhang, Y.,X. Cheng, J. Wang, Y. Wang, B. Shi, C. Huang, X. Yang, and T. Liu. 2006. Novel chitosan/collagen
scaffold containing transforming growth factor-β1 DNA for periodontal tissue engineering. *Biochemical
and Biophysical Research Communications* 344:362–369.

Zhang, C., Q. N. Ping, Y. Ding, Y. Cheng, and J. Shen. 2004. Synthesis, characterization and microspheres
formation of galactosylated chitosan. *Journal of Applied Polymer Science* 91:659–665.

Zhang H. and S. H. Neau. 2002. *In vitro* degradation of chitosan by bacterial enzymes from rat cecal and
colonic contents. *Biomaterials* 23:2761–2766.

Zhang, C., G. Qu, Y. Sun, X. Wu, Z. Yao, Q. Guo, Q. Ding, S. Yuan, Z. Shen, Q. Ping, and H. Zhou. 2008b.
Pharmacokinetics, biodistribution, efficacy and safety of N-octyl-O-sulfate chitosan micelles loaded with
paclitaxel. *Biomaterials* 29:1233–1241.

Zhang, Y., Y. Wang, B. Shi, and X. Cheng. 2007. A platelet-derived growth factor releasing chitosan/coral composite scaffold for periodontal tissue engineering. *Biomaterials* 28:1515–1522.

Zhao, X., L. Yin, J. Ding, C. Tang, S. H. Gu, and C. Yin. 2010. Thiolated trimethyl chitosan nanocomplexes
as gene carriers with high *in vitro* and *in vivo* transfection efficiency. *Journal of Controlled Release*
144:46–54.

Zheng, F., X. W. Shi, G. F. Yang, L. L. Gong, H. Y. Yuan, and Y. J. Cui. 2007. Chitosan nanoparticle as gene
therapy vector via gastrointestinal mucosa administration: Results of an *in vitro* and *in vivo* study. *Life
Science* 80:388–396.

Zheng, H., X. Zhang, Y. Yin, F. Xiong, X. Gong, Z. Zhu, B. Lu, and P. Xu. 2011. *In vitro* characterization, and *in
vivo* studies of crosslinked lactosaminated carboxymethyl chitosan nanoparticles. Carbohydrate Polymer
84:1048–1053.

Zhong, Z., R. Chen, and R. Xing. 2007. Synthesis and antifungal properties of sulfanilamide derivatives of
chitosan. *Carbohydrate Research* 342:2390–2395.

Zhu, G. and S. P. Schwendeman. 2000. Stabilization of proteins encapsulated in cylindrical poly(lactide-*co*-gly-
colide) implants: Mechanism of stabilization by basic additives. *Pharmaceutical Research* 17:351–357.

Ziani, K., I. Ferńandez-Pan, M. Royo, and J. I. Mate. 2009. Antifungal activity of films and solutions based on
chitosan against typical seed fungi. *Food Hydrocolloid* 23:2309–2314.

Zoldners, J. T. Kiseleva, and I. Kaiminsh 2005. Influence of ascorbic acid on the stability of chitosan solutions.
Carbohydrate Polymer 60: 215–218.

16 Modifications of Chitosan for Gene Delivery

Bijay Singh, Sushila Maharjan, Sang-Ki Kang, Yun-Jaie Choi,
Myung-Haing Cho, and Chong-Su Cho

CONTENTS

16.1 INTRODUCTION

Gene therapy is a method of treatment for several genetic and acquired diseases by replacing defective genes, substituting missing genes, or silencing unwanted gene expression (Kabanov and Kabanov 1995). Although gene therapy is regarded as a promising therapeutic approach, the method still lacks an efficient and safe delivery system or vector that delivers therapeutic genes to a specific target tissue or organ selectively and efficiently with minimal toxicity (Li and Huang 2000). Although viral vectors have been commonly used in the majority of clinical trials due to their high transfection efficiency and capability of inducing long-term gene expression, they have several drawbacks such as immunogenicity, potential infectivity, complicated production, and inflammation (Somia and Verma 2000).

Therefore, nonviral vectors have been extensively used as alternatives for gene delivery because of low immune response, low toxicity, high flexibility of size of the delivered gene, and easy production (Putnam 2006). Among nonviral vectors, cationic synthetic polymers such as poly(L-lysine) (Kadlecova et al. 2012), polyethylenimine (PEI) (Lai 2011), poly(ester amine) (Islam et al. 2011), and polyamidoamine dendrimer (Wen et al. 2012) have been studied owing to their well-defined chemistries, effective condensation of DNA, stability of polyplexes, and low immunogenicity.

Among natural polymers, chitosan and chitosan derivatives have been extensively studied as gene carriers because they are biocompatible, less immunogenic, and less toxic. Chitosan is a linear, cationic polysaccharide of $\beta(1 \rightarrow 4)$-linked D-glucosamine and N-acetyl-D-glucosamine, which can be obtained by deacetylation of chitin. Although chitosan could be used as potential gene carriers due to its biodegradability, biocompatibility, and strong DNA binding ability, it has several defects, such as poor solubility in physiological pH, low transfection efficiency, and low cell specificity

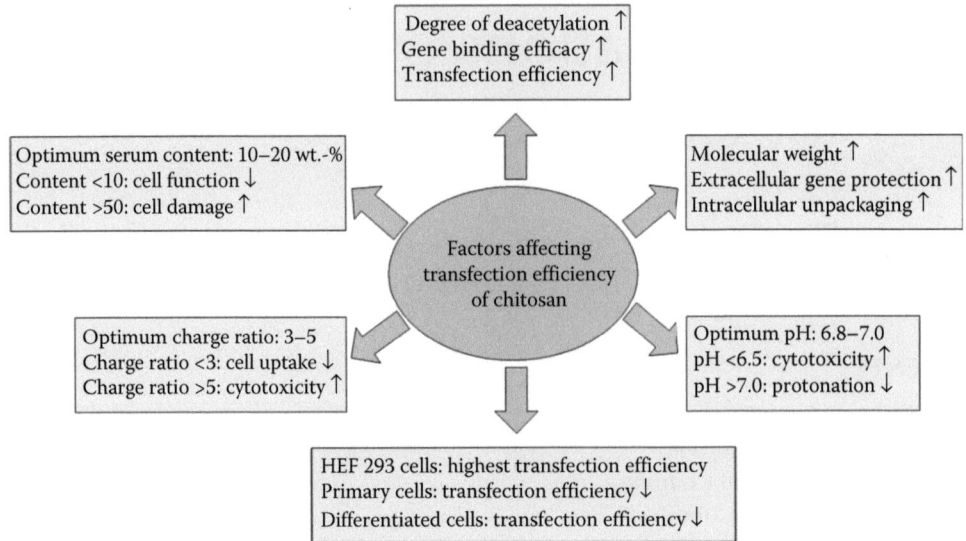

FIGURE 16.1 Schematic illustration of the factors affecting transfection efficiency of chitosan as a gene carrier. (Modified from Kim, T. H. et al., *Progress in Polymer Science* 32: 726–753. 2007. With permission.)

(Park et al. 2003). Other factors that affect the transfection efficiency of chitosan are summarized in Figure 16.1 (Kim et al. 2007).

To overcome the low transfection efficiency and low cell specificity of chitosan, it has been modified by hydrophilic, hydrophobic, pH-sensitive, thermosensitive, and cell-specific ligand groups. In this review, only pH-sensitive modification of chitosan will be discussed as the chemical modification of chitosan by pH-sensitive groups, such as PEI, amino acid (or amino acid derivatives), and spermine, critically affect the transfection efficiency of DNA or siRNA. Besides, conjugation of specific ligand into pH-sensitive-modified chitosan for cell specificity will be explained.

16.2 pH-SENSITIVE MODIFICATION

Chemical modification is a powerful method to improve the physical and chemical properties of chitosan. Generally, chitosan is modified by grafting small molecules or polymer chains onto the chitosan backbone, and sometimes by quaternization of the amino groups. There are three attractive reactive sites, two hydroxyl groups (primary or secondary) and one primary amine, in chitosan chains for chemical modification. Therefore, several changes can be made in chitosan chains to improve its property for various applications, which are discussed below.

16.2.1 PEI-MODIFIED CHITOSAN

To overcome the low transfection efficiency of chitosan/DNA complexes owing to the inefficient release of DNA in the complexes from endosomes, low molecular weight (MW) of PEI is introduced to the chitosan because it facilitates endosomal escape to the cytoplasm due to its high buffering capacity and because it is less cytotoxic compared to high MW PEI. PEI-graft-chitosan was initially prepared by the polymerization of aziridine with low MW chitosan to evaluate it as a DNA carrier (Wong et al. 2006). The PEI-graft-chitosan showed higher transfection efficiency with lower cytotoxicity than PEI 25 K both *in vitro* and *in vivo* due to the combination of the buffering capacity of low MW PEI and the biocompatible nature of chitosan. Another chitosan-graft-PEI was prepared by imine reaction between periodate-oxidized chitosan and low MW PEI as shown in Figure 16.2

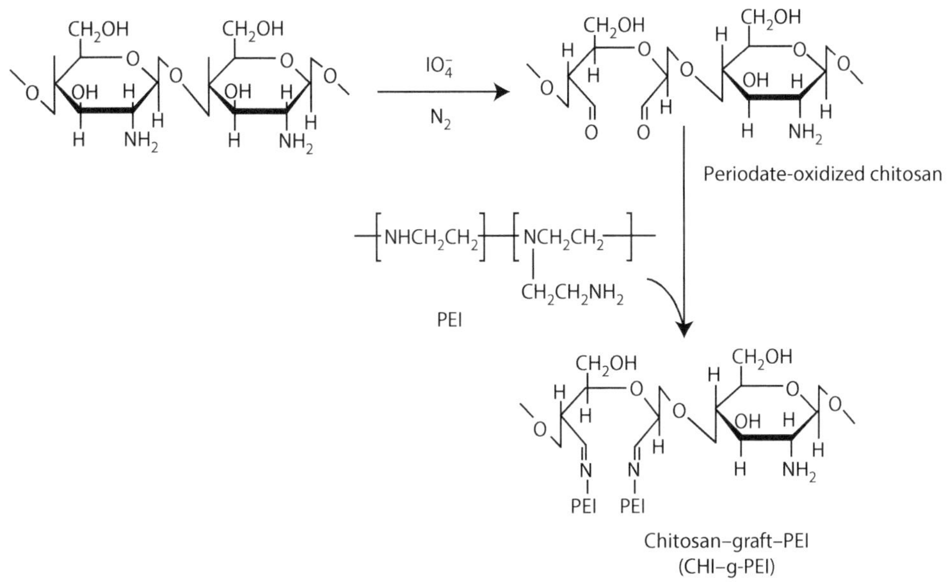

FIGURE 16.2 Proposed reaction scheme for the synthesis of chitosan-graft-PEI.

(Jiang et al. 2007a). The chitosan-graft-PEI also showed higher transfection efficiency in three different cell lines, such as HeLa, 293T, and HepG2, than PEI 25 K with lower cytotoxicity due to the buffering effect of PEI.

Similarly, N-maleated chitosan (NMC)-graft-PEI (Lu et al. 2008) and N-succinyl chitosan (NSC)-graft-PEI (Lu et al. 2009) showed higher transfection efficiency than PEI 25 K in HeLa and 293T cells with lower cytotoxicity. Moreover, the transfection efficiency of the NSC-graft-PEI was not affected by the presence of serum because of the hydrophilic succinyl group in the NSC-graft-PEI. Likewise, chitosan was grafted with low MW PEI (600 Da) through a short poly(ethylene glycol) with terminal epoxide rings to increase water solubility and transfection efficiency of chitosan (Lou et al. 2009). The transfection efficiency of the PEI-graft-chitosan was significantly higher than only chitosan with lower cytotoxicity even at higher weight ratios.

Next, 1,1'-carbonyl-diimidazole was used as a linking agent to link chitosan with low MW PEI for preparing chitosan-graft-PEI (Gao et al. 2010). The transfection efficiency of the chitosan-graft-PEI in three different cancer cells was higher than PEI with lower cytotoxicity. Also, the tumor growth rate of mice inoculated with CCL22 gene by the chitosan-graft-PEI was much more suppressed because of the rapid escape of the gene from endosome due to PEI in the copolymer. In another study, polyethylene glycol (PEG) was grafted to chitosan-graft-PEI to increase its solubility in water (Zhang et al. 2010). The transfection efficiency of the PEG-graft-chitosan-graft-PEI was higher than that of PEI 25 K with lower cytotoxicity, and serum did not affect the transfection efficiency because of the PEG in the copolymer.

Moreover, the effect of the degree of grafting of PEI in chitosan-graft-PEI on cytotoxicity and transfection efficiency was also studied (Pezzoli et al. 2012). It was found that intermediate degree of grafting of 2.7% resulted in low cytotoxicity and higher transfection efficiency compared to lower and higher degree of grafting of PEI revealing the significance of structure–activity relationship. Likewise, 27.7% grafting of PEI in the linear PEI-graft-chitosan prepared by the reaction of chlorohydrin chitosan with linear low MW PEI (2.5 kD) demonstrated 2–24-fold higher transfection efficiency compared to chitosan, PEI (25 kD), and Lipofectamine (Tripathi et al. 2012a).

Recently, PEI-conjugated stearic acid (SA)-graft-chitosan was synthesized for effective antitumor gene therapy because PEI has buffering capacity and stearic acid has faster internalization ability into tumor cells (Hu et al. 2013). The carrier showed comparable transfection efficiency with

Lipofectamine 2000 in HeLa and MCF-7 cells and the transfection efficiency was enhanced in the presence of serum. Also, chitosan-SA-graft-PEI complexed with pigment epithelium-derived factor effectively suppressed the tumor growth (above 60% tumor suppression) without systematic toxicity against mice after intravenous (IV) injection.

16.2.2 Amino Acid-Modified Chitosan

Several amino acids or amino acid derivatives have been used to modify chitosan for enhancing solubility and gene transfection. Among them, imidazole or imidazole derivatives were mostly used to modify chitosan as imidazole side chain of histidine has a suitable pK_a of approximately 6 that imparted buffering capacity to the carrier and enhanced the release of polyplexes into the cytoplasm following adsorptive pinocytosis (Wang et al. 2008).

Likewise, the imidazole group in urocanic acid (UA) when introduced into water-soluble chitosan (Figure 16.3) was reported to have buffering capacity (Kim et al. 2003). Therefore, the UA-coupled chitosan (UAC) mediated about 1000-fold higher transfection efficiency than chitosan itself in 293T cells. Programmed cell death 4 (PDCD4) as a therapeutic gene was also delivered by the UAC to a K-ras null lung cancer model mouse via inhalation (Jin et al. 2006). The UAC/PDCD4 suppressed tumor angiogenesis and facilitated apoptosis in comparison to UAC itself. Furthermore, phosphatase and tensin homolog deleted on chromosome 10 (PTEN) as another therapeutic gene was delivered into a K-ras lung cancer model mouse via inhalation (Jin et al. 2008). The UAC/PTEN suppressed lung tumors through Akt-related signal pathway and cell cycle arrest mechanism. Similarly, the UAC was also used to deliver p53 gene into HepG2 and BALB/c nude mice via intratumoral injection (Wang et al. 2008). The UAC/p53 facilitated apoptosis by inhibition of HepG2 growth *in vitro* and suppressed tumor growth *in vivo*.

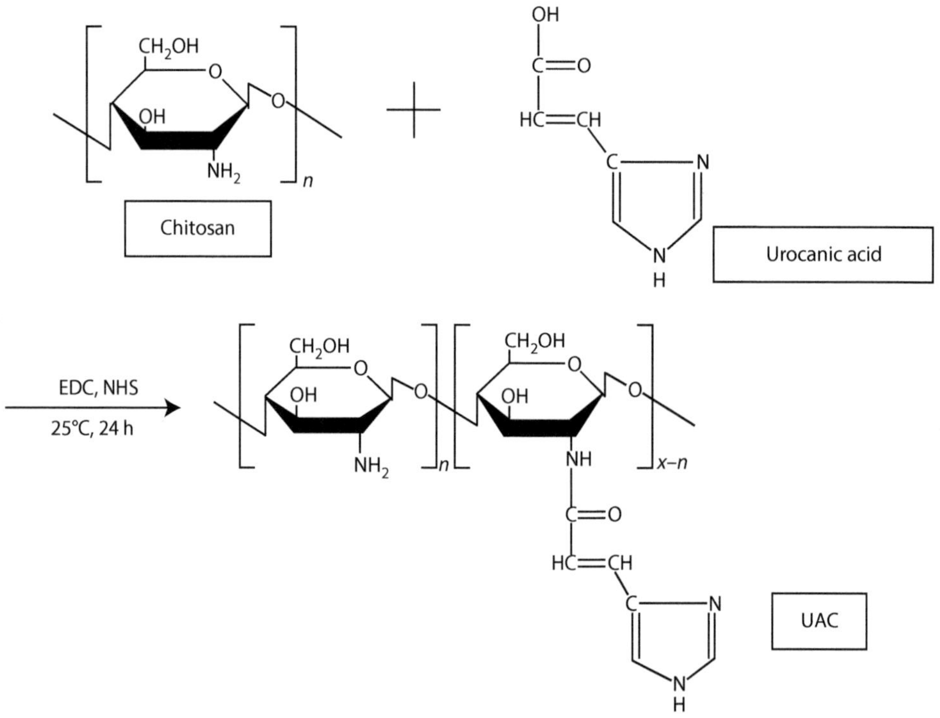

FIGURE 16.3 Synthetic scheme of urocanic acid-coupled chitosan.

Several compounds that contain imidazole ring have been incorporated into chitosan to induce the buffering capacity of chitosan. As a proof of concept, imidazole acetic acid (IAA)-modified chitosan showed about 100-fold increase of transfection efficiency than chitosan (Ghosn et al. 2008). The transfection efficiency also depends on the degree of substitution (DS) of IAA in the modified chitosan. Hence, IAA-modified chitosan with 22.1 mol % of DS showed highest β-galactosidase expression in HepG2 and 293T cells (Moreira et al. 2009). Moreover, imidazole-grafted chitosan has been used as a tool for tuning the expression of a gene delivered in the context of regenerative medicine (Pires et al. 2011). While the transfection mediated by imidazole-grafted chitosan was time-dependent, the expression of the delivered gene occurred during a minimum 7-day period.

Recently, carboxyl and imidazolyl groups were introduced into chitosan to increase its water solubility (Shi et al. 2012). *N*-Imidazolyl-*O*-carboxymethyl chitosan not only showed low cytotoxicity against HEK293T cells but also demonstrated high transfection efficiency, which is dependent on the degree of imidazolyl substitution. Histidine itself was introduced into chitosan through disulfide bond between histidine–cysteine and iminothiolane-modified chitosan (Chang et al. 2010). The histidine-modified chitosan showed improved transfection efficiency than chitosan. In addition, lysine–histidine dendron modified chitosan was developed to increase the gene transfection efficiency of chitosan by promoting its endosomal escape property (Chang et al. 2011). Hence, the lysine–histidine dendron modified chitosan showed higher gene transfection efficiency in HEK 293 cells than chitosan.

It is well known that arginine residues in the cell penetrating peptides (CPPs) delivered molecules efficiently by intracellular translocation because of the membrane permeability of arginine residues (Eguchi et al. 2001). Accordingly, arginine-modified chitosan showed higher transfection efficiency in HEK 293 and COS-7 cells with lower cytotoxicity (Gao et al. 2008). The *in vitro* transfection efficiency was also found to be pH-dependent and the highest transfection efficiency was obtained at pH 7.2. To understand the cellular uptake mechanism, Alexa Fluor 488-conjugated arginine-modified chitosan was treated with A10 cells in the presence of a variety of inhibitors of different endocytic pathways (Zhang et al. 2011). The study revealed that conjugation of arginine into chitosan enhanced cellular uptake and hence transfection efficiency probably due to the preference of internalization of arginine-modified chitosan/DNA complexes by caveolin-mediated endocytosis, thereby avoiding the lysosomal degradation when compared to chitosan/DNA ones. Remarkably, the introduction of an arginine moiety in chitosan did not show any negative impact on the inflammatory activities of the macrophages and differentiation of the naïve CD4+ T cells (Liu et al. 2011). In another study, an arginine moiety was introduced into chitosan through disulfide linkage for reversible thiol-disulfide exchange reactions in cytoplasm in the presence of glutathione to facilitate DNA release (Ho et al. 2011). Although the arginine-modified chitosan had a 48-h delay of green fluorescent protein (GFP) expression in HEK 293 cells, it showed higher and more sustainable GFP expression than its unmodified chitosan.

16.2.3 SPERMINE-MODIFIED CHITOSAN

Spermine is a tetraamine, with two primary and two secondary amines, involved in cellular metabolism found in all eukaryotic cells. Spermine has not only DNA condensing capacity but also high buffering capacity owing to the secondary amines in it (Hosseinkhani et al. 2004). Owing to these advantages, spermine was grafted into chitosan (Figure 16.4) to improve the transfection efficiency of chitosan (Jiang et al. 2011). The spermine-modified chitosan showed good DNA-binding ability and protected the DNA from nuclease degradation. Therefore, the DNA transfection in A549, WI-38, and HepG2 cells by spermine-modified chitosan was higher than chitosan. In addition, aerosol delivery of GFP with spermine-modified chitosan showed higher GFP expression in mouse lung without cytotoxicity when compared to chitosan.

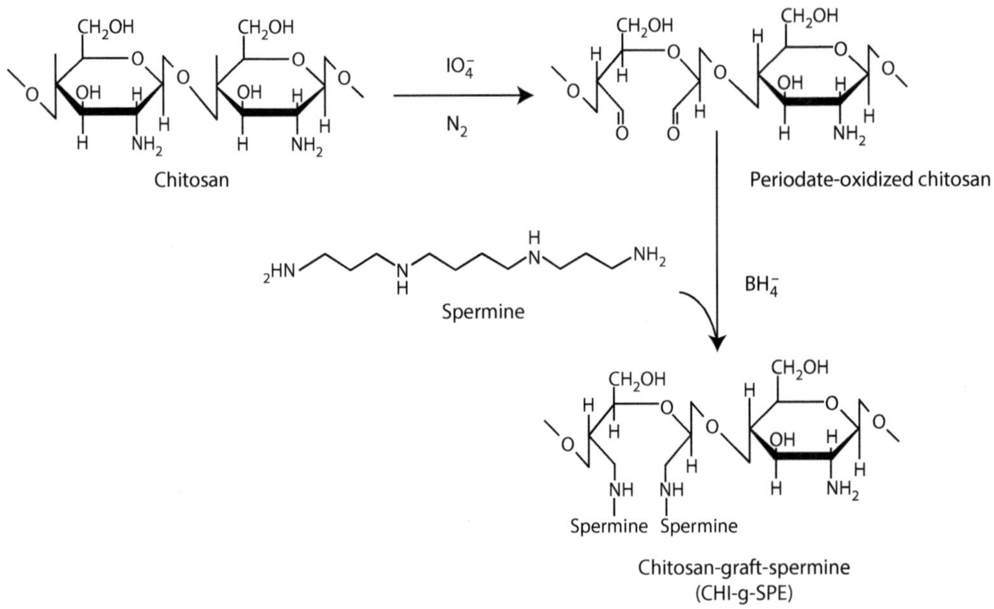

FIGURE 16.4 Proposed reaction scheme for the synthesis of chitosan-graft-spermine.

16.3 siRNA DELIVERY BY MODIFIED CHITOSAN

After the discovery of RNA interference, considerable efforts have been directed toward siRNA-mediated gene silencing using chitosan as gene carrier. One successful application is the delivery of Akt1 siRNA by chitosan-graft-PEI to A549 cells (Jere et al. 2009). The chitosan-graft-PEI carrier efficiently silenced Akt1 protein inducing apoptosis in A549 cells. The carrier also decreased lung cancer proliferation malignancy and metastatsis of the cancer cells. Similarly, low MW PEI-modified β-cyclodextrin (β-CD)-coupled oxidized chitosan could successfully deliver siRNA in HEK293, L929, and COS7 cell lines (Ping et al. 2011). The chitosan-graft-PEI-β-CD showed higher gene silencing in HEK 293 and L929 cell lines with lower cytotoxicity.

IAA-modified chitosan was also used to deliver glyceraldehyde-3-phosphate dehydrogenase (GAPDH) siRNA either via IV or via intranasal (IN) routes to the mouse for *in vivo* study (Ghosn et al. 2010). The PEGylated IAA-modified chitosan siRNA nanoparticles showed significant silencing of GAPDH in both lung and liver of the mouse as low as 1 mg/kg siRNA dose through IV administration, whereas significant silencing of GAPDH was obtained in the lungs with only 0.5 mg/kg/day siRNA delivered over 3 consecutive days through IN delivery. Moreover, the efficient and dose-dependent silencing of apolipoprotein B in the liver was seen.

Among repeating units of arginine to promote cell internalization, nine repeating units of arginine was reported to be highly effective for enhancing cellular uptake of the gene (Fuchs and Raines 2004). Strikingly, nona-arginine tethered PEGylated chitosan showed efficient delivery of siRNA in Hepal-6, A549, and VK cells (Noh et al. 2010). Besides, intratumoral administration of the modified chitosan polyplexes significantly silenced the expression of red fluorescent protein (RFP) in tumor tissues *in vivo*. A similar *in vitro* study of siRNA delivery by nona-arginine-modified chitosan demonstrated enhanced cellular association and gene silencing with lower cytotoxicity (Park et al. 2013). Recently, histidine-conjugated glycol chitosan was developed to deliver siRNA along with quantum dots to trace the delivery of siRNA into the target site (Azari et al. 2011). The study revealed that the localization of siRNA loaded into the histidine-modified glycol chitosan was increased in cytoplasm suggesting that histidine promoted their dissociation from the endosomal membranes. Hence, there was a marked reduction (64%) of MDM2 protein (negative regulator of p53 tumor suppressor) expression 24 h after transfection in MCF-7 cells.

16.4 SPECIFIC LIGAND IN MODIFIED CHITOSAN

There are several extracellular and intracellular barriers that hinder the delivery of DNA or siRNA into the nucleus or cytosol of the cells. One of the usual ways to overcome the extracellular barrier in the nonviral vector system is to couple specific ligand with the gene carrier, enabling specific binding to the cell receptors that enhance the cellular uptake of the carrier. In this section, several ligands that target specific receptors will be discussed.

16.4.1 GALACTOSE LIGAND

Liver is an attractive target tissue for gene therapy because of its high metabolic capacity and rich blood supply that is very useful for the delivery of genes to the liver and the distribution of genes from the liver and to the systemic circulation (Kim et al. 2007). Therefore, hepatocytes, the parenchymal liver cells, are useful target cells as they possess large numbers of high-affinity cell-surface receptors (asialoglycoprotein) that recognize and bind molecules with exposed galactose moieties in the vector (Wu and Wu 1998). For this reason, galactose was introduced into the chitosan-graft-PEI (Figure 16.5) for hepatocyte specificity (Jiang et al. 2007b). The galactosylated chitosan-graft-PEI (GC-g-PEI) demonstrated DNA-binding ability and protection of DNA from nuclease. The transfection efficiency of the GC-g-PEI in HepG2 cells was higher than in HeLa cells due to the presence of asialoglycoprotein receptors in HepG2 cells. In addition, the GC-g-PEI transfected liver cells in mice more effectively than PEI 25 K after intraperitoneal administration.

Moreover, galactosylated PEG-chitosan-graft-PEI (Gal-PEG-CH-g-PEI) (Figure 16.6) was prepared to decrease aggregation, cytotoxicity, and opsonization of the polyplexes with serum proteins in the blood stream (Jiang et al. 2008). The Gal-PEG-CH-g-PEI transfected HepG2 cells much more than HeLa cells with lower cytotoxicity. The higher distribution of ^{99m}Tc radiolabeled-Gal-PEG-CH-g-PEI/DNA complexes than ^{99m}Tc-PEI/DNA and ^{99m}Tc-chitosan-graft-PEI/DNA complexes in liver after IV injection in mice further indicated the recognition of galactose group

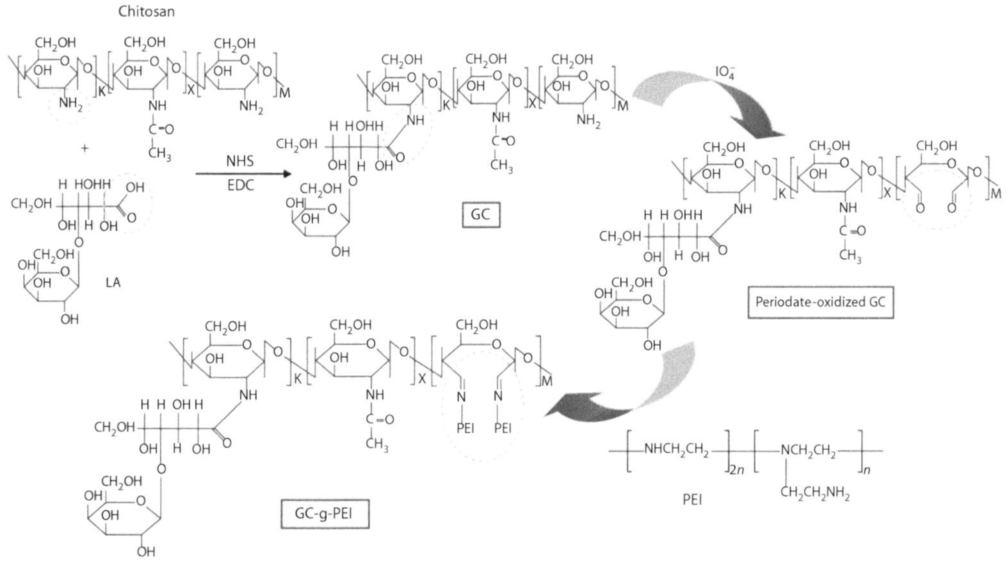

FIGURE 16.5 Proposed reaction scheme for the synthesis of galactosylated chitosan-graft-PEI.

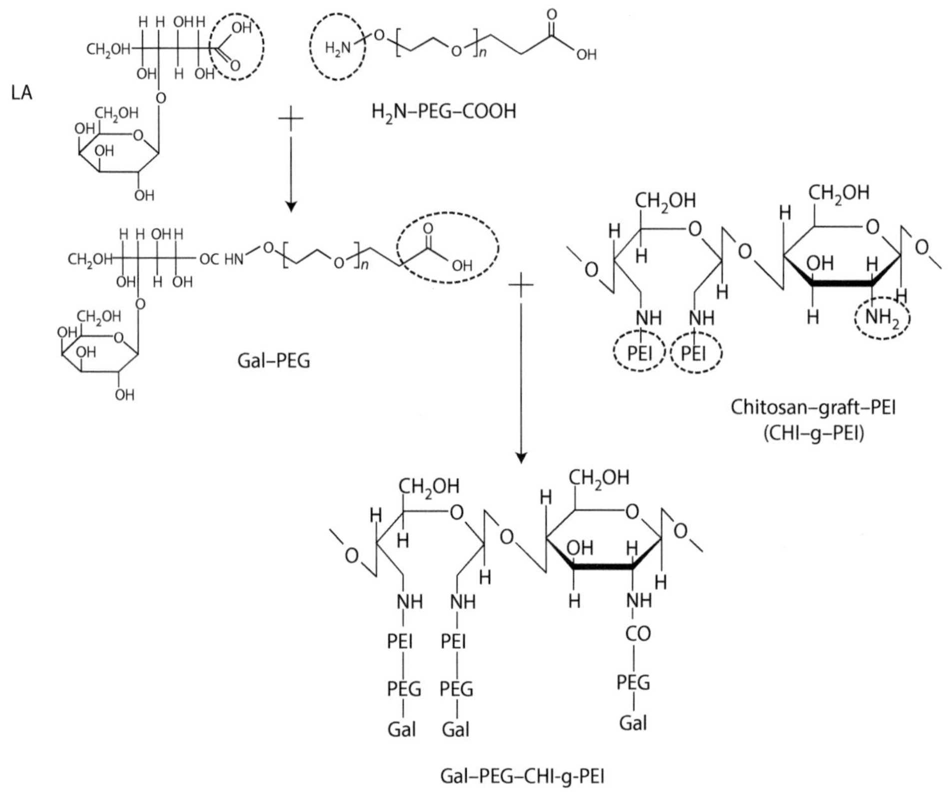

FIGURE 16.6 Proposed reaction scheme for the synthesis of galactosylated-PEG-chitosan-graft-PEI.

of Gal-PEG-CH-g-PEI by asialoglycoprotein receptors in liver cells. Likewise, galactosylated chitosan-graft-spermine (GCS) was prepared for selective transfection in liver cells (Alex et al. 2011). The GCS showed higher transfection efficiency in HepG2 cells with lower cytotoxicity and the transfection efficiency was decreased by the treatment of cellular uptake inhibitors, suggesting the receptor-mediated internalization of polyplexes. Similar GCS, prepared by a different method, also demonstrated specific delivery of gene into hepatocytes *in vitro* as well as *in vivo* (Kim et al. 2012a).

Moreover, galactosylated PEG-chitosan-graft-spermine (GPCS) (Figure 16.7) was prepared to enhance the interactions between PEG-containing polyplexes and hepatocytes in the liver (Kim et al. 2012b). GPCS not only had lower cytotoxicity in HepG2, HeLa, and A549 cells than PEI 25 K but also showed good hepatocyte-targeting ability. When compared to PEG-chitosan-graft-spermine, the GPCS showed higher GFP expression in the mice liver after IV administration without remarkable fibriosis, inflammation, lipidosis, or necrosis. Furthermore, IV injection of the GPCS/PDCD4 complexes significantly suppressed tumor growth, proliferation, and angiogenesis in liver tumor-bearing H-ras 12V mice, indicating GPCS as a hepatocyte-targeting gene carrier.

16.4.2 Folate Ligand

Folate receptors are usually overexpressed on the surfaces of human cancer cells, and folic acid (FA) has high affinity to the folate receptors. Therefore, FA was introduced into PEGylated trimethylated chitosan grafted with arginine for cancer cell targeting (Morris and Sharma 2010a).

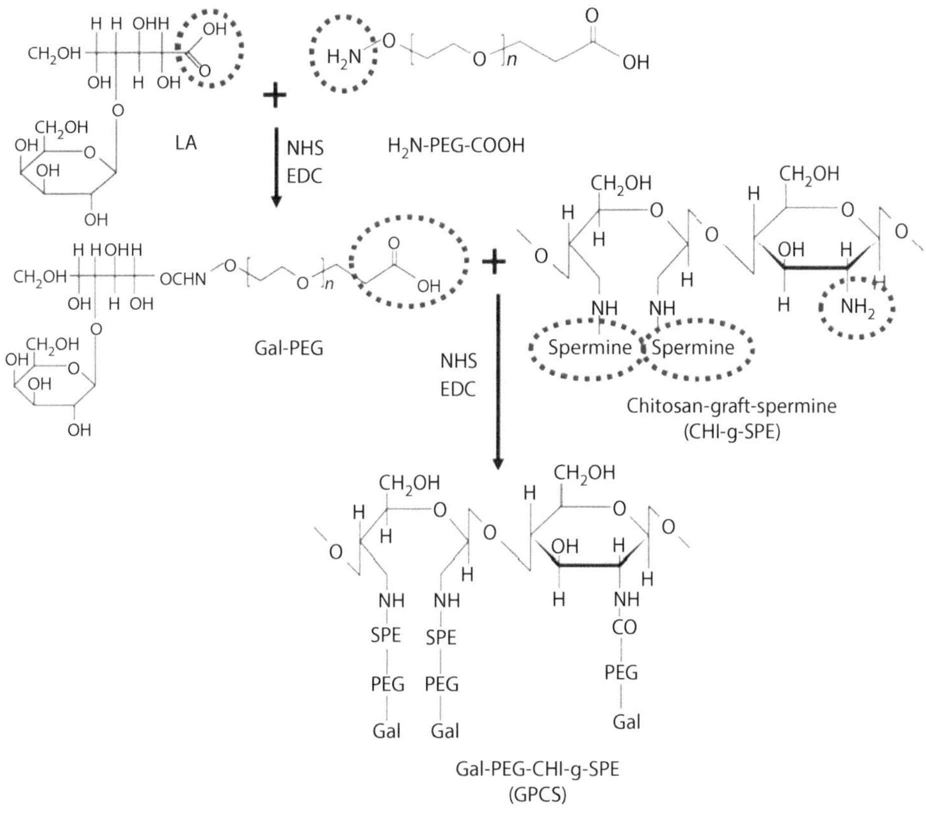

FIGURE 16.7 Proposed reaction scheme for the synthesis of galactosylated-PEG-chitosan-graft-spermine.

The carrier showed buffering capacity, blood compatibility, and protection of DNA degradation from both DNAse I and plasma proteins. Also, the transfection efficiency of carrier was comparable to PEI 25 K in KB cells expressing folate receptors. In a similar study, PEG-FA was introduced to the histidine-conjugated trimethylated chitosan (Morris and Sharma 2010b). Both cellular and nuclear uptake of the carrier was enhanced due to conjugation of FA and histidine with trimethylated chitosan. The carrier not only showed comparable transfection efficiency with PEI 25 K in KB cells but also demonstrated good blood compatibility.

In a similar manner, FA-conjugated chitosan-graft-PEI (Figure 16.8) was used to deliver Akt1 small hairpin RNA (shRNA) (Jiang et al. 2009a). The FA-chitosan-graft-PEI showed good shRNA condensation ability and high protection of shRNA from nuclease attack. The carrier had good targeting ability in KB cells with lower cytotoxicity. In addition, aerosol delivery of FA-chitosan-graft-PEI/Akt1 shRNA complexes suppressed tumorigenesis in a urethane-induced lung cancer model mouse via Akt signaling pathway.

16.4.3 MACROPHAGE TARGETING

Macrophages, one of antigen-presenting cells, express high levels of mannose receptors that are used for endocytosis and phagocytosis of a variety of antigens (Jiang et al. 1995). Therefore, mannose was introduced into chitosan-graft-PEI (Figure 16.9) for macrophage cell-targeted gene delivery (Jiang et al. 2009b). The mannose-chitosan-graft-PEI exhibited good DNA-binding ability and high DNA protection against nuclease attack. The transfection efficiency of mannose-chitosan-graft-PEI

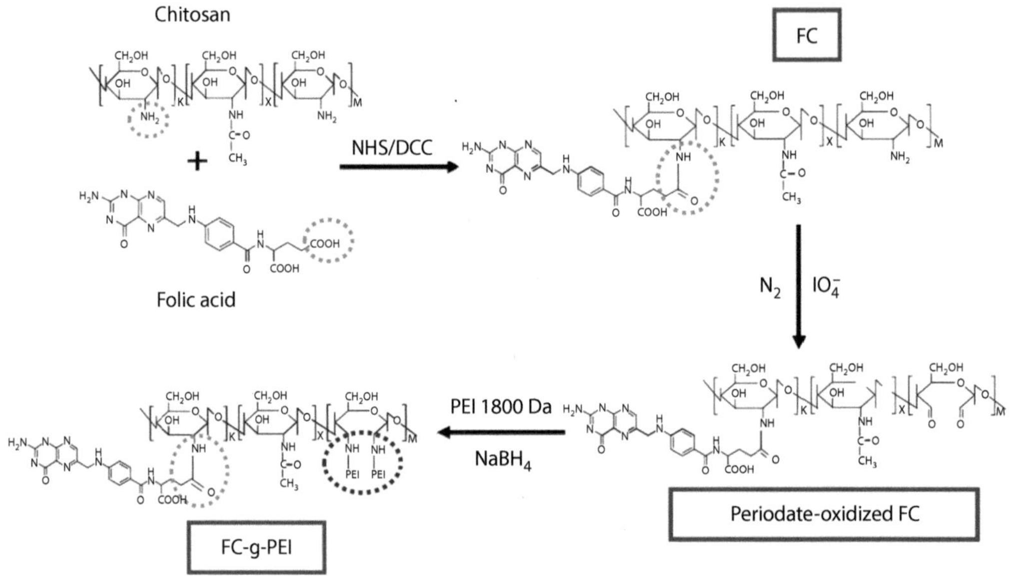

FIGURE 16.8 Proposed reaction scheme for the synthesis of folic acid-conjugated chitosan-graft-PEI.

in RAW 264.7 cells was higher than chitosan-graft-PEI. Recently, tuftsin, a known macrophage targeting molecule, was conjugated with chitosan-graft-PEI to develop it as a gene carrier for macrophage targeting (Tripathi et al. 2012b). The tuftsin-chitosan-graft-PEI/DNA complexes showed higher transfection efficiency than chitosan-graft-PEI/DNA in cultured mouse peritoneal macrophages without any cytotoxicity.

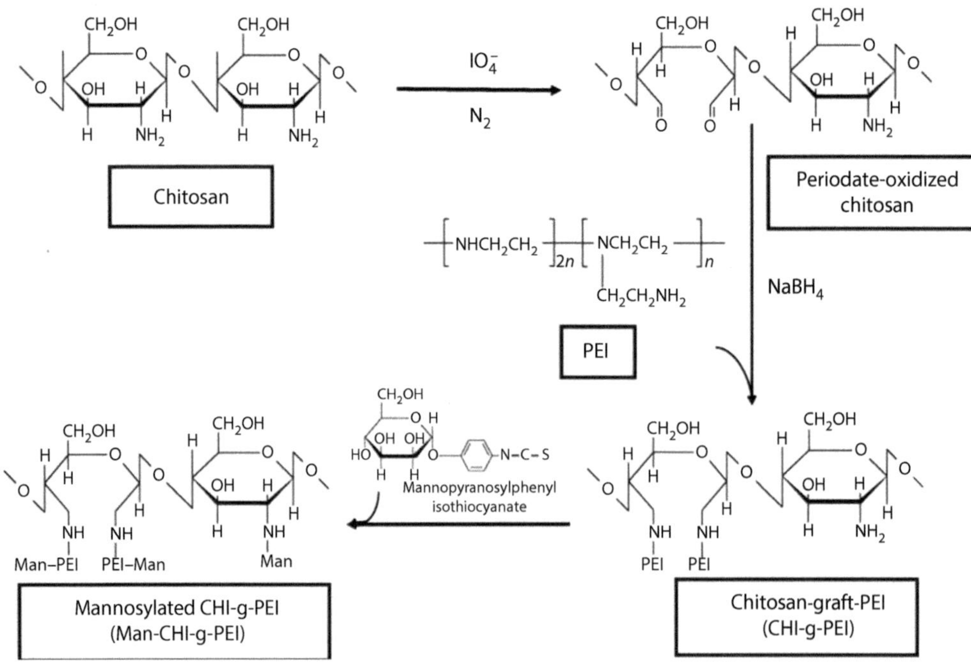

FIGURE 16.9 Proposed reaction scheme for the synthesis of mannose-conjugated chitosan-graft-PEI.

16.5 SUMMARY

In this chapter, the modification of chitosan and conjugation of specific ligand into chitosan have been reviewed as nonviral vectors for gene delivery system. Although their use in these studies seems convincing, the relationship between the structure and function of chemically modified chitosan in cells should be studied in detail. In addition, studies on cellular processes *in vitro* and *in vivo* as well as quality control of modified chitosan are necessary to develop chitosan derivatives as gene carriers. Furthermore, preclinical studies of modified chitosan need to be performed for future clinical application.

REFERENCES

Alex, S. M., M. R. Rekha, and C. P. Sharma. 2011. Spermine grafted galactosylated chitosan for improved nanoparticle mediated gene delivery. *International Journal of Pharmaceutics* 410:125–137.

Azari, F., M. G. Sandros, and M. Tabrizian. 2011. Self-assembled multifunctional nanoplexes for gene inhibitory therapy. *Nanomedicine (London)* 6:669–680.

Chang, K. L., Y. Higuchi, S. Kawakami, F. Yamashita, and M. Hashida. 2010. Efficient gene transfection by histidine-modified chitosan through enhancement of endosomal escape. *Bioconjugate Chemistry* 21:1087–1095.

Chang, K. L., Y. Higuchi, S. Kawakami, F. Yamashita, and M. Hashida. 2011. Development of lysine-histidine dendron modified chitosan for improving transfection efficiency in HEK293 cells. *Journal of Controlled Release* 156:195–202.

Eguchi, A., T. Akuta, H. Okuyama, T. Senda, H. Yokoi, H. Inokuchi, S. Fujita et al. 2001. Protein transduction domain of HIV-1 Tat protein promotes efficient delivery of DNA into mammalian cells. *Journal of Biological Chemistry* 276:26204–26210.

Fuchs, S. M., and R. T. Raines. 2004. Pathway for polyarginine entry into mammalian cells. *Biochemistry* 43:2438–2444.

Gao, J. Q., Q. Q. Zhao, T. F. Lv, W. P. Shuai, J. Zhou, G. P. Tang, W. Q. Liang, Y. Tabata, and Y. L. Hu. 2010. Gene-carried chitosan-linked-PEI induced high gene transfection efficiency with low toxicity and significant tumor-suppressive activity. *International Journal of Pharmaceutics* 387:286–294.

Gao, Y., Z. Xu, S. Chen, W. Gu, L. Chen, and Y. Li. 2008. Arginine-chitosan/DNA self-assemble nanoparticles for gene delivery: *In vitro* characteristics and transfection efficiency. *International Journal of Pharmaceutics* 359:241–246.

Ghosn, B., S. P. Kasturi, and K. Roy. 2008. Enhancing polysaccharide-mediated delivery of nucleic acids through functionalization with secondary and tertiary amines. *Current Topics in Medicinal Chemistry* 8:331–340.

Ghosn, B., A. Singh, M. Li, A. V. Vlassov, C. Burnett, N. Puri, and K. Roy. 2010. Efficient gene silencing in lungs and liver using imidazole-modified chitosan as a nanocarrier for small interfering RNA. *Oligonucleotides* 20:163–172.

Ho, Y. C., Z. X. Liao, N. Panda, D. W. Tang, S. H. Yu, F. L. Mi, and H. W. Sung. 2011. Self-organized nanoparticles prepared by guanidine- and disulfide-modified chitosan as a gene delivery carrier. *Journal of Materials Chemistry* 21:16918–16927.

Hosseinkhani, H., T. Azzam, Y. Tabata, and A. J. Domb. 2004. Dextran-spermine polycation: An efficient nonviral vector for *in vitro* and *in vivo* gene transfection. *Gene Therapy* 11:194–203.

Hu, F. Q., W. W. Chen, M. D. Zhao, H. Yuan, and Y. Z. Du. 2013. Effective antitumor gene therapy delivered by polyethylenimine-conjugated stearic acid-g-chitosan oligosaccharide micelles. *Gene Therapy* 20:597–606.

Islam, M. A., C. H. Yun, Y. J. Choi, J. Y. Shin, R. Arote, H. L. Jiang, S. K. Kang et al. 2011. Accelerated gene transfer through a polysorbitol-based transporter mechanism. *Biomaterials* 32:9908–9924.

Jere, D., H. L. Jiang, Y. K. Kim, R. Arote, Y. J. Choi, C. H. Yun, M. H. Cho, and C. S. Cho. 2009. Chitosan-graft-polyethylenimine for Akt1 siRNA delivery to lung cancer cells. *International Journal of Pharmaceutics* 378:194–200.

Jiang, H. L., Y. K. Kim, R. Arote, J. W. Nah, M. H. Cho, Y. J. Choi, T. Akaike, and C. S. Cho. 2007a. Chitosan-graft-polyethylenimine as a gene carrier. *Journal of Controlled Release* 117:273–280.

Jiang, H. L., J. T. Kwon, Y. K. Kim, E. M. Kim, R. Arote, H. J. Jeong, J. W. Nah et al. 2007b. Galactosylated chitosan-graft-polyethylenimine as a gene carrier for hepatocyte targeting. *Gene Therapy* 14:1389–1398.

Jiang, H. L., J. T. Kwon, E. M. Kim, Y. K. Kim, R. Arote, D. Jere, H. J. Jeong et al. 2008. Galactosylated poly(ethylene glycol)-chitosan-graft-polyethylenimine as a gene carrier for hepatocyte-targeting. *Journal of Controlled Release* 131:150–157.

Jiang, H. L., C. X. Xu, Y. K. Kim, R. Arote, D. Jere, H. T. Lim, M. H. Cho, and C. S. Cho. 2009a. The suppression of lung tumorigenesis by aerosol-delivered folate-chitosan-graft-polyethylenimine/Akt1 shRNA complexes through the Akt signaling pathway. *Biomaterials* 30:5844–5852.

Jiang, H. L., Y. K. Kim, R. Arote, D. Jere, J. S. Quan, J. H. Yu, Y. J. Choi, J. W. Nah, M. H. Cho, and C. S. Cho. 2009b. Mannosylated chitosan-graft-polyethylenimine as a gene carrier for Raw 264.7 cell targeting. *International Journal of Pharmaceutics* 375:133–139.

Jiang, H. L., H. T. Lim, Y. K. Kim, R. Arote, J. Y. Shin, J. T. Kwon, J. E. Kim et al. 2011. Chitosan-graft-spermine as a gene carrier *in vitro* and *in vivo*. *European Journal of Pharmaceutics and Biopharmaceutics* 77:36–42.

Jiang, W., W. J. Swiggard, C. Heufler, M. Peng, A. Mirza, R. M. Steinman, and M. C. Nussenzweig. 1995. The receptor DEC-205 expressed by dendritic cells and thymic epithelial cells is involved in antigen processing. *Nature* 375:151–155.

Jin, H., T. H. Kim, S. K. Hwang, S. H. Chang, H. W. Kim, H. K. Anderson, H. W. Lee et al. 2006. Aerosol delivery of urocanic acid-modified chitosan/programmed cell death 4 complex regulated apoptosis, cell cycle, and angiogenesis in lungs of K-ras null mice. *Molecular Cancer Therapeutics* 5:1041–1049.

Jin, H., C. X. Xu, H. W. Kim, Y. S. Chung, J. Y. Shin, S. H. Chang, S. J. Park et al. 2008. Urocanic acid-modified chitosan-mediated PTEN delivery via aerosol suppressed lung tumorigenesis in K-ras(LA1) mice. *Cancer Gene Therapy* 15:275–283.

Kabanov, A. V., and V. A. Kabanov. 1995. DNA complexes with polycations for the delivery of genetic material into cells. *Bioconjugate Chemistry* 6:7–20.

Kadlecova, Z., L. Baldi, D. Hacker, F. M. Wurm, and H. A. Klok. 2012. Comparative study on the *in vitro* cytotoxicity of linear, dendritic, and hyperbranched polylysine analogues. *Biomacromolecules* 13:3127–3137.

Kim, T. H., J. E. Ihm, Y. J. Choi, J. W. Nah, and C. S. Cho. 2003. Efficient gene delivery by urocanic acid-modified chitosan. *Journal of Controlled Release* 93:389–402.

Kim, T. H., H. L. Jiang, D. Jere, I. K. Park, M. H. Cho, J. W. Nah, Y. J. Choi, T. Akaike, and C. S. Cho. 2007. Chemical modification of chitosan as a gene carrier *in vitro* and *in vivo*. *Progress in Polymer Science* 32:726–753.

Kim, J. H., Y. K. Kim, M. T. Arash, S. H. Hong, J. H. Lee, B. N. Kang, Y. B. Bang et al. 2012a. Galactosylation of chitosan-graft-spermine as a gene carrier for hepatocyte targeting *in vitro* and *in vivo*. *Journal of Nanoscience and Nanotechnology* 12:5178–5184.

Kim, J. H., A. Minai-Tehrani, Y. K. Kim, J. Y. Shin, S. H. Hong, H. J. Kim, H. D. Lee et al. 2012b. Suppression of tumor growth in H-ras12V liver cancer mice by delivery of programmed cell death protein 4 using galactosylated poly(ethylene glycol)-chitosan-graft-spermine. *Biomaterials* 33:1894–1902.

Lai, W. F. 2011. *In vivo* nucleic acid delivery with PEI and its derivatives: Current status and perspectives. *Expert Review of Medical Devices* 8:173–185.

Li, S., and L. Huang. 2000. Nonviral gene therapy: Promises and challenges. *Gene Therapy* 7:31–34.

Liu, L., Y. Bai, D. Zhu, L. Song, H. Wang, X. Dong, H. Zhang, and X. Leng. 2011. Evaluation of the impact of arginine-chitosan/DNA nanoparticles on human naive CD4+ T cells. *Journal of Biomedical Materials Research Part A* 96:170–176.

Lou, Y. L., Y. S. Peng, B. H. Chen, L. F. Wang, and K. W. Leong. 2009. Poly(ethylene imine)-g-chitosan using EX-810 as a spacer for nonviral gene delivery vectors. *Journal of Biomedical Materials Research Part A* 88:1058–1068.

Lu, B., X. D. Xu, X. Z. Zhang, S. X. Cheng, and R. X. Zhuo. 2008. Low molecular weight polyethylenimine grafted N-maleated chitosan for gene delivery: Properties and *in vitro* transfection studies. *Biomacromolecules* 9:2594–2600.

Lu, B., Y. X. Sun, Y. Q. Li, X. Z. Zhang, and R. X. Zhuo. 2009. N-Succinyl-chitosan grafted with low molecular weight polyethylenimine as a serum-resistant gene vector. *Molecular Biosystems* 5:629–637.

Moreira, C., H. Oliveira, L. R. Pires, S. Simoes, M. A. Barbosa, and A. P. Pego. 2009. Improving chitosan-mediated gene transfer by the introduction of intracellular buffering moieties into the chitosan backbone. *Acta Biomaterialia* 5:2995–3006.

Morris, V. B., and C. P. Sharma. 2010a. Folate mediated *in vitro* targeting of depolymerised trimethylated chitosan having arginine functionality. *Journal of Colloid and Interface Science* 348:360–368.

Morris, V. B., and C. P. Sharma. 2010b. Folate mediated histidine derivative of quaternised chitosan as a gene delivery vector. *International Journal of Pharmaceutics* 389:176–185.

Noh, S. M., M. O. Park, G. Shim, S. E. Han, H. Y. Lee, J. H. Huh, M. S. Kim et al. 2010. Pegylated poly-l-arginine derivatives of chitosan for effective delivery of siRNA. *Journal of Controlled Release* 145:159–164.

Park, I. K., T. H. Kim, S. I. Kim, Y. H. Park, W. J. Kim, T. Akaike, and C. S. Cho. 2003. Visualization of transfection of hepatocytes by galactosylated chitosan-graft-poly(ethylene glycol)/DNA complexes by confocal laser scanning microscopy. *International Journal of Pharmaceutics* 257:103–110.

Park, S., E. J. Jeong, J. Lee, T. Rhim, S. K. Lee, and K. Y. Lee. 2013. Preparation and characterization of nonaarginine-modified chitosan nanoparticles for siRNA delivery. *Carbohydrate Polymers* 92:57–62.

Pezzoli, D., F. Olimpieri, C. Malloggi, S. Bertini, A. Volonterio, and G. Candiani. 2012. Chitosan-graft-branched polyethylenimine copolymers: Influence of degree of grafting on transfection behavior. *Plos One* 7:e34711.

Ping, Y., C. Liu, Z. Zhang, K. L. Liu, J. Chen, and J. Li. 2011. Chitosan-graft-(PEI-beta-cyclodextrin) copolymers and their supramolecular PEGylation for DNA and siRNA delivery. *Biomaterials* 32:8328–8341.

Pires, L. R., H. Oliveira, C. C. Barrias, P. Sampaio, A. J. Pereira, H. Maiato, S. Simoes, and A. P. Pego. 2011. Imidazole-grafted chitosan-mediated gene delivery: *In vitro* study on transfection, intracellular trafficking and degradation. *Nanomedicine (London)* 6:1499–1512.

Putnam, D. 2006. Polymers for gene delivery across length scales. *Nature Materials* 5:439–451.

Shi, B., Z. Shen, H. Zhang, J. Bi, and S. Dai. 2012. Exploring *N*-imidazolyl-*O*-carboxymethyl chitosan for high performance gene delivery. *Biomacromolecules* 13:146–153.

Somia, N., and I. M. Verma. 2000. Gene therapy: Trials and tribulations. *Nature Reviews Genetics* 1:91–99.

Tripathi, S. K., R. Goyal, P. Kumar, and K. C. Gupta. 2012a. Linear polyethylenimine-graft-chitosan copolymers as efficient DNA/siRNA delivery vectors *in vitro* and *in vivo*. *Nanomedicine* 8:337–345.

Tripathi, S. K., R. Goyal, M. P. Kashyap, A. B. Pant, W. Haq, P. Kumar, and K. C. Gupta. 2012b. Depolymerized chitosans functionalized with bPEI as carriers of nucleic acids and tuftsin-tethered conjugate for macrophage targeting. *Biomaterials* 33:4204–4219.

Wang, W., J. Yao, J. P. Zhou, Y. Lu, Y. Wang, L. Tao, and Y. P. Li. 2008. Urocanic acid-modified chitosan-mediated p53 gene delivery inducing apoptosis of human hepatocellular carcinoma cell line HepG2 is involved in its antitumor effect *in vitro* and *in vivo*. *Biochemical and Biophysical Research Communications* 377:567–572.

Wen, Y., Z. Guo, Z. Du, R. Fang, H. Wu, X. Zeng, C. Wang, M. Feng, and S. Pan. 2012. Serum tolerance and endosomal escape capacity of histidine-modified pDNA-loaded complexes based on polyamidoamine dendrimer derivatives. *Biomaterials* 33:8111–8121.

Wong, K., G. Sun, X. Zhang, H. Dai, Y. Liu, C. He, and K. W. Leong. 2006. PEI-g-chitosan, a novel gene delivery system with transfection efficiency comparable to polyethylenimine *in vitro* and after liver administration *in vivo*. *Bioconjugate Chemistry* 17:152–158.

Wu, C. H., and G. Y. Wu. 1998. Receptor-mediated delivery of foreign genes to hepatocytes. *Advanced Drug Delivery Reviews* 29:243–248.

Zhang, H., D. Zhu, L. Song, L. Liu, X. Dong, Z. Liu, and X. Leng. 2011. Arginine conjugation affects the endocytic pathways of chitosan/DNA nanoparticles. *Journal of Biomedical Materials Research Part A* 98:296–302.

Zhang, W., S. Pan, Y. Wen, X. Luo, and X. Zhang. 2010. Synthesis of poly(ethylene glycol)-g-chitosan-g-poly(ethylene imine) co-polymer and *in vitro* study of its suitability as a gene-delivery vector. *Journal of Biomaterials Science, Polymer Edition* 21:741–758.

17 Prospective Corollary of Ophthalmic Nanomedicine

A Concept Shift toward Chitosan-Based Mucoadhesive Nanomedicine

Farhan Jalees Ahmad, Sohail Akhter, Mohammad Zaki Ahmad, Farshad Ramazani, Mohammad Samim, Musarrat Husain Warsi, Mohammad Anwar, and Ziyaur Rahman

CONTENTS

17.1 INTRODUCTION

The development of drug delivery for the transportation of drug in a bioavailable and safe manner to the target site is now becoming an exceedingly important area of pharmaceutics research. A large number of novel drug delivery technologies surface every year and targeting every body part as potential target sites. As a result, various smart drug delivery technologies with significant outcomes have been reported for biopharmaceutical classification system (BCS) class-II and class-IV drugs, peptides, proteins, and so on. Further, among novel drug delivery technologies, advancement of nanotechnology in formulation development for biodegradable nanoparticles, nanoemulsions (NEs), vesicular systems, implants, bioadhesive systems, and so on are currently under intensive exploratory studies (Akhter et al., 2011, 2012). Apart from that, ocular drug delivery has remained one of the most challenging topic of research as it requires a series of specified characteristics for effective drug delivery (Tiffany, 1991; Van Ooteghem, 1987). Eye is a unique and challenging organ for therapeutic delivery on to the surface as well as in its interior structure (Sieg and Robinson, 1997; Shedden et al., 2001; Rupenthal et al., 2011). Most of its anatomical and physiological architecture interferes with the fate of the administered drug (Diepold et al., 1989; Hanrahan et al., 2012). Tears permanently wash the surface of the eye and enhance anti-infectious activity by the lysozyme and immunoglobulins (Tian et al., 2012). In addition, the drug may bind to tear proteins and conjunctival mucin (Davies, 2000). Moreover, corneal and conjunctival epithelia, along with the tear film, build a complex barrier that prevents the intraocular absorption of topically applied drugs and results in low ocular bioavailability and unwanted systemic side effects (Lang, 1995; Bourlais et al., 1998; Jain et al., 2011a,b). Topical eye drops are the most desirable dosage form for ophthalmic disease and account for nearly 90% of the currently available ophthalmic formulations owing to their simplicity and preference in use by patients. However, conventional eye drops, most of which are present in solution form, usually have quite a limited therapeutic efficacy due to the low bioavailability (Davies et al., 1997; Jain et al., 2011a,b). In addition, topical ocular drug delivery is associated with rapid and extensive precorneal loss as a major drawback caused by drainage in the extraocular area and high tear fluid turnover (Chrai and Robinson, 1974; Zignani et al., 1995; Zimmer et al., 1994; Bourlais et al., 1998; Wei et al., 2002). Typically, less than 5% of the drug applied penetrates the cornea/sclera and reaches the inner tissues, with the foremost part of the installed dose often absorbed systemically through the conjunctiva and nasolacrimal duct

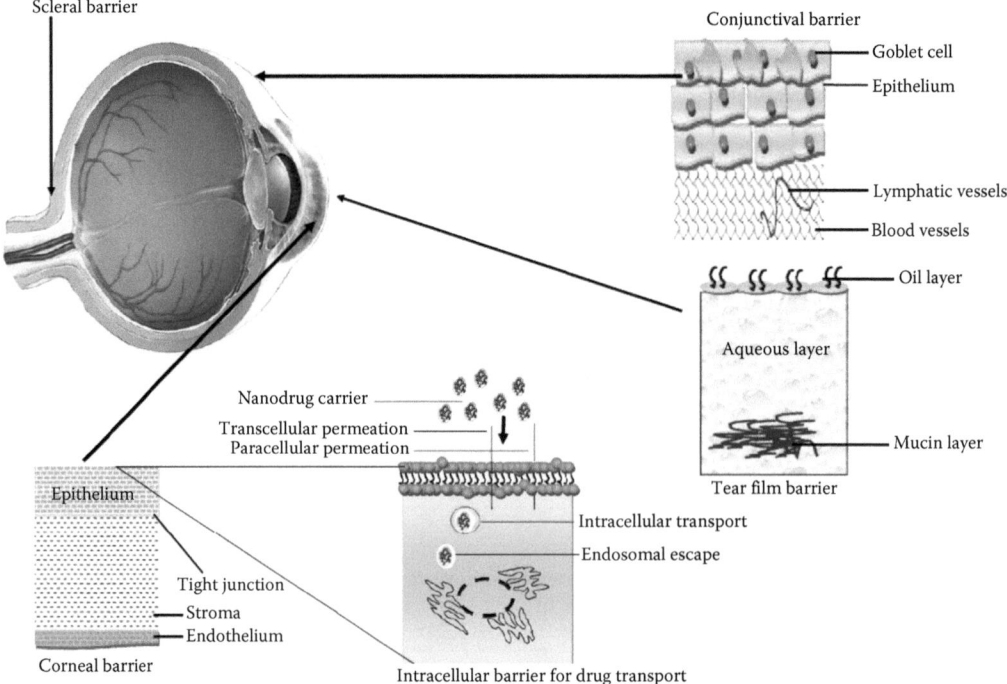

FIGURE 17.1 Illustration of ocular barriers against ocular drug/drug carriers in different areas of the eyes and possible mean of drug transport on topical application.

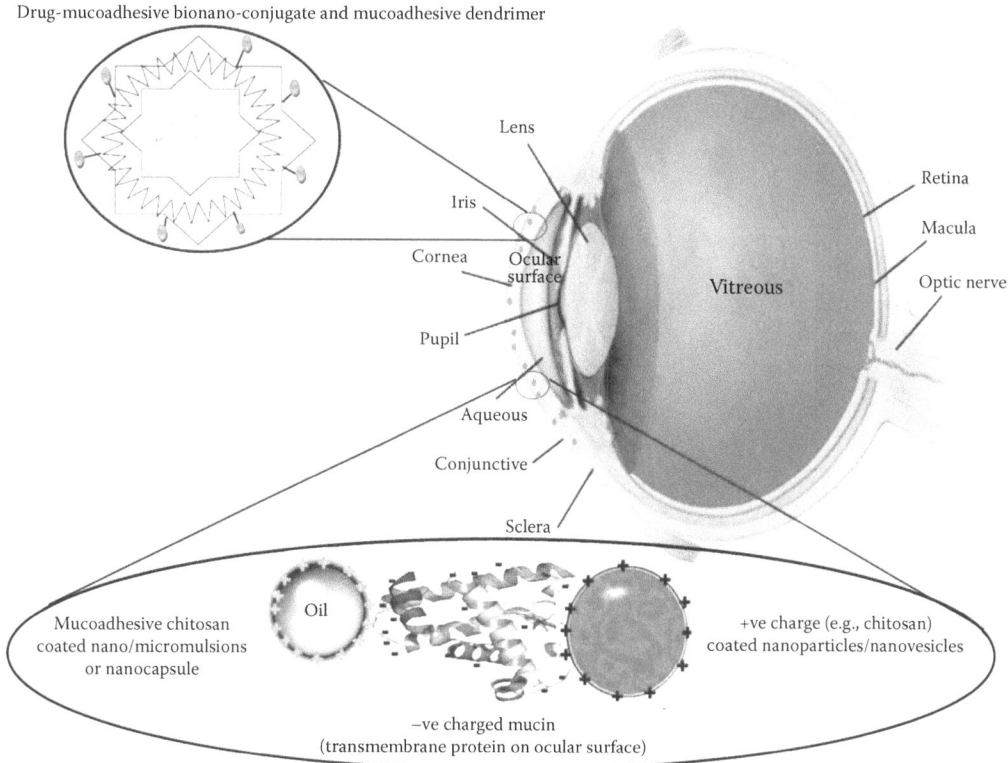

FIGURE 17.2 Design showing the conceptualization of different mucoadhesive nanocarriers that may enhance the ocular bioavailability of drug on topical application.

(Kupferman et al., 1974; MacKeen, 1980). For easy understanding, the ocular barriers and drug disposition in the eyes are illustrated in Figures 17.1 and 17.2, respectively.

On topical ocular delivery of drug, slow removal of drug from the absorption site and the improvement of paracellular and intracellular pathways of epithelial cells would be of great benefit to reduce the dose, its frequent instillation, and consequently, the side effects inherent to applied drugs. Therefore, ophthalmic drug delivery is always a subject of interest to researchers working in the multidisciplinary areas pertaining to the eye. In general, the major problem in ocular therapeutics is to maintain an effective drug concentration in ocular tissue for a significant period, to achieve the expected therapeutic response (Akhter et al., 2011).

Thus, for effective topical delivery of therapeutic agents to the eyes, an optimum delivery system would be one that can be delivered in the form of eye drops without blurred vision or irritability with low dosing frequency and better patient compliance (Sahoo et al., 2008).

Consequently, the development of drug delivery system with improved bioavailability of drug onto the ocular surface would be a promising step toward the management of external ophthalmic disorders (Krauland et al., 2003). Ophthalmic drug delivery, probably more than any other route of administration, may benefit from the uniqueness of nanotechnology-based drug delivery (Sahoo and Labhasetwar, 2003). Thus, the application of nanotechnology provides attractive replacements for the conventional topical ocular delivery because of its capacity to protect the encapsulated molecule, along with the drug transport to the intraocular compartments (Losa et al., 1993). Additionally, nanotechnology offers prolonged and controlled drug delivery, thereby being an attractive vehicle for the treatment of ophthalmic disorders such as glaucoma, cytomegalovirus (CMV) retinitis, retinal neovascularization, and dry eye disease (Alonso, 2004; Mainardes and Evangelista, 2005; Sahoo et al., 2008). The versatility of nanomaterials allowed us to carry the hydrophilic and lipophilic drugs to the inaccessible area of the eye with better tolerance (Bourlais et al., 1998; De Campos et al., 2001). Furthermore, nanomaterials are currently critical to exploit the emerging gene delivery platform (Bochot et al., 1998; Joossand and Chirmule, 2003). However, eye is a delicate organ and their neural tissue is extremely sensitive to toxicity (Furrer et al., 2000). Therefore, toxicity concern with the ocular nanomaterials is necessary and great care must be taken to ensure that the ophthalmic nanocarriers are safe. Thus, taking into consideration the effect of nanomaterials and their inherent components, toxicity in the eye is an underrepresented prospect that is unfortunately lacking in terms of scientific data.

In this review, we presented the application of various nano-approaches in the field of topical ophthalmic drug delivery attempted by investigators during the last decade. Moreover, this review also enlightens the detailed account on mucoadhesive nanotechnology for effective topical ocular delivery, advantage, and limitations with different nanocarriers, and regulatory aspects of nanomedicines.

17.2 NANOMEDICINE: INSIGHT INTO THE CURRENT R&D TREND IN OPHTHALMIC TOPICAL DELIVERY

Conventional drug delivery systems, such as solutions, emulsions, suspensions, and ointments, account for almost 90% of the ophthalmic formulations in the market (Lang, 1995; Bourlais et al., 1998; Qian et al., 2010). Although they offer some advantages such as ease of administration and have better patient compliance, ease of preparation, and low production costs, these conventional dosage forms have significant disadvantages such as very short contact time with the ocular surface and fast nasolacrimal drainage that subsequently leads to poor bioavailability of installed drugs (Patton and Robinson, 1975; Olejnik, 1993; Ooya et al., 2003; Nanjwade et al., 2012).

The use of nano-approaches, such as nanosuspensions, nanoparticles, NE, niosomes, and liposomes, has led to the solution of various solubility- and permeability-related problems of poorly soluble drugs, such as dexamethasone, cyclosporine, dorzolamide, gancyclovir, and many more (Sahoo et al., 2008). Drugs can also be targeted to ocular tissue to allow region-specific delivery and minimize side effects to other organs (Wood et al., 1985). Besides this, depending on their surface properties and hydrophobicity, nanoparticles can be designed for successfully overcoming corneal barriers. In addition to these, encapsulation of drug in nanoparticles, nanospheres, liposomes, and so on can also provide stability to the drug along with prolonged exposure of the drug by controlled release behavior (Sahoo et al., 2008). Nano-approaches-based drug delivery may demonstrate the best drug delivery tools for chronic ocular diseases, in which regular drug administration is required, for instance, in ophthalmic diseases, such as chronic CMV retinitis and dry eye disease.

17.3 NANOPARTICLES/NANOSPHERES/NANOCAPSULES

Nanoparticles are a submicroscopic, colloidal system (vary in size from 10 to 1000 nm) representing promising drug carriers for ophthalmic application. The drug may be dissolved, entrapped, adsorbed, attached, or encapsulated into the nanoparticle. Based on the method of preparation, nanoparticles, nanospheres, or nanocapsules can be obtained with different release profiles for the encapsulated drugs (Sahoo and Labhasetwar, 2003; Vandervoort and Ludwig, 2007). After optimal binding to these particles, the drug absorption in the eye is enhanced significantly in comparison to eye drop solutions owing to the much slower ocular elimination rate. Smaller particles are better abided by the patients than larger ones; therefore, nanoparticles may comfortably be used for ophthalmic delivery systems (Calvo et al., 1996a). Additionally, surface-modified nanoparticulate carriers may be used to accommodate a wide variety of activities. It helps in the ocular retention of the drug with increased corneal permeation. Although several synthetic methods and drug loading techniques are reported to be safe and reproducible, the major developmental issues, including formulation stability, particle size uniformity, and large-scale manufacture of sterile preparations, remain to be address (Mainardes et al., 2005). Nanoparticles used in ophthalmology may be made up of biodegradable and nonbiodegradable polymers (Gurny, 1986; Ludwig et al., 1992a). In such particle suspension on topical administration, particles reside at the delivery site and the drug is released from the polymer matrix through diffusion or erosion, or combinations thereof. The risk of toxicity due to the intracellular overloading of nondegradable polymers would be a limitation of their systemic administration, making these materials more suitable for removable, topical eye drop, inserts, or implants (Lee et al., 1986). In the past decade, biodegradable polymers were extensively used for particulate system development because of their controlled degradation time and minimum inflammatory or nontoxic nature.

17.3.1 CHITOSAN NANOPARTICLES

Chitosan (CS) nanocarriers have attracted a great deal of attention because of their unique properties, such as acceptable biocompatibility, biodegradability, and carrying positive charge (Calvo et al., 1997a; Genta et al., 1997). CS is a cationic polysaccharide that is able to convert to gel or particle-like structure when it comes in contact with multivalent polyanions, such as sodium tripolyphosphate (TPP). CS nanoparticles are spontaneously formed on mixing of CS and TPP solution, through the formation of inter- and intramolecular linkage between the phosphate group of TPP and the amino groups of CS (Calvo et al., 1997a). Using this technique, it has been possible to efficiently associate hydrophilic compounds, such as small molecules, peptides, and proteins, and genes can easily be loaded with this particulate system. The establishment of electrostatic interactions either with the positively charged CS or with the negative polyanion TPP is the main mechanism behind the entrapment of drugs (de la Fuente et al., 2008). Moreover, hydrophobic molecules can also be encapsulated with CS particles with some modifications in the nanoparticle preparation technique. Entrapment of the hydrophobic polypeptide cyclosporine (CyA) was achieved by a prior dissolution of the peptide in an acetonitrile:water mixture, and further precipitation into nanoparticles (De Campos et al., 2001). *In vivo* evaluation of these nanoparticles in rabbits evidenced that these nanoparticles provide a selective and prolonged delivery of CyA to the cornea and conjunctiva (De Campos et al., 2001). Using a very similar technological approach, other authors also reported the success in the development of lipophilic molecules loaded CS nanoparticles (Badawi et al., 2008). Indomethacin entrapment into CS nanoparticles was achieved by a previous dissolution in dichloromethane (DCM) and a further nanoprecipitation into the CS nanoparticles (Badawi et al., 2008). Another strategy for the loading of lipophilic drugs within CS nanoparticles has involved the chemical modification of the CS with hydrophobic residues (Uchegbu et al., 2001; Qu et al., 2006). For example, quaternary ammonium palmitoyl glycol CS aggregated into a hierarchically organized micellar structure to enable the entrapment of hydrophobic compounds, such as prednisolone (Uchegbu et al., 2001; Qu et al., 2006). On topical ocular administration, it was found that these new nanostructures have a capacity for overcoming the ocular barriers (Qu et al., 2006).

17.3.2 ALBUMIN NANOPARTICLES

Merodio et al. developed Ganciclovir (GCV)-loaded albumin nanoparticles for the treatment of CMV retinitis. Its *in vitro* studies indicated a burst release of the drug in 1 h, which continued in a sustained manner for 5 days and continued for almost 30 days. The result demonstrated its controlled delivery for a longer period (Merodio et al., 2001). Zimmer et al. developed Pilocarpine-loaded albumin nanoparticles for the treatment of glaucoma. Albumin particles increased the bioavailability of pilocarpine by about 50–90% (meiosis) and 50–70% (intraocular pressure, IOP), respectively, compared to a pilocarpine solution (Zimmer et al., 1994).

17.3.3 GELATIN AND ALGINATE NANOPARTICLES

Owing to its biocompatible and biodegradable nature, gelatin nanoparticles were studied as nano-sized ocular drug carrier. The developed formulations were studied for various characteristics such as particle size, zeta potential, encapsulation efficiency, and *in vitro* drug release. Pilocarpine HCl and hydrocortisone-loaded gelatin nanoparticles were obtained in the range 300–500 and 110–220 nm, respectively (Vandervoort and Ludwig, 2004). The potential of sodium alginate nanoparticles has been explored as a novel vehicle for the prolonged topical ophthalmic delivery (Motwani et al., 2008). Gatifloxacin-loaded alginate nanoparticles studied by Motwani et al. revealed a fast release during the first hour followed by a more gradual drug release during 24 h by a non-Fickian diffusion mechanism (Motwani et al., 2008).

17.3.4 Polymethylmethacrylate Nanoparticles

Polymethylmethacrylate (PMMA) nanoparticles are made up of *in situ* emulsion polymerization technique (Gurny, 1983). Briefly, monomeric methylmethacrylate is dissolved in water or phosphate-buffered saline or a solution or suspension of drugs or antigens in a concentration range of 0.1–1.5% (Kreuter, 1992). Nanoparticles made of polyacrylamide or PMMA do not degrade either biologically or enzymatically, which makes them less attractive for topical nanodispersion, but act as good drug delivery carrier for contact lenses and hydrogels (Gurny, 1983).

17.3.5 Polyacryl-Cyanoacrylate Nanoparticles

Polyacryl-cyanoacrylate (PACA) particles possess properties of biodegradation and bioadhesion that make them interesting drug carriers for controlled ocular drug delivery and drug targeting. Wood et al. showed that PACA nanoparticles were able to adhere to the corneal and conjunctival surfaces, due to the ability to entangle in the mucin matrix and form a noncovalent or ionic bond with the mucin layer of the conjunctiva, which represent their mucoadhesion property (Wood et al., 1985). PACA nanoparticles and nanocapsules have been shown to improve and prolong the corneal penetration of hydrophilic and lipophilic drugs such as Betaxolol and amikacin sulfate. Despite these attractive results, the potential of the PACA nanoparticles is limited due to the disruption to the corneal epithelium cell membrane (Zimmer et al., 1991).

17.3.6 Poly ε caprolactone Nanoparticle/Nanocapsules

Poly ε caprolactone (PECL) nanocapsules are considered as one of the excellent polymeric carrier systems for ocular drug delivery system due to their biodegradable and biocompatible nature. Marchal-Heussler et al. successfully demonstrated infusion of nanoparticles prepared by using PECL nanoparticles (Marchal-Heussler et al., 1992). It was shown that PECL nanoparticles yielded the highest pharmacological effect. This was believed to be due to the agglomeration of these nanoparticles in the conjunctival sac. Nanocapsules of PECL were also tried for the topical ocular delivery of cyclosporin A (CyA) (Calvo et al., 1996). It was found that PECL coating increased the corneal levels of the drug by five times compared to the oily solution of the drug when administered to the cul-de-sac of fully awake rabbits (Calvo et al., 1996). Moreover, PECL nanocapsules also showed significant improvement in ocular bioavailability of antiglaucomal drugs, such as metipranolol and betaxolol (Marchal-Heussler et al., 1992). It was also found to be safe, as investigators ascribed that PECL nanocapsules are endocytocyged by the corneal epithelium without damaging the cellular membrane (Calvo et al., 1996).

17.3.7 PLGA Nanoparticles

Polylactide and polylactide-*co*-glycolide biopolymer-based nano- and microparticles were extensively studied for topical as well as intravitreal administration (Contia et al., 1997). Poly(lactic-co-glycolic acid) (PLGA) is regarded as a safe drug carrier in ophthalmology owing to its biodegradable and biocompatible nature. PLGA is hydrolytically degraded into nontoxic oligomer and monomer such as lactic acid and glycolic acid (Gupta et al., 2010). Agnihotri and Vavia evaluated diclofenac sodium-loaded PLGA nanoparticles for ocular use and found good biocompatibility with the eye (Agnihotri and Vavia, 2009). Flurbiprofen-PLGA nanoparticles were also studied extensively on various aspects for ocular inflammation and proved to have good stability and ocular tolerance (Pignatello et al., 2002a). Recently, PLGA nanoparticles for sparfloxacin ophthalmic delivery were developed to improve precorneal residence time and ocular bioavailability (Gupta et al., 2010). Satisfactory results in relation to corneal permeation, retention, and ocular bioavailability were obtained when compared to the marketed preparation (Gupta et al., 2010).

17.3.8 Inorganic Nanoparticles

Chen et al. developed a new system for the local delivery of methazolamide to the eye based on calcium phosphate (CaP) nanoparticles. The methazolamide-loaded CaP nanoparticles were prepared through the formation of an inorganic core of CaP and further adsorption of the methazolamide (Agnihotri and Vavia, 2009; Chen et al., 2010). *In vivo* data showed that the intraocular pressure-lowering effect of the inorganic nanoparticles eye drops lasted for 18 h as compared to the effect of 1% brinzolamide eye drops (6 h) (Agnihotri and Vavia, 2009). Chen et al. and Sheikpranbabu et al. developed silver nanoparticles (Ag-NP) to determine the effects on vascular endothelial growth factor (VEGF)- and interleukin-1 beta (IL-1β)-induced vascular permeability (Chen et al., 2010; Sheikpranbabu et al., 2009). They found that VEGF and IL-1β increased the flux of dextran across a cultured porcine retinal endothelial cell (PREC) monolayer, and Ag-NP blocked the solute flux induced by VEGF and IL-1β. They reported that VEGF and IL-1β stimulated endothelial permeability via Src-dependent pathway by increasing the Src phosphorylation, and Ag-NP blocks the VEGF- and IL-1β-induced Src phosphorylation at Y419. These results demonstrate that Ag-NP may inhibit the VEGF- and IL-1β-induced permeability through the inactivation of Src kinase pathway that may represent a potential therapeutic target to inhibit ocular diseases such as diabetic retinopathy (Sheikpranbabu et al., 2009).

17.3.9 Microemulsion/Nanoemulsion

NEs are defined as the dispersions of water and oil in the presence of combination of surfactant and cosurfactant (Smix) in such a manner as to reduce interfacial tension. On the basis of the nature of dispersion and disperse phase, NEs were classified as o/w, w/o, and bicontinuous type (Akhter et al., 2008). These systems are usually characterized by clear appearance, higher thermodynamic stability, small droplet size (200 nm), high drug solubility, and drug reservoir for lipophilic and hydrophilic drugs (Ansari et al., 2008). Moreover, they achieve sustained release of a drug applied to the cornea and higher penetration into the deeper layers of the ocular structure than the native drug. These systems offer additional advantages, including low viscosity, greater ability, and as absorption promoters (Ansari et al., 2008). Most notably, ME/NE possesses low surface tension and therefore exhibits good wetting and spreading that is considered an important property for the topical ocular drug delivery. Furthermore, the possibility of prolonged release of drugs in NEs makes these vehicles attractive for ocular delivery that can greatly decrease the frequency of application of eye drops. Selection and optimization of oily phase and surfactant/cosurfactant (Smix) systems are critical parameters that can affect the stability of the NE system drug stability (Vandamme, 2002). While the presence of surfactants is advantageous due to an increase in cellular membrane permeability, which facilitates drug absorption and bioavailability, caution needs to be taken in terms of the amount of surfactant, as high concentrations can lead to ocular toxicity (Bagwe et al., 2001). In general, nonionic surfactants are preferred over ionic ones, which are generally too toxic to the ocular tissues. Surfactants most frequently utilized for the preparation of MEs include poloxamers, polysorbates, and PEG derivatives (Attwood, 1994). As with the other colloidal delivery systems discussed above, it was hypothesized by numerous research teams that a positive charge would increase the ocular residence time of the formulation due to electrostatic interactions with the negatively charged mucin residues (Benita and Levy, 1993). Toxicological studies, however, contradicted this assumption, and so far there has been no publication showing a distinct beneficial effect for the addition of charged surfactants. MEs can be classified into three different types depending on their microstructure: oil-in-water (o/w ME), water-in-oil (w/o ME), and bicontinuous ME. Water-in-oil MEs composed of water, Crodamol EO, Crill 1, and Crillet 4 were investigated as potential ocular delivery systems by Alany et al. (Alany et al., 2006). It was hypothesized that w/o ME undergo phase transition into lamellar liquid crystals on aqueous dilution by the tears, prolonging the precorneal retention time due to an increase in the formulation's viscosity. Methylprednisolone acetate (MPA) was suspended in a copolymer of poly(ethylacrylate, methyl-methacrylate, and chlorotrimethyl-ammonioethyl methacrylate) and examined for its anti-inflammatory effect in rabbits that having endotoxin-induced uveitis (EIU) (Adibkia et al., 2007a). Similar studies were carried out using piroxicam NS in eudragit RS100 and significant improvement in anti-inflammatory effect was found as compared to the microsuspensions (Adibkia et al., 2007b). In another approach, three different glucocorticoids, hydrocortisone, prednisolone, and dexamethasone, were formulated in nanosuspension. *In vivo* study in rabbits suggested that the nanosuspensions significantly enhanced the ocular absorption of these glucocorticoids (Kassem et al., 2007). Cloricromene (AD6) was formulated in nanosuspensions by using eudragit RS100 and RL100. AD6-loaded eudragit retarded nanosuspension offered a significant edge in enhancing the shelf life and bioavailability of the drug (Pignatello et al., 2006). Recently, oil-in-water-type NE system consisting of pilocarpine as drug and using lecithin, propylene glycol, PEG 200 as surfactant/cosurfactants, and isopropyl myristate as the oil phase has been designed, which is nonirritating to the rabbit animal model (Hasse and Keipert, 1997). These studies showed that pilocarpine-loaded NE system lowers the frequency of administration to two times as compared to four times with conventional eye drops in a day. Another report in timolol NE system was laden in a 2-hydroxyethyl methacrylate (HEMA) gels, which was studied to modulate its transport across the gel (Li et al., 2007). In another attempt to deliver timolol, a stable o/w and w/o emulsion was formulated and a highly lipophilic drug with aqueous solubility of 2.6 μg/mL was formulated in NE system, which could hold 1 mg of drug in the system with excellent stability and tolerability (Buech et al., 2007). An alcohol-free, NE-based formulation consisting of chloramphenicol was developed. This formulation exhibited excellent stability compared to commercially available formulation. Although NEs have excellent advantages, limitations in the selection of surfactant/cosurfactant system and potential toxicity associated with higher concentrations of surfactant/cosurfactant often restrict its use (Kaur and Kanwar, 2002a). Based on the principle of mucoadhesion imparted by CS, Badawi et al. investigated the CS-based NE loaded with indomethacin intended to be used in the management of postoperative inflammation and intraocular irritation (Badawi et al., 2008). NE system (220–690 nm) showed 86% of indomethacin releases over 24 h with initial fast release. Most importantly, NE system significantly improved the corneal chemical ulcer with moderate effective inhibition of polymorph nuclear leukocytic infiltration. The enhanced therapeutic response attribute to the enhanced concentration of drug at the site of action for prolonged period. Moreover, the wound-healing effect of the CS may impart the synergistic effect. More recently, Akhter et al. investigated the CS-based triacetin/Tween 20/Transcutol P NE system for the effective delivery of GCV (Akhter et al., 2011). Because of unique physiochemical features, many more valuable findings have now surfaced based on ME/NE system.

17.3.10 Liposomes

Liposomes are microscopic vesicles composed of one or more concentric lipidic bilayers, separated by water or aqueous buffer with a diameter ranging from 100 nm to 10 μm (Wadhwa et al., 2009; Davis et al., 2004). Liposomes offer advantages over most ophthalmic delivery systems in being biodegradable and nontoxic. Another potential advantage of liposomes is their ability to come in intimate contact with the corneal and conjunctival surfaces, thereby increasing drug absorption. In recent years, liposome has been extensively studied in ophthalmology (Fitzgerald et al., 1987;

Lajavardi et al., 2007; Shen and Tu, 2007). GCV-loaded liposomal formulation was formulated by a reversed-phase evaporation method and *in vivo* pharmacokinetic evaluation was performed in rabbit. Transcorneal permeability and bioavailability of GCV from liposome was found to be 3.9- and 1.7-fold higher than the solution (Shen and Tu, 2007). The antibiotic drug Ciprofloxacin was also formulated in multilamellar liposomal (MLVs) formulation, which lowered tear-driven dilution in the conjunctival sac. This approach produced sustained release of the drug depending on the nature of the lipid composition selected (Budai et al., 2007). Acetazolamide was encapsulated in liposomal formulation for topical delivery; its effectiveness was studied *in vivo* by measuring intraocular pressure. MLVs exhibited prolonged efficacy than liposomes prepared by reverse phase evaporation (REV) (Hathout et al., 2007). In another study, cationic liposomes containing herpes simplex virus type 1 (HSV-1), glycoprotein B (gB1s), or DTK1 and DTK2 (polylysine-rich peptides) were prepared and examined in the rabbit ocular model of HSV-1 keratitis. It was found that peptide liposomal formulation has the potential to act as very effective anti-HSV vaccine (Nagarsenker et al., 1999). Recently, positively charged liposomal formulation of tropicamide and oligolamellar liposomes of ACV were generated. Both formulations resulted in significantly higher penetration across the cornea (Cortesi et al., 2006). Antisense oligonucleotides (ODNs) encapsulated in liposomes have been studied for their potential to treat ocular disorders. Results revealed that liposomes may not be able to release the entire payload of high molecular weight ODNs relative to a free solution form (Fattal and Bochot, 2006). PEGylated liposome containing ODNs results in a higher percentage of intact ODNs (30% of the total dose) after 2 weeks. Some researchers have formulated liposomes coated with an envelope of inactivated hemagglutinating virus of Japan (HVJ) to treat choroidal neovascularization (CNV) in rats. They successfully delivered phosphorothioate ODNs to inhibit VEGF (Ogata et al., 1999). Liposomes are potentially valuable as ocular drug delivery systems due to the easiness of its preparation and versatility in physical characteristics. However, their use in ocular drug delivery is limited by instability (due to hydrolysis of the phospholipids), limited drug loading capacity, technical difficulties in obtaining sterile preparations, and blurred vision due to their size (Ansari et al., 2008; Araujo et al., 2009). In addition, liposomes are subject to the same rapid precorneal clearance as conventional ocular solutions, especially the ones with a negative or no surface charge (Vandamme, 2002). however, Some researchers have reported, however, that positively charged liposomes have a prolonged precorneal retention due to electrostatic interactions with the negative sialic acid residues of the mucin layer (Hasse and Keipert, 1997; Kaur and Smitha, 2002b; Vandamme, 2002; Budai et al., 2007; Lajavardi et al., 2007; Li et al., 2007; Shen and Tu, 2007; Kakkar and Kaur, 2011).

17.3.11 NIOSOMES

Like liposomes, in recent years, niosomes have been successfully studied for ophthalmic drug delivery as vesicular systems to provide controlled drug delivery, prolonged drug precorneal residence time, and enhanced ocular bioavailability and prevention of metabolism of the drug by enzymes present at the tear/corneal surface (Kaur et al., 2004). The drug enclosed in the vesicles allows for an improved partitioning and transport through the cornea. Moreover, vesicles offer a promising avenue to fulfill the need for an ophthalmic drug delivery system that has the convenience of a drop, but will localize and maintain drug activity at its site of action. Niosomes in topical ocular delivery may be preferred over liposomal vesicular system as they are chemically stable than liposomes, incur lower production cost, and are composed of biodegradable and nonimmunogenic materials (Kaur et al., 2004). Unlike phospholipids, niosomes do not require expensive handling (storage at freezers and preparation under nitrogen gas). Moreover, they handle surfactants with no special precautions or conditions; they can improve the performance of the drug via better availability and controlled delivery at a particular site; they are biodegradable, biocompatible, and nonimmunogenic. The enhanced corneal/conjunctival deposition and transport by means of lipophilic vesicular system can mechanistically be explored through various hypothesis, such as liposomal and niosomal vesicular system can be specifically or nonspecifically adsorbed onto the corneal and conjunctival surface or can be fused with corneal cell membranes, and release encapsulated drug inside the cell (Mishra et al., 2011; Shimazaki et al., 2011). The lipophilic vesicular system may itself internalize into the deeper corneal tissues by endocytocis process (Abdelkader et al., 2010). Negatively charged liposomal and niosomal systems have been found to be more efficient than their neutral complimentary for internalization into the cells by the endocytocis process (Kaur et al., 2010). It is also proposed that during adsorption there may be an additive possibility of drug absorption mechanism; lipo- or niosomal vesicles may release drug in front of the membrane, and the released drug can enter cell via micropinocytosis (Abdelkader et al., 2010). They may adopt the transcellular and paracellular pathways of corneal and conjunctival transport (Hosny, 2009; Akhter et al., 2011). In addition, particularly in niosomes, the presence of surfactant may facilitate the permeation (Akhter et al., 2011). However, the untailored formulated niosomes are normally negative and neutral charged that may only improve the ocular bioavailability to some extent due to corneal permeation enhancement owing to their rapid clearance such as conventional eye drops.

17.3.12 Dendrimers

Dendrimers are unique nanostructured drug carriers that have a series of branches around a central core (Sahoo et al., 2008). Their nanosize, ease of preparation, functionalization, and the possibility to attach multiple surface groups render them suitable alternative vehicles for ophthalmic drug delivery (Ihre et al., 2002, 104–105; Quintana et al., 2002; Sahoo et al., 2008). Dendrimers are liquid or semisolid polymers and contain amine, carboxylic, and hydroxyl surface groups, which keep on increasing as the generation number increases (G0, G1, G2, etc.). Dendrimers based on poly (amido-amine) (PAMAM) have been widely employed in drug delivery. This system of branched polymers represents unique architecture, and can entrap both hydrophilic and lipophilic drugs into their structure (Sahoo et al., 2008). The selection of the functional group on the surface (amine, carboxylate, and hydroxyl) and the size and molecular weight of the dendrimer are important parameters to be considered in designing a delivery system. Dendrimer-based approach has shown better bioavailability of pilocarpine and tropicamide, when DG1.5 and DG4.0 (OH) dendrimers with carboxylate and hydroxyl surface groups were instilled (Vandamme and Brobeck, 2005). Some of the scientists have developed water-soluble conjugates of D(+)-glucosamine and D(+)-glucosamine 6-sulfate to obtain synergistic immunomodulatory and antiangiogenic effect by using anionic PANAM (G3.5) dendrimers (Shaunak et al., 2004). In a study, cell adhesion peptides were attached to dendrimers that were later used to link with collagen scaffolds (Duanand and Sheardown, 2006). Stratification of corneal epithelial cell was studied using the above-mentioned collagen scaffolds. All peptides have shown a promoted effect on stratification. The same investigators have actively developed biocompatible conjugates of dendrimers with collagen scaffolds to obtain better mechanical and adhesion ability. In another work, the dendrimer-based approach was used to deliver anti-VEGF ODN and was successfully tested in rats to treat CMV (Marano et al., 2005).

17.3.13 Solid Lipid Nanoparticles

Solid lipid nanoparticles (SLN) are characteristically spherical particles with an average diameter between 50 and 100 nm (Muller and Keck, 2004). SLNs are particularly advantageous in ocular drug delivery as they have the ability to enhance the ocular bioavailability of both hydrophilic and lipophilic drugs. Furthermore, they can be easily autoclaved for sterilization, which is an important aspect of ocular administration for drug formulation (Seyfoddin et al., 2010). Recently, Li et al. reported the utility of SLN in the management of glaucoma and concluded that higher therapeutic efficacy retarded the occurrence of maximum action, and a more prolonged effect of methazolamide was found in comparison of drug solution and commercial product (Li et al., 2011). In the same year, Liu et al. demonstrated that baicalin-loaded SLN had a better and desirable pharmacokinetic profile than baicalin solution (Liu et al., 2011). Similarly, an enhanced topical ophthalmic efficacy of cyclosporine A was found when it was incorporated into SLNs (Gokce et al., 2009). The study on the same drug revealed release profiles that were not decreasing during 48 h, representing controlled and prolonged release of the drug from SLN formulations. In the same study, similarities in drug concentration data exhibited that interindividual variance did not affect the ocular penetration of cyclosporine A when formulated as SLN (Basaran et al., 2010).

17.4 CURRENT CONCEPT: A PARADIGM SHIFT TOWARD MUCOADHESIVE NANOSYSTEMS

For the last 20 years, efforts have been directed in the rational design of ocular drug delivery consisting of mucoadhesive nanocarriers. Thereafter, the plan was directed to generate nanocarriers with a hydrophilic coating with the idea of improving their stability and their interaction with the mucosa (Calvo et al., 1997b; Schipper et al., 1997; Alonso and Sanchez, 2003; De Campos et al., 2003; Ludwig, 2005). It was proficient via optimization of nanocarriers' ocular drug delivery to obtain long-lasting bioadhesion/residence time by the so-called mucoadhesive property based on entrapment of particles in the ocular mucus layer and interaction of bioadhesive polymer chains with mucins (Du Toit et al., 2011). Maintenance of the designed nanocarriers in the ocular delivery following topical application is thus decisive to accomplish unremitting drug release and prolonged therapeutic activity. Therefore, the manufacture of nanocarriers from mucoadhesive materials is crucial for effective retention in ocular cul-de-sac (du Toit et al., 2011). Ocular mucoadhesion, exclusively, refers to the capability of certain polymers to hold on to the mucus layer casing the conjunctival and corneal surfaces of the eye by noncovalent bonds (Round et al., 2002). The washout time of the mucoadhesive polymeric system is reduced, since this depends on mucus turnover rate rather than lachrymal discharge turnover rate. It is expected that positively charged nanocarriers may enhance the drug corneal retention, permeation, and subsequently the ocular bioavailability than the neutral and negatively charged systems owing to the result of interaction of positively charged vesicles with the polyanionic corneal and conjunctival surfaces due to the presence of mucin (Fresta et al., 1999; Tian et al., 2012). Figure 17.2 illustrated here show the conceptualization of mucoadhesive nanocarriers' enhanced bioavailable topical ocular drug delivery. Mucoadhesive polymer with plentiful hydrophilic functional group, namely, sulfate, hydroxyl, carboxyl, and amide have found a fundamental role in ocular drug delivery system owing to their adhesion property with precorneal/conjunctival mucin layer via noncovalent bonds, and remaining in place for as long as the mucin is available there. Taking into account this theory, it was thought that the use of mucoadhesive cationic CS polymer is potentially worthy of tailoring the nanosystem to positively charge-coated niosomes using cationic lipid such as stearylamine as a positively charged substance that may lead to irritation and potential toxic effect to the eye (Taniguchi et al., 1988; Tian et al., 2012). It is reported from many studies that the negatively charged mucin and CS interaction induced enhanced concentration and residence time of the associated drug (Illum, 1998). CS has unique properties such as acceptable biocompatibility and biodegradability with low toxicity and high charge density (Dornish et al., 1997; Xu and Du, 2003). Moreover, CS exhibits interesting physicochemical characteristic with a good potential for ocular drug delivery such as bioadhesion prolonging the corneal residence time (Felt et al., 1999; De Campos et al., 2001; De Campos et al., 2003) and penetration-enhancing properties, which were initially attributed to the modulation of the tight junction barrier between epithelial cells (Koch et al., 1998; Mainardes et al., 2005). It was found that CS increases the cell permeability by affecting both paracellular and intracellular pathways of epithelial cells in a reversible manner without affecting cell viability or causing membrane wounds (Artusson et al., 1994; Dodane et al., 1999; Tian et al., 2012). Moreover, CS may impart favorable rheological behavior, where its solutions have shown pseudoplastic and viscoelastic properties. This behavior is particularly important in ophthalmic formulations since it facilitates the retention while it permits the easy spreading of the formulation due to the blinking of the eye (Wang et al., 2011).

During the selection of the bioadhesive polymer intended for ophthalmic drug delivery, the viscosity and wetting properties of the polymer are considered. Viscosity measures the resistance to flow, which depends on its molecular mass, concentration, temperature, and shear stress. In Newtonian system, above a certain range of viscosity, there is no real improvement of bioavailability and no further increase of residence contact time and blinking becomes a panic (Schoenwald and Ward, 1978; Ludwig et al., 1992a; Ludwig and van Ooteghem, 1992b; Van Ooteghem, 1995). However, polymer showing nonNewtonian behavior, when incorporated in formulation possessing pseudoplastic behavior in which viscosity decreases with increasing shear rate (due to blinking and ocular movement), results in significantly less resistance to blinking and demonstrates greater acceptance as compared to formulation possessing polymer exhibiting Newtonian flow (Greaves et al., 1992, 1993; Van Ooteghem, 1995). The mucoadhesive properties of polyacrylic acid hydrogels and their ability to penetrate the mucin at the surface of the eye have been investigated extensively (Ponchel et al., 1987; Park and Robinson, 1987; Duchfine et al., 1988; Slovin and Robinson, 1993). In the mean time, several other synthetic polymers have been examined for the fabrication of mucoadhesive nanocarriers for ocular delivery, for example, Pignatello et al. reported the formulation and evaluation of nanocarriers composed of Eudragit-RL100 with good ocular tolerance, and no inflammation or discomfort in the rabbit eye (Pignatello et al., 2002b). According to Dillen et al. (2006), positively charged nanoparticles could also be prepared when Eudragit RL100 was combined with PLGA (Dillen et al., 2006). Barbault-Foucher et al. (2002) reported hyaluronic acid (HA) to be a natural, nonirritating polysaccharide showing pseudoplastic behavior with desirable ocular mucoadhesive properties (Barbault-Foucher et al., 2002). This group designed the novel ocular drug delivery system based on biodegradable nanospheres coated with a mucoadhesive polymer. The system was composed of a core of PECL coated by corona of the bioadhesive HA molecule. In this investigation, this group proposed the noncovalent attachment of unmodified HA to the exterior of the nanoparticles. They use three approaches, namely, (i) coating the PECL core during particle formation by chain entanglement with HA; (ii) coating of preformed PECL nanosystems by HA adsorption; and (iii) coating of PECL nanosystem by electrostatic interactions between negatively charged HA and a cationic surfactant used in the formulation (i.e., a cationic lipid, stearylamine, and a preservative usually used in ophthalmic formulation and absorption enhancer, benzalkonium chloride). The results made known that HA was robustly attached to nanospheres that had been conferred with a positive charge by cationic surfactant, resulting in intact HA-coated nanospheres. However, like CS, it does not possess permeability-enhancing properties (Lemarchand et al., 2004); in addition, the toxicity of stearylamine must be taken into consideration (du Toit et al., 2011) Vandervoort and Ludwig (2004) reported the use of gelatin nanoparticle based on its biocompatibility and biodegradability, since it was derived from the collagen obtained from the stroma of the eye, and has been used extensively in the ocular drug delivery (Sintzel et al., 1996; Friess, 1998). Diagrammatically, the mucoadhesive/positively charged mucoadhesive nanosystem and its interaction mechanism over ocular surface are presented in Figure 17.3.

Tailoring of nanosystems with positive surface functionalization was extensively explored in the recent research trend of ocular delivery technology. Guo et al. reported the synthesis and evaluation of positively charged phospholipids and cholesterols as components for liposomes (Guo et al., 1989). This group reported that some liposome preparations containing these synthetic lipid materials were found to be noncytotoxic. Further, they observe that insertion of the positively charged lipid derivatives into the liposomes appreciably improved the ocular withholding compared with neutral or negatively charged liposomes in an unanesthetized rabbit eye model due to molecular association with polyanionic corneal and conjunctival surface mucoglycoproteins.

17.5 REGULATORY CONCERN

The positive impact on the impaired quality of life in life-threatening diseases and the progressive result of clinical trials in nanomedicines are well evident. However, involving these R&D

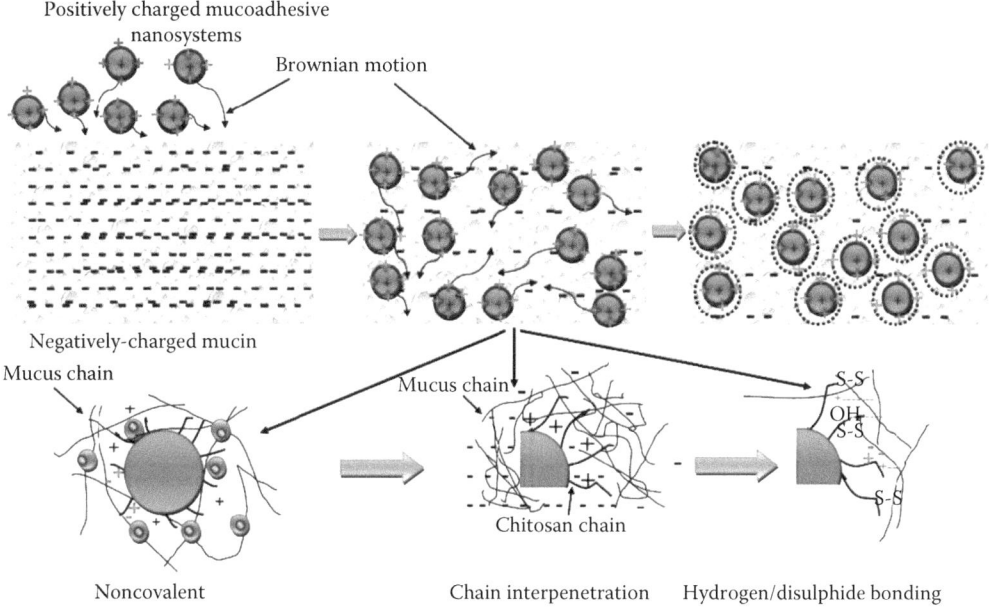

Positively charged mucoadhesive
nanosystems

Brownian motion

Negatively-charged mucin

Mucus chain

Mucus chain

Chitosan chain

S-S
OH
S-S

S-S

Noncovalent   Chain interpenetration   Hydrogen/disulphide bonding

FIGURE 17.3 Probable interaction mechanism describing mucoadhesion over the corneal surface by muco-adhesive positively charged nano-systems. Negatively-charged mucin (due to presence of sialic acid and sulfated sugars) interacts with positively charged mucoadhesive system.

outcomes to the clinical practice still need to be rationalized and carefully controlled in terms of risk-to-benefit ratio. These steps and clinical practice as pharmacovigilance is carefully overseen by regulatory authorities such as Food and Drug Administration (FDA), Medicines and Healthcare products Regulatory Agency (MHRA), European Medicines Agency (EMEA), and so on to ensure that the new medicines, diagonistic agents, and devices are safe for clinical practice in terms of the risk-to-benefit ratio. For any such products, the agencies have a set of protocols that basically works for the assessment of three parameters: quality, safety, and efficacy. The first generation of nanomedicines is already in clinical practice and some of them are now in the phase of generic development. Both generic manufacturers and regulatory agencies are struggling to finalize the studies that required to demonstrate the bioequivalency of nanometric generic medicine compared to the innovator and the developed generic nanomedicine have the same physicochemical properties and are safe and effective. This setback was clearly visualized with several unproductive efforts in the generic development of nab-paclitaxel formulation. The claimed biosimilar of nab-paclitaxel, when evaluated, did not fit to reproduce particle size, stability, potency, and other physicochemical attributes of nab-paclitaxel. Moreover, unlike nab-paclitaxel, the reconstituted nanomedicine exhibited poor stability when evaluated for accelerated stability testing and formed aggregates even within 24 h of the study period. Pharmaceutical/chemical and/or bioequivalence may not sufficiently symbolize the function of the nanomedicine at the site of action, as assumed for standard preparations. In addition, several liposomal formulations available in the market, such as those containing amphotericin B and doxorubicin, have recently gone off patent. The lack of critical information regarding composition, dimensional configuration of components, and critical parameters that are essential for the function of nanomedicine, raise the concern of risk that "generic" may be approved based on the guidelines compiled with conventional manufacturing and bioequivalence standards for generic drug approvals but that may result in substandard products in the market. This exemplifies how generic nanomedicine manufacturers and regulatory authorities are going to face troubles in their development and approval. Therefore, it is crucial that a comprehensive physicochemical-based understanding of nanomedicine and recognition of critical parameters that affect their functions be conducted early in development stage to put down a defined rule for such generics in the near future. Indeed, FDA has recently begun to consider relevant approval standards for generic copies of nanomedicines. A recently issued guideline in case of doxorubicin-loaded liposome (Doxil®) is an example that is reliable with this approach. The current scenario illustrates that ample nanomedicines have defined the clinical pharmacokinetics; considering this, it is an imperative need to improve the pharmacokinetic models for such nanometric medicines. Considerations such as pathophysiology, target tissue, absorption, distribution, metabolism, and excretion (ADME), impact of their size and surface characteristics on organ, tissue, and cellular localization, and better understanding of pharmacokinetics and pharmacodynamics (PK–PD) correlations need to be address in experimental models. Moreover, quantitative techniques need to be strengthened for biodistribution study of polymers and metals. In the context of polymeric and metallic nanomedicine therapeutics, new carrier systems are emerging with newer complex architectures, and such carrier systems may be administered by different route such as pulmonary, intravenous (iv), and organ-directed injection. With the increased complexity of the architecture, induction of multifunctionality in single carrier system frequently falls into a gap between medicines to medical devices regulation. It is noted that biological and medical devices assessment guidelines are based on general and nonspecific standards. Still the standard validation specifications for nanomedicines are lacking in current formats of regulatory guidelines and such complications will probably push the introduction of a completely new set of regulatory guidelines. Moreover, a regulatory guideline addressing new nanometric devices and different category of medicines harmonized for the global regulatory will be highly productive.

17.6 CONCLUSION

Overall, the promising results illustrated in the literatures point toward the acceptance of nanotechnology-based drug delivery system as future nanomedicine for topical ocular administration bioactive. Additionally, mucoadhesive polymer-based nanocarriers offer improved ocular residency and enhanced corneal permeation of drugs with prolonged action.

REFERENCES

Abdelkader, H., S. Ismail, A. Kamal, and R.G. Alany. 2010. Preparation of niosomes as an ocular delivery system for naltrexone hydrochloride: Physicochemical characterization. *Pharmazie* 65:811–817.

Adibkia, K., Y. Omidi, M.R. Siahi, A.R. Javadzadeh, M. Barzegar-Jalali, J. Barar, N. Maleki, G. Mohammadi and A. Nokhodchi. 2007a. Inhibition of endotoxin-induced uveitis by methylprednisolone acetate nanosuspension in rabbits. *Journal of Ocular Pharmacology and Therapeutics* 23:421–432.

Adibkia, K., M.R. Siahi Shadbad, A. Nokhodchi, A. Javadzedeh, M. Barzegar-Jalali, J. Barar, G. Mohammadi, and Y. Omidi. 2007b. Piroxicam nanoparticles for ocular delivery: Physicochemical characterization and implementation in endotoxin-induced uveitis. *Journal of Drug Targeting* 15:407–416.

Agnihotri, S.M., and P.R. Vavia. 2009. Diclofenac-loaded biopolymeric nanosuspensions for ophthalmic application. *Nanomedicine* 5:90–95.

Akhter, S., G. K. Jain, F. J. Ahmad, R. K. Khar, N. Jain, Z. I. Khan, and S. Talegaonkar. 2008. Investigation of nanoemulsion system for transdermal delivery of domperidone: *Ex-vivo* and *in vivo* studies. *Current Nanoscience* 4:381–390.

Akhter, S., S. Kushwaha, M.H. Warsi, M. Anwar, M.Z. Ahmad, I. Ahmad, S. Talegaonkar, Z.I. Khan, R.K. Khar and F.J Ahmad. 2012. Development and evaluation of nanosized niosomal dispersion for oral delivery of Ganciclovir. *Drug Development and Industrial Pharmacy* 38:84–92.

Akhter, S., S. Talegaonkar, Z.I. Khan, G.K. Jain, R.K. Khar, and F.J. Ahmad. 2011. Assessment of ocular pharmacokinetics and safety of Ganciclovir loaded nanoformulation. *Journal of Biomedical Nanotechnology* 7:144–145.

Alany, R.G., T. Rades, J. Nicoll, I.G. Tucker, and N.M. Davies. 2006. W/O microemulsions for ocular delivery: Evaluation of ocular irritation and precorneal retention. *Journal of Controlled Release* 111:145–152.

Alonso, M.J. 2004. Nanomedicine for overcoming biological barriers. *Biomedicine & Pharmacotherapy* 58:168–172.

Alonso, M.J. and A. Sanchez. 2003. The potential of chitosan in ocular drug delivery. *Journal of Pharmacy and Pharmacology* 55:1451–1463.

Ansari, M.J., K. Kohli, and N. Dixit. 2008. Microemulsions as potential drug delivery systems: A review. *PDA Journal of Pharmaceutical Science and Technology* 62:66–79.

Araujo, J.E. Gonzalez, M.A. Egea, M.L. Garcia, and E.B. Souto. 2009. Nanomedicines for ocular NSAIDs: Safety on drug delivery. *Nanomedicine* 5:394–401.

Artusson, P., T. Lindmark, S.S. Davis, and L. Illum. 1994. Effect of chitosan on the permeability of monolayers of intestinal epithelial cells (caco-2). *Pharmaceutical Research* 11:1358–1361.

Attwood, D. 1994. Microemulsions. In *Colloidal drug delivery Systems*. ed. J. Kreuter, Marcel Dekker, 31–71. New York.

Badawi, A.A., H.M. El-.Laithy, R.K. El-Qidra, H. El-Mofty, and M. El-dally. 2008. Chitosan based nanocarriers for indomethacin ocular delivery. *Archives of Pharmacal Research* 31:1040–1049.

Bagwe, R.P., J.R. Kanicky, B.J. Palla, P.K. Patanjali, and Shah DO. 2001. Improved drug delivery using microemulsions: Rationale, recent progress, and new horizons. *Critical Reviews in Therapeutic Drug Carrier Systems* 18:77–140.

Barbault-Foucher, S., R. Gref, P. Russo, J. Guechot, and A. Bochot. 2002. Design of poly–caprolactone nanospheres coated with bioadhesive hyaluronic acid for ocular delivery. *Journal of Controlled Release* 83:365–375.

Basaran, E., M. Demirel, B. Sirmagul, and Y. Yazan. 2010. Cyclosporine-A incorporated cationic solid lipid nanoparticles for ocular delivery. *Journal of Microencapsulation* 27:37–47.

Benita, S. and M.Y. Levy. 1993. Submicron emulsions as colloidal drug carriers for intravenous administration: Comprehensive physicochemical characterization. *Journal of Pharmaceutical Sciences* 82: 1069–1079.

Bochot, A., B. Mashhour, F. Puisieux, P. Couvreur, and E. Fattal.1998. Comparison of the ocular distribution of a model oligonucleotide after topical instillation in rabbits of conventional and new dosage forms. *Journal of Drug Targeting* 6:309–313.

Bourlais, C.L., L. Acar, H. Zia, P.A. Sado, T. Needham, and R. Leverge. 1998. Ophthalmic drug delivery systems-recent advances. *Progress in Retinal and Eye Research* 17:33–58.

Budai, L., M. Hajdú, M. Budai, P. Gróf, S. Béni, B. Noszál, I. Klebovich, and I. Antal. 2007. Gels and liposomes in optimized ocular drug delivery: Studies on ciprofloxacin formulations. *International Journal of Pharmaceutics* 343:34–40.

Buech, G., E. Bertelmann, U. Pleyer, I. Siebenbrodt, and H.H. Borchert. 2007. Formulation of sirolimus eye drops and corneal permeation studies. *Journal of Ocular Pharmacology and Therapeutics* 23:292–303.

Calvo, P., C. Remunan-Lopez, J.L Vila-Jato, and M.J. Alonso. 1997a. Chitosan and chitosan/ethylene oxide-propylene oxide block copolymer nanoparticles as novel carriers for proteins and vaccines. *Pharmaceutical Research* 14:1431–1436.

Calvo, P., A. Sanchez, J. Martinez, M.I. Lopez, M. Calonge, J.C. Pastor, and M.J. Alonso. 1996a. Polyester Nanocapsules as new topical ocular delivery systems for cyclosporin A *Pharmaceutical Research* 13:311–315.

Calvo, P., J.L. Vila-Jato, and M.J. Alonso. 1996b. Comparative *in vitro* evaluation of several colloidal systems, nanoparticles, nanocapsules, and nanoemulsions, as ocular drug carriers. *Journal of Pharmaceutical Sciences* 85:530–536.

Calvo, P., J.L. Vila-Jato, and M.J. Alonso. 1997b. Evaluation of cationic polymer-coated nanocapsules as ocular drug carriers. *International Journal of Pharmaceutics* 153:41–50.

Chen, R., Y. Qian, R. Li, Q. Zhang, D. Liu, M. Wang, and Q. Xu. 2010. Methazolamide Calcium Phosphate Nanoparticles in an Ocular Drug Delivery System. *Yakugaku Zasshi* 130:419–424.

Chrai, S.S. and J.R. Robinson. 1974. Ocular evaluation of methylcellulose vehicle in albino rabbits. *Journal of Pharmaceutical Sciences* 63:1218–1223.

Contia, B., C. Bucolob, C. Giannavolac, G. Puglisic, P. Giunchedia, and U. Contea. 1997. Biodegradable microspheres for the intravitreal administration of acyclovir: *In vitro in vivo* evaluation. *European Journal of Pharmaceutical Sciences* 5:287–293.

Cortesi, R., R. Argnani, E. Esposito, A Dalpiaz, A. Scatturin, F. Bortolotti, M. Lufino et al. 2006. Cationic liposomes as potential carriers for ocular administration of peptides with anti-herpetic activity. *International Journal of Pharmaceutics* 317:90–100.

Davies, N.M. 2000. Biopharmaceutical Considerations in Topical Ocular Drug Delivery. *Clinical and Experimental Pharmacology and Physiology* 27:558–562.

Davies, N.M., G. Wang, and I.G. Tucker. 1997. Evaluation of a hydrocortisone/hydroxypropyl-β-cyclodextrin solution for ocular drug delivery. *International Journal of Pharmaceutics* 156:201–209.

Davis, J.L., B.C. Gilger, and M.R. Robinson. 2004. Novel approaches to ocular drug delivery. *Current Opinion in Molecular Therapeutics.* 6, 195–205.

De Campos, A.M, A. Sanchez, and M.J. Alonso. 2001. Chitosan nanoparticles: A new vehicle for the improvement of the delivery of drugs to the ocular surface. Application to cyclosporine A. *International Journal of Pharmaceutics* 224:159–168.

De Campos, A.M, A. Sanchez, R. Gref, P. Calvo, and M.J. Alonso. 2003. The effect of a PEG versus a chitosan coating on the interaction of drug colloidal carriers with the ocular mucosa. *European Journal of Pharmaceutical Sciences* 20:73–81.

de la Fuente, M., N. Csaba, M. Garcia-Fuentes, and M.J. Alonso. 2008. Nanoparticles as protein and gene carriers to mucosal surfaces. *Nanomedicine (Lond).* 3:845–857.

Diepold, R., J. Kreuter, J. Himber, R. Gurny, V.H. Lee, J.R. Robinson, M.F. Saettone, and O.E. Schnaudigel. 1989. Comparison of different models for the testing of pilocarpine eyedrops using conventional eyedrops as a novel depot formulation (nanoparticles). Graefe's Arch. *Clinical and Experimental Ophthalmology.* 227:188–193.

Dillen, K., J. Vandervoort, G. Van den Mooter, and A. Ludwig. 2006. Evaluation of ciprofloxacin-loaded Eudragit RS100 or RL100/PLGA nanoparticles. *International Journal of Pharmaceutics* 314:72–82.

Dodane, V., M.A. Khan, and J.R. Merwin. 1999. Effect of chitosan on epithelium permeability and structure. *International Journal of Pharmaceutics* 182:21–32.

Dornish, M., A. Hagan, E. Hansson, C. Pecheur, F. Verdier, and Q. Skaugrud. 1997. Safety of protasan: Ultrapure chitosan salts for biomedical and pharmaceutical use. *Advances in Chitin Science* 2:664–670.

du Toit, L.C., V. Pillay, Y.E. Choonara, T. Govender, and T. Carmichael. 2011. Ocular drug delivery—a look towards nanobioadhesives. *Expert Opinion on Drug Delivery* 8:71–94.

Duanand, X. and H. Sheardown. 2006. Dendrimer crosslinked collagen as a corneal tissue engineering scaffold: Mechanical properties and corneal epithelial cell interactions. *Biomaterials* 27:4608–4617.

Duchfine, D., F. Touchard, and N.A. Peppas. 1988. Pharmaceutical and medical aspects of bioadhesive systems of drug administration. *Drug Development and Industrial Pharmacy* 14:283–318.

Fattal, E. and A. Bochot. 2006. Ocular delivery of nucleic acids: Antisense oligonucleotides, aptamers and siRNA. *Advanced Drug Delivery Reviews* 58:1203–1223.

Felt, O., P. Furrer, J.M. Mayer, B. Plazonnet, P. Buri, and R. Gurny. 1999. Topical use of chitosan in ophthalmology: Tolerance assessment and evaluation of pre-corneal retention. *International Journal of Pharmaceutics* 180:185–193.

Fitzgerald, P., J. Hadgraft, J. Kreuter, and C.G. Wilson. 1987. A γ-scintigraphic evaluation of microparticulate ophthalmic delivery systems: Liposomes and nanoparticles. *International Journal of Pharmaceutics* 40:81–84.

Fresta, M., A.M. Panico, C. Bucolo, C. Giannavola, and G. Puglisi. 1999. Characterization and in-vivo ocular absorption of liposome-encapsulated acyclovir. *Journal of Pharmacy and Pharmacology* 51:565–576.

Friess, W. 1998. Collagen-material for drug delivery. *European Journal of Pharmaceutics and Biopharmaceutics* 45:113–136.

Furrer, P., B. Plazonnet, J.M. Mayer, and R. Gurny. 2000. Application of *in vivo* confocal microscopy to the objective evaluation of ocular irritation induced by surfactants. *International Journal of Pharmaceutics* 2007:89–98.

Genta, I., B. Conti, P. Perugini, F. Pavaneto, A. Spadaro, and G. Puglisi. 1997. Bioadhesive microspheres for ophthalmic administration of acyclovir. *Journal of Pharmacy and Pharmacology* 49:737–742.

Gokce, E.H., G. Sandri, S. Egrilmez, M.C. Bonferoni, T. Guneri, and C. Caramella. 2009. Cyclosporine A-loaded solid lipid nanoparticles: Ocular tolerance and *in vivo* drug release in rabbit eyes. *Current Eye Research* 34:996–1003.

Greaves, J.L., O. Olejnik, and C.G. Wilson. 1992. Polymers and the precorneal tear film. *STP Pharma Sciences* 2:13–33.

Greaves, J.L., C.G. Wilson, and A.T. Birmingham.1993. Assessment of the precorneal residence of an ophthalmic ointment in healthy subjects. *British Journal of Clinical Pharmacology* 35:188–192.

Guo, L.S.S., A.M. Sarris, and M.D. Levy. 1989. A safe bioadhesive liposomal formulation for ophthalmic applications. *Investigative Ophthalmology & Visual Science* 29:Suppl-439.

Gupta, H., M. Aqil, R.K. Khar, A. Ali, A. Bhatnagar, and G. Mittal. 2010. Sparfloxacin-loaded PLGA nanoparticles for sustained ocular drug delivery. *Nanomedicine* 6:324–333.

Gurny, R. 1983. Latex systems. In Topics in Pharmaceutical Sciences. ed. D.D, Breimer, and P. Speiser, 277–288. Elsevier Science Publishers: Amsterdam.

Gurny, R. 1986. Ocular therapy with nanoparticles. In Polymeric Nanoparticles and Microspheres. ed. P. Guiot, and P. Couvreur, 127–136. CRC Press: Boca Raton.

Hanrahan, F., M. Campbell, A.T. Nguyen, M. Suzuki, A.S. Kiang, L.C. Tam, O.L. Gobbo et al. 2012. On Further Development of Barrier Modulation as a Technique for Systemic Ocular Drug Delivery. *Advances in Experimental Medicine and Biology* 723:155–159.

Hasse A. and S. Keipert. 1997. Development and characterization of microemulsions for ocular application. *European Journal of Pharmaceutics and Biopharmaceutics* 43:179–183.

Hathout, R.M., S. Mansour, N.D. Mortada, and A.S. Guinedi. 2007. Liposomes as an ocular delivery system for acetazolamide: *In vitro* and *in vivo* studies. *American Association of Pharmaceutical Scientists* (AAPS) 8:1.

Hosny, K.M. 2009. Preparation and evaluation of thermosensitive liposomal hydrogel for enhanced transcorneal permeation of ofloxacin. *American Association of Pharmaceutical Scientists* (AAPS) 10:1336–1342.

Ihre, H.R., O.L. Padilla, D. Jesus, F.C. Szoka Jr, and J.M. Frechet. 2002. Polyester dendritic systems for drug delivery applications: Design, synthesis, and characterization. *Bioconjugate Chemistry* 13:443–452.

Illum, L. 1998. Chitosan and its uses as a pharmaceutical excipient. *Pharmaceutical Research* 15:1326–1331.

Jain, N., S. Akhter, G.K. Jain, Z.I. Khan, R.K. Khar, and F.J. Ahmad. 2011a. Antiepileptic intranasal Amiloride loaded mucoadhesive nanoemulsion: Development and safety assessment. *Journal of Biomedical Nanotechnology* 7:142–143.

Jain, G.K., S.A. Pathan, S. Akhter, N. Jayabalan, S.Talegaonkar, R.K. Khar and F.J. Ahmad. 2011b. Microscopic and spectroscopic evaluation of novel PLGA-chitosan nanoplexes as an ocular delivery system. *Colloids and Surfaces B: Biointerfaces* 82:397–403.

Joossand, K. and N. Chirmule. 2003. Immunity to adenovirus and adeno-associated viral vectors: Implications for gene therapy. *Gene Therapy* 10:955–963.

Kakkar, S. and I.P. Kaur. 2011. Spanlastics—A novel nanovesicular carrier system for ocular delivery. *International Journal of Pharmaceutics* 413:202–210.

Kassem, M.A., R.A.A. Abdel, M.M. Ghorab, M.B. Ahmed, and R.M. Khalil. 2007. Nanosuspension as an ophthalmic delivery system for certain glucocorticoid drugs. *International Journal of Pharmaceutics* 340:126–133.

Kaur, I.P., D. Aggarwal, H. Singh, and S. Kakkar. 2010. Improved ocular absorption kinetics of timolol maleate loaded into a bioadhesive niosomal delivery system. *Graefes Archive for Clinical and Experimental Ophthalmology* 248:1467–1472.

Kaur, I.P., A. Garg, A.K. Singla, and D. Aggarwal. 2004. Vesicular systems in ocular drug delivery: An overview. *International Journal of Pharmaceutics* 269:1–14.

Kaur, I.P. and M. Kanwar. 2002a. Ocular preparations: The formulation approach. *Drug Development and Industrial Pharmacy* 28:473–493.

Kaur, I.P. and R. Smitha. 2002b. Penetration enhancers and ocular bioadhesives: Two new avenues for ophthalmic drug delivery. *Drug Development and Industrial Pharmacy*. 28:353–369.

Koch, M.A., V. Dodane, M.A. Khan, and J.R. Merwin. 1998. Chitosan induced effects on epithelial morphology as seen by confocal scanning microscopy. *Scanning* 20:262–263.

Krauland, A.H., V.M. Leitner, and A. Bernkop-Schnurch. 2003. Improvement in the *in situ* gelling properties of deacetylated gellan gum by the immobilization of thiol groups. *Journal of Pharmaceutical Sciences* 92:1234–1241.

Kreuter, J. 1992. Nanoparticles-Preparation and applications. In Microcapsules and Nanocapsules in Medicine and Pharmacy. ed. M. Donbrow, 126–143. CRC Press Inc: Boca Raton.

Kupferman, A., M.V. Pratt, Suckewer K, and H.M. Leibowitz. 1974. Topically applied steroids in corneal disease, the role of drug derivative in stromal absorption of Dexamethasone. *Archives of Ophthalmology* 91:373–376.

Lajavardi, L., A. Bochot, S. Camelo, B. Goldenberg, M.C. Naud, F. Behar-Cohen, E. Fattal, and Y. de Kozak. 2007. Downregulation of endotoxin-induced uveitis by intravitreal injection of vasoactive intestinal peptide encapsulated in liposomes. *Investigative Ophthalmology & Visual Science* 48:3230–3238.

Lang, J.C. 1995. Ocular drug delivery conventional ocular formulations. *Advanced Drug Delivery Reviews* 16:39–43.

Lee, V.H.K., R.W. Wood, J. Kreuter, T. Harima, and J.R. Robinson. 1986. Ocular drug delivery of progesterone using nanoparticles. *Journal of Microencapsulation* 3:213–218.

Lemarchand, C., R. Gref, and P. Couvreur. 2004. Polysaccharide-decorated nanoparticles. *European Journal of Pharmaceutics and Biopharmaceutics* 58:327–341.

Li, C.C., M. Abrahamson, Y. Kapoor, and A. Chauhan. 2007. Timolol transport from microemulsions trapped in HEMA gels. *Journal of Colloid and Interface Science* 315:297–306.

Li, R., S. Jiang, D. Liu, X. Bi, F. Wang, Q. Zhang, and Q. Xu. 2011. A potential new therapeutic system for glaucoma: Solid lipid nanoparticles containing Methazolamide. *Journal of Microencapsulation* 28:134–141.

Liu, Z., X. Zhang, H. Wu, J. Li, L. Shu L, R. Liu, L. Li, and N. Li. 2011. Preparation and evaluation of solid lipid nanoparticles of baicalin for ocular drug delivery system *in vitro* and in vivo. *Drug Development and Industrial Pharmacy* 37:475–481.

Losa, C., L. Marchal-Heussler, F. Orallo, J.L. Vila Jato, and M.J. Alonso. 1993. Design of new formulations for topical ocular administration: Polymeric nanocapsules containing metipranolol. *Pharmaceutical Research* 10:80–87.

Ludwig, A. 2005. The use of mucoadhesive polymers in ocular drug delivery. *Advanced Drug Delivery Reviews* 57:1595–1639.

Ludwig, A., N.J. van Haeringen, V.M. Bodelier, and R. Van Ooteghem. 1992a. Relationship between precorneal retention of viscous eye drops and tear fluid composition. *International Ophthalmology* 16:23–26.

Ludwig, A. and M. van Ooteghem. 1992b. Influence of viscoslysers on the residence of ophthalmic solutions evaluated by slip lamp fluorophotometry. *STP Pharma Sciences* 2:81–87.

MacKeen, D.L. 1980. Aqueous formulations and ointments. *International Ophthalmology Clinics* 20:79–92.

Mainardes, R.M. and R.C. Evangelista. 2005. PLGA nanoparticles containing praziquantel: Effect of formulation variables on size distribution. *International Journal of Pharmaceutics* 290:137–144.

Marano, R.J., I. Toth, N. Wimmer, M. Brankov, and P.E. Rakoczy. 2005. Dendrimer delivery of an anti-VEGF oligonucleotide into the eye: A long-term study into inhibition of laser-induced CNV, distribution, uptake and toxicity. *Gene therapy* 12:1544–1550.

Marchal-Heussler, L., H. Fessi, J.P. Devissaguet, M. Hoffman, and P. Maincent. 1992. Colloidal drug delivery systems for the eye. A comparison of the efficacy of three different polymers: Polyisobutylcyanoacrylate, polylactic-coglycolic acid, poly-epsilon-caprolactone. *Journal of Pharmaceutical Sciences* 2:98.

Merodio, M., A. Arnedo, M.J. Renedo and J.M. Irache. 2001. Ganciclovir-loaded albumin nanoparticles: Characterization and *in vitro* release properties. *European Journal of Pharmaceutical Sciences* 12:251–259.

Mishra, G.P., M. Bagui, V. Tamboli, and A.K. Mitra. 2011. Recent applications of liposomes in ophthalmic drug delivery. *Journal of Drug Delivery* doi:10.1155/2011/863734.

Motwani, S.K., S. Chopra, S. Talegaonkar, K. Kohli, F.J. Ahmad, and R.K. Khar. 2008. Chitosan–sodium alginate nanoparticles as submicroscopic reservoirs for ocular delivery: Formulation, optimisation and *in vitro* characterization. *European Journal of Pharmaceutics and Biopharmaceutics* 68:513–525.

Muller, R.H. and C.M. Keck. 2004. Challenges and solutions for the delivery of biotech drugs—a review of drug nanocrystal technology and lipid nanoparticles. *Journal of Biotechnology* 113:151–170.

Nagarsenker, M.S., V.Y. Londhe, and G.D. Nadkarni. 1999. Preparation and evaluation of liposomal formulations of tropicamide for ocular delivery. *International Journal of Pharmaceutics* 190:63–71.

Nanjwade, B.K., R.V. Deshmukh, K.R. Gaikwad, K.A. Parikh, and F.V. Manvi. 2012. Formulation and evaluation of micro hydrogel of Moxifloxacin hydrochloride. *European Journal of Drug Metabolism and Pharmacokinetics* 37:117–123.

Ogata, N., T. Otsuji, M. Matsushima, T. Kimoto, R. Yamanaka, K. Takahashi, M. Wada, M. Uyama, and Y. Kaneda. 1999. Phosphorothioate oligonucleotides induction into experimental choroidal neovascularization by HVJ-liposome system. *Current Eye Research* 18:261–269.

Olejnik, O. 1993. Conventional Systems in Ophthalmic Drug Delivery. In Ophthalmic Drug Delivery Systems. ed. Mitra, A.K. 177–198. Marcel Dekker; New York.

Ooya, T., J. Lee, and K. Park. 2003. Effects of ethylene glycol-based graft, star-shaped, and dendritic polymers on solubilization and controlled release of paclitaxel. *Journal of Controlled Release* 93:121–127.

Park, H. and J.R. Robinson. 1987. Mechanisms of mucoadhesion of poly(acrylic acid) hydrogels. *Pharmaceutical Research* 4:457–464.

Patton, T.F. and J.R. Robinson. 1975. Ocular evaluation of polyvinyl alcohol vehicle in rabbits. *Journal of Pharmaceutical Sciences* 64:1312–1316.

Paul, W. and C. Sharma. 2000. Chitosan, a drug carrier for the 21st century. *S.T.P. Pharma Sciences* 10:5–22.

Pignatello, R., C. Bucolo, P. Ferrara, A. Maltese, A. Puleo, and G. Puglisi. 2002a. Eudragit RS100® nanosuspensions for the ophthalmic controlled delivery of ibuprofen. *European Journal of Pharmaceutical Sciences* 16:53–61.

Pignatello, R., C. Bucolo, and G. Puglisi. 2002b. Ocular tolerability of Eudragit RS[(R)] 100 and RL[(R)] 100 nanosuspensions as carriers for ophthalmic controlled drug delivery. *Journal of Pharmaceutical Sciences*. 91:2636–2641.

Pignatello, R., N. Ricupero, C. Bucolo, F. Maugeri, A. Maltese, and G. Puglisi. 2006. Preparation and characterization of eudragit retard nanosuspensions for the ocular delivery of cloricromene. *American Association of Pharmaceutical Scientists (AAPS)* 7:E27.

Ponchel, G., F. Touchard, D. Duchfine, and N.A. Peppas. 1987. Bioadhesive analysis of controlled-release systems. I. Fracture and interpenetration analysis in poly (acrylic acid)-containing systems. *Journal of Controlled Release* 5:129–141.

Qian, Y., F. Wang, R. Li, Q. Zhang, and Q. Xu. 2010. Preparation and evaluation of *in situ* gelling ophthalmic drug delivery system for methazolamide. *Drug Development and Industrial Pharmacy* 36:1340–1347.

Qu. X., V. V. Khutoryanskiy, A. Stewart, S. Rahman, B. Papahadjopoulos-Sternberg, C. Dufes, D. McCarthy et al. 2006. Carbohydrate-based micelle clusters which enhance hydrophobic drug bioavailability by up to 1 order of magnitude. *Biomacromolecules* 7:3452–3459.

Quintana, A., E. Raczka, L. Piehler, I. Lee, A. Myc, I. Majoros, A.K. Patri, T. Thomas, J. Mulé, and J.R. Baker. 2002. Design and function of a dendrimer-based therapeutic nanodevice targeted to tumor cells through the folate receptor. *Pharmaceutical Research* 19:1310–1316.

Round, A.N., M. Berry, T.J. McMaster, S. Stoll, D. Gowers, A.P. Corfield, and M.J. Miles. 2002. Heterogeneity and persistence length in human ocular mucins. *Biophysical Journal* 83:1661–1670.

Rupenthal, I.D., C.R. Green, and R.G. Alany. 2011. Comparison of ion-activated *in situ* gelling systems for ocular drug delivery. Part 2: Precorneal retention and *in vivo* pharmacodynamic study. *International Journal of Pharmaceutics* 411:78–85.

Sahoo, S.K., F. Dilnawaz, and S. Krishnakumar. 2008. Nanotechnology in ocular drug delivery. *Drug Discovery Today* 13:144–151.

Sahoo, S.K. and V. Labhasetwar. 2003. Nanotech approaches to drug delivery and imaging. *Drug Discovery Today* 8:1112–1120.

Schipper, N.G., S. Olsson, J.A. Hoogstraate, A.G. deBoer, K.M. Vårum, and P. Artursson. 1997. Chitosans as absorption enhancers for poorly absorbable drugs 2: Mechanism of absorption enhancement. *Pharmaceutical Research* 14:923–929.

Schoenwald, R.D. and P. Stewart. 1980. Effect of particle size on ophthalmic bioavailability of dexamethasone suspensions in rabbits. *Journal of Pharmaceutical Sciences* 69:391–394.

Schoenwald, R.D. and R. Ward. 1978. Relationship between steroid permeability across excised rabbit cornea and octanol-water partition coefficients. *Journal of Pharmaceutical Sciences* 67:786–788.

Seyfoddin, A., J. Shaw, and R. Al-Kassas. 2010. Solid lipid nanoparticles for ocular drug delivery. *Drug Delivery* 17:467–489.

Shaunak, S., S. Thomas, E. Gianasi, A. Godwin, E. Jones, I. Teo, K. Mireskandari et al. 2004. Polyvalent dendrimer glucosamine conjugates prevent scar tissue formation. *Nature Biotechnology* 22:977–984.

Shedden, A.H., J. Laurence, A. Barrish, and T.V. Olah. 2001. Plasma timolol concentrations of timolol maleate: Timolol gel-forming solution (TIMOPTICXE) once daily versus timolol maleate ophthalmic solution twice daily. *Documenta Ophthalmologica* 103:73–79.

Sheikpranbabu, S., K. Kalishwaralal, D. Venkataraman, S.H. Eom, J. Park, and S. Gurunathan. 2009. Silver nanoparticles inhibit vegf- and il-1-induced vascular permeability via src dependent pathway in porcine retinal endothelial cells. *Journal of Nanobiotechnology* 30:7:8.

Shen, Y. and J. Tu. 2007. Preparation and ocular pharmacokinetics of ganciclovir liposomes, *AAPS Journal* 9:E371–377.

Shimazaki, H., K. Hironaka, T. Fujisawa, K. Tsuruma, Y. Tozuka, M. Shimazawa, H. Takeuchi, and H. Hara. 2011. Edaravone-loaded liposome eyedrops protect against light-induced retinal damage in mice. *Investigative Ophthalmology & Visual Science* 52:7289–7297.

Sieg, J.W. and J.R. Robinson. 1997. Vehicle effects on ocular drug bioavailability II: Evaluation of pilocarpine. *Journal of Pharmaceutical Sciences* 66:1222–1228.

Sintzel, M.B., S.F. Bernatchez, C. Tabatabay, and R. Gurny. 1996. Biomaterials in ophthalmic drug delivery. *European Journal of Pharmceutics and Biopharmaceutics* 42:358–374.

Slovin, E.M. and J.R. Robinson.1993. Bioadhesives in ocular drug delivery. In *Biopharmaceutics of Ocular Drug Delivery*. ed. Edman, P. 145–157. CRC Press: Boca Raton, Florida.

Taniguchi, K., Y. Yamamoto, K. Itakura, H. Miichi, and S. Hayashi. 1988. Assessment of ocular irritability of liposome preparations. *Journal Pharmacobiodyn* 11:607–611.

Tian, B., Q. Luo, S. Song, D. Liu, H. Pan, W. Zhang, L. He, S. Ma, X. Yang, and W. Pan. 2012. Novel surface-modified nanostructured lipid carriers with partially deacetylated water-soluble chitosan for efficient ocular delivery. *Journal of Pharmaceutical Sciences* 101:1040–1049.

Tiffany, J.M. 1991. The viscosity of human tears. *International Ophthalmology* 15:371–376.

Uchegbu, I.F., L. Sadiq, M. Arastoo, A.I. Gray, W. Wang, R.D. Waigh, and A.G. Schatzlein. 2001. Quaternary ammonium palmitoyl glycol chitosan—A new polysoap for drug delivery. *International Journal of Pharmaceutics* 224:185–199.

Van Ooteghem, M. 1987. Factors influencing the retention of ophthalmic solutions on the eye surface. In *Ophthalmic Drug Delivery. Biopharmaceutical, Technological and Clinical Aspects.* ed. Saettone, M., F, Bucci, and M.P. Speiser. 7–17. Livinia Press. (Vol. 11): Padova.

Van Ooteghem, M. 1995. Preparations ophtalmiques. In *Technique and Documentation.* ed. Galenica, Lavoisier; Paris.

Vandamme, T.F. 2002. Microemulsions as ocular drug delivery systems: Recent developments and future challenges. *Progress in Retinal and Eye Research* 21:15–34.

Vandamme, T.F. and L. Brobeck. 2005. Poly(amidoamine) dendrimers as ophthalmic vehicles for ocular delivery of pilocarpine nitrate and tropicamide. *Journal of Controlled Release.* 102: 23–38.

Vandervoort, J. and A. Ludwig. 2004. Preparation and evaluation of drug-loaded gelatin nanoparticles for topical ophthalmic use. *European Journal of Pharmaceutics and Biopharmaceutics* 57:251–261.

Vandervoort, J. and A. Ludwig. 2007. Ocular drug delivery: Nanomedicine applications. *Nanomedicine (Lond)* 11–21.

Wadhwa, S., R. Paliwal, S.R. Paliwal, and S.P. Vyas. 2009. Nanocarriers in ocular drug delivery: An update review. *Current Pharmaceutical Design* 15:2724–2750.

Wang, S., J. Zhang, T. Jiang, L. Zheng, Z. Wang, J. Zhang, and P. Yu. 2011. Protective effect of Coenzyme Q (10) against oxidative damage in human lens epithelial cells by novel ocular drug carriers. *International Journal of Pharmaceutics* 403:219–229.

Wei, G., H. Xu, P.T. Ding, S.M. Li, and J.M. Zheng. 2002. Thermosetting gels with modulated gelation temperature for ophthalmic use: The rheological and gamma scintigraphic studies. *Journal of Controlled Release* 83:65–74.

Wood, R.W., V.H.K. Li, J. Kreuter, and J.R. Robinson. 1985. Ocular disposition of poly-hexyl-2-cyano (3–14C) acrylate nanoparticles in the albino rabbit. *International Journal of Pharmaceutics* 23:175–183.

Xu, Y. and Y. Du. 2003. Effect of molecular structure of chitosan on protein delivery properties of chitosan nanoparticles. *International Journal of Pharmaceutics* 250:215–226.

Zignani, M., C. Tabatabay, and R. Gurny. 1995. Topical semi-solid drug delivery: Kinetics and tolerance of ophthalmic hydrogels. *Advanced Drug Delivery Reviews* 16:51–60.

Zimmer, A., J. Kreuter, and J.R. Robinson. 1991. Studies on the transport pathway of PBCA nanoparticles in ocular tissues. *Journal of Microencapsulation* 8:497–504.

Zimmer, A., E. Mutschler, G. Lambrecht, D. Mayer, and J. Kreuter. 1994. Pharmacokinetic and pharmacodynamic aspects of an ophthalmic pilocarpine nanoparticle-delivery-system. *Pharmaceutical Research* 11:1435–1442.

Zimmer, A.K., H. Zerbe, and J. Kreuter. 1994. Evaluation of pilocarpine-loaded albumin particles as drug delivery systems for controlled delivery in the eye I. *In vitro* and *in vivo* characterisation. *Journal of Controlled Release* 32:57–70.

18 Biomedical Applications of Chitin and Chitosan Derivatives

Sougata Jana, Arijit Gandhi, Kalyan Kumar Sen, and Sanat Kumar Basu

CONTENTS

18.1 INTRODUCTION

Biopolymers have received attention as tissue engineering (TE) substrates, with several studies examining materials such as alginate, chitosan, and gelatin as cell scaffolds for both two-dimensional (2D) and three-dimensional (3D) cell culture (Pan et al. 2005, Park et al. 2005). In recent years, carbohydrate polymers have been extensively used in biomedical and pharmaceutical applications because of their biocompatibility and biodegradability (Tabata and Ikada 1998). Polysaccharides represent one of the most abundant industrial raw materials and have been the subject of intensive research owing to their sustainability, biodegradability, and biosafety. Currently, the uses of chitin and chitosan have been postulated in numerous areas of biopharmaceutical research, such as mucoadhesion, permeation enhancement, vaccine technology, gene therapy, and wound healing. Chitin is a known biodegradable natural polymer based on polysaccharides, which is obtained from crustacean shell (e.g., crab, shrimp, and lobster), some insects (e.g., true fly and sulfur butterfly), and fungi such as yeasts and plants. It principally occurs in animals of the phylum Arthopoda (Zheng et al. 2001). Historically, in 1811, Professor Henri Braconnot, isolated fibrous substances from mushroom and found them to be insoluble in aqueous acidic solution. A decade later in 1823, Ojer named it "chitin" from Greek "khiton" meaning "envelope" present in certain insects. In 1894, Hope Seyle named it as "chitosan." From 1930 to 1940, this biopolymer of glucosamine gained much interest in the field of medicine. Chitosan obtained from partial deacetylation of chitin is a polysaccharide comprising copolymers of glucosamine and N-acetylglucosamine. Chitosan is commercially available in several types and grades that vary in molecular weight between 10,000 and 1,000,000 and vary in degree of deacetylation and viscosity. Technically speaking, Chitosan is a naturally occurring substance that is chemically similar to cellulose, which is a plant fiber. Like plant fibers, chitosan possesses many of the same properties as fiber; however, unlike plant fiber, it has the ability to significantly bind fat, acting like a "fat sponge" in the digestive tract. It attracts the biohazardous substances, such as greases, oils, heavy metals, and other potentially toxic substances from the water and it is used for detoxifying water (Miyazaki et al. 1981).

18.2 CHEMICAL STRUCTURE

Chitin is similar to cellulose both in chemical structure and in biological function as a structural polymer. The crystalline structure of chitin has been shown to be similar to cellulose in the arrangements

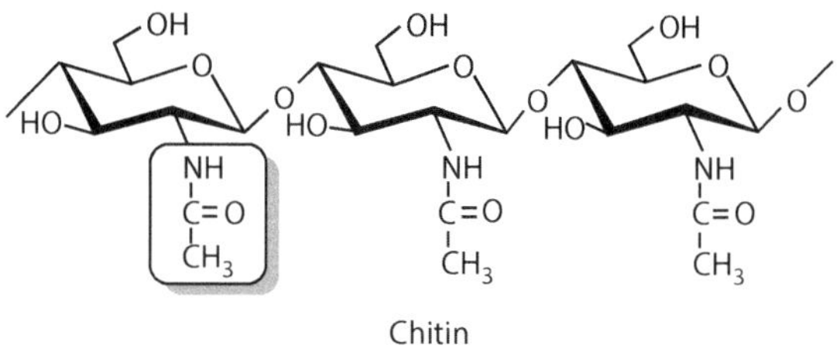

FIGURE 18.1 Structure of the chitin molecule, showing two of the *N*-acetylglucosamine units that repeat to form long chains in β-1,4 linkages.

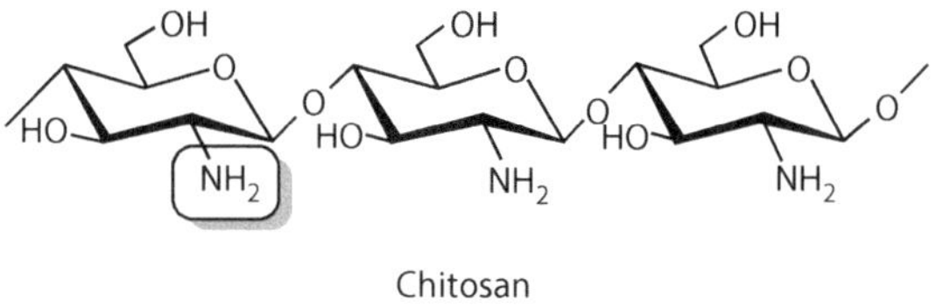

FIGURE 18.2 Chitosan is a linear polysaccharide composed of randomly distributed β-(1–4)-linked D-glucosamine (deacetylated unit) and *N*-acetyl-D-glucosamine (acetylated unit).

of inter- and intrachain hydrogen bonding. Chitosan is made by alkaline *N*-deacetylation of chitin. The term "chitosan" refers to a family of copolymers with various fractions of acetylated units (Figures 18.1 and 18.2). It consists of two types of monomers: chitin monomers and chitosan monomers. Chitin is a linear polysaccharide consisting of (1-4)-linked 2-acetamido-2-deoxy-β-D-glucopyranose. Chitosan is a linear polysaccharide consisting of (1-4)-linked 2-amino-2-deoxy-β-D-glucopyranose (Yazdani-Pedram and Retuert 1997, Sugimoto et al. 1998).

18.3 METHOD OF MANUFACTURE

It is manufactured by chemically treating the shells of crustaceans, such as shrimps and crabs. The process involves the separation of proteins by treating with alkali and minerals, such as calcium carbonate and calcium phosphate, and followed by treatment with acid (Gupta and Ravikumar 1986). Initially, the shells are deproteinized by treatment with (3–5%) aqueous sodium hydroxide solution. The resulting product is neutralized and calcium is removed by treatment with (3–5%) aqueous hydrochloric acid solution at room temperature to precipitate chitin. The chitin is dried and deacetylated to give chitosan. This can be achieved by treatment with 40–45% aqueous sodium hydroxide solution at moderate temperature (110°C) and the precipitate is washed with water. The crude sample is dissolved in 2% acetic acid and the insoluble material is removed. The resulting clear supernatant solution is neutralized with aqueous sodium hydroxide to give a white precipitate of chitosan. It can be further purified and ground to a fine uniform powder or granules (Niederhofer and Maller, 2004).

18.4 PROPERTIES OF CHITOSAN

The word "chitosan" refers to a large number of polymers, which differ in their degree of *N*-deacetylation (40–98%) and molecular weight (50,000–2,000,000 Da). These two characteristics are very important to the physicochemical properties of the chitosans and hence, they have a major effect on the biological properties (Singla and Chawla 2001).

The chemical properties of chitosan are as follows:

- Cationic polyamine
- High charge density at pH 6.5
- Adheres to negatively charged surfaces
- Forms gels with polyanions
- High-molecular-weight linear polyelectrolyte
- Viscosity is <5 cps
- Chelates certain transitional metals
- Amiable to chemical modification
- Reactive amino/hydroxyl groups
- The molecular weight of chitosan ranges from 1×10^5 to 3×10^5, whereas the average molecular weight ranges from 3.8×10^3 to 2000×10^3
- The degree of acetylation of chitosan ranges from 66 to 99.8%
- Moisture content is >10%

The biological properties of chitosan are as follows:

- Biocompatibility
- Natural polymer
- Biodegradable to normal body constituents
- Safe and nontoxic
- Hemostatic, bacteriostatic, and fungistatic
- Spermicidal
- Anticancerogen
- Anticholesteremic
- Reasonable cost
- Versatile

The overview of the biological properties of chitosan is shown in Figure 18.3.

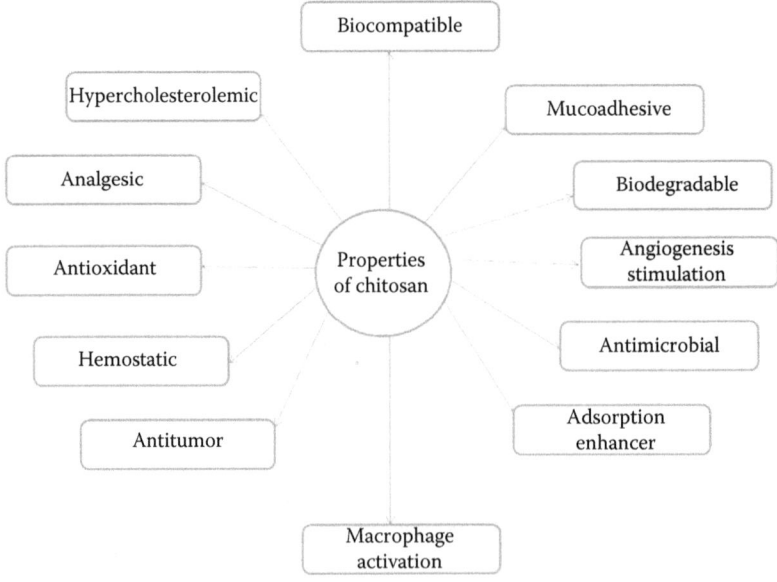

FIGURE 18.3   Biological properties of chitosan.

18.5 NOVEL CHITOSAN DERIVATIVES

Various chemical procedures have been employed to enhance the aqueous solubility of chitosan, including carboxymethylation (Lu et al. 2007), carboxyethylation (Jiang et al. 2005), reductive amination with phosphorylcholine glyceraldehydes (Tiera et al. 2006), sulfation (Zhang et al 2008), *N*- or *O*-acylation (Badawy et al. 2004), alkylation (Kang et al. 2006), and quaternization (Huang et al. 2007). Many functional derivatives have been prepared by Michael addition reaction to increase the solubility of chitosan in water.

18.5.1 CARBOXYMETHYL CHITOSAN

The synthesis of carboxymethyl chitosan has been depicted in Figure 18.4 (Chen et al. 2004). *N*-Acetyl chitosan (10 g) was suspended in 50% (w/w) NaOH and kept at –20°C overnight. The frozen alkali chitosan was transferred to 100 mL of 2-propanol, and monochloroacetic acid (ClCH₂COOH) was added in portions. After stirring at room temperature for 2 h, heat was applied to bring the reaction mixture to a temperature of 60°C for another 2 h. Following dialyzing against deionized water, the product was vacuum dried at room temperature.

Carboxymethyl chitosan aids in absorption by enhancing paracellular transport through the controlled, transient, and reversible opening of intestinal tight junctions without being absorbed or interacting with charges of the lipid bilayer of the cell membrane.

18.5.2 *N*-TRIMETHYL CHITOSAN

N-Trimethyl chitosan is a soluble cationic methylated derivative of chitosan, which can widen the paracellular route for the passage of hydrophilic and macromolecular drugs after mucosal administration (Dung et al. 1994). The charge density and the structural features of chitosan salts and *N*-trimethyl chitosan chloride (TMC) are important factors that determine their potential use as absorption enhancers. Other TMC studies demonstrated that TMC is an absorption enhancer over a wide pH range and that TMC polymers with high degrees of quaternization are less mucoadhesive compared to those with lower degrees of quaternization and this was attributed to changes in the flexibility of the polymer chain. The same study revealed that TMC had no toxic effects (Kotzé et al. 2002). TMC reduced the transepithelial electrical resistance (TEER) of Caco-2 cell monolayers in a slightly acidic environment (pH 6.2). TMC polymers with higher degrees of quaternization were found to have a higher number of positive charges available for more electrostatic interactions between TMC and the cell membranes, resulting in a greater number of tight junctions that are opened to allow for paracellular movement of ions and thereby reduction of TEER. It was reported that Schiff's base reacted with methyl iodide and *N,N,N*-trimethyl chitosan, *N,N*-propyl-*N,N*-dimethyl chitosan, and *N*-furfuryl-*N,N*-dimethyl chitosan were obtained depending on the reaction conditions (Jia et al. 2001). The productivity, degree of quaternization, and water solubility of quaternized chitosan were influenced by the molecular weight of the chitosan sample.

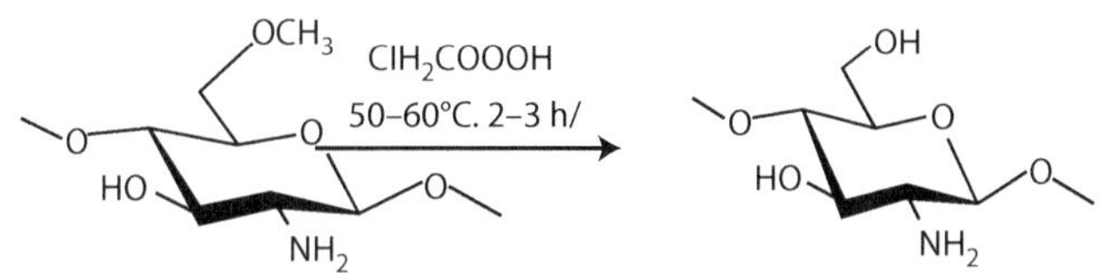

FIGURE 18.4 Synthesis of carboxymethyl chitosan.

18.5.3 *N*-Butyl Chitosan/*N*-Cetyl Chitosan

D,L-Lactic acid modified by *N*-butyl chitosan and *N*-cetyl chitosan had a significant difference in cell viability. D,L-Lactic acid modified by *N*-butyl chitosan could strengthen the survival ability of osteoblast. Compared with *N*-cetyl chitosan, *N*-butyl chitosan has a good biocompatibility (Cai et al. 2002).

18.5.4 *N*-(2-Hydroxyl) Propyl-3-Trimethyl Ammonium Chitosan

N-(2-hydroxyl)propyl-3-trimethyl ammonium chitosan chloride (HTCC) was synthesized by the reaction between glycidyl-trimethyl-ammonium chloride and chitosan. The HTCC nanoparticles were formed based on the ionic gelation process of HTCC and sodium tripolyphosphate (TPP). A model protein drug, bovine serum albumin (BSA), was incorporated into the HTCC nanoparticles (Xu et al. 2003).

18.5.5 *N*-Octyl-*O*-Sulfate Chitosan

N-Octyl-*O*-sulfate chitosan (OCS) contains long-chain alkyl groups as hydrophobic moieties and sulfated groups as hydrophilic moieties. Taxol was dissolved in the polymeric micelle by physical entrapment higher than that in water. Therefore, OCS micelle may be useful as a carrier for taxol (Zhang et al. 2003).

18.5.6 *N*-Alkyl-*N*-Methylene Phosphonic Chitosan

N-Methylene phosphonic chitosan has not only the hydrophilic group but also the hydrophobic group. Lauryl was connected with an amino group to obtain a new kind of chitosan derivate of *N*-lauryl-*N*-methylene phosphonic chitosan. The derivate is a compound of amphiphilic properties, such as surface activity typical for surfactants; this derivative opens new perspectives in the pharmaceutical and cosmetic fields (Ramos et al. 2003).

18.5.7 *N*-Monophosphonomethylated, *N,N*-Diphosphonomethylated, and *N*-Methylene Phosphonic Chitosan

A water-soluble chitosan derivative carrying phosphonic groups was synthesized by a one-step reaction, including *N*-monophosphonomethylated, *N,N*-diphosphonomethylated, and *N*-methylene phosphonic chitosan. These new derivatives take advantage of the known chelating ability of the phosphonic groups, especially for calcium, and opens new perspectives as biomaterial (Heras et al. 2001).

18.5.8 *N*-(2-Carboxyethyl) Chitosan

Low-molecular-weight chitosan, obtained from coarse ground crab, was treated with aqueous 3-chloropropionic acid in the presence of sodium hydrocarbonate, and the desired product was isolated from the system (Skorik et al. 2000).

18.5.9 5-Methylpyrrolidinone-Chitosan

In the synthesis of 5-methylpyrrolidinone-chitosan (MPCS), the amino groups of glucosamine units of the polysaccharide backbone were partially replaced by methylpyrrolidinone at position 5 (Gavini et al. 2008), and were loaded with the antibacterial metronidazole (MET).

FIGURE 18.5 Synthesis of thiolated chitosan.

18.5.10 THIOLATED CHITOSAN

The preparation method of thiolated chitosan is given in Figure 18.5. One gram of chitosan was dissolved in a 1% v/v acetic acid solution. The pH was then adjusted to pH 5 using 5 M NaOH. Two hundred milligram of 2-iminothiolane HCl was added to the chitosan solution and the reaction mixture was incubated for 6 h at room temperature under continuous stirring. To eliminate the residual 2-iminothiolane HCl and to isolate the polymer conjugate, the reaction mixture was dialyzed in tubings membrane, Sigma, for 3 days at 10°C in the dark against 5 mM HCl, then twice against the same medium but containing 1% NaCl. Therefore, the samples were exhaustively dialyzed against 1 mM HCl to adjust the pH of the polymer to 4. The polymer conjugates were lyophilized at –30°C and stored at 4°C (Bernkop-Schnurch et al. 2003). Polymers with thiol groups provide much higher adhesive properties than those polymers generally considered to be mucoadhesive. The enhancement of mucoadhesion can be explained by the formation of covalent bonds between the polymer and the mucus layer, which are stronger than noncovalent bonds. These thiolated polymers, or the so-called thiomers, are supposed to interact with cysteine-rich subdomains of mucus glycoproteins via disulfide exchange reactions. Various cationic thiomers (chitosan–cysteine, chitosan–thiobutylamidine, and chitosan–thioglycolic acid) and anionic thiomers (poly acylic acid–cysteine, poly(acrylic acid)–cysteamine, carboxymethyl cellulose–cysteine, and alginate–cysteine) have been studied so far (Jayakumar et al. 2007). Owing to the immobilization of thiol groups on these polymers, their mucoadhesive properties have been improved. The chemical modification is easy due to the primary amino group at the C-2 position and the carboxyl group at the C-6 position of each polymer subunit.

18.5.11 PHOSPHORYLATED CHITIN AND CHITOSAN

Recently, several methods have been reported to obtain phosphorylated derivatives of chitin and chitosan because of their interesting biological and chemical properties (Granja et al. 2000). They can exhibit bactericidal, biocompatible, bioabsorbable, osteoinductive, and metal chelating properties (Jung et al 1999). Phosphorylated chitin can be prepared using chitin with $H_3PO_4/Et_3PO_4/P_2O_5$/hexanol. Phosphorylated chitosan can also be prepared by reaction with $H_3PO_4/Et_3PO_4/P_2O_5$/hexanol as reported previously (Jayakumar et al. 2008). Phosphorylated chitosan is synthesized by graft copolymerization technique using 2-carboxethylphosphonic acid, chitosan with 1-ethyl-3-(3-dimethylaminopropyl) carbodiimide (EDC) catalyst. The reaction scheme is shown in Figure 18.6.

18.6 CHITIN AND CHITOSAN IN DIFFERENT FORMS

18.6.1 NANOPARTICLES

Different methods have been used to prepare chitosan nanoparticles. Selection of any of the methods depends on various factors such as particle size requirement, thermal and chemical stability of the active agent, reproducibility of the release kinetic profiles, stability of the final product, and residual toxicity associated with the final product. The methods are as follows.

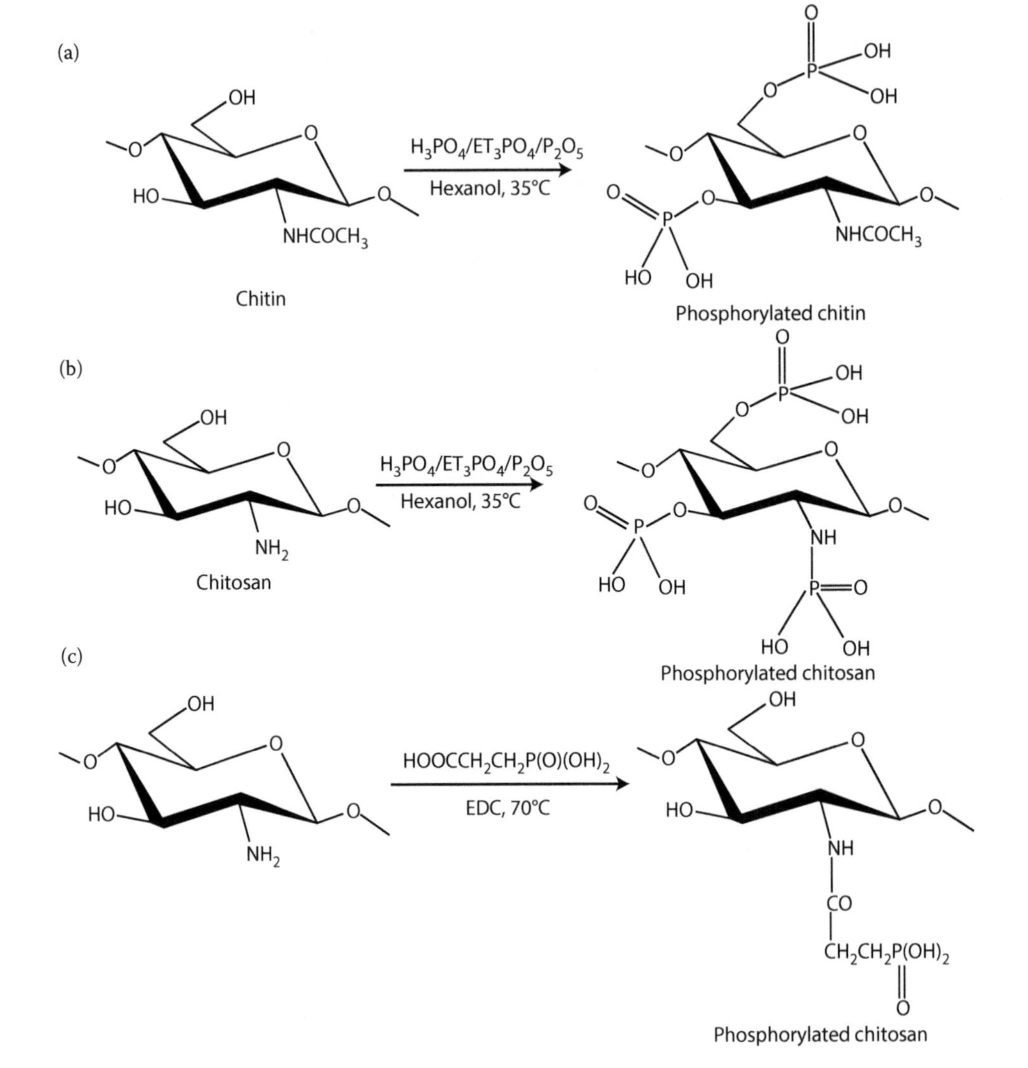

FIGURE 18.6 Synthesis of phosphorylated (a) chitin; (b) chitosan by $H_3PO_4/Et_3PO_4/P_2O_5$/hexanol; and (c) phosphorylated chitosan by grafting method.

18.6.1.1 Emulsion Cross-Linking

This method utilizes the reactive functional amine group of chitosan to cross-link with aldehyde groups of the cross-linking agent. In this method, water-in-oil (w/o) emulsion is prepared by emulsifying the chitosan aqueous solution in the oil phase. Aqueous droplets are stabilized using a suitable surfactant. The stable emulsion is cross-linked by using an appropriate cross-linking agent, such as glutaraldehyde, to harden the droplets. Nanoparticles are filtered and washed repeatedly with *n*-hexane followed by alcohol and then dried. Using this method, the size of the particles can be controlled by controlling the size of aqueous droplets. However, the particle size of the final product depends on the extent of cross-linking agent used while hardening in addition to the speed of stirring during the formation of emulsion (Akbuga and Durmaz 1994).

18.6.1.2 Coacervation/Precipitation

This method utilizes the physicochemical property of chitosan because it is insoluble in alkaline pH medium, but precipitates/coacervates when it comes in contact with an alkaline solution. Particles are produced by blowing the chitosan solution into an alkaline solution, such as sodium hydroxide, NaOH–methanol or ethanediamine, using a compressed air nozzle to form coacervate droplets. Separation and purification of particles was done by filtration/centrifugation followed by successive washing with hot and cold water. Varying compressed air pressure or spray-nozzle diameter controlled the size of the particles and subsequently using a cross-linking agent to harden particles controlled the drug release. In another technique (Berthod and Kreuter 1996), sodium sulfate solution was added dropwise to an aqueous acidic solution of chitosan containing a surfactant under stirring and ultrasonication for 30 min. Particles were purified by centrifugation and resuspended in demineralized water. Particles were cross-linked with glutaraldehyde. Particles produced by this method have better acid stability than observed by other methods.

18.6.1.3 Spray Drying

Drug is first dispersed in an aqueous acidic solution of chitosan. The next step is the addition of a suitable cross-linking agent. This solution is then atomized in a stream of hot air. This leads to the formation of small droplets from which solvent evaporates, leading to the formation of free flowing particles (He et al. 1999).

18.6.1.4 Emulsion Droplet Coalescence Method

This method utilizes the principle of emulsion cross-linking and precipitation. In the first step, a stable emulsion containing aqueous solution of chitosan along with the drug is produced in liquid paraffin oil. In the second step, another stable emulsion containing aqueous solution of chitosan in NaOH is prepared. Both emulsions are then mixed under high speed stirring, whereby the droplets of each emulsion collide at random and coalesce, thereby precipitating chitosan droplets to give small-sized particles (Tokumitsu et al. 1999).

18.6.1.5 Ionic Gelation

An aqueous acidic solution of chitosan is added dropwise under constant stirring to sodium TPP solution. Chitosan undergoes gelation due to the complexation between oppositely charged species and precipitates to form spherical particles (Polk et al. 1994).

18.6.1.6 Reverse Micellar Method

The surfactant is first dissolved in an organic solvent to produce reverse micelles. To this, an aqueous solution of chitosan and drug are added with constant vortexing to avoid any turbidity. The aqueous solution is kept in such a way as to keep the entire mixture in an optically transparent microemulsion phase. Additional amount of water may be added to obtain nanoparticles of larger size. To this solution, a cross-linking agent is added and the mixture kept overnight under constant stirring. The organic solvent is then evaporated to obtain the transparent dry mass. The material is dispersed in water, followed by the addition of a suitable salt, which helps to precipitate the surfactant. It is then centrifuged and the supernatant decanted, which contains the drug-loaded nanoparticles. The aqueous dispersion is immediately dialyzed through dialysis membrane for about 1 h and the liquid is lyophilized to dry powder (Leong and Candau 1982).

18.6.2 Microspheres

As a drug carrier, chitosan helps to overcome certain adverse characteristics of drugs such as insolubility and hydrophobicity, but the semicrystalline powder does not lend itself to direct compression. Nevertheless, chitosan powders have been evaluated in direct compression tests, and the formulations so far developed include excipients to facilitate compression. For example, commercial chitosan tablets for weight control contain magnesium stearate as a binder with negative consequences on the efficacy of chitosan.

18.6.2.1 Preparation of Microspheres by Spray Drying

Spray drying includes four sequential stages: atomization through a spray nozzle, contact of the sprayed feed with warm air, drying of the droplets, and collection of the solid chitosan. Chitosan solutions with drug can be fed to a spray drier at a slightly acidic pH. The size of the particles is influenced by various process parameters such as size of the nozzle, rate of feeding, and inlet air temperature. One point concerning the exposure of the sample to hot air needs to be explained. The inlet air temperature is measured prior to flowing into the drying chamber and may be set at 160 °C or higher; however, the gradient between the wet surface and unsaturated gas actually leads to evaporation at much lower temperatures (Liu et al. 2008).

18.6.2.2 Preparation of Microspheres by Multiple Emulsion/Solvent Evaporation and Coacervation Methods

This multiple emulsion technique includes three steps: (1) preparation of a primary o/w emulsion in which the oil-dispersed phase constitutes CH_2Cl_2 and the aqueous continuous phase is a mixture of 2% v/v acetic acid solution:methanol (4/1 v/v) containing chitosan (1.6%) and Tween (1.6 w/v); (2) multiple emulsion formation with mineral oil (oily outer phase) containing Span 20 (2% w/v); and (3) evaporation of aqueous solvents under reduced pressure. Details can be obtained from various publications. Chemical cross-linking is an option of this method if the cross-linking agent is added just after the emulsion formation; enzymatic cross-linking can also be performed. Physical cross-linking may take place to a certain extent if chitosan is exposed to high temperature. The emulsion technique is convenient when the drug is particularly sensitive to certain parameters connected to the spray drying. The emulsion technique may be associated to cross-linking or other treatments of the microspheres (Il'ina and Varlamov 2005).

One of the examples is the preparation of microspheres of polyacrylamide-grafted-chitosan to encapsulate indomethacin, a nonsteroidal anti-inflammatory drug (NSAID). The microspheres were produced by the water-in-oil emulsion technique and cross-linking with glutaraldehyde. The release of indomethacin is dependent on the degree of cross-linking and on the amount of drug loading. This was further supported by the calculation of drug-diffusion coefficients (Prabaharan and Mano 2005).

18.6.3 HYDROGELS

Gel materials are utilized in a variety of technological applications and are currently investigated for advanced exploitations such as the formulation of "intelligent gels" and the synthesis of "molecularly imprinted polymers." Chitosan hydrogels have been prepared in a number of different shapes, geometries, and formulations that include liquid gels, powders, beads, films, tablets, capsules, microspheres, microparticles, sponges, nanofibrils, textile fibers, and inorganic composites (Denkbas and Ottenbrite 2006). In each preparation, chitosan is either physically associated or chemically cross-linked to form the hydrogel. Our discussion below will focus on these two distinct hydrogel engineering approaches.

18.6.3.1 Physical Association Networks

To satisfy the requisite features of a hydrogel, the chitosan polymer network must satisfy two conditions: (1) interchain interactions must be sufficiently strong to form semipermanent junction points in the molecular network and (2) the network should promote the access and residence of water molecules inside the polymer network. Gels that meet these demands may be prepared by noncovalent strategies that capitalize on electrostatic, hydrophobic, and hydrogen bonding forces between polymer chains (Berger et al. 2004). Figure 18.7 shows the schematics of four major physical interactions (i.e., ionic, polyelectrolyte, interpolymer complex, and hydrophobic associations) that lead to the gelation of a chitosan solution. Because the network formation by all of these interactions is purely physical, gel formation can be reversed.

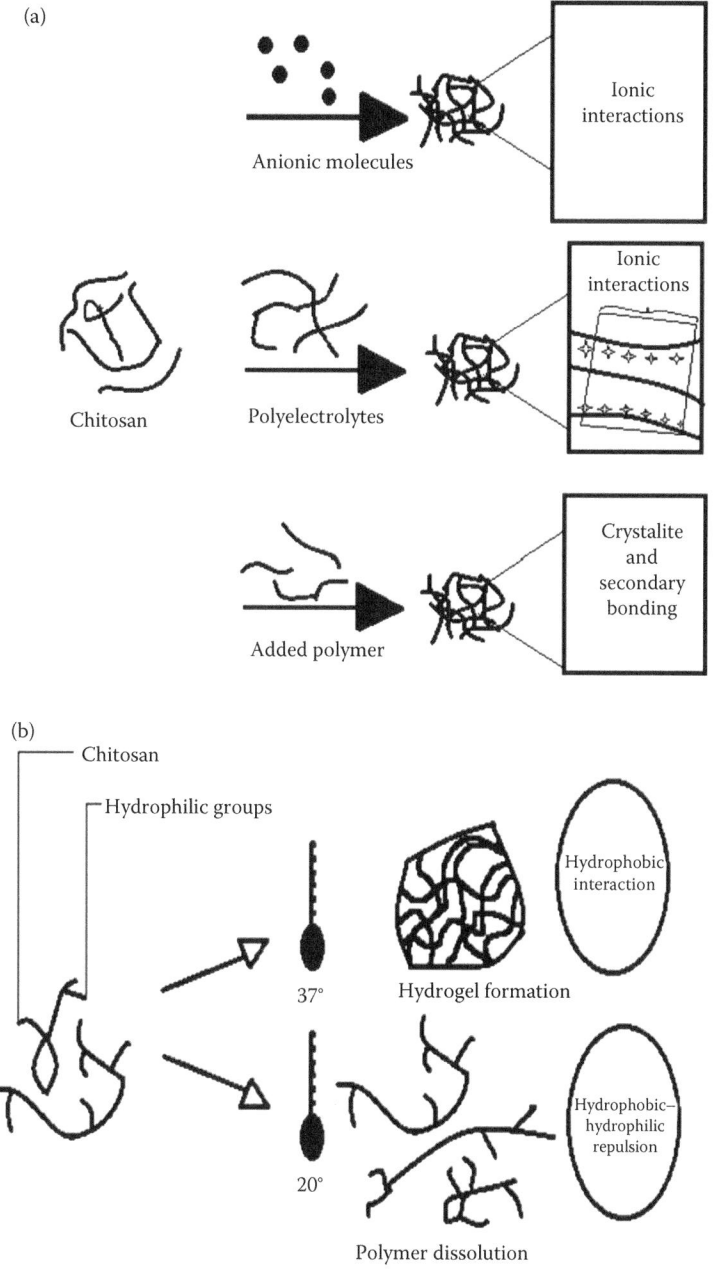

FIGURE 18.7 Schematic representation of chitosan-based hydrogel networks derived from different physical associations: (a) networks of chitosan formed with ionic molecules; polyelectrolyte polymer and neutral polymers; (b) thermoreversible networks of chitosan graft copolymer resulting in semisolid gel at body temperature and liquid below room temperature.

18.6.3.2 Cross-Linked Networks

Although physically bonded hydrogels have the advantage of gel formation without the use of cross-linking entities, they have limitations. It is also difficult to precisely control the physical gel pore size, chemical functionalization, and degradation or dissolution, leading to inconsistent performance *in vivo*. Alternatively, robust chitosan hydrogels can be produced using irreversible networks.

Polymeric chains of these hydrogels are covalently bonded together by using small cross-linker molecules, secondary polymerizations, or irradiation chemistry. Most of these linker molecules react with the primary amines of chitosan and form irreversible inter- or intramolecular bridges among the chitosan chains. Covalently cross-linked hydrogels are also obtained by attaching photoreactive or enzyme-sensitive molecules on the chitosan, followed by their subsequent exposure to UV or sensitive enzymes, respectively. The properties of cross-linked hydrogels depend mainly on their cross-linking density and the ratio of moles of cross-linker molecules to the moles of polymer repeating units (Hennink and Van Nostrum 2002).

18.6.4 FILMS

The choice of biomaterials suitable for forming the carrier film matrix and the barrier films were dictated by various factors, namely, (a) compatibility with the gastric environment, (b) stability during the time of drug delivery, (c) adequate mechanical properties, (d) ease of fabrication and cost, and (e) no appreciable swelling in water and softening point above 37°C. Several polymers were found to fill all or most of the above criteria and used to prepare carrier films. Often, some plasticizers and excipients were used along with the base polymer to impart a suitable degree of flexibility and facilitate the diffusion of the drug from the films, respectively. Chitosan possesses good film-forming properties. *In vitro* and *in vivo* degradation profiles of chitin and chitosan films by solution casting method using specimens of varied deacetylation degrees, namely, 68.8%, 73.3%, 84.0%, 90.1%, and 100% was reported; the thickness of the films was 150 μm (Hoffman 2002). The equilibrated water content of the films decreased with increase in deacetylation, while the tensile strength of the water-swollen films increased with increasing deacetylation.

Recently, there has been a growing interest in grafting of vinyl monomers onto chitosan for biomedical and industrial applications. This chemical combination of natural and synthetic polymers yields new materials that could have desirable properties, including biodegradability. Chitosan films grafted by 2-hydroxyethyl methacrylate (HEMA) under Co γ-irradiation was prepared. It was observed that the tensile properties were largely affected by the graft levels (Qiu and Park 2001). Design and evaluation of chitosan/ethylcellulose mucoadhesive bilayered devices for buccal drug delivery was also carried out (Peppas et al. 2000). The investigations highlight preparation of devices comprising a drug-containing mucoadhesive layer and a drug-free backing layer (Figure 18.8) by two different methods. Bilaminated films were produced by casting/solvent evaporation technique, and bilayered tablets were obtained by direct compression. The mucoadhesive layer was composed of a mixture of drug and chitosan with or without an anionic cross-linking polymer (polycarbophil, sodium alginate, and gellan gum), and the backing layer was made of ethylcellulose. It was clearly

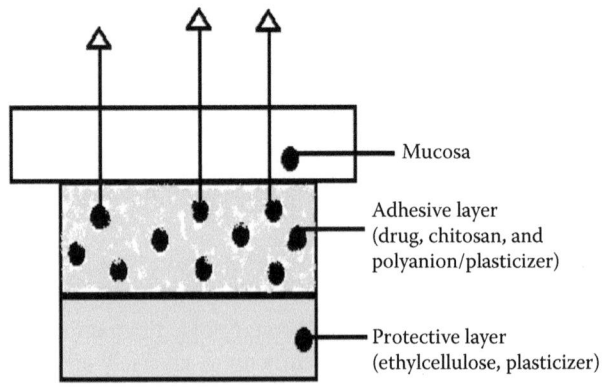

FIGURE 18.8 Schematic explanation of the structure of mucoadhesive bilayers.

evident that hydrophilic chitosan was easily laminated onto hydrophobic ethylcellulose and that a perfect binding between the mucoadhesive and the backing layers was achieved.

18.6.5 FIBERS

The subject of fiber production from chitin and chitosan has been reviewed from various viewpoints because of inherent interest in the exploitation of chitin as a textile material. Fiber production, however, has appeared since the beginning to be a challenging and difficult task. As it became established for cellulose, the xanthate process was developed for chitin, but the chitin fiber had poor tensile strength (Ganta et al. 2008).

Nontoxic and noncorrosive solvent systems can be used when chitosan is considered instead of chitin for the manufacture of fibers. Many articles and patents address the spinning of chitosan solutions. Although there was interest in taking advantage of the liquid crystalline nature of chitosan, no great improvement of the mechanical properties was obtained. The same can be said for alkyl chitin fibers (Majeti and Kumar 2000). Cross-linking agents have been proposed for improvement of chitin fibers in the wet state. Epichlorohydrin is a convenient base-catalyzed cross-linker to be used in 0.067 M NaOH (pH 10) at 40°C. The wet strength of the fibers was considerably improved, whereas cross-linking had a negligible effect on the dry fiber properties (Tamura et al. 2002). Of course, the more extended the chemical modification, the more unpredictable the biochemical characteristics and effects *in vivo*. Every modified chitin or modified chitosan fiber should be studied in terms of biocompatibility, biodegradability, and overall effects on the wounded tissues. The present trend is to coat other polysaccharide fibers with a film of chitosan or modified chitosan to impart novel characteristics to the textile. For example, a detailed study has been made on alginate fibers coated with various chitosans that actually form polyelectrolyte complexes (PECs) and, under certain conditions, improve the general characteristics of the fiber while keeping the biochemical significance of chitosan (Knilla et al. 2004). Another study described wet-spun chitosan-collagen fibers, their chemical modification, and their blood compatibility.

The summary of different types of chitosan-based drug delivery systems are given in Figure 18.9.

18.7 PHARMACEUTICAL AND DRUG DELIVERY APPLICATIONS OF CHITOSAN AND ITS DERIVATIVES

18.7.1 DIRECT COMPRESSIBLE EXCIPIENTS AND AS BINDER

Chitosan has an excellent property as excipients for direct compression of tablets where the addition of 50% chitosan result in rapid disintegration. The degree of deacetylation determines the extent of moisture absorption. Chitosan, higher than 5% was superior to corn starch and microcrystalline cellulose as a disintegrant. The efficiency was dependent on chitosan crystallinity, degree of deacetylation, molecular weight, and particle size (Ritthidej et al. 1994). Chitosan is found to be an excellent tablet binder as compared to other excipients with the rank order correlation for binder efficiency. Hydroxypropyl methyl cellulose > chitosan > methyl cellulose > sodium carboxymethyl cellulose (Upadrashta et al. 1992).

18.7.2 CONTROLLED RELEASE DOSAGE FORMS

Chitosan and chitosan derivatives, in combination with other excipients, are used to give zero-order release profile. The rate of release of drug from tablet was found to some degree to be directly related to the amount and type of chitosan used. It was found that the addition of sodium alginate to the tablet preparation gave the tablet an extended release property (Miyazaki et al. 1990). Similar results had been obtained that suggested that citric acid can gel the chitosan and thereby impart sustained release properties. Film coating of theophylline granules with PEC of chitosan and sodium

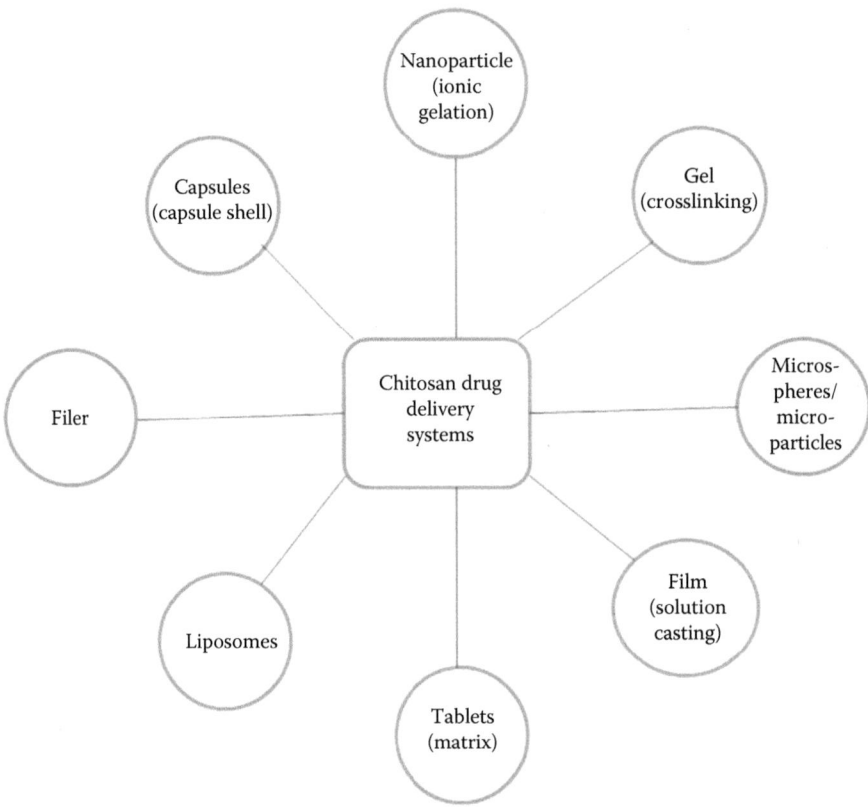

FIGURE 18.9 Different types of chitosan-based drug delivery systems.

TPP has produced a controlled release system. The rate of release of drug could be controlled by pH. At low pH value, the reduced charge of the anionic TPP reduces the electrostatic interaction in the complex and the film network loosens. The influence of physicochemical characteristics of drugs on their release characteristics from chitosan maleate matrix tablet was studied and it was found that the drug solubility, degree of ionization, and the molecular weight of the drug are features of importance (Akbuga 1993).

Thiolated chitosan has excellent cohesive properties. The reduced thio functions on the chitosan backbone enable thiolated chitosan not only to form disulfide bond with mucus glycoproteins but also to form inter- as well as intramolecular disulfide bonds. Such a cross-linking of the polymeric chains results in high stability of the drug carrier system based on thiolated chitosan. To improve therapeutic efficacy, clotrimazole matrix tablets based on chitosan–thioglycolic acid conjugate and chitosan–4-thiobutylamidine (chitosan–TBA) conjugates were quantified. Both the thiolated chitosan tablets remained stable during the whole experiment (6 h) and no disintegration could be observed. However, only the chitosan–TBA conjugate was able to guarantee a significant delay in the drug release in comparison to unmodified chitosan, leading to sustained release over a much longer time period (Kast et al. 2002). The release profile of salmon calcitonin matrix tablets based on chitosan–TBA conjugate was determined. A pseudo-zero-order release profile of salmon calcitonin during the first 8 h was observed in an artificial intestinal fluid. During the experiment, the tablet swelled continuously, maintaining a good cohesiveness and releasing an active agent via a controlled diffusion process (Guggi et al. 2003). These release studies, in which a peptide drug was liberated, formed a thiolated chitosan matrix system, permitting information concerning the chemical events within the formulation to be gained. Strong unintended interaction between the polymeric matrix system and the peptide drug could be excluded according to controlled and sustained release profile. The possibility of using mixtures and/or PECs from both chitosan-alginate and chitosan-carrageenan system have evaluated as prolonged drug release systems. It has been found that the chitosan-alginate system is better than the chitosan-carrageenan system as prolonged drug release matrix because the drug release is controlled at low percentage in the formulation, the mean dissolution time is high, and a different dissolution profile could be obtained by changing the mode of inclusion of the polymers (Tapia et al. 2004). Good arrangement between t_d and k_f/k_r values for the chitosan-alginate system was found, which means that the swelling behavior of the polymer controlled the drug release from the matrix. In case of the chitosan-carrageenan system, the high capacity of carrageenan promotes the entry of water into the tablet and therefore the main mechanism of drug release would be disintegration instead of swelling of the matrix.

18.7.3 OPHTHALMIC DRUG DELIVERY

Because of its low toxicity, chitosan exhibits favorable biological behavior, such as bioadhesion, permeability-enhancing properties, and interesting physicochemical characteristics, which make it a unique material for the design of ocular drug delivery vehicles. The potential of chitosan-based systems is for improving the retention and biodistribution of drugs applied topically onto the eye. Chitosan-based formulations for ophthalmic drug delivery are chitosan gels, chitosan-coated colloidal systems, and chitosan nanoparticles. Chitosan-based colloidal systems are found to work as transmucosal drug carriers, either facilitating the transport of drugs to the inner eye (chitosan-coated colloidal systems containing indomethacin) or their accumulation into the corneal/conjunctival epithelia (chitosan nanoparticles containing cyclosporin) (Liu et al. 2001). The microparticulate drug carrier (microspheres) is a promising means of the topical administration of acyclovir to the eye (Gental et al. 1997).

18.7.4 NASAL DRUG DELIVERY

The nasal mucosa presents an ideal site for bioadhesive drug delivery systems. Chitosan-based drug delivery systems, such as microspheres, liposomes, and gels, have been demonstrated to have good bioadhesive characteristics and swell easily when in contact with the nasal mucosa, increasing the bioavailability and residence time of the drugs to the nasal route. Various chitosan salts such as chitosan lactate, chitosan aspartate, chitosan glutamate, and chitosan hydrochloride are good candidates for nasal sustained release of vancomycin hydrochloride (Thanou et al. 2007). Nasal administration of diphtheria toxoid (DT) incorporated into chitosan microparticles results in a protective systemic and local immune response against DT with enhanced IgG production. Nasal formulations have induced significant serum IgG responses similar to secretory IgA levels, which are superior to parenteral administration of the vaccine (Hamman et al. 2003). Research showed that bioadhesive chitosan microspheres of pentazocine for intranasal systemic delivery significantly improved the bioavailability with sustained and controlled blood level profiles compared to intravenous and oral administration (Kean et al. 2005). Nasal absorption of insulin after administration into chitosan powder were found to be the most effective formulation for nasal drug delivery of insulin in sheep compared to chitosan nanoparticles and chitosan solution.

18.7.5 BUCCAL DRUG DELIVERY

An ideal buccal delivery system should stay in the oral cavity for few hours and release the drug in a unidirectional way toward the mucosa in a controlled or sustained-release fashion. Mucoadhesive polymers prolong the residence time of the device in the oral cavity, while bilayered devices ensure that the release of the drug occurs in a unidirectional way. Chitosan is an excellent polymer to be used for buccal delivery because it has muco/bioadhesive properties and can act as an absorption enhancer (Guggi and Bernkop 2003). Directly compressible bioadhesive tablets of ketoprofen containing chitosan and sodium alginate in the weight ratio of 1:4 showed sustained release 3 h after intraoral (into sublingual site of rabbits) drug administration. Buccal tablets based on chitosan microspheres containing chlorhexidine diacetate gives prolonged release of the drug in the buccal cavity, improving the antimicrobial activity of the drug. Chitosan microparticles with no drug incorporated have antimicrobial activity due to the chitosan. The buccal bilayered devices (bilaminated films, palavered tablets) using a mixture of drugs (nifedipine and propranolol hydrochloride) and chitosan, with or without anionic cross-linking polymers (polycarbophil, sodium alginate, and gellan gum) has promising potential for use in controlled delivery in the oral cavity. Bioadhesive tablets of nicotine containing 0–50% w/w glycol chitosan gives good adhesion (Sandri et al. 2005).

18.7.6 Vaginal Drug Delivery

Chitosan-based vaginal tablets containing metronidazole were prepared by directly compressing the polymer, loosely cross-linked with glutaraldehyde, together with sodium alginate with or without microcrystalline cellulose. Tablets showed adequate release properties in pH 4.8 and 7 and gave low values of swelling index. These tablets also showed comparatively good adhesive properties (El-Kamel et al. 2002). In another study, the mucoadhesive and permeation enhancing properties of four different chitosan derivatives: MPCS, two low-molecular-mass chitosans, and a partially reacetylated chitosan were evaluated via the vaginal and buccal mucosa using acyclovir as model drug (Sandri et al. 2004). Methylpyrrolidinone-chitosan showed the highest mucoadhesive and permeation enhancing properties in both vaginal and buccal environments. The capability of enhancing the permeation/penetration of acyclovir was decreased by partial depolymerization of chitosan and disappeared after partial reacetylation. The antimicrobial properties of chitosan, however, might have a negative impact on the vaginal microflora (Raafat and Sahl 2009). Its vaginal use for treatment of chronic diseases has therefore to be seen with caution.

18.7.7 Periodontal Drug Delivery

Local delivery of drugs and other bioactive agents directly into the periodontal pocket has received increased attention. For moderate to severe periodontal diseases, antimicrobial agents are used to eradicate and/or suppress the plaque bacteria. However, systemic administration of these drugs requires frequent dosing to maintain the drug concentrations at the therapeutic level in the plasma causing poor patient compliance. Chitosan itself possesses antibacterial activity. Muco/bioadhesive polymers increase the residence time of the formulation in the oral cavity. This enhances drug penetration, localizes the drug for local therapy, targets the diseased tissue, and improves efficacy and acceptability. The antibacterial activity of chitosan is because of the electrostatic interactions between the amine groups of chitosan and the anionic sites on the bacterial cell wall because of the presence of carboxylic acid residues and phospholipids. Chitosan inhibits the adhesion of *Candida albicans* to buccal cells showing antifungal activity. Research shows that a monolayer and multilayered film of chitosan/PLGA containing ipriflavone prolongs drug release for 20 days *in vitro*.

18.7.8 Transdermal Delivery

Chitosan has good film-forming properties. The drug release from the devices is affected by the membrane thickness and cross-linking of the film (Sakkinen et al. 2004). Chitosan-alginate PEC has been prepared *in situ* in beads and microspheres for potential applications in packaging, controlled release systems, and wound dressings. Chitosan gel beads are a promising biocompatible and biodegradable vehicle for the treatment of local inflammation for drugs, such as prednisolone, which showed sustained release action improving therapeutic efficacy. The rate of drug release was found to be dependent on the type of membrane used (Tozaki et al. 1997). A combination of chitosan membrane and chitosan hydrogel containing lidocaine hydrochloride, a local anesthetic, is a good transparent system for controlled drug delivery and release kinetics.

18.7.9 PARENTERAL DRUG DELIVERY

As highly purified low-molecular-mass chitosan was shown neither to be toxic nor has hemolytic property when given intravenously, its use in injectable preparations received considerable attention in the last few years. In controlled release technology, biodegradable polymeric carriers offer potential advantages for prolonged release of low-molecular-weight compounds to macromolecular drugs. The susceptibility of chitosan to lysozyme makes it biodegradable (Nordtveit et al. 1994). Pharmacokinetic and tissue-distribution studies were performed in mice using fluorescent glycol chitosan and N-succinyl-chitosan. Both chitosans demonstrated a good retention in blood circulation and a slight accumulation in tissues, suggesting that chitosan is an effective carrier for drugs that are excreted rapidly. In another study, hydrophobically modified glycol chitosan was synthesized with 5-β-cholanic acid groups, which were subsequently used to prepare nanoparticles with incorporated RGD peptide (Arg–Gly–Asp). Intratumoral administration of RGD-loaded chitosan nanoparticles demonstrated a substantially decreased tumor growth as compared to the RGD peptide administered intravenously.

18.7.10 INTRAVESICAL DRUG DELIVERY

Investigations of mucoadhesive properties of chitosan and thiolated chitosan nanoparticles on the intravesical mucosa were carried out to prolong the residence time of instilled drugs in urinary bladders (Kim et al. 2008). It was shown that thiolated chitosan nanoparticles might be a useful tool for local intravesical drug delivery, which increases the residence time of the drug and enables sustainable delivery for an extended period of time. Evidence for this theory could at least be provided by *in vivo* studies in rats.

18.7.11 VACCINE DELIVERY

Although chitosan was already shown in the early 1990s to stimulate IgM production of human lymphocytes in serum-free culture (Barthelmes et al. 2011), it took several years until the first vaccine delivery systems were developed. Jabbal-Gill and coworkers were likely the first to provide evidence for a successful nasal immunization against *Bordetella pertussis* in mice combining different antigens with chitosan. In the following, various chitosan-based carriers have been designed and evaluated for nasal delivery of antigens showing valid levels of both systemic and local immune response (Garmise et al. 2007). Utilizing trimethylated chitosan nanoparticles for nasal administration even led to an increased immune response that could be further improved by the incorporation of certain immunopotentiators (Borges et al. 2008). In the following, oral vaccine delivery systems based on chitosan were also developed. The uptake of chitosan microparticles by the epithelium of the murine Peyer's patches was also demonstrated (Bal et al. 2012). Similar to nasal vaccine delivery systems, also in case of oral delivery trimethylated-chitosan exhibiting in contrast to unmodified chitosan, an intrinsic adjuvant effect on dendritic cells showed comparatively greater potential (Slütter et al. 2009).

18.7.12 PERORAL DRUG DELIVERY

As chitosan and most of its derivatives have a mucoadhesive property, a presystemic metabolism of peptides can be strongly reduced, leading to a strongly improved bioavailability of many perorally given peptide drugs, such as insulin, calcitonin, and buserelin. Unmodified chitosan has a permeation-enhancing effect for peptide drugs. A protective effect for polymer-embedded peptides toward degradation by intestinal peptidases can be achieved by the immobilization of enzyme inhibitors on the polymer. The mucoadhesive property of chitosan gel can be enhanced threefold to sevenfold by admixing chitosan-glyceryl mono-oleate. Drug release from the gel followed a matrix diffusion controlled mechanism. Nifedipine embedded in a chitosan matrix in the form of beads have prolonged release of drug compared to granules (Kose-Ozkan et al. 2010).

18.7.13 GENE DELIVERY

Chitosan is a good candidate for gene delivery system because positively charged chitosan can be complexed with negatively charged DNA (Richardson et al. 1999). Chitosan can effectively bind DNA and protect it from nuclease degradation (Cui and Mumper 2001). It has the advantage of not necessitating sonication and organic solvents for its preparation, thereby minimizing possible damage to DNA during complexation. DNA-loaded chitosan microparticles were found to be stable during storage. The application of DNA–chitosan nanospheres has advanced *in vitro* DNA transfection research and the data have showed their usefulness for gene delivery (Corsi et al. 2003).

Chitosan and lactosylated chitosan vectors were investigated for their transfection efficiency of in HeLa cells in the presence of 10% fetal salt serum and was found comparable to that of another cationic polymer, polyethyleneimine (Erbacher et al. 1998). It was shown that these vectors were poorly efficient in transfecting galactose-specific membrane lectin (HepG2 cells). This was probably caused by the aggregation of the complexes due to decrease of the zeta potential after lactocylation, accompanied by a lower affinity of the lactosylated polymer to DNA. The presence of chloroquine, a weak base, prevents lysosome acidification, not improving transfection efficiency (Pouton and Seymour 1998). Galactocylated chitosan graft-polyethylene glycol is utilized as a DNA vector (Park et al. 2001). DNA complexed with galactocylated chitosan graft-polyethylene glycol (GCP) is stable

TABLE 18.1
Application Summary of Chitosan

Sl. No.	Delivery System	Dosage Form	Drugs
1	Opthalmic drug delivery system	Chitosan gel Chitosan microparticle Chitosan nanoparticles Chitosan ophthalmic film	Indomethacin Acyclovir Cyclosporine
2	Nasal drug delivery system	Microparticles Powder Chitosan nanoparticles Chitosan solution	Diphtheria toxoid (DT), pentazocine, insulin
3	Buccal drug delivery	Mucoadhesive tablets Bilaminated films Palavered tablets	Ketoprofen, nicotine, nifedipine, and propranolol hydrochloride
4	Peridontal drug delivery	Monolayer and multilayered film	Ipriflavon
5	Gastrointestinal drug delivery	Microspheres Microcapsules Chitosan granules	Metoclopramide and glipizide Melatonin Prednisolone
6	Peroral drug delivery	Chitosan beads Chitosan granules	Peptide drugs, such as insulin, calcitonin, buserelin, and nifedipine
7	Intestinal drug delivery	Microspheres Microcapsules	Insulin
8	Colon drug delivery	Microspheres Microcapsules	5-Fluorouracil and insulin
9	Transdermal delivery	Chitosan gel beads	Prednisolone
10	Vaginal drug delivery	Vaginal tablets	Metronidazole, acriflavine

TABLE 18.2
Modification of Chitosan and Their Applications

Sl. No.	Chitosan Modification	Drug Delivery and Biomedical Applications	References
1	Chitosan blends with hydrophilic polymers, including polyvinyl alcohol (PVA), polyethylene oxide (PEO), and polyvinyl pyrrolidone (PVP)	Oral gingival delivery	Cynthia et al. (2003)
2	Ionic cross-linking of chitosan with hydroxypropyl methyl cellulose phthalate (HPMCP) as a pH-sensitive polymer	Oral insulin delivery	Makhlof et al. (2011)
3	Carboxymethyl chitosans (CMC)	Oral delivery of protein drugs	Chen et al. (2004)
4	Chitosan–TBA (chitosan–4-thiobutylamidine) conjugates	Nasal peptide drug delivery	Kraulan et al. (2006)
5	5-Methylpyrrolidinone-chitosan	Delivery of metronidazole	Perioli et al. (2008)
6	Cholesterol-modified chitosan conjugate	Novel carrier of epirubicin for anticancer therapy	Wang et al. (2007)
7	Methoxy poly(ethylene glycol)-grafted-chitosan (mPEG-g-CS) conjugates	Sustained release carrier of methotrexate	Aral and Akbuga (1998)
8	Methylated N-(4-N,N-dimethylaminobenzyl) chitosan	Absorption enhancer and it showed good paracellular permeability through Caco-2 cell	Kowapradit et al. (2008)
9	Albumin-chitosan microparticles	Ocular drug delivery	Addo et al. (2010)
10	N-Trimethyl chitosan–mono-N-carboxymethyl chitosan	Mucosal delivery of vaccines	Sayin et al. (2009)
11	Alginate and N,O-carboxymethyl chitosan conjugate	Site specific protein delivery in the intestine	Chen and Tan (2006)
12	Chitosan/polyguluronate nanoparticles	siRNA delivery	Lee et al. (2009)
13	Chitosan–sodium deoxycholate nanoparticles	Plasmid delivery system	Cadete et al. (2012)
14	N-Octyl-O-sulfate chitosan-modified liposomes	Delivery of docetaxel for tumor targeting	Guowei et al. (2012)
15	Quaternary ammonium–chitosan conjugates	Enhanced intestinal drug permeation	Zambito et al. (2008)

and protected against enzyme degradation with DNA. However, the transfection efficiency using GCP/DNA complexes is very low, mainly because of the interaction with plasma, leading to the dissociation of GCP/DNA complexes. This study was challenged by the data of Gao et al. in different cell lines, which supported the positive use of low-molecular-weight chitosan as a modified vector for gene therapy (Gao et al. 2003). Another approach to increase the transfection rate was to use chitosan as a vector, consisting of preparing trimethylated chitosan oligomers through quaternization of oligomeric chitosan (Thanou et al. 2002). Chitosan/DNA/ligand complexes are used to increase the transfection rate. This system is based on transferring receptor-mediated endocytosis to carry exogenous DNA into the cells to yield a higher transfection rate. Compared to nonmodified chitosan, the method results in fourfold enhanced transfection efficiency in HeLa cells and several fold in H3K293 (Sato et al. 2001).

Yau and Liu have proposed a DNA/N-dodecylated chitosan complex and salt-induced gene delivery (CS-12) from dodecyl bromide and chitosan (DNA-CS-12-PEC). Incorporating dodecylated hydrolyze by DNase can be broken into fragments. However, DNA dissociated from the complex is well protected and remains intact due to the protection from DNase offered by alkylated chitosan (Liu et al. 2001). A hydrophobic modification of chitosan ($Mv = 7.0 \times 10^4$, degree of deacetylation 80%) was carried out with deoxycholic acid (Li et al. 2002). Deoxycholic acid is a main component of bile acid, which is biologically the most detergent-like molecule in the body. Since bile acid can assemble in water, the deoxycholic acid–modified chitosan also self-associates to form missiles of a mean diameter of 160 nm. The transfection of COS-1 cells (Monkey kidney) with chitosan self-aggregated DNA complexes were examined using the plasmid encoding chlormphenicol acetyl transferase. The transfection efficiency of the complex is enhanced compared to that achieved by naked DNA but is lower than that achieved by lipofectamine (Liu and Yao 2002).

18.7.14 Wound-Healing Properties

The efficacy of chitosan in the promotion of wound healing was first reported in 1978. Chitosan acetate films, which were tough and protective, had the advantage of good oxygen permeability, high water absorptivity, and slow enzymatic degradation, thereby avoiding the need for repeated application. Treatment of various dock tissues with chitosan solutions resulted in the inhibition of fibroplasias with enhanced tissue regeneration (Weng et al. 2008).

18.7.15 Tissue Engineering

γ-Poly(glutamic acid) (γ-PGA), a hydrophilic and biodegradable polymer, was chosen to modify chitosan matrices to produce a γ-PGA/chitosan composite biomaterial. This has hydrophilic, cytocompatible, and mechanical properties, and are very promising biomaterials for TE applications (Hsieh et al. 2005). Chitosan/glycerophosphate salt (GP) hydrogels scaffolds have been reported for neural TE (Crompton et al. 2007). The "Application Summary of Chitosan" and "Modification of Chitosan and Their Applications" were given in Tables 18.1 and 18.2, respectively.

18.8 CONCLUSIONS

Chitosan has prompted the continuous movement for the development of safe and effective drug delivery systems because of its unique physicochemical and biological characteristics. The primary hydroxyl and amine groups located on the backbone of chitosan allow for chemical modification to control its physical properties. When the hydrophobic moiety is conjugated to a chitosan molecule, the resulting amphiphile may form self-assembled nanoparticles that can encapsulate a quantity of drugs and deliver them to a specific site of action. Chemical attachment of the drug to the chitosan throughout the functional linker may produce useful prodrugs, exhibiting the appropriate biological activity at the target site. Mucoadhesive and absorption enhancement properties of chitosan increase the *in vivo* residence time of the dosage form in the gastrointestinal tract and improve the bioavailability of various drugs. As a result of the physical, chemical, and biological properties, chitosan has been used in different formulations for drug and gene delivery in the GI tract. We anticipate that more uses of chitosan will be forthcoming as additional derivatives are synthesized and newer formulations are developed.

REFERENCES

Addo, R.T., A. Siddig, R. Siwale et al. 2010. Formulation, characterization and testing of tetracaine hydrochloride-loaded albumin-chitosan microparticles for ocular drug delivery. *J. Microencapsul.* 27:95–104.

Akbuga, J. 1993. The effect of physiochemical properties of a drug on its release from chitosan maleate tablets. *Int. J. Pharm.* 100:257–261.

Akbuga, J. and G. Durmaz. 1994. Preparation and evaluation of crosslinked chitosan microspheres containing furosemide. *Int. J. Pharm.* 11:217–222.

Aral, C. and J. Akbuga. 1998. Alternative approach to the preparation of chitosan beads. *Int. J. Pharm.* 168:9–15.

Badawy, M.E.I., E.I. Rabea, T.M. Rogge, C.V. Stevens, G. Smagghe, and W. Steurbaut. 2004. Synthesis and fungicidal activity of new *N,O*-acyl chitosan derivatives. *Biomacromolecules* 5:589–595.

Bal, S.M., B. Slütter, R. Verheul, J.A. Bouwstra, and W. Jiskoot. 2012. Adjuvanted, antigen loaded *N*-trimethyl chitosan nanoparticles for nasal and intradermal vaccination: Adjuvant- and site-dependent immunogenicity in mice. *Eur. J. Pharm. Sci.* 45:475–481.

Barthelmes, J., G. Perera, J. Hombach, S. Dünnhaupt, and A. Bernkop-Schnürch. 2011. Development of a mucoadhesive nanoparticulate drug delivery system for a targeted drug release in the bladder. *Int. J. Pharm.* 416:339–345.

Berger, J., M. Reist, J.M. Mayer, O. Felt, and R. Gurny. 2004. Structure and interactions in chitosan hydrogels formed by complexation or aggregation for biomedical applications. *Eur. J. Pharm. Biopharm.* 57:35–52.

Bernkop-Schnurch, A., M. Hornof, and T. Zoidl. 2003. Thiolated polymers-thiomers: Synthesis and *in vitro* evaluation of chitosan-2-iminothiolane conjugates. *Int. J. Pharm.* 260:229–237.

Berthod, A. and J. Kreuter. 1996. Chitosan microspheres–improved acid stability and change in physicochemical properties by cross-linking. *Proc. Int. Symp. Control. Release Bioact. Mater.* 23: 369–370.

Borges, O., A. Cordeiro-da-Silva, J. Tavares et al. 2008. Immune response by nasal delivery of hepatitis B surface antigen and codelivery of a CpG ODN in alginate coated chitosan nanoparticles. *Eur. J. Pharm. Biopharm.* 69:405–416.

Cadete, A., L. Figueiredo, R. Lopes, C.C.R. Calado, A.J. Almeida, and L.M.D. Gonçalves. 2012. Development and characterization of a new plasmid delivery system based on chitosan–sodium deoxycholate nanoparticles. *Eur. J. Pharm. Sci.* 45: 451–458.

Cai, K., W. Liu, and F. Li. 2002. Modulation of osteoblast function using poly(D,L-lactic acid) surfaces modified with alkylation derivative of chitosan. *J. Biomater. Sci. Polym. Ed.* 13:53–66.

Chen, Y. and H. Tan. 2006. Crosslinked carboxymethylchitosan-g-poly(acrylic acid) copolymer as a novel super absorbent polymer. *Carbohydr. Res.* 341:887–896.

Chen, L., Z. Tian, and Y. Du. 2004. Synthesis and pH sensitivity of carboxymethyl chitosan-based polyampholyte hydrogels for protein carrier matrices. *Biomaterials* 25:3725–3732.

Corsi, K., F. Chellat, L. Yahia, and J.C. Fernandes. 2003. Mesenchymal stem cells, MG63 and HEK293 transfection using chitosan-DNA nano particles. *Biomaterials* 24:1255–1264.

Crompton, K..E., J.D. Goud, R.V. Bellamkonda, T.R. Gengenbach, and D.I. Finkelstein. 2007. Polylysine-functionalised thermoresponsive chitosan hydrogel for neural tissue engineering. *Biomaterials* 28:441–449.

Cui, Z. and R.J. Mumper. 2001. Chitosan-based nanoparticles for topical genetic immunization. *J. Control. Release* 75:409–419.

Cynthia G.L., S. Frantzicha, A. Rosinskia, M. Sjöströmb, and J. Hoogstraate. 2003. Oral gingival delivery systems from chitosan blends with hydrophilic polymers. *Eur. J. Pharm. Biopharm.* 55:47–56.

Denkbas, E.B. and R.M. Ottenbrite. 2006. Perspectives on: Chitosan drug delivery systems based on their geometries. *J. Bioact. Compat. Polym.* 21:351–368.

Dung, L., P. Milas, and M.M.J. Rinaudo. 1994. Water soluble derivatives obtained by controlled chemical modifications of chitosan. *Carbohydr. Polym.* 24:209–214.

El-Kamel, A., M. Sokar, V. Naggar, and S. Al Gamal. 2002. Chitosan and sodium alginate-based bioadhesive vaginal tablets. *AAPS J.* 4:224–230.

Erbacher, P., S. Zou, T. Bettinger, A.M. Steffen, and J.S. Remy. 1998. Chitosan based vector/DNA complexes for gene delivery: Biophysical characteristics and transfection ability. *Pharm. Res.* 15:1332–1339.

Ganta, S., H. Devalapally, A. Shahiwala, and M. Amiji. 2008. A review of stimuli-responsive nanocarriers for drug and gene delivery. *J. Control. Release* 126:187–204.

Gao, S., J. Chen, X. Xu et al. 2003. Galactosylated low molecular weight chitosan as DNA carrier for hepatocyte-targeting. *Int. J. Pharm.* 255:57–68.

Garmise, R., H. Staats, and A. Hickey. 2007. Novel dry powder preparations of whole inactivated influenza virus for nasal vaccination. *AAPS PharmSciTech* 8: 2–10.

Gavini, E., G. Rassu, C. Muzzarelli, M. Cossu, and P. Giunchedi. 2008. Spray-dried microspheres based on methylpyrrolidinone chitosan as new carrier for nasal administration of metoclopramide. *Eur. J. Pharm. Biopharm.* 68:245–252.

Gental C.B., P. Perugini, F. Pavantto, A. Spadaro, and G. Puglisi. 1997. Bioadhesive micro spheres for ophthalmic administration of Acyclovir. *Pharm. Pharmacol.* 49:737–742.

Granja, P.L., L. Pouysgu, M. Ptraud, B.D. Jso, C. Baquey, and M.A. Barbosa. 2000. Cellulose phosphates as biomaterials. II. Surface chemical modification of regenerated cellulose hydrogels. *J. Appl. Polym. Sci.* 8:3354–3365.

Guggi, D. and S.A. Bernkop. 2003. *in vitro* evaluation of polymeric excipients protecting calcitonin against degradation by intestinal serine proteases. *Int. J. Pharm.* 252:187–196.

Guggi, D., A.H. Kkrauland, and A. Bernkop-schnurch. 2003. Systemic peptide delivery via the stomach: *In vivo* evaluation of an oral dosage form for salmon calcitonin. *J. Control. Release* 92:125–135.

Guowei, Q., X. Wu, L. Yin, and C. Zhang. 2012. N-octyl-O-sulfate chitosan-modified liposomes for delivery of docetaxel: Preparation, characterization, and pharmacokinetics. *Biomed. Pharmacother.* 66:46–51.

Gupta, K.C. and M.N.V. Ravikumar. 1986. An overview on chitin and chitosan applications with an emphasis on controlled drug release formulations. *Sci. Rev. Macromol. Chem. Phys.* 4:273.

Hamman, J.H., C.M. Schultz, and A.M. Kotze. 2003. *N*-trimethyl chitosan chloride: Optimum degree of quaternization for drug absorption enhancement across epithelial cells. *Drug Dev. Ind. Pharm.* 29: 161–172.

He, P., S.S. Davis, and L. Illum, L. 1999. Chitosan microspheres prepared by spray drying. *Int. J. Pharm.* 187:53–65.

Hennink, W.E. and C.F. Van Nostrum. 2002. Novel crosslinking methods to design hydrogels. *Adv. Drug Deliv. Rev.* 54:13–36.

Heras, A., N.M. Rodríguez, and V.M. Ramos. 2001. *N*-methylene phosphonic chitosan: A novel soluble derivative. *Carbohydr. Polym.* 44:1–8.

Hoffman, A.S. 2002. Hydrogels for biomedical applications. *Adv. Drug Deliv. Rev.* 54:3–12.

Hsieh, C.Y., S.P. Tsai, D.M. Wang, Y.N. Chang, and H.J. Hsieh. 2005. Preparation of γ-PGA/chitosan composite tissue engineering matrices. *Biomaterials* 26:5617–5623.

Huang, R.H., G.H. Chen, M.K. Sun, Y.M. Hu, and C.J. Gao. 2007. Preparation and characterization of quaternized chitosan/poly(acrylonitrile) composite nanofiltration membrane from epichlorohydrin crosslinking. *Carbohydr. Polym.* 70:318–323.

Il'ina, A.V. and V.P. Varlamov. 2005. Chitosan-based polyelectrolyte complexes: A review. *Prikl. Biokhim. Mikrobiol.* 41:9–16.

Jayakumar, R., H. Nagahama, T. Furuike, and H. Tamura. 2008. Synthesis of phosphorylated chitosan by novel method and its characterization. *Int. J. Biol. Macromol.* 42:335–339.

Jayakumar, R., R.L. Reis, and J.F. Mano. 2007. Synthesis and characterization of pH-sensitive thiol-containing chitosan beads for controlled drug delivery applications. *Drug Deliv.* 14:9–17.

Jia, Z., D. Shen, and W. Xu. 2001. Synthesis and antibacterial activities of quaternary ammonium salt of chitosan. *Carbohydr. Res.* 333:1–6.

Jiang, H., Y. Wang, Q. Huang, Y. Li, C. Xu, and K. Zhu. 2005. Biodegradable hyaluronic acid/N-carboxyethyl chitosan/protein ternary complexes as implantable carriers for controlled protein release. *Macromol. Biosci.* 5:1226–1233.

Jung, B.O., C.H. Kim, K.S. Choi, Y.M. Lee, and J.J. Kim. 1999. Preparation of amphiphilic chitosan and their antimicrobial activities. *J. Appl. Polym. Sci.* 72:1713–1719.

Kang, H.M., Y.L. Cai, J.J Deng, H.J. Zhang, Y.F. Liu, and P.S Tang. 2006. Synthesis and aqueous solution behavior of phosphonate functionalized chitosans. *Eur. Polym. J.* 42:2678–2685.

Kast, C.E., C. Valenta, M. Leopold, and A. Bernkop-schnurch. 2002. Design and *in vitro* evaluation of a novel bioadhesive vaginal drug delivery system for clotrimazole. *J. Control. Release* 81:347–354.

Kean, T., S. Roth, and T. Thanou. 2005. Trimethylated chitosans as non-viral gene delivery vectors: Cytotoxicity and transfection efficiency. *J. Control. Release* 103: 643–653.

Kim, J.H., Y.S. Kim, K. Park et al. 2008. Self-assembled glycol chitosan nanoparticles for the sustained and prolonged delivery of antiangiogenic small peptide drugs in cancer therapy. *Biomaterials* 29 2008: 1920–1930.

Knilla C.J., J.F. Kennedya, J. Mistrya et al. 2004. Alginate fibres modified with unhydrolysed and hydrolysed chitosans for wound dressings. *Carbohydr. Polym.* 55:65–76.

Kose-Ozkan, C., A. Savaser, C. Tas, and Y. Ozkan. 2010. The development and *in vitro* evaluation of sustained release tablet formulations of benzydamine hydrochloride and its determination. *Comb. Chem. High Throughput Screening* 13: 683–689.

Kotzé, A.F., J.H. Hamman, D. Snyman, C. Jonker, and M. Stander. 2002. Mucoadhesive and absorption enhancing properties of *N*-trimethyl chitosan chloride. In *Chitosan in Pharmacy and Chemistry*, eds. R.A.A. Muzzarelli and C. Muzzarelli, 31–39. Italy: Atec.

Kowapradit, J., P. Opanasopit, T. Ngawhiranpat et al. 2008. Methylated *N*-(4-*N*,*N*-dimethylaminobenzyl) chitosan, a novel chitosan derivative, enhances paracellular permeability across intestinal epithelial cells (Caco-2). *AAPS PharmSciTech* 9:1143–1152.

Lee, D.W., K.S. Yun, H.S. Ban, W. Choe, S.K. Lee, and K.Y. Lee. 2009. Preparation and characterization of chitosan/polyguluronate nanoparticles for siRNA delivery. *J. Control. Release* 139: 146–152.

Leong, Y.S. and F. Candau. 1982. Inverse microemulsion polymerization. *J. Phys. Chem.* 86: 2269–2271.

Li, F., W.G. Liu, and K.D. Yao. 2002. Preparation of oxidized glucose-crosslinked N-alkylated chitosan membrane and *in vitro* studies of pH-sensitive drug delivery behavior. *Biomaterials* 23:343–347.

Liu, W.G. and K.D. Yao. 2002. Chitosan and its derivatives—A promising non-viral vector for gene transfection. *J. Control. Release* 83:1–11.

Liu, W.G., K.D. Yao, and Q.G. Liu. 2001. Formation of a DNA/N-dodecylated chitosan complex and salt induced gene delivery. *J. Appl. Polym. Sci.* 82:3391–3395.

Liu, X.F., Y.L. Guan, D.Z. Yang, Z. Li, and K.D. Yao. 2001. Antibacterial action of chitosan and carboxymethylated Chitosan. *J. Appl. Polym. Sci.* 79:1324–1335.

Liu, Z., Y. Jiao, Y. Wang, C. Zhou, and Z. Zhang. 2008. Polysaccharides-based nanoparticles as drug delivery systems. *Adv. Drug Deliv. Rev.* 60:1650–1662.

Lu, G., L. Kong, B. Sheng, G. Wang, Y. Gong, and X. Zhang. 2007. Degradation of covalently cross-linked carboxymethyl chitosan and its potential application for peripheral nerve regeneration. *Eur. Polym. J.* 43:3807–3818.

Majeti, N.V. and R. Kumar. 2000. A review of chitin and chitosan applications. *React. Funct. Polym.* 46:1–27.

Makhlof, A., Y. Tozuka, and H. Takeuchi. 2011. Design and evaluation of novel pH-sensitive chitosan nanoparticles for oral insulin delivery. *Eur. J. Pharm. Sci.* 42: 445–451.

Miyazaki, S., K. Ishii, and T. Nadai. 1981. The use of chitin and chitosan as drug carriers. *Chem. Pharm. Bull.* 29:3067–3069.

Miyazaki, T., T. Komuro, C. Yomota, and S. Okada. 1990. Usage of chitosan as a pharmaceutical material: Effectiveness as an additional additive of sodium alginate. *Eisei Shikenjo ho Koku* 108:95–97.

Niederhofer, A. and B.W. Maller. 2004. A method for direct preparation of chitosan with low molecular weight from fungi. *Eur. J. Pharm. Biopharm.* 57:101–105.

Nordtveit R.J., K.M. Vårum, and O. Smidsrød. 1994. Degradation of fully water-soluble, partially N-acetylated chitosans with lysozyme. *Carbohydr. Polym.* 23: 253–260.

Pan, J.L., Z.M. Bao, J.L. Li et al. 2005. Chitosan-based scaffolds for hepatocyte culture, In: ASBM6. *Adv. Biomater.* 6:91–94.

Park, I.K., T.H. Kim, Y.H. Park et al. 2001. Galactosylated chitosan–graft-poly(ethelene glycol) as hypatocyte targeting DNA carrier. *J. Control. Release* 76:349–362.

Park, Y., M. Sugimoto, A. Watrin, M. Chiquet, and E.B. Hunziker. 2005. BMP-2 Induces the expression of chondrocyte-specific genes in bovine synovium-derived progenitor cells cultured in three-dimensional alginate hydrogel. *Osteoarthr. Cartilage.* 13:527–536.

Peppas, N.A., P. Bures, W. Leobandung, and H. Ichikawa. 2000. Hydrogels in pharmaceutical formulations. *Eur. J. Pharm. Biopharm.* 50:27–46.

Perioli, L., V. Ambrogi, L. Venezia, C. Pagano, M. Ricci, and C. Rossi. 2008. Chitosan and a modified chitosan as agents to improve performances of mucoadhesive vaginal gels. *Colloids Surf. B Biointerfaces* 66:141–145.

Polk, A., B. Amsden, K.D. Yao, T. Peng, and M.F.A. Goosen. 1994. Controlled release of albumin from chitosan-alginate microcapsules. *J. Pharm. Sci.* 83:178–185.

Pouton, C.W. and L.W. Seymour. 1998. Key issues in non-viral gene delivery. *Adv. Drug Deliv. Rev.* 34:3–19.

Prabaharan, M. and J.F. Mano. 2005. Chitosan-based particles as controlled drug delivery systems. *Drug Deliv.* 12:41–57.

Qiu, Y. and K. Park. 2001. Environment-sensitive hydrogels for drug delivery. *Adv. Drug Deliv. Rev.* 53:321–339.

Raafat, D. and H.G. Sahl. 2009. Chitosan and its antimicrobial potential—A critical literature survey. *Microb. Biotechnol.* 2: 186–201.

Ramos, V.M., M.S. Rodríguez, and A. Rodríguez. 2003. Modified chitosan carrying phosphonic and alkyl groups. *Carbohydr. Polym.* 51:425–429.

Richardson, S.C.W., H.V.J. Kolbe, and R. Duncan. 1999. Potential of low molecular mass chitosan as a DNA delivery system: Biocompatibility, body distribution and ability to complex and protect DNA. *Int. J. Pharm.* 178:231–243.

Ritthidej, G.C., P. Chemto, S. Pummangura, and P. Menasveta. 1994. Chitin and chitosan as disintegrant in paracetamol tablets. *Drug Dev. Ind. Pharm.* 20:2109–2134.

Sakkinen, M., J. Marvola, H. Kanerva et al. 2004. Gamma scintigraphic evaluation of the fate of microcrystalline chitosan granules in human stomach. *Eur. J. Pharm. Biopharm.* 57:133–143.

Sandri, G., S. Rossi, F. Ferrari, M.C. Bonferoni, C. Muzzarelli, and C. Caramella. 2004. Assessment of chitosan derivatives as buccal and vaginal penetration enhancers. *Eur. J. Pharm. Sci.* 21:351–359.

Sandri, G.,S. Rossi, M.C. Bonferoni et al. 2005. Buccal penetration enhancement properties of *N*-trimethyl chitosan: Influence of quaternization degree on absorption of a high molecular weight molecule. *Int. J. Pharm.* 297:146–155.

Sato, T., T. Ishii, and Y. Okahata. 2001. *In vitro* gene delivery mediated by chitosan. Effect of pH, serum, and molecular mass of chitosan on the transfection efficiency. *Biomaterials* 22:2075–2080.

Sayin, B., S. Somavarapu, X.W. Li et al. 2009. TMC–MCC (N-trimethyl chitosan–mono-N-carboxymethyl chitosan) nanocomplexes for mucosal delivery of vaccines. *Eur. J. Pharm. Sci.* 38: 362–369.

Singla, A.K. and M. Chawla. 2001. Chitosan. Some pharmaceutical and biological aspects—An update. *J. Pharm. Pharmacol.* 53:1047–1067.

Skorik, Y.A., C.A.R. Gomes, M.S.T.D. Vasconcelos, and Y.G. Yatluk. 2000. *N*-(2-Carboxyethyl)chitosans: Regioselective synthesis, characterization and protolytic equilibria. *Carbohydr. Res.* 338: 271.

Slütter, B., L. Plapied, V. Fievez et al. 2009. Mechanistic study of the adjuvant effect of biodegradable nanoparticles in mucosal vaccination. *J. Control. Release* 138:113–121.

Sugimoto, M., M. Morimoto, H. Sashiwa, H. Saimoto, and Y. Shigemasa. 1998. Preparation and characterization of water-soluble chitin and chitosan derivatives. *Carbohydr. Polym.* 36:49–59.

Tabata, Y. and Y. Ikada. 1998. Protein release from gelatin matrices. *Adv. Drug Deliv. Rev.* 31:287–301.

Tamura, H., Y. Tsuruta and S. Tokura. 2002. Preparation of chitosan-coated alginate filament. *Mater. Sci. Eng. C* 20:143–147.

Tapia, C., Z. Escobar, E. Costa et al. 2004. Comparative studies on polyelectrolyte complexes and mixtures of chitosan-alginate and chitosan-carrageenan as prolonged diltiazem chlorhydrate release system. *Eur. J. Pharm. Biopharm.* 57:65–75.

Thanou, M., B.I. Florea, M. Geldof, H.E. Junginger, and G. Borchard. 2002. Quaternized chitosan oligomers as novel gene delivery vector in epithelial cell lines. *Biomaterials* 23:153–159.

Thanou, M., S. Henderson, A. Kydonieus, and C. Elson. 2007. *N*-sulfonato-*N,O*-carboxymethylchitosan: A novel polymeric absorption enhancer for the oral delivery of macromolecules. *J. Control. Release* 117:171–178.

Tiera, M.J., X.P. Qiu, S. Bechaouch, Q. Shi, J.C. Fernandes, and F.M. Winnik. 2006. Synthesis and characterization of phosphorylcholine-substituted chitosans soluble in physiological pH conditions. *Biomacromolecules* 7:3151–3156.

Tokumitsu, H., H. Ichikawa, and Y. Fukumori. 1999. Chitosan–gadopentetic acid complex nanoparticles for gadolinium neutron capture therapy of cancer: Preparation by novel emulsion droplet coalescence technique and characterization. *Pharm. Res.* 16:1830–1835.

Tozaki, H., J. Komoike, C. Tada et al. 1997. Chitosan capsules for colon-specific drug delivery: Improvement of insulin absorption from the rat colon. *J. Pharm. Sci.* 86:1016–1021.

Upadrashta, S.M., P.R. Katikaneni, and N.O. Nuessle. 1992. Chitosan as a tablet binder. *Drug Dev. Ind. Pharm.* 18:1701–1708.

Wang, Y.S., L.R. Liu, Q. Jiang, and Q.Q. Zhang. 2007. Self aggregated nanoparticles of cholesterol-modified chitosan conjugate as a novel carrier of epirubicin. *Eur. Polym. J.* 43:43–51.

Weng, L., A. Romanov, J. Rooney, and W. Chen. 2008. Noncytotoxic, in situ gelable hydrogels composed of *N*-carboxyethyl chitosan and oxidized dextran. *Biomaterials* 29:3905–3913.

Xu, Y., Y. Du, and R. Huang. 2003. Preparation and modification of *N*-(2-hydroxyl) propyl-3-trimethyl ammonium chitosan chloride nanoparticle as a protein carrier. *Biomaterials* 24:5015–5022.

Yazdani-Pedram, M. and J. Retuert. 1997. Homogeneous grafting reaction of vinyl pyrrolidone onto chitosan. *J. Appl. Polym. Sci.* 63:1321–1326.

Zambito, Y., C. Zaino, G. Uccello-Barretta, F. Balzano, and G. Di Colo. 2008. Improved synthesis of quaternary ammonium-chitosan conjugates (N⁺-Ch) for enhanced intestinal drug permeation. *Eur. J. Pharm. Sci.* 33: 343–350.

Zhang, C., G. Qu, Y. Sun, X. Wu, Z. Yao, and Q. Guo. 2008. Pharmacokinetics, biodistribution, efficacy and safety of *N*-octyl-*O*-sulfate chitosan micelles loaded with paclitaxel. *Biomaterials* 29:1233–1241.

Zhang, C., Q. Ping, and H.J. Zhang. 2003. Preparation of N-alkyl-O-sulfate chitosan derivatives and micellar solubilisation of taxol. *Carbohydr. Polym.* 54:137–141.

Zheng, H., Y. Du, J. Yu, R. Huang, and L. Zhang. 2001. Preparation and characterization of chitosan/poly(vinyl alcohol) blend fibers. *J. Appl. Polym. Sci.* 80:2558–2565.

19 Chitosan Derivatives for Oral Delivery of Insulin

Anumita Chaudhury and Surajit Das

CONTENTS

19.1   INTRODUCTION

Chitosan is a polymer of *N*-acetylglucosamine and glucosamine, obtained naturally from the shells of crustaceans (Illum, 1998). It has shown pharmacological activity in wound healing, against microbes, for treatment of ulcers, and also as a cholesterol-reducing agent (Azad et al., 2004; Felt et al., 1998; Sugano et al., 1988). However, being a nontoxic and nonimmunogenic polymer, it has been widely used as a drug delivery matrix. Because it is obtained from natural sources, chitosan is a cost-effective polymer. It is also available in different molecular weights and degree of deacetylation (Illum, 1998). Chitosans have been used to prepare gels, capsule shells, solid lipid nanoparticles, nanoparticles, and microparticles for the delivery of various small therapeutic agents, peptides, vaccines, DNA, and genes (Chaudhury and Das, 2011; Illum, 1998). However, the solubility profile of chitosan has raised concern for oral peptide delivery. This is because chitosan is easily soluble in dilute acidic conditions while it is insoluble in neutral pH, water, and organic solvents (George and Abraham, 2006). As a result, oral drug delivery systems (DDS) made of chitosan were incapable of protecting protein and peptide drugs. This problem was solved by the derivatization of chitosan molecule by reacting with different functional groups or with other polymers or peptides. These newly formed chitosan derivatives showed superior stability in acidic pH conditions and could slowly release drugs in neutral or intestinal pH. Such modifications paved a new way for the delivery of macromolecules through the oral route. Among them, insulin delivery has been the most attempted and researched topic among scientists.

There are several reasons for attempting insulin delivery via the oral route. Diabetes is marked by the lack of insulin production by the body (type I) or when human body is unable to utilize the insulin produced by it (type II) (WHO, 2007). It is one of the leading causes of death among the world's population. An estimated 246 million people are affected by this disease and the number is expected to cross 380 million by 2025 (Science, 2010). This chronic disease can affect multiple organs of the body if not properly treated in time. Although several drugs and supplementary therapy can help to control type II diabetes, the same cannot be said for the treatment of type I diabetes. The only way to treat diabetes effectively is by providing adequate amounts of insulin to the body from external sources (WHO, 2007). However, insulin being a peptide drug, its delivery via the oral route is impeded due to the degradation of the molecule by protease enzymes and acidic condition of the intestine. Till today, the main delivery route of insulin is through subcutaneous injections. It is extremely painful for the patients as the treatment requires administering insulin daily with multiple dosing. Although various other routes of insulin delivery have been attempted in the past, only few of them have met with limited success (Iyer et al., 2011; Krauland et al., 2006). This provides a huge opportunity for investigation and probably can solve the biggest patient compliance issue existing in today's diabetes therapy. Chitosan, being the naturally occurring biodegradable polymer, has the potential to be a suitable delivery matrix for oral insulin delivery. This chapter mainly focuses on the suitability of chitosan for oral insulin delivery, the different delivery matrices explored in recent research, and the pharmacological effects shown by such delivery systems.

19.2   SUITABILITY OF CHITOSAN AS A DRUG DELIVERY SYSTEM

19.2.1   SAFETY OF CHITOSAN

Whenever a new polymer is used as a DDS, the question that arises is on its safety and pharmacological effects. Chitosan is a naturally occurring biocompatible polymer producing no toxic or immunogenic reaction within the animal body (Baldrick, 2011; Zhang et al., 2010a). Chitosan has a GRAS (generally regarded as safe) status from the U.S. Food and Drug Administration (Boateng et al., 2008). Its safety has been assessed extensively in human, mice, and rats through oral and parenteral administration. Chitosan has an LD_{50} of 16 g/kg body weight when orally administered to mice. This LD_{50} value is almost equivalent to LD_{50} of household sugar or salt (Kean and Thanou, 2011). In fact, continual intake of chitosan for 3 months at a dose of 15 g/kg/day also did not show any signs of toxic effect on mice and rats (Baldrick, 2011). Similarly, healthy human volunteers did not show any significant clinical signs when orally fed with chitosan up to a dose of 6.75 g/day (Tapola et al., 2008). Not only that, very recently, chitosan is also used as a dietary supplement to prevent fat absorption (Boateng et al., 2008).

However, the safety of chitosan should not be confused with the safety of chitosan derivatives. This is because chemical modifications can considerably change the inherent properties of chitosan and this may result in toxicity as well (Chaudhury and Das, 2011). But currently, several studies have assessed the cytotoxicity of various chitosan derivatives in cell lines and animal models (Kean and Thanou, 2011).

19.2.2 BIODEGRADABILITY

Several enzymes present in human physiological fluid and mucosa are reported to be involved in the biodegradation of chitosan. Some commonly known enzymes are lysozymes, chititriosidase, di-*N*-acetylchitobiase, *N*-acetylglucosaminidase, and so on (Kean and Thanou, 2011; Muzzarelli, 1997). Moreover, it is reported that chitosanase enzymes secreted by microorganisms in the intestine helps in the digestion of chitosan after oral administration (Nagpal et al., 2010).

19.2.3 VERSATILITY FOR REACTIONS

Chitosan is available in different molecular weights. Most importantly, the molecule has two functional groups (amino and hydroxyl) that can be readily substituted with a suitable ligand to prepare new substituted chitosan derivatives (Werle et al., 2009). The highly reactive amino and hydroxyl groups also allow reacting with other active molecules or polymer to form newer complexes. These properties enable the polymer to be tailored into a suitable platform for the delivery of a variety of molecules.

19.2.4 SOLUBILITY

Chitosan itself is hydrophilic in nature; however, its solubility is strongly dependent on the number of acetylations and its distribution in the chain. As mentioned before, various substitutions in the amino or hydroxyl group of chitosan can give birth to derivatives with different solubility profiles. This makes chitosan a versatile polymer as it can be used for the delivery of both cationic and anionic drugs (Chaudhury and Das, 2011). Previously, the high solubility of chitosan in lower pH conditions of the intestine had impeded the delivery of macromolecules through the oral route. However, the recent development of chitosan derivatives having high stability in acidic pH has opened the possibility of protein delivery via the oral route (George and Abraham, 2006).

19.2.5 MUCOADHESIVE PROPERTY OF CHITOSAN

Chitosan itself has a cationic property due to the presence of the primary amino group in its structure. It is also known that the mucus layer of the intestine is anionic in nature due to the presence of sialic acid and sulfonic acid in its structure (Dhawan et al., 2004). This gives rise to an ionic interaction between the cationic chitosan and anionic mucus layer on oral delivery of chitosan (Dhawan et al., 2004). In addition, hydrophobic interactions also take place that enhance the mucoadhesive property of chitosan. Substitution with thiolated groups has shown to further improve the mucoadhesive property of chitosan (Bernkop-Schnurch and Dunnhaupt, 2011).

19.2.6 Controlled Drug Delivery System

Owing to its versatile chemical nature, chitosan can form stable ionic complexes with a variety of drugs (Bernkop-Schnurch and Dunnhaupt, 2011). Drugs are then slowly released from these complexes by the process of diffusion and erosion. In addition, chitosan can be complexed with other polymers such as alginate, pectin, hyaluronic acid, and acrylates to form highly stable complexes of high density and molecular weight (Bernkop-Schnurch and Dunnhaupt, 2011). Such systems provide an excellent platform for the controlled release of a variety of drugs.

19.2.7 Permeation Enhancer Effect

Various *in vitro* and *in vivo* studies have indicated the permeation enhancing effect of chitosan and its derivatives. The positively charged amino group of the chitosan molecule can interact with the proteins of the cell membrane to loosen the tight junctions and thereby enhance the permeation of the molecule (Bravo-Osuna et al., 2007). Chitosan also demonstrates increasing permeation enhancing property with increasing molecular weight or higher degree of acetylations (Bonferoni et al., 2009; Bravo-Osuna et al., 2007). Moreover, chitosan has demonstrated additive or synergistic effect on permeation enhancement property when coadministered with other polysaccharides, such as sodium dodecylsulfate or cyclodextrin (Shah et al., 2007; Trapani et al., 2010). In fact, about 30-fold improvements in permeation enhancing effect were observed when chitosan was thiolated (Langoth et al., 2006).

19.2.8 Efflux Pump Inhibition

Although not pronounced, the thiolated derivative of chitosan has shown inhibitory properties toward multidrug efflux transporters (Foger et al., 2006). This has been demonstrated by the improved oral uptake of intestinal P-glycoprotein (P-gp) substrate rhodamine-123 (Foger et al., 2006). This property can be better utilized for the improved delivery of efflux pump substrate drugs in future.

19.3 ROUTES OF DELIVERY

All of the above-mentioned properties of chitosan have enabled it to be administered via various routes of administration. The section below gives a quick overview on the different routes of administration possible via chitosan-based DDS.

19.3.1 Oral Drug Delivery

The oral route of drug delivery is most preferred due to various reasons but most convincingly because it has maximum patient compliance. Among the different oral dosage forms, tablets are the most convenient because they are easy to manufacture, transport, handle, and also because they are most accurate, stable, and patient compliant. As a result, chitosans have also been used to prepare sustained released tablets, mostly by homogenously mixing with the drugs followed by compression (Mura et al., 2003). Moreover, as mentioned before, the mucoadhesive property of chitosan had enabled to prepare a more controlled oral delivery system for drugs such as pramipexole hydrochloride (Papadimitriou et al., 2008). In addition, nanoparticles or microparticles of chitosan and its derivatives have been prepared for oral delivery of several other drugs, such as ammonium glycyrrhizinate, ciprofloxacin hydrochloride, streptomycin, gentamicin, tobramycin, and catechin (Chaudhury and Das, 2011). Such oral delivery systems made of chitosan has shown improved pharmacological and pharmacokinetic property for the encapsulated drug (Chaudhury and Das, 2011).

19.3.2 PARENTERAL DRUG DELIVERY

Low-molecular-weight chitosan does not show any hemolytic property on intravenous injection. In addition, its nontoxic property and biodegradability has made it possible to utilize this polymeric carrier for parenteral delivery of both small molecular drugs and macromolecules as well (Bernkop-Schnurch and Dunnhaupt, 2011). DDS prepared with chitosan derivatives, such as glycol chitosan and *N*-succinyl chitosan, have demonstrated long circulation lifetime and minimal accumulation in tissues (Bernkop-Schnurch and Dunnhaupt, 2011). In another study with nanoparticles prepared with hydrophobically modified glycol, chitosan has shown substantial tumor accumulation of the arginyl glycyl aspartic acid (RGD) peptide on intravenous administration (Bernkop-Schnurch and Dunnhaupt, 2011).

19.3.3 BUCCAL DRUG DELIVERY

Mucoadhesive polymers such as chitosan are probably ideal candidates for preparing DDS for buccal delivery. Chitosan provides a controlled buccal DDS that can bypass the first pass metabolism as well as intestinal degradation by enzymes (Sandri et al., 2005). Drug such as nifedipine and propranolol have been successfully delivered through the buccal route through bilayered devices prepared from chitosan. Moreover, trimethylated chitosans and thiolated chitosan have shown promising results for the delivery of macromolecules (Sandri et al., 2005). Tablets prepared with thiolated chitosan have shown to provide a long-term therapeutic level of peptide drugs (Langoth et al., 2006).

19.3.4 OCULAR DRUG DELIVERY

Chitosan has an *in situ* gelling property on change of pH and when combined with other polymers or drugs (Gupta and Vyas, 2011). Chitosan-based hydrogels, nanoparticles, microparticles, and colloidal systems have shown potential for ocular drug delivery due to its nontoxicity, prolonged ocular retention time, quick gel-forming ability, and good biodistribution of the drug (De Campos et al., 2001). Drugs such as cyclosporine A, acyclovir, indomethacine, and so on have shown significantly improved ocular drug release when embedded in a chitosan-based ocular DDS (De Campos et al., 2001).

19.3.5 NASAL DRUG DELIVERY

To prepare an effective nasal delivery system, it is important to have a long residence time and effective absorption from the nasal cavity. Owing to its effective mucoadhesive property and permeation enhancing effect, chitosan is a suitable candidate for being an effective adjuvant in nasal DDS (Zhang et al., 2008). Chitosan nanoparticles encapsulating insulin have shown to be more effective in the nasal delivery of the molecule when compared to chitosan solutions (Zhang et al., 2008). Nasal sprays using chitosan and other polymers have been prepared for the nasal delivery of fentanyl and results have shown to significantly improve the systemic exposure of the drug in human subjects (Fisher et al., 2011).

19.3.6 VAGINAL DRUG DELIVERY

Low-molecular-weight chitosan, 5-methyl-pyrrolidinone chitosan, and partially reacetylated chitosan have been extensively explored for the vaginal delivery of acyclovir based on the enhanced mucoadhesive properties of such derivatives. Other than that, metronidazole has also been delivered through this route via compressed tablets made of chitosan, microcrystalline cellulose, and other ingredients (Sandri et al., 2004).

19.3.7 COLON-SPECIFIC DELIVERY

As any other polysaccharide, chitosan is naturally degraded in the colon. Owing to this property, chitosan has been utilized for coating tablets intended for colon-specific drug delivery. In addition, chitosan matrix has also been used for colon-specific delivery of anionic drug. Insulin nanoparticles prepared with chitosan have shown improved hypoglycemic effect when targeted via colon after delivery through the oral route.

19.4 PREPARATION OF CHITOSAN NANOPARTICLES

Chitosan and its derivatives have been quite ideal candidates for the preparation of nanoparticles or microparticles due to their quick and easy preparation methods. Below is a short description of the four different methods by which nano- or microsized DDS of chitosan are widely prepared.

19.4.1 POLYELECTROLYTE COMPLEXATION METHOD

In polyelectrolyte complexation method the polyanionic alginate or dextran molecules are complexed with the cationic amine group of chitosan to spontaneously form nanoparticles (Tiyaboonchai and Limpeanchob, 2007). These polyelectrolyte complexes are generally of various sizes ranging from 100 to 700 nm (Tiyaboonchai and Limpeanchob, 2007). This method provides the easiest way to prepare the nanoparticles in a mild condition.

19.4.2 IONOTROPIC GELATION METHOD

In this method, the positively charged amino group of chitosan is complexed with the negatively charged phosphate group of tripolyphosphate (Nagpal et al., 2010). The electrostatic attraction between the groups causes ionic gelation and chitosan nanoparticles or microparticles are formed. The molecular weight of chitosan, degree of deacetylation, and concentration of chitosan solutions determines the size and release mechanism of such nanoparticles (Nagpal et al., 2010). The biggest advantage of this method is that it is carried out in aqueous medium with minimum preparation time and effort (Nagpal et al., 2010).

19.4.3 EMULSION CROSS-LINKING METHOD

In this method, a w/o emulsion is prepared by dissolving aqueous solution of chitosan in oil, and thereafter stabilizing the system with a surfactant (Nagpal et al., 2010). A cross-linking agent is then used to further harden the microemulsion droplets to form microparticles. Similar cross-linking has been carried out with glutaraldehyde to prepare microparticles of narrow size distribution (Nagpal et al., 2010). However, one disadvantage of this method is the use of organic solvents and longer preparation time.

19.4.4 COMPLEX COACERVATION METHOD

This method is generally used to prepare chitosan nanoparticles encapsulating DNA, proteins, bovine serum albumin, and rhodamine. Here, the positively charged amino group of chitosan is complexed with the negatively charged sulfate group of dextran sulfate polymer to form nanoparticles. This method can be entirely performed under aqueous conditions; however, the entrapment efficiency is often poor (Makhlof et al., 2011). In addition, stability issues have been found with such nanoparticles and often need the use of toxic chemicals such as glutaraldehyde for cross-linking of the complex (Makhlof et al., 2011).

19.5 CHITOSAN NANOPARTICLES FOR ORAL DELIVERY OF INSULIN

Chitosan and its derivatives have been widely used for the oral delivery of insulin. Chitosan has also been combined with other peptides or molecules to prepare nanoparticulate matrix or microparticles to encapsulate insulin (Table 19.1). Below is a short description of each type of chitosan derivative or combination used for this purpose.

19.5.1 INSULIN DELIVERY VIA UNMODIFIED CHITOSAN

Chitosan itself has been used to prepare nanoparticles of insulin by the ionotropic gelation method as described above. It was found that the pH of the formulation had significant impact on the insulin loading efficiency and release rates (Ma et al., 2005; Pan et al., 2002). Such nanoparticles have shown significantly improved intestinal absorption of insulin in rats when compared with CS solution. Two different studies showed that the hypoglycemic effect of insulin could last till 11 or 15 h (Ma et al., 2005; Pan et al., 2002). However, the bioavailability of insulin through this delivery system was only 14.9% (Pan et al., 2002).

19.5.2 INSULIN DELIVERY VIA MODIFIED AMINO GROUP OF CHITOSAN

As mentioned earlier, the versatile nature of chitosan molecule provides the opportunity to prepare several modified chitosan derivatives. Mostly, the amino group of chitosan has been substituted with alky groups to prepare trimethyl chitosan (TMC), triethyl chitosan, diethyl-methyl chitosan, dimethyl-ethyl chitosan, and N-TMC. All these different alkyl-substituted derivatives of chitosan have been used to prepare insulin nanoparticles for oral delivery.

The polyelectrolyte complexation (PEC) method was used to prepare insulin complexes with TMC (Jintapattanakit et al., 2007). Such complexes showed good insulin entrapment efficiency and improved stability in the presence of proteolytic enzymes. This further led to the preparation of insulin complexes with PEG (polyethylene glycol)-grafted TMC derivatives (Figure 19.1) (Jintapattanakit et al., 2007). These complexes also exhibited superior stability in stimulated intestinal fluid. However, when nanoparticles were prepared using the same derivatives, it showed lower stability and encapsulation efficiency *in vitro* (Jintapattanakit et al., 2007).

Other quaternized derivatives of chitosan, such as TMC and diethyl-methyl chitosan, have been used to prepare nanoparticles using both ionotropic gelation method and PEC method (Sadeghi et al., 2008). Nanoparticles prepared by the PEC method had better insulin loading capability. The nanoparticles formed were able to protect insulin from the harsh intestinal pH and enzymes. These newly prepared quaternized polymers also showed enhanced antibacterial property against Gram-positive bacteria when used in the free form. Such positively charged polymers are also known for their enhanced penetration property as they are able to loosen the tight junctions of the intestine in neutral and alkaline pH conditions (Sadeghi et al., 2008).

Two other quaternized chitosan derivatives have been used for the colon-specific delivery of insulin via the oral route. Nanoparticles of insulin were prepared by triethyl chitosan and dimethyl-ethyl chitosan using the PEC method (Bayat et al., 2008a,b). Such nanoparticles showed 80% loading efficiency and sustained release effect when measured in *in vitro* conditions. *Ex vivo* studies in rat colon showed faster transport of insulin across the colonic sites when compared to nanoparticles prepared with chitosan alone (Bayat et al., 2008a,b). Prolonged hypoglycemic effect in rats revealed better *in vivo* absorption of tri-ethyl chitosan and dimethyl-ethyl chitosan nanoparticles when compared to free insulin (Bayat et al., 2008a,b).

N-TMC has also been exploited to prepare nanoparticles of insulin owing to its better mucoadhesive and permeation enhancing property (Sandri et al., 2010). *In vitro* studies with Caco-2 cell line showed that nanoparticles made of N-TMC improved the absorption of insulin by the process of endocytosis, whereas nanoparticles prepared with unmodified chitosan helps in permeation by

TABLE 19.1
Chitosan and Chitosan Derivatives for Oral Delivery of Insulin

Chitosan Derivatives	Type of DDS	Method of Preparation	Main Features of the Formulation or Pharmacological Effect	References
Unmodified chitosan	Nanoparticles	Ionotropic gelation	Improved *in vivo* absorption in rats; Bioavailability up to 14.9%	Pan et al. (2002)
Trimethyl chitosan	Insulin complexes	PEC	Good insulin entrapment efficiency; Improved stability in presence of proteolytic enzymes	Jintapattanakit et al. (2007)
PEG grafted trimethyl chitosan	Insulin complexes	PEC	Superior stability in stimulated intestinal fluid	Jintapattanakit et al. (2007)
Trimethyl chitosan	Nanoparticles	PEC and ionotropic gelation	Good insulin loading capability (by PEC); Superior stability in intestinal pH (by PEC)	Sadeghi et al. (2008)
Diethyl-methyl chitosan	Nanoparticles	PEC and ionotropic gelation	Good insulin loading capability (by PEC); Superior stability in intestinal pH (by PEC)	Sadeghi et al. (2008)
Triethyl chitosan	Nanoparticles	PEC	Good insulin loading capability (>80%); Enhanced colonic absorption of insulin	Bayat et al. (2008a,b)
Dimethyl-ethyl chitosan	Nanoparticles	PEC	Good insulin loading capability (>80%); Enhanced colonic absorption of insulin	Bayat et al. (2008a,b)
N-trimethyl chitosan	Nanoparticles	Ionotropic gelation	High mucoadhesive property in jejunum; Better penetration into duodenum and jejunum by endocytosis	Sandri et al. (2010); Sandri et al. (2007)
N-(2-hydroxypropyl)-3-trimethyl ammonium chitosan	Microparticles	Ionotropic gelation	Superior mucoadhesive property; Enhanced intestinal permeation	Sonia and Sharma (2011)
Alginate chitosan	Nanoparticles	Ionotropic gelation and PEC	High association efficiency; Prolonged hypoglycemic effect (~24 h)	Sarmento et al. (2007a,b)

Derivative	Type	Method	Properties	References
Alginate chitosan	Microspheres	Membrane emulsification technique	Sustained release of insulin in blood (14 days) Hypoglycemic effect up to 60 h	Zhang et al. (2011)
Alginate chitosan and cationic β-cyclodextrin	Nanoparticles	Complexation and PEC	Good insulin loading capability (>80%) Superior stability in intestinal pH	Zhang et al. (2010)
Chitosan and poly(γ-glutamic acid)	Nanoparticles	Ionotropic gelation	Increased permeability Hypoglycemic effect up to 10 h	Lin et al. (2007, 2008); Sonaje et al. (2007, 2010)
Chitosan and methyl methacrylate	Nanoparticles	Free radical polymerization	High encapsulation efficiency up to 100% Controlled drug release	Qian et al. (2006)
Chitosan and poly (methyl methacrylate)	Nanoparticles	Graft polymerization	pH-dependent release	Cui et al. (2009)
Chitosan and carboxymethyl-β-cyclodextrin	Nanoparticles	Ionotropic gelation	High encapsulation efficiency up to 90%	Krauland and Alonso (2007)
Lauryl succinyl chitosan	Nanoparticles	Cross-linking	Superior mucoadhesive property Enhanced intestinal permeation	Rekha and Sharma (2009)
Chitosan and Arabic gum	Nanoparticles	Ionotropic gelation	Lower encapsulation than CS-NPs	Avadi et al. (2011)
Chitosan and hydroxypropyl methylcellulose phthalate	Nanoparticles	Ionic cross-linking	Superior stability in acidic pH	Makhlof et al. (2011)
Chitosan and Eudragit L100-55	Nanoparticles	Complex coacervation method	Protection in acidic pH	Jelvehgari et al. (2011)
Chitosan and glutaraldehyde	Microspheres	Emulsion cross-linking	Good encapsulation efficiency up to 70% Significantly lower blood sugar level *in vivo*	Jose et al. (2011)

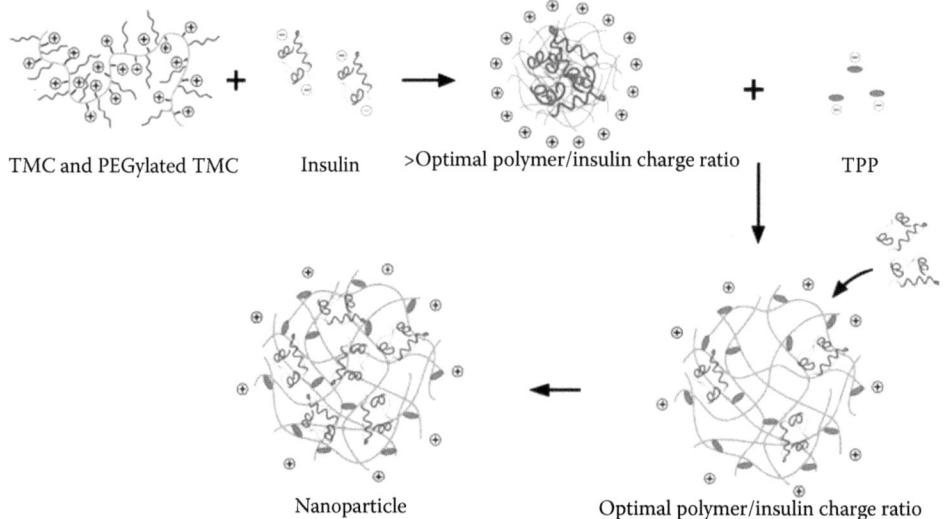

TMC and PEGylated TMC   Insulin   >Optimal polymer/insulin charge ratio   TPP

Nanoparticle   Optimal polymer/insulin charge ratio

FIGURE 19.1 Schematic representation of insulin nanoparticle formation. (From Jintapattanakit, A. et al. 2007. *Int J Pharm* 342:240–249. With permission.)

opening tight junctions of the intestinal membrane. Confocal images confirmed such observation. In addition, *ex vivo* studies with excised rat duodenum, jejunum, and ileum showed that nanoparticles made with N-TMC had good mucoadhesive property in jejunum (Sandri et al., 2007, 2010).

In a recent study, another quaternized derivative of chitosan, namely, *N*-(2-hydroxy)propyl-3-trimethyl ammonium chitosan was used for the preparation of insulin microparticles. Microparticles of insulin were prepared by the ionotropic gelation method and it showed controlled release property under *in vitro* conditions. *Ex vivo* studies with freshly excised rat intestine showed improved mucoadhesion of the microparticles on oral administration. The microparticles could also improve the absorption of insulin by opening the tight intestinal junctions, as understood from confocal imaging. However, no bioavailability studies were performed with these microparticles (Sonia and Sharma, 2011).

19.5.3 INSULIN DELIVERY VIA ALGINATE–CHITOSAN COMPLEX

Alginate is a polyanionic polymer widely used for drug delivery because of its biocompatible and biodegradable features, just like chitosan (George and Abraham, 2006). Alginate is soluble in high pH and insoluble in low pH, in contrast to chitosan. This enables to complex chitosan and alginate to prepare a stable complex that is able to protect insulin during transit through the intestine (George and Abraham, 2006). Mainly, the amino groups of chitosan react with the carboxyl groups in alginic acid via electrostatic interaction to form a polyelectrolyte complex.

Insulin nanoparticles have been prepared using this polymer interaction. Initially, ionotropic gelation method was used to prepare insulin–alginate nanoparticles and then reacted with chitosan via PEC method to finally prepare chitosan–alginate–insulin nanoparticles (Sarmento et al., 2007a, 2007b). These nanoparticles showed controlled release of insulin for up to 24 h under *in vitro* gastric pH conditions. Formulation conditions and molecular mass/weight ratios of chitosan and alginate affected the release characteristics and insulin encapsulation efficiency (Sarmento et al., 2007a, 2007b). Prolonged hypoglycemic effect was observed up to 24 h in diabetic mice on administration of these nanoparticles at a dose of 50 IU/kg of insulin. This hypoglycemic effect was found to be dose dependent up to 50 IU/kg but did not show any further improvement on increasing dose to 100 IU/kg (Sarmento et al., 2007a, 2007b).

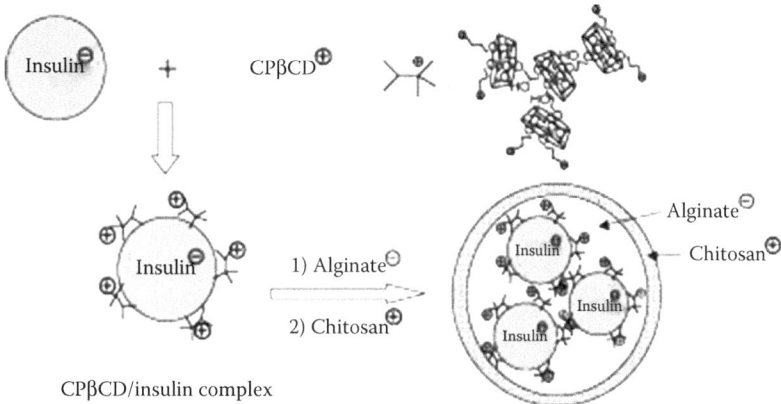

FIGURE 19.2 Schematic representation of the complexation and protection of insulin by CPβCDs in alginate/chitosan nanoparticles. (From Zhang, N. et al., 2010b. *Int J Pharm* 393:212–218. With permission.)

In another study, the alginate–chitosan complex was used as the main DDS but insulin was further complexed with a derivative of cyclodextrin to provide more stability during gastric transit (Zhang et al., 2010b). Insulin was first entrapped in cationic β-cyclodextrin (CP-β-CD) polymer, which is a hemocompatible and water-soluble derivative of cyclodextrin (Figure 19.2). Following it, this complex was included in the alginate–chitosan matrix to form nanoparticles of 150–350 nm size range. Nanoparticles prepared without cationic β-cyclodextrin showed significantly faster insulin release in simulated gastric fluid indicating the effectiveness of the inclusion complex to withhold insulin in the core of the nanoparticles (Zhang et al., 2010b).

Recently, alginate–chitosan microparticles encapsulating insulin have been prepared for oral administration. These microparticles were prepared by membrane emulsification technique and subsequently polymer solidification approach was used to improve insulin loading efficiency (up to 55%) (Zhang et al., 2011). The microparticles showed good stability under *in vitro* stimulated gastric and intestinal conditions releasing only 32% of entrapped insulin. When similar experiment was conducted in pH condition of blood, it showed extremely slow release of insulin, which continued up to 14 days (Zhang et al., 2011). However, the most significant observation made with such microparticles was the long-term reduction in blood glucose level in orally fed diabetic rats. The hypoglycemic effect could last up to 60 h showing promise as an oral insulin delivery system (Zhang et al., 2011).

19.5.4 INSULIN DELIVERY VIA CHITOSAN–PEPTIDE COMPLEX

Poly(γ-glutamic acid), a biodegradable, relatively nonimmunogenic anionic peptide has been combined with chitosan to prepare insulin nanoparticles for oral delivery. These nanoparticles could improve the permeability of insulin molecules in Caco-2 cells by opening the tight junctions of the membranes (Lin et al., 2007, 2008). This resulted in an enhanced bioavailability of insulin as understood from *in vivo* studies conducted in diabetic rats. However, this formulation was unstable in higher pH and released insulin suddenly at pH 7.4. When these nanoparticles were delivered via an enteric-coated capsule after freeze-drying, it showed a controlled release of insulin in the intestinal pH. The process of freeze drying did not affect the functional molecule as understood from small-angle x-ray scattering results (Sonaje et al., 2007, 2010).

19.5.5 INSULIN DELIVERY VIA CHITOSAN AND METHYL METHACRYLATE DERIVATIVES

Insulin nanoparticles have been prepared with the combination of chitosan and derivatives of methyl methacrylate such as *N*-dimethylaminoethyl methacrylate hydrochloride and *N*-trimethylaminoethyl methacrylate chloride (Qian et al., 2006). The free radical polymerization method was used to prepare these functionalized graft copolymer nanoparticles of insulin, without the use of any organic solvents. This enabled them to prepare nanoparticles showing up to 100% of loading efficiency. Such nanoparticles showed a controlled release of insulin in neutral pH for about 4 days; however, no *in vitro* data was shown for acidic or alkaline pH. When the plasma glucose level of diabetic rats was measured on oral administration of these nanoparticles, it showed an improved oral absorption and bioavailability of insulin in comparison to insulin solution (Qian et al., 2006).

The polymer poly(methyl methacrylate) has also been reacted with carboxylated chitosan to prepare insulin nanoparticles by the method of graft polymerization (Cui et al., 2009). Such nanoparticles were pH sensitive in nature and showed controlled release of insulin in stimulated gastric fluid, whereas rapid release was observed in stimulated intestinal fluid. *In vivo* studies indicated that bioavailability of insulin on oral administration was around 9.7% (Cui et al., 2009).

19.5.6 INSULIN DELIVERY VIA CHITOSAN AND OTHER MOLECULES

Carboxymethyl-β-cyclodextrin, an anionic derivative of the inclusion compound cyclodextrin, has also been combined with chitosan for the preparation of insulin nanoparticles (Krauland and Alonso, 2007). High encapsulation efficiency of up to 93% was found for insulin-containing nanoparticles. However, insulin was released within 15 min under *in vitro* conditions from these nanoparticles, which made it a suitable delivery system for oral absorption (Krauland and Alonso, 2007).

Sodium laurate has been combined with chitosan to synthesize lauryl succinyl to enhance the paracellular permeability of the molecule (Rekha and Sharma, 2009). Insulin nanoparticles were prepared by simple cross-linking with sodium tripolyphosphate. Such nanoparticles had both a hydrophilic and a hydrophobic internal core due to the presence of lauryl (phobic) and succinyl (philic) moiety. Such property helped in the retention of insulin in the core for a longer time, thereby showing controlled release property. Moreover, the inherent mucoadhesive property of the molecule and the ability to open the tight junctions of the intestinal membrane resulted in an enhanced cellular uptake of insulin. Thus, hypoglycemic effect was observed in diabetic rats upon oral administration, although for a short period of time (Rekha and Sharma, 2009).

Acacia, a commonly used pharmaceutical excipient has been combined with chitosan to prepare insulin nanoparticles. The negatively charged acacia molecule was reacted with the positively charged chitosan by the ionotropic gelation method to form the nanoparticles (Avadi et al., 2011). However, such nanoparticles did not show any significant benefit over the normal chitosan insulin nanoparticles with respect to insulin loading and release properties.

Hydroxypropyl methylcellulose phthalate (HPMCP), widely used in pharmaceutical formulations as a pH-sensitive polymer, have been recently combined with chitosan to form insulin nanoparticles with better stability in acidic pH (Makhlof et al., 2011). Results from quantitative fluorescence analysis and confocal microscopy showed that these nanoparticles had superior mucoadhesion and intestinal penetration property when compared to chitosan nanoparticles. This enhanced absorption of insulin was also evident in *in vivo* studies with diabetic rats, where it showed a nine times improvement in hypoglycemic effect in comparison to insulin solution. Moreover, the reduction in blood sugar level was three times more in rats fed with HPMCP–chitosan nanoparticles when compared to rats fed with chitosan nanoparticles (Makhlof et al., 2011).

Eudragit L100-55 polymer has been used to prepare chitosan–insulin nanoparticles with an enteric coating property. Different types of nanoparticles were prepared by varying the molecular weight of chitosan and utilizing complex coacervation method technique (Jelvehgari et al., 2011). These nanoparticles were stable at acidic pH and showed controlled release of insulin at neutral pH. However, the loading efficiency of insulin was quite low (30%) and it did not show any additional benefit as compared to other chitosan nanoparticles (Jelvehgari et al., 2011).

Cross-linking agent glutaraldehyde has been recently used to prepare chitosan–insulin microspheres for oral delivery of insulin. Different formulations were prepared by varying the concentration of chitosan and glutaraldehyde and cross-linking time (Jose et al., 2011). The optimized formulation showed controlled release properties and an insulin entrapment efficiency of about 70%. Importantly, *in vivo* studies in diabetic rats revealed that the formulation was able to significantly reduce blood glucose level when compared to subcutaneous injection of insulin (Jose et al., 2011).

19.6 FUTURE DIRECTION: A WORD OF CAUTION?

All of the above studies have shown that chitosan-based DDS are able to effectively deliver insulin via the oral route. Several derivatives of chitosan showed promising results in increasing the mucoadhesion of nanoparticles in the intestine and then increasing the permeation by opening the tight junctions of the membrane. Most importantly, most of these DDS were stable in gastric pH and enzymatic conditions and thereby could protect insulin from intestinal degradation. The nanoparticles and microparticles also served as a controlled DDS in many instances. Overall, it can be said that almost all chitosan-based delivery systems showed promising results *in vitro* for oral delivery of insulin.

However, future medical use of such systems would require intensive clinical studies in human subjects and promising supporting data. Thus far, to the best of our knowledge, none of these systems has entered clinical trial for oral delivery of insulin (Chaudhury and Das, 2011). Only preclinical data in diabetic rat and sometimes in diabetic mice model is available for few nanoparticulate and microparticulate systems. Most of these systems showed very low bioavailability of insulin (10–20%) *in vivo*; as a result, the dose used is not sufficient to replace insulin subcutaneous injection for daily use in human. In addition, very few studies have been conducted to evaluate the toxicity of these newly developed chitosan derivative-based DDS (Chaudhury and Das, 2011). As mentioned before, although chitosan is safe for human use up to very high dose, its toxicity and safety profile may not remain the same when modified or combined with other molecules (Chaudhury and Das, 2011). As such, lack of in-depth *in vivo* study or clinical data makes it difficult to predict the future of each of these delivery systems.

19.7 CONCLUSION

On the basis of the above discussion, it can be seen that chitosan and its multiple derivatives or chitosan combined with other molecules have been successfully used for the oral delivery of insulin. Although, few particular studies have shown promising results, it cannot be confirmed that these delivery systems can see the light of clinical usage in future. However, further modifications of the delivery systems and in-depth clinical investigations should be continued in future for effective oral delivery of insulin via chitosan matrix.

REFERENCES

Avadi, M.R., Sadeghi, A.M., Mohammadpour, N., Abedin, S., Atyabi, F., Dinarvand, R., and Rafiee-Tehrani, M., 2011. Preparation and characterization of insulin nanoparticles using chitosan and Arabic gum with ionic gelation method. *Nanomedicine* 6:58–63.

Azad, A.K., Sermsintham, N., Chandrkrachang, S., and Stevens, W.F., 2004. Chitosan membrane as a wound-healing dressing: Characterization and clinical application. *J Biomed Mater Res B Appl Biomater* 69: 216–222.

Baldrick, P., 2011. The safety of chitosan as a pharmaceutical excipient. *Regul Toxicol Pharmacol* 56:290–299.

Bayat, A., Dorkoosh, F.A., Dehpour, A.R., Moezi, L., Larijani, B., Junginger, H.E., and Rafiee-Tehrani, M., 2008a. Nanoparticles of quaternized chitosan derivatives as a carrier for colon delivery of insulin: *Ex vivo* and *in vivo*studies. *Int J Pharm* 356:259–266.

Bayat, A., Larijani, B., Ahmadian, S., Junginger, H.E., and Rafiee-Tehrani, M., 2008b. Preparation and characterization of insulin nanoparticles using chitosan and its quaternized derivatives. *Nanomedicine* 4:115–120.

Bernkop-Schnurch, A. and Dunnhaupt, S., 2011. Chitosan-based drug delivery systems. *Eur J Pharm Biopharm* 81:463–469.

Boateng, J.S., Matthews, K.H., Stevens, H.N., and Eccleston, G.M., 2008. Wound healing dressings and drug delivery systems: A review. *J Pharm Sci* 97:2892–2923.

Bonferoni, M.C., Sandri, G., Rossi, S., Ferrari, F., and Caramella, C., 2009. Chitosan and its salts for mucosal and transmucosal delivery. *Expert Opin Drug Deliv* 6:923–939.

Bravo-Osuna, I., Vauthier, C., Farabollini, A., Palmieri, G.F., and Ponchel, G., 2007. Mucoadhesion mechanism of chitosan and thiolated chitosan-poly(isobutyl cyanoacrylate) core-shell nanoparticles. *Biomaterials* 28:2233–2243.

Chaudhury, A. and Das, S., 2011. Recent advancement of chitosan-based nanoparticles for oral controlled delivery of insulin and other therapeutic agents. *AAPS Pharm Sci Tech* 12:10–20.

Cui, F., Qian, F., Zhao, Z., Yin, L., Tang, C., and Yin, C., 2009. Preparation, characterization, and oral delivery of insulin loaded carboxylated chitosan grafted poly(methyl methacrylate) nanoparticles. *Biomacromolecules* 10:1253–1258.

De Campos, A.M., Sanchez, A., and Alonso, M.J., 2001. Chitosan nanoparticles: A new vehicle for the improvement of the delivery of drugs to the ocular surface. Application to cyclosporin A. *Int J Pharm* 224:159–168.

Dhawan, S., Singla, A.K., and Sinha, V.R., 2004. Evaluation of mucoadhesive properties of chitosan microspheres prepared by different methods. *AAPS PharmSciTech* 5:122–128.

Felt, O., Buri, P., and Gurny, R., 1998. Chitosan: A unique polysaccharide for drug delivery. *Drug Dev Ind Pharm* 24:979–993.

Fisher, A., Watling, M., Smith, A., and Knight, A., 2011. Pharmacokinetic comparisons of three nasal fentanyl formulations; pectin, chitosan and chitosan-poloxamer 188. *Int J Clin Pharmacol Ther* 48:138–145.

Foger, F., Schmitz, T., and Bernkop-Schnurch, A., 2006. *In vivo* evaluation of an oral delivery system for P-gp substrates based on thiolated chitosan. *Biomaterials* 27:4250–4255.

George, M. and Abraham, T.E., 2006. Polyionic hydrocolloids for the intestinal delivery of protein drugs: Alginate and chitosan—A review. *J Control Release* 114:1–14.

Gupta, S. and Vyas, S.P., 2011. Carbopol/chitosan based pH triggered *in situ* gelling system for ocular delivery of timolol maleate. *Sci Pharm* 78:959–976.

Illum, L., 1998. Chitosan and its use as a pharmaceutical excipient. *Pharm Res* 15:1326–1331.

Iyer, H., Khedkar, A., and Verma, M., 2011. Oral insulin—A review of current status. *Diabetes Obes Metab* 12:179–185.

Jelvehgari, M., Zakeri-Milani, P., Siahi-Shadbad, M.R., Loveymi, B.D., Nokhodchi, A., Azari, Z., and Valizadeh, H., 2011. Development of pH-sensitive insulin nanoparticles using Eudragit L100-55 and chitosan with different molecular weights. *AAPS PharmSciTech* 11:1237–1242.

Jintapattanakit, A., Junyaprasert, V.B., Mao, S., Sitterberg, J., Bakowsky, U., and Kissel, T., 2007. Peroral delivery of insulin using chitosan derivatives: A comparative study of polyelectrolyte nanocomplexes and nanoparticles. *Int J Pharm* 342:240–249.

Jose, S., Fangueiro, J.F., Smitha, J., Cinu, T.A., Chacko, A.J., Premaletha, K., and Souto, E.B., 2011. Cross-linked chitosan microspheres for oral delivery of insulin: Taguchi design and *in vivo* testing. *Colloids Surf B Biointerfaces* 92:175–179.

Kean, T. and Thanou, M., 2011. Biodegradation, biodistribution and toxicity of chitosan. *Adv Drug Deliv Rev* 62:3–11.

Krauland, A.H. and Alonso, M.J., 2007. Chitosan/cyclodextrin nanoparticles as macromolecular drug delivery system. *Int J Pharm* 340:134–142.

Krauland, A.H., Leitner, V.M., Grabovac, V., and Bernkop-Schnurch, A., 2006. *In vivo* evaluation of a nasal insulin delivery system based on thiolated chitosan. *J Pharm Sci* 95:2463–2472.

Langoth, N., Kahlbacher, H., Schoffmann, G., Schmerold, I., Schuh, M., Franz, S., Kurka, P., and Bernkop-Schnurch, A., 2006. Thiolated chitosans: Design and *in vivo* evaluation of a mucoadhesive buccal peptide drug delivery system. *Pharm Res* 23:573–579.

Lin, Y.H., Mi, F.L., Chen, C.T., Chang, W.C., Peng, S.F., Liang, H.F., and Sung, H.W., 2007. Preparation and characterization of nanoparticles shelled with chitosan for oral insulin delivery. *Biomacromolecules* 8:146–152.

Lin, Y.H., Sonaje, K., Lin, K.M., Juang, J.H., Mi, F.L., Yang, H.W., and Sung, H.W., 2008. Multi-ion-crosslinked nanoparticles with pH-responsive characteristics for oral delivery of protein drugs. *J Control Release* 132:141–149.

Ma, Z., Lim, T.M., and Lim, L.Y., 2005. Pharmacological activity of peroral chitosan-insulin nanoparticles in diabetic rats. *Int J Pharm* 293:271–280.

Makhlof, A., Tozuka, Y., and Takeuchi, H., 2011. Design and evaluation of novel pH-sensitive chitosan nanoparticles for oral insulin delivery. *Eur J Pharm Sci* 42:445–451.

Mura, P., Zerrouk, N., Mennini, N., Maestrelli, F., and Chemtob, C., 2003. Development and characterization of naproxen-chitosan solid systems with improved drug dissolution properties. *Eur J Pharm Sci* 19:67–75.

Muzzarelli, R.A., 1997. Human enzymatic activities related to the therapeutic administration of chitin derivatives. *Cell Mol Life Sci* 53:131–140.

Nagpal, K., Singh, S.K., and Mishra, D.N., 2010. Chitosan nanoparticles: A promising system in novel drug delivery. *Chem Pharm Bull* 58:1423–1430.

Pan, Y., Li, Y.J., Zhao, H.Y., Zheng, J.M., Xu, H., Wei, G., Hao, J.S., and Cui, F.D., 2002. Bioadhesive polysaccharide in protein delivery system: Chitosan nanoparticles improve the intestinal absorption of insulin *in vivo*. *Int J Pharm* 249:139–147.

Papadimitriou, S., Bikiaris, D., Avgoustakis, K., Karavas, E., and Georgarakis, M., 2008. Chitosan nanoparticles loaded with dorzolamide and pramipexole. *Carbohydr Polymer* 73:44–54.

Qian, F., Cui, F., Ding, J., Tang, C., and Yin, C., 2006. Chitosan graft copolymer nanoparticles for oral protein drug delivery: Preparation and characterization. *Biomacromolecules* 7:2722–2727.

Rekha, M.R. and Sharma, C.P., 2009. Synthesis and evaluation of lauryl succinyl chitosan particles towards oral insulin delivery and absorption. *J Control Release* 135:144–151.

Sadeghi, A.M., Dorkoosh, F.A., Avadi, M.R., Saadat, P., Rafiee-Tehrani, M., and Junginger, H.E., 2008. Preparation, characterization and antibacterial activities of chitosan, N-trimethyl chitosan (TMC) and N-diethylmethyl chitosan (DEMC) nanoparticles loaded with insulin using both the ionotropic gelation and polyelectrolyte complexation methods. *Int J Pharm* 355:299–306.

Sandri, G., Bonferoni, M.C., Rossi, S., Ferrari, F., Boselli, C., and Caramella, C., 2010. Insulin-loaded nanoparticles based on N-trimethyl chitosan: *In vitro* (Caco-2 model) and *ex vivo* (excised rat jejunum, duodenum, and ileum) evaluation of penetration enhancement properties. *AAPS PharmSciTech* 11:362–371.

Sandri, G., Bonferoni, M.C., Rossi, S., Ferrari, F., Gibin, S., Zambito, Y., Di Colo, G., and Caramella, C., 2007. Nanoparticles based on N-trimethylchitosan: Evaluation of absorption properties using *in vitro* (Caco-2 cells) and *ex vivo* (excised rat jejunum) models. *Eur J Pharm Biopharm* 65:68–77.

Sandri, G., Rossi, S., Bonferoni, M.C., Ferrari, F., Zambito, Y., Di Colo, G., and Caramella, C., 2005. Buccal penetration enhancement properties of N-trimethyl chitosan: Influence of quaternization degree on absorption of a high molecular weight molecule. *Int J Pharm* 297:146–155.

Sandri, G., Rossi, S., Ferrari, F., Bonferoni, M.C., Muzzarelli, C., and Caramella, C., 2004. Assessment of chitosan derivatives as buccal and vaginal penetration enhancers. *Eur J Pharm Sci* 21:351–359.

Sarmento, B., Ribeiro, A., Veiga, F., Sampaio, P., Neufeld, R., and Ferreira, D., 2007a. Alginate/chitosan nanoparticles are effective for oral insulin delivery. *Pharm Res* 24:2198–2206.

Sarmento, B., Ribeiro, A.J., Veiga, F., Ferreira, D.C., and Neufeld, R.J., 2007b. Insulin-loaded nanoparticles are prepared by alginate ionotropic pre-gelation followed by chitosan polyelectrolyte complexation. *J Nanosci Nanotechnol* 7:2833–2841.

Science, W.H., 2010. http://www.worldhealthsciences.com/diabetes-statistics-in-developed-countries.html#ixzz2CMztCnTQ.

Shah, P., Jogani, V., Mishra, P., Mishra, A.K., Bagchi, T., and Misra, A., 2007. Modulation of ganciclovir intestinal absorption in presence of absorption enhancers. *J Pharm Sci* 96:2710–2722.

Sonaje, K., Chen, Y.J., Chen, H.L., Wey, S.P., Juang, J.H., Nguyen, H.N., Hsu, C.W., Lin, K.J., and Sung, H.W., 2010. Enteric-coated capsules filled with freeze-dried chitosan/poly(gamma-glutamic acid) nanoparticles for oral insulin delivery. *Biomaterials* 31:3384–3394.

Sonaje, K., Italia, J.L., Sharma, G., Bhardwaj, V., Tikoo, K., and Kumar, M.N., 2007. Development of biodegradable nanoparticles for oral delivery of ellagic acid and evaluation of their antioxidant efficacy against cyclosporine A-induced nephrotoxicity in rats. *Pharm Res* 24:899–908.

Sonia, T.A. and Sharma, P.J., 2011. *In vitro* evaluation of N-(2-hydroxy) propyl-3-trimethyl ammonium chitosan for oral insulin delivery. *Carbohydr Polymer* 84:103–109.

Sugano, M., Watanabe, S., Kishi, A., Izume, M., and Ohtakara, A., 1988. Hypocholesterolemic action of chitosans with different viscosity in rats. *Lipids* 23:187–191.

Tapola, N.S., Lyyra, M.L., Kolehmainen, R.M., Sarkkinen, E.S., and Schauss, A.G., 2008. Safety aspects and cholesterol-lowering efficacy of chitosan tablets. *J Am Coll Nutr* 27:22–30.

Tiyaboonchai, W. and Limpeanchob, N., 2007. Formulation and characterization of amphotericin B-chitosan-dextran sulfate nanoparticles. *Int J Pharm* 329:142–149.

Trapani, A., Lopedota, A., Franco, M., Cioffi, N., Ieva, E., Garcia-Fuentes, M., and Alonso, M.J., 2010. A comparative study of chitosan and chitosan/cyclodextrin nanoparticles as potential carriers for the oral delivery of small peptides. *Eur J Pharm Biopharm* 75:26–32.

Werle, M., Takeuchi, H., and Bernkop-Schnurch, A., 2009. Modified chitosans for oral drug delivery. *J Pharm Sci* 98:1643–1656.

WHO, 2007. Definition, diagnosis and classification of diabetes mellitus and its complications: Report of a WHO Consultation. Part 1. Diagnosis and classification of diabetes mellitus.

Zhang, J., Xia, W., Liu, P., Cheng, Q., Tahirou, T., Gu, W., and Li, B., 2010a. Chitosan modification and pharmaceutical/biomedical applications. *Mar Drugs* 8:1962–1987.

Zhang, N., Li, J., Jiang, W., Ren, C., Li, J., Xin, J., and Li, K., 2010b. Effective protection and controlled release of insulin by cationic beta-cyclodextrin polymers from alginate/chitosan nanoparticles. *Int J Pharm* 393:212–218.

Zhang, X., Zhang, H., Wu, Z., Wang, Z., Niu, H., and Li, C., 2008. Nasal absorption enhancement of insulin using PEG-grafted chitosan nanoparticles. *Eur J Pharm Biopharm* 68:526–534.

Zhang, Y., Wei, W., Lv, P., Wang, L., and Ma, G., 2011. Preparation and evaluation of alginate-chitosan microspheres for oral delivery of insulin. *Eur J Pharm Biopharm* 77:11–19.

20 Chitosan and Chitosan Derivatives for Mucosal Delivery of Biopharmaceutical Drugs

Fernanda Andrade, Francisca Araújo, and Bruno Sarmento

CONTENTS

20.1 INTRODUCTION

Biopharmaceutical medicines are increasingly becoming promising tools for modern therapeutics. This new class of therapeutics is quite heterogeneous and includes different molecules such as proteins, peptides, vaccines, and nucleic acids, among others, with use in virtually all therapeutic fields (e.g., cancer and infectious diseases treatment, vaccination, and metabolic dysfunctions) and diagnostics. Since at the moment most of biopharmaceuticals are only active after parenteral administration, methods to develop successful delivery systems that may explore the mucosal pathway to exert local effect or enter into the bloodstream are emerging.

Mucosal drug delivery is a multistep process comprising targeting of the delivery system at a specific region or cell type in a mucosa, retention of that delivery system where it is anchored, drug release from the delivery system in a predetermined pattern that is not necessarily constant, and access of the drug to the drug-transport machinery in the epithelial cells (Lee 2001). Mucus can not only be seen as the first barrier for drug absorption but also as a target for delivery systems docking as prolonged retention will increase the likelihood of delivering its payload to the underlying mucosa (das Neves et al. 2011).

The interaction between mucosal tissues and chitosan has been demonstrated to benefit the absorption of drugs in special biopharmaceutical molecules because of the absorption-enhancing properties that chitosan exerts on the epithelial cell layer and its mucoadhesiveness, providing retention of the active molecules on the potential sites of absorption. Chitosan possesses a direct effect on the tight junctions of epithelial cells, acting on the expression of claudin-4 protein and its redistribution from the cell membrane to the cytosol, associated with its degradation in lysosomes and a decrease in tight junction strength, thus resulting in an increase in paracellular permeability (Yeh et al. 2011). Reconstruction of tight junctions during recovery was shown to be dependent on claudin-4 synthesis. The interaction of the positively charged amino group at the C-2 position of chitosan with the opening mechanism of the tight junctions has also been demonstrated by a decrease in ZO-1 proteins and the change in the cytoskeletal protein F-actin from a filamentous to a globular structure (Cano-Cebrian et al. 2005).

Biopharmaceutical drugs have thus been attempted to result in better bioavailability outcomes, taking into account the symbiotic association of chitosan-based delivery systems and mucosal administration.

20.2 MUCOSAL DELIVERY OF BIOPHARMACEUTICAL DRUGS: MECHANISMS OF DRUG ABSORPTION THROUGH EPITHELIAL TISSUES AND CELLS

Several mucosal routes can be found in the human organism, presenting the main benefit of being noninvasive. However, they are still not an easy pathway to drug delivery because of certain difficulties while permeating across biological barriers. Biopharmaceutical drug's absorption is a complex and dynamic process in which drugs need to cross several functional pathways in parallel to reach the systemic circulation. Thus, the achieved bioavailability mainly depends on the properties of both the drug and delivery systems and the pathophysiological state of the mucosa. These are not impenetrable barriers, but rather are cell assemblies that control cross-talk between the lumen and the lamina propria using multiple strategies. Understanding the characteristics and function of physiological barriers as well as the knowledge of the affecting factors and the mechanisms through which drugs can pass through the epithelial tissues and cells is of utmost importance.

Although each mucosal route has different features in cellular organization, both in the cell layer number of the epithelium and the several cells types that affects in different ways of the absorption, the mechanisms through which drugs pass the epithelial tissues and cells can be generally divided in two different pathways: the transcellular and the paracellular transport (Sudhakar et al. 2006; Campisi et al. 2010; Sarmento et al. 2012). In general, drugs may use both routes simultaneously to cross epithelial tissues and cells in the process of absorption, but one route is predominant over the other, depending on its physicochemical properties.

The transcellular transport of drugs occurs across the apical cell membrane, through the cytoplasm and across the basolateral membrane, either passively, by facilitated diffusion or carrier-mediated by specific transporters. Such permeation depends on drug-specific physicochemical properties, such as lipophilicity, molecular size, and shape. As the cell membrane is lipophilic in nature, hydrophilic solutes will have difficulty in permeate through the cell membrane. In general, lipophilic drugs cross the biological membrane through the transcellular pathway, this being the pathway that comprises the majority of drugs that passively cross any epithelium (Griffin and O'Driscoll 2011). Such permeation occurs more frequently in mucosa with a single cell layer epithelium, such as pulmonary mucosa at the level of alveoli and the gastrointestinal tract. This layer is constituted by several types of cells that are tightly sealed by intercellular junctions, thereby ensuring that the intercellular space is minimum essentially making it impermeable for macromolecules. Moreover, the antigen-sampling membranous cells, M-cells, have the capacity to internalize and pass drugs and particles to the underlying lymphocytes and antigen-presenting cells, being responsible for the uptake and transcytosis of some macromolecules. However, drug degradation from intracellular organelles, existing in all cell types such as lysosomes, as well as the existence of efflux transporters may prevent drug from reaching the bloodstream (Mahato et al. 2003; Griffin and O'Driscoll 2011).

The paracellular absorption occurs by passive diffusion that is concentration gradient dependent, between cells across the cell junctions. Distinct from the transcellular, this route does not depend on lipophilicity but rather in the size of proteins and its ionic charge. Because the intercellular spaces are hydrophilic in character, lipophilic compounds would have low solubility in this environment; being the hydrophilic drugs, the main drugs to cross the epithelium paracellularly. Owing to the smaller area and the resistance of the tight junctions present between adjacent cells and by the resistance of the lateral intercellular space, this route is reduced, being generally restricted to small hydrophilic molecules. Nevertheless, studies suggest that this is the major pathway through which large molecules pass across stratified epithelium, such as the buccal and vagina mucosa, since in this epithelium, the junctions are looser than in epitheliums with a single layer (Gorodeski 2001; Sudhakar et al. 2006; Campisi et al. 2010). It has the great advantage of not having proteolytic activity. In this absorption pathway, molecules such as junctional modulators may enhance the permeability of drugs and offer possibilities as absorption enhancers (Salamat-Miller and Johnston 2005; Rosenthal et al. 2012).

20.3 CHALLENGES IN MUCOSAL DELIVERY OF BIOPHARMACEUTICAL DRUGS BASED ON ADMINISTRATION ROUTE

Being based on noninvasive routes, mucosal delivery of drugs enjoys greater compliance by patients, specially the oral one, when compared to the parenteral administration. However, mucosal administration of biopharmaceuticals is challenging, as stated before, mainly due to their high molecular weight, hydrophilicity, charged nature, and immunogenic potential with consequent low bioavailability and physical/chemical instability (Morishita and Peppas 2006; Antosova et al. 2009).

There are a variety of factors related to the administration route that influences the therapeutic index of the delivered drugs, namely, the pH of organs' fluids and the presence of mucus layer, proteolytic enzymes, and macrophages. Of utmost importance is the fact that mucus is produced by mucosal cells and is continuously secreted into the lumen of the cavities it coats, being shed or digested subsequently. Is essentially composed of water (90% or more) and mucins (~5%), which are responsible for the viscoelastic gel-like properties of this fluid, although other components such as electrolytes, lipids, proteins, enzymes, immunoglobulins, nucleic acids, and cells or cell debris are also present (das Neves et al. 2011).

Although the oral route is largely preferred in the clinic for the administration of drugs (Hearnden et al. 2012), the intestinal epithelium (>200 m^2 of surface area) (DeSesso and Jacobson 2001) is almost impermeable to biopharmaceuticals, which associated to the thicker mucus layer (300, 150–400, and 800–900 μm in stomach, small intestine, and distal colon, respectively) (Derrien et al. 2010), the acidic gastric pH (1–3), and the existence of digestive proteases (DeSesso and Jacobson 2001) limit in large scale its absorption. Only low molecular weight and more hydrophobic peptidic drugs present the capacity to permeate the enterocyte membrane via the paracellular route (Salamat-Miller and Johnston 2005). The peroral bioavailability of biological drugs is also negatively affected by the hepatic first pass metabolism, which is absent in both buccal and colonic routes.

Taking into account the physiological characteristics of the oral cavity, some researchers have proposed the buccal delivery as an alternative to the peroral administration. In fact, some proteins present higher bioavailability when administered buccally, explaining the existence of a marketed buccal insulin formulation (Oral-lyn™, Generex Biotechnology, Toronto, Canada), instead of oral administration. In contrast to the gastric environment, the pH of the oral cavity is close to neutrality (5.8–7.6) and the enzymatic activity is only moderate (Patel et al. 2011). However, the "saliva wash out" phenomenon, related to the constant production and renovation of the salivary film limiting the residence time of the drugs in the oral cavity, the small surface area of absorption (≈100 cm^2), the presence of

TABLE 20.1

Marketed Nasal Formulations of Biopharmaceuticals

Trade Name(®)	Biopharmaceutical Drug
Miacalcin	Salmon calcitonin
Fortical	Salmon calcitonin
Kryptocur	Luteinizing-hormone-releasing hormone
Suprefact	Buserelin
Suprecur	Buserelin
FluMist	Influenza virus live attenuated

mucus (40–300 μm), the low fluid volume, and the existence of membrane coating granules are the major obstacles in the buccal administration of drugs (Patel et al. 2011; Sudhakar et al. 2006).

Besides the formulation/delivery device characteristics that affect the efficacy of aerosolized drugs, physiological factors of lungs and airways have a great impact on the success of inhaled medicines. Although the respiratory system presents a high surface area of absorption (100–140 m²), thin epithelium, almost neutral pH ($\approx$ 6.7), low enzymatic activity, and the absence of hepatic first-pass metabolism, the existence of a mucociliary clearance and alveolar macrophages play a vital role as barriers to the absorption of drugs (Effros and Chinard 1969; Patton 1996; Labiris and Dolovich 2003; Hastings et al. 2004). Nevertheless, the pulmonary absorption of peptidic drugs is 10- to 200-fold higher than other noninvasive routes (Andrade et al. 2011c).

However, on the respiratory system, the nasal cavity has been used for local and systemic administration of drugs and vaccination, including proteins such as calcitonin, insulin, and oxytocin (Illum 2002). Currently, there are some marketed formulations of biopharmaceuticals by nasal delivery (Table 20.1) (Kammona and Kiparissides 2012). The main limitations of nasal delivery are the small area for absorption, well-developed mucociliary clearance, and the presence of efflux systems (Rahisuddin et al. 2011).

Regarding the ocular route, the main barriers to drug absorption beyond the corneal, conjunctival, and scleral epithelium are enzymatic degradation and the quick vascular and nasolacrimal drainage of the eye (El Sanharawi et al. 2010).

To overcome the limitations of the transmucosal delivery of biopharmaceuticals, several strategies have been proposed to enhance their bioavailability, including coadministration of enzyme inhibitors and mucolytic compounds, use of absorption enhancers and mucoadhesive/mucus-penetrating systems, drug chemical modifications, and the development of novel formulation strategies (nano- and microparticles) (Gentilucci et al. 2010; Andrade et al. 2011a; Pernot et al. 2011).

20.4 CHITOSAN AND CHITOSAN DERIVATIVES-BASED FORMULATIONS FOR MUCOSAL DELIVERY OF BIOPHARMACEUTICALS

During the last few decades, biocompatible and biodegradable polymers such as chitosan, alginate, poly(lactic acid) (PLA), poly(glycolic acid) (PGA), and poly(lactic-*co*-glycolic acid) (PLGA) have been explored and used in the development of a variety of drug delivery formulations. Owing to its mucoadhesiveness and the capacity to enhance the permeability of mucosal barriers by transiently opening the tight junctions (Shi and Huang 2009), chitosan and its derivatives have been proposed as good candidates for mucosal delivery (Andrade et al. 2011a), as also recently reviewed (Silva et al. 2011; Sarmento and das Neves 2012). In addition, characteristics such as hydrophilicity, the capacity to protect the drugs from pH and enzymatic degradation, and efflux pump inhibitory properties make chitosan a good choice for the delivery of biopharmaceuticals (Bernkop-Schnürch 2000; Amidi et al. 2010; Kammona and Kiparissides 2012). Chitosan can also be used to increase the entrapment efficiency of biopharmaceutics and to control the drug release properties of other systems, by electrostatic interactions between chitosan and the peptidic drugs (Zheng et al. 2004, 2010; Kang and Song 2008). Regarding gene delivery, chitosan has gained attention as a nonviral vector due to its capacity to form polyplexes through the condensation of its positive charge with the negatively charged genetic material, being less toxic than other cationic polymers such as polyethyleneimine, polylysine, or polyarginine (Bernkop-Schnürch and Dünnhaupt 2012).

20.4.1 ORAL DELIVERY

Chitosan and its derivatives have been used in large scale in the last couple of decades as drug delivery systems for oral administration (Prego et al. 2005). Owing to its biological properties of being biocompatible and biodegradable and its low oral toxicity (LD of 16 g/kg in rats), chitosan became an attractive polymer in this pharmaceutical field. Moreover, its mucoadhesive properties improve the residence time of the drug in the small intestine, the absorption site, maintaining its integrity all the way through the gastrointestinal tract. However, owing to the basic nature of chitosan ($pK_a \approx 6.5$), it is only soluble in the acidic media (pH below 6.5), which makes it incapable of enhancing small intestinal absorption.

To overcome this, trimethyl chitosan (TMC) was developed. TMC is a partially quaternized derivative of chitosan, which is soluble under neutral and basic conditions and it is the chitosan derivative that is used the most (Hejazi and Amiji 2003; van der Merwe et al. 2004; Thompson and Ibie 2011; Vasconcelos et al. 2012). Many studies using TMC were conducted and demonstrated that it is a good enhancer in the intestinal absorption of peptides such as buserelin, both *in vitro* and *in vivo*, presenting a bioavailability between 6% and 13% in contrast to buserelin alone (0.8% bioavailability) (Luessen et al. 1996; Thanou et al. 2000a); octreotide (Sandostatin®), increasing the *in vitro* permeation by 34- to 121-fold and the *in vivo* experiments resulting in a fivefold increase in bioavailability when compared with controls (Thanou et al. 2000b); salmon calcitonin, increasing the area above the blood calcium concentration time curves (AAC) from 3.13 ± 20.50 to 448.84 ± 103.56 when compared to the calcitonin solution (Huang et al. 2011); curcumin, which exhibited enhanced bioavailability when liposomes were TMC coated (AUC = 416.58 µg/L h) than with liposomes without coating (AUC = 263.77 µg/L h) or curcumin in solution (AUC = 244.77 µg/L h) (Chen et al. 2012); and with heparin that showed that particles had higher uptake by gastrointestinal epithelium than with the chitosan and also showed that the heparin oral bioavailability had a significant increase ($p < 0.05$) in comparison with both chitosan and heparin solution (Paliwal et al. 2012).

Thiolated chitosans, another kind of chitosan derivative, formed by the derivatization of the primary amino groups of chitosan with coupling reagents bearing thiol functions, are multifunctional polymers that exhibit improved mucoadhesive, cohesive, and permeation-enhancing as well as efflux-pump-inhibitory properties. Many studies using this chitosan derivative were also done in the last few years (Werle and Bernkop-Schnürch 2008; Saremi et al. 2011; Dünnhaupt et al. 2012; Iqbal et al. 2012).

With the attempt to combine the mucoadhesion and the permeation-enhancing effects of TMC and thiolated polymers, a trimethyl chitosan–cysteine conjugate (TMC-Cys) was developed. TMC-Cys-insulin nanoparticles (TMC-Cys NP) showed a 2.1- to 4.7-fold increase in mucoadhesion compared to TMC-insulin nanoparticles (TMC NP), which might be partly attributed to disulfide formation between TMC-Cys NP and mucin. Compared to insulin solution and TMC NP, TMC-Cys NP induced increased insulin transport through rat intestine by 3.3–11.7- and 1.7–2.6-folds, promoted Caco-2 cell internalization by 7.5–12.7- and 1.7–3.0-folds, and augmented uptake in Peyer's patches by 14.7–20.9- and 1.7–5.0-folds, respectively (Figure 20.1) (Yin et al. 2009).

Insulin has been by far the peptide of choice for nanoparticle-based oral drug delivery. Many other studies trying to achieve high bioavailability of insulin using chitosan-based carriers were conducted, such as chitosan cross-linked with hydroxypropyl methylcellulose phthalate (Makhlof

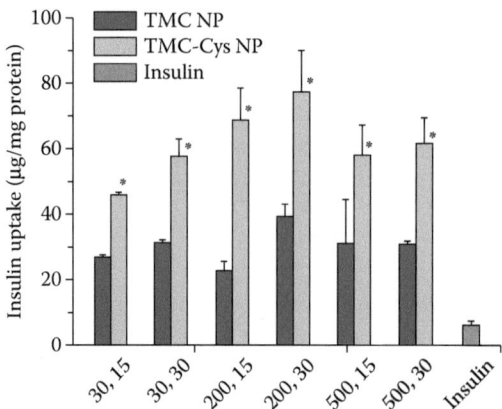

FIGURE 20.1 Uptake of TMC NP and TMC-Cys NP in Caco-2 cells at 37°C and Peyer's patches in rat intestine. The nomenclature of m and n in the x-axis refers to MW (kDa) and DQ (%) of TMC and TMC-Cys. Indicated values were means S.D. of three experiments. Significant difference from TMC NP: $p < 0.05$. (Reprinted with permission from Yin, L. et al. 2009. *Biomaterials* 30 (29):5691–700.)

et al. 2011), complexation of chitosan and dextran sulfate (Sarmento et al. 2006, 2007c), chitosan and poly(gamma-glutamic) acid (Lin et al. 2007; Sonaje et al. 2010), alginate-coated chitosan nanoparticles (Sarmento et al. 2007a,d), solid-lipid nanoparticles coated with chitosan (Sarmento et al. 2007b; Fonte et al. 2011, 2012), lecithin–chitosan nanoparticles (Hafner et al. 2009), and lauryl succinyl chitosan (Rekha and Sharma 2009). These studies presented the evidence of a significant absorption of insulin with pharmacological effects lasting between 7 and 24 h and some of them have estimated an insulin pharmacological bioavailability around 15–20% as compared to a subcutaneous injection.

Other derivatives also based on chemically modified chitosans were developed to improve the properties of chitosan for oral drug delivery, diminish the toxicity of therapeutic biomolecules, and improve their bioavailability. They are described in more detail elsewhere (Werle et al. 2009; Wang et al. 2011; Kammona and Kiparissides 2012). Some examples are formulations based on chitosan half-acetylated derivatives (Sogias et al. 2012), with poly(ethylene glycol) (Kulkarni et al. 2005; Prego et al. 2006; Casettari et al. 2012), and polyelectrolyte complexes between enoxaparin with different kinds of chitosan derivatives (Sun et al. 2010).

Chitosan and its derivatives have also been used both for immunization through oral vaccines and for gene therapy because of their properties of promoting adhesion and absorption across a mucus surface (Dass and Choong 2008). Proteins such as diphtheria toxoid, tetanus toxoid (van der Lubben et al. 2001a), and model proteins such as ovalbumin and bovine serum albumin (van der Lubben et al. 2001a,b) were loaded into chitosan microparticles and were successfully delivered to Payer's patches. Despite the increase in genetic material being delivery to enterocytes, Peyer's patches, and mesenteric lymph nodes by oral route, as shown by MacLaughlin in female New Zealand white rabbits, the levels of expression are low, which makes the transfection efficiency of chitosan very low for clinical application (MacLaughlin et al. 1998). Chitosan-based formulation parameters, such as the chitosan molecular weight (MW), its degree of deacetylation (DDA), the charge of chitosan:DNA, the genetic material concentration, and the preparation techniques of chitosan–nucleic acid particles are some features that influence the transfection efficiency (Chen et al. 2007; Wang et al. 2011). Lavertu et al. (2006) found that the maximum transfection efficiency occurred when the ratio DDA:MW moves from high DDA:low MW to low DDA:high MW. Zhen and coworkers confirmed that transfection efficiency is better in *in vitro* and *in vivo* experiments for quaternized chitosan −60% trimethylated chitosan oligomer (TMCO-60%) followed by chitosan (43–45 kDa, 87%), while the lowest transfection efficiency was obtained with chitosan (230 kDa, 90%) (Figure 20.2) (Zheng et al. 2007). Together with other polymers, it also acts prolonging gene

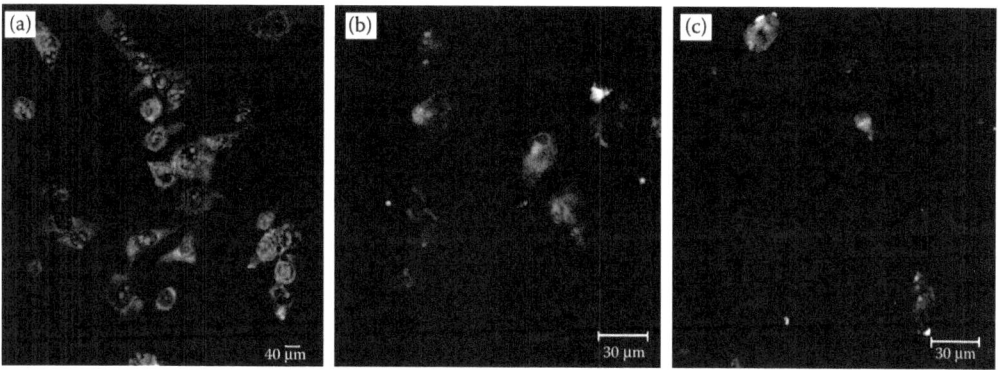

FIGURE 20.2 GFP expression after three types of chitosan/pDNA complexes transfected HCCLM6. (a) TMCO-60% as vector (×400; bar = 40 μm); (b) C(43–45 kDa, 87%) as vector (×800; bar = 30 μm); (c) C(230 kDa, 90%) as vector (×800; bar = 30 μm). (Reprinted with permission from Zheng, F. et al. 2007. *Life Sci* 80 (4):388–96.)

transfer and increasing transfection efficiency (Chen et al. 2007). A more detailed review highlighting the potential of chitosan as a component of oral gene delivery system can be found elsewhere (Bowman and Leong 2006).

20.4.2 Buccal Delivery

Chitosan is also an excellent candidate for buccal mucosa drug delivery due to its mucoadhesive properties, acting as a bioadhesive and an absorption enhancer, overcoming the quite low bioavailability characteristic of this region. It can also reduce application frequency and the drug amount necessary to obtain a pharmacological effect.

Many films and tablets based on chitosan have been developed in the last few years showing promising results for nonbiopharmaceutical drugs. Park et al. (2012) developed films consisting of ondansetron-loaded mucoadhesive gels, containing 1%, 2%, or 3% chitosan and impermeable backing layer that presented a half-life, mean residence time, and AUC (0–24 h) of about 1.7-, 1.4-, and 2.0-fold higher than those of the oral solution, respectively. Rossi and coworkers investigated the delivery of acyclovir through films based on chitosan hydrochloride (HCS) and polyacrylic acid sodium salt (PAA). All the films promoted permeation across porcine cheek epithelium when compared with acyclovir in suspension and the commercial cream. However, the penetration enhancement properties were affected by the mixing ratio of the two polymers. The film based on 1/1.3 HCS/PAA weight ratio, besides possessing the best resilience properties on the mucosa, was also characterized by the highest permeation profile and, therefore, represents a promising formulation for buccal delivery of acyclovir (Rossi et al. 2003). Lyophilized wafers from chitosan were also tested for buccal delivery. The optimized formulation were loaded with bovine serum albumin, as a protein model, with or without annealing and the cumulative percentage of drug release after 7 h was of 91.5% and 80.1% for the annealed and nonannealed wafers, respectively (Ayensu et al. 2012).

Biopharmaceutical drugs such as transforming growth factor-β (TGF-β) (Senel et al. 2000), pituitary adenylate cyclase-activating polypeptide (Langoth et al. 2005, 2006), and insulin (Portero et al. 2007) were also tested for buccal delivery with chitosan-based carriers and showed higher solubility and permeability.

20.4.3 Pulmonary Delivery

The use of chitosan and derivatives as pulmonary delivery systems of drugs was extensively reviewed elsewhere (Andrade et al. 2011b). Chitosan have shown *in vivo* the capacity to enhance the pulmonary absorption and the bioavailability of peptidic drugs such as insulin (Grenha et al. 2005; Al-Qadi et al. 2012), α-interferon (Yamada et al. 2005), and calcitonin (Makhlof et al. 2010), without toxicity, thereby being the highest bioavailability obtained using chitosan hexamers (Yamada et al. 2005). Comparing chitosan with some derivatives, TMC strongly enhances the *in vitro* permeation through Calu-3 monolayers and the bioavailability of octreotide, probably because of its higher cationic nature (Florea et al. 2006). Owing to its mucoadhesiveness, chitosan–PLGA nanospheres increased the retention time of elcatonin in lung tissue and reduced the blood calcium levels to 80% of the initial calcium concentration, prolonging its pharmacological action (Yamamoto et al. 2005). Regarding insulin, chitosan nanoparticles with different molecular weights and protein:polymer ratios presenting good encapsulation efficiencies (~50–80%) were spray-dried, and the resulting dry powder possesses adequate aerodynamic properties for deposition in deep lungs (Grenha et al. 2005; Al-Qadi et al. 2012). Moreover, besides the compatibility with lung epithelial cells (Grenha et al. 2007), nanoparticles improved the hypoglycemic effect of insulin after intratracheal instillation to rats (Figure 20.3) (Al-Qadi et al. 2012). The therapeutic effect of insulin was not dependent on the chitosan molecular weight (Al-Qadi et al. 2012). In another study, TMC nanoparticles prepared by supercritical fluid drying processes presenting characteristics suitable for inhalation were developed (Amidi et al. 2008).

In the last few decades, gene delivery to the airways has been developed for both therapeutic and vaccination proposes. The molecular weight of chitosan has influenced the transfection efficiency of the genetic material (Köping-Höggård et al. 2004; Teijeiro-Osorio et al. 2009a). Low-molecular-weight chitosan presents a higher transfection efficiency of pDNA encoding secreted alkaline phosphatase in Calu-3 cells compared to medium-molecular-weight chitosan (Teijeiro-Osorio et al. 2009a). Similarly, chitosan oligomers had a higher gene expression *in vivo* after intratracheal administration to mouse than high-molecular-weight chitosan, being the gene expression values comparable to polyethyleneimine (Köping-Höggård et al. 2004). The transfection efficiency is also dependent on the medium pH and the gene:chitosan ratio. At optimized conditions, chitosan was more efficient than Lipofectin®, mainly due to the resistence of chitosan to serum, contrary to Lipofectin (Sato et al. 2001). Aerosolized chitosan/siRNA nanoparticles administered intratracheally to mice were able to silence enhanced green fluorescent protein (EGFP) expression with 68% reduction compared to the mismatch group (Nielsen et al. 2010).

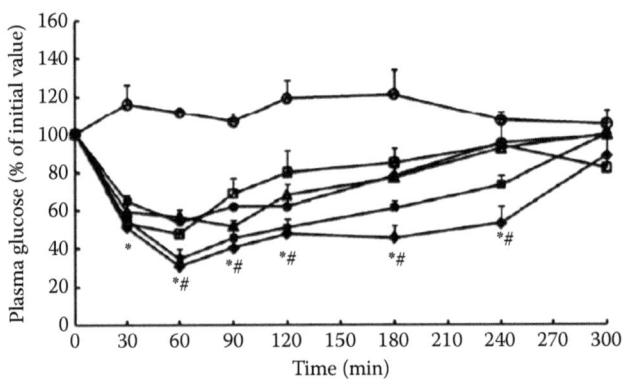

FIGURE 20.3 Hypoglycemic profiles following intratracheal administration to rats of microencapsulated insulin-loaded chitosan nanoparticles (INS-loaded CS NPs) prepared using chitosans of different MW (CS 113 and CS 213), and control formulations (mean ± SD, $n \geq 3$): (♦) microencapsulated INS-loaded CS NPs—CS 113; (■) microencapsulated INS-loaded CS NPs—CS 213; (○) microencapsulated blank (without insulin) CS NPs—CS 113; (□) mannitol microspheres containing INS; (Δ) suspension of INS-loaded CS NPs—CS 113; (●) INS solution in PBS pH 7.4; *statistically significant differences from microencapsulated blank CS NPs ($p < 0.05$); #statistically significant differences from INS solution ($p < 0.05$). (Reprinted with permission from Al-Qadi, S. et al. 2012. *J Control Release* 157 (3):383–90.)

20.4.4 Nasal Delivery

The first report of chitosan as an absorption enhancer of proteins through the nasal mucosa was published by Illum et al. in 1994 (Illum et al. 1994). Since then, many chitosan-based formulations have been proposed for nasal administration of proteins, especially insulin, and antigens for vaccination. *In vivo* studies in rats showed that insulin in 1% chitosan solution effectively reduced blood glucose levels, the therapeutic efficacy being increased by the synergic absorption enhancement of chitosan and 5% hydroxypropyl-β-cyclodextrin (Yu et al. 2004). Also, chitosan-based nanoparticles were found to be good vehicles for nasal protein administration. Insulin-loaded chitosan (Fernández-Urrusuno et al. 1999), chitosan-*N*-acetyl-L-cysteine (Wang et al. 2009), and chitosan-g-PEG (Zhang et al. 2008), chitosan/alginate (Goycoolea et al. 2009), and chitosan/cyclodextrin (Krauland and Alonso 2007; Teijeiro-Osorio et al. 2009b) nanoparticles were found to improve the mucosal absorption and/or the therapeutic index of insulin by blood glucose levels after nasal administration. The molecular weight of chitosan did not have a significant impact on the therapeutic effect of insulin (Fernández-Urrusuno et al. 1999), as referred previously for pulmonary delivery. Insulin-loaded chitosan-*N*-acetyl-L-cysteine (thiolated chitosan) nanoparticles improved the absorption of insulin across the nasal epithelium as compared to nonthiolated chitosan nanoparticles as well as chitosan solution (Wang et al. 2009). Other thiolated chitosans, chitosan-4-thiobutylamidine, and chitosan-thioglycolic acid, also presented higher *in vivo* bioavailability of insulin and leuprolide, respectively, when compared to normal chitosan and drug solutions (Krauland et al. 2006; Shahnaz et al. 2012). Thiolated nanoparticles had a 6.9-fold increase in area under the curve and a relative nasal bioavailability (versus subcutaneous injection) of 19.6% as compared to leuprolide solution (2.8% relative nasal bioavailability) (Shahnaz et al. 2012). Chitosan/cyclodextrin nanoparticles used to incorporate heparin (Krauland and Alonso 2007) and TMC nanoparticles were also proposed as vehicles for proteins, in the way that they were able to be taken up by rat nasal epithelial cells (Amidi et al. 2006).

Chitosan and its derivatives have also been extensively studied as adjuvants of intranasal vaccination because they have shown to enhance the immune response to antigens, including influenza (Wang et al. 2012), diphtheria (McNeela et al. 2000, 2004), tetanus toxoid (Vila et al. 2004; Sayin et al. 2007, 2008), hepatitis B (Borges et al. 2008; Khatri et al. 2008; Mangal et al. 2011), *Bordetella bronchiseptica* dermonecrotoxin (Kang et al. 2006, 2008), and cholera toxin (Nagamoto et al. 2004). For example, a chitosan in a 0.2 M sodium glutamate solution-based vaccine provided protection to mice against a challenge with the homologous and heterologous influenza virus, after intranasal immunization by increasing the levels of antigen-specific antibodies and the population of IFN-γ-secreting T cells (Wang et al. 2012). In another study, a diphtheria toxin vaccine based on chitosan glutamate enhanced the levels of antigen-specific IL-5 after intranasal administration to human volunteers, when compared to the vaccine without chitosan and the vaccine administered by intramuscular injection. In addition, higher levels of toxin-neutralizing antibodies in serum were found in the formulation having chitosan (McNeela et al. 2004). The complexation of TMC with mono-*N*-carboxymethyl chitosan (MCC) increased the serum levels of IgG antibody against tetanus toxoid after intranasal administration as compared to chitosan, TMC, and MCC alone (Figure 20.4), and induced both the mucosal and systemic immune responses (Sayin et al. 2008, 2009), proving the beneficial effects of polymer combination in the development of improved delivery systems. Similar result was obtained with the incorporation of Pluronic F127 in chitosan microspheres for antigen delivery (Kang et al. 2007). The decoration of chitosan particles with targeting moieties enhances the efficacy of the system. Mannosylated chitosan microspheres present the capacity to target the macrophage mannose receptors and consequently enhance the immune-stimulating activity against *Bordetella bronchiseptica* (Jiang et al. 2008).

A chitosan–DNA vaccine complex intended for respiratory syncytial virus immunization via nasal mucosa was developed by Iqbal et al. The immune response obtained after intranasal immunization was comparable to the intradermal immunization. In addition, the formulation

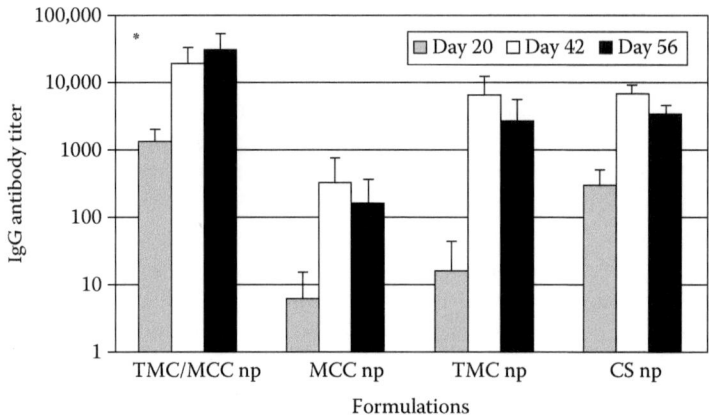

FIGURE 20.4 Comparison of tetanus toxoid (TT)-specific serum IgG antibody titers following intranasal delivery of *N*-trimethyl chitosan (TMC), mono-*N*-carboxymethyl chitosan (MCC), chitosan (CS), and TMC/MCC nanoparticles ($n = 5$, mean ± S.D.). Booster dose administration: day 22, intranasal application dose: 5 Lf TT/mice dose administered in 25 μL, *significantly higher than other formulations ($p < 0.01$). (Reprinted with permission from Sayin, B. et al. 2009. *Eur J Pharm Sci* 38 (4):362–9.)

significantly reduced the viral load in the mice lungs (Iqbal et al. 2003). In another study, a chitosan–DNA vaccine induced immunization against *Campylobacter jejuni* infection in chickens (Huang et al. 2010).

Chitosan nanoparticles can also be used for delivery of siRNA. EGFP knockdown was observed after nasal administration to transgenic EGFP mouse model (43% compared to untreated mice control) without the development of adverse effects (Howard et al. 2006).

20.4.5 OCULAR DELIVERY

Gene therapy appears as a promising therapeutic approach to treat ocular diseases that can lead to blindness such as corneal neovascularization, corneal haze, and corneal dystrophies; however, successful gene therapy remains challenging (Hao et al. 2010). Some research groups have been developing different strategies to ocular gene delivery, including the use of chitosan formulations. de la Fuente et al. developed hyaluronic acid–chitosan nanoparticles for DNA delivery. The particles, in particular, those made of low-molecular-weight chitosan presented high transfection levels to the corneal and conjunctival epithelium both *in vitro* and *in vivo* using rabbits (de la Fuente et al. 2008a,b). The formulation did not show signs of cytoxicity, ocular discomfort, or irritation (de la Fuente et al. 2008a,b; Contreras-Ruiz et al. 2010). This formulation was internalized by an active transport mechanism mediated mainly by a caveolin-dependent endocytic pathway (Contreras-Ruiz et al. 2011). DNA/chitosan oligomer complexes also enhanced the gene expression in rat corneas 5.4 times greater than polyethyleneimine-DNA (Klausner et al. 2010). However, this system was administered by injection into the corneal stroma (Klausner et al. 2010, 2012). The therapeutic efficacy of local administration in the eye needs to be evaluated. Nevertheless, these results reveal the potential of chitosan as an efficient nonviral vector for gene delivery to the eye.

20.5 CONCLUSIONS AND FUTURE PERSPECTIVES

Biopharmaceuticals are now widespread in therapy to treat or to provide relief from symptomatology related to many metabolic and oncological diseases. Advances in the field of pharmaceutical biotechnology were fundamental for these achievements. However, the delivery of biopharmaceuticals is still a major drawback against their maximum pharmacodynamic due to their physicochemical properties, poor stability, permeability, and biodistribution. In this regard, chitosan exhibits several favorable biological properties, such as biocompatibility, biodegradability, nontoxicity, and muco-adhesiveness, which made it a promising candidate for the formulation of biopharmaceutical drugs.

The development of novel biopharmaceutical delivery systems based on chitosan is a rising subject irrespective of the intended route of administration. As widely demonstrated in the past, chitosan may be fundamental to overcome the intrinsic poor mucosal bioavailability of biopharmaceutical drugs, but despite the fundamental evidences of its role, conclusive and regulatory concepts must be taken into account. The evaluation of chitosan potential for biopharmaceutical drug delivery requires clinical trials studies, as well as chitosan must pass through the quality demands of regulatory agencies to be accepted as inert excipient. In fact, chitosan has the potential to be a safe pharmaceutical excipient for nonparenteral, nonblood contact use as shown by publicly available data (Baldrick 2010), but, still, the GRAS status has had no regulatory impact on regulatory submissions for the use of chitosan in biomedical applications. To date, no chitosan products have been registered as GRAS (Kean and Thanou 2010), nor have they been granted the GRAS status by the FDA, which may be an important milestone for its industrialized use.

ACKNOWLEDGMENTS

The authors acknowledge national Portuguese funding through FCT-Fundação para a Ciência e a Tecnologia, grants SFRH/BD/73062/2010, SFRH/BD/87016/2012 and SFRH/BPD/35996/2007.

REFERENCES

Al-Qadi, S., A. Grenha, D. Carrión-Recio, B. Seijo, and C. Remuñán-López. 2012. Microencapsulated chitosan nanoparticles for pulmonary protein delivery: In vivo evaluation of insulin-loaded formulations. *J Control Release* 157 (3):383–90.

Amidi, M., E. Mastrobattista, W. Jiskoot, and W. E. Hennink. 2010. Chitosan-based delivery systems for protein therapeutics and antigens. *Adv Drug Deliv Rev* 62 (1):59–82.

Amidi, M., H. C. Pellikaan, A. H. de Boer, D. J. Crommelin, W. E. Hennink, and W. Jiskoot. 2008. Preparation and physicochemical characterization of supercritically dried insulin-loaded microparticles for pulmonary delivery. *Eur J Pharm Biopharm* 68 (2):191–200.

Amidi, M., S. G. Romeijn, G. Borchard, H. E. Junginger, W. E. Hennink, and W. Jiskoot. 2006. Preparation and characterization of protein-loaded N-trimethyl chitosan nanoparticles as nasal delivery system. *J Control Release* 111:107–16.

Andrade, F., F. Antunes, A. V. Nascimento et al. 2011a. Chitosan formulations as carriers for therapeutic proteins. *Curr Drug Discov Technol* 8 (3):157–72.

Andrade, F., F. Goycoolea, D. A. Chiappetta, J. das Neves, Alejandro S., and B. Sarmento. 2011b. Chitosan-grafted copolymers and chitosan-ligand conjugates as matrices for pulmonary drug delivery. *Int J Carbohydr Chem* 2011:1–14.

Andrade, F., M. Videira, D. Ferreira, and B. Sarmento. 2011c. Nanocarriers for pulmonary administration of peptides and therapeutic proteins. *Nanomedicine (Lond)* 6 (1):123–41.

Antosova, Z, M. Mackova, V Kral, and T Macek. 2009. Therapeutic application of peptides and proteins: Parenteral forever? *Trends Biotechnol* 27 (11):628–35.

Ayensu, I., J. C. Mitchell, and J. S. Boateng. 2012. Development and physico-mechanical characterisation of lyophilised chitosan wafers as potential protein drug delivery systems via the buccal mucosa. *Colloids Surf B Biointerfaces* 91:258–65.

Baldrick, P. 2010. The safety of chitosan as a pharmaceutical excipient. *Regul Toxicol Pharmacol* 56:290–99.

Bernkop-Schnürch, A. 2000. Chitosan and its derivatives: Potential excipients for peroral peptide delivery systems. *Int J Pharm* 194 (1):1–13.

Bernkop-Schnürch, A., and S. Dünnhaupt. 2012. Chitosan-based drug delivery systems. *Eur J Pharm Biopharm* 81 (3):463–9.

Borges, O., A. Cordeiro-da-Silva, J. Tavares et al. 2008. Immune response by nasal delivery of hepatitis B surface antigen and codelivery of a CpG ODN in alginate coated chitosan nanoparticles. *Eur J Pharm Biopharm* 69 (2):405–16.

Bowman, K., and K. W. Leong. 2006. Chitosan nanoparticles for oral drug and gene delivery. *Int J Nanomedicine* 1 (2):117–28.

Campisi, G., C. Paderni, R. Saccone, O. Di Fede, A. Wolff, and L. I. Giannola. 2010. Human buccal mucosa as an innovative site of drug delivery. *Curr Pharm Des* 16 (6):641–52.

Cano-Cebrian, M. J., T. Zornoza, L. Granero, and A. Polache. 2005. Intestinal absorption enhancement via the paracellular route by fatty acids, chitosans and others: A target for drug delivery. *Curr Drug Deliv* 2 (1):9–22.

Casettari, L., D. Vllasaliu, E. Castagnino, S. Stolnik, S. Howdle, and L. Illum. 2012. PEGylated chitosan derivatives: Synthesis, characterizations and pharmaceutical applications. *Prog Polym Sci* 37 (5):659–85.

Chen, H., J. Wu, M. Sun et al. 2012. N-trimethyl chitosan chloride-coated liposomes for the oral delivery of curcumin. *J Liposome Res* 22 (2):100–9.

Chen, J., B. Tian, X. Yin et al. 2007. Preparation, characterization and transfection efficiency of cationic PEGylated PLA nanoparticles as gene delivery systems. *J Biotechnol* 130 (2):107–13.

Contreras-Ruiz, L., M. de la Fuente, C. García-Vázquez et al. 2010. Ocular tolerance to a topical formulation of hyaluronic acid and chitosan-based nanoparticles. *Cornea* 29 (5):550–8.

Contreras-Ruiz, L., M. de la Fuente, J. E. Párraga et al. 2011. Intracellular trafficking of hyaluronic acid-chitosan oligomer-based nanoparticles in cultured human ocular surface cells. *Mol Vis* 17:279–90.

das Neves, J., M. F. Bahia, M. M. Amiji, and B. Sarmento. 2011. Mucoadhesive nanomedicines: Characterization and modulation of mucoadhesion at the nanoscale. *Expert Opin Drug Deliv* 8 (8):1085–1104.

Dass, C. R., and P. F. Choong. 2008. Chitosan-mediated orally delivered nucleic acids: A gutful of gene therapy. *J Drug Target* 16 (4):257–61.

de la Fuente, M., B. Seijo, and M. J. Alonso. 2008a. Bioadhesive hyaluronan-chitosan nanoparticles can transport genes across the ocular mucosa and transfect ocular tissue. *Gene Ther* 15 (9):668–76.

de la Fuente, M., B. Seijo, and M. J. Alonso. 2008b. Novel hyaluronic acid-chitosan nanoparticles for ocular gene therapy. *Invest Ophthalmol Vis Sci* 49 (5):2016–24.

Derrien, M., M. W. van Passel, J. H. van de Bovenkamp, R. G. Schipper, W. M. de Vos, and J. Dekker. 2010. Mucin-bacterial interactions in the human oral cavity and digestive tract. *Gut Microbes* 1 (4):254–68.

DeSesso, J. M., and C. F. Jacobson. 2001. Anatomical and physiological parameters affecting gastrointestinal absorption in humans and rats. *Food Chem Toxicol* 39 (3):209–28.

Dünnhaupt, S., J. Barthelmes, J. Iqbal et al. 2012. In vivo evaluation of an oral drug delivery system for peptides based on S-protected thiolated chitosan. *J Control Release* 160 (3):477–85.

Effros, R. M., and F. P. Chinard. 1969. The *in vivo* pH of the extravascular space of the lung. *J Clin Invest* 48 (11):1983–96.

El Sanharawi, M., L. Kowalczuk, E. Touchard, S. Omri, Y. de Kozak, and F. Behar-Cohen. 2010. Protein delivery for retinal diseases: From basic considerations to clinical applications. *Prog Retin Eye Res* 29 (6):443–65.

Fernández-Urrusuno, R., P. Calvo, C. Remuñán-López, J. L. Vila-Jato, and M. J. Alonso. 1999. Enhancement of nasal absorption of insulin using chitosan nanoparticles. *Pharm Res* 16 (10):1576–81.

Florea, B. I., M. Thanou, H. E. Junginger, and G. Borchard. 2006. Enhancement of bronchial octreotide absorption by chitosan and N-trimethyl chitosan shows linear in vitro/in vivo correlation. *J Control Release* 110 (2):353–61.

Fonte, P., F. Andrade, F. Araujo, C. Andrade, J. d. Neves, and B. Sarmento. 2012. Chitosan-coated solid lipid nanoparticles for insulin delivery. *Methods Enzymol* 508:295–14.

Fonte, P., T. Nogueira, C. Gehm, D. Ferreira, and B. Sarmento. 2011. Chitosan-coated solid lipid nanoparticles enhance the oral absorption of insulin. *Drug Deliv Transl Res* 1 (4):299–308.

Gentilucci, L., R. De Marco, and L. Cerisoli. 2010. Chemical modifications designed to improve peptide stability: Incorporation of non-natural amino acids, pseudo-peptide bonds, and cyclization. *Curr Pharm Des* 16 (28):3185–203.

Gorodeski, G. I. 2001. Vaginal-cervical epithelial permeability decreases after menopause. *Fertil Steril* 76 (4):753–61.

Goycoolea, F. M., G. Lollo, C. Remuñán-López, F. Quaglia, and M. J. Alonso. 2009. Chitosan-alginate blended nanoparticles as carriers for the transmucosal delivery of macromolecules. *Biomacromolecules* 10 (7):1736–43.

Grenha, A, C. I. Grainger, L. A. Dailey et al. 2007. Chitosan nanoparticles are compatible with respiratory epithelial cells in vitro. *Eur J Pharm Sci* 31 (2):73–84.

Grenha, A, B. Seijo, and C. Remuñán-López. 2005. Microencapsulated chitosan nanoparticles for lung protein delivery. *Eur J Pharm Sci* 25 (4–5):427–37.

Griffin, B. T., and C. M. O'Driscoll. 2011. Opportunities and challenges for oral delivery of hydrophobic versus hydrophilic peptide and protein-like drugs using lipid-based technologies. *Ther Deliv* 2 (12):1633–53.

Hafner, A., J. Lovric, D. Voinovich, and J. Filipovic-Grcic. 2009. Melatonin-loaded lecithin/chitosan nanoparticles: Physicochemical characterisation and permeability through Caco-2 cell monolayers. *Int J Pharm* 381 (2):205–13.

Hao, J., S. K. Li, W. W. Kao, and C. Y. Liu. 2010. Gene delivery to cornea. *Brain Res Bull* 81 (2–3):256–61.

Hastings, R. H., H. G. Folkesson, and M. A. Matthay. 2004. Mechanisms of alveolar protein clearance in the intact lung. *Am J Physiol Lung Cell Mol Physiol* 286 (4):L679–89.

Hearnden, V., V. Sankar, K. Hull et al. 2012. New developments and opportunities in oral mucosal drug delivery for local and systemic disease. *Adv Drug Deliv Rev* 64 (1):16–28.

Hejazi, R., and M. Amiji. 2003. Chitosan-based gastrointestinal delivery systems. *J Control Release* 89 (2): 151–65.

Howard, K. A., U. L. Rahbek, X. Liu et al. 2006. RNA interference *in vitro* and *in vivo* using a novel chitosan/siRNA nanoparticle system. *Mol Ther* 14 (4):476–84.

Huang, A., A. Makhlof, Q. Ping, Y. Tozuka, and H. Takeuchi. 2011. N-trimethyl chitosan-modified liposomes as carriers for oral delivery of salmon calcitonin. *Drug Deliv* 18 (8):562–9.

Huang, J. L., Y. X. Yin, Z. M. Pan et al. 2010. Intranasal immunization with chitosan/pCAGGS-flaA nanoparticles inhibits *Campylobacter jejuni* in a White Leghorn model. *J Biomed Biotechnol* 2010.

Illum, L. 2002. Nasal drug delivery: New developments and strategies. *Drug Discov Today* 7 (23):1184–89.

Illum, L., N. F. Farraj, and S. S. Davis. 1994. Chitosan as a novel nasal delivery system for peptide drugs. *Pharm Res* 11 (8):1186–9.

Iqbal, J., G. Shahnaz, G. Perera, F. Hintzen, F. Sarti, and A. Bernkop-Schnurch. 2012. Thiolated chitosan: Development and *in vivo* evaluation of an oral delivery system for leuprolide. *Eur J Pharm Biopharm* 80 (1):95–102.

Iqbal, M., W. Lin, I. Jabbal-Gill, S. S. Davis, M. W. Steward, and L. Illum. 2003. Nasal delivery of chitosan-DNA plasmid expressing epitopes of respiratory syncytial virus (RSV) induces protective CTL responses in BALB/c mice. *Vaccine* 21 (13–14):1478–85.

Jiang, H. L., M. L. Kang, J. S. Quan et al. 2008. The potential of mannosylated chitosan microspheres to target macrophage mannose receptors in an adjuvant-delivery system for intranasal immunization. *Biomaterials* 29 (12):1931–9.

Kammona, O., and C. Kiparissides. 2012. Recent advances in nanocarrier-based mucosal delivery of biomolecules. *J Control Release* 161 (3):781–94.

Kang, G. D., and S. C. Song. 2008. Effect of chitosan on the release of protein from thermosensitive poly(organophosphazene) hydrogels. *Int J Pharm* 349 (1–2):188–95.

Kang, M. L., H. L. Jiang, S. G. Kang et al. 2007. Pluronic F127 enhances the effect as an adjuvant of chitosan microspheres in the intranasal delivery of Bordetella bronchiseptica antigens containing dermonecrotoxin. *Vaccine* 25 (23):4602–10.

Kang, M. L., S. G. Kang, H. L. Jiang et al. 2008. Chitosan microspheres containing Bordetella bronchiseptica antigens as novel vaccine against atrophic rhinitis in pigs. *J Microbiol Biotechnol* 18 (6):1179–85.

Kang, M. L., S. G. Kang, H. L. Jiang, et al.. 2006. *In vivo* induction of mucosal immune responses by intranasal administration of chitosan microspheres containing Bordetella bronchiseptica DNT. *Eur J Pharm Biopharm* 63 (2):215–20.

Kean, T., and M. Thanou. 2010. Biodegradation, biodistribution and toxicity of chitosan. *Adv Drug Deliv Rev* 62 (1):3–11.

Khatri, K., A. K. Goyal, P. N. Gupta, N. Mishra, and S. P. Vyas. 2008. Plasmid DNA loaded chitosan nanoparticles for nasal mucosal immunization against hepatitis B. *Int J Pharm* 354 (1–2):235–41.

Klausner, E. A., Z. Zhang, R. L. Chapman, R. F. Multack, and M. V. Volin. 2010. Ultrapure chitosan oligomers as carriers for corneal gene transfer. *Biomaterials* 31 (7):1814–20.

Klausner, E. A., Z. Zhang, S. P. Wong, R. L. Chapman, M. V. Volin, and R. P. Harbottle. 2012. Corneal gene delivery: Chitosan oligomer as a carrier of CpG rich, CpG free or S/MAR plasmid DNA. *J Gene Med* 14 (2):100–8.

Krauland, A. H., and M. J. Alonso. 2007. Chitosan/cyclodextrin nanoparticles as macromolecular drug delivery system. *Int J Pharm* 340 (1–2):134–42.

Krauland, A. H., D. Guggi, and A. Bernkop-Schnürch. 2006. Thiolated chitosan microparticles: A vehicle for nasal peptide drug delivery. *Int J Pharm* 307 (2):270–7.

Kulkarni, A. R., V. I. Hukkeri, H. W. Sung, and H. F. Liang. 2005. A novel method for the synthesis of the PEG-crosslinked chitosan with a pH-independent swelling behavior. *Macromol Biosci* 5 (10):925–8.

Köping-Höggård, M., K. M. Vårum, M. Issa et al. 2004. Improved chitosan-mediated gene delivery based on easily dissociated chitosan polyplexes of highly defined chitosan oligomers. *Gene Ther* 11 (19): 1441–52.

Labiris, N. R., and M. B. Dolovich. 2003. Pulmonary drug delivery. Part I: Physiological factors affecting therapeutic effectiveness of aerosolized medications. *Br J Clin Pharmacol* 56 (6):588–99.

Langoth, N., H. Kahlbacher, G. Schoffmann et al. 2006. Thiolated chitosans: Design and *in vivo* evaluation of a mucoadhesive buccal peptide drug delivery system. *Pharm Res* 23 (3):573–9.

Langoth, N., J. Kalbe, and A. Bernkop-Schnurch. 2005. Development of a mucoadhesive and permeation enhancing buccal delivery system for PACAP (pituitary adenylate cyclase-activating polypeptide). *Int J Pharm* 296 (1–2):103–11.

Lavertu, M., S. Methot, N. Tran-Khanh, and M. D. Buschmann. 2006. High efficiency gene transfer using chitosan/DNA nanoparticles with specific combinations of molecular weight and degree of deacetylation. *Biomaterials* 27 (27):4815–24.

Lee, Vincent H. L. 2001. Mucosal Drug Delivery. *JNCI Monographs* 2001 (29):41–44.

Lin, Y. H., F. L. Mi, C. T. Chen et al. 2007. Preparation and characterization of nanoparticles shelled with chitosan for oral insulin delivery. *Biomacromolecules* 8 (1):146–52.

Luessen, H. L., B. J. de Leeuw, M. W. Langemeyer, A. B. de Boer, J. C. Verhoef, and H. E. Junginger. 1996. Mucoadhesive polymers in peroral peptide drug delivery. VI. Carbomer and chitosan improve the intestinal absorption of the peptide drug buserelin in vivo. *Pharm Res* 13 (11):1668–72.

MacLaughlin, F. C., R. J. Mumper, J. Wang et al. 1998. Chitosan and depolymerized chitosan oligomers as condensing carriers for *in vivo* plasmid delivery. *J Control Release* 56 (1–3):259–72.

Mahato, R. I., A. S. Narang, L. Thoma, and D. D. Miller. 2003. Emerging trends in oral delivery of peptide and protein drugs. *Crit Rev Ther Drug Carrier Syst* 20 (2–3):153–214.

Makhlof, A., Y. Tozuka, and H. Takeuchi. 2011. Design and evaluation of novel pH-sensitive chitosan nanoparticles for oral insulin delivery. *Eur J Pharm Sci* 42 (5):445–51.

Makhlof, A., M. Werle, Y. Tozuka, and H. Takeuchi. 2010. Nanoparticles ofglycolchitosananditsthiolatedderivativesignificantly improved thepulmonarydeliveryofcalcitonin. *Int J Pharm* 15;397(1–2):92–95.

Mangal, S., D. Pawar, N. K. Garg et al. 2011. Pharmaceutical and immunological evaluation of mucoadhesive nanoparticles based delivery system(s) administered intranasally. *Vaccine* 29 (31):4953–62.

McNeela, E. A., I. Jabbal-Gill, L. Illum et al. 2004. Intranasal immunization with genetically detoxified diphtheria toxin induces T cell responses in humans: Enhancement of Th2 responses and toxin-neutralizing antibodies by formulation with chitosan. *Vaccine* 22 (8):909–14.

McNeela, E. A., D. O'Connor, I Jabbal-Gill et al. 2000. A mucosal vaccine against diphtheria: Formulation of cross reacting material (CRM(197)) of diphtheria toxin with chitosan enhances local and systemic antibody and Th2 responses following nasal delivery. *Vaccine* 19 (9–10):1188–98.

Morishita, M., and N. A. Peppas. 2006. Is the oral route possible for peptide and protein drug delivery? *Drug Discov Today* 11 (19–20):905–10.

Nagamoto, T., Y. Hattori, K. Takayama, and Y. Maitani. 2004. Novel chitosan particles and chitosan-coated emulsions inducing immune response via intranasal vaccine delivery. *Pharm Res* 21 (4):671–4.

Nielsen, E. J., J. M. Nielsen, D. Becker et al. 2010. Pulmonary gene silencing in transgenic EGFP mice using aerosolised chitosan/siRNA nanoparticles. *Pharm Res* 27 (12):2520–7.

Paliwal, R., S. R. Paliwal, G. P. Agrawal, and S. P. Vyas. 2012. Chitosan nanoconstructs for improved oral delivery of low molecular weight heparin: *In vitro* and *in vivo* evaluation. *Int J Pharm* 422 (1–2):179–84.

Park, D. M., Y. K. Song, J. P. Jee, H. T. Kim, and C. K. Kim. 2012. Development of chitosan-based ondansetron buccal delivery system for the treatment of emesis. *Drug Dev Ind Pharm* 38 (9):1077–83.

Patel, V. F., F. Liu, and M. B. Brown. 2011. Advances in oral transmucosal drug delivery. *J Control Release* 153 (2):106–16.

Patton, J. S. 1996. Mechanisms of macromolecule absorption by the lungs. *Adv Drug Deliv Rev* 19 (1):3–36.

Pernot, M., R. Vanderesse, C. Frochot, F. Guillemin, and M. Barberi-Heyob. 2011. Stability of peptides and therapeutic success in cancer. *Expert Opin Drug Metab Toxicol* 7 (7):793–802.

Portero, A., D. Teijeiro-Osorio, M. J. Alonso, and C. Remuñán-López. 2007. Development of chitosan sponges for buccal administration of insulin. *Carbohydr Polym* 68 (4):617–25.

Prego, C., D. Torres, and M. J. Alonso. 2005. The potential of chitosan for the oral administration of peptides. *Expert Opin Drug Deliv* 2 (5):843–54.

Prego, C., D. Torres, E. Fernandez-Megia, R. Novoa-Carballal, E. Quiñoá, and M. J. Alonso. 2006. Chitosan-PEG nanocapsules as new carriers for oral peptide delivery. Effect of chitosan pegylation degree. *J Control Release* 111 (3):299–308.

Rahisuddin, P. k S., G. Garg, and M. Salim. 2011. Review on nasal drug delivery system with recent advancemnt. *Int J Pharm Pharm Sci* 3:6–11.

Rekha, M. R., and C. P. Sharma. 2009. Synthesis and evaluation of lauryl succinyl chitosan particles towards oral insulin delivery and absorption. *J Control Release* 135 (2):144–51.

Rosenthal, R., D. Gunzel, C. Finger et al. 2012. The effect of chitosan on transcellular and paracellular mechanisms in the intestinal epithelial barrier. *Biomaterials* 33 (9):2791–800.

Rossi, S., G. Sandri, F. Ferrari, M. C. Bonferoni, and C. Caramella. 2003. Buccal delivery of acyclovir from films based on chitosan and polyacrylic acid. *Pharm Dev Technol* 8 (2):199–208.

Salamat-Miller, N., and T. P. Johnston. 2005. Current strategies used to enhance the paracellular transport of therapeutic polypeptides across the intestinal epithelium. *Int J Pharm* 294 (1–2):201–16.

Saremi, S., F. Atyabi, S. P. Akhlaghi, S. N. Ostad, and R. Dinarvand. 2011. Thiolated chitosan nanoparticles for enhancing oral absorption of docetaxel: Preparation, *in vitro* and ex vivo evaluation. *Int J Nanomedicine* 6:119–28.

Sarmento, B., F. Andrade, S. B. da Silva, F. Rodrigues, J. d. Neves, and D. Ferreira. 2012. Cell-based *in vitro* models for predicting drug permeability. *Expert Opin Drug Metab Toxicol* 8 (5):607–21.

Sarmento, B., D. C. Ferreira, L. Jorgensen, and M. van de Weert. 2007a. Probing insulin's secondary structure after entrapment into alginate/chitosan nanoparticles. *Eur J Pharm Biopharm* 65 (1):10–7.

Sarmento, B., S. Martins, D. Ferreira, and E. B. Souto. 2007b. Oral insulin delivery by means of solid lipid nanoparticles. *Int J Nanomedicine* 2 (4):743–9.

Sarmento, B., A. Ribeiro, F. Veiga, and D. Ferreira. 2006. Development and characterization of new insulin containing polysaccharide nanoparticles. *Colloids Surf B Biointerfaces* 53 (2):193–202.

Sarmento, B., A. Ribeiro, F. Veiga, D. Ferreira, and R. Neufeld. 2007c. Oral bioavailability of insulin contained in polysaccharide nanoparticles. *Biomacromolecules* 8 (10):3054–60.

Sarmento, B., A. Ribeiro, F. Veiga, P. Sampaio, R. Neufeld, and D. Ferreira. 2007d. Alginate/chitosan nanoparticles are effective for oral insulin delivery. *Pharm Res* 24 (12):2198–206.

Sarmento, B., and das Neves. 2012. *Chitosan-Based Systems for Biopharmaceuticals: Delivery, Targeting and Polymer Therapeutics*. Chichester, UK: John Wiley & Sons.

Sato, T., T. Ishii, and Y. Okahata. 2001. *In vitro* gene delivery mediated by chitosan. effect of pH, serum, and molecular mass of chitosan on the transfection efficiency. *Biomaterials* 22 (15):2075–80.

Sayin, B., S. Somavarapu, X. W. Li, D. Sesardic, S. Senel, and O. H. Alpar. 2009. TMC-MCC (N-trimethyl chitosan-mono-N-carboxymethyl chitosan) nanocomplexes for mucosal delivery of vaccines. *Eur J Pharm Sci* 38 (4):362–9.

Sayin, B., S. Somavarapu, X. W. Li et al. 2008. Mono-N-carboxymethyl chitosan (MCC) and N-trimethyl chitosan (TMC) nanoparticles for non-invasive vaccine delivery. *Int J Pharm* 363 (1–2):139–48.

Sayin, B., S. Somavarapu, X. G. Li, O. Alpar, and S. Senel. 2007. A new vaccine delivery system with chitosan derivatives for intranasal immunization: In vitro/in vivo evaluations. In *Advances in Chiton Science—Vol. X*, edited by S. Senel, K. M. Varum, M. M. Sumnu, and A. A. Hincal. Ankara: Alp Ofset.

Senel, S., M. J. Kremer, S. Kas, P. W. Wertz, A. A. Hincal, and C. A. Squier. 2000. Enhancing effect of chitosan on peptide drug delivery across buccal mucosa. *Biomaterials* 21 (20):2067–71.

Shahnaz, G., A. Vetter, J. Barthelmes et al. 2012. Thiolated chitosan nanoparticles for the nasal administration of leuprolide: Bioavailability and pharmacokinetic characterization. *Int J Pharm* 428 (1–2):164–70.

Shi, Y., and G. Huang. 2009. Recent developments of biodegradable and biocompatible materials based micro/nanoparticles for delivering macromolecular therapeutics. *Crit Rev Ther Drug Carrier Syst* 26 (1): 29–84.

Silva, S. B. da, J. Fernandes, F. Tavira, M. Pintado, and B. Sarmento. 2011. The Potential of Chitosan in Drug Delivery Systems. In *Focus on Chitosan Research*, edited by A. N. Ferguson, and A. G. O'Neill: Nova Publisher.

Sogias, I. A., A. C. Williams, and V. V. Khutoryanskiy. 2012. Chitosan-based mucoadhesive tablets for oral delivery of ibuprofen. *Int J Pharm* 436 (1–2):602–10.

Sonaje, K., Y. J. Chen, H. L. Chen et al. 2010. Enteric-coated capsules filled with freeze-dried chitosan/poly(gamma-glutamic acid) nanoparticles for oral insulin delivery. *Biomaterials* 31 (12):3384–94.

Sudhakar, Y., K. Kuotsu, and A. K. Bandyopadhyay. 2006. Buccal bioadhesive drug delivery—A promising option for orally less efficient drugs. *J Control Release* 114 (1):15–40.

Sun, W., S. Mao, Y. Wang et al. 2010. Bioadhesion and oral absorption of enoxaparin nanocomplexes. *Int J Pharm* 386 (1–2):275–81.

Teijeiro-Osorio, D., C. Remuñán-López, and M. J. Alonso. 2009a. Chitosan/cyclodextrin nanoparticles can efficiently transfect the airway epithelium in vitro. *Eur J Pharm Biopharm* 71 (2):257–63.

Teijeiro-Osorio, D., C. Remuñán-López, and M. J. Alonso. 2009b. New generation of hybrid poly/oligosaccharide nanoparticles as carriers for the nasal delivery of macromolecules. *Biomacromolecules* 10 (2): 243–9.

Thanou, M., B. I. Florea, M. W. Langemeyer, J. C. Verhoef, and H. E. Junginger. 2000a. N-trimethylated chitosan chloride (TMC) improves the intestinal permeation of the peptide drug buserelin *in vitro* (Caco-2 cells) and *in vivo* (rats). *Pharm Res* 17 (1):27–31.

Thanou, M., J. C. Verhoef, P. Marbach, and H. E. Junginger. 2000b. Intestinal absorption of octreotide: N-trimethyl chitosan chloride (TMC) ameliorates the permeability and absorption properties of the somatostatin analogue *in vitro* and in vivo. *J Pharm Sci* 89 (7):951–7.

Thompson, C., and C. Ibie. 2011. The oral delivery of proteins using interpolymer polyelectrolyte complexes. *Ther Deliv* 2 (12):1611–31.

van der Lubben, I. M., J. C. Verhoef, G. Borchard, and H. E. Junginger. 2001a. Chitosan for mucosal vaccination. *Adv Drug Deliv Rev* 52 (2):139–44.

van der Lubben, I. M., J. C. Verhoef, A. C. van Aelst, G. Borchard, and H. E. Junginger. 2001b. Chitosan microparticles for oral vaccination: Preparation, characterization and preliminary *in vivo* uptake studies in murine Peyer's patches. *Biomaterials* 22 (7):687–94.

van der Merwe, S. M., J. C. Verhoef, J. H. Verheijden, A. F. Kotzé, and H. E. Junginger. 2004. Trimethylated chitosan as polymeric absorption enhancer for improved peroral delivery of peptide drugs. *Eur J Pharm Biopharm* 52 (2):225–35.

Vasconcelos, T., P. Barrocas, and R. Cerdeira. 2012. Use of Chitosan and Derivatives in Conventional Biopharmaceutical Dosage Forms Formulation. In *Chitosan-Based Systems for Biopharmaceuticals*, edited by B. Sarmento, and J. d. Neves: John Wiley & Sons, Ltd.

Vila, A., A. Sánchez, K. Janes et al. 2004. Low molecular weight chitosan nanoparticles as new carriers for nasal vaccine delivery in mice. *Eur J Pharm Biopharm* 57 (1):123–31.

Wang, J. J., Z. W. Zeng, R. Z. Xiao et al. 2011. Recent advances of chitosan nanoparticles as drug carriers. *Int J Nanomedicine* 6:765–74.

Wang, X., W. Zhang, F. Liu et al. 2012. Intranasal immunization with live attenuated influenza vaccine plus chitosan as an adjuvant protects mice against homologous and heterologous virus challenge. *Arch Virol* 157 (8):1451–61.

Wang, X., C. Zheng, Z. M. Wu et al. 2009. Chitosan–NAC nanoparticles as a vehicle for nasal absorption enhancement of insulin. *J Biomed Mater Res B Appl Biomater* 88B:150–61.

Werle, M., and A. Bernkop-Schnürch. 2008. Thiolated chitosans: Useful excipients for oral drug delivery. *J Pharm Pharmacol* 60 (3):273–81.

Werle, M., H. Takeuchi, and A. Bernkop-Schnürch. 2009. Modified chitosans for oral drug delivery. *J Pharm Sci* 98 (5):1643–56.

Yamada, K, M. Odomi, N. Okada, T. Fujita, and A. Yamamoto. 2005. Chitosan oligomers as potential and safe absorption enhancers for improving the pulmonary absorption of interferon-alpha in rats. *J Pharm Sci* 94 (11):2432–40.

Yamamoto, H., Y. Kuno, S. Sugimoto, H. Takeuchi, and Y. Kawashima. 2005. Surface-modified PLGA nanosphere with chitosan improved pulmonary delivery of calcitonin by mucoadhesion and opening of the intercellular tight junctions. *J Control Release* 102:373–81.

Yeh, T.-H., L.-W. Hsu, M. T. Tseng et al. 2011. Mechanism and consequence of chitosan-mediated reversible epithelial tight junction opening. *Biomaterials* 32 (26):6164–73.

Yin, L., J. Ding, C. He, L. Cui, C. Tang, and C. Yin. 2009. Drug permeability and mucoadhesion properties of thiolated trimethyl chitosan nanoparticles in oral insulin delivery. *Biomaterials* 30 (29):5691–700.

Yu, S, Y Zhao, F Wu et al. 2004. Nasal insulin delivery in the chitosan solution: *In vitro* and *in vivo* studies. *Int J Pharm* 281 (1–2):11–23.

Zhang, X., H. Zhang, Z. Wu, Z. Wang, H. Niu, and C. Li. 2008. Nasal absorption enhancement of insulin using PEG-grafted chitosan nanoparticles. *Eur J Pharm Biopharm* 68 (3):526–34.

Zheng, C. H., J. Q. Gao, Y. P. Zhang, and W. Q. Liang. 2004. A protein delivery system: Biodegradable alginate-chitosan-poly(lactic-*co*-glycolic acid) composite microspheres. *Biochem Biophys Res Commun* 323 (4):1321–7.

Zheng, F., X. W. Shi, G. F. Yang et al. 2007. Chitosan nanoparticle as gene therapy vector via gastrointestinal mucosa administration: Results of an *in vitro* and *in vivo* study. *Life Sci* 80 (4):388–96.

Zheng, X., Y. Huang, C. Zheng, S. Dong, and W. Liang. 2010. Alginate-chitosan-PLGA composite microspheres enabling single-shot hepatitis B vaccination. *AAPS J* 12 (4):519–24.

21 Chitosans as Delivery Vectors in Biomedical Applications

Vandana Patravale and Swati Vyas

CONTENTS

21.1 INTRODUCTION

A rare, cationic mucopolysaccharide, chitosan (CS) is composed of D-glucosamine and *N*-acetyl-D-glucosamine monomeric units. Produced from chitin, which is simply *N*-acetylated CS, it is nature's most plentiful biopolymer after cellulose. Chitin and CS are present in cell walls of microbes such as fungi and form the basic skeletal material of invertebrates (Hirano 2002), thereby supporting numerous living organisms (Van et al. 2006). CS polymers have an assortment of important qualities such as biodegradability, compatibility, pH sensitivity, nonantigenicity, and nonallergenicity that make them significant for biomedical applications.

CS is polycationic, attributed by the occurrence of an amino functionality at the 2-position as a substitute of the hydroxyl group usually seen in sugars; this sets it apart from other natural polysaccharides and thus, CS lends itself to structural alterations (Ravi Kumar 2000). Several efforts have been made to prepare functional derivatives of CS with enhanced properties to generate CSs that are precise for delivery applications. The features indispensable for each application dictate the type of modifications that have been conducted and which may impart suitable function. For instance, in gene delivery, the amino group in CS should remain protonated at physiological pH to maintain the stability of the CS–plasmid complex held by electrostatic bonds (Mei et al. 2009). This criterion is fulfilled by quaternizing the amino group that then remains positively charged at all pH values (Thanou et al. 2000). Similarly, conjunction with sugars such as galactose, which are recognized by certain cell surface receptors fortifies the resulting derivative with cell specificity and therefore, enhanced drug-targeting ability (Jiang et al. 2008b). Many workers have successfully constructed conjugates of CS with functionally important molecules. These conjugates couple properties of CS with those of the biomolecule attached to it. The combined effect either overcomes the limitations of the biomolecule or leads to a functionally better candidate. Similarly, poor solubility of CS at physiological pH is surpassed by altering it to yield more water-soluble derivatives such as N-succinyl chitosan (Aiedeh ad Taha 2001).

Some significant uses of CS include gene transport (Park et al. 2010), drug conveyance (Mura et al. 2012), vaccine delivery (Gunbeyaz et al. 2010), and as adjuvant for imaging in diagnostics (Willman et al. 2008). Recent literature reports insightful developments taking place with respect to different chemical modifications of CS that impart desirable biological properties to it (Dutta et al. 2004). This chapter views each area mentioned above, discusses recently reported strategies to modulate CS toward achieving derivatives with enhanced delivery function that surpass unmodified CS. The focal point is *en route* to understanding the development of CS polymers by optimization of their properties toward applicability in the above fields.

The chapter will also look at the strengths, limitations, and challenges of gene, drug and vaccine delivery, molecular imaging, as well as applications of CS nanoparticles in these subject areas. The various techniques for generation of CS nano- and microparticles are also discussed. In the last part of this chapter, the versatility of dosing CS-based formulations by different routes of administration is described in detail. In summary, this chapter highlights the potential role and application of CS and derivatives as an important delivery mediator and constituent in biomedicine.

21.2 BIOMEDICAL APPLICATIONS OF CHITOSAN AS A NANOSIZED DELIVERY VECTOR

The inherent properties of CS confer upon it key attributes necessary for delivery vectors to convey drugs and even genes, to various endogenous sites. Interestingly, CS structurally provides avenues for covalent modification, which have been used by several researchers to direct the delivery of CS-based nanocarriers to specific tissues and tumors.

21.2.1 CHITOSAN: A NONVIRAL VECTOR IN GENE TRANSPORT

The introduction of nucleic acids into cells as a universal remedy for genetic disorders has been extensively researched by various workers. Cell transfection with genetic material is a useful line of therapy for treating disorders such as cancer that arise due to defective genes. This involves either transfection of cells directly with naked plasmid (pDNA, siRNA, miRNA, and antisense RNA) (Kim et al. 2008) or a more efficient transfection using viral or nonviral carrier–plasmid complexes (Park et al. 2010), which transport the therapeutic gene specifically to afflicted cells. Viral carriers, however, carry the risk of inducing adverse immune responses (Figure 21.1); therefore, biocompatible, nonimmunogenic CS vectors have a strategic advantage in gene delivery. Nevertheless, nucleic acid therapy based on carrier–plasmid complexes, in addition, has potential uses in immune response modulation, regenerative medicine, deoxyribonucleic acid (DNA) vaccination, and bacterial transformation.

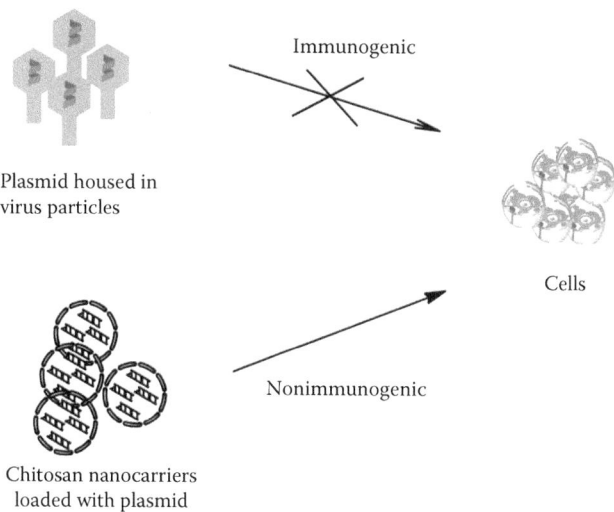

FIGURE 21.1 Viral versus nonviral carriers for gene delivery.

There are some key points that influence gene delivery to the target cell. First, gene transport requires the genes to reach the cell nucleus intact. This is often difficult as certain physical, chemical, and biological factors govern the stability of the gene–carrier complex. The complex is susceptible to dissociation *per se* during storage or once it is exposed to physiological pH in the biological environment. Transmembrane transport of plasmid–vector complexes occurs via endocytosis (Dutta et al. 2004) and the extent to which the complexes are endocytosed affects transfection efficiency (Pillai et al. 2009). Other complications comprise intracellular transport to the nucleus, nuclear localization, and failure of the carrier–plasmid complex to dissociate once inside the nucleus. Degradative action by endolysosomes is yet another determinant for successful gene delivery (Ravi Kumar 2000).

CS offers multiple facets to enhance gene transfection. To begin with, the presence of an amino group in its structure (Figure 21.2) upon protonation allows CS to form stable ionic complexes with negatively charged DNA. Second, CS protects the plasmid from nuclease degradation. Apart from being biodegradable, nonallergenic, biocompatible, and nonimmunogenic, CS is easily amenable for chemical modifications. A direct advantage of this is cell-specific gene transport using tailor-made CS polymers (Jayakumar et al. 2010).

CS-based gene transfer is influenced by ionic as well as nonionic interactions occurring between the carbohydrate backbone of CS and surface proteins of transfected cells. Being a membrane perturbant due to its cationicity, CS can penetrate the membrane lipid bilayer. Another route of cellular

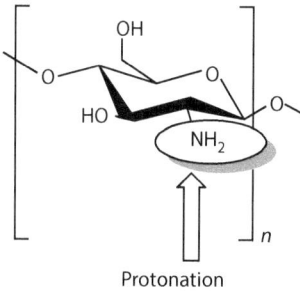

FIGURE 21.2 Structure of CS. The amino species in CS are the most reactive group that lends itself readily to chemical conjugations with various biomolecules.

entry involves endocytosis and incorporation within the endosome. In the normal course of events, this endosome fuses with lysosome forming the endolysosome, which then triggers degradation of its contents. However, CS–plasmid complex escapes the endosomal sac by virtue of its membrane perturbant activity and localizes at the nucleus. Within the nucleus, the complex dissociates and the plasmid is released, which is then free to execute its function (Gill et al. 1997). In addition, it is reported that CS protects the plasmid from degradation by the nucleases present in the vicinity (Ozbas-Turan et al. 2003).

The focal downside of plain CS is its low transfection efficiency. Unmodified CS shows poor gene transfection efficiency because it exhibits a transfection mechanism that relies on pH. Because the –NH$_2$ functionality in CS protonates in an acidic atmosphere, the stability of the CS–DNA electrostatic complex with DNA is confined to an environment with pH below 6. At neutral pH, the plasmid is likely to be released before nuclear transfection is achieved. Furthermore, protonation of amino groups in CS is essential for endosomal escape (Erbacher et al. 1998) by the carrier–plasmid complex so as to prevent degradation by lysosome, which is also pH dependent (Mao et al. 2001). Consequently, derivatives of CS have been developed to improve transfection efficiency.

Nevertheless, for the remedial gene to exert its therapeutic effect, it is crucial for the plasmid–carrier complex to dissociate within the nuclear surroundings. So far, studies have indicated that although the therapeutic effects due to introduction of CS–plasmid complex are evident, the mechanism behind dissociation of the plasmid from CS or its derivatives within the nucleus is unclear (Figure 21.3). More investigations are being conducted to determine gene release profiles in view of gene therapy.

Enhancement of transfection efficiency of CS consists of alterations in the structure of CS that encompass functional group modifications, such as quaternization or chemical conjugation, with ligand moieties such as transferrin that are recognized by cell surface receptors or copolymerization (e.g., with polyethylene glycol [PEG]). The molecular weight, degree of deacetylation, pH of the transfection medium, and cell type affect the transfection efficiency of CS. As the molecular weight increases, transfection efficiency also improves as does a moderate degree of deacetylation (Zhang and Neau 2001a). High levels of transfection can be achieved at optimal pH values between 6.8 and 7.0 (Huang et al. 2005). Higher gene transfer efficiency was achieved in human embryonic kidney 293 (HEK 293) cells as compared to other cell types (Mao et al. 2001). These design strategies have resulted in derivatives that are more stable, resistant to enzymatic degradation, and effectively surpass physiological barriers, thereby facilitating gene transport inside the nuclear environment.

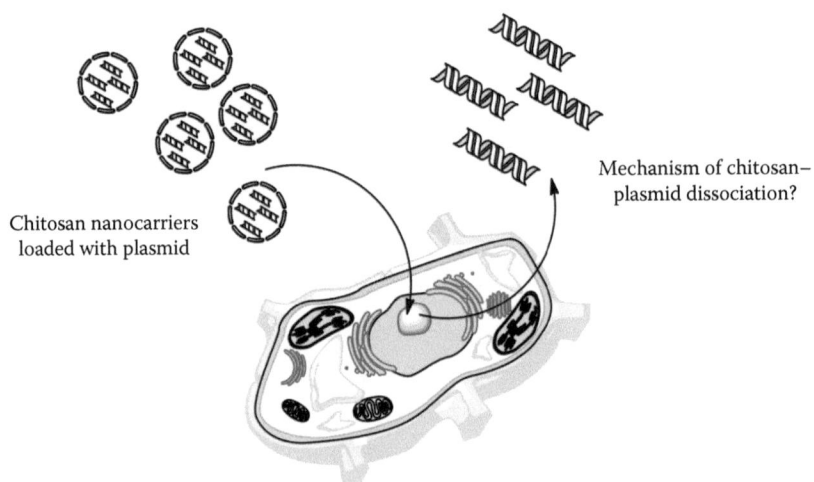

Mechanism of chitosan–plasmid dissociation?

Chitosan nanocarriers loaded with plasmid

FIGURE 21.3 Plasmid release from CS–plasmid complex within the nucleus.

Several covalent modifications have been achieved by many groups to increase transfection efficiency, for instance, quaternization (Lu et al. 2009), deoxycholic acid modification (Kim et al. 2001), galactosylation (Park et al. 2001), polyethyleneimine (PEI) grafting (Erbacher et al. 1998), and thiolation (Zhao et al. 2010). Table 21.1 describes the derivatization strategies that have been carried out for increasing transfection efficiency. Table 21.1 highlights the results produced by bringing about such chemical modifications of CS.

21.2.2 Drug Delivery: Passive and Active Targeting

Interestingly, CS appositely fulfills the criteria for ideal nanocarriers; its biodegradability, nontoxicity, and affordability make it a well-suited targeting adjuvant. It can be formulated as a nanoparticulate carrier for active and passive targeting. The 6-hydroxy and 2-amino groups of CS show high chemical reactivity and can be covalently conjugated to ligands recognized by receptors on the apical membranes of specific cell surfaces. A range of CS polymers thus produced have been fabricated into nanosized formulations specific for a given tissue predicament.

CS nanoparticles, just like any other xenobiotic, are recognized by the endogenous reticuloendothelial system (RES), which brings about rapid clearance of foreign particles. They can, therefore, be used to target RES agents such as macrophages with immunotherapeutic drugs. However, this also poses a major difficulty in passive targeting if the drug is meant to bypass the RES and reach other tissues (Fahmy et al. 2005). To circumvent this problem, the CS nanocarrier should have sufficient hydrophilicity to escape being detected by the RES, which preferably recognizes hydrophobic substances. This has been accomplished by chemical conjugation of PEG with CS. The PEG portion of the conjugate (Figure 21.4) imparts hydrophilicity attributed to the presence of multiple hydroxyl groups to the nanocarrier; thus, the nanocarrier exhibits reduced RES clearance (Vasir et al. 2005).

In contrast, CS has been hydrophobically modified using cholanic acid and this derivative (Figure 21.5) can passively deliver paclitaxel to tumor-bearing tissues. This is facilitated by enhanced permeability retention (EPR) effect, where increased vascular permeability of blood vessels present in neoplastic tissues allows nanoparticles to concentrate in these areas (Kim et al. 2006). The branching of cholanic acid was achieved through 1-ethyl-3-(3-dimethylaminopropyl) carbodiimide (EDC)-mediated coupling of glycol CS. N-Hydroxysuccinimide (NHS) forms an activated ester with cholanic acid that eases coupling with amino functionality of CS.

The grafting of succinic acid onto CS via an amide bond to bring about an increase in the pK_a of the resulting derivative has been proposed. The resulting N-succinylated derivative (Figure 21.6) of CS is soluble at basic values of pH due to the anionic, free carboxyl group, but is insoluble in acidic media; as a nanocarrier, it can effect the release of drugs in the colon where the pH is alkaline (Aiedeh and Taha 1999). This makes it suitable for colon-specific targeting of drugs, such as 5-aminosalicylic acid and diclofenac (Mura et al. 2012).

Because CS cannot be used as such for cell-specific active targeting, it is chemically attached to receptor-specific ligand molecules, such as galactose (Sashiwa et al. 2000a), folic acid (Lu and Low 2003), and transferrin (Wagner et al. 1990) to generate nanocarriers that transport drugs to cells selectively.

Several receptors are overexpressed on the cell surfaces of cancerous tissues due to genetic manipulations. These receptors include folate and transferrin receptors that become abundant on the apical surface of cancerous cells and, therefore, are good anticancer targets (Kwon 2008).

CS has been widely used in pharmaceutical and biomedical areas due to a plethora of valuable biological properties, such as biodegradability, biocompatibility, low toxicity, hemostatic, bacteriostatic, fungistatic, antineoplastic, and anticholesterolemic properties, as well as low cost (Liu et al. 2001b). The advantages of using CS as a vector include the ability to administer repeatedly with minimal host immune response, targetability through bioconjugation, stability in storage, and easy production in large quantities (Chen et al. 2007).

TABLE 21.1
Derivatives of CS for Gene Delivery

Derivative	Structure	Significance	Cell Line Tested	Transfection Efficiency
Quaternary ammonium derivatives: trimethyl chitosan chloride (Thanou et al. 2002)		pH-independent transfection mechanism Positive quaternary nitrogen: stable complex formation with plasmid	COS-1 cells Caco-2 cells	500 times higher No increase
CS–EDTA conjugates (Loretz and Bernkop-Schnurch 2006)		Metal-ion chelation: protection from enzymatic degradation Cross-linking: CS–DNA complex stabilization	Caco-2 cells	35% higher
CS–DNA–ligand complexes Galactosylated CS (Park et al. 2003)		Galactose recognition: liver-specific delivery	HepG2 cells	300 times higher

Name/Reference	Structure	Function	Cells	Enhancement
Transferrin–CS conjugate (Wagner et al. 1990)		Transferrin-mediated endocytosis: anticancer targeting in gene therapy	HEK293 cells HeLa cells	40 times increase 50% higher
Deoxycholic acid-modified CS (Kim et al. 2001)		Amphiphilic nature: self-assembly in water, thereby protecting hydrophobic DNA	COS-1 cells	8 times higher
Dodecylated CS (Liu et al. 2001a)		Long-chain dodecyl group: thermal stability enhancement of DNA	HEK293 cells	Higher
CS–PEG conjugate (Mao et al. 2001)		Hydrophilic PEG: RES escape Erythrocyte aggregation is prevented Bacterial surface growth inhibition	HEK293 cells	25 times higher

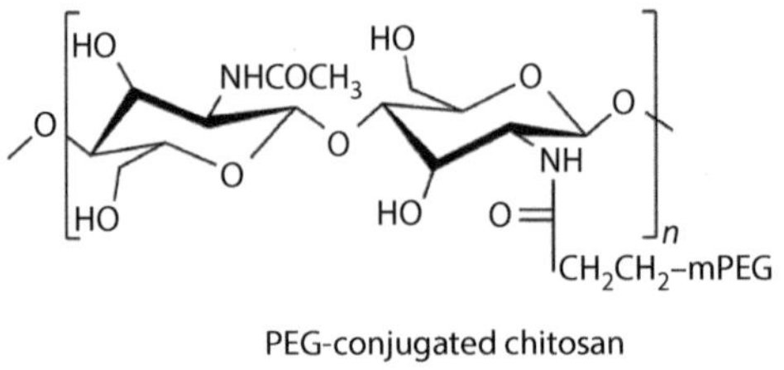

PEG-conjugated chitosan

FIGURE 21.4 PEGylated CS–hydrophilic derivative that escapes RES recognition.

Polycationic carriers that incorporate cell-specific ligands such as galactose have been shown to improve delivery owing to its recognition by the asialoglycoprotein receptors present on the hepatocyte surface. Chemical linkage of galactose is brought about by esterification of galactose to lactobionic acid–conjugated CS. Lactobionic acid serves the purpose of a 4-carbon linker, incorporated in the conjugate to mediate the attachment of galactose to CS. Galactosylated CS (Figure 21.7) shows good hepatocyte targeting due to galactose recognition; thus, the nanocarriers obtained from this derivative are specifically endocytosed by liver cells. This derivative is useful in targeting drugs for treating hepatic clinical conditions, such as liver metastasis (Kato et al. 2001) and acute rejection following xenotransplantation (Sashiwa et al. 2000a). The results of the study show the liver-focused targeting capability of galactosylated CS in comparison with the passive, nonspecific targeting ability of plain CS.

Many mammalian cells possess transferrin receptors, which function to bring essential molecules such as iron into the cells. Small-molecular-weight drugs, bioactive macromolecules, and liposomes have been efficiently transported into cells by transferrin-recognizing receptors via endocytosis (Wagner et al. 1990); plasmid DNA and oligonucleotides were delivered by this mechanism. In addition, oligonucleotides or polycations attached to transferrin or antibodies against transferrin receptor were complexed with plasmid DNA (pDNA) (Cheng 1996). The import and the expression of genetic material in cells affluent with transferrin receptors on their surface are facilitated by these delivery systems.

CS has been linked to transferrin molecule by two methods—reversible conjugation and Schiff's base formation with periodate-oxidized transferrin. The researchers have reported that transferrin-bound CS on treatment with a solution of DNA formed polyelectrolyte complexes to yield DNA-containing

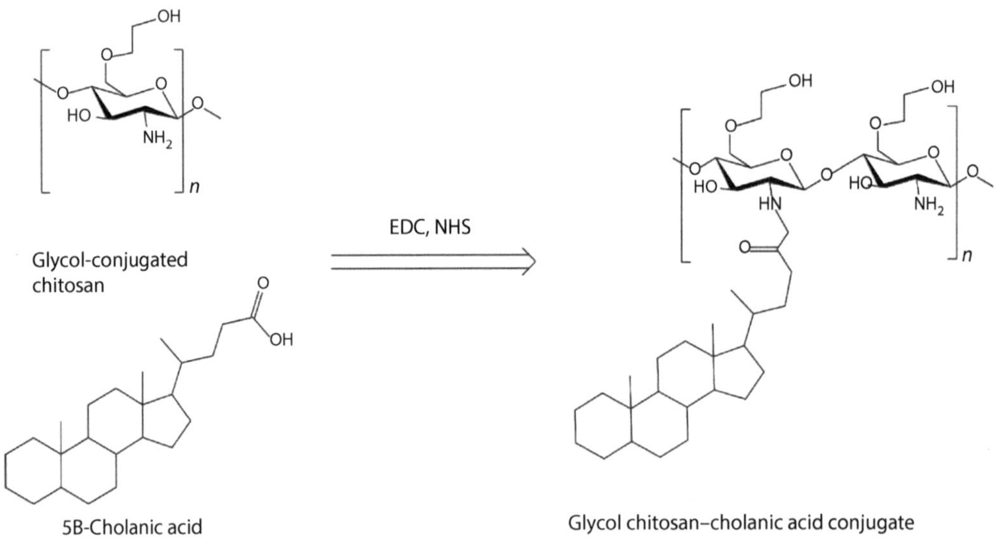

Glycol-conjugated chitosan

EDC, NHS

5B-Cholanic acid

Glycol chitosan–cholanic acid conjugate

FIGURE 21.5 Cholanic acid arms integrated with CS—an excellent passive carrier for antineoplastic agents.

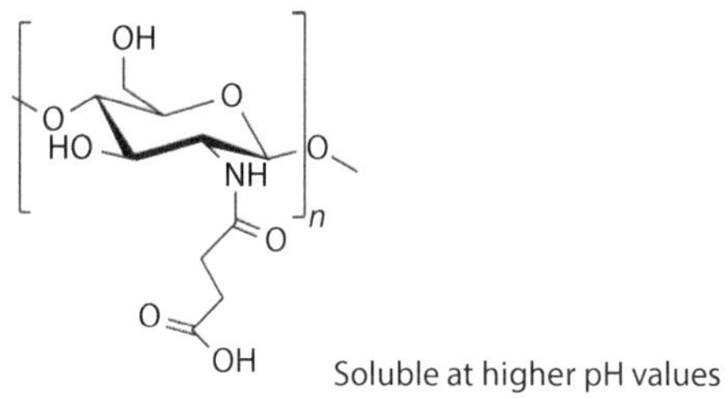

FIGURE 21.6 *N*-Succinyl chitosan–colon-specific drug carrier.

nanoparticles. Transferrin-conjugated nanoparticles (Figure 21.8) have shown DNA transfection efficiency around fourfold higher than nanoparticles without transferrin in HEK 293 cells and around 50% higher in HeLa cells (Wagner et al. 1990).

In addition, C-terminal globular domain of transferrin (KNOB) conjugated to CS via disulfide linkages resulted in a significant 130-fold increase in gene expression (Wagner et al. 1990).

Overexpression of folate receptors localized to apical surfaces of polarized epithelia in various cancer cells is a known phenomenon (Zhao and Lee 2004). Because folate receptors are seldom seen on the apical surfaces of normal cells, the use of folic acid as a tumor-selective ligand has been extensively applied to anticancer cell targeting (Lu et al. 2003). Covalent linkage of folic acid molecules to polymers brings about internalization of the folate–polymer conjugates into cells via receptor-mediated endocytosis (Dube et al. 2002). Therefore, tumor cells that express folate-binding receptors are targeted using folate as a ligand for directing proteins (Ward et al. 2000), liposomes (Gabizon et al. 2004), and other molecules (Kalber et al. 2011). In addition, recycling of folate receptors in target cells can lead to further accumulation of folate conjugates.

Structurally, the folic acid molecule contains two carboxyl groups at α- and γ-positions (Figure 21.9). Carbodiimide-mediated coupling of folic acid to amino groups at the γ- position yields a bioconjugate that exhibits a much higher reactivity. Furthermore, folate linked via its γ-carboxyl group retains a strong affinity toward its receptor (Wang et al. 1997).

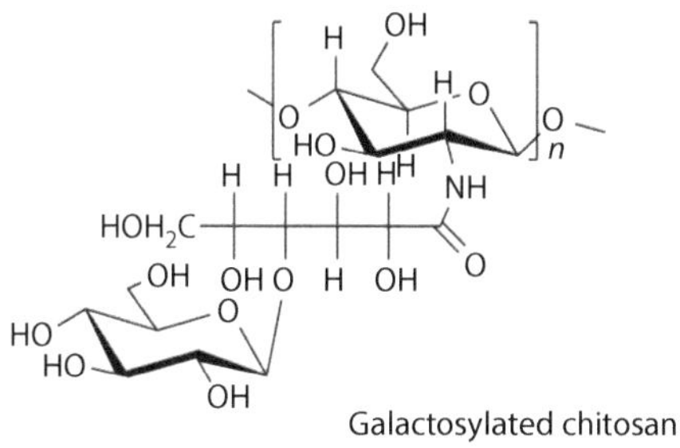

Galactosylated chitosan

FIGURE 21.7 Asialoglycoprotein receptor-targeted CS derivative.

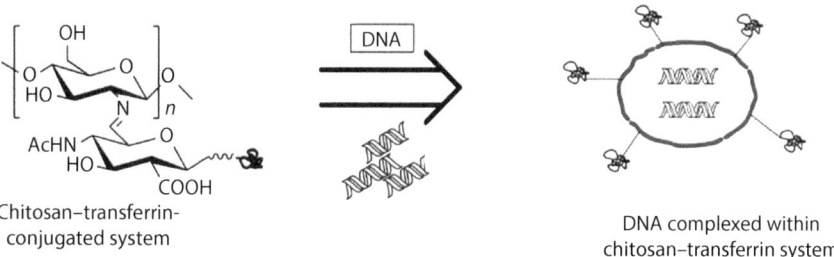

Chitosan–transferrin-conjugated system

DNA complexed within chitosan–transferrin system

FIGURE 21.8 Complex formation of CS–transferrin bioconjugate with DNA.

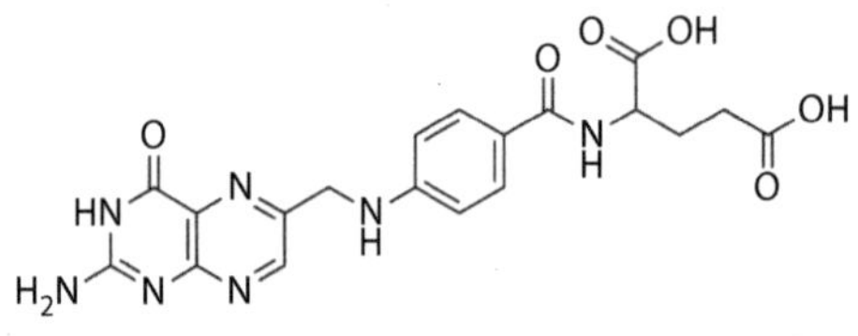

Folic acid–folate receptor-specific
ligand

FIGURE 21.9 Targeting overexpressed folate receptors on apical surfaces in tumors.

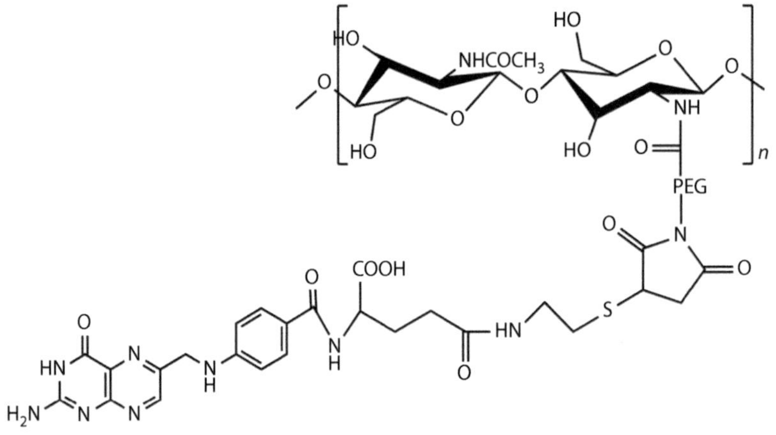

FIGURE 21.10 More efficient folic acid–PEG–CS system.

Folate conjugates, such as protein toxins, anti-T-cell receptor antibodies, radioimaging agents, chemotherapeutic agents, and gene transfer vectors, have demonstrated receptor-specific delivery properties as well (Lu et al. 2003). Also, polylysine, poly(dimethylaminomethyl methacrylate), and polyethylenimine attached to folate via a PEG linker (Figure 21.10) have exhibited both long systemic circulation time and efficient folate-selective gene delivery (Chan et al. 2007).

On the same lines, macromolecules containing CS as the main chain connected to folate via a PEG spacer have been designed, synthesized, and characterized (Chan et al. 2007). The prospective of using this CS-based carrier for tumor-targeted gene delivery has been investigated by determining water solubility, DNA-binding ability, and cytotoxicity of the modified CS. The investigators of this study, Chan et al., showed that water solubility of CS increased as a function of PEG grafting degree, although the water solubility of folic acid–PEG–CS was slightly lower than that of PEG–CS due to the hydrophobicity imparted by folic acid. The introduction of PEG and folic acid–PEG did not affect the DNA-binding ability of CS. In addition, the presence of PEG and folic acid–PEG did not cause cytotoxicity to CS, when evaluated in HEK 293 cells as compared to the commercial carriers such as Lipofectamine™2000 and TransFast™ transfection reagents. Thus, CS-grafted PEG–folate bioconjugate can be a promising nonviral vector for targeted gene delivery to tumors.

Apart from these, other sugar-based ligands have also been integrated with adjuvant CS-based vectors and include mannosylated (Jiang et al. 2008a) and peptide- (Jayaraman et al. 2012) conjugated CSs (Hansson et al. 2012) toward targeting drugs to neoplastic tissues.

21.3 VACCINATION

Adjuvants are included in vaccines to enhance the immune response and contribute toward reducing the dose levels of antigen required to elicit a strong immune response and in turn lower costs for the purposes of immunization (Gupta and Siber 1995). The favorable attributes of CS have been applied to vaccine development using CS as an ancillary carrier support (Ghendon et al. 2008, 2009). Vaccine adjuvants, such as CS, can act as repositories that control the rate of antigen presentation and may direct the antigen to selective sites (Banzhoff et al. 2009). CS produces immunogenic reactions by increasing cytokine production, attracting polymorphonuclear cells, and stimulating antibody formation, and has been used to formulate vaccines comprising antigenic epitopes and cellular components responsible for various viral and bacterial diseases. CS nanoparticles and nanocomposites of antigenic components have shown immunostimulatory potential for these diseases and have been customized to deliver antigens orally, intradermally, and via the nasal route. CS aptly fits the space of an ideal adjuvant by inducing both a strong and sustained humoral response with elevated antibody titers in conjunction with a cellular response with memory cells (Fraser et al. 2007).

21.3.1 HUMAN INFLUENZA

Vaccines coupled with a good immunostimulatory adjuvant often show better induction of immune response most desirable for a vaccine. Likewise, CS with a 50% degree of deacetylation, termed Viscogen developed by Andersson et al., enhances the immunogenic capability of vaccine against *Haemophilus influenzae* type b (Act-HIB) in murine models *in vivo* that demonstrates quick infiltration by granulocytes and stronger cellular and humoral responses (Andersson et al. 2011). The vaccine is administered as a hydrogel (Viscogel) and Viscogen easily forms the hydrogel due to better water solubility at physiological pH conditions, compared to its higher deacetylated counterparts. In a recent study by Gupta et al., projected toward achieving better immunomodulatory activity for preventing influenza, CS has been employed as a coating agent for poly-(ε-caprolactone) nanoparticles bearing the influenza antigen, recombinant influenza A virus, and H1N1 hemagglutinin (HA) protein. This nasally administered vaccine shows promising results by displaying humoral, cellular, and mucosal immunity in murine models *in vivo* by enhancing immunoglobulin G1 (IgG1), immunoglobulin G2a (IgG2a), and immunoglobulin A (IgA) responses, and γ-interferon (IFN-γ) levels (Gupta et al. 2011).

N-Alkylated derivatives, more hydrophobic than CS per se, have further demonstrated strong immunostimulation on incorporation as adjuvants in vaccine delivery systems. In concordance with this, *N*-trimethyl chitosan (TMC) nanoparticles loaded with ovalbumin (OVA) as a model subunit antigen increased the immune response after nasal and intradermal administration (Bal et al. 2012). Intranasally administered TMC nanoparticles have shown similar accruement of immune response when loaded with influenza subunit antigens as well.

HA present on the surface of the influenza virion is recognized by sialic acid receptors on the cell surface resulting in its endocytosis. It is a widely accepted fact that inhibition of hemagglutination of erythrocytes by influenza viruses is brought about by polymers bearing sialic acid. As an extension, CS–sialic acid conjugate (Figure 21.11) has been prepared as a possible antiviral agent via reductive *N*-alkylation using *p*-formylphenyl-α-sialoside (Sashiwa et al. 2000b).

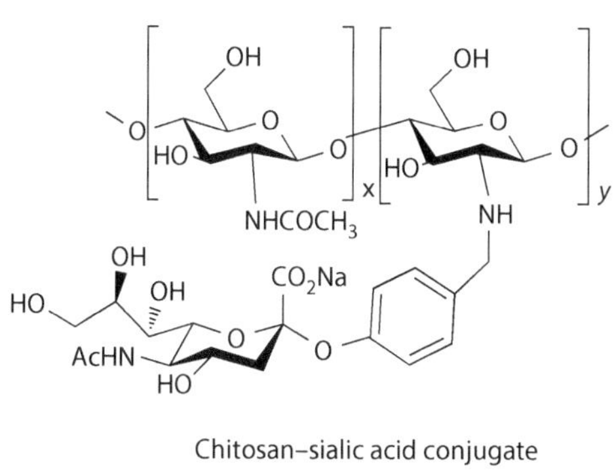

Chitosan–sialic acid conjugate

FIGURE 21.11 Antiviral CS derivative.

21.3.2 Hepatitis B

Hepatitis affects approximately 400 million people worldwide and problems associated with current vaccination methods generally encompass safety, stability, and cost. CS in its capacity as an adjuvant provides cheaper and better delivery by boosting the cellular immune response and oral administration. CS microspheres loaded with hepatitis B surface antigen can be effectively administered perorally, are stable, and provide better fortification by immunostimulation. Hepatitis B surface antigen-loaded CS microspheres with bacitracin as protease inhibitor showed better protective levels of immunity after oral administration compared with aprotinin as protease inhibitor and good stability at room temperature up to a period of 4 months (Premaletha et al. 2012).

21.3.3 Herpes

A vaccine adjuvant should enhance the immunological response to the antigen, without being toxic or eliciting a response toward itself. The vaccine adjuvant capacity of CS has been extended to develop vaccines for cattle, to prevent infection with bovine herpes virus (BHV-1) that affects animal health and productivity. CS, a biodegradable, biocompatible, and bioadhesive natural polysaccharide, has been developed into a mucosal vaccine, promising both as a vaccine delivery system and immunomodulator. Stable gel formulations of CS microparticles (<10 μm) of differential molecular weights, with an inherent positive surface charge and prepared for mucosal delivery show effective induction of immunization against BHV-1 (Gunbeyaz et al. 2010).

21.3.4 Polio

A disease more relevant to the subtropical regions, this debilitating disorder affects younger populations. The reduction on the dose of poliomyelitis antigen was studied in the context of investigating salt derivatives of CS for effective production of immune cells. CS as a carrier protects the antigen against rapid clearance. Ghendon et al. prepared CS sulfate microparticles composed of poliovirus type 1, 2, and 3 strains that can amplify immunogenicity by inducing high neutralizing antibody titers with reduced antigen doses and fewer immunizations. CS glutamate solutions containing poliovirus particles can correspondingly elicit stronger immune responses, contributed by the immunogenic induction of CS glutamate (Ghendon et al. 2011).

21.4 DIAGNOSTICS: MOLECULAR IMAGING

Biomedical imaging research has flourished in recent years due to major advancements in electronics, biotechnology, information technology, and nanomedicine. Molecular imaging typically takes advantage of molecular probes as well as inherent tissue characteristics on the basis of image contrast, and offers the prospective for understanding integrative biology, prior detection and depiction of disease, and estimation of treatment (Willmann et al. 2008). Molecular imaging could also assist judgments in choosing potential drug candidates that can have a good therapeutic effect and screen out drug entities that might have toxicological implications (Willmann et al. 2008). The objective of molecular imaging is to envisage biological processes at the tissue and cellular levels by noninvasive procedures (Agrawal et al. 2010).

Molecular imaging is an imperative tool in tackling challenges presented during interpretation and characterization of biological mechanisms at the cellular level. Diagnostic methods involving imaging applied for routine clinical use are predeveloped in an experimental research setting. Hence, designing and conducting preclinical imaging experiments defines later clinical procedures and additionally provides mechanistic understanding of the observed biological response. Although contrast imaging of biological specimens both *ex vivo* and *in vivo* has been by and large dependent on optical microscopy involving fluorescence and luminescence imaging, advanced noninvasive imaging technologies have taken precedence in the recent decade and include scanning procedures by computed tomography (CT), magnetic resonance (MR) (Hogemann et al. 2001), positron emission tomography (PET) (Pomper and Hammoud 2004), single-photon emission computed tomography (SPECT), ultrasound (US) (Sharma et al. 2006), and optical imaging (OI) (Mahmood and Weissleder 2002), including their variations (Sharma et al. 2006).

Imaging modalities are classified into two groups, morphological or anatomical and molecular or functional imaging techniques. The high spatial resolution, morphological imaging technologies include CT, magnetic resonance imaging (MRI), and US; however, they cannot detect diseases until changes at the structural level within the tissue, such as growth of a tumor or the extent of inflammation is large enough to be morphologically identified. In contrast, molecular imaging methods, comprising OI, PET, and SPECT, present the possibility to detect molecular and cellular alterations caused by a disorder at the structural level, for instance, before the tumor is large enough to cause morphological changes. However, these molecular modalities possess a low spatial resolution capacity with currently available equipment and do not present morphological insights (Agrawal et al. 2010). There are excellent reviews that discuss these aspects in the subject matter of molecular imaging in detail (Tremoleda et al. 2011).

Molecular imaging technologies, including OI, PET, and SPECT present the prospective to identify molecular and cellular alterations of disorders, but they have reduced spatial resolution. MRI, by contrast, has higher spatial resolution, but exhibits low sensitivity toward recognition of contrast agents and this is a major obstacle in its use as a molecular imaging modality (Sharma et al. 2006). In preclinical evaluation, the imaging method requires both sufficient spatial resolutions in the range of tens of millimeters to micrometers and appreciable sensitivity to distinguish biochemical changes, as well as capability to detect minute, clinically relevant alterations with the progression of time. Designing hybrid systems where the combined strengths of morphological and molecular imaging modalities are utilized permits detection of pathophysiological events in early disease stages at high sensitivity using, for example, PET and with high spatial resolution, using CT or MRI (Agrawal et al. 2010).

CS composites integrated with supplementary contrast materials bring forth a new class of biomaterials that have mechanical, physicochemical, and functional characteristics that are difficult to observe with pristine CS per se or the incorporated contrast agent alone. Further, amalgamation of imaging agents such as magnetite for MRI into self-associated superparamagnetic iron oxide nanoparticles (SPION)-loaded nanoparticles can boost targeted imaging applications (Lee et al. 2009b) and the particle can serve as MR molecular imaging agent. An assortment of inorganic materials can be integrated in the CS composite preparations and their collective properties have been demonstrated as beneficial for many biomedical purposes. Composites of CS–polyion complex can be generated by contact of CS with natural and synthetic anionic macromolecules (Hein et al. 2008). Moreover, fluorescent CS quantum dot composites embedded with drugs can enable the combination of therapy in terms of targeted drug and gene delivery with OI (Koping-Hoggard et al. 2001).

Nonalcoholic steatohepatitis (NASH) can be diagnosed with MRI, an efficient contrast probe, by a stable suspension of superparamagnetic Fe_3O_4 nanoparticles (Luo et al. 2012). The negatively charged Fe_3O_4 nanoparticles coated with positive CS are assembled with poly(vinyl acetate–methylacrylic acid) (P(VAc–MAA). Poly(vinyl acetate–methylacrylic acid)/chitosan/magnetite (P(VAc–MAA)/CS/Fe_3O_4) nanoparticles had a spherical or ellipsoidal morphology with an average diameter in the range of 14–20 nm shown by transmission electron microscope and dynamic light scattering. The superparamagnetic property and spinel structure of the Fe_3O_4 nanoparticles was conserved due to the protection by P(VAc–MAA)/CS layers on the surface of the Fe_3O_4 nanoparticles. The *in vivo* murine experiments confirmed that the P(VAc–MAA)/CS/Fe_3O_4 nanoparticles were an efficient imaging probe for MRI to diagnose NASH.

The rising significance of MRI in the diagnostic estimation of intestinal disorders, with description of mucosal environment was obtained using conventional intravenous contrast agents. Regular clinical use of contrast agents has been carried out by intravenous injection for mucosal imaging. Imaging modalities that specifically localize within the intestinal mucosa are therefore required to enhance clinical imaging of the mucosal surface. A novel contrast agent consisting of gadopentetic acid (Gd-DTPA)-loaded CS nanoparticles was developed by a research group and investigated for the absorption of the nanoparticles in the colon wall of healthy murine models by MRI (Cheng et al. 2012). The cytotoxicity of gadolinium–CS nanoparticles was evaluated by an 3-(4,5-Dimethylthiazol-2-yl)-2,5-diphenyltetrazolium bromide (MTT) assay that indicated low cytotoxic potential. Gadolinium–CS nanoparticles were administered to the colon mucosa of healthy rats by rectal administration, and MRI scans *in vivo* were carried out with a 3.0 T imaging scanner. The prepared gadolinium–CS nanoparticles were ~420 nm in diameter with a 74.4% Gd-DTPA content that showed retention in both the stratum submucosum and epithelial cells of the colon for almost 80 min. Gadolinium–CS nanoparticles were localized inside the mucosal cells or intercellular space. Owing to the infusion of gadolinium–CS nanoparticles, the MR signal intensity of colon mucosa increased from about 6% to 35% and the contrast enhancement was highest at 20 min after administration. Gadolinium–CS nanoparticles with high Gd-DTPA content can be successfully developed for use as a novel MRI contrast agent and thus, rectally administered gadolinium–CS nanoparticles have the potential for MRI diagnosis of colon mucosal disease.

CS composites can also be made to contain contrast agents for imaging purposes. Lee et al. investigated the ability of CS microspheres loaded with superparamagnetic iron oxide nanoparticles (SPIOs) to aid in anticancer embolotherapy. For these applications, SPIO-based contrast agents are dispersed in aqueous milieu and exhibit better stability and a narrow size distribution (Lutz et al. 2006). CS nanoparticles with prolonged residence time in blood are advantageous for *in vivo* biomedical applications. Lee et al. have prepared novel self-assembling nanoparticles composed of amphipathic water-soluble chitosan–linoleic acid (WSC–LA) conjugates for entrapment of SPIOs as a contrast agent to target hepatocytes (Lee et al. 2009a,b). The potential of CS as a delivery vehicle for biocompatible MRI contrast agent has been researched by several groups (Hari et al. 1996a,b; Takahashi et al. 1990), in particular, in SPION nanoparticles (Lee et al. 2005). Shi et al. developed biocompatible carboxymethyl chitosan (CMCS)-coated SPIO that have potential for use in envisaging human mesenchymal stem cells (hMSCs) by MRI (Shi et al. 2009). The carboxymethylation of CS increases its solubility in water, and hence, the CMCS-coated SPIO easily suspend in aqueous medium.

21.5 METHODS OF PREPARING NANO- AND MICROPARTICLES OF CHITOSAN

Optimal drug loading and encapsulation are significant and challenging to achieve without deterring biological and pharmacological profiles, especially in cases of labile protein and peptide drugs that can undergo degradative processes in hostile pH conditions, hydrolytic and deamidation stresses, and aggregate formation by adverse temperatures and shear stresses. Drug adherence to polymers by means of adsorption can result in conformational disruptions by triggering protein unfolding and eventual loss of the drug. Strong shear forces and mechanical stressors can quickly bring about deformation of biotherapeutics, most of which are proteinaceous (Green et al. 1999) and cause reduction in activity, altered activity or nondesirable immunogenic rejection, and adverse reactions (Jorgensen et al. 2006).

The formulation of CS nanoparticulate systems embedded with therapeutics has been thoroughly researched in recent times by various groups. In this perspective, the peculiar nature of CS as a hydrophilic polymer characterized by cationic charge allowed the authors to develop relatively mild manufacturing methods for obtaining micro- and nanoparticulate carriers, mostly involving ionic gelation or polyelectrolyte cross-linking, which can be successfully applied for loading these systems with biopharmaceutical drugs (Janes et al. 2001). Various methods for the preparation of CS nano/microparticles are emulsification/solvent evaporation, spray drying, ionotropic gelation and coacervation, emulsion cross-linking, and sieving. The following section describes preparative methods for CS-based micro- and nanoparticles on the basis of the principles of techniques that are currently in use.

21.5.1 Emulsion Cross-Linking

Emulsion-based preparation techniques are largely employed toward generation of polymer-based nanoparticles. For drug delivery systems of hydrophilic polymers, such as CS, the polymer is incorporated in an aqueous environment mainly having low pH, mediated by weakly acidic agents, and more commonly, acetic acid. Protein and peptide therapeutics are also contained within the aqueous phase, which is emulsified with oils, such as liquid paraffin, and the micro- or nanosized emulsion droplets are stabilized with suitable surfactants.

Degradation of peptide and protein drugs at times occurs during the emulsification process due to mechanical stress in homogenization (Jorgensen et al. 2006). Further, contact of protein biotherapeutics with water–oil surfaces can bring about degradation due to interfacial tension at the emulsion globule surfaces (Dill 1990). A few approaches can stabilize peptide and protein drugs and protect them from degradation during emulsification. CS being a hydrophilic polymer can significantly reduce the free energy of water-soluble proteins and offer protein protection by causing protein stabilization (Wu et al. 2008). The high viscosity imparted by CS increases the energy barrier for intramolecular movement and intermolecular interaction of proteins. A universal stabilization strategy can comprise introduction of bulky surface-active agents for inhabiting more interfacial area to control protein adsorption of the proteins, leading to prevention of fractional dose reduction and undesirable protein denaturation and unfolding at the emulsion droplet interface (Jorgensen et al. 2006).

The isolation and purification of CS particles is brought about usually by simple filtration techniques and washings, for instance, with n-hexane. The nanoparticles are strengthened by a cross-linking step essential to impart rigidity to aqueous droplets of CS solution. Conventional cross-linkers such as glutaraldehyde mostly react with CS amino groups. However, generally, aldehyde-based cross-linking agents are not appropriate for peptide drugs (Mao et al. 2006) contained in nanoparticles because they are not only difficult to remove completely, leading to possible toxicity issues, but can also cross-react with amino substituents of peptides such as insulin causing agglomeration (Ubaidulla et al. 2007). Less toxic ascorbyl palmitate and dehydroascorbyl palmitate are excellent substitutes of glutaraldehyde (Bugamelli et al. 1998). Ascorbyl palmitate or dehydroascorbyl palmitate localizes at the globule surface of the emulsion and protects the protein from denaturation (Varshosaz et al. 2004).

21.5.2 Coacervation/Precipitation

CS nanoparticles and microparticles are formed by coacervation in two ways. The first technique involves introduction of sodium sulfate solution by dropwise addition to an aqueous, low-pH CS solution in the presence or absence of a surfactant, where coacervation occurs by the desolvation phenomenon. Cross-linkers such as glutaraldehyde are sometimes incorporated. This technique is often employed for synthesizing CS-based nanoparticles in gene delivery for incorporation of pDNA, microRNA (miRNA), and small interfering ribonucleic acid (siRNA) components, which occurs by "polyplex" formation. Moreover, proteins such as recombinant interleukin-2 (IL-2) have been loaded in CS microspheres using this method by Ozbas-Turan et al. (2002). The other technique exploits the lack of the solubilizing capacity of CS in higher, basic pH conditions, where aqueous CS solution generally of low pH is brought in by means of an atomizer into basic solutions comprising sodium hydroxide, sodium hydroxide–methanol, or ethylenediamine solutions. On contact with higher pH conditions, CS precipitates as nano- or microparticles.

21.5.3 Drying Methods

Diverse drying methods developed for embedding biotherapeutics in CS particulate delivery systems, offer ease and integrity with drug stability, simple scale-up, and the opportunity to manufacture nano- and microformulations with distinct sizes. These techniques are discussed in this section.

Spray drying is an important method for the preparation of powders, granules, for tablet manufacture, or agglomerates from a homogenized blend of drug and excipients. In fact, this process of drug entrapment is not as much reliant as other methodologies on the aqueous solubility of drug and adjuvant polymer and is uncomplicated, reproducible, and simple toward scale-up. Spray drying is achieved by a spray-drying equipment, composing an air nozzle and a drying chamber. Solutions and suspensions of nanoparticles are atomized to fine droplets and a flow of hot air causes rapid removal of solvent from the droplets in the drying compartment, leading to dry powders of nanoparticles (Giunchedi and Conte 1995). The droplets are subjected to lower temperatures as hot air temperatures are lowered by evaporative cooling. The addition of a suitable cross-linking agent is done during preatomization in the CS solution for particle integrity. The spray-drying process leads to the formation of dry, free-flowing particles (Amidi et al. 2006).

Electrospraying technique can create particles by utilizing an even electrohydrodynamic force to rupture solutions into thin jets. It is a promising process for quick and efficient manufacture of nano- to microsized systems (Wu and Clark 2007; Wu and Jin 2008). Electrospraying and electrospinning are effortless electrohydrodynamic production methods that create micro- or nanospheres or fibers in a single stage. A few reviews have discussed different techniques involved in electrospraying in detail (Chew et al. 2006; Rutledge and Fridrikh 2007; Moghe and Gupta 2008; Sill and von Recum 2008).

Controlled release kinetics can be achieved in nanoparticulate forms fabricated using this technique and the adaptability of the delivery systems acquired is augmented by facilitating a near-zero-order release mechanism and preventing burst discharge of the drug. Critical factors such as voltage, conductivity, surface tension, flow rate, density, and viscosity of sprayed liquid affect and bring about changes in particle diameter and morphology. Electrospraying avoids utilization of an exterior dispersion medium that frequently incorporates components unsuitable for biopharmaceutical use. Some researchers have worked toward the characterization and optimization of differentially sized CS-based nanosystems (Kuo et al. 2004) prepared by electrospraying technique (Pancholi et al. 2009). Poly(lactide) and CS–tripolyphosphate (TPP) particles prepared by electrospraying and encapsulated with bovine serum albumin (BSA) were investigated by Xu and Hanna (2007). BSA release kinetics from CS nanoparticles revealed steady-state achievement in 24-h duration in contrast to poly(lactide) nanoparticles that took longer. Yoo et al. investigated CS, polycaprolactone, and PEG as entrapment vector materials for proteins such as albumin, processed using electrospraying and efficient nanoencapsulation of albumin was easily attained by electrospraying (Yoo et al. 2007). The focal biomedical applications of CS micro- and nanoparticles obtained by electrospraying comprise delivery of drugs, proteins, and cells.

21.5.4 Ionic Gelation

This technique of preparing chitin and CS nanoparticles and microparticles uses interactive, opposite ionic forces between positively charged CS and anions leading to stable cross-linking by virtue of complexation and eventually particle formation. Generally, an aqueous acidic solution of CS is gradually introduced by aid of stirring to a solution containing anions, which may be low-molecular-weight molecules such as cyclodextrin derivatives and sodium TPP, or large molecules such as dextran sulfate, sodium alginate, poly-g-glutamic acid, and hyaluronic acid. Spherical particles are generated because of the precipitation triggered by complexation between oppositely charged cationic and anionic species (Sandri et al. 2010). In preparing CS nanoparticles and microparticles by this exclusively aqueous system-based methodology, the use of organic solvents or elevated temperatures is avoided and thus, thermolabile biopharmaceuticals can be flexibly incorporated with efficiency.

21.5.5 Reverse Micellar Method

In this method, reverse micelles of surfactant are formed in a low-vapor-pressure organic solvent and aqueous CS–drug solution is introduced with high-speed stirring, to form a microemulsion.

The size of nanoparticles is controlled by the quantity of water; more water addition results in larger nanoparticles. To impart rigidity, cross-linkers are added and subsequently, the organic solvent is removed by evaporation. The material is redispersed in water and the surfactant is salted out of the dispersion and removed by centrifugation. The supernatant containing the drug-loaded CS nanoparticles is purified by dialysis and nanoparticles are obtained as a dry powder postlyophilization (Mitra et al. 2001).

21.5.6 SIEVING METHOD

Sieving is a technique that offers particle size control to already-formed larger particles of CS, a result of direct cross-linking. Specifically, a solution of CS is cross-linked using glutaraldehyde and cross-linked CS particles are then forced through a sieve with an appropriate mesh dimension to achieve microparticles. The microparticles are subsequently subjected to washings with 0.1 N sodium hydroxide solutions to remove excess glutaraldehyde and are dried at 40°C (Agnihotri et al. 2004).

21.6 CHITOSAN AND DERIVATIVES IN CONVENTIONAL DOSAGE FORM APPLICATIONS

The mode of delivery of various dosage form types significantly determines the extent and duration of therapeutic response. Nanoformulations of CS comprising biotherapeutics have been administered by many of the existing delivery routes for treating disorders.

21.6.1 ORAL ROUTE

Novel therapeutic delivery systems for oral administration have resulted due to amalgamation of technological developments and biomaterials design for peptide delivery and other hydrophilic small- and large-molecule drugs. An ideal mode of delivery for such molecules, the oral route offers improved patient compliance and convenience. Proteins and peptide drugs exhibit poor bioavailability profiles, largely because of lack of stability in the gastrointestinal tract, where poor permeability, peptidases, and low stomach pH cause premature degradation and loss, a result of high molecular weight and hydrophilicity. CS exhibits mucoadhesion and, mainly when pooled with enzyme inhibitors, has received ample attention toward incorporation of orally administered peptide and protein delivery therapeutics. Enzyme inhibitory agents prevent proteolytic enzymes from destroying peptide and protein therapeutics.

The restriction of enzyme inhibitors to a nonabsorbable mucoadhesive carrier template lowers luminal denaturation and degradation of orally administered peptide and protein drugs and yet, the long-term toxicological and metabolic effects of these drugs are questionable. In contrast, macromolecular mucoadhesive polymers confine the nanoformulation at the absorption site, thereby increasing the barrier between the released drug and the luminal tissue region, resulting in low drug degradation by proteolytic enzymes. CS and chitin derivatives can additionally enhance the intestinal absorption of macromolecular compounds (Thanou et al. 2001a). The CS–ethylene diamine tetraacetic acid (EDTA) complex, for instance, has strong affinity toward scavenging divalent cations, such as Ca^{2+}, which might widen gaps between tight junctions in mucosal tissues allowing improved oral absorption of peptide and protein drugs (Luessen et al. 1996). CS and its derivatives are recognized as strong absorption enhancers (Thanou et al. 2001b; Cano-Cebrian 2005; van der Merwe 2004). CS in its pristine form is, however, soluble only at low pH values and hence is less competent at improving the absorption in the small intestine. The positively charged, quaternized derivative of CS, TMC obtained by reductive methylation of CS, is soluble in neutral and basic pH environments and can appreciably lessen the transepithelial resistance, directly causing opening of the tight junctions between epithelial cells.

Gemcitabine is a wide-spectrum, cytotoxic agent employed in anticancer therapy for treating various neoplastic disorders of the lung, ovary, breast, and bladder cancers and, also in the treatment of pancreatic cancer, non-small-cell lung cancer (NSCLC), and leukemia. Gemcitabine acts at the level of DNA and incorporates into DNA causing inhibition of cell proliferation. However, gemcitabine is rapidly cleared by metabolism of cytidine–deaminase enzyme considerably limiting its efficacy. Owing to extensive deamination by intestinal cells, its oral administration as such has very low bioavailability. CS nanoparticles were prepared by the ionic gelation method using TPP to introduce a novel oral formulation of the drug with improved physicochemical properties (Derakhshandeh and Fathi 2012). Physicochemical properties such as particle size and shape, loading efficiency and release rate, and oral absorption of both free and nanoparticle-loaded drugs were measured using the rat intestinal sac model. Gemcitabine-loaded CS nanoparticles, spherical with a mean size of 95 ± 8 nm and high drug loading (63%), showed controlled release pattern characterized by a fast initial release during the first 8 h, followed by slower and continuous release. The absorption study showed that gemcitabine intestinal transport increased 3–5-folds by loading in CS nanocarrier and displayed the potential of CS nanocarrier systems as efficient oral delivery vectors.

21.6.2 BUCCAL MODE OF DELIVERY

Buccal delivery systems have interestingly been explored for a variety of conditions, although these have less preference over the oral and other routes for drug delivery. Park et al. investigated mucoadhesive ondansetron buccal films for the treatment of emesis using CS as a mucoadhesive polymer (Park et al. 2012). CS can enhance drug absorption and lower dosing frequency and the drug dose per se. A different research group developed mucoadhesive CS-based films, embedded with insulin-loaded nanoparticles made of poly(ethylene glycol) methyl ether-*block*-polylactide (PEG-*b*-PLA) (Giovino et al. 2012). Blank nanoparticles were prepared by double-emulsion solvent evaporation methodology with varying concentrations of the copolymer (5% and 10%, w/v) and optimized before loading with insulin (model protein) at preliminary loadings of 2%, 5%, and 10% with respect to the weight of the copolymer. The *in vitro* release profiles of both formulations displayed a biphasic sustained release of protein during 5 weeks that was influenced by the pH of the release medium. Optimized CS films containing 3 mg of insulin-loaded nanoparticles prepared by solvent casting with uniform nanoparticle distribution in the mucoadhesive template exhibited exceptional physicomechanical properties. The drug delivery system presents a novel platform for prospective buccal delivery of macromolecular biotherapeutics and protein and peptide drugs.

21.6.3 NASAL ADMINISTRATION

Current research has focused on nasally administered vaccines in the recent decade. However, several factors such as the quick clearance of dosage forms by mucociliary cells (Donovan et al. 1990) limited the diffusion of large-molecule antigens across the mucosal barrier (Pereswetoff-Morath 1998) and enzymatic degradation (Schipper et al. 1991) has restricted the antigen delivery via the nasal mucosa frequently resulting in a low immune response (Czerkinsky et al. 1999). Nevertheless, distinctive approaches can be employed to circumvent these limitations, by aiding the administration of antigens with mucoadhesive polymeric adjuvants as different nanosized drug delivery systems (Sarkar 1992). CS nanoparticles can considerably improve the immune response of antigens for vaccines such as those for hepatitis B, influenza, pertussis, and diphtheria by stimulating blood immunoglobulin, IgG, and IgA (Illum 2003). CS-based nanoformulations loaded with antigens can augment the efficiency of the immune response by possibly enhancing the permeation of the antigens across the nasal mucosa.

Low-molecular-weight CSs and derivatives can activate long-term humoral and cell-mediated immunity. One research group recently reported studies in murine models, which showed that TMC, a quaternary alkylamine and cationic derivative of CS, exhibited a considerably affirmative influence on the immunological behavioral response postnasal administration of hepatitis B surface antigen-based vaccine (Illum et al. 2001). Another research group synthesized and optimized CS–dextran sulfate nanoparticles for the integration of pertussis toxin (PTS) and a possible targeting ligand, IgA (Sharma et al. 2012). *In vitro* characterization and *in vivo* uptake studies were performed for the evaluation of developed nanoparticles. The ratio of CS to dextran sulfate, the order of mixing, and pH of nanoparticle suspension were influential formulation factors regulating the size and zeta potential of nanoparticles loaded with PTX and IgA. The *in vivo* uptake of IgA-loaded nanoparticles in mice showed a favored uptake of nanoparticles by nasal membranes following intranasal administration and show promise as a nasal vaccine delivery system.

21.6.4 Pulmonary Delivery Pathway

Parkinson's disease is a neuronal debilitating disorder and levodopa, (*S*)-2-amino-3-(3,4-dihydroxyphenyl) propanoic acid, is still a prominent and preferred drug in therapy today. However, oral levodopa is known to display poor pharmacokinetics and its efficiency becomes diminished with the progression of the disease. Pulmonary delivery by means of adjuvant polymeric carriers such as CS is a promising tactic to surmount this challenge. A stability-indicating liquid chromatography method for the quantitative determination of levodopa microparticles for pulmonary delivery was developed by Pareira et al. as well as its photodegradation kinetics in solution. The developed and validated method was applied for the analyses of the novel formulation as well as for protocols of stability studies (Pereira et al. 2012).

siRNA-based strategies seem to be an exciting new approach for the treatment of respiratory diseases. To extrapolate siRNA-mediated interventions from bench to bedside in this area, several aspects have to be jointly considered, including a safe and efficient gene carrier with pulmonary deposition efficiency, as well as *in vivo* method for siRNA/nanoparticles delivery. CS-derived nanodelivery vehicle, that is, siRNA-loaded guanidinylated chitosan (SGCS) in ribonucleic acid (RNA) interference therapy for lung diseases via aerosol inhalation was prepared and studied by Xu et al. In this research paper, a nonviral DNA vector, guanidinylated CS was developed and tested for siRNA delivery. It was demonstrated that guanidinylated chitosan (GCS) could entirely condense siRNA at weight ratio 40:1, forming nanoparticles of diameter 100 nm, 15 mV in surface potential. Guanidinylation of CS reduced cytotoxicity and also mediated cellular engulfment of siRNA nanoparticles, resulting in higher gene-silencing efficiency than pure CS nanoparticles. Guanidinylated CS could protect both the pDNA and siRNA against harsh shear forces produced by mesh-based nebulizers. Aerosol treatment enhanced nanoparticle size distribution and thus, transfection efficiency (Xu et al. 2011).

21.6.5 Transdermal Means of Administration

The effectiveness of CS and its derivatives for transdermal delivery as for mediating permeation across skin barriers has been investigated in some research papers (He et al. 2009; Huang et al. 2009; Lee et al. 2008; Wong 2009). Changes in cell membrane potential and fluidity and, secondary structure changes of keratin and water content in the stratum corneum, brought about by CS-based vectors are thought to facilitate skin penetration (He et al. 2009). Li et al. prepared covalently crosslinked hydrogel composing *N,O*-CMCS and oxidized alginate for intraperitoneal drug delivery. *In vitro/vivo* cytocompatibility and biocompatibility of the developed hydrogel were preliminary evaluated, where *in vitro* cytocompatibility test revealed good cytocompatibility against murine embryonic fibroblasts cell line (NH_3T3) cells after 3-day incubation of the developed hydrogel. Acute toxicity tests indicated total absence of cytotoxicity for major organs during the period of 21-day intraperitoneal administration and did not induce any cutaneous reaction within 72 h of subcutaneous injection. Further, the developed Hydrogel showed 0% of hemolysis ratio. *N,O*-CMCS/ oxidized alginate hydrogel with noncytotoxicity and good biocompatibility might be promising for drug delivery applications (Li et al. 2012). In another study, the main objective was to formulate warfarin-β-cyclodextrin (WAF-β-CD)-loaded CS nanoparticles for transdermal delivery (Khalil et al. 2012). CS is a hydrophilic carrier and, therefore, hydrophobic WAF was incorporated into CS nanoparticles, by prior complexation with β-CD. The nanoparticles were made by ionotropic pregelation using TPP and optimized. Nanoparticles prepared with 3:1 (CS:TPP) weight ratio and 2 mg/mL final CS concentration had spherical particles (35 ± 12 nm diameter) with narrow size distribution (PDI = 0.364) and 94% entrapment efficiency. The *in vitro* release and permeation profiles of WAF-β-CD from the selected nanoparticle formulation were studied and the nanoparticles followed Higuchi release, whereas *ex vivo* permeation (at pH 7.4) followed a zero-order permeation profile. These WAF-β-CD-loaded CS carriers can potentially present programmable delivery of WAF transdermally.

21.6.6 OCULAR DELIVERY

Topical ophthalmic drops are the first choice of medical practitioners and the patients in terms of compliance. Despite its simplicity, however, eye drops present issues, which include low ocular drug bioavailability, pulsated drug entry after topical administration, and frequent administration leading to ocular irritation (Ding 1998). CS and its derivatives have recently been reported as materials with high potential to extend the corneal residence time of drugs and CS-coated nanoparticles are more efficient for increasing the intraocular penetration of drugs (Felt et al. 1999). CS-coated sodium alginate–CS nanoparticles loaded with 5-fluorouracil (5-FU) developed for ophthalmic delivery by ionic gelation technique using sodium alginate and CS and then suspended in CS solution presented a sustained release of 5-FU compared to the 5-FU solution with high burst release. *In vivo* study in rabbit eye showed higher levels of 5-FU in aqueous humor compared to plain 5-FU solution. Optimized formulation was nonirritant and tolerable when tested by modified Draize test in rabbit eye (Nagarwal et al. 2012). Mucoadhesiveness imparted by CS nanoparticles improved bioavailability significantly, showing that CS-based systems have great potential in ocular- based therapy.

21.7 CONCLUSION

In summary, the relationship of the chemical structure of CS with its biological function has served as the blueprint for producing derivatives with enhanced potentials. The research strategies adopted for chemical modifications of CS circle mostly around the reactive group of CS. CS polymers display distinctiveness in view of their excellent characteristics that make them remarkably safe and optimal for biomedical applications. Over the years, they have been well customized in the pursuit of obtaining polymers that are precise for each pharmaceutical purpose. To this end, CSs have been chemically modified by different methods to achieve different outcomes such as altered physicochemical properties and conjugates with various functional ligands, without compromising biological compatibility. Recently, CSs devised with augmented antimicrobial properties are being investigated as novel agents that could be potentially used for treating microbial infections, which cause complexities in therapy because of the ever-prevailing predicament of resistance development. CSs, in addition, continue to be widely researched as excipients in drug and gene therapy. CS is extensively available and can offer low-cost therapy. All these properties combined, CS polymers themselves form a unique class of new-age biotherapeutics that has excellent scope for future development.

The growing attention of CS in biomedical purposes has produced new arenas for the manufacture of available marketed products such as injectable thermoresponsive CS hydrogel for cartilage tissue engineering (Biosyntech Inc., Canada) and CS oligomers for gene delivery (FMC Biopolymer AS, Norway). Further advancements have led to the commercialization of ecofriendly CS-based antimicrobial textiles such as Crabyon from Tec Services (Italy) and Chitopoly from Fuji Spinning Co. Ltd (Japan). With respect to more conventional CS-based products available in the commercial sector, CS composites are comparatively still quite novel and are yet to appear in the market.

SYMBOLS

–NH$_2$ Amino group
Ca^{2+} Calcium

REFERENCES

Aiedeh, K., and M. O. Taha. 1999. Synthesis of chitosan succinate and chitosan phthalate and their evaluation as suggested matrices in orally administered, colon-specific drug delivery systems. *Archiv der Pharmazie* 332:103–107.

Aiedeh, K., and M. O. Taha. 2001. Synthesis of iron-crosslinked chitosan succinate and iron-crosslinked hydroxamated chitosan succinate and their *in vitro* evaluation as potential matrix materials for oral theophylline sustained-release beads. *European Journal of Pharmaceutical Sciences* 13:159–168.

Agnihotri, S. A., N. N. Mallikarjuna, and T. M. Aminabhavi. 2004. Recent advances on chitosan-based micro- and nanoparticles in drug delivery. *Journal of Controlled Release* 100:5–28.

Agrawal, P., G. J. Strijkers, and K. Nicolay. 2010. Chitosan-based systems for molecular imaging. *Advanced Drug Delivery Reviews* 62:42–58.

Amidi, M., S. G. Romeijn, G. Borchard, H. E. Junginger, W. E. Hennink, and W. Jiskoot. 2006. Preparation and characterization of protein-loaded *N*-trimethyl chitosan nanoparticles as nasal delivery system. *Journal of Controlled Release* 111:107–116.

Andersson, T. N., A. C. Hällgren, M. Andersson, J. Langebäck, L. Zettergren, J. Nilsen-Nygaard, K. I. Draget et al. 2011. Improved immune responses in mice using the novel chitosan adjuvant Viscogel with a *Haemophilus influenzae* type b glycoconjugate vaccine. *Vaccine* 29:8965–8973.

Bal, S. M., B. Slutter, R. Verheul, J. A. Bouwstra, and W. Jiskoot. 2012. Adjuvanted, antigen loaded *N*-trimethyl chitosan nanoparticles for nasal and intradermal vaccination: Adjuvant- and site-dependent immunogenicity in mice. *European Journal of Pharmaceutical Sciences* 45:475–481.

Banzhoff, A., R. Gasparini, F. Laghi-Pasini, T. Staniscia, P. Durando, E. Montomoli, P. L. Capecchi et al. 2009. MF59-adjuvanted H5N1 vaccine induces immunologic memory and heterotypic antibody responses in non-elderly and elderly adults. *PLoS One* 4:e4384.

Bugamelli, F., M. A. Raggi, I. Orienti, and V. Zecchi. 1998. Controlled insulin release from chitosan microparticles. *Archiv der Pharmazie* 331:133–138.

Cano-Cebrian, M. J., T. Zornoza, L. Granero, and A. Polache. 2005. Intestinal absorption enhancement via the paracellular route by fatty acids, chitosans and others: A target for drug delivery. *Current Drug Delivery* 2:9–22.

Chan, P., M. Kurisawa, J. E. Chung, and Y. Y. Yang. 2007. Synthesis and characterization of chitosan-*g*-poly(ethylene glycol)-folate as a non-viral carrier for tumor-targeted gene delivery. *Biomaterials* 28:540–549.

Chen, Y., V. J. Mohanraj, F. Wang, and H. A. Benson. 2007. Designing chitosan–dextran sulfate nanoparticles using charge ratios. *AAPS PharmSciTech* 8:131–139.

Cheng, P. W. 1996. Receptor ligand-facilitated gene transfer: Enhancement of liposome-mediated gene transfer and expression by transferrin. *Human Gene Therapy* 7:275–282.

Cheng, J. J., J. Zhu, X. S. Liu, D. N. He, J. R. Xu, L. M. Wu, J. Zhou, and Q. Feng. 2012. Gadolinium–chitosan nanoparticles as a novel contrast agent for potential use in clinical bowel-targeted MRI: A feasibility study in healthy rats. *Acta Radiologica* 53:900–907.

Chew, S. Y., Y. Wen, Y. Dzenis, and K. W. Leong. 2006. The role of electrospinning in the emerging field of nanomedicine. *Current Pharmaceutical Design* 12:4751–4770.

Czerkinsky, C., F. Anjuere, J. R. McGhee, A. George-Chandy, J. Holmgren, M. P. Kieny, K. Fujiyashi et al. 1999. Mucosal immunity and tolerance: Relevance to vaccine development. *Immunological Reviews* 170:197–222.

Derakhshandeh, K., and S. Fathi. 2012. Role of chitosan nanoparticles in the oral absorption of gemcitabine. *International Journal of Pharmaceutics* 437:172–177.

Dill, K. A. 1990. Dominant forces in protein folding. *Biochemistry* 29:7133–7155.

Ding, S. 1998. Recent developments in ophthalmic drug delivery. *Pharmaceutical Science and Technology Today* 1:328–335.

Donovan, M. D., G. L. Flynn, and G. L. Amidon. 1990. Absorption of polyethylene glycols 600 through 2000: The molecular weight dependence of gastrointestinal and nasal absorption. *Pharmaceutical Research* 7:863–868.

Dube, D., M. Francis, J. C. Leroux, and F. M. Winnik. 2002. Preparation and tumor cell uptake of poly(*N*-isopropylacrylamide) folate conjugates. *Bioconjugate Chemistry* 13:685–692.

Dutta, P. K., J. Dutta, and V. S. Tripathi. 2004. Chitin and chitosan: Chemistry, properties and applications. *Journal of Scienctific and Industrial Research* 63:20–31.

Erbacher, P., S. M. Zou, T. Bettinger, A. M. Steffan, and J. S. Remy. 1998. Chitosan-based vector/DNA complexes for gene delivery: Biophysical characteristics and transfection ability. *Pharmaceutical Research* 15:1332–1339.

Fahmy, T. M., P. M. Fong, A. Goyal, and W. M. Saltzman. 2005. Targeted nanoparticles for drug delivery. *Materials Today (Nano Today)* 8:18–26.

Felt, O., P. Furrer, J. M. Mayer, B. Plazonnet, P. Buri, and R. Gurny. 1999. Topical use of chitosan in ophthalmology: Tolerance assessment and evaluation of precorneal retention. *International Journal of Pharmaceutics* 180:185–193.

Fraser, C., K. Diener, M. Brown, and J. Hayball. 2007. Improving vaccines by incorporating immunological coadjuvants. *Expert Review of Vaccines* 6:559–578.

Gabizon, A., H. Shmeeda, A. T. Horowitz, and S. Zalipsky. 2004. Tumor cell targeting of liposome-entrapped drugs with phospholipids-anchored folic acid–PEG conjugates. *Advanced Drug Delivery Reviews* 56:1177–1192.

Ghendon, Y., S. Markushin, G. Krivtsov, and I. Akopova. 2008. Chitosan as an adjuvant for parenterally administered inactivated influenza vaccines. *Archives of Virology* 153:831–837.

Ghendon, Y., S. Markushin, Y. Vasiliev, I. Akopova, I. Koptiaeva, G. Krivtsov, O. Borisova et al. 2009. Evaluation of properties of chitosan as an adjuvant for inactivated influenza vaccines administered parenterally. *Journal of Medical Virology* 81: 494–506.

Ghendon, Y., S. Markushin, I. Akopova, I. Koptiaeva, and G. Krivtsov. 2011. Chitosan as an adjuvant for poliovaccine. *Journal of Medical Virology* 83:847–852.

Gill, D. R., K. A. Southern, K. A. Mofford, T. Seddon, H. F. Sorgi, A. Thompson, L. J. Macvinish et al. 1997. A placebo controlled study of liposome-mediated gene transfer to the nasal epithelium of patients with cystic fibrosis. *Gene Therapy* 4:199–209.

Giovino, C., I. Ayensu, J. Tetteh, and J. S. Boateng. 2012. Development and characterisation of chitosan films impregnated with insulin loaded PEG-*b*-PLA nanoparticles (NPs): A potential approach for buccal delivery of macromolecules. *International Journal of Pharmaceutics* 428:143–151.

Giunchedi, P., and U. Conte. 1995. Spray-drying as a preparation method of microparticulate drug delivery systems: An overview. *STP Pharma Pratiques* 5:276–290.

Green, R. J., I. Hopkinson, and R. A. L. Jones. 1999. Unfolding and intermolecular association in globular proteins adsorbed at interfaces. *Langmuir* 15:5102–5110.

Gunbeyaz, M., A. Faraji, A. Ozkul, N. Purali, and S. Senel. 2010. Chitosan based delivery systems for mucosal immunization against bovine herpesvirus 1 (BHV-1). *European Journal of Pharmaceutical Sciences* 41:531–545.

Gupta, R. K., and G. R. Siber. 1995. Adjuvants for human vaccines—Current status, problems and future prospects. *Vaccine* 13:1263–1276.

Gupta, N. K., P. Tomar, V. Sharma, and V. K. Dixit. 2011. Development and characterization of chitosan coated poly-(caprolactone) nanoparticulate system for effective immunization against influenza. *Vaccine* 29:9026–9037.

Hansson, A., N. Hashom, F. Falson, P. Rousselle, O. Jordan, and G. Borchard. 2012. *In vitro* evaluation of an RGD-functionalized chitosan derivative for enhanced cell adhesion. *Carbohydrate Polymers* 90: 1494–1500.

Hari, P. R., T. Chandy, and C. P. Sharma. 1996a. Chitosan/calcium–alginate beads for oral delivery of insulin. *Journal of Applied Polymer Science* 59:1795–1801.

Hari, P. R., T. Chandy, and C. P. Sharma. 1996b. Chitosan/calcium alginate microcapsules for intestinal delivery of nitrofurantoin. *Journal of Microencapsulation* 13:319–329.

He, W., X. Guo, L. Xiao, and M. Feng. 2009. Study on the mechanisms of chitosan and its derivatives used as transdermal penetration enhancers. *International Journal of Pharmaceutics* 382:234–243.

Hein, S., K. Wang, W. F. Stevens, and J. Kjems. 2008. Chitosan composites for biomedical applications: Status, challenges and perspectives. *Materials Science and Technology* 24:1053–1061.

Hirano, S. 2002. Chitin and chitosan. *Ullmann's Encyclopedia of Industrial Chemistry*. Weinham: Wiley-VCH Verlag GmbH & Co. KGaA.

Hogemann, D., J. P. Basilion, and R. Weissleder. 2001. Molecular imaging techniques in magnetic resonance imaging and nuclear imaging. *Radiologe* 41:116–120.

Huang, M., C. W. Fong, E. Khor, and L. Y. Lim. 2005. Transfection efficiency of chitosan vectors: Effect of polymer molecular weight and degree of deacetylation. *Journal of Controlled Release* 106:391–406.

Huang, H. N., T. L. Li, Y. L. Chan, C. L. Chen, and C. J. Wu. 2009. Transdermal immunization with low-pressure-gene–gun mediated chitosan-based DNA vaccines against Japanese encephalitis virus. *Biomaterials* 30:6017–6025.

Illum, L., I. Jabbal-Gill, M. Hinchcliffe, A. N. Fisher, and S. S. Davis. 2001. Chitosan as a novel nasal delivery system for vaccines. *Advanced Drug Delivery Reviews* 51:81–96.

Illum, L. 2003. Nasal drug delivery—Possibilities, problems and solutions. *Journal of Controlled Release* 87:187–198.

Janes, K. A., P. Calvo, and M. J. Alonso. 2001. Polysaccharide colloidal particles as delivery systems for macromolecules. *Advanced Drug Delivery Reviews* 47:83–97.

Jayakumar, R., D. Menon, K. Manzoor, S. V. Nair, and H. Tamura. 2010. Biomedical applications of chitin and chitosan based nanomaterials—A short review. *Carbohydrate Polymers* 82:227–232.

Jayaraman, M. S., D. J. Bharali, T. Sudha, and S. A. Mousa. 2012. Nano chitosan peptide as a potential therapeutic carrier for retinal delivery to treat age-related macular degeneration. *Molecular Vision* 18:2300–2308.

Jiang, H. L., M. L. Kang, J. S. Quan, S. G. Kang, T. Akaike, H. S. Yoo, and C. S. Cho. 2008a. The potential of mannosylated chitosan microspheres to target macrophage mannose receptors in an adjuvant-delivery system for intranasal immunization. *Biomaterials* 29:1931–1939.

Jiang, H. L., J. T. Kwon, E. M. Kim, Y. K. Kim, R. Arote, D. Jere, H. J. Jeong et al. 2008b. Galactosylated poly(ethylene glycol)–chitosan–graft–polyethylenimine as a gene carrier for hepatocyte-targeting. *Journal of Controlled Release* 131:150–157.

Jorgensen, L., E. H. Moeller, M. van deWeert, H. M. Nielsen, and S. Frokjaer. 2006. Preparing and evaluating delivery systems for proteins. *European Journal of Pharmaceutical Sciences* 29:174–182.

Kalber, T. L., N. Kamaly, P. W. So, J. A. Pugh, J. Bunch, C. W. McLeod, M. R. Jorgensen, A. D. Miller, and J. D. Bell. 2011. A low molecular weight folate receptor targeted contrast agent for magnetic resonance tumor imaging. *Molecular Imaging and Biology* 13:653–662.

Khalil, S. K., G. S. El-Feky, S. T. El-Banna, and W. A. Khalil. 2012. Preparation and evaluation of warfarin-β-cyclodextrin loaded chitosan nanoparticles for transdermal delivery. *Carbohydrate Polymers* 90: 1244–1253.

Kim, I. Y., S. J. Seo, H. S. Moon, M. K. Yoo, I. Y. Park, B. C. Kim, and C. S. Cho. 2008. Chitosan and its derivatives for tissue engineering applications. *Biotechnology Advances* 26:1–21.

Kim, J. H., Y. S. Kim, S. Kim, J. H. Park, K. Kim, K. Choi, H. Chung et al. 2006. Hydrophobically modified glycol chitosan nanoparticles as carriers for paclitaxel. *Journal of Controlled Release* 111:228–234.

Kim, Y. H., S. H. Gihm, C. R. Park, K. Y. Lee, T. W. Kim, I. C. Kwon, H. Chung, and S. Y. Jeong. 2001. Structural characteristics of size-controlled self-aggregates of deoxycholic acid-modified chitosan and their application as a DNA delivery carrier. *Bioconjugate Chemistry* 12:932–938.

Koping-Hoggard, M., K. M. Varum, M. Issa, S. Danielsen, B. E. Christensen, B. T. Stokke, and P. Artursson. 2004. Improved chitosan-mediated gene delivery based on easily dissociated chitosan polyplexes of highly defined chitosan oligomers. *Gene Therapy* 11:1441–1452.

Kuo, S. M., G. C. Niu, S. J. Chang, C. H. Kuo, and M. S. Bair. 2004. A one-step method for fabricating chitosan microspheres. *Journal of Applied Polymer Science* 94:2150–2157.

Kwon, I. C. 2008. Chitosan-based nanoparticles for cancer therapy: tumor specificity and enhanced therapeutic efficacy in tumor-bearing mice. *Journal of Controlled Release* 132:e69–e70.

Lee, C. M., H. J. Jeong, S. L. Kim, E. M. Kim, D. W. Kim, S. T. Lim, K. Y. Jang, Y. Y. Jeong, J. W. Nah, and M. H. Sohn. 2009a. SPION-loaded chitosan–linoleic acid nanoparticles to target hepatocytes. *International Journal of Pharmaceutics* 371:163–169.

Lee, C. M., H. J. Jeong, S. L. Kim, E. M. Kim, S. T. Lim, H. T. Kim, I. K. Park, Y. Y. Jeong, J. W. Kim, and M. H. Sohn. 2009b. Superparamagnetic iron oxide nanoparticles as a dual imaging probe for targeting hepatocytes *in vivo*. *Magnetic Resonance in Medicine* 62:1440–1446.

Lee, H. S., E. H. Kim, H. Shao, and B. K. Kwak. 2005. Synthesis of SPIO–chitosan microspheres for MRI-detectable embolotherapy. *Journal of Magnetism and Magnetic Materials* 293:102–105.

Lee, P. W., S. F. Peng, C. J. Su, F. L. Mi, H. L. Chen, M. C. Wei, H. J. Lin, and H. W. Sung. 2008. The use of biodegradable polymeric nanoparticles in combination with a low-pressure gene gun for transdermal DNA delivery. *Biomaterials* 29:742–751.

Li, X., X. Kong, Z. Zhang, K. Nan, L. Li, X. Wang, and H. Chen. 2012. Cytotoxicity and biocompatibility evaluation of *N,O*-carboxymethyl chitosan/oxidized alginate hydrogel for drug delivery application. *International Journal of Biological Macromolecules* 50:1299–1305.

Liu, W. G., K. D. Yao, and Q. C. Liu. 2001a. Formation of a DNA/*N*-dodecylated chitosan complex and salt-induced gene delivery. *Journal of Applied Polymer Science* 83:3391–3395.

Liu, X. F., Y. L. Guan, D. Z. Yang, Z. Li, and K. D. Yao. 2001b. Antibacterial action of chitosan and carboxymethylated chitosan. *Journal of Applied Polymer Science* 79:1324–1335.

Loretz, B., and A. Bernkop-Schnurch. 2006. *In vitro* evaluation of chitosan–EDTA conjugate polyplexes as a nanoparticulate gene delivery system. *AAPS Journal* 8:E756–E764.

Lu, B., C. F. Wang, D. Q. Wu, C. Li, X. Z. Zhang, and R. X. Zhuo. 2009. Chitosan based oligoamine polymers: Synthesis, characterization, and gene delivery. *Journal of Controlled Release* 137:54–62.

Lu, Y., and P. S. Low. 2003. Immunotherapy of folate receptor-expressing tumors: Review of recent advances and future prospects. *Journal of Controlled Release* 91:17–29.

Luessen, H. L., B. J. de Leeuw, M. W. Langemeyer, A. B. de Boer, J. C. Verhoef, and H. E. Junginger. 1996. Mucoadhesive polymers in peroral peptide drug delivery. VI. Carbomer and chitosan improve the intestinal absorption of the peptide drug buserelin *in vivo*. *Pharmaceutical Research* 13:1668–1672.

Luo, X. L., J. J. Xu, Y. Du, and H. Y. Chen. 2004. A glucose biosensor based on chitosan–glucose oxidase–gold nanoparticles biocomposite formed by one-step electrodeposition. *Analytical Biochemistry* 334:284–289.

Lutz, J. F., S. Stiller, A. Hoth, L. Kaufner, U. Pison, and R. Cartier. 2006. One-pot synthesis of PEGylated ultrasmall iron-oxide nanoparticles and their *in vivo* evaluation as magnetic resonance imaging contrast agents. *Biomacromolecules* 7:3132–3138.

Mahmood, U., and R. Weissleder. 2002. Some tools for molecular imaging. *Academic Radiology* 9:629–631.

Mao, H. Q., K. Roy, V. L. Troung-Le, K. A. Janes, K. Y. Lin, Y. Wang, J. T. August, and K. W. Leong. 2001. Chitosan–DNA nanoparticles as gene carriers: Synthesis, characterization and transfection efficiency. *Journal of Controlled Release* 70:399–421.

Mao, S., U. Bakowsky, A. Jintapattanakit, and T. Kissel. 2006. Self-assembled polyelectrolyte nanocomplexes between chitosan derivatives and insulin. *Journal of Pharmaceutical Sciences* 95:1035–1048.

Mitra, S., U. Gaur, P. C. Ghosh, and A. N. Maitra. 2001. Tumour targeted delivery of encapsulated dextran–doxorubicin conjugate using chitosan nanoparticles as carrier. *Journal of Controlled Release* 74:317–323.

Moghe, A. K., and B. S. Gupta. 2008. Co-axial electrospinning for nanofiber structures: Preparation and applications. *Polymer Reviews* 48:353–377.

Mura, C., A. Nácher, V. Merino, M. Merino-Sanjuán, M. Manconi, G. Loy, A. M. Fadda, and O. Díez-Sales. 2012. Design, characterization and *in vitro* evaluation of 5-aminosalicylic acid loaded *N*-succinyl–chitosan microparticles for colon specific delivery. *Colloids and Surfaces B: Biointerfaces* 94:199–205.

Nagarwal, R. C., R. Kumar, and J. K. Pandit. 2012. Chitosan coated sodium alginate–chitosan nanoparticles loaded with 5-FU for ocular delivery: *In vitro* characterization and *in vivo* study in rabbit eye. *European Journal of Pharmaceutical Sciences* 47:678–685.

Ozbas-Turan, S., J. Akbuga, and C. Aral. 2002. Controlled release of interleukin-2 from chitosan microspheres. *Journal of Pharmaceutical Sciences* 91:1245–1251.

Ozbas-Turan, S., C. Aral, L. Kabasakal, and M. Keyer-Uysal. 2003. Co-encapsulation of two plasmids in chitosan microspheres as a non-viral gene delivery vehicle. *Journal of Pharmacy and Pharmaceutical Sciences* 1:27–32.

Pancholi, K., N. Ahras, E. Stride, and M. Edirisinghe. 2009. Novel electrohydrodynamic preparation of porous chitosan particles for drug delivery. *Journal of Materials Sciences: Materials in Medicine* 20:917–923.

Pandey, M., M. Batzloff, and M. Good. 2009. Mechanism of protection induced by group A *Streptococcus* vaccine candidate J8-DT: Contribution of B and T-cells towards protection. *PLoS One* 4:e5147.

Park, D. M., Y. K. Song, J. P. Jee, H. T. Kim, and C. K. Kim. 2012. Development of chitosan-based ondansetron buccal delivery system for the treatment of emesis. *Drug Development and Industrial Pharmacy* 38:1077–1083.

Park, I. K., T. H. Kim, Y. H. Park, B. A. Shin, E. S. Choi, E. H. Chowdhury, T. Akaike, and C. S. Cho. 2001. Galactosylated chitosan–graft–poly(ethylene glycol) as hepatocyte-targeting DNA carrier. *Journal of Controlled Release* 76:349–362.

Park, I. K., J. Yang, H. J. Jeong, H. S. Bom, I. Harada, T. Akaike, S. I. Kim, and C. S. Cho. 2003. Galactosylated chitosan as a synthetic extracellular matrix for hepatocytes attachment. *Biomaterials* 24:2331–2337.

Park, J. H., G. Saravanakumar, K. Kim, and I. C. Kwon. 2010. Targeted delivery of low molecular drugs using chitosan and its derivatives. *Advanced Drug Delivery Reviews* 62:28–24.

Pereira, R. L., C. S. Paim, A. B. Barth, R. P. Raffin, S. S. Guterres, and E. E. Schapoval. 2012. Levodopa microparticles for pulmonary delivery: Photodegradation kinetics and LC stability-indicating method. *Pharmazie* 67:605–610.

Pereswetoff-Morath, L. 1998. Microspheres as nasal drug delivery systems. *Advanced Drug Delivery Reviews* 29:185–194.

Pillai, C. K. S., W. Paul, and C. P. Sharma. 2009. Chitin and chitosan polymers: Chemistry, solubility and fiber formation. *Progress in Polymer Science* 34:641–678.

Pomper, M. G., and D. A. Hammoud. 2004. Positron emission tomography in molecular imaging. *IEEE Engineering in Medicine and Biology Magazine* 23:28–37.

Premaletha, K., C. D. Licy, S. Jose, A. Saraladevi, A. Shirwaikar, and A. Shirwaikar. 2012. Formulation, characterization and optimization of hepatitis B surface antigen (HBsAg)-loaded chitosan microspheres for oral delivery. *Pharmaceutical Development and Technology* 17:251–258.

Ralston, G. B., M. V. Tracey, and P. M. Wrench. 1964. The inhibition of fermentation in baker's yeast by chitosan. *Biochimica et Biophysica Acta* 93:652–655.

Ravi Kumar, M. N. V. 2000. A review of chitin and chitosan applications. *Reactive and Functional Polymers* 46:1–7.

Rutledge, G. C., and S. V. Fridrikh. 2007. Formation of fibers by electrospinning. *Advanced Drug Delivery Reviews* 59:1384–1391.

Sandri, G., M. C. Bonferoni, S. Rossi, F. Ferrari, C. Boselli, and C. Caramella. 2010. Insulin-loaded nanoparticles based on *N*-trimethyl chitosan: *In vitro* (Caco-2 model) and *ex vivo* (excised rat jejunum, duodenum, and ileum) evaluation of penetration enhancement properties. *AAPS PharmSciTech* 11:362–371.

Sarkar, M. A. 1992. Drug metabolism in the nasal mucosa. *Pharmaceutical Research* 9:1–9.

Sashiwa, H., J. M. Thompson, S. K. Das, Y. Shigemasa, S. Tripathy, and R. Roy. 2000a. Chemical modification of chitosan: Preparation and lectin binding properties of α-galactosyl–chitosan conjugates. Potential inhibitors in acute rejection following xenotransplantation. *Biomacromolecules* 1:303–305.

Schipper, N. G., J. C. Verhoef, and F. W. Merkus. 1991. The nasal mucociliary clearance: Relevance to nasal drug delivery. *Pharmaceutical Research* 8:807–814.

Sharma, P., S. Brown, G. Walter, S. Santra, and B. Moudgil. 2006. Nanoparticles for bioimaging. *Advances in Colloid and Interface Science* 123:471–485.

Sharma, S., T. K. Mukkur, H. A. Benson, and Y. Chen. 2012. Enhanced immune response against pertussis toxoid by IgA-loaded chitosan–dextran sulfate nanoparticles. *Journal of Pharmaceutical Sciences* 101: 233–244.

Shi, Z., K. G. Neoh, E. T. Kang, B. Shuter, S. C. Wang, C. Poh, and W. Wang. 2009. Carboxymethyl chitosan-modified superparamagnetic iron oxide nanoparticles for magnetic resonance imaging of stem cells. *ACS Applied Materials and Interfaces* 1:328–335.

Sill, T. J., and H. A. von Recum. 2008. Electrospinning: Applications in drug delivery and tissue engineering. *Biomaterials* 29:1989–2006.

Takahashi, T., K. Takayama, Y. Machida, and T. Nagai. 1990. Characteristics of polyion complexes of chitosan with sodium alginate and sodium polyacrylate. *International Journal of Pharmaceutics* 61:35–41.

Thanou, M. M., A. F. Kotze, T. Scharringhausen, H. L. Leussen, A. G. de Boer, J. C. Verhoef, and H. E. Junginger. 2000. Effect of degree of quaternization of *N*-trimethyl chitosan chloride for enhanced transport of hydrophilic compounds across intestinal Caco-2 cell monolayers. *Journal of Controlled Release* 64:15–25.

Thanou, M., J. C. Verhoef, and H. E. Junginger. 2001a. Chitosan and its derivatives as intestinal absorption enhancers. *Advanced Drug Delivery Reviews* 50:S91–S101.

Thanou, M., J. C. Verhoef, and H. E. Junginger. 2001b. Oral drug absorption enhancement by chitosan and its derivatives. *Advanced Drug Delivery Reviews* 52:117–126.

Thanou, M. M., B. I. Florea, M. Geldof, H. E. Junginger, and G. Borchard. 2002. Quaternized chitosan oligomers as novel gene delivery vectors in epithelial cell lines. *Biomaterials* 23:153–159.

Tremoleda, J. L., M. Khalil, L. L. Gompels, M. Wylezinska-Arridge, T. Vincent, and W. Gsell. 2011. Imaging technologies for preclinical models of bone and joint disorders. *EJNMMI Research* 1:11.

Ubaidulla, U., R. K. Khar, F. J. Ahmad, Y. Sultana, and A. K. Panda. 2007. Development and characterization of chitosan succinate microspheres for the improved oral bioavailability of insulin. *Journal of Pharmaceutical Sciences* 96:3010–3023.

Van, T. N., C. H. Ng, K. N. Aye, T. S. Trang, and W. F. Stevens. 2006. Production of high quality chitin and chitosan from preconditioned shrimp shells. *Journal of Chemical Technology and Biotechnology* 81:1113–1118.

van der Merwe, S. M., J. C. Verhoef, J. H. Verheijden, A. F. Kotzé, and H. E. Junginger. 2004. Trimethylated chitosan as polymeric absorption enhancer for improved peroral delivery of peptide drugs. *European Journal of Pharmaceutics and Biopharmaceutics* 58:225–235.

Varshosaz, J., H. Sadrai, and R. Alinagari. 2004. Nasal delivery of insulin using chitosan microspheres. *Journal of Microencapsulation* 21:761–774.

Vasir, J. K., M. K. Reddy, and V. D. Labhasetwar. 2005. Nanosystems in drug targeting: Opportunities and challenges. *Current Nanoscience* 1:47–64.

Verheul, R. J., N. Hagenaars, T. van Es, E. V. van Gaal, P. H. de Jong, S. Bruijns, E. Mastrobattista et al. 2012. A step-by-step approach to study the influence of *N*-acetylation on the adjuvanticity of *N,N,N*-trimethyl chitosan (TMC) in an intranasal nanoparticulate influenza virus vaccine. *European Journal of Pharmaceutical Sciences* 45:467–474.

Wagner, E., M. Zenke, M. Cotten, H. Beug, and M. L. Birnstiel. 1990. Transferrin–polycation conjugates as carriers for DNA uptake into cells. *Proceedings of the National Academy of Sciences* 87:3410–3414.

Wang, S., J. Luo, D. A. Lantrip, D. J. Waters, C. J. Mathias, and M. A. Green. 1997. Design and synthesis of [111In] DTPA–folate for use as tumor-targeted radiopharmaceutical. *Bioconjugate Chemistry* 8:673–679.

Ward, C., N. Acheson, and L. Seymour. 2000. Folic acid targeting of protein conjugates into ascites tumour cells from ovarian cancer patients. *Journal of Drug Targeting* 8:119–123.

Willmann, J. K., N. van Bruggen, L. M. Dinkelborg, and S. S. Gambhir. 2008. Molecular imaging in drug development. *Nature Reviews Drug Discovery* 7:591–607.

Wong, T. W. 2009. Chitosan and its use in design of insulin delivery system. *Recent Patents on Drug Delivery and Formulation* 3:8–25.

Wu, Y., and R. L. Clark. 2007. Controllable porous polymer particles generated by electrospraying. *Journal of Colloid and Interface Science* 310:529–535.

Wu, F., and T. Jin. 2008. Polymer-based sustained-release dosage forms for protein drugs, challenges, and recent advances. *AAPS PharmSciTech* 9:1218–1229.

Xu, J., W. Dai, Z. Wang, B. Chen, Z. Li, and X. Fan. 2011. Intranasal vaccination with chitosan–DNA nanoparticles expressing pneumococcal surface antigen A protects mice against nasopharyngeal colonization by *Streptococcus pneumoniae*. *Clinical and Vaccine Immunology* 18:75–81.

Xu, Y., and M. A. Hanna. 2007. Electrosprayed bovine serum albumin-loaded tripolyphosphate cross-linked chitosan capsules: Synthesis and characterization. *Journal of Microencapsulation* 24:143–151.

Yoo, J. Y., M. Kim, and J. Lee. 2007. Electrospraying of micro/nano particles for protein drug delivery. *Polymer* 31:215–220.

Zhao, X. B., and R. J. Lee. 2004. Tumor-selective targeted delivery of genes and antisense oligodeoxyribonucleotides via the folate receptor. *Advanced Drug Delivery Reviews* 56:1193–1204.

Zhao, X., L. Yin, J. Ding, C. Tang, S. Gu, C. Yin, and Y. Mao. 2010. Thiolated trimethyl chitosan nanocomplexes as gene carriers with high *in vitro* and *in vivo* transfection efficiency. *Journal of Controlled Release* 144:46–54.

Zhang, H., and S. H. Neau. 2001a. *In vitro* degradation of chitosan by a commercial enzyme preparation: Effect of molecular weight and degree of deacetylation. *Biomaterials* 22:1653–1658.

FURTHER READING

Aguilar, J. C., and E. G. Rodríguez. 2007. Vaccine adjuvants revisited. *Vaccine* 25:3752–3762.

Atmar, R. L., and M. K. Estes. 2006. The epidemiologic and clinical importance of norovirus infection. *Gastroenterology Clinics of North America* 35:275–290.

Bae, Y. H. 2009. Drug targeting and tumor heterogenicity. *Journal of Controlled Release* 133:2–3.

Bal, S. M., S. Hortensius, Z. Ding, W. Jiskoot, and J. A. Bouwstra. 2011. Co-encapsulation of antigen and toll-like receptor ligand in cationic liposomes affects the quality of the immune response in mice after intradermal vaccination. *Vaccine* 29:1045–1052.

Bernkop-Schnürch, A., and S. Steininger. 2000. Synthesis and characterization of mucoadhesive thiolated polymers. *International Journal of Pharmaceutics* 194:1–13.

Bremer, S., A. P. Worth, M. Paparella, K. Bigot, E. Kolossov, B. K. Fleischmann, J. Hescheler, and M. Balls. 2001. Establishment of an *in vitro* reporter gene assay for developmental cardiac toxicity. *Toxicology In Vitro* 15:215–223.

Cerundolo, V., J. Silk, S. Masri, and M. Salio. 2009. Harnessing invariant NKT cells in vaccination strategies. *Nature Reviews Immunology* 9:28–38.

Chandrasekar, D., R. Sistla, F. J. Ahmad, R. K. Khar, and P. V. Diwan. 2007. Folate coupled poly(ethylene glycol) conjugates of anionic poly(amidoamine) dendrimer for inflammatory tissue specific drug delivery. *Journal of Biomedical Materials Research A* 82:92–103.

Chen, C. S., W. Y. Liau, and G. J. Tsai. 1998. Antibacterial effects of *N*-sulfonated and *N*-sulfobenzoyl chitosan and application to oyster preservation. *Journal of Food Protection* 61:1124–1128.

Chen, X., S. Ding, G. Qu, and C. Zhang. 2008. Synthesis of novel chitosan derivatives for micellar solubilization of cyclosporine A. *Journal of Bioactive and Compatible Polymers* 23:563–578.

Cho, Y.W, Y. N. Cho, S. H. Chung, G. Yoo, and S. K. Ko. 1999. Water-soluble chitin as a wound healing accelerator. *Biomaterials* 20:2139–2145.

Chow, K. S., and E. Khor. 2002. New fluorinated chitin derivatives: Synthesis, characterization and cytotoxicity assessment. *Carbohydrate Polymers* 47:357–363.

Chung, T. W., J. Yang, T. Akaike, K. Y. Cho, J. W. Nah, S. Kim, and C. S. Cho. 2002. Preparation of alginate/galactosylated chitosan scaffold for hepatocyte attachment. *Biomaterials* 23:2827–2834.

Danesh-Bahreini, M. A., J. Shokri, A. Samiei, E. Kamali-Sarvestani, M. Barzegar-Jalali, and S. Mohammadi-Samani. 2011. Nanovaccine for leishmaniasis: Preparation of chitosan nanoparticles containing *Leishmania* superoxide dismutase and evaluation of its immunogenicity in BALB/c mice. *International Journal of Nanomedicine* 6:835–842.

Dautry-Varsat, A. 1986. Receptor-mediated endocytosis: The intracellular journey of transferrin and its receptor. *Biochimie* 68:375–381.

Defaye, J., A. Gadelle, and C. Pederson. 1994. A convenient access to β-(1 → 4)-linked 2-amino-2-deoxy-D-glucopyranosyl fluoride oligosaccharides and β-(1 → 4)-linked 2-amino-2-deoxy-D-glucopyranosyl oligosaccharides by fluorolysis and fluorohydrolysis of chitosan. *Carbohydrate Research* 261:267–277.

Gamian, A., M. Chomik, C. A. Laferriere, and R. Roy. 1991. Inhibition of influenza A virus hemagglutinin and induction of interferon by synthetic sialylated glycoconjugates. *Canadian Journal of Microbiology* 37:233–237.

Garnett, M. C. 2001. Targeted drug conjugates: Principles and progress. *Advanced Drug Delivery Reviews* 53:171–216.

Hashem, F. M., S. A. Fahmy, A. M. El-Sayed, and M. M. Al-Sawahli. 2011. Development and evaluation of chitosan microspheres for tetanus, diphtheria and divalent vaccines: A comparative study of subcutaneous and intranasal administration in mice. *Pharmaceutical Development Technology* 0:1–11.

Heck, D. E., D. L. Laskin, C. R. Gardner, and J. D. Laskin. 1992. Epidermal growth factor suppresses nitric oxide production by keratinocytes. *Journal of Biological Chemistry* 267:21277–21280.

Hejazi, R., and M. J. Amiji. 2003. Chitosan-based gastrointestinal delivery systems. *Journal of Controlled Release* 89:151–165.

Hogemann, D., and J. P. Basilion. 2002. Seeing inside the body: MR imaging of gene expression. *European Journal of Nuclear Medicine* 29:400–408.

Hwang, H. Y., I. S. Kim, I. C. Kwon, and Y. H. Kim. 2008. Tumor targetability and anti-tumor effect of docetaxel-loaded hydrophobically modified glycol chitosan nanoparticles. *Journal of Controlled Release* 128:23–31.

Il'ina, A. V., and V. P. Varlamov. 2007. Galactosylated derivatives of low-molecular-weight chitosan: Production and properties. *Prikladnaia Biokhimiia Mikrobiologiya* 43:82–87.

Ishihara, M., K. Nakanishi, K. Ono, M. Sato, M. Kikuchi, Y. Saito, H. Yura et al. 2001. Photocrosslinkable chitosan as a dressing for wound occlusion and accelerator in healing process. *Biomaterials* 23:833–840.

Ishii, T., Y. Okahata, and T. Sato. 2001. Mechanism of cell transfection with plasmid/chitosan complexes. *Biochimica et Biophysica Acta-Biomembranes* 1514:51–64.

Ivanov, I. I., B. S. McKenzie, L. Zhou, C. E. Tadokoro, A. Lepelley, J. J. Lafaille, D. J. Cua, and D. R. Littman. 2006. The orphan nuclear receptor RORgammat directs the differentiation program of proinflammatory IL-17+ T helper cells. *Cell* 126:1121–1133.

Jia, Z., D. Shen, and W. Xu. 2001. Synthesis and antibacterial activities of quaternary ammonium salt of chitosan. *Carbohydrate Research* 333:1–6.

Jiang, H. L., Y. K. Kim, R. Arote, D. Jere, J. S. Quan, J. H. Yu, Y. J. Choi, J. W. Nah, M. H. Cho, and C. S. Cho. 2009. Mannosylated chitosan–graft–polyethylenimine as a gene carrier for Raw 264.7 cell targeting. *International Journal of Pharmaceutics* 375:133–139.

Jiang, X., H. Dai, K. W. Leong, S. H. Goh, H. Q. Mao, and Y. Y. Yang. 2006. Chitosan-g-PEG/DNA complexes deliver gene to the rat liver via intrabiliary and intraportal infusions. *Journal of Gene Medicine* 8:477–487.

Juliano, R. L. 1998. Factors affecting the clearance kinetics and tissue distribution of liposomes, microspheres and emulsions. *Advanced Drug Delivery Reviews* 2:31–54.

Kang, M. L., C. S. Cho, and H. S. Yoo. 2009. Application of chitosan microspheres for nasal delivery of vaccines. *Biotechnology Advances* 27:857–865.

Kast, C. E., and A. Bernkop-Schnürch. 2001. Thiolated polymers–thiomers: Development and *in vitro* evaluation of chitosan thioglycolic acid conjugates. *Biomaterials* 22:2345–2352.

Kato, Y., H. Onishi, and Y. Mashida. 2001. Lactosaminated and intact *N*-succinyl chitosans as drug carriers in liver metastasis. *Journal of Controlled Release* 70:295–307.

Kim, T. H., S. I. Kim, T. Akaike, and C. S. Cho. 2005. Synergistic effect of poly(ethylenimine) on the transfection efficiency of galactosylated chitosan/DNA complexes. *Journal of Controlled Release* 105:354–366.

Kim, T. H., J. W. Nah, M. H. Cho, T. G. Park, and C. S. Cho. 2006. Receptor-mediated gene delivery into antigen presenting cells using mannosylated chitosan/DNA nanoparticles. *Journal of Nanoscience and Nanotechnology* 6:2796–2803.

Kong, M., X. G. Chen, C. S. Liu, C. G. Liu, X. H. Meng, and L. J. Yu. 2008. Antibacterial mechanism of chitosan microspheres in a solid dispersing system against *E. coli*. *Colloids and Surfaces B: Biointerfaces* 65:197–202.

Koping-Hoggard, M., I. Tubulekas, H. Guan, K. Edwards, M. Nilsson, K. M. Varum, and P. Artursson. 2001. Chitosan as a nonviral gene delivery system. Structure–property relationships and characteristics compared with polyethylenimine *in vitro* and after lung administration *in vivo*. *Gene Therapy* 8: 1108–1121.

Lai, W. F., and M. C. Lin. 2009. Nucleic acid delivery with chitosan and its derivatives. *Journal of Controlled Release* 134:158–168.

Lentacker, I., R. E. Vandenbroucke, B. Lucas, J. Demeester, S. C. De Smedt, and N. N. Sanders. 2008. New strategies for nucleic acid delivery to conquer cellular and nuclear membranes. *Journal of Controlled Release* 132:279–288.

Leroux-Roels, I., E. Vets, R. Freese, M. Seiberling, F. Weber, C. Salamand, and G. Leroux-Roels. 2008. Seasonal influenza vaccine delivered by intradermal microinjection: A randomised controlled safety and immunogenicity trial in adults. *Vaccine* 26:6614–6619.

Li, F., W. G. Liu, and K. D. Yao. 2002. Preparation of oxidized glucose-crosslinked *N*-alkylated chitosan membrane and *in vitro* studies of pH-sensitive drug delivery behavior. *Biomaterials* 23:343–347.

Luo, X., X. Song, A. Zhu, Y. Si, L. Ji, Z. Ma, Z. Jiao, and J. Wu. 2012. Development of poly(vinyl acetate–methylacrylic acid)/chitosan/Fe(3)O(4) nanoparticles for the diagnosis of non-alcoholic steatohepatitis with magnetic resonance imaging. *Journal of Materials Science: Materials in Medicine* 23:3075–3082.

Maeda, H., J. Wu, T. Sawa, Y. Matsumura, and K. Hori. 2000. Tumor vascular permeability and the EPR effect in macromolecular therapeutics. *Journal of Controlled Release* 65:271–284.

Mahmood, U., C. H. Tung, A. Jr. Bogdanov, and R. Weissleder. 1999. Near-infrared optical imaging of protease activity for tumor detection. *Radiology* 213:866–870.

Mahmood, U. 2004. Near infrared optical applications in molecular imaging. *IEEE Engineering in Medicine and Biology Magazine* 23:58–66.

Mao, S., X. Shuai, F. Unger, M. Simon, D. Bi, and T. Kissel. 2004. The depolymerization of chitosan: Effects on physicochemical and biological properties. *International Journal of Pharmaceutics* 281:45–54.

Mattioli-Belmonte, M., N. Nicoli-Aldini, A. DeBenedittis, G. Sgarbi, S. Amati, M. Fini, G. Biagini, and R. A. Muzzarelli. 1999. Morphological study of bone regeneration in presence of 6-oxychitin. *Carbohydrate Polymers* 40:23–27.

Mei Lin Tan, P., F. M. Chong, and C. R. Dass. 2009. Chitosan nanotechnology and cancer therapy. *Journal of Pharmacy and Pharmacology* 61:3–21.

Mia, S., T. Tina, J. Heidi, V. Peep, and M. Martti. 2003. Evaluation of microcrystalline chitosans for gastro-retentive drug delivery. *European Journal of Pharmaceutical Sciences* 19:345–353.

Mori, T., M. Okumura, M. Matsuura, K. Ueno, S. Tokura, Y. Okamoto, S. Minami, and T. Fujinaga. 1997. Effects of chitin and its derivatives on the proliferation and cytokine production of fibroblasts *in vitro*. *Biomaterials* 18:947–951.

Muzzarelli, R. A. A., C. Muzzarelli, A. Cosani, and M. Terbojevich. 1999. 6-Oxychitins, novel hyaluronan-like regiospecifically carboxylated chitins. *Carbohydrate Polymers* 39:361–367.

Obara, K., M. Ishihara, T. Ishizuka, M. Fujita, Y. Ozeki, T. Maehara, Y. Saito et al. 2003. Photocrosslinkable chitosan hydrogel containing fibroblast growth factor-2 stimulates wound healing in healing–impaired db/db mice. *Biomaterials* 24:3437–3444.

Ohya T., R. Cai, H. Nishizawa, K. Hara, and T. Ouchi. 2000. Preparation of PEG-grafted chitosan nanoparticles as peptide drug carriers. *STP Pharma Sciences* 10:77–82.

Park, Y. K., Y. H. Park, B. A. Shin, E. S. Choi, Y. R. Park, T. Akaike, and C. S. Cho. 2000. Galactosylated chitosan–graft–dextran as hepatocyte-targeting DNA carrier. *Journal of Controlled Release* 9:97–108.

Prabaharan, M. 2008. Chitosan derivatives as promising materials for controlled drug delivery. *Journal of Biomaterials Applications* 23:5–36.

Prabaharan, M., and J. F. Mano. 2006. Chitosan derivatives bearing cyclodextrin cavities as novel adsorbent matrices. *Carbohydrate Polymers* 63:153–166.

Quong, D., and R. J. Nuefeld. 1998. DNA protection from extracapsular nucleases within chitosan or poly-L-lysine-coated alginated beads. *Biotechnology and Bioengineering* 60:124–134.

Ren, D., H. Yi, W. Wang, and X. Ma. 2005. The enzymatic degradation and swelling properties of chitosan matrices with different degrees of *N*-acetylation. *Carbohydrate Research* 340:2403–2410.

Rinaudo, M. 2006. Chitin and chitosan: Properties and applications. *Progress in Polymer Science* 31:603–632.

Roy, R., and C. A. Laferriere. 1988. Synthesis of antigenic copolymers of *N*-acetylneuraminic acid binding to wheat germ agglutinin and antibodies. *Carbohydrate Research* 177:C1–C4.

Roy, R. 1996. Blue-prints synthesis and applications of glycopolymers. *Trends in Glycoscience and Glycotechnology* 8:79–99.

Sashiwa, H., Y. Makimura, Y. Shigemasa, and R. Roy. 2000b. Chemical modification of chitosan: Preparation of chitosan–sialic acid branched polysaccharide hybrids. *Chemical Communications* 11:909–910.

Sashiwa, H., M. Yamamori, Y. Ichinose, J. Sunamoto, and S. Aiba. 2003. Michael reaction of chitosan with various acryl reagents in water. *Biomacromolecules* 4:1250–1254.

Sashiwa, H., and S. Aiba. 2004. Chemically modified chitin and chitosan as biomaterials. *Progress in Polymer Science* 29:887–908.

Sieval, A. B., M. Thanou, A. F. Kotze, J. C. Verhoef, I. Brossee, and H. E. Junginger. 1998. Preparation and NMR characterization of highly substituted *N*-trimethyl chitosan chloride. *Carbohydrate Polymers* 36:157–165.

Sukarto, A., C. Yu, L. E. Flynn, and B. G. Amsden. 2012. Co-delivery of adipose-derived stem cells and growth factor-loaded microspheres in RGD-grafted *N*-methacrylate glycol chitosan gels for focal chondral repair. *Biomacromolecules* 13:2490–2502.

Sudarshan, N. R., D. G. Hoover, and D. Knorr. 1992. Antibacterial action of chitosan. *Food Biotechnology* 6:257–272.

Tenover, F. C., M. V. Lancaster, B. C. Hill, C. D. Steward, S. A. Stocker, G. A. Hancock, C. M. O'Hara, N. C. Clark, and K. J. Hiramtsu. 1998. Characterization of staphylococci with reduced susceptibilities to vancomycin and other glycopeptides. *Journal of Clinical Microbiology* 36:1020–1027.

Torchilin, V. P. 2000. Drug targeting. *European Journal of Pharmaceutical Sciences* 11:S81–S91.

van der Lubben, I. M., J. C. Verhoef, G. Borchard, and H. E. Junginger. 2001a. Chitosan for mucosal vaccination. *Advanced Drug Delivery Reviews* 52:139–144.

van der Lubben, I. M., J. C. Verhoef, G. Borchard, and H. E. Junginger. 2001b. Chitosan and its derivatives in mucosal drug and vaccine delivery. *European Journal of Pharmaceutical Sciences* 14:201–207.

van Steenis, J. H., E. M. van Maarseveen, F. J. Verbaan, R. Verrijk, D. J. Crommelin, G. Storm, and W. E. Hennink. 2003. Preparation and characterization of folate-targeted PEG-coated pDMAEMA-based polyplexes. *Journal of Controlled Release* 87:167–176.

Werb, Z., and S. Gordon. 1975a. Elastase secretion by stimulated macrophages. Characterization and regulation. *Journal of Experimental Medicine* 142:361–377.

Werb, Z., and S. Gordon. 1975b. Secretion of a specific collagenase by stimulated macrophages. *Journal of Experimental Medicine* 142:346–360.

Yalpani, M., and L. D. Hall. 1984. Some chemical and analytical aspects of polysaccharide formation. Formation of branched-chain soluble chitosan derivatives. *Macromolecules* 17:272–281.

Young, D. H., H. Kohle, and H. Kauss. 1982. Effect of chitosan on membrane permeability of suspension-cultured glycine max and *Phaseolus vulgaris* cells. *Plant Physiology* 70:1499–1554.

Young, D. H., and H. Kauss. 1983. Release of calcium from suspension-cultured glycine max cells by chitosan, other polycations, and polyamines in relation to effects on membrane permeability. *Plant Physiology* 73:698–702.

Zhang, H., I. A. Alsarra, and S. H. Neau. 2001b. An *in vitro* evaluation of a chitosan-containing multiparticulate system for macromolecule delivery to the colon. *International Journal of Pharmaceutics* 239:197–205.

Zhang, X. Q., X. L. Wang, P. C. Zhang, Z. L. Liu, R. X. Zhuo, H. Q. Mao, and K. W. Leong. 2005. Galactosylated ternary DNA/polyphosphoramidate nanoparticles mediate high gene transfection efficiency in hepatocytes. *Journal of Controlled Release* 102:749–763.

Zuo, G. Y., G. C. Wang, Y. B. Zhao, G. L. Xu, X. Y. Hao, J. Han, and Q. J. Zhao. 2008. Screening of Chinese medicinal plants for inhibition against clinical isolates of methicillin-resistant *Staphylococcus aureus* (MRSA). *Journal of Ethnopharmacology* 120:287–290.

22 Chitosan as Coating Material for Titanium Implants

Ramona Lieder and Olafur E. Sigurjonsson

CONTENTS

22.1 INTRODUCTION

Clinical treatment of orthopedic tissue injuries often requires fixation via bone implant material. Implants produced from titanium and titanium alloys have been the gold standard in load-bearing orthopedic applications since many years because of their favorable biological and mechanical properties. The advantages of titanium and its alloys for biomedical devices include the following: (1) high corrosion resistance, (2) biocompatibility owing to the spontaneous formation of oxide layers, (3) high specific strength, and (4) lack of toxicity (Elias et al. 2008). Nevertheless, there is still room for improvement, particularly concerning the stabilization of the implant and the integration at the bone–biomaterial interface (Goodacre et al. 1999).

Successful integration and stabilization of the implant critically depend on surface characteristics, that is, surface chemistry, roughness, topography, and wettability. Increasing the surface roughness and modifying the surface architecture by sandblasting, plasma spraying, or acid etching has been extensively used to enhance initial stabilization of the implant and promote bone formation at the peri-implant site (Bagno and Di Bello 2004). Osseo-integration is another factor that critically deter-mines the lifetime of the implant and describes the direct interaction of the implant with the bone tissue, thereby resulting in bone growth on the implant surface. During implantation, damage to the bone environment and the direct contact of the implant with body fluids can promote the formation of a fibrous tissue capsule, preventing osteoblastic cell attachment to the implant surface. This may ultimately lead to the loosening of the implant and decrease the patient's quality of life (Puleo and Nanci 1999).

Titanium has high affinity for oxygen, which makes the implant surface sufficiently reactive for chemical modification, including the deposition of bioactive coatings to enhance bioactivity, osseo-integration, and implant stabilization (Hench and Polak 2002). One of the materials being investigated as bioactive coating for titanium and its alloys is chitosan. The direct use of chitosan in combination with ample amounts of different biomaterials and the easy molding abilities, make this polymer an attractive tool for tissue engineering applications (Shahidi and Abuzaytoun 2005).

22.2   CHITOSAN AS BIOACTIVE COATING

The prerequisite for the preparation of coatings and membranes is the cationic nature of chitosan, as it is the premise for the solubility in dilute aqueous acids (Yi et al. 2005). After dissolution, membranes can be cast on virtually any substrate and the resulting coating becomes insoluble in aqueous solutions after a simple neutralization step (Yi et al. 2005). For easy handling, the valida-tion of a new coating procedure or starting material and the evaluation of bioactivity are generally performed on tissue culture plastic rather than the final substrate. This allows for straightforward initial characterization of surface properties and biological performance.

The biological properties of chitosan are strongly dependent on the number of charged groups (degree of deacetylation—DD), the molecular weight or its distribution (polydispersity index), and the successive order of acetylated and deacetylated residues in the chain (Harish Prashanth and Tharanathan 2007). Inconsistency in reports describing the biological performance of chitosan coatings are frequently because of the lack of detailed information on source, sample preparation, and chemical properties of the polymer that strongly affects the comparability of results from dif-ferent studies (Hamilton et al. 2006).

22.2.1   Biological Performance

Cellular behavior of any cell type is crucially affected by the surface characteristics and DD, with distinct effects depending on the cell type (Chatelet et al. 2001). Generally, cell attachment and proliferation are superior on membranes prepared from higher DD chitosan, yet lower DD chitosan coatings hold the promising capacity to induce healing without formation of scar tissue (Prasitsilp et al. 2000).

The cationic charge of the polymer is indicated in mediating the attachment of osteoblasts, rather than fibroblasts, offering a remarkably useful strategy for the prevention of fibrous tissue capsule formation around medical implants (Fakhry et al. 2004). Negatively charged cytokines and growth factors readily interact with positively charged chitosan, which in turn promotes the migration of polymorphonuclear cells to aid in the tissue regeneration process at the implantation site (Di Martino et al. 2005). However, the cationic charge also provokes interaction with the negatively charged phosphate groups of lipopolysaccharides, the toxicity inducing factor in endotoxins (Naberezhnykh et al. 2008). The affinity for endotoxins is in fact so strong that cross-linked chitosan microfiltration membranes have been used for the removal of endotoxin contamination from medical prepara-tions (Lamb Machado et al. 2006). However, chitosan polymers have also been indicated in reduc-ing lipopolysaccharide-mediated toxicity by competitive inhibition of endotoxin receptor binding (Yermak et al. 2006). On the basis of our research with mesenchymal stem cells and chitosan poly-mers, we suggest that the reduction of endotoxin toxicity is in addition dependent on the presence of soluble versus membrane-bound CD14 receptors (Lieder et al. 2013). CD14 engages in the transfer of lipopolysaccharide monomers to downstream extracellular adaptor proteins (Wright et al. 1990).

Chitosan membranes and coatings as biological substrates have been studied extensively for a variety of cell types and coating protocols, including varying concentrations of chitosan solution, different neutralization and sterilization agents, and starting materials from diverse sources. As a general rule, osteoblastic cell attachment and proliferation are favored on high DD chitosan membranes, which aid in the differentiation process and stimulate the secretion of extracellular matrix proteins (Bumgardner et al. 2003a). More specifically, Amaral et al. (2007) demonstrated that MG-63 human osteosarcoma osteoblast-like cells require chitosan membranes with a DD higher than 87% for spreading and the proliferation behavior strongly correlates with increases in the DD. Furthermore, human osteoblasts grown on 90% DD chitosan membranes sustained a spherical morphology as compared to spindle-shaped cells on tissue culture plastic, and preserved collagen type I expression during short-term culture (Lahiji et al. 2000).

However, the detailed mechanism of how chitosan membranes affect bone cell attachment and proliferation is still insufficiently understood and particularly attachment rates to lower DD chitosan coatings remain unsatisfactorily low for clinical use in orthopedic applications (Zheng et al. 2009). Our group has recently developed a standardized solution casting protocol that can be applied for the preparation of chitosan membranes with a wide range of DD (at least 47–96%). Following this protocol, cell attachment and proliferation of the mouse preosteoblastic cell line MC3T3-E1 can be sustained for extended periods of time. This is achieved by a simple stepwise procedure for the casting process, which applies adequate considerations for the underlying substrate, implements an intricate understanding of the elementary chemical mechanism, and challenges traditional cell culture methods (Lieder et al. 2012).

Concerning the attachment and differentiation of mesenchymal stem cell precursors into osteoblasts, more elaborate considerations are required than for the culture of preosteoblastic cell lines. Uygun et al. (2010) demonstrated that attachment is strongly dependent on the thickness of the membranes, where increased width of the coating correlates to more homogeneous cell spreading. Chitosan membranes may promote the differentiation of osteoprogenitor cells and aid in bone formation, but a DD of at least 96% is required (Amaral et al. 2005).

Furthermore, several non-osteoblastic cell lines have been evaluated for their adherence and proliferation potential on chitosan membranes. The mouse myoblastic cell line C2C12 attached and proliferated on 84% DD chitosan membranes, yet the downregulation of genes associated with differentiation was observed, while genes involved in cell cycle regulation were increased (Abarrategi et al. 2009). Keratinocytes can adhere to chitosan membranes prepared from 53% to 97.5% DD; however, the proliferation potential is directly correlated with increases in the DD (Chatelet et al. 2001). Furthermore, BHK21(C13) hamster kidney cells only attach to high DD chitosan membranes, whereas L920 mouse fibroblast-like cells can adhere to a wide range in DD. In the latter, the adhesion strength increases with the DD and was shown to negatively correlate to the proliferation abilities, effectively preventing cell cycle entry (Prasitsilp et al. 2000; Mao et al. 2004).

22.2.2 Chitosan Derivatives as Coating Material

Literature on chitosan derivatives used as bioactive membrane on tissue culture plastic is limited. This might be because of increased difficulties in sample preparation and the increased water solubility of many commonly used chitosan derivatives. Nevertheless, a few interesting reports have been published on how derivatization may affect membrane surface characteristics and bioactivity.

N-acylation of chitosan was shown to result in fibrillar membrane morphology, whereas *N*-arylidene modified chitosan derivatives displayed smooth surfaces. Cell attachment and bioactivity of these derivatives was not evaluated, yet solubility was shown to depend on the length of the acyl chain (Hirano et al. 1981). Cai et al. (2002) evaluated *N*-butyl and *N*-cetyl chitosan derivatives as coating material and the effect on neonatal rat calvaria osteoblasts. *N*-butyl chitosan membranes promoted proliferation and alkaline phosphatase activity, while *N*-cetylchitosan derivatives performed similar to control substrates apart from increased cell viability (Cai et al. 2002). We have recently evaluated the effect of *N*-lauroyl chitosan derivatives with a degree of substitution (DS) of 0.02–0.20 on attachment of the mouse preosteoblastic cell line MC3T3-E1. The incentive of this study was based on decreasing the solubility of low DD chitosan membranes in aqueous solutions without the use of cross-linking reagents. Unfortunately, only derivatives with a DS of 0.02 were soluble under the general conditions applied during solution casting methods. Although cells initially attached to these derivatized membranes, attachment could not be sustained more than a few days, likely because of the inhomogeneous distribution of the membrane (R. Lieder and P. Sahariha, unpublished observations).

Chitin is rarely used as coating material, yet derivatization can increase its solubility and facilitate application. Highly sulfated chitin membranes with a DS of 1.55 sustained MC3T3-E1 mouse preosteoblastic cell attachment, adsorbed high levels of fibronectin, and promoted osteogenic differentiation. In the same experimental set-up, carboxymethylchitin with a DS of 1.0 was shown to significantly increase osteogenic differentiation at early time points, accompanied by enhanced calcification and alkaline phosphatase activity (Abe et al. 2010).

22.3 FACTORS INFLUENCING THE BIOACTIVITY OF CHITOSAN MEMBRANES

Surface characteristics and the biological microenvironment surrounding any biomaterial, strongly influence the interactions with approaching cells and determine bioactivity (Marquis et al. 2009). Surface considerations generally not only include the surface chemistry, topography, and wettability but also the influence of these material properties on polymer crystallinity, matrix protein adsorption, and degradation rate (Puleo and Nanci 1999). The surface characteristics critically determine the initial amount and the conformation of proteins that are adsorbed onto the biomaterial surface. This outermost atomic layer of adhesive proteins is the primary interaction site for approaching cells and fundamentally affects integrin signaling, cell responses, and regeneration processes (Boyan et al. 1996). Although the general parameters affecting cell responses and protein adsorption are known, the optimal surface characteristics and the detailed mechanisms at the bone–biomaterial interface are only partially understood (Puleo and Nanci 1999).

22.3.1 HOMOGENEITY

Homogeneity of chitosan membranes is a factor seldom discussed in the literature, yet we know from our experience that it may strongly affect cell attachment and proliferation. During the solution casting process, undissolved chitosan particles need to be removed either by centrifugation or filtration, as cells generally do not attach onto these residues when present at the surface of the membrane. Furthermore, because un-cross-linked chitosan membranes are clear, the homogeneity of the casting process needs to be confirmed regularly to assure the quality of membrane preparation. Defects in the membrane crucially affect attachment and compromise the results of proliferation studies. Owing to the cationic charge of the chitosan polymer, the membrane can be easily stained with an acidic anionic dye, for example, Alizarin Red. As for chitosan derivatives in membrane casting, the choice of dye depends on the chemical modification used because the engagement of the amino group in derivatization may abrogate the anionic dye-binding capacity. Fatty acid modified chitosan derivatives, for example, can be stained by crystal violet even with a DS as low as 0.05 because the fatty acid residues retain the triarylmethane structure in the dye (R. Lieder et al. unpublished observations).

22.3.2 SURFACE TOPOGRAPHY

One of the parameters strongly affecting initial protein adsorption and cell interactions is the geometrical architecture of the surface (Esposito et al. 2005). Rougher surfaces are associated with enhanced osteoblastic differentiation and cell attachment, whereas fibroblastic cells prefer smooth surfaces (Anselme et al. 2000). The geometrical architecture on the micrometer and the millimeter scales is important for the successful integration of a biomaterial, whereas the nano surface roughness mediates cell-specific interactions (Junker et al. 2009).

Generally, the surface roughness of chitosan membranes decreases with an increase in DD. Yet, we have shown that the surface roughness is also strongly dependent on the source of chitosan used in the membrane casting process. Shrimp shell chitosan displays a surface roughness similar to tissue culture plastic, whereas crab shell chitosan of the same DD is slightly rougher, most probably as a result of hydrolysis of the polysaccharide backbone and the subsequent decrease in molecular weight due to the requirement of longer deacetylation processes to obtain the same DD (Rhazi et al. 2000; Lieder et al. 2012). Topographically, chitosan membranes produce valley and hill-like structures, whereas tissue culture plastic is characterized by evenly distributed long fibers (Lieder et al. 2012).

Surface roughness and topography can be determined on a nano-scale by atomic force microscopy (AFM), enabling the visualization of distinct surface structures and consecutive calculation of average surface roughness (Siedlecki and Marchant 1998). Other suitable methods include light interference microscopy and SEM, yet the latter cannot deduce nanometer surface structures (Ma et al. 2007; Abarrategi et al. 2009).

22.3.3 Wettability

Water contact angle measurements are used as an estimation of the hydrophilicity/hydrophobicity of a material surface. Wettability can be determined in a goniometer applying the sessile drop method and surfaces with water contact angles that are smaller than 90° are generally considered as hydrophilic (Wenling et al. 2005; Hu et al. 2010). Low water contact angles are associated with superior cell responses, yet matrix protein adsorption may be thermodynamically undesirable. Hydrophobic surfaces attract high levels of matrix proteins to their surface but the strong adsorption may damage the protein conformation and hence reduce bioactivity (Ma et al. 2007).

Owing to its functional groups, chitosan is a hydrophilic polymer, however, reports on water contact angles on chitosan membranes are controversial (Wenling et al. 2005). A decrease in water contact angle with increase in DD was shown, while other groups did not find significant differences in hydrophilicity between 70% DD and 95% DD chitosan membranes (Amaral et al. 2005; Wenling et al. 2005). In our studies, we demonstrated that chitosan membranes prepared from 87% DD and 94% DD are statistically more hydrophobic than 47% DD and 68% DD, yet both surfaces are still considered hydrophilic (water contact angle <90°) (Lieder et al. 2012).

22.3.4 Surface Chemistry

The role of surface charge in modulating cellular functions is controversial, as both positively and negatively charged surfaces have been shown to increase bone formation (Hamamoto et al. 1995). A strong positive charge at the biomaterial surface may induce unnaturally strong focal adhesion and integrin binding, but since the cell membrane is negatively charged, electrostatic interactions with positively charged surfaces can be similarly favored (Shelton et al. 1988). Additionally, the osteoblast-mediated secretion of a mineralized matrix was shown to be particularly enhanced on cationic surfaces (Healy et al. 1996). At physiologic pH, the DD of chitosan is not correlated to the surface charge because of the pKa value (6.2–7.0) of the amino group (Wenling et al. 2005). Because the zeta potential of chitosan is close to zero based on the alkaline pH of the amino group, the electrostatic interaction of chitosan with approaching cells might be negligible (Tomihata and Ikada 1997).

The effect of surface charge is closely related to the chemical groups present on the biomaterial surface. Amino functionalities at the surface are especially desirable as they are associated with: (1) increased cell adhesion and proliferation, (2) high adsorption of matrix proteins, (3) stimulation of integrin binding, (4) moderate hydrophilicity, and (5) enhanced mineralization and osteogenic gene expression (Faucheux et al. 2004). Considerations concerning surface chemistry become more important when chitosan derivatives are used for the preparation of membranes, as the chemical groups present on the surface will strongly affect cell responses.

22.3.5 CRYSTALLINITY

Crystallinity describes the structural organization of a biomaterial and high crystallinity is correlated to decreased protein adsorption and subsequently lower cell attachment (Uygun et al. 2010). Chitosan powder is a highly crystalline polymer with the potential for extensive intra- and intermolecular hydrogen bonding (Ogawa et al. 2004). Hundred percent DD chitosan and 0% DD chitin display high chemical regularity, leading to high crystallinity, as opposed to any other chitosan preparations (Jaworska et al. 2003). This can be explained by the decrease in molecular weight and the occurrence of defects in the polymer structure after deacetylation (Tomihata and Ikada 1997). Higher DD is associated with higher inter-chain flexibility, which favors hydrogen bonds and establishes a superior crystalline structure (Anthonsen et al. 1993).

The membrane casting process strongly affects the crystallinity of the polymer, depending on the underlying substrate, the casting methods, and the evaporation conditions (Reje and Block 1999). Both increase and decrease of crystallinity have been reported with increase in DD, which can most likely be attributed to differences in material processing (Li et al. 1997; Wenling et al. 2005). X-ray diffraction analysis is used for the determination of crystallinity (Aranaz et al. 2009).

22.3.6 ADSORPTION OF MATRIX PROTEINS

The interaction of a biomaterial with approaching cells is mediated by the unspecific adsorption of matrix proteins (Boyan et al. 1996). Fibronectin is one of the main adhesion proteins at the bone–biomaterial interface and contains an integrin-specific binding sequence (RGD-sequence). This protein is particularly important for the attachment and differentiation of osteoblastic cells (Stephansson et al. 2002). The conformation of fibronectin at the biomaterial surface strongly depends on the surface wettability, with generally higher bioactivity on hydrophilic surfaces because of decreased denaturation of the secondary structure (Culp and Sukenik 1998). This negatively charged protein is frequently used to coat biomaterials to promote cell attachment *in vitro*, and it was shown that structural changes upon surface interaction mediate the protein's bioactivity (Steele et al. 1993).

The interaction of fibronectin with chitosan membranes is based on the hydrophilicity and the cationic charge, which plays an important role in decreasing denaturation processes (Conti et al. 2002). We have shown that higher DD chitosan membranes can absorb similar or even higher amounts of fibronectin to their surface than tissue culture plastic, whereas low DD chitosan membranes show significantly reduced fibronectin retention (Lieder et al. 2012). These results are in agreement with previously published studies, showing a direct correlation of fibronectin with increases in the DD. Chitosan might even be more specific for the adsorption of fibronectin than tissue culture plastic, as membranes with 85% DD or higher maintain the protein conformation (Amaral et al. 2005). The amount of matrix protein adsorption can be determined by in-house ELISA and coating chitosan membranes with fibronectin to improve initial cell attachment does not affect subsequent gene expression studies (Lieder et al. 2012).

22.3.7 DEGRADATION RATE

The degradation rate describes the reduction of chitosan molecular weight based on chain scission by enzymes, hydrolysis, or action of free radicals. Lysozyme is the main enzyme in the human body interacting with chitosan and is frequently used in the determination of chitosan degradation rates depending on the chemical properties of the polymer (Dash et al. 2011). The enzymatic mechanism of lysozyme degradation is based on the hydrolysis of the bond between two adjacent *N*-acetyl glucosamine residues in the chitosan polymer chain (Aiba 1992). Several parameters are known to decrease the degradation rate, including: (1) increases in DD, (2) increases in degree of crystallinity, (3) higher molecular weight, (4) use of acetic acid rather than formic acid as solvent in the casting process, and (5) the more homogeneous distribution of *N*-acetyl glucosamine residues in the chain (Dash et al. 2011).

Furthermore, chitosan membranes with DD close to 50% are particularly soluble in aqueous solutions and require cross-linking to improve stability. The solubility of these materials is attributed to the degradation of the secondary structure and the subsequent increase in hydrophilicity (Sannan et al. 1976). The degradation rate can be determined after incubation with lysozyme by measuring the reduction in molecular weight with viscosimetric or chromatographic methods (Bumgardner et al. 2003b). Alternatively, the changes in molecular weight can be related to membrane weight loss and morphological changes observed by SEM (Cao et al. 2005). The mechanisms involved in the *in vivo* degradation of chitosan are currently insufficiently understood (Dash et al. 2011).

Cross-linking methods are commonly used in biomaterials research to prolong the stability and improve the properties of biomaterials, that is, degradation, chemical resistance, porosity, and mechanical properties (Feng et al. 2005). A soluble biomaterial can be either cross-linked before the casting process by straightforward addition of the cross-linking agent into the solution (bulk cross-linking), or surface-modified by immersion after the drying process (surface cross-linking) (Wan et al. 2003). Frequently used agents in the cross-linking of chitosan solutions and membranes include glutaraldehyde, genipin, and tripolyphosphate (Silva et al. 2004). Residual traces of the cross-linking agent or leaking during cell experiments cause decreases in cell viability and can be toxic *in vivo* (Phaechamud 2008). Glutaraldehyde is by far the most widely used cross-linking reagent for chitosan membranes and acts via the formation of an imine bond between the primary amino group of chitosan and the aldehyde group, turning the final membranes slightly yellowish/ orange. A low degree of cross-linking does not negatively affect cell survival and proliferation and provides a convenient balance between bioactivity and degradation of chitosan membranes prepared from materials with a DD close to 50% (Lieder et al. 2012).

22.3.8 STERILIZATION

Before application in cell culture experiments and animal models, chitosan membranes require sterilization. Special consideration should be given to the choice of sterilization technique, as some of the commonly applied methods can induce chemical modifications in the polymer structure and ultimately affect the biological performance (Rao and Sharma 1995). Autoclaving and dry heat were shown to decrease molecular weight and negatively affect aqueous solubility (Lim et al. 1999). Sterilization via gamma irradiation or ethylene oxide is associated with chain scission, resulting in decreased mechanical properties and increased degradation susceptibility (Lim et al. 1998). In our experience, sterilization with 70% ethanol followed by UV irradiation for 30 min is suitable for chitosan membranes prepared from different sources and with a wide DD (Lieder et al. 2012; R. Lieder et al. unpublished work).

22.4 COATING METHODS FOR TITANIUM IMPLANTS

Currently, there are four main techniques to coat titanium implants with bioactive chitosan membranes, that is, solution casting, silanization, layer-by-layer self-assembly, and electrophoretic deposition. In this section, we will briefly discuss the incentive, advantages, and disadvantages for each of these methods.

22.4.1 Solution Casting

Solution casting is a simple dipping method based on electrostatic interactions between positively charged chitosan and the oxide layer on titanium surfaces (Figure 22.1) (Zhang et al. 2011). Chitosan deposition can be improved by surface pretreatment of the titanium implants, that is, sandblasting or acid etching to increase surface roughness, ultrasonic cleaning to remove residues from the machining process, and sterilization before coating (Park et al. 2012a). Solution casting methods are particularly interesting as the process is cheap, fast, and does not require special equipment. Modification of coating solution pH and salt conditions can improve the coating procedure (Park et al. 2012a). Yet, the bonding strength of the coating to the medical device is insufficiently low for clinical applications and successful coverage is limited to noncomplex implant shapes, that is, disks rather than porous implants (Bumgardner et al. 2003b).

22.4.2 Surface Grafting

Silanization is a chemical bonding process based on the reaction of an amino silane with the titanium oxide layer, resulting in covalent bonds. In a second step, glutaraldehyde reacts with the surface-bound silane by secondary ketimine formation and simultaneously presents aldehyde groups for interaction with the chitosan polymer (Bumgardner et al. 2003a). This final cross-linking step is mediated by the formation of imine bonds between the primary amino group of chitosan and the aldehyde group (Figure 22.2). This process is independent of the DD, results in significantly superior bonding strength as compared to solution casting methods, and allows for direct control of the coating thickness (Bumgardner et al. 2003a; Yuan et al. 2008). However, residual traces of glutaraldehyde could be released during biodegradation of the coating and damage the surrounding tissue (Phaechamud 2008). Following the same principle, a dopamine linker can be used instead of an amino silane to covalently bond chitosan coatings to titanium implants (Shi et al. 2008).

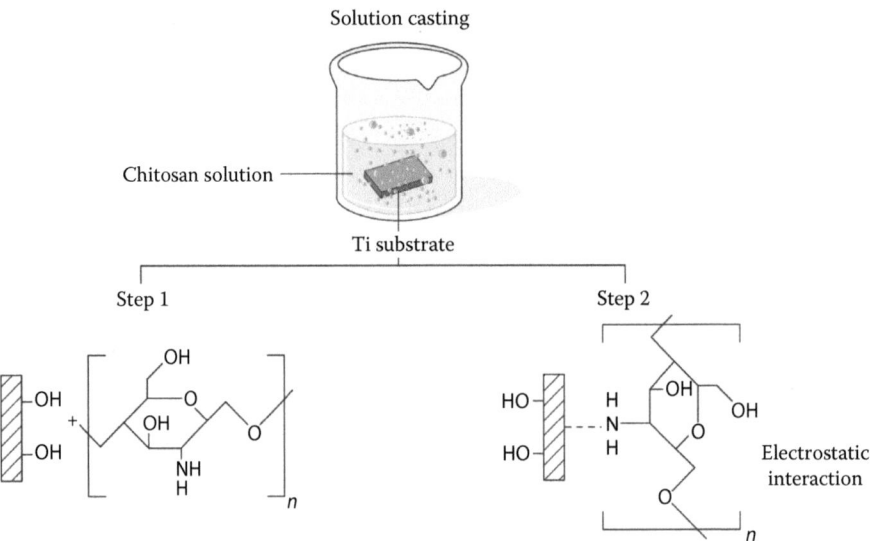

FIGURE 22.1 Solution casting of bioactive chitosan coatings. Electrostatic interactions between cationic chitosan and the oxide layer on titanium implant surfaces are employed in a simple dipping method. After immersion in the chitosan solution, noncomplex implant shapes are rapidly coated, yet the bonding strength of the coating remains insufficiently low for clinical applications.

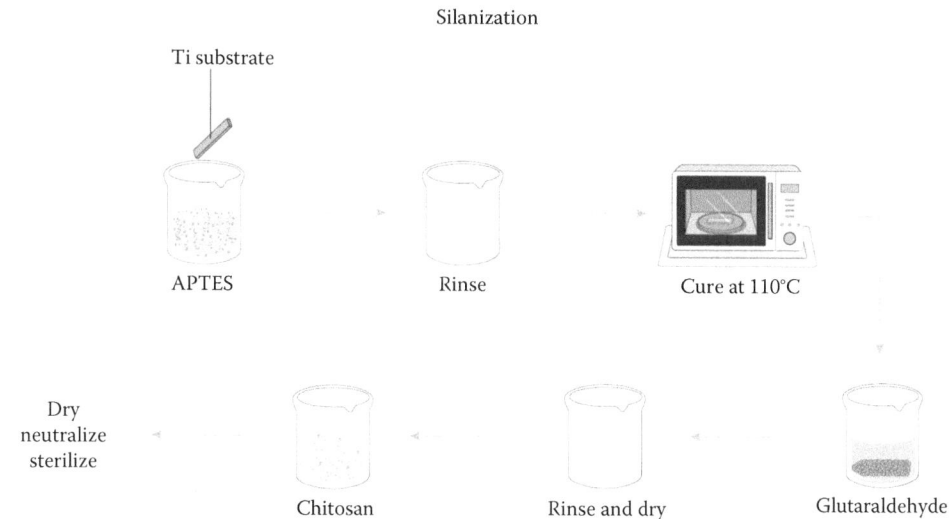

Silanization

Ti substrate

APTES   Rinse   Cure at 110°C

Dry
neutralize
sterilize

Chitosan   Rinse and dry   Glutaraldehyde

FIGURE 22.2 Bonding of chitosan membranes to titanium implants via silanization. Pretreated titanium implants are immersed in amino silane solution, for example, APTES (3-aminopropyl-triethoxysilane), rinsed to remove unreacted reagent residues, and then cured to induce covalent bond formation. Following the immersion in glutaraldehyde solution, the primary amine group of the surface-bound amino silane reacts with one of the aldehyde groups of glutaraldehyde. Via the second aldehyde group, chitosan can be covalently linked to the surface. In a final step, the coated implant surface is dried, neutralized, and sterilized for subsequent use.

22.4.3 LAYER-BY-LAYER SELF-ASSEMBLY

Self-assembled coatings can be generated by alternate solution casting using a negatively charged biomaterial solution followed by dipping into a positively charged chitosan solution. Similar to traditional solution casting, this method is based on the electrostatic interaction between the positively charged chitosan and a negatively charged biomaterial solution of choice (Figure 22.3) (Cai et al. 2005). The resulting coating layers are relatively stable and uniform, permitting the use of implants with complex shape. Layer-by-layer self-assembly is a cheap and simple method that does not require the use of special equipment or potentially toxic chemicals (Channasanon et al. 2007). Thickness and surface topography can be flexibly modulated by the process parameters, resulting in layers of bioactive coatings with alternating properties (Cai et al. 2008). Yet, this method ultimately relies on the combination of two biomaterials, as well as initial optimization of salt conditions for optimal deposition (Lvov et al. 1998).

22.4.4 ELECTROPHORETIC DEPOSITION

Electrophoretic deposition is based on the principles of electrophoresis, that is, the mobility of charged particles toward an electrode because of an electrical stimulus (Boccaccini et al. 2010). At the electrode, the release of hydrogen gas locally raises the pH, resulting in the deposition of a thin chitosan film (Figure 22.4). This colloidal process is generally performed in a two-electrode cell (Simchi et al. 2009). Electrophoretic deposition is an inexpensive, fast, and easily scalable technique for biomaterial processing. Thickness and surface topography can be controlled by the process parameters, allowing for the deposition of homogeneous and highly pure coatings at ambient temperatures (Sharma et al. 2009). However, because the process parameters strongly affect the final coating, time-consuming optimization is necessary, which is presently inefficiently accomplished by trial-and-error methods (Pishbin et al. 2011). Furthermore, because the system is based on aqueous solutions, gas entrapment is unavoidable and can interfere with the deposition process (Besra et al. 2008).

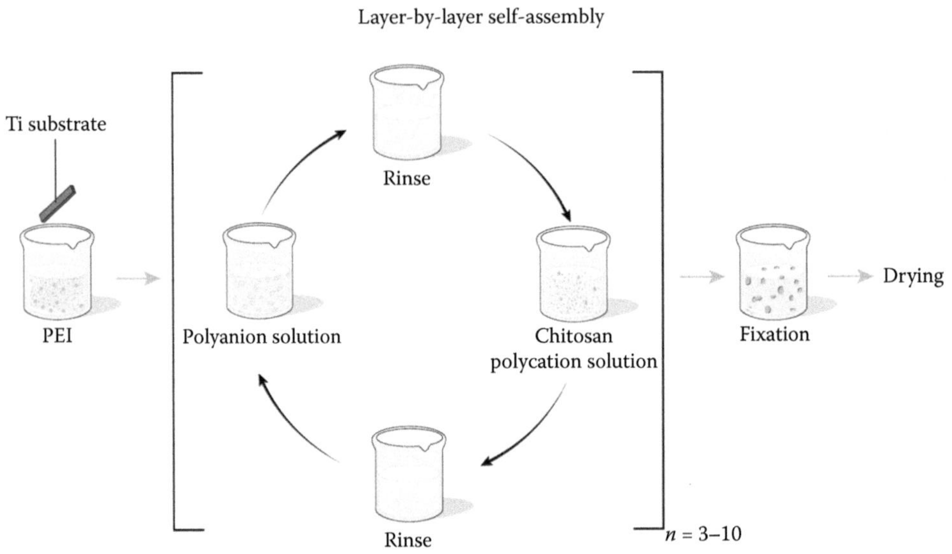

Layer-by-layer self-assembly

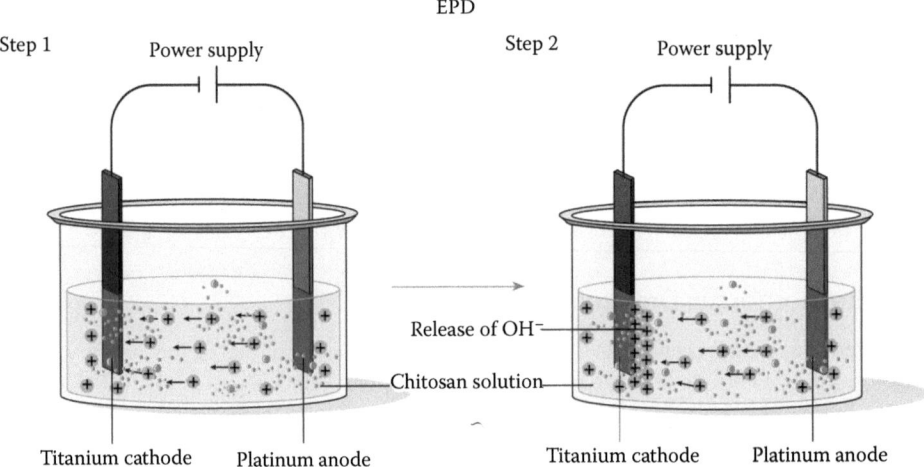

FIGURE 22.3 Layer-by-layer self-assembly methods. A stable positive charge is introduced to the titanium implant surface by immersion in polyethylenimine (PEI) solution. This positive charge allows for electrostatic interactions with a negatively charged polyanion solution. The now reversed surface charge can be employed for reactions with positively charged chitosan solution. This cycle is repeated until the desired amount of bilayers is deposited on the implant surface. Finally, the multilayer coated implant is immersed in fixation solution and dried.

EPD

FIGURE 22.4 Electrophoretic deposition of chitosan coatings. The titanium implant surface is used as electrode in a two-electrode cell system immersed in positively charged chitosan solution. When an electrical stimulus is applied, the positively charged particles move toward the electrode. Due to the release of hydrogen gas at the electrode, the pH locally increases, resulting in the deposition of a thin chitosan layer on the titanium surface.

22.5 *IN VITRO* AND *IN VIVO* BIOACTIVITY OF CHITOSAN-COATED TITANIUM IMPLANTS

Chitosan has been under investigation as bioactive coating for titanium implants based on the following desirable properties: (1) easy molding into desired shape and combination with other biomaterials, (2) bacteriostatic, (3) biodegradable, (4) biocompatible, (5) stimulation of wound healing, and (6) favorable effects on osteogenesis *in vitro* and *in vivo* (Bumgardner et al. 2007; Avila et al. 2009). However, reports on the evaluation of chitosan-coated titanium and even more so, the use of chitosan derivatives as coating material is scarcely described in the literature (Martin et al. 2007). In this section, we highlight some of the outstanding work focused on the surface coating of titanium.

22.5.1 UNMODIFIED CHITOSAN AS BIOACTIVE COATINGS

The *in vitro* evaluation of 80–89% DD chitosan membranes on titanium grade 2 disks and their effect on MG-63 human osteoblast-like cells showed increased cell attachment and decreased osteocalcin expression for coated samples, whereas this was effectively reversed, when titanium disks were pretreated with oxygen plasma before solution casting (Park et al. 2012a,b).

Bonding of 91.2% DD and 92.3% DD chitosan via silanization to implant quality titanium presented increased bonding strength over solution casted membranes, minimal degradation over 8 weeks, and significantly higher cell attachment of UMR-106 osteosarcoma cells as compared to control samples (Bumgardner et al. 2003a; Yuan et al. 2008). No significant differences in bioactivity could be attributed to dopamine-anchored 84% DD chitosan on Ti-6Al-4V in the presence of bone marrow-derived mesenchymal stem cells, whereas the same experimental set-up with MC3T3-E1 mouse pre-osteoblastic cells showed higher cell attachment, enhanced alkaline phosphatase activity, and bacteriostatic properties of the coating (Shi et al. 2008; Lim et al. 2009). Similarly, dopamine anchoring of 82% DD chitosan onto titanium foils displayed significantly increased cell attachment of MC3T3-E1 cells over uncoated titanium (Chua et al. 2008a).

Alternating layers of 82% DD chitosan and hydroxyapatite coated onto titanium foils via layer-by-layer self-assembly demonstrated regular multilayer deposition and bacteriostatic properties; however, cell attachment was lower than that of the uncoated control samples (Chua et al. 2008a,b).

Although electrophoretic deposition of 95% DD chitosan onto commercially pure titanium grade 2 displayed high bonding strength, attachment of MG-63 osteoblast-like cells was insufficiently low and the coating was porous because of gas entrapment (Jiang et al. 2010).

In vivo, the implantation of silane-bonded 80.7% DD chitosan on titanium grade 2 loaded with antibiotics into the muscle of Sprague–Dawley rats, increased the inflammatory reaction but sustained rapid release of antibiotics (Norowski et al. 2011). However, bonding of 92.3% DD to titanium pins via silanization and subsequent implantation into the tibia of New Zealand white rabbits improved osseointegration at the bone–biomaterial interface and did not induce inflammation (Bumgardner et al. 2007).

22.5.2 TITANIUM IMPLANTS COATED WITH CHITOSAN DERIVATIVES

Commercially pure titanium has been coated with alternating layers of sulfated chitosan and collagen and evaluated for its effect on platelets. Because the sulfated chitosan derivative is increasingly more soluble in aqueous solutions, the outermost layer of the multilayer coating has to be collagen. Platelet adhesion and activation was shown to be decreased (Li et al. 2009).

Dopamine anchoring of carboxymethylchitosan to titanium foils displayed bacteriostatic properties and minimal dissolution in aqueous medium, yet attachment of MC3T3-E1 mouse pre-osteoblastic cells was insignificant (Hu et al. 2010). In a similar set-up, C6-carboxymethylchitosan dopamine-linked to Ti-6Al-4V was shown to sustain MC3T3-E1 attachment, spreading, and osteogenic differentiation, accompanied by high antimicrobial activity. The same trend was shown for human mesenchymal stem cells on the same substrate, but cell seeding densities were unusually high (Shi et al. 2009).

To the best of our knowledge there are no reports on the *in vivo* evaluation of titanium implants coated with membranes prepared from chitosan derivatives.

22.6 FUTURE CHALLENGES AND PROSPECTS

Much research has focused attention on the use of chitosan membranes as bioactive coatings, yet the opportunities in tissue engineering applications are far from being exhausted. Numerous challenges remain before successful translation into the clinics: (1) batch-to-batch variability of chitosan starting material complicate the prediction of the clinical outcome, (2) bonding strength of titanium-coated implants often remains insufficiently low to sustain the pressure during implantation, (3) cell attachment and biocompatibility need further improvement, (4) effects of long-term storage of coated titanium implants before implantation remain unknown, and (5) the application of electrophoretic deposition techniques—however promising—requires a more profound understanding of the relationship between processing parameters and structure of the final coating (Khor and Lim 2003; Corni et al. 2008; Yuan et al. 2008).

What we would like to see in the next few years is an increased understanding and integration of the parameters that are involved in the design of chitosan membranes as bioactive coatings for titanium implants (Figure 22.5). The optimization of bioactive chitosan coatings requires the intricate knowledge of the mechanisms influencing bioactivity, surface properties, and bonding strength to titanium implants. Additionally, more extensive *in vivo* evaluation of chitosan-coated titanium implants is necessary before even considering clinical applications. The evaluation of chitosan derivatives with their promising properties could revolutionize the use of bioactive coatings prepared from chitosan. However, certain standards need to be established for publishing reports on the use of chitosan, which should at least include the DD, molecular weight, source, detailed processing parameters, and potential contaminants.

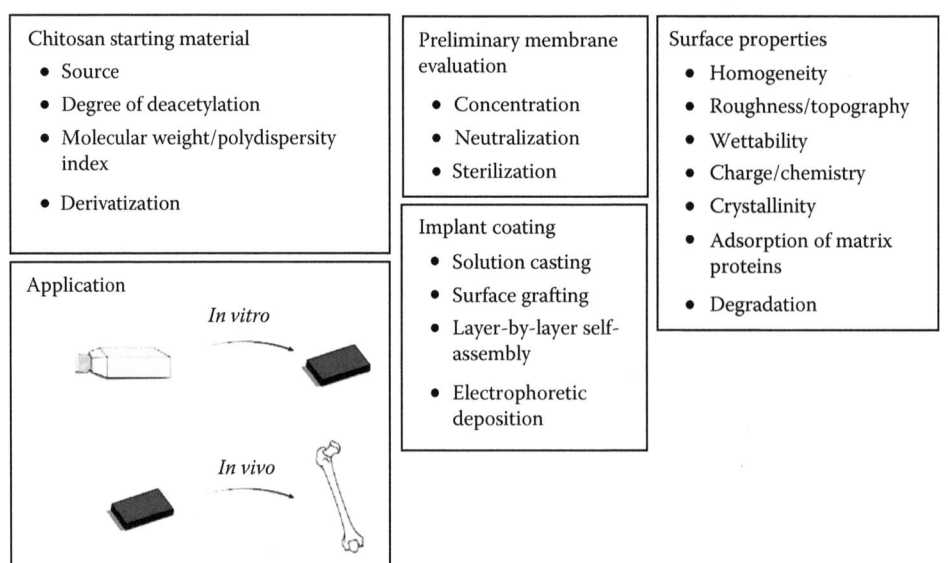

FIGURE 22.5 Essential considerations involved in the design of chitosan membranes as bioactive coatings for titanium implants.

Chitosan is one of the most promising natural substances used in biomaterials research and provides several essential key properties for the use in tissue engineering applications (nontoxic, biodegradable, biocompatible, bacteriostatic, etc.) (Khor and Lim 2003). The polymer can be easily combined with other biomaterials for virtually any application, rapidly processed into bioactive coatings, and offers the cost-effective possibility of delivering growth factors and drugs to the implantation site (Kurita 2006). An increase of implant biocompatibility and early strength development at the bone–biomaterial interface could significantly improve the existing practices in the clinical treatment of orthopedic tissue injuries.

REFERENCES

Abarrategi, A., J. García-Cantalejo, C. Moreno-Vicente et al. 2009. Gene expression profile on chitosan/rhBMP-2 films: A novel osteoinductive coating for implantable materials. *Acta Biomater* 5 (7):2633–46.

Abe, K., M. Nagahata, and A. Teramoto. 2010. Regeneration of tissue in the living body. In: *Material Science of Chitin and Chitosan*, edited by T. Uragami, and S. Tokura. Japan: Springer.

Aiba, S. 1992. Studies on chitosan: 4. Lysozymic hydrolysis of partially N-acetylated chitosans. *Int J Biol Macromol* 14 (4):225–8.

Amaral, I. F., A. L. Cordeiro, P. Sampaio, and M. A. Barbosa. 2007. Attachment, spreading and short-term proliferation of human osteoblastic cells cultured on chitosan films with different degrees of acetylation. *J Biomater Sci Polym Ed* 18 (4):469–85.

Amaral, I. F., M. Lamghari, S. R. Sousa, P. Sampaio, and M. A. Barbosa. 2005. Rat bone marrow stromal cell osteogenic differentiation and fibronectin adsorption on chitosan membranes: The effect of the degree of acetylation. *J Biomed Mater Res A* 75 (2):387–97.

Anselme, K., M. Bigerelle, B. Noel et al. 2000. Qualitative and quantitative study of human osteoblast adhesion on materials with various surface roughnesses. *J Biomed Mater Res* 49 (2):155–66.

Anthonsen, W. M., K. M. Varum, and O. Smidsrod. 1993. Solution properties of chitosan: Conformation and chain stiffness of chitosans with different degress of N-acetylation. *Carbohydr Polym* 22:193–201.

Aranaz, I., M. Mengíbar, R. Harris et al. 2009. Functional characterization of chitin and chitosan. *Curr Chem Biol* 3:203–30.

Avila, G., K. Misch, P. Galindo-Moreno, and H. L. Wang. 2009. Implant surface treatment using biomimetic agents. *Implant Dent* 18 (1):17–26.

Bagno, A., and C. Di Bello. 2004. Surface treatments and roughness properties of Ti-based biomaterials. *J Mater Sci Mater Med* 15 (9):935–49.

Besra, L., T. Uchikoshi, T. S. Suzuki, and Y. Sakka. 2008. Bubble-free aqueous electrophoretic deposition (EPD) by pulse-potential application. *J Am Ceram Soc* 91 (10):3154–9.

Boccaccini, A. R., S. Keim, R. Ma, Y. Li, and I. Zhitomirsky. 2010. Electrophoretic deposition of biomaterials. *J R Soc Interface* 7 (Suppl 5):S581–613.

Boyan, B. D., T. W. Hummert, D. D. Dean, and Z. Schwartz. 1996. Role of material surfaces in regulating bone and cartilage cell response. *Biomaterials* 17 (2):137–46.

Bumgardner, J. D., B. M. Chesnutt, Y. Yuan et al. 2007. The integration of chitosan-coated titanium in bone: An *in vivo* study in rabbits. *Implant Dent* 16 (1):66–79.

Bumgardner, J. D., R. Wiser, S. H. Elder, R. Jouett, Y. Yang, and J. L. Ong. 2003a. Contact angle, protein adsorption and osteoblast precursor cell attachment to chitosan coatings bonded to titanium. *J Biomater Sci Polym Ed* 14 (12):1401–9.

Bumgardner, J. D., R. Wiser, P. D. Gerard et al. 2003b. Chitosan: Potential use as a bioactive coating for orthopaedic and craniofacial/dental implants. *J Biomater Sci Polym Ed* 14 (5):423–38.

Cai, K., Y. Hu, K. D. Jandt, and Y. Wang. 2008. Surface modification of titanium thin film with chitosan via electrostatic self-assembly technique and its influence on osteoblast growth behavior. *J Mater Sci Mater Med* 19 (2):499–506.

Cai, K., W. Liu, F. Li et al. 2002. Modulation of osteoblast function using poly(D,L-lactic acid) surfaces modified with alkylation derivative of chitosan. *J Biomater Sci Polym Ed* 13 (1):53–66.

Cai, K., A. Rechtenbach, J. Hao, J. Bossert, and K. D. Jandt. 2005. Polysaccharide-protein surface modification of titanium via a layer-by-layer technique: Characterization and cell behaviour aspects. *Biomaterials* 26 (30):5960–71.

Cao, W., M. Cheng, Q. Ao, Y. Gong, N. Zhao, and X. Zhang. 2005. Physical, mechanical and degrada-tion properties, and Schwann cell affinity of cross-linked chitosan films. *J Biomater Sci Polym Ed* 16 (6):791–807.

Channasanon, S., W. Graisuwan, S. Kiatkamjornwong, and V. P. Hoven. 2007. Alternating bioactivity of multi-layer thin films assembled from charged derivatives of chitosan. *J Colloid Interface Sci* 316 (2):331–43.

Chatelet, C., O. Damour, and A. Domard. 2001. Influence of the degree of acetylation on some biological prop-erties of chitosan films. *Biomaterials* 22 (3):261–8.

Chua, P. H., K. G. Neoh, E. T. Kang, and W. Wang. 2008a. Surface functionalization of titanium with hyal-uronic acid/chitosan polyelectrolyte multilayers and RGD for promoting osteoblast functions and inhibit-ing bacterial adhesion. *Biomaterials* 29 (10):1412–21.

Chua, P. H., K. G. Neoh, Z. Shi, and E. T. Kang. 2008b. Structural stability and bioapplicability assessment of hyaluronic acid-chitosan polyelectrolyte multilayers on titanium substrates. *J Biomed Mater Res A* 87 (4):1061–74.

Conti, M., G. Donati, G. Cianciolo, S. Stefoni, and B. Samorì. 2002. Force spectroscopy study of the adhesion of plasma proteins to the surface of a dialysis membrane: Role of the nanoscale surface hydrophobicity and topography. *J Biomed Mater Res* 61 (3):370–9.

Corni, Ilaria, Mary P. Ryan, and Aldo R. Boccaccini. 2008. Electrophoretic deposition: From traditional ceram-ics to nanotechnology. *Journal of the European Ceramic Society* 28:1353–67.

Culp, L. A., and C. N. Sukenik. 1998. Cell type-specific modulation of fibronectin adhesion functions on chemically-derivatized self-assembled monolayers. *J Biomater Sci Polym Ed* 9 (11):1161–76.

Dash, M., F. Chiellini, R. M. Ottenbrite, and E. Chiellini. 2011. Chitosan—A versatile semi-synthetic polymer in biomedical applications. *Progr Polymer Sci* 36:981–1014.

Di Martino, A., M. Sittinger, and M. V. Risbud. 2005. Chitosan: A versatile biopolymer for orthopaedic tissue-engineering. *Biomaterials* 26 (30):5983–90.

Elias, C. N., J. H. C. Lima, R. Valiev, and M. A. Meyers. 2008. Biomedical applications of titanium and its alloys. *Biol Mater Sci* 60:46–9.

Esposito, M., P. Coulthard, P. Thomsen, and H. V. Worthington. 2005. The role of implant surface modifi-cations, shape and material on the success of osseointegrated dental implants. A Cochrane systematic review. *Eur J Prosthodont Restor Dent* 13 (1):15–31.

Fakhry, A., G. B. Schneider, R. Zaharias, and S. Senel. 2004. Chitosan supports the initial attachment and spreading of osteoblasts preferentially over fibroblasts. *Biomaterials* 25 (11):2075–9.

Faucheux, N., R. Schweiss, K. Lützow, C. Werner, and T. Groth. 2004. Self-assembled monolayers with different terminating groups as model substrates for cell adhesion studies. *Biomaterials* 25 (14):2721–30.

Feng, F., Y. Liu, B. Zhao, and Ke'ao Hu. 2005. *In vitro* biomineralization of glutaraldehyde crosslinked chito-san films. *J Wuhan Univ Technol—Mater Sci Ed* 20:20–23.

Goodacre, C. J., J. Y. Kan, and K. Rungcharassaeng. 1999. Clinical complications of osseointegrated implants. *J Prosthet Dent* 81 (5):537–52.

Hamamoto, N., Y. Hamamoto, T. Nakajima, and H. Ozawa. 1995. Histological, histocytochemical and ultra-structural study on the effects of surface charge on bone formation in the rabbit mandible. *Arch Oral Biol* 40 (2):97–106.

Hamilton, V., Y. Yuan, D. A. Rigney et al. 2006. Characterization of chitosan films and effects on fibroblast cell attachment and proliferation. *J Mater Sci Mater Med* 17 (12):1373–81.

Harish Prashanth, K. V., and R. N. Tharanathan. 2007. Chitin/chitosan: Modifications and their unlimited appli-cation potential—An overview. *Trends Food Sci Technol* 18:117–31.

Healy, K. E., C. H. Thomas, A. Rezania et al. 1996. Kinetics of bone cell organization and mineralization on materials with patterned surface chemistry. *Biomaterials* 17 (2):195–208.

Hench, L. L., and J. M. Polak. 2002. Third-generation biomedical materials. *Science* 295 (5557):1014–7.

Hirano, S., K. Tobetto, and Y. Noishiki. 1981. SEM ultrastructure studies of *N*-acyl- and *N*-benzylidene-chitosan and chitosan membranes. *J Biomed Mater Res* 15 (6):903–11.

Hu, X., K. G. Neoh, Z. Shi, E. T. Kang, C. Poh, and W. Wang. 2010. An *in vitro* assessment of titanium func-tionalized with polysaccharides conjugated with vascular endothelial growth factor for enhanced osseo-integration and inhibition of bacterial adhesion. *Biomaterials* 31 (34):8854–63.

Jaworska, M, K Sakurai, P Gaudon, and E Guibal. 2003. Influence of chitosan characteristics on polymer prop-erties. I: Crystallographic properties. *Polym Int* 52:198–205.

Jiang, T., Z. Zhang, Y. Zhou et al. 2010. Surface functionalization of titanium with chitosan/gelatin via electro-phoretic deposition: Characterization and cell behavior. *Biomacromolecules* 11 (5):1254–60.

Junker, R., A. Dimakis, M. Thoneick, and J. A. Jansen. 2009. Effects of implant surface coatings and composition on bone integration: A systematic review. *Clin Oral Implants Res* 20 (Suppl 4):185–206.

Khor, E., and L. Y. Lim. 2003. Implantable applications of chitin and chitosan. *Biomaterials* 24 (13):2339–49.

Kurita, K. 2006. Chitin and chitosan: Functional biopolymers from marine crustaceans. *Mar Biotechnol (NY)* 8 (3):203–26.

Lahiji, A., A. Sohrabi, D. S. Hungerford, and C. G. Frondoza. 2000. Chitosan supports the expression of extracellular matrix proteins in human osteoblasts and chondrocytes. *J Biomed Mater Res* 51 (4):586–95.

Lamb Machado, R., Eduardo José de Arruda, C. Costapinto Santana, and S. Maria Alves Bueno. 2006. Evaluation of a chitosan membrane for removal of endotoxin from human IgG solutions. *Process Biochem* 41:2252–57.

Li, J., J. F. Revol, and R. H. Marchessault. 1997. Effect of degree of deacetylation of chitin on the properties of chitin crystallites. *J Appl Poly Sci* 65:373–380.

Li, Q. L., N. Huang, J. Chen et al. 2009. Anticoagulant surface modification of titanium via layer-by-layer assembly of collagen and sulfated chitosan multilayers. *J Biomed Mater Res A* 89 (3):575–84.

Lieder, R., M. Darai, M. B. Thor et al. 2012. *In vitro* bioactivity of different degree of deacetylation chitosan, a potential coating material for titanium implants. *J Biomed Mater Res A.* 100:3392–3399.

Lieder, R., M. Darai, G. Örlygsson, and O. E. Sigurjonsson. Unpublished work. Solution casting of chitosan membranes for *in vitro* evaluation of bioactivity.

Lieder, R., V. S. Gaware, F. Thormodsson et al. 2013. Endotoxins affect bioactivity of chitosan derivatives in cultures of bone marrow-derived human mesenchymal stem cells. *Acta Biomater* 9 (1):4771–8.

Lim, L. Y., E. Khor, and O. Koo. 1998. Gamma irradiation of chitosan. *J Biomed Mater Res* 43 (3):282–90.

Lim, L. Y., E. Khor, and C. E. Ling. 1999. Effects of dry heat and saturated steam on the physical properties of chitosan. *J Biomed Mater Res* 48 (2):111–6.

Lim, T. Y., W. Wang, Z. Shi, C. K. Poh, and K. G. Neoh. 2009. Human bone marrow-derived mesenchymal stem cells and osteoblast differentiation on titanium with surface-grafted chitosan and immobilized bone morphogenetic protein-2. *J Mater Sci Mater Med* 20 (1):1–10.

Lvov, Y., M. Onda, K. Ariga, and T. Kunitake. 1998. Ultrathin films of charged polysaccharides assembled alternately with linear polyions. *J Biomater Sci Polym Ed* 9 (4):345–55.

Ma, Z., Z. Mao, and C. Gao. 2007. Surface modification and property analysis of biomedical polymers used for tissue engineering. *Colloids Surf B Biointerfaces* 60 (2):137–57.

Mao, J. S., Y. L. Cui, X. H. Wang et al. 2004. A preliminary study on chitosan and gelatin polyelectrolyte complex cytocompatibility by cell cycle and apoptosis analysis. *Biomaterials* 25 (18):3973–81.

Marquis, M. E., E. Lord, E. Bergeron et al. 2009. Bone cells–biomaterials interactions. *Front Biosci* 14:1023–67.

Martin, H. J., K. H. Schulz, J. D. Bumgardner, and K. B. Walters. 2007. XPS study on the use of 3-aminopropyltriethoxysilane to bond chitosan to a titanium surface. *Langmuir* 23 (12):6645–51.

Naberezhnykh, G. A., V. I. Gorbach, G. N. Likhatskaya, V. N. Davidova, and T. F. Solov'eva. 2008. Interaction of chitosans and their N-acylated derivatives with lipopolysaccharide of gram-negative bacteria. *Biochemistry (Mosc)* 73 (4):432–41.

Norowski, P. A., H. S. Courtney, J. Babu, W. O. Haggard, and J. D. Bumgardner. 2011. Chitosan coatings deliver antimicrobials from titanium implants: A preliminary study. *Implant Dent* 20 (1):56–67.

Ogawa, K., T. Yui, and K. Okuyama. 2004. Three D structures of chitosan. *Int J Biol Macromol* 34 (1–2):1–8.

Park, J. H., R. Olivares-Navarrete, C. E. Wasilewski, B. D. Boyan, R. Tannenbaum, and Z. Schwartz. 2012a. Use of polyelectrolyte thin films to modulate osteoblast response to microstructured titanium surfaces. *Biomaterials* 33 (21):5267–77.

Park, J. H., C. E. Wasilewski, N. Almodovar et al. 2012b. The responses to surface wettability gradients induced by chitosan nanofilms on microtextured titanium mediated by specific integrin receptors. *Biomaterials* 33 (30):7386–93.

Phaechamud, T. 2008. Hydrophobically modified chitosans and their pharmaceutical applications. *Int J Pharm Sci Tech* 1 (1):2–9.

Pishbin, F., A. Simchi, M. P. Ryan, and A. R. Boccaccini. 2011. Electrophoretic deposition of chitosan/45S5 Bioglass composite coatings for orthopaedic applications. *Surf Coat Technol* 205:5260–8.

Prasitsilp, M., R. Jenwithisuk, K. Kongsuwan, N. Damrongchai, and P. Watts. 2000. Cellular responses to chitosan in vitro: The importance of deacetylation. *J Mater Sci Mater Med* 11 (12):773–8.

Puleo, D. A., and A. Nanci. 1999. Understanding and controlling the bone–implant interface. *Biomaterials* 20 (23–24):2311–21.

Rao, S. B., and C. P. Sharma. 1995. Sterilization of chitosan: Implications. *J Biomater Appl* 10 (2):136–43.

Reje, P. R., and L. H. Block. 1999. Chitosan processing: Influence of process paramters during acidic and alkaline hydrolysis and effect of the processing sequence on the resultant chitosan's properties. *Carbohydr Res* 321:235–45.

Rhazi, M, J Desbrières, A Tolaimate, A Alagui, and P Vottero. 2000. Investigation of different natural sources of chitin: Influence of the source and deactylation process on the physicochemical characteristics of chitosan. *Polym Int* 49:337–344.

Sannan, T., K. Kurita, and Y. Iwakura. 1976. Studies on chitin, 2. Effect of deacetylation on solubility. *Die Makromol Chem* 177:3589–3600.

Shahidi, F., and R. Abuzaytoun. 2005. Chitin, chitosan, and co-products: Chemistry, production, applications, and health effects. *Adv Food Nutr Res* 49:93–135.

Sharma, S., V. P. Soni, and J. R. Bellare. 2009. Electrophoretic deposition of nanobiocomposites for orthopedic applications: Influence of current density and coating duration. *J Mater Sci Mater Med* 20 (Suppl 1):S93–100.

Shelton, R. M., A. C. Rasmussen, and J. E. Davies. 1988. Protein adsorption at the interface between charged polymer substrata and migrating osteoblasts. *Biomaterials* 9 (1):24–9.

Shi, Z., K. G. Neoh, E. T. Kang, C. K. Poh, and W. Wang. 2009. Surface functionalization of titanium with carboxymethyl chitosan and immobilized bone morphogenetic protein-2 for enhanced osseointegration. *Biomacromolecules* 10 (6):1603–11.

Shi, Z., K. G. Neoh, E. T. Kang, C. Poh, and W. Wang. 2008. Bacterial adhesion and osteoblast function on titanium with surface-grafted chitosan and immobilized RGD peptide. *J Biomed Mater Res A* 86 (4):865–72.

Siedlecki, C. A., and R. E. Marchant. 1998. Atomic force microscopy for characterization of the biomaterial interface. *Biomaterials* 19 (4–5):441–54.

Silva, R. M., G. A. Silva, O. P. Coutinho, J. F. Mano, and R. L. Reis. 2004. Preparation and characterisation in simulated body conditions of glutaraldehyde crosslinked chitosan membranes. *J Mater Sci Mater Med* 15 (10):1105–12.

Simchi, A., F. Pishbin, and A. R. Boccaccini. 2009. Electrophoretic deposition of chitosan. *Mater Lett* 63:2253–6.

Steele, J. G., C. McFarland, B. A. Dalton et al. 1993. Attachment of human bone cells to tissue culture polystyrene and to unmodified polystyrene: The effect of surface chemistry upon initial cell attachment. *J Biomater Sci Polym Ed* 5 (3):245–57.

Stephansson, S. N., B. A. Byers, and A. J. García. 2002. Enhanced expression of the osteoblastic phenotype on substrates that modulate fibronectin conformation and integrin receptor binding. *Biomaterials* 23 (12):2527–34.

Tomihata, K., and Y. Ikada. 1997. *In vitro* and *in vivo* degradation of films of chitin and its deacetylated derivatives. *Biomaterials* 18 (7):567–75.

Uygun, B. E., T. Bou-Akl, M. Albanna, and H. W. Matthew. 2010. Membrane thickness is an important variable in membrane scaffolds: Influence of chitosan membrane structure on the behavior of cells. *Acta Biomater* 6 (6):2126–31.

Wan, Y., K. A. M. Creber, B. Peppley, and V. T. Bui. 2003. Ionic conductivity and related properties of crosslinked chitosan membranes. *J Appl Polym Sci* 89 (2):306–17.

Wenling, C., J. Duohui, L. Jiamou, G. Yandao, Z. Nanming, and Z. Xiufang. 2005. Effects of the degree of deacetylation on the physicochemical properties and Schwann cell affinity of chitosan films. *J Biomater Appl* 20 (2):157–77.

Wright, S. D., R. A. Ramos, P. S. Tobias, R. J. Ulevitch, and J. C. Mathison. 1990. CD14, a receptor for complexes of lipopolysaccharide (LPS) and LPS binding protein. *Science* 249 4975:1431–3.

Yermak, I. M., V. N. Davidova, V. I. Gorbach et al. 2006. Forming and immunological properties of some lipopolysaccharide–chitosan complexes. *Biochimie* 88 (1):23–30.

Yi, H., L. Q. Wu, W. E. Bentley et al. 2005. Biofabrication with chitosan. *Biomacromolecules* 6 (6):2881–94.

Yuan, Y., B. M. Chesnutt, L. Wright, W. O. Haggard, and J. D. Bumgardner. 2008. Mechanical property, degradation rate, and bone cell growth of chitosan coated titanium influenced by degree of deacetylation of chitosan. *J Biomed Mater Res B Appl Biomater* 86 (1):245–52.

Zhang, Z., T. Jiang, K. Ma, X. Cai, Y. Zhou, and Y. Wang. 2011. Low temperature electrophoretic deposition of porous chitosan/silk fibroin composite coating for titanium biofunctionalization. *J Mater Chem* 21:7705–13.

Zheng, Z., L. Zhang, L. Kong, A. Wang, Y. Gong, and X. Zhang. 2009. The behaviour of MC3T3-E1 cells on chitosan/poly-L-lysine composite films: Effect of nanotopography, surface chemistry and wettability. *J Biomed Mater Res* 89A:453–65.

23 Current Scenario and Future Prospects of Chitosan in Dentistry

Manal Farea, Ahmad Sukari Halim, and Nor Shamsuria Omar

CONTENTS

23.1 INTRODUCTION

A paradigm shift is taking place in the application of tissue engineering techniques in the repair or replacement of impaired tissues. Stem cell biology, an emerging field of research, provides promising methods that make speculation about a future application in human dentistry reasonable.

The three key components believed to be essential for tissue regeneration are stem cells, scaffold, and morphogens or growth factors (as illustrated in Figure 23.1). The scaffold should provide a three-dimensional porous and fully interpenetrable space for tissue ingrowth, accelerate the formation of tissue structure, and ultimately be replaced by a new extracellular matrix (ECM) to form completely natural tissues (Hubbell 1995; Ma et al. 2008). The scaffold may be implanted alone or in combination with cells and growth factors. After implantation, the scaffold allows for cell migration and organization. Ideally, a scaffold should be porous (to allow for placement of cells and growth factors), biocompatible with the host tissues, biodegradable, leave no toxic by-products, and have the correct shape and form to allow for replacement of the lost tissues (Young et al. 2002; Murray et al. 2007; Prescott et al. 2008). The rate at which degradation occurs has to coincide as much as possible with the rate of tissue formation; this means that while cells are fabricating their own natural matrix structure around themselves, the scaffold is able to provide structural integrity within the body, and eventually break down, leaving a newly formed tissue that will take over the mechanical load (Freed et al. 1994). Currently, two types of scaffolds are available: synthetic or natural scaffolds. Synthetic scaffolds are made from polyester materials, such as polylactic acid (PLA), polyglycolic acid (PGA), and polycaprolactone (PCL) that degrade within the human body. Synthetic scaffolds offer improved control over the degradation rate along with the ability to support the growth of different stem cell types (Taylor et al. 1994; Sharma and Elisseeff 2004). However, the difficulties in obtaining high porosity and regular pore size are the disadvantages of these scaffolds that have led researchers to concentrate their efforts on engineering scaffolds at the nanostructural level to modify cellular interactions with the scaffold (Tuzlakoglu et al. 2005). Natural scaffolds are made from collagen and polysaccharidic materials, such as chitosan or glycosaminoglycans (GAGs) (Griffon et al. 2005; Guo et al. 2006; Prescott et al. 2008).

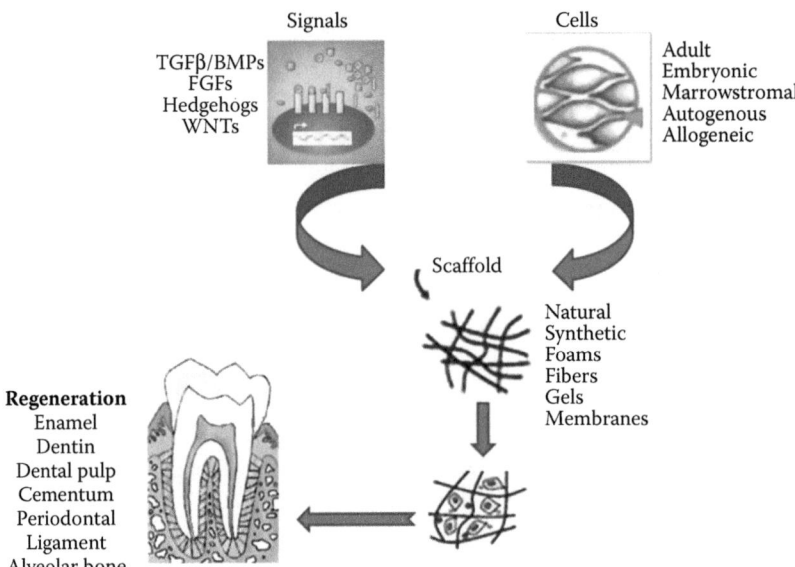

FIGURE 23.1 The elements (stem cells, scaffold, and signals) of tissue engineering triad and regeneration of dental tissues.

23.2 CHITOSAN

Chitosan was first discovered in 1811 by Henri Braconnot, a French chemist and pharmacist. Braconnot observed that certain substance (chitin) found in mushrooms did not dissolve in sulfuric acid. Later in the century, chitin was found in crustaceans (such as crabs, lobsters, shellfish, and shrimp), the indigestible outer skeleton of insects and the material from which the cell walls of the mycelial fungi are made. Chitin was also found in the radulae of mollusks, and the beaks of cephalopods (Labrude and Becq 2003; Dai et al. 2011).

Chitin (β-1,4-D-linked polymer of N-acetylglucosamine) is a naturally abundant mucopolysaccharide and is second to cellulose in terms of the amount produced annually by biosynthesis (Lim and Halim 2010). It is visually characterized as a white, hard, inelastic, nitrogenous polysaccharide, and approximately 1 billion tons are synthesized each year (Peter 1997). Chitin is a common constituent of the exoskeleton in animals, particularly in crustaceans, mollusks, and insects. Commercially sold chitin is usually extracted from shellfish waste (Skjak-Braek et al. 1989; Goosen 1997). Chitin is structurally similar to cellulose; however, less attention has been paid to chitin than cellulose, primarily because of its inertness. Hence, it remains an essentially underutilized resource. Deacetylation of chitin yields chitosan, which is a relatively reactive compound and is produced in numerous forms, such as powder, paste, film, and fiber. Chitosan is a poly(β-1,4-D-glucosamine) derived from the N-deacetylation of chitin. Chitosan is a partially deacetylated derivative of natural chitin, a primary structural polymer in arthropod exoskeletons. One of the current trends in biomedical research requires the use of materials that are derived from nature as natural materials have been shown to exhibit better biocompatibility with humans. For example, chitosan's monomeric unit, N-acetylglucosamine, occurs in hyaluronic acid, an extracellular macromolecule that is important in wound repair. Chitosan has been considered a candidate natural polymer for scaffold because it shows several advantages over other synthetic biomaterials, and stable porous structure is easily manipulated by lyophilization. Studies on chitosan have been intensified since 1990 because of its low cytotoxicity (Lim et al. 2007; Mohammad et al. 2010), antimicrobial (Tanigawa et al. 1992; Tokura et al. 1997), immunological (Nishimura et al. 1984; Mori et al. 1997) and wound-healing activities (Khnor and Lim 2003; Kweon et al. 2003; Okamoto et al. 1993), and excellent biodegradable properties in the human body (Sashiwa et al. 1990; Shigemasa et al. 1994), where these properties have contributed to the better biocompatibility of chitosan (Li et al. 1992; Shigemasa et al. 1994; Sashiwa et al. 1990; Prabaharan and Mano 2006; Van et al. 2006; Lim et al. 2007).

Additionally, the *N*-acetylglucosamine moiety in chitosan is structurally similar to GAGs, heparin, chondroitin sulfate, and hyaluronic acid, which are biocompatible, and hold specific interactions with various growth factors, receptors, and adhesion proteins besides being biologically important mucopolysaccharides in all mammals. Therefore, the analogous structure in chitosan may also exert similar bioactivity and biocompatibility (Li et al. 1992; Suh and Matthew 2000). Therefore, chitosan attracts considerable interest in the dental field because of its favorable properties, such as low cost, nontoxicity, biocompatibility, biodegradability, antimicrobial activity, wound healing ability, and hemostatic and tissue regenerative properties.

For improving the mechanical or biological properties of chitosan, collagen is added to accessorize the Arg–Gly–Asp (RGD)-like sequence to promote cell adhesion and migration (Mizuno et al. 2000). The chitosan/collagen composite has demonstrated its usability in tissue engineering applications. Collagen shares similar chemical and biological characteristics with those of natural tissues, and it has a low antigenicity. Collagen type 1 is the predominant component of the dentin matrix as well as the dental pulp. The presence of collagen type 1 in dentin is considered to offer initiation sites for calcification. Collagen also allows the arrangement of preodontoblasts, and it also binds newly formed odontoblasts to pulp tissue, supporting a reparative dentinogenesis framework (Kitasako et al. 2002). It has been used as a capping material to achieve regeneration of the dental–pulp complex (Pieper et al. 1999; Zhang et al. 2006a).

In 2012, Miranda et al. studied the effects of a chitosan–gelatin (C/G) scaffold seeded with bone marrow mesenchymal stem cells (BMMSCs) in the healing process of tooth sockets in rats. BMMSCs were isolated from transgenic rats and then expanded and seeded on a C/G scaffold; the immunohistochemical results showed that BMMSCs seeded on the C/G scaffold contributed to bone, epithelial, and vascular repair. A composite membrane made of C/G is also beneficial for facilitating the transfer of human adipose stem cells (ASCs) in spheroids. A blend film containing 75% chitosan and 25% gelatin showed promising results to serve as a biomaterial for human ASC-based cell therapy (Cheng et al. 2012).

Recently, increased efforts have been focused on combining growth factors with scaffold materials to modify cell behavior. A growth factor delivery system helps the polypeptide to reach the desired defect site within the requisite time period and further controls the time course of the release of the growth factor. Bone morphogenetic proteins (BMP) are a family of pleiotropic signaling molecules critically involved at various stages in the formation of a variety of tissues and organs, including bones and teeth (Thesleff and Sharpe 1997; Nakashima and Reddi 2003; Nakashima 2005). BMP-7 is present in developing teeth, and its expression may be developmentally regulated. Some studies have indicated that recombinant BMP-7, when applied to freshly cut dentin in teeth, stimulated reparative dentin formation (Helder et al. 1998; Rutherford and Gu 2000; Sloan et al. 2000; Six et al. 2002).

Chitosan (and its preparations) has also been widely used as a vehicle to deliver growth factors into wounds to facilitate healing. Growth factors are involved in cell division, migration, differentiation, protein expression, and enzyme production. Growth factors participate in the inflammatory, proliferative, and migratory phases of wound healing. A variety of growth factors, such as EGF, PDGF, FGF, TGF-β1, IGF-1, human growth hormone, and granulocyte–macrophage colony-stimulating factor (Dai et al. 2011), which participate in the process of wound healing, have been reported. In 2003, Mizuno et al. studied the stability of basic fibroblast growth factor (bFGF) incorporated into a chitosan film as a delivery vehicle for providing sustained release of bFGF. The results showed that the wounds were smaller on day 20 in the bFGF-chitosan group than in the chitosan-alone group. Proliferation of fibroblasts with an increase in the number of capillaries was observed in both groups, but granulation tissue was more abundant in the bFGF-chitosan group.

The investigators suggested that chitosan itself facilitates wound repair, and bFGF incorporated into chitosan film is a stable delivery vehicle for accelerating wound healing.

In a similar study, chitosan scaffolds loaded with bFGF contained in gelatin microparticles were developed and tested for treating pressure ulcers in an aged mouse model, mimicking the situations in an elderly population. It was demonstrated that both chitosan and chitosan-bFGF scaffolds significantly accelerated wound closure compared with gauze control. By day 10, all wounds achieved similar closure. Delivery and angiogenic function of bFGF was verified through ELISA and histology. Elevated neutrophil levels were observed in chitosan and chitosan-bFGF groups. *In vitro*, chitosan inhibited elastase activity. *In vivo*, elastase protein levels in wounds were reduced with chitosan-bFGF scaffolds by day 10. These results suggest that chitosan is an effective material for growth factor delivery and can help to heal chronic ulcers (Park et al. 2009; Dai et al. 2011). Moreover, in 2011, Jun et al. investigated a hybrid material composed of a silica xerogel and chitosan that was coated on Ti for the delivery of growth factors. Fibroblast growth factor (FGF) and green fluorescence protein were incorporated into the coatings for hard tissue engineering. The authors found that the hybrid coating containing FGF showed significantly improved osteoblast cell responses compared to the pure xerogel coating with FGF or the hybrid coating without FGF. These results indicate that the hybrid coating is potentially very useful in enhancing the bioactivity of metallic implants by delivering growth factors in a controlled manner.

In 2011, Yang et al. evaluated the adhesion of human dental pulp stem cells (DPSCs) on chitosan/collagen scaffolds that were loaded with the plasmid vector encoding human bone morphogenetic protein-7 (BMP-7) gene at day 4 after cell seeding, and found that cells adhered and spread well on the surfaces of the scaffolds.

In 2010, Lim and Halim evaluated the adhesion and proliferation of human dermal fibroblasts on chitosan porous skin regenerating template (CPSRT), and found that the cells grew and proliferated on the surface of chitosan. The reason for this occurrence is that chitosan is positively charged, and it binds electrostatically to the negatively charged cell surface (Figure 23.2).

In contrast, in 2006b, Zhang et al. prepared chitosan/collagen composites combined with plasmid and adenoviral vector encoding human transforming growth factor-b1 (Ad TGF-b1) gene using a freeze-drying method and evaluated the cytocompatibility through the seeding of human periodontal ligament cells (HPLCs) onto the scaffold *in vitro*. They found that the scaffold containing Ad-TGF-b1 exhibited the highest proliferation rate and the expression of type I and type III collagen was upregulated in the Ad-TGF-b1 containing scaffold. The authors concluded that the potential of chitosan/collagen scaffold combined with Ad-TGF-b1 is a good substrate candidate for periodontal tissue engineering.

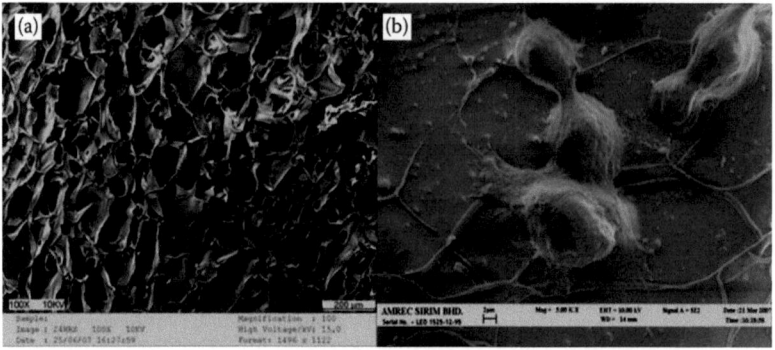

FIGURE 23.2 Scanning electron micrographs. (a) Porous structures of a CPSRT. (b) Proliferating cells in the CPSRT. (From Lim, C. K., and Halim, A. S. 2010. Biomedical-grade chitosan in wound management and its biocompatibility *in vitro*. *Biopolymers* 19–35. With permission.)

Furthermore, chitosans and its derivatives have been studied for formulations that enhance the absorption of macromolecular biotherapeutics (peptides, protein therapeutics, and antigens, as well as plasmid DNA) and for the preparation of particulate drug-targeting systems (Muzzarelli 2012). Silica xerogel–chitosan hybrids containing vancomycin were fabricated by the sol–gel process at room temperature and their potential as a drug eluting bone replacement was evaluated in terms of their mechanical properties and drug release behaviors. The hybrids with more than 30% chitosan could release the vancomycin for an extended period of time in a controlled manner (Lee et al. 2010).

Therefore, it seems that chitosan derivatives can be used as carriers for a variety of drugs for controlled release applications. Chitosan is useful as a bio-scaffold for tissue engineering especially for skin and bone. Chitosan in the form of a porous structure can serve as a template for bone osteoblasts or skin fibroblasts and keratinocytes where the cells can grow within the scaffold and take the shape of the scaffold. For example, the chitosan porous structure in a thin sheet can represent a scaffold for skin where skin fibroblasts and keratinocytes grow to form a basic skin substitute *in vitro*. Furthermore, chitosan possesses all the characteristics required for an ideal contact lens, such as optical clarity, mechanical stability, sufficient optical correction, gas permeability, wettability, and immunological compatibility. Recently, the transdermal absorption promoting characteristics of chitosan have been explored for nasal and oral delivery of polar drugs to include peptides and proteins and for vaccine delivery. These properties have enabled the chitosan together with appropriate derivatives to be a promising biomaterial for the pharmaceutical industry and other applications (Lim and Halim 2009).

23.3 DENTAL APPLICATION OF CHITOSAN

Gram-negative anaerobic bacteria are the primary cause of chronic periodontitis (CP), and conventional periodontal therapy is either surgical or nonsurgical. However, these procedures are not always successful, and hence, the use of antimicrobial agents to suppress or eradicate pathogenic microbiota is widespread. However, systemic administration of antibiotics is associated with sev-eral drawbacks, such as the development of antibiotic-resistant bacteria, low patient compliance, requirement of frequent dosing to maintain therapeutic drug concentrations in the serum and gingival crevicular fluid, hypersensitivity reactions, gastrointestinal discomfort, central nervous system disturbances, and development of super infections (Genco 1981; Golomb et al. 1984; Slots and Rams 1990; Slots 1996; Mariotti and Monroe 1998). Therefore, to overcome all these limitations, the use of antimicrobials in local delivery systems has received considerable attention. A biode-gradable delivery system has several advantages, including maintaining an effective drug release rate in the periodontal pocket while simultaneously eroding throughout the duration of treatment. Additionally, a second visit is not required for the removal of a biodegradable delivery system. Bioadhesive delivery systems play an important role in retaining a formulation on the periodontal hard and soft tissue surfaces and in increasing the retention time of the formulation (Langer 1990; Smart 1991; Needleman et al. 1997, 1998). In 2002, Ikinci et al. studied the antimicrobial activity of chitosan formulations either in gel or in film forms at different molecular weights and deacetylation degree in the absence or presence of chlorhexidine gluconate (Chx) against a periodontal pathogen, *Porphyromonas gingivalis*. The combination of chitosan with Chx showed a higher activity when compared with that of Chx alone. Chitosan showed similar effectiveness in the treatment of CP and when applied as an adjunctive to scaling and root planing (SRP), and also when used alone or in combination with metronidazole (Akıncıbay et al. 2006). Thus, chitosan is a promising delivery system when used either alone or with antibacterial drugs for periodontal therapy.

Injectable thermosensitive hydrogel (CS–HTCC/-,--GP) was designed using chitosan (CS), quaternized chitosan (HTCC), and a,b-glycerophosphate (-,--GP) without any additional chemical stimulus. The gelation point of CS–HTCC/-,--GP can be set at a temperature close to normal body temperature or at other temperatures above 25°C. In addition, CS–HTCC/-,--GP thermosensitive hydrogels exhibited stronger antibacterial activity toward two periodontal pathogens (*Porphyromonas gingivalis*, *P.g*, and *Prevotella intermedia*, *P.i*). CS–HTCC/-,--GP thermosensitive hydrogels can be considered a local drug delivery system for periodontal treatment (Ji et al. 2009).

Oral mucositis is defined as inflammation and ulceration of the oral mucosa that occurs as a result of chemo/radiotherapy for cancer (Sonis 2004). This pathology is often accompanied by bacterial and fungal infections. It has been reported that chitosan is an excellent candidate for the treatment of oral mucositis; it provides an extended retention time on the oral mucosa (Miyazaki et al. 1994; Senel et al. 2000).

In addition to its bioadhesive properties, chitosan inhibits the adhesion of *Candida albicans* to human buccal cells and prevents the development of mycosis (Knapczyk et al. 1992). In 2004, Aksungur et al. prepared gel and film formulations using chitosans at different molecular weights and in different solvents incorporating nystatin, which is considered a prophylactic agent for oral mucositis, to develop an occlusive bioadhesive system for prophylaxis and/or treatment of oral mucositis. The effect of these formulations was investigated *in vivo* in hamsters with chemotherapy- induced mucositis. The authors found that the topical application of both chitosan gel and a suspen-sion incorporating nystatin on the oral mucosa significantly reduced the severity and the incidence of oral mucositis, reduced weight loss and increased survival, and resulted in significantly faster healing.

Moreover, in 2020, Rossi et al. developed a thermally sensitive mucoadhesive gel based on chitosan derivatives for the treatment of oral mucositis; trimethyl chitosan (TMC) and methyl-pyrrolidinone chitosan (MPC) were mixed with glycerophosphate (GP). The best properties (gelation and mucoadhesion) were shown by TMC/GP, and then, this mixture was loaded with benzydamine hydrochloride, an antiinflammatory drug with antimicrobial properties used in the treatment of oral mucositis. The mucoadhesive properties of the mixtures were also assessed using porcine buccal mucosa. The formulation based on TMC/GP mixture was able to prolong drug release and to withstand the removal of physiological mechanisms. Also, in the absence of drug, the TMC/GP mixture showed antimicrobial properties, indicating that the TMC/GP mixture is a promising candidate for buccal administration of benzydamine hydrochloride in the treatment of oral mucositis.

Furthermore, films of lidocaine prepared with three different molecular weights and concentrations of chitosan and cross-linked with tripolyphosphate pentasodium salt were investigated *in vitro* to evaluate the effect of chitosan on lidocaine release. It was found that increasing the concentration and molecular weight of chitosan caused an increase in both the rate and extent of lidocaine release, which is expected to prolong the anesthetic effect of lidocaine in the oral cavity during the treatment of oral mucositis (Varshosa and Karimzadeh 2007).

Another *in vitro* study was carried out to evaluate the suitability of using chitosan, poly(lactide-*co*-glycolide) (PLGA), and polymethyl methacrylate (PMMA) as a coating material for the absorbent paper point to control the release of (Chx) root canal disinfection; it was found that chitosan showed better results than PLGA and PMMA (Lee et al. 2005), which may be attributed to the high water sensitivity of chitosan, causing it to swell in water and rupture, thereby releasing Chx from the chitosan-coated paper points.

Denture retention is considered a real challenge in dentistry; therefore, chitosan has been evaluated for its ability to serve as a denture adhesive. It has been suggested that chitosan with high viscosity may improve the retention and masticatory efficiency of complete dentures (Cheng et al. 2004). Moreover, various denture base resin materials were evaluated for their effects on bacterial adherence, and it was found that the denture base resin materials with chitosan have a rougher surface and lower bacteria adherence than those without chitosan (Chung et al. 2002).

Caries are infections that cause demineralization of the hard tissues (enamel, dentin, and cementum) and destruction of the organic matter of the tooth, usually through the production of acid by hydrolysis of the food debris accumulated on the tooth surface, and it remains one of the most com-mon diseases. The two groups of bacteria that are responsible for initiating caries are *Streptococcus mutans* and *Lactobacillus*. Fluoride is still widely used to control the caries disease. The amino groups of chitosan capture hydrogen ions in acid media, which in turn results in an overall positive charge that imparts bioadhesive ability to negatively charged surfaces, such as tooth enamel, soft tissue, cell membrane, among others (Pawtlowska 1997; Sezer et al. 2008). Owing to its functional properties, chitosan has been evaluated for its influence on the process of tooth enamel deremineralization while assessing its ability as biomaterial with anticariogenic activity. An *in vitro* study was carried out to evaluate the effect of chitosan treatment on enamel deremineralization behavior on a pH cycling assay, and it was found that chitosan acts as a barrier against acid penetration and interferes with the process of demineralization of the tooth enamel inhibiting the release of phosphorus (Arnaud et al. 2010).

An *in vivo* study was carried out at Nagasaki University School of Dentistry in Japan to evaluate whether chewing gum containing chitosan can effectively suppress the growth of oral bacteria in saliva. They found that the amount of oral bacteria was significantly decreased in the chitosan group (Hayashi et al. 2007), and this finding suggests that a supplementation of chitosan in gum is an effective method for controlling the number of cariogenic bacteria.

It has been found that water-soluble chitosan directly suppresses the growth of the typical cariogenic bacterium *Streptococcus mutans* even at pH 6.5, without causing demineralization of the tooth surface (Fujiwara et al. 2004). Three types of chitosan, polymer, oligomer, and monomer, were used at 4% (w/v) as well as three different levels of pH: 6.0, 6.5, and 7.4; the bactericidal activity was calculated by the growth ratio. The authors reported that all three types of chitosan strongly inhibited bacterial growth at pH 6.0. Furthermore, nearly complete inhibition was obtained with 2% (w/v) chitosan solution at constant pH 6.5.

In another study, a group of researchers examined the effects of a newly developed water-soluble reduced chitosan on *S. mutans*, plaque regrowth, and biofilm vitality. A 1.0%, water-soluble reduced chitosan, with pH ranging from 6.0 to 6.5, molecular weights between 3000 and 5000 Da, and 70% degree of deacetylation, was used. To determine the antibacterial and antiplaque potency of chitosan, minimal inhibitory concentrations (MICs) for *S. mutans* and *S. sanguinis* (formerly *S. sanguinis*), short-term exposure to *S. mutans*, and a clinical trial of plaque regrowth and biofilm vitality were conducted. The chitosan solution reduced the plaque index and the vitality of the plaque flora significantly when compared with distilled water, but this effect was less than the reductions found with the positive control of 0.1% chlorhexidine solution. The water-soluble reduced chitosan exhibited potent antibacterial effect on *S. mutans* and displayed significant antibacterial and plaque-reducing action during the 4-day plaque regrowth (Bae et al. 2006).

In 2012, Uraz et al. evaluated the microbiological and clinical effects of a chitosan (CH) mouth rinse on plaque inhibition. Thirty-six healthy participants were recruited. The following clinical data were recorded: a plaque index (PI), gingival index (GI), Quickley–Hein plaque index (QPI), and probing depth (PD). Volunteers were given oral hygiene (OH) instruction and trained on scaling and professional tooth cleaning (PTC). Group A used rinsed 2% CH mouth rinse, group B used 0.2% chlorhexidine digluconate (CHX) mouth rinse, and group C used 2% CH þ 0.2% CHX mouth rinse. Plaque samples were collected and assayed for *Streptococcus mutans*, *Candida albicans*, and enterococci. After a nonbrushing period, the full-mouth PI and QPI values between the CH and CHX þ CH groups differed significantly. A higher PI score at sampling sites was observed in the CH group, but no significant differences were observed between groups. The *S. mutans* and *C. albicans* levels were statistically significant in each group on days 0 and 4. Therefore, the authors concluded that further investigations are needed to evaluate the potential value of CH as an effective antiplaque mouth rinse.

Bone defects may develop in various systemic and dental disorders. Osteolysis in periodontal diseases accounts for most cases of need for bone repair. The conventional methods of bone repair that are commonly used, such as autografts and allografts, have their own shortcomings. Autografts are limited in terms of availability of materials and may result in donor site morbidity (Wang et al. 2002). The use of allografts may be more desirable in some cases, but the possible immune reaction and infection transmission limit their application. To overcome these limitations, various synthetic bone substitutes made of metal, ceramics, polymers, and various composite structures have been introduced to accelerate and improve the process of bone regeneration; although their safety and efficacy remain uncertain (Damien and Parsons 1991). Currently, owing to an increase in the rate of invasive surgical procedures, especially in the fields of orthopedics and dentistry, bone repair techniques using new materials are becoming more popular. The use of these new materials should help to reduce the operation time, scar size, postoperative pain, and also improve patient recovery (Liu et al. 2006; Song et al. 2009). One of the best materials that can fulfill these requirements is chitosan (Madihally and Matthew 1999; Khan et al. 2002; Nascimento et al. 2009). Moreover, chitosan/hydroxyapatite quick hardening paste has been used as a bone substitute material for dental treatment such as the augmentation of edentulous alveolar ridges and implants (Ito 1991).

Tricalcium phosphate (TCP)/chitosan composite microgranules were developed as bone substitutes and tissue engineering scaffolds with the aim of obtaining a high bone forming efficacy. In addition, the transforming growth factor-beta 1(TGF-β1) was added to the microgranules to improve bone healing efficacy. The authors found that the TCP/chitosan microgranules were effective in achieving control release of TGF-1 for 28 days. The microgranules supported the proliferation of seeded osteoblasts as well as their differentiation, as indicated by the ALPase activity and osteocalcin content (Lee et al. 2004). In 2012, Kim et al. investigated the effect of alendronate released from chitosan scaffolds on enhancement of osteoblast functions and inhibition of osteoclast differentiation *in vitro*. For evaluating osteoblast functions in MG-63 cells, cell proliferation, alkaline phosphatase (ALP) activity, and calcium deposition were investigated. To evaluate the inhibition of osteoclast differentiation in RAW 264.7 cells, tartrate-resistant acid phosphatase (TRAP) activity, TRAP staining, and gene expressions were investigated. The authors reported that alendronate-eluting chitosan substrates are promising materials for enhancing osteoblast functions and inhibiting osteoclast differentiation in orthopedic and dental fields.

In dentistry, the barrier properties of dentin provide better protection for dental pulp compared with any artificial restorative material. Dental pulp-derived adult stem cells (DPSCs) are especially attractive because they have shown potential for odontogenic differentiation and the ability to form dentin. In addition to cells, scaffold material is often critical to recapitulate the *in vivo* milieu and to support cell proliferation and function. Various materials, such as ceramic, titanium, and collagen, have been used for dental tissue engineering. Although these materials can support cell growth and differentiation, their low flexibility has limited their use for the regeneration of the dentin–pulp complex via dental pulp capping procedure. Therefore, in 2010, Guan et al. concluded that chitosan scaffolds fabricated through centrifugation, freeze-drying, and stabilization were nontoxic, and could promote the growth of DPSC and stem cells from human exfoliated deciduous teeth (SHED). Moreover, in 2010, Liao et al. reported that β-tricalcium phosphate/chitosan composite scaffold could promote the differentiation of HPLCs toward osteoblast and cementoblast phenotypes.

Therefore, chitosan in different physical forms, such as solutions, films, porous structure scaffolds, pastes, sheets, hydrogels, fibers, tablets, nanoparticles, and microspheres, is a promising biocompatible dental material that has shown bioadhesive, biodegradable, permeabilizing, and antimicrobial properties. Moreover, chitosan plays an important role in dental tissue regeneration, and is an excellent candidate for the treatment of oral mucositis, offering not only the palliative effects of an occlusive dressing but also the potential for delivering therapeutic compounds. More detailed *in vivo* experiments are needed before any clinical investigations are carried out for assessing the dental pulp, dentin, and cementum regeneration ability of chitosan when used in combination with stem cells. This will contribute more to the field of regenerative dentistry and help in increasing the chances of saving many teeth that would otherwise have a poor to hopeless prognosis. Regenerative therapy will provide promising methods that make speculation about a future application in human dentistry reasonable. However, experiments need to be carried out in human teeth, as the outcome may differ from that in animal teeth.

REFERENCES

Akıncıbay, H., Senel, S., and Yetkin, Z. A. 2006. Application of chitosan gel in the treatment of chronic periodontitis. *Journal of Biomedical Materials Research Part B: Applied Biomaterials* 80:290–296.

Aksungur, P., Sungur, A., Unal, S., Iskit, A. B., Squier, C. A., and Senel, S. 2004. Chitosan delivery systems for the treatment of oral mucositis: In vitro and in vivo studies. *Journal of Controlled Release* 98:269–279.

Arnaud, T. M. S., Neto, B. B., and Diniz, F. B. 2010. Chitosan effect on dental enamel de-remineralization: An in vitro evaluation. *Journal of Dentistry* 38:848–852.

Bae, K., Jun, E. J., Lee, S. M., Paik, D. I., and Kim, J. B. 2006. Effect of water-soluble reduced chitosan on *Streptococcus mutans*, plaque regrowth and biofilm vitality. *Clinical Oral Investigations* 10:102–107.

Cheng, J. H., Xiang-bin, L. I., Lei, Z., and Xing-qiang, W. 2004. A study on the ability of chitosan with high viscosity to serve as denture adhesive. *Chinese Journal of Prosthodontics* 03.

Cheng, N. C., Chang, H. H., Tu, Y. K., and Young, T. H. 2012. Efficient transfer of human adipose-derived stem cells by chitosan/gelatin blend films. *Journal of Biomedical Materials Research Part B: Applied Biomaterials* 100:1369–1377.

Chung, S. H., Vang, M. S., and Park, H. O. 2002. Adherence of oral bacteria on chitosan: Added denture base materials *in vitro*. *Journal of Korean Academy of Prosthodontics* 40:525–535.

Damien, C. J., and Parsons, J. R. 1991. Bone graft and bone graft substitutes: A review of current technology and applications. *Journal of Applied Biomaterials* 2:187–208.

Dai, T., Tanaka, M., Huang, Y. Y., and Hamblin, R. H. 2011. Chitosan preparations for wounds and burns: Antimicrobial and wound-healing effects. 2011. *Expert Review of Anti-Infective Therapy* 9:857–879.

Freed, L. E., Vunjak-Novakovic, G., Biron, R. J., Eagles, D. B., Lesnoy, D. C., Barlow, S. K., and Langer, R. 1994. Biodegradable polymer scaffolds for tissue engineering. *Biotechnology (New York)* 12:689–693.

Fujiwara, M., Hayashi, Y., and Ohara, N. 2004. Inhibitory effect of water-soluble chitosan on growth of Streptococcus mutans. *New Microbiologica* 27:83–86.

Genco, R. J. 1981. Antibiotics in the treatment of human periodontal diseases. *Journal of Periodontology* 52:545–558.

Golomb, G., Friedman, M., Soskolne, A., Stabholz, A., and Sela, M. N. 1984. Sustained release device containing metronidazole for periodontal use. *Journal of Dental Research* 63:1149–1153.

Goosen, M. F. A. 1997. *Application of Chitin and Chitosan*, Technomic Publishing, Lancaster, USA, pp.320.

Griffon, D. J., Sedighi, M. R., Sendemir-Urkmez, A., Stewart, A. A., and Jamison, R. 2005. Evaluation of vacuum and dynamic cell seeding of polyglycolic acid and chitosan scaffolds for cartilage engineering. *American Journal of Veterinary Research* 66:599–605.

Guan, Z., Kunming, Y., China, S., and SHI, S. 2010. Chitosan as scaffolds for DPSC and SHED. Poster, The Preliminary Program for IADR General Session (July14–17).

Guo, T., Zhao, J., Chang, J., Ding, Z., Hong, H., Chen, J., and Zhang, J. 2006. Porous chitosan-gelatin scaffold containing plasmid DNA encoding transforming growth factor-beta1 for chondrocytes proliferation. *Biomaterials* 27:1095–1103.

Hayashi, Y., Ohara, N., Ganno, T., Yamaguchi, K., Ishizaki, T., and Nakamura, T. 2007. Chewing chitosan-containing effectively inhibits the growth cariogenic bacteria. *Archives of Oral Biology* 52:290–294.

Helder, M., Karg H., Bervoets, T., Vukicevic, S., Burger, E., D'souza, R., Wltgens, J., Karsenty, G., and Bronckers, A. 1998. Bone morphogenetic protein-7 (osteogenic protein-1, OP-1) and tooth development. *Journal of Dental Research* 77:545–554.

Hubbell, J. A. 1995. Biomaterials in tissue engineering. *Biotechnology (New York)* 13:565–576.

Ikinci, G. I., Senel, S., Akıncıbay, H., Kas, S., Ercis, S., Wilson, C. G., and Hıncal, A. A. 2002. Effect of chitosan on a periodontal pathogen *Porphyromonas gingivalis*. *International Journal of Pharmaceutics* 235:121–127.

Ito, M. 1991. In vitro properties of a chitosan-bonded hydroxyapatite bone-filling paste. *Biomaterials* 12:41–45.

Ji, Q. X., Chen, X. G., Zhao, Q. S., Liu, C. S., Cheng, X. J., and Wang, L. C. 2009. Injectable thermosensitive hydrogel based on chitosan and quaternized chitosan and the biomedical properties. *Journal of Materials Science: Materials in Medicine* 20:1603–1610.

Jun, S. H., Lee, E. J., Kim, H. E., Jang, J. H., and Koh, Y. H. 2011. Chitosan hybrid coating on Ti for controlled release of growth factors. *Journal of Materials Science: Materials in Medicine* 22:2757–2764.

Khan, T. A., Peh, K. K., and Ch'ng, H. S. 2002. Reporting degree of deacetylation values of chitosan: The influence of analytical methods. *Journal of Pharmacy and Pharmaceutical Sciences* 5:205–212.

Khnor, E., and Lim, L. 2003. Implantated applications of chitin and chitosan. *Biomaterials* 24:2339–2349.

Kim, S. E., Suh, D. H., Yun, Y. P., Lee, J. Y., Park, K. S., Chung, J. Y., and Lee, D. W. 2012. Local delivery of alendronate eluting chitosan scaffold can effectively increase osteoblast functions and inhibit osteoclast differentiation. *Journal of Materials Science: Materials in Medicine* 23:2739–2749.

Kitasako, Y., Shibata, S., Cox, C. F., and Tagami, J. 2002. Location, arrangement and possible function of interodontoblastic collagen fibres in association with calcium hydroxide-induced hard tissue bridges. *International Endodontic Journal* 35:996–1004.

Knapczyk, J. Macura, A. B., and Pawlik, B. 1992. Simple tests demonstrating the antimycotic effect of chitosan. *International Journal of Pharmaceutics* 80:33–38.

Kweon, D. K., Song, S. B., and Park, Y. Y. 2003. Preparation of water-soluble chitosan/heparin complex and its application as wound healing accelerator. *Biomaterials* 24:1595–1601.

Labrude, P., and Becq, C. 2003. Pharmacist and chemist Henri Braconnot. *Review History of Pharmacy (Paris)* 51:61–78.

Langer, R. 1990. New methods of drug delivery. *Science* 249:1527–1533.

Lee, D. Y., Spangberg, L. S. W., Bok, Y. B., Lee, C. Y., and Kum, K. W. 2005. The sustaining effect of three polymers on the release of chlorhexidine from a controlled release drug device for root canal disinfection. *Oral Surgery Oral Medicine Oral Pathology Oral Radiology and Endodontology* 100:105–111.

Lee, E. J., Jun, S. H., Kim, H. E., Kim, H. W., Koh, Y. H., and Jang, J. H. 2010. Silicaxerogel-chitosan nano-hybrids for use as drug eluting bone replacement. *Journal of Materials Science: Materials in Medicine* 21:207–214.

Lee, J. Y., Seol, Y. J., Kim, K. H., Lee, Y. M., Park, Y. J., Rhyu, I. C., Chung, C. P., and Lee, S. J. 2004. Transforming growth factor (TGF)-1 releasing tricalcium phosphate/chitosan microgranules as bone substitutes. *Pharmaceutical Research* 21(10):1790–1796.

Liao, F., Chen, Y., Li, Z., Wang, Y., Shi, B., Gong, Z., and Cheng, X. 2010. A novel bioactive three-dimensional beta-tricalcium phosphate/chitosan scaffold for periodontal tissue engineering. *Journal of Materials Science: Materials in Medicine* 21:489–496.

Li, O., Grandmaison, E. W., Goosen, M. F. A., and Dunn, E. T. 1992. Applications and properties of chitosan. *Journal of Bioactive and Compatible Polymers* 7:370–397.

Lim, C. K., and Halim, A. S. 2009. In vitro models in biocompatibility assessment for biomedical-grade chitosan derivatives in wound management. *International Journal of Molecular Sciences* 10:1300–1313.

Lim, C. K., and Halim, A. S. 2010. Biomedical-grade chitosan in wound management and its biocompatibility *in vitro*. In: M. Elnashar, (Ed.), *Biopolymers*. Sciyo Publishing, New York, pp. 19–35. ISBN: 978-953-307-109-1.

Lim, C. K., Halim, A. S., Lau, H. Y., Ujang, Z., and Hazri, A. 2007. In vitro cytotoxicology model of oligo-chitosan and n,o-carboxymethyl chitosan using primary normal human epidermal keratinocyte cultures. *Journal of Applied Biomaterials* 5:82–87.

Liu, H., Li, H., Cheng, W., Yang, Y., Zhu, M., and Zhou, C. 2006. Novel injectable calcium phosphate/chitosan composites for bone substitute materials. *Acta Biomaterialia* 2:557–565.

Mizuno, K., Yamamura, K. Yano, K., Osada, T., Saeki, S., Takimoto, N., Sakurai, T., and Nimura, Y. 2003. Effect of chitosan film containing basic fibroblast growth factor on wound healing in genetically diabetic mice. *Journal of Biomedical Materials Research Part A* 64:177–181.

Ma, Z., Li, S., Song, Y., Tang, L., Ma, D., Liu, B., and Jin, Y. 2008. The biological effect of dentin noncollagenous proteins (DNCPs) on the human periodontal ligament stem cells (HPDLSCs) in vitro and in vivo. *Tissue Engineering Part A* 14:2059–2068.

Madihally, S. V., and Matthew, H. W. 1999. Porous chitosan scaffolds for tissue engineering. *Biomaterials* 20:1133–1142.

Mariotti, A., and Monroe, P. J. 1998. Pharmacologic management of periodontal diseases using systemically administered agents. *Dental Clinics of North America* 42:245–261.

Miranda, C. C. S., Silva, A. B G., Mendes, R. M., Abreu, F. a. M., Caliari, M. V., Alves, J. B., and Goes, A. M. 2012. Mesenchymal stem cells associated with porous chitosan–gelatin scaffold: A potential strategy for alveolar bone regeneration. *Journal of Biomedical Materials Research Part A* 100A:2775–2786.

Miyazaki, S., Nakayama, A., Oda, M., Takada, M., and. Attwood, D. 1994. Chitosan and sodium alginate based bioadhesive tablets for intraoral delivery. *Biological and Pharmaceutical Bulletin* 17:745–747.

Mizuno, M., Imai, T., Fujisawa, R., Tani, H., and Kuboki, Y. 2000. Bone sialoprotein (BSP) is a crucial factor for the expression of osteoblastic phenotypes of bone marrow cells cultured on type I collagen matrix. *Calcified Tissue International* 66:388–396.

Mohammad, S. B. A. R., Ahmad, S. H., Kamaruddin, H., Ahmad, H. A. R., Norimah, Y., and Shaharum, S. 2010. In vitro evaluation of novel chitosan derivatives sheet and paste cytocompatibility on human dermal fibroblasts. *Carbohydrate Polymers* 79:1094–1100

Mori, T., Okumura, M., Matsuura, M., Ueno, K., Tokura, S., and Okamoto, Y. 1997. Effects of chitin and its derivatives on the proliferation and cytokine production of fibroblasts in vitro. *Biomaterials* 18:947–951.

Murray, P. E., Garcia-Godoy, F., and Hargreaves, K. M. 2007. Regenerative endodontics: A review of current status and a call for action. *Journal of Endodontics* 33:377–390.

Muzzarelli, A. A. R. 2012. Chitosan-based systems for biopharmaceuticals: Delivery, targeting and polymer therapeutics, First edition. Edited by Bruno Sarmento, *José das Neves.* John Wiley & Sons, Ltd.

Nakashima, M., and Reddi, A. H. 2003. The application of bone morphogenetic proteins to dental tissue engineering. *Nature Biotechnology* 21:1025–1032.

Nakashima, M. 2005. Bone morphogenetic proteins in dentin regeneration for potential use in endodontic therapy. *Cytokine Growth Factor Review* 16:369–376.

Nascimento, E. G., Sampaio, T. B., Medeiros, A. C., and Azevedo, E. P. 2009. Evaluation of chitosan gel with 1% silver sulfa-diazine as an alternative for burn wound treatment in rats. *Acta Cirurgica Brasileira* 24:460–465.

Needleman, I. G., Smales, F. C., and Martin, G. P. 1997. An investigation of bioadhesion for mucosal and oral mucosal drug delivery. *Journal of Clinical Periodontology* 24:394–400.

Needleman, I. G., Martin, G. P., and Smales, F. C. 1998. Characterization of bioadhesives for periodontal and oral mucosal drug delivery. *Journal of Clinical Periodontoogy* 25:74–82.

Nishimura, K., Nishimura, S., Nishi, N., Saiki, I., Tokura, S., and Azuma, I. 1984. Immunological activity of chitin and its derivatives. *Vaccine* 2:93–99.

Okamoto, Y., Minami, S., Matsuhashi, A., Sashiwa, H., Saimoto, H., and Shigemasa, Y. 1993. Polymeric *N*-acetyl-D-glucosamine (chitin) induces histrionic activation in dogs. *Journal of Veterinary Medical Science* 55:739–742.

Park, C. J., Clark, S. G., Lichtensteiger, C. A., Jamison, R. D., and Johnson, A. J. 2009. Accelerated wound closure of pressure ulcers in aged mice by chitosan scaffolds with and without bFGF. *Acta Biomaterialia* 5:1926–1936.

Pawtlowska, E. 1997. The assessment of influence of chitosan on the dental pulp in rats. In: A. Domard, G. A. F. Roberts, and K. M. Varum, (Eds.), *Advances in Chitin Science.* Jacques. Andre. Publishers, Lyon, pp. 705–10.

Peter, M. G. 1997. Introductory remarks. *Carbohydrates in Europe* 19:9–15.

Pieper, J. S., Oosterhof, A., Dijkstra, P. J., Veerkamp, J. H., and vanKuppevelt, T. H. 1999. Preparation and characterization of porous crosslinked collagenous matrices containing bioavailable chondroitin sulphate. *Biomaterials* 20:847–858.

Prabaharan, M., and Mano, J. F. 2006. Chitosan derivatives bearing cyclodextrin cavities as novel adsorbent matrices. *Carbohydrate Polymers* 63:153–166.

Prescott, R. S., Alsanea, R., Fayad, M. I., Johnson, B. R., Wenckus, C. S., Hao, J., John, A. S., and George, A. 2008. In vivo generation of dental pulp-like tissue by using dental pulp stem cells, a collagen scaffold, and dentin matrix protein 1 after subcutaneous transplantation in mice. *Journal of Endodontics* 34:421–426.

Rossi, S., Marciello, M., Bonferoni, M. C., Ferrari, F., Sandri, G., Dacarro, C., Grisoli, P., and Caramella, C. 2010. Thermally sensitive gels based on chitosan derivatives for the treatment of oral mucositis. *European Journal of Pharmaceutics and Biopharmaceutics* 74:248–254.

Rutherford, R. B., and Gu, K. 2000. Treatment of inflamed ferret dental pulps with recombinant bone morphogenetic protein-7. *European Journal of Oral Science* 108:202–206.

Sashiwa, H., Saimoto, H., Shigemasa, Y., Ogawa, R., and Tokura, S. 1990. Lysozyme susceptibility of partially deacetylated chitin. *International Journal of Biological Macromolecules* 12:295–296.

Sharma, B., and Elisseeff, J. H. 2004. Engineering structurally organized cartilage and bone tissues. *Annals of Biomedical Engineering* 32:148–159.

Shigemasa, Y., Saito, K., Sashiwa, H., and Saimoto, H. 1994. Enzymatic degradation of chitins and partially deacetylated chitins. *International Journal of Biological Macromolecules* 16:43–49.

Six, N., Lasfargues, J. J., and Goldberg, M. 2002. Differential repair responses in the coronal and radicular areas of the exposed rat molar pulp induced by recombinant human bone morphogenetic protein 7 (osteogenic protein 1). *Archive of Oral Biology* 47:177–187.

Sloan, A., Rutherford, R., and Smith, A. 2000. Stimulation of the rat dentin–pulp complex by bone morphogenetic protein-7 in vitro. *Archive of Oral Biology* 45:173–177.

Slots, J. 1996. Position paper: Systemic antibiotics in periodontics. *Journal of Periodontology* 67:831–838.

Slots, J., and Rams, T. E. 1990. Antibiotics in periodontal therapy: Advantages and disadvantages. *Journal of Clinical Periodontology* 17:479–493.

Senel, S. Kas, H. S., and Squier, C. A. 2000. Application of chitosan in dental drug delivery and therapy, In: A. A. R. Muzzarelli (Ed.), *Chitosan per Os: From Dietary Supplement to Drug Carrier,* Atec, Grottammare, Italy, pp. 241–256.

Sezer, A. D., Cevher, E., Hatipoglu, F., Ogurtan, Z., Bas, A. L., and Akbuga, J. 2008. Preparation of fucoidan–chitosan hydrogel and its application as burn healing accelerator on rabbits. *Biological and Pharmaceutical Bulletin* 31:2333–2362.

Smart, J. D. 1991. An in vitro assessment of some mucoadhesive dosage forms. *International Journal of Pharmaceutics* 73:69–74.

Skjak-Braek, G., Anthonsen, T., and Sandford, P. 1989. *Chitin and Chitosan*, Elsevier Applied Science, London, pp. 560.

Song, H. Y., Esfakur Rahman, A. H., and Lee, B. T. 2009. Fabrication of calcium phosphate- calcium sulfate injectable bone substitute using chitosan and citric acid. *Journal of Material Science: Materials in Medicine* 20:935–941.

Sonis, S. T. 2004. Oral mucositis in cancer therapy. *Journal of Supportive Oncology* 2:3–8.

Suh, F. J. K., and Matthew, H. W. T. 2000. Applications of chitosan-based polysaccharide biomaterials in cartilage tissue engineering: A review. *Biomaterials* 21:2589–2598.

Tanigawa, T., Tanaka, Y., Sashiwa, H., Saimoto, H., and Shigemasa, Y. 1992. Various biological effects of chitin derivatives. In: C. J. Brine, P. A. Sandford, and J. P. Zikakis (Eds.), *Advances in Chitin and Chitosan* (pp. 206–215). Elsevier Science Publisher, London.

Taylor, M. S., Daniels, A. U., Andriano, K. P., and Heller, J. 1994. Six bioabsorbable polymers: In vitro acute toxicity of accumulated degradation products. *Journal of Applied Biomaterials* 5:151–157.

Thesleff, I., and Sharpe, P. 1997. Signalling networks regulating dental development. *Mechanisms of Development* 67:111–123.

Tokura, S., Ueno, K., Miyazaki, S., and Nishi, N. 1997. Molecular weight dependent antimicrobial activity by chitosan. *Macromolecular Symposia* 120:1–9.

Tuzlakoglu, K., Bolgen, N., Salgado, A. J., Gomes, M. E., Piskin, E., and Reis, R. L. 2005. Nano- and microfiber combined scaffolds: A new architecture for bone tissue engineering. *Journal of Material Science: Materials in Medicine* 16:1099–1104.

Uraz, A., Boynuegri, D., Ozcan, G., Karaduman, B., Uc, D., Senel, S., Pehlivan, S., Ogus, E., and Sultan, N. 2012. Two percent chitosan mouthwash: A microbiological and clinical comparative study. *Journal of Dental Sciences* 7:342–349.

Van, T. N., Ng, C. H., Aye, K. N., Trang, T. S., and Stevens, W. F. 2006. Production of high-quality chitin and chitosan from preconditioned shrimp shells. *Journal of Chemical Technology and Biotechnology* 81:1113–1118.

Varshosa, J., and Karimzadeh, S. 2007. Development of cross-linked chitosan films for oral mucosal delivery of lidocaine. *Research in Pharmaceutical Sciences* 2:43–52.

Wang, X., Ma, J., Wang, Y., and He, B. 2002. Bone repair in radii and tibias of rabbits with phosphorylated chitosan reinforced calcium phosphate cements. *Biomaterials* 23:4167–4176.

Yang, X., Han, G., Pang, X., and Fan, M. 2011. Chitosan/collagen scaffold containing bone morphogenetic protein-7 DNA supports dental pulp stem cell differentiation in vitro and in vivo. *Journal of Biomedical Materials Research Part A.* 00A:1–8.

Young, C. S., Terada, S., Vacanti, J. P., Honda, M., Bartlett, J. D., and Yelick, P. C. 2002. Tissue engineering of complex tooth structures on biodegradable polymer scaffolds. *Journal of Dental Research* 81:695–700.

Zhang, W., Walboomers, X. F., van Kuppevelt, T. H., Daamen, W. F., Bian, Z., and Jansen, J. A. 2006a. The performance of human dental pulp stem cells on different three- dimensional scaffold materials. *Biomaterials* 27:5658–5668.

Zhang, Y., Cheng, X., Wang, J., Wang, Y., Shi, B., Huang, C., Yang, X., and Liu, T. 2006b. Novel chitosan/collagen scaffold containing transforming growth factor-b1 DNA for periodontal tissue engineering. *Biochemical and Biophysical Research Communications* 344:362–369.

24 Bionanocomposites of Chitosan for Multitissue Engineering Applications

P. N. Sudha, S. Aisverya, Maximas H. Rose, Jayachandran Venkatesan, and Se-Kwon Kim

CONTENTS

24.1 INTRODUCTION

Tissue engineering requires the cells to have regenerative potential when seeded on three-dimensional networks that are implanted into the defect. Various synthetic and other biopolymers are used to construct scaffolds for tissue engineering applications (Bartold et al. 2006). The emerging field of regenerative medicine tries to find solutions for the incomplete regeneration of tissues in the human body. It employs living cells, biomaterials, soluble mediators of tissue regeneration, or a combination of these to recapitulate normal tissue structure and function (Nerem 1991; Marler et al. 1998; Vacanti and Langer 1999). The National Institute of Biomedical Imaging and Bioengineering (NIBIB) (http://www.nibib1.nih.gov/) defines tissue engineering as "a rapidly growing area that seeks to create, repair and/or replace tissues and organs by using combinations of cells, biomaterials, and/or biologically active molecules." Various techniques, such as phase separation, self-assembly, and electrospinning have been developed to fabricate nanofibrous scaffolds with unique properties. Among these techniques, electrospinning technology has become popular for the fabrication of tissue engineering scaffolds in recent years because it is a simple, rapid, efficient, and inexpensive method for producing nanofibers by applying a high voltage to electrically charged liquid (Huang et al. 2003; Zhang et al. 2007; Teo and Ramakrishna 2009).

The present generation of tissue engineering research is based on the seeding of cells onto porous biodegradable polymer matrixes. A primary factor is the availability of good biomaterials to serve as the temporary matrix. These biomaterials must be capable of being prepared in porous forms to offer a channel for the migration of host cells into the matrix, thereby permitting growth into complete tissue analogs and thereafter be biodegradable into nontoxic products once they have served their function *in vivo*. The scaffold material has an essential function concerning cell anchorage, proliferation, and tissue formation in three dimensions. Performance of these properties usually demands a porous scaffold structure, with the porosity characteristics being application specific. Recently, chitosan and its derivatives have been reported as attractive candidates for scaffolding materials because they degrade as the new tissues are formed, eventually without inflammatory reactions or toxic degradation (Kim et al. 2003; Tuzlakoglu et al. 2004; Baran et al. 2004). Cells are often implanted or "seeded" into an artificial structure capable of supporting three-dimensional tissue formation of these structures, typically called scaffolds. The ideal scaffold must satisfy various often-conflicting demands: (1) appropriate levels and sizes of porosity allowing cell migration; (2) sufficient surface area and a variety of surface chemistries that encourage cell adhesion, growth, migration, and differentiation; and (3) a degradation rate that closely matches the regeneration rate of the desired natural tissue.

Scaffolds developed from natural and synthetic polymers are commonly used in tissue engineering. These polymers mimic the chemical and physical properties of natural extracellular matrix (ECM) (Kim et al. 2011; Liu et al. 2008). The scaffold should be biodegradable and elastic for contractile tissues such as blood vessels and heart valves. It is therefore necessary to develop tissue such as matrix with interconnected network, to act as templates to guide cell growth transfer of nutrients, oxygen, and waste products (Kathuria et al. 2009). The aim of this chapter is to review the recent developments on the chemically modified chitosan derivatives that are specially designed for multitissue engineering applications.

24.2 CHITOSAN

Partial deacetylation of chitin results in the production of chitosan, which is a polysaccharide comprising copolymers of glucosamine and *N*-acetylglucosamine (NAG). Indeed, chitosan is a derivative of natural chitin, the second most abundant polysaccharide in nature after cellulose (Nishimura et al. 1991). The properties of materials are governed by their structure. Chitosan is a polysaccharide primarily formed by the repeating units of β-(1 → 4)-2-amino-2-deoxy-D-glucose (or D-glucosamine) (Figure 24.1). This polycationic biopolymer is generally obtained by alkaline deacetylaton of chitin, which is the main component of the exoskeleton of crustaceans, such as shrimps (Muzzarelli 1973). Most commercial and laboratory-created chitosan contains NAG and glucosamine repeating units linked by β-glycosidic bonds (Shen et al. 2010; Aam et al. 2010; Chen et al. 2011). Chitin and chitosan are of great commercial interest because of their high percentage of nitrogen (6.89%) compared to synthetically substituted cellulose (1.25%). Chitosan can be processed in a sort of devices varying in shape and size, such as membranes, nanoparticles, microparticles/microspheres, gels, tablets, and capsules (Pillai and Panchagnula 2001; Khor and Lim 2003). Chitosan is a nontoxic, semicrystalline biodegradable and biocompatible, linear polysaccharide of randomly distributed NAG and glucosamine units. However, chitosan can only be soluble in few dilute acid solutions, which limits its wide applications. Recently, there has been a growing interest in the chemical modification of chitosan to improve its solubility and widen its applications (Sashiwa and Shigemasa 1999). Chitosan can be easily modified into various forms such as films, fibers, beads, sponges, and more complex shapes for orthopedic treatment (Di Martino et al. 2005). The cationic nature of chitosan is responsible for attracting various negatively charged proteoglycans. Although the chemical modification of chitosan modifies its properties, it is possible to maintain some interesting characteristics such as mucoadhesivity, biocompatibility, and biodegradability (Jayakumar et al. 2005). Chitosan has been combined with a variety of materials such as HAp, alginate, hyaluronic acid (HA), calcium phosphate, poly(methyl methacrylate), poly-L-lactic acid, and growth factors for potential application in orthopedics (Kast and Bernkop-Schnürch 2001; Kast et al. 2003).

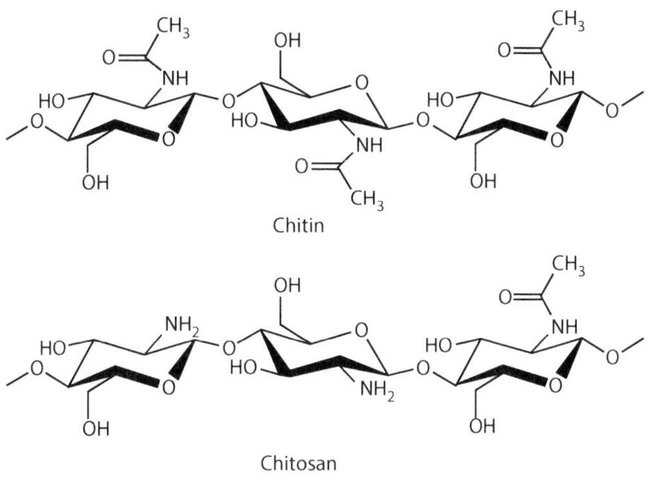

FIGURE 24.1 Structure of chitin and chitosan.

24.3 CHITOSAN AND CHITOSAN DERIVATIVES IN TISSUE ENGINEERING

Chitosan has three types of reactive functional groups that allow modifications of chitosan to produce various useful hydrogels for tissue engineering applications (Kim et al. 2008). Recently, nanofibers based on chitosan derivatives have also been prepared for biomedical applications. Carboxymethyl, carboxyethyl, and hexanoyl chitosan (Neamnark et al. 2006; Peesan et al. 2006; Mincheva et al. 2007; Vondran 2007) derivatives were used for the preparation of nanofibers. Hexanoyl chitosan was used for medical applications because it was proven to be resistant to hydrolysis by lysosome (Lee et al. 1995) and is antithrombogenic (Hirano and Noishiki 1985). An important chemical modification method is carboxymethylation. Polymers with carboxylate functionality have been found to elicit a broad range of biological activities.

The important carboxymethyl derivatives are *O*-carboxymethyl-*N*-chitosan (O-CMC), *N,O*-carboxymethyl chitosan (N,O-CMC), *N*-carboxymethyl chitosan (N-CMC), and *N*-succinyl chitosan.

Chitin (Shalumon et al. 2009), chitosan (Sajeev et al. 2008), alginate, and so on are used for tissue regeneration due to their nontoxicity, enhanced biocompatibility, cell adhesion, and proliferation. Because natural polymers have certain limitations, such as low stability and the release of toxic degradation products that can be harmful to cells, they are often blended with synthetic polymers (Yingshan et al. 2008; Ying Wun et al. 2008) to enhance the mechanical properties, degradation stability, and enhanced affinity to cellular components. The hydroxyapatite/carboxymethyl chitosan composite scaffolds showed degradability and bioactivity (Oliveira et al. 2009).

Polymeric fibers having diameters ranging from a few nanometers up to some microns can be successfully used for a wide variety of applications such as reinforcements in nanocomposites (Huang et al. 2003), as nanowires, nanotubes (Hu et al. 1999), and in tissue engineering (Deitzel et al. 2002; Khil et al. 2004; Xu et al. 2004). Chitosan and HA are two polysaccharides that can form hydrogels and are widely exploited with various modifications for its use as scaffold for tissue engineering (Huaping Tan et al. 2009). Chitosan is considered one of the most valuable polymer for biomedical and pharmaceutical applications due to its biodegradability, biocompatibility, antimicrobial property, nontoxicity, and antitumor properties. Nanoparticles, microspheres, hydrogels, films, and fibers are typical chitosan-based forms for biomedical and pharmaceutical applications. Examples of such applications include nasal, ocular, oral, parenteral, and transdermal drug delivery

(Kumar 2000). Tissue engineering investigations continue for the repair of different tissues such as bone, cornea, skin, liver, and nerve.

24.3.1 CHITOSAN BIONANOCOMPOSITES IN MULTITISSUE ENGINEERING

Biopolymers have an advantage of being biodegradable and these materials contain structural groups similar to natural extracellular components. Nanosurface and nanoparticles are known to influence cell behavior, including attachment and spreading. The cells' attachment and proliferation were found to be good on micro- and nanostructured materials (Laurencin et al. 1999; Teixeira et al. 2003). Scaffolds are important components in tissue engineering as they are artificial structures on which cells are seeded and support the three-dimensional tissue formation.

The ideal properties of a scaffold can be summarized as follows (Gomes et al. 2002; Barnes et al. 2007; Hollister and Lin 2007; Hong et al. 2007; Liu et al. 2007; Malafaya et al. 2008):

- Provides anatomical shape and volume.
- Should be compatible with the surrounding biological fluids and tissues, to minimize the immunological response.
- The scaffold itself and its degradation products should be nontoxic, and should be prepared by a biodegradable polymer with an adequate degradation rate (if the degradation is faster than the cell proliferation, the scaffold might degrade before the tissue construction; if the degradation of the scaffold is slower than the cell proliferation, cell death can be observed).
- The mechanical properties of the scaffold should provide temporary mechanical support at the site of implantation. It should be capable of carrying and delivering cells and/or biosignals.
- Should have the favorable surface properties for cell attachment and differentiation.
- Should be porous to increase the surface area for cell attachment.
- Pores of the scaffold should be interconnected not only for cell migration but also for the diffusion of gases, nutrients, and metabolic wastes; otherwise, cell death can be observed.

24.3.1.1 Bone Regeneration

Bone tissue engineering is gaining popularity as an alternative method for the treatment of osseous defects. Various biodegradable polymers have been explored for tissue engineering purposes. These include synthetic polymers such as poly(caprolactone), poly(lactic-*co*-glycolic acid), poly(ethylene glycol), poly(vinyl alcohol), and natural polymers like alginate, collagen, gelatin, chitin, and chitosan (Hirano et al. 1990). Bone is made up of seven hierarchical structures and consists of hydroxyapatite and collagen as major constituents (Weiner and Wagner 1998; Venkatesan and Kim 2010a,b; Venkatesan et al. 2011a,b). Chitosan favors cell adhesion due to its chemical backbone, which resembles glycosaminoglycans (GAGs), a major component of bone and cartilage.

Several studies have focused on the use of chitosan as a component in calcium-based cements in the development of bone substitutes. Yokoyama et al. (2002) used chitosan as a component of the liquid phase that included citric acid and glucose in combination with --TCP (tricalcium phos-phate) and tetra-calcium phosphate to produce an easily moldable cement. Novel poly(L-lactic acid) (PLLA)–chitosan hybrid scaffolds were prepared as tissue engineering scaffolds and simultane-ously as drug release carriers (Prabaharan et al. 2007).

Hydrogels derived from natural polymers are used for tissue engineering applications as they resemble the native ECM of the tissue (Tan and Marra 2010). Natural polymers such as alginate, chitin, and chitosan enhance osteogenesis. Therefore, it can be used as a bone substitute for bone repair and reconstruction (Kumar 2000). As a biomedical material, calcium phosphate has bioactive and osteoconductive properties (Cleries et al. 2000). The most widely occurring biological calcium phosphate is HAp. It is the main inorganic component of bones and teeth having biocompatibility for biomedical applications (Oliveira et al. 2006). Ideally, the scaffolds for bone tissue engineering should provide high structural integrity, high surface area for cell–material interaction, and slow degradation at a rate that commensurate with the elucidation of new bone tissue formation (Joshua et al. 2009). Moreover, the mechanical properties of the scaffolds should also match with the moduli of bone since it has to transmit mechanical forces and manage mineralization requirements (Vunjak-Novakovic and Goldstein, 2005). HAp is also a highly osteoconductive material facilitating better bonding of the biomaterial with bone (Jarcho et al. 1977). Biopolymer–HAp composites have been synthesized by many methods, such as blending (Furuzono et al. 2000), biomimetic process using simulated body fluid (SBF) (Zhang et al. 2004; Jayakumar and Tamura 2006; Maeda et al. 2008), *in situ* precipitation (Chen et al. 2002; Beppu and Santana 2003), and electrochemical deposition (Andrew et al. 1998; Manara et al. 2008). Bioactive glass ceramic (BGC) are a group of osteoconductive silicate-based materials used for bone repair. Bioglass was developed by Hench (1991) as a biomaterial to repair bone defects and is widely used in orthopedics and dentistry. BGC coatings on the surface of titanium are superior to HAp in their ability for osteointegration (Wheeler et al. 2000). Moreover, BGC can also bond to soft and hard tissues (Verrier et al. 2004).

As mentioned above, synthetic and natural polymers (e.g., polyglycolic acid (PGA), poly(lactic-*co*-glycolic acid) (PLGA), PLLA, polylactic acid (PLA), gelatin, collagen, and chitosan) are excellent candidates for bone/cartilage tissue engineering applications because of their biodegradability and ease of fabrication. It has been observed that the combination of carbon nanotube (CNT) with CTS (chitosan) leads to an enormous increase in the mechanical strength of the composite (Wang et al. 2005). In addition, owing to their superior cytocompatible, mechanical, and electrical properties, CNTs/carbon nanofibers (CNFs) are ideal scaffold candidates for bone tissue engineering applications (Venkatesan et al. 2011a,b).

24.3.1.2 Cartilage Tissue Regeneration

Cartilage is a connective tissue that has complex mechanical properties. There are four major types of cartilage, which can be distinguished by their specific constitutive components: hyaline cartilage, fibrocartilage, elastic cartilage, and costochondral cartilage (Bonassar 2002). Hyaline articular cartilage is the most abundant type in the body, composed of one cell type (the chondrocyte) dispersed in an abundant ECM. ECM is composed mainly of collagen type II and a large proteoglycan, aggrecan (Bonassar 2002; Nesic et al. 2006). ECM provides most of the functional properties associated with hyaline cartilage, including resistance to compression and provision of low friction articulating surfaces in the joints. Injuries to cartilage are often painful and may severely affect movement. For cartilaginous tissue, GAGs, consisting of HA, chondroitin sulfate, keratin sulfate, dermatan sulfate, and heparan sulfate are the major components of the extracellular proteoglycan (Martin et al. 1999). The side-chain electrostatic bonds of the collagen–GAG complex enhance the hydrophilic ECM to become a structure with high water contents (Fransson 1987). For tissue-engineered cartilage, scaffold substrates play the most significant role in their biomimetic function as ECM for chondrocyte multiplication. Unfortunately, articular cartilage has a relatively poor capacity for self-repair (related to the lack of a direct blood supply), and cartilage injury is frequently associated with the onset of chronic problems, including osteoarthritis (Bonassar 2002; Magne et al. 2005). Currently, there is no agreed method of restoring fully damaged cartilage (Steinert et al. 2007). Current therapies include abrasion arthroplasty, subchondral drilling, prosthetic joint replacement, and, ultimately, transplantation of autologous chondrocytes or tissues (Steinert et al. 2007; Martin et al. 2007). In cartilage engineering, chitosan and HA can be blended together to create a three-dimensional scaffold that has a subchondral (bone layer) and cartilage layer. By creating a bilayered scaffold, it eliminates donor site morbidity associated with traditional autografts. Oliveira et al. (2006) created a bilayer scaffold utilizing the freeze-dry method and pouring a 3% chitosan solution onto a sintered HA scaffold. The scaffolds demonstrated a high connectivity, adequate water uptake and porosity, good mechanical properties, and cellular adhesion. Yamane et al. (2005) also indicated that chitosan-based hyaluronan hybrid polymer fibers show a great potential as a desirable biomaterial for cartilaginous tissue scaffolds due to their excellent characters of cell adhesion, proliferation, and matrix secretion. Because of the similarity in the framework of chitosan and glucosaminoglycons in cartilage, chitosan has been tested as a scaffold for cartilage repair (Lee et al. 2004). A chitosan hydrogel in the form of a scaffold was prepared for chondrocyte cells to reconstruct tissue-engineered cartilage and repair articular cartilage defects in the sheep model as reported by Hao et al. (2010). Chitosan-based scaffolds deliver growth factors in a controlled fashion to promote the in-growth and biosynthetic ability of chondrocytes. Lee et al. (2004) reported porous collagen/chitosan/GAG scaffolds loaded with transforming growth factor (TGF)-β1. This scaffold exhibited controlled release of TGF-β1 and promoted cartilage regeneration. The addition of chitosan to the collagen scaffold improved the mechanical properties and the stability of the collagen network by inhibiting the action of collagenases (Taravel and Domard 1996). Kim et al. (2003) used a porous freeze-dried chitosan scaffold incorporating TGF-β1-containing microspheres for the treatment of cartilage defects.

24.3.1.3 Nerve Regeneration

The goal of repairing nerve lesions is to direct the regenerating nerve fibers into the proper endoneurial tubes (Thurman et al. 1999). The current strategies can be classified into two categories: (a) bridging, using grafting and tubulization techniques; and (b) end-to-end suturing of the nerve stumps. The former technique seems to be more effective as it avoids tension across the repair site (Dash et al. 2011). Generally, the nervous system can be divided into two main parts: the central nervous system (CNS) (including the brain and the spinal cord) and the peripheral nervous system (PNS) (including the spinal and autonomic nerves). These two systems have two different repair procedures after injury (Bahr and Bonhoeffer 1994; Zhang et al. 2005; Huang and Huang 2006). For the PNS, the damaged axons usually regenerate and recover via proliferating Schwann cells, phagocytosing myelin by macrophages or monocytes, forming bands of Bünger by the bundling of Schwann cells, and sprouting axons in the distal segment (Evans 2001). In another report, electrospun polycaprolactone (PCL)/chitosan nanofiber scaffolds exhibited improved mechanical properties compared to chitosan (Prabhakaran et al. 2008). Schwann cells also proliferated well on this PCL/chitosan nanofiber scaffold. Hydralazine-encapsulated chitosan nanoparticles are expected to yield dual benefits after acrolein-mediated cellular damage: first, by mediating the sealing of damaged membrane and second, by accomplishing neuroprotection by inhibiting the generation of reactive oxygen species (ROS) and lipid peroxidation or LPO. Consistent with previous reports, the hydralazine entrapped inside chitosan nanoparticle exhibited enhanced cytoprotective effects by interfering with the generation of ROS and LPO after acrolein-mediated challenge (Cho et al. 2010a through e). Chitosan is suitable for nerve regeneration based on its biocompatibility and biodegradability. Haipeng et al. (2000) reported that neurons cultured on the chitosan membrane can grow well and that chitosan tube can promote repair of the PNS. Yuan et al. (2004) found that chitosan fibers supported the adhesion, migration, and proliferation of SCs, which provide a similar guide for regenerating axons to Büngner bands in the nervous system. A novel biomaterial for nerve regeneration through immobilization of laminin peptide in molecularly aligned chitosan by covalent bonding was designed by Matsuda et al. (2005). Progesterone delivered from chitosan prostheses were reported by Chavez-Delgado et al. (2003) to provide better facial nerve regenerative response of the rabbits than chitosan prostheses without progesterone.

24.3.1.4 Skin Regeneration

Skin repair is an important field of tissue engineering, especially in the case of extended third-degree burns, where the current treatments are ineffective in promoting satisfying skin regeneration. Bioinspired bilayered physical hydrogels that are only constituted of chitosan and water were processed and applied in the treatment of full-thickness burn injuries (Boucard et al. 2007). An ideal scaffold used for skin tissue engineering should possess the characteristics of excellent biocompatibility, suitable microstructure such as 100–200 mm mean pore size and porosity above 90%, controllable biodegradability, and suitable mechanical property (Dagalakis et al. 1980; Radhika et al. 1999; Chen et al. 2000; Freyman et al. 2001). The potential use of the CECS/PVA (carboxyethyl chitosan/poly(vinyl alcohol)) electrospun fiber mats as scaffolding materials for skin regeneration was evaluated *in vitro* using mouse fibroblasts (L929) as reference cell lines. Indirect cytotoxicity assessment of the fiber mats indicated that the CECS/PVA electrospun mat was non-toxic to the L929 cell. Cell culture results showed that fibrous mats were good in promoting the L929 cell attachment and proliferation (Zhou et al., 2008). This novel electrospun matrix would be used as potential wound dressing for skin regeneration. The results indicated that gold colloid/ chitosan film scaffold was nontoxic to keratinocytes, and was a good candidate for wound dressing in skin tissue engineering (Zhang et al. 2009). Further, Ma et al. (2003) showed that collagen/ chitosan scaffolds were potential candidates for the dermal equivalent, with enhanced biostability and good biocompatibility.

24.3.2 OTHER TISSUE REGENERATION APPLICATIONS OF CHITOSAN BIONANOCOMPOSITES

Many tissue analogs, including cartilage, bone, liver, bladder, cardiovascular, and nerve have been prepared using this engineering technology. The combination of chitosan with another material appears to be a common theme in various reports. Risbud et al. (2001a) have studied the interaction of chitosan–gelatin hydrogels and their cell interactions. Extending this approach, chitosan–poly-vinyl pyrrolidone hydrogels have also been evaluated and found not to have significant interactions with endothelial cells and therefore suitable as immunoisolation materials (Risbud et al. 2001b). This membrane was used to demonstrate the possibility to support respiratory epithelial cells for a possible tissue-engineered trachea. Chitosan has also been shown to exert a strong influence on nerve cell attachment and proliferation (Haipeng et al. 2000). The chitosan system performed better when coated with ECM materials such as laminin. Liver tissue engineering requires a per-fect ECM for primary hepatocytes culture to maintain a high level of liver-specific functions and desirable mechanical stability (Feng et al. 2009). Lee et al. (2010) developed a microfluidic-based pure chitosan microfibers for liver tissue engineering applications without the use of any chemical additives. Periodontal regeneration is of utmost importance in the field of dentistry that essentially reconstitutes and replaces the lost tooth supporting structures. Alveolar bone loss is a common finding associated with periodontal degeneration. Various treatment modalities have been used to regenerate or fill bony defects using different biomaterials such as bioglass and hydroxyapatite. Biodegradable materials are most extensively used in cardiovascular tissue engineering. Polymers such as collagen, gelatin, fibrin, HA, alginate, and decellularized matrices (Zimmermann et al. 2004; Christman and Lee 2006) can be produced from biological sources and no toxic degradation or inflammatory reactions are expected (Kong et al. 2004). Cardiac patches of silk fibroin (SF) combined with microparticles of chitosan or HA was fabricated by Yang et al. (2009). Chitosan/ collagen scaffold combined TGF-β1 as a good substrate candidate in periodontal tissue engineer-ing. Chitosan/coral sponge with platelet-derived growth factor B (PDGFB) encoding pDNA was prepared to construct periodontal tissue. Increased expression of PDGFB and significant cell proliferation was observed *in vitro*, and increased expression of PDGFB and new vascular tissue growth were observed *in vivo* (Zhang et al. 2007).

24.4 FUTURE PROSPECTS

Chitosan has a great potential in a variety of biomedical applications and chitosan's physicochemi-cal and mechanical properties utilized in fabricating particles and films can be modulated for spe-cific purposes. In addition, chitosan, having both amine and hydroxyl groups, can be modified with various ligands. Chitosan derivatives can be promising candidates as wound healing and supporting material for tissue engineering applications. Future biomaterials must simultaneously enhance tissue regeneration while minimizing immune responses and inhibiting infection. While promoters of tissue engineering promised to develop materials that can trigger tissue regeneration for the entire body, such promises have yet not become reality.

24.5 CONCLUSION

In the past few decades, substrates for tissue engineering have significantly progressed from being initially biologically inert structural support to multifunctional systems capable of orchestrating the formation and regeneration of complex tissue architectures. Chitosan and its derivatives are versatile and promising biodegradable polymers with a wide range of applications and various contributions in environmental, pharmaceutical, and biomedical applications. There are many parameters that are unexplored in terms of chitosan scaffold design and more research is needed.

ACKNOWLEDGMENTS

The authors are grateful to authorities of DKM College for Women and Thiruvalluvar University, Vellore, Tamil Nadu, India for the support. Thanks are also due to the editor Dr. Se-Kwon Kim, Marine Bio Process Research Center, Pukyoung National University, South Korea for the opportunity to review such an innovating field.

REFERENCES

Andrew, A.C., E. Wan, E. Khor, and G.W. Hastings. 1998. Preparation of a chitin-apatite composite by *in situ* precipitation onto porous chitin scaffolds. *J. Biomed. Mater. Res.* 41:541–548.

Aam, B.B., E.B. Heggset, A.L. Norberg et al. 2010. Production of chitooligosaccharides and their potential applications in medicine. *Mar. Drugs* 8:1482–1517.

Bahr, M. and F. Bonhoeffer. 1994. Perspectives on axonal regeneration in the mammalian CNS. *Trends Neurosci.* 17:473–479.

Baran, E.T., K. Tuzlakoglu, A.J. Salgado, and R.L. Reis. 2004. Multichannel mould processing of 3D structures from porous corallinehydroxy apatitegranules and chitosan support material for tissue regeneration/ engineering. *J. Mater. Sci. Mater. Med.* 2:161–165.

Barnes, C.P., S.A. Sell, E.D. Boland, D.G. Simpson, and G.L. Bowlin. 2007. Nanofiber technology: Designing the next generation of tissue engineering scaffolds. *Adv. Drug Deliv. Rev.* 59:1413–1433.

Bartold, P.M., Y. Xiao, S.P. Lyngstaadas, M.L. Paine, and M.L. Snead. 2006. Principles and applications of cell delivery systems for periodontal regeneration. *Periodontology* 41:123–135.

Beppu, M.M. and C.C. Santana. 2003. PAA influence on chitosan membrane calcification. *Mater. Sci. Eng.* 23:651–658.

Bonassar, L.J. 2002. Cartilage reconstruction. In: A. Atala and R. Lanza (Eds.), *Methods of Tissue Engineering.* San Diego: Academic Press. 1027–1039.

Boucard, N., C. Vitona, D. Agayb et al. 2007. The use of physical hydrogels of chitosan for skin regeneration following third-degree burns. *Biomaterials* 28:3478–3488.

Chavez-Delgado, M.E., J. Mora-Galindo, U. Gomez-Pinedo et al. 2003. Facial nerve regeneration through progesterone-loaded chitosan prosthesis. A preliminary report. *J. Biomed. Mater. Res. B Appl. Biomater.* 67:702–711.

Chen, H.J., R.T. Pires, and S.C. Tseng. 2000. Amniotic membrane transplantation for severe neurotrophic corneal ulcers. *Br. J. Ophthalmol.* 84:826–833.

Chen, G.P., Y. Ushida, and T. Tateishi. 2002. Scaffold design for tissue engineering. *Macromol. Bio. Sci.* 2:67–77.

Chen, J.K., C.R. Shen, C.H. Yeh et al. 2011. N-acetyl glucosamine obtained from chitin by chitin degrading factors in *Chitinbacter tainanesis. Int. J. Mol. Sci.* 12:1187–1195.

Cho, Y., R. Shi, and R. Ben Borgens. 2010a. Chitosan nanoparticle-based neuronal membrane sealing and neuroprotection following acrolein-induced cell injury. *J. Biol. Eng.* 4:2.

Cho, Y., R. Shi, and R.B. Borgens. 2010b. Chitosan nanoparticle-based neuronal membrane sealing and neuroprotection following acrolein-induced cell injury. *J. Bionic. Eng.* 29:2.

Cho, Y., R. Shi, and R.B. Borgens. 2010c. Chitosan produces potent neuroprotection and physiological recovery following traumatic spinal cord injury. *J. Exp. Biol.* 213:1513–1520.

Cho, Y., R. Shi, A. Ivanisevic, and R.B. Borgens. 2010d. Functional silica nanoparticle-mediated neuronal membrane sealing following traumatic spinal cord injury. *J. Neurosci. Res.* 88:1433–1444.

Cho, Y., R. Shi, A. Ivanisevic, and R.B. Borgens. 2010e. A mesoporous silica nanosphere-based drug delivery system using an electrically conducting polymer. *Nanotechnology* 20:275102–275114.

Christman, K.L. and R.J. Lee. 2006. Biomaterials for the treatment of myocardial infarction, *J. Am. Coll. Cardiol.* 5:907–913.

Cleries, L., J.M. Fernandez-Pradas, and J.L. Morenza. 2000. Bone growth, and resorption of, calcium phosphate coatings obtained by pulsed laser deposition. *J. Biomed. Mater. Res.* 49:43–52.

Dagalakis, N., J. Flink, P. Stasikelis, J.F. Burke, and I.V. Yannas. 1980. Design of an artificial skin III. *Control Pore Structure* 14:511–528.

Dash, M., F. Chiellini, R.M. Ottenbrite, and E. Chiellini. 2011. Chitosan—A versatile semi-synthetic polymer in biomedical Applications. *Progr. Polym. Sci.* 36:981–1014.

Deitzel, J.M., W. Kosik, S.H. McKnight et al. 2002. Electrospinning of polymer nanofibers with specific surface chemistry. *Polymer* 43:1025–1029.

Di Martino, A., M. Sittinger, and M.V. Risbud. 2005. Chitosan: A versatile biopolymer for orthopaedic tissue-engineering. *Biomaterials* 26:5983–5990.

Evans, G.R.D. 2001. Peripheral nerve injury: A review and approach to engineered constructs. *Anat. Rec.* 263:396–404.

Feng, Z.Q., X. Chu, N.P. Huang et al. 2009. The effect of nanofibrous galactosylated chitosan scaffolds on the formation of rat primary hepatocyte aggregates and the maintenance of liver function. *Biomaterials* 30:2753–2763.

Fransson, L.-Å. 1987. Structure and function of cell-associated proteoglycans. *Trends Biochem. Sci.* 12:406–411.

Freyman, T.M., I.V. Yannas, and L.J. Gibson. 2001. Cellular materials as porous scaffolds for tissue engineering. *Progr. Mater. Sci.* 46:273–282.

Furuzono, T., T. Tagushi, A. Kishida, M. Akashi, and Y. Tamada. 2000. Preparation and characterization of apatite deposited on silk fabric using an alternate soaking process. *J. Biomed. Mater. Res.* 50:344–352.

Gomes, M.E., J.S. Godinho, D. Tchalamov, A.M. Cunha, and R.L. Reis. 2002. Alternative tissue engineering scaffolds based on starch: Processing methodologies, morphology, degradation and mechanical properties. *Mater. Sci. Eng. C* 20:19–26.

Haipeng, G., Y. Zhong, J. Li, Y. Gong, N. Zhao, and X. Zhang. 2000. Studies on nerve cell affinity of chitosan-derived materials. *J. Biomed. Mater. Res.* 52:285–295.

Hao, T., N. Wen, J.K. Cao et al. 2010. The support of matrix accumulation and the promotion of sheep articular cartilage defects repair *in vivo* by chitosan hydrogels. *Osteoarthr. Cartil.* 18:257–265.

Hench, L.L. 1991. From concept to clinic. *Bioceramics J. Am. Ceram. Soc.* 74:1487–1510.

Hirano, S., and Y. Noishiki. 1985. The blood compatibility of chitosan and N-acylchitosans. *J. Biomed. Mater. Res.* 19:413–417.

Hirano, S., C. Itakura, H. Seino et al. 1990. Chitosan as an ingredient for domestic animal feeds. *J. Agric. Food Chem.* 38:1214–1217.

Hollister, S.J., and C.Y. Lin. 2007. Computational design of tissue engineering scaffolds. *Comput. Methods Appl. Mech. Engrg.* 196:2991–2998.

Hong, Y., H. Song, Y. Gong et al. 2007. Covalently crosslinked chitosan hydrogel: Properties of *in vitro* degradation and chondrocyte encapsulation. *Acta Biomater.* 3:23–31.

Hu, J., T.W. Odom, and C.M. Lieber. 1999. Chemistry and physics in one dimension: Synthesis and properties of nanowires and nanotubes. *Acc. Chem. Res.* 32:435–445.

Huang, Y.C. and Y.Y. Huang. 2006. Biomaterials and strategies for nerve regeneration. *Artif. Organs* 30:514–522.

Huang, Z.M., Y.Z. Zhang, M. Kotaki, and S. Ramakrishna. 2003. A review on polymer nanofibers by electro spinning and their applications in nanocomposites. *Comput. Sci. Technol.* 63:2223–2253.

Huang, Z.M., Y.Z. Zhang, M. Kotaki, and S. Ramakrishna. 2003. A review on polymer nanofibers by electrospinning and their applications in nanocomposites. *Compos. Sci. Technol.* 63:2223–2253.

Huaping Tan, H.P., C.R. Chu, K.A. Payne, and K.J. Marra. 2009. Injectable *in situ* forming biodegradable chitosan–hyaluronic acid based hydrogels for cartilage tissue engineering. *Biomaterials* 30:2499–2506.

Jarcho, M., J.F. Kay, K.I. Gumaer, R.H. Doremus, and H.P. Drobeck. 1977. Tissue, cellular and subcellular events at a bone-ceramic hydroxyapatite interface. *J. Bioeng.* 1:79–92.

Jayakumar, R., M. Prabaharan, R.L. Reis, and J.F. Mano. 2005. Graft copolymerized chitosan—Present status and applications. *Carbohyd. Polym.* 62:142–158.

Jayakumar, R. and H. Tamura.2006. Apatite forming ability of N-carboxymethyl chitosan gels in a simulated body fluid. *Asian Chitin J.* 2:91–96.

Joshua, R.P., T. Timothy, C.P. Ketul et al. 2009. Bone tissue engineering: A review in bone biomimetics and drug delivery strategies. *Biotechnol. Progr.* 25:1539–1560.

Kast, C.E. and A. Bernkop-Schnürch. 2001. Thiolated polymers: Development and *in vitro* evaluation of chitosan-thioglycolic acid conjugates. *Biomaterials* 22:2345–2352.

Kast, C.E., W. Frick, U. Losert, and A.B. Schnürch. 2003. Chitosan-thioglycolic acid conjugate: A new scaffold material for tissue engineering. *Int. J. Pharm.* 256:183–189.

Kathuria, N., A. Tripathi, K.K. Kar, and A. Kumar. 2009. Synthesis and characterization of elastic and macroporous chitosan–gelatin cryogels for tissue engineering. *Acta Biomater.* 1:406–418.

Khil, M.S., H.Y. Kim, M.S. Kim, S.Y. Park, and D.R. Lee. 2004. Nanofibrous mats of poly(trimethylene terephthalate) via electrospinning. *Polymer* 45:295–301.

Khor, E. and L.Y. Lim. 2003. Implantable applications of chitin and chitosan. *Biomaterials* 24:2339–2349.

Kim, B.S., I.K. Park, T. Hoshiba et al. 2011. Design of artificial extracellular matrices for tissue engineering. *Progr. Polym. Sci.* 36:238–268.

Kim, I.Y., S.J. Seo, H.S. Moon et al. 2008. Chitosan and its derivatives for tissue engineering applications. *Biotechnol. Adv.* 26:1–21.

Kim, S.E., J.H. Park, Y.W. Cho et al. 2003. Porous chitosan scaffold containing microspheres loaded with transforming growth factor-beta1: Implications for cartilage tissue engineering. *J. Control. Release* 91, 365–374.

Kong, H.J., E. Alsberg, D. Kaigler, K.Y. Lee, and D.L. Mooney. 2004. Controlling degradation of hydrogels via the size of cross-linked junctions. *Adv. Mater.* 16:1917–1921.

Kumar, M.N.V.R. 2000. A review of chitin and chitosan applications. *Reactive Functional Polym.* 46:1–27.

Laurencin, C.T., A.M.A. Ambrosio, M.D. Borden, and J.A. Cooper. 1999. Issue engineering: Orthopedic applications. *Ann. Rev. Biomed. Eng.* 1:19–46.

Lee J.E., K.E. Kim, I.C. Kwon et al. 2004. Effects of the controlled-released TGF-beta 1 from chitosan microspheres on chondrocytes cultured in a collagen/chitosan/glycosaminoglycan scaffold. *Biomaterials* 25:4163–4173.

Lee, K.H., S.J. Shin, C.B. Kim et al. 2010. Microfluidic synthesis of pure chitosan microfibers for bio-artificial liver chip. *Lab. Chip.* 10:1328–1334.

Lee, K.Y., W.S. Ha, and W.H. Park. 1995. Blood compatibility and biodegradability of partially N-acylated chitosan derivatives. *Biomaterials* 16:1211–1216.

Liu, A., Z. Hong, X. Zhuang et al. 2008. Surface modification of bioactive glass nanoparticles and the mechanical and biological properties of poly(L-lactide) composites. *Acta Biomater.* 4:1005–1015.

Liu, C., Z. Xia, and J.T. Czernuszka. 2007. Design and development of three-dimensional scaffolds for tissue engineering. *Chem. Eng. Res. Design* 85:1051–1064.

Ma, L., C. Gao, Z. Mao et al. 2003. Collagen/chitosan porous scaffolds with improved biostability for skin tissue engineering. *Biomaterials* 24:4833–4841.

Maeda, Y., R. Jayakumar, H. Nagahama, T. Furuike, and H. Tamura. 2008. Synthesis, characterization and bioactive studies of novel β-chitin scaffolds for tissue engineering applications. *Int. J. Biol. Macromol.* 42:463–467.

Magne, D., C. Vinatier, M. Julien, P. Weiss, and J. Guicheux. 2005. Mesenchymal stem cell therapy to rebuild cartilage. *Trends Mol. Med.* 11:519–526.

Malafaya, P.B., T.C. Santos, M. van Griensven, and R.L. Reis. 2008. Morphology, mechanical characterization and *in vivo* neo-vascularization of chitosan particle aggregated scaffolds architectures. *Biomaterials* 29:3914–3926.

Manara, S., F. Paolucci, B. Palazzo et al. 2008. Electrochemically assisted deposition of biomimetic hydroxyapatite-collagen coatings on titanium plate. *Inorg. Chim. Acta* 361:1634–1645.

Marler, J.J., J. Upton, R. Langer, and J.P. Vacanti. 1998. Transplantation of cells in matrices for tissue regeneration. *Adv. Drug Deliv. Rev.* 33:165–182.

Martin, I., S. Miot, A. Barbero, M. Jakob, and D. Wendt. 2007. Osteochondral tissue engineering. *J. Biomech.* 40:750–765.

Martin, I., G. Vunjak-Novakovic, J. Yang, R. Langer, and L.E. Freed. 1999. Mammalian chondrocytes expanded in the presence of Fibroblast growth factor 2 maintain the ability to differentiate and regenerate three-dimensional cartilaginous tissue. *Exp. Cell. Res.* 253:681–688.

Matsuda, A., H. Kobayashi, S. Itoh, K. Kataoka, and J. Tanaka. 2005. Immobilization of laminin peptide in molecularly aligned chitosan by covalent bonding. *Biomaterials* 26:2273–2279.

Mincheva, R., N. Manolova, and I. Rashkov. 2007. Biocomponent aligned nanofibers of *N*-carboxyethyl chitosan and poly (vinyl alcohol). *Eur. Polym. J.* 43:2809–2818.

Muzzarelli, R. 1973. Natural chelating polymers. In R. Muzzarelli, Ed., *Chitosan*. Oxford, UK: Pergamon Press 144–176.

Neamnark, A., R. Rujiravanit, and P. Supaphol. 2006. Electrospinning of hexanoyl chitosan. *Carbohydr. Polym.* 66:298–305.

Nerem, R.M. 1991. Cellular engineering. *Ann. Biomed. Eng.* 19:529–545.

Nesic, D., R. Whiteside, M. Brittberg et al. 2006. Cartilage tissue engineering for degenerative joint disease. *Adv. Drug. Deliv. Rev.* 58:300–322.

Nishimura, S., O. Kohgo, K. Kurita, H. Kuzuhara. 1991. Chemospecific manipulations of a rigid polysaccharide synthesis of novel chitosan derivatives with excellent solubility in common organic solvents by regioselective chemical modifications. *Macromolecules* 24:4745–8.

Oliveira, J.M., M.T. Rodrigues, S.S. Silva et al. 2006. Novel hydroxyapatite/chitosan bilayered scaffold for osteochondral tissue engineering applications: Scaffold Design and its performance when seeded with goat bone marrow stromal cells. *Biomaterials* 276:123–6137.

Oliveira, J.M., S.A. Costa, I.B. Leonor et al. 2009. Novel hydroxyapatite/carboxymethylchitosan composite scaffolds prepared through an innovative autocatalytic electroless coprecipitation route. *J. Biomed. Mater. Res.* 88A:470–480.

Peesan, M., R. Rujiravanit, and P. Supaphol. 2006. Electrospinning of hexanoyl chitosan/polylactide blends. *J. Biomater. Sci. Polym.* 17:547–565.

Pillai, O., and R. Panchagnula. 2001. Polymers in drug delivery. *Curr. Opin. Chem. Biol.* 5:447–451.

Prabaharan, M., M.A. Rodriguez-Perez, J.A. de Saja, and J.F. Mano. 2007. Preparation and characterization of poly (L-lactic acid)-chitosan hybrid scaffolds with drug release capability. *J. Biomed. Mater. Res. Part B: Appl. Biomater.* 81:427–434.

Prabhakaran, M.P., J.R. Venugopal, T.T. Chyan et al. 2008. Electrospun biocomposite nanofibrous scaffolds for neural tissue engineering. *Tissue Eng. Part A* 14:1787–1797.

Radhika, M., B. Mary, and P.K. Sehgal. 1999. Cellular proliferation on desamidated collagen matrices. Comparative. *Biochem. Physiol. Part C* 124:131–1319.

Risbud, M., M. Endres, J. Ringe, R. Bhonde, and M. Sittinger. 2001a. Biocompatible hydrogel supports the growth of respiratory epithelial cells: Possibilities in tracheal tissue engineering. *J Biomed. Mater. Res.* 56:120–127.

Risbud, M.V., M.R. Bhonde, and R.R. Bhonde. 2001b. Effect of chitosanpolyvinyl pyrrolidone hydrogel on proliferation and cytokine expression of endothelial cells: Implications in islet immunoisolation. *J. Biomed. Mater. Res.* 57:300–305.

Sajeev, U.S., A.K. Anoop, M. Deepthy, and S.V. Nair. 2008. Control of nanostructures in PVA, PVA/chitosan blends and PCL polymers through electrospinning. *Bull. Mater. Sci.* 31:343–351.

Sashiwa, H., and Y. Shigemasa. 1999. Chemical modification of chitin and chitosan 2: Preparation and water soluble property of N-acylated or N-alkylated partially deacetylated chitins. *Carbohydr. Polym.* 39:127–138.

Shalumon, K.T., N.S. Binulal, N. Selvamurugan et al. 2009. Electrospinning of carboxymethyl chitin/poly (vinyl alcohol) nanofibrous scaffolds for tissue engineering applications. *Carbohydr. Polym.* 77:863–869.

Shen, C.R., Y.S. Chen, C.J. Yang, J.K. Chen, and C.L. Liu. 2010. Colloid chitin azure is a dispersible, low-cost substrate for chitinase measurements in a sensitive, fast, reproducible assay. *J. Biomol. Screen.* 15:213–217.

Steinert A.F., S.C. Ghivizzani, A. Rethwilm et al. 2007. Major biological obstacles for persistent cell-based regeneration of articular cartilage. *Arthritis Res. Ther.* 9:213.

Tan, H. and K.G. Marra. 2010. Injectable, biodegradable hydrogels for tissue engineering applications. *Materials*, 3:1746–1767.

Taravel, M.N. and A. Domard. 1996. Collagen and its interactions with chitosan, III some biological and mechanical properties. *Biomaterials* 17:451–455.

Teixeira, A.I., G.A. Abrams, P.J. Bertics, C.J. Murphy, and P.F. Nealey.2003. Epithelial contact guidance on well-defined micro- and nanostructured substrates. *J. Cell Sci.* 116:1881–1892.

Teo, W.E. and S. Ramakrishna. 2009. Electrospun nanofibers as a platform for multifunctional, hierarchically organized nanocomposite. *Compos. Sci. Technol.* 69:1804–1817.

Thurman, D.J., C. Alverson, K.A. Dunn, J. Guerrero, and J.E. Sniezek. 1999. Traumatic brain injury in the United States: A public health perspective. *J. Head Trauma Rehabil.* 14:602–615.

Tuzlakoglu, K., C.M. Alves, J.F. Mano, and R.L. Reis. 2004. Production and characterization of chitosan fibers and 3D fiber meshscaf folds fort issue engineering applications. *Macromol. Biosci.* 4:811–819.

Vacanti, J.P. and R. Langer. 1999. Tissue engineering: The design and fabrication of living replacement devices for surgical reconstruction and transplantation. *Lancet* 254:SI32–SI43 (http://www.nibib1.nih.gov/).

Venkatesan, J. and S.K. Kim. 2010a. Chitosan composites for bone tissue engineering—An overview. *Mar. Drugs* 8:2252–2266.

Venkatesan, J. and S.K. Kim. 2010b. Effect of temperature on isolation and characterization of hydroxyapatite from tuna (*Thunnus obesus*) bone. *Materials* 3:4761–4772.

Venkatesan, J., Z.-J. Qian, B. Ryu, N. Ashok Kumar, and S.-K. Kim. 2011a. Preparation and characterization of carbon nanotube-grafted-chitosan-natural hydroxyapatite composite for bone tissue engineering. *Carbohydr. Polym.* 83:569–577.

Venkatesan, J., Z.-J. Qian, B. Ryu, V.T. Noel, and S.-K. Kim. 2011b. A comparative study of thermal calcinations and an alkaline hydrolysis method in the isolation of hydroxyapatite from *Thunnus obesus* bone. *Biomed. Mater.* 6:035003.

Verrier, S., J.J. Blaker, M. Maquet, L.L. Hench, and R.A. Boccaccinia. 2004. PDLLA/Bioglasss composites for soft-tissue and hard-tissue engineering: An *in vitro* cell biology assessment. *Biomaterials* 25:3013–3021.

Vondran, J.L. 2007. Fabrication, optimization, and characterization of carboxymethylated chitosan nanofiber mats for cartilage regeneration applications. Materials science and engineering Philadelphia: Drexel University 123.

Vunjak-Novakovic, G., and S.A. Goldstein. 2005. In V.C. Mow, and R. Huiskes (Eds.), *Basic Orthopaedic Biomechanics and Mechano-Biology*. Philadelphia: Lippincott Williams and Wilkins. 343–408.

Wang, S.F., L. Shen, W.D. Zhang, and Y.J. Tong. 2005. Preparation and mechanical properties of chitosan/carbon nanotubes composites. *Biomacromolecules* 6:3067–3072.

Weiner, S., and H.D. Wagner. 1998. The materials bone: Structure-mechanical function relations. *Annu. Rev. Mater. Sci.* 28:271–298.

Wheeler, D.L., M.J. Montfort, and S.W. McLoughlin. 2000. Differential healing response of bone adjacent to porous implant coated with hydroxyapatite and bioactive glass. *J. Biomed. Mater. Res.* 55:603–612.

Xu, C.Y., R. Inai, M. Kotaki, and S. Ramakrishna. 2004. Aligned biodegradable nanofibrous structure: A potential scaffold for blood vessel engineering. *Biomaterials* 25:877–886.

Yamane, S., N. Iwasaki, T. Majima et al. 2005. Feasibility of chitosan-based hyaluronic acid hybrid biomaterial for a novel scaffold in cartilage tissue engineering. *Biomaterials* 26:611–619.

Yang, M.C., S.S. Wang, N.K. Chou et al. 2009. The cardiomyogenic differentiation of rat mesenchymal stem cells on silk fibroin-polysaccharide cardiac patches in vitro. *Biomaterials* 30:3757–3765.

Ying Wun, H.W., C. Xiaoying, and D. Siqin. 2008. Compressive mechanical properties and biodegradability of porous poly(caprolactone)/chitosan scaffolds. *Polym. Degrad. Stabil.* 93:1736–1741.

Yingshan, Z., Y. Dongzhi, C. Xiangmei et al. 2008. Electrospun water-soluble carboxyethyl chitosan/poly (vinyl alcohol) nanofibrous membrane as potential wound dressing for skin regeneration. *Biomacromolecules* 9:349–354.

Yokoyama, A., S. Yamamoto, T. Kawasaki, T. Kohgo, and M. Nakasu. 2002. Development of calcium phosphate cement using chitosan and citric acid for bone substitute materials. *Biomaterials* 23:1091–1101.

Yuan, Y., P. Zhang, Y. Yang, X. Wang, and X. Gu. 2004. The interaction of Schwann cells with chitosan membranes and fibers *in vitro*. *Biomaterials* 25:4273–4278.

Zhang Y, H. He, W.J. Gao et al. 2009. Rapid adhesion and proliferation of keratinocytes on the gold colloid/chitosan film scaffold. *Mater. Sci. Eng.* 29:908–912.

Zhang, L-J., X-S. Feng, H-G. Liu et al. 2004. Hydroxyapatite/collagen composite materials formation in simulated body fluid environment. *Mater. Lett.* 58:719–722.

Zhang, N., H. Yan, and X. Wen. 2005. Tissue- engineering approaches for axonal guidance. *Brain Res. Brain Res. Rev.* 49:48–64.

Zhang, Y.Z., B. Su, J. Venugopal, S. Ramakrishna, and C.T. Lim. 2007. Biomimetic and bioactive nanofibrous scaffolds from electrospun composite nanofibers. *Int. J. Nanomedicine* 2:623–638.

Zhang,Y., Y. Wang, B. Shi, and X. Cheng. 2007. A platelet derived growth factor releasing chitosan/coral composite scaffold for periodontal tissue engineering. *Biomaterials* 28:1515–1522.

Zhou, Y., D. Yang, X. Chen, Q. Xu, F. Lu, and J. Nie. 2008. Electrospun water-soluble carboxyethyl chitosan/poly (vinyl alcohol) nanofibrous membrane as potential wound dressing for skin regeneration. *Biomacromolecules* 9:349–354.

Zimmermann, W.H., I. Melnychenko, and T. Eschenhagen. 2004. Engineered heart tissue for regeneration of diseased hearts. *Biomaterials* 25:1639–1647.

25 Recent Advancements in Research on Chitin and Chitosan Derivatives for Drug Delivery Application

P.N. Sudha, T. Gomathi, K. Nasreen, Jayachandran Venkatesan, and Se-Kwon Kim

CONTENTS

25.1 INTRODUCTION

Natural polymers such as proteins and polysaccharides are mainly used in the pharmaceutical field. Among them, a polysaccharide analog chitin is the most abundantly used, second only to cellulose (Cohen 1993). It is mostly derived from the exoskeleton of crustaceans and has a wide range of application in various fields (Inmaculada Aranaz et al. 2009). Chitosan, a derivative of chitin, is

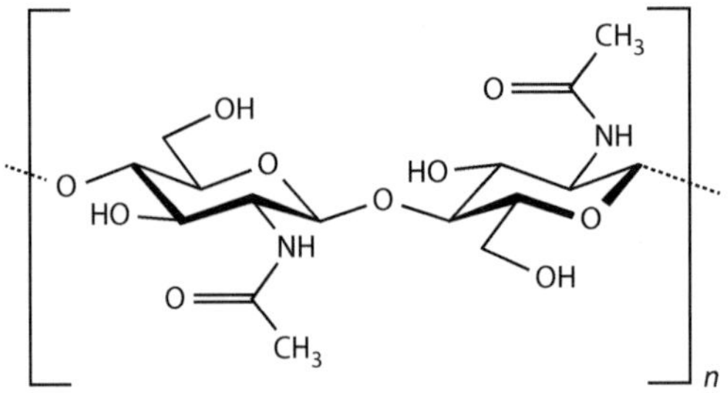

FIGURE 25.1 Structure of chitin.

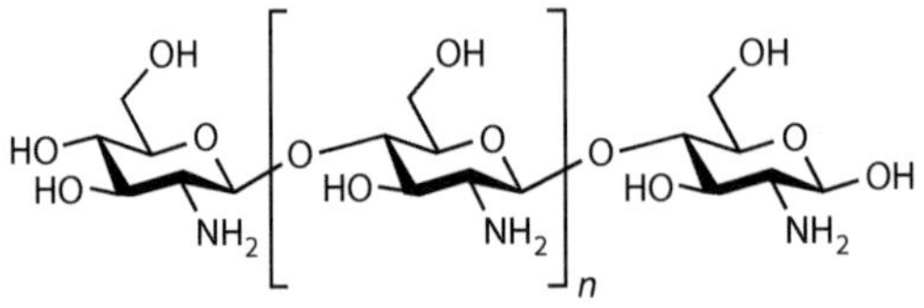

FIGURE 25.2 Structure of chitosan.

obtained by partial deacetylation under alkaline conditions or by enzymatic hydrolysis in the presence of chitin deacetylase.

Chitin and chitosan are described as a family of linear polysaccharides consisting of varying amounts of β(1–4)-linked residues of N-acetyl-2-amino-2-deoxy-D-glucose (glucosamine) and 2-amino-2-deoxy-D-glucose (N-acetylglucosamine) (Kast et al. 2003; Bernkop-Schnürch et al. 2004) (Figures 25.1 and 25.2). Chitin/chitosan possess good antimicrobial property, nontoxicity, biocompatibility, physiological inertness, hydrophilicity, adsorption property, gel-forming properties, and affinity for proteins (Krajewska 2004), and is used in various fields such as medicine, pharmaceuticals, cosmetics, textile, paper, agriculture, water treatment, and food industries. Compared to many other natural polymers, chitin and chitosan are used extensively in pharmaceuticals mainly in the field of drug delivery (Dodane and Vilivalam 1998; Felt et al. 1998; Kas 1997; Illum 1998).

However, chitosan has certain drawbacks, which can be rectified by chemical modification. Chemical modification expands the applications of chitin and enables it to contribute to certain specific functions (Mincea et al. 2012). It has a large number of applications in pharmaceutical dosage form; its further application can be exploited by the modifications of its basic structure to obtain polymers with a range of properties. This modification can be carried out by following one of the several approaches, such as chemically, mechanically, as well as enzymatically (Vipin Bansal et al. 2011). Almost all works concerning chitosan application refer to the polymer as a nontoxic, biologically compatible material, and thus suitable to be used in carrier production in the field of drug delivery (Susana Rodrigues et al. 2012).

25.2 SYNTHESIS OF CHITIN AND CHITOSAN

Chitin is easily obtained from the shells of crustaceans (crabs, prawns, lobsters, and shrimps), crawfish, the exoskeletons of insects, and the cell walls of various fungi (Dodane and Vilivalam 1998; Kumar 2000; Khor and Lim 2003; Krajewska 2004; Sinha et al. 2004). Crustaceans and crawfish are the richest sources of chitin. They are composed of protein (30–40%), inorganic salts (30–50%), and chitin (20–30%) on the dry weight basis (Johnson and Peniston 1982) and lipids as main structural components. Chitin represents one-third of the shell composition, and is highly hydrophobic and insoluble in water and most organic solvents.

Chitin and chitosan can be synthesized by two methods: chemical method and enzymatic method. The chemical method results in environmental pollution and it also consumes large amount of energy and wastes a lot of concentrated alkaline solution. However, the chemical method is the traditional method that consists of three basic steps with standard procedures (No and Meyers 1989): (1) demineralization, (2) deproteinization, and (3) decolorization. In the first step, calcium carbonate and calcium phosphate are separated followed by protein separation and then the removal of pigments. Chitin can also be produced from shell waste by fermentation with microorganisms or with the aid of enzymes (Rao et al. 2000). Enzymatic deacetylation of chitin to chitosan has been accomplished at the laboratory scale, but has not yet been attempted on an industrial scale (Win et al. 2000; Win and Stevens 2001).

Deacetylation leads to the more advantageous material chitosan, which overcomes the difficulties of chitin application. When the degree of deacetylation (DD) is 75% or above, then chitin is known as chitosan (Knaul et al. 1999). The determination of the DD of chitosan includes ninhydrin test, linear potentiometric titration, near-infrared spectroscopy, nuclear magnetic resonance spectroscopy, hydrogen bromide titrimetry, infrared spectroscopy, and first derivative ultraviolet–visible (UV) spectrophotometry (Khan et al. 2002).

25.3 PROPERTIES OF CHITIN AND CHITOSAN

25.3.1 PHYSICOCHEMICAL PROPERTIES

The physicochemical characteristics of chitin and chitosan control their functional properties, such as solubility, chemical reactivity, biological activities (Al Sagheer et al. 2009), and biodegradability (Sato et al. 1998; Sorlier et al. 2001), which are different based on the source of the crustacean species and preparation method followed (No et al. 2003). A wide range of natural polysaccharides, for example, cellulose, carrageenans, dextran, pectin, agar-agar, agarose, heparin, and alginic acid, are neutral or acidic in nature, but chitin and chitosan differ from other polysaccharides and show a highly basic nature due to the presence of an amine group. Molecular weight (MW) and DD are very important factors that affect the properties of chitosan. During the processing of chitosan, parameters such as concentration of reagents, time, and temperature employed can affect the physical characteristics of chitosan. The MW of the chitosan depends on viscosity, solubility, elasticity, and shear strength. In neutral and basic pH, chitosan is insoluble in aqueous media, but in acidic pH, the amino group of chitosan gets protonated and makes it soluble in aqueous medium. It forms salts with most of the inorganic and organic acids.

Chemical modification of various reactive (amino and hydroxyl) groups present on the chitosan employs a powerful means to promote new biological activities and to modify the mechanical properties. The viscosity of chitosan (CS) solution increases with an increase in chitosan concentration and decreases with increase in temperature, and influences the biological properties such as wound healing and biodegradation by lysozyme. The hydrophilic nature of chitosan tends to form gels at acidic pH. This type of gels can be used as a carrier in sustained drug delivery system. Cross-linking of chitosan decreases its solubility as well as it swelling property (Chen 2008; Dutta et al. 2004).

25.3.2 BIOLOGICAL PROPERTIES

Nontoxicity and biodegradability of chitosan make it a safe carrier in drug formulations. The bioadhesive property makes it adhere to hard and soft tissues and this has been used in dentistry, orthopedics, ophthalmology, and surgical procedures. It adheres to epithelial tissues and to the mucus coat present on the surface of the tissues. The favorable properties of chitosan toward biology makes it as a superior molecule in biological applications (Figure 25.3) (Dutta et al. 2004).

25.4 CHARACTERIZATION OF CHITIN AND CHITOSAN

The structural details of the heteropolymer chitosan tells that neither random nor block orientation is meant to be implied. The main parameters influencing the characteristics of chitosan are its

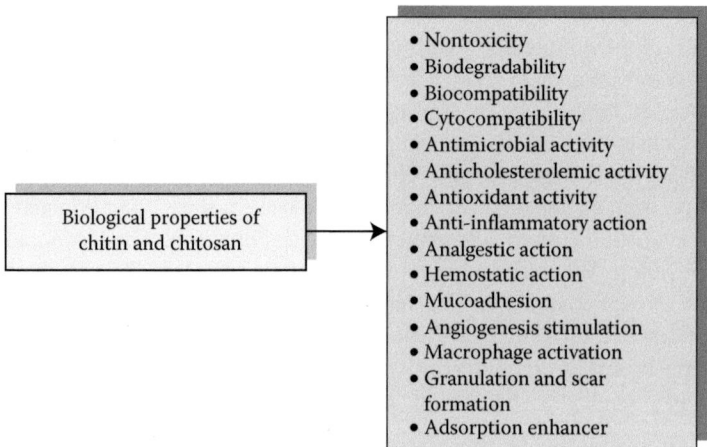

FIGURE 25.3   Biological properties of chitin and chitosan.

MW and DD, which affect the solubility and rheological and physical properties (Wang et al. 2003; Nunthanid et al. 2001; Chen et al. 1996). Various measuring techniques for MW and DD determinations are shown in Table 25.1.

The DD, defined as the mole fraction of deacetylated units in the polymorph chain (Zhang et al. 2005), is one of the most important factors influencing the properties of chitin and chitosan, such as solubility, flexibility, polymer conformation, and viscosity (Crini et al. 2009; Dash et al. 2011). The degree of acetylation (DA) can be measured by employing Fourier transform infrared spectroscopy (FT–IR), UV, nuclear magnetic resonance spectroscopy (NMR), and differential scanning calorimetry (DSC) techniques.

25.5   MODIFICATION OF CHITIN AND CHITOSAN

Modification of chitin and chitosan provides a powerful means to promote new biological activities and to modify their mechanical properties. The major limiting factors of chitin and chitosan for

TABLE 25.1
Physicochemical Characteristics and Their Methods of Determination

Physicochemical Characteristics	Determination Method
Deacetylation degree	Infrared spectroscopy (Brugnerotto et al. 2001; Kassai 2008)
	UV spectrophotometry (Wu and Zivanovic 2008)
	Nuclear magnetic resonance spectroscopy (^{1}HNMR) and (^{13}CNMR) (Varum et al. 1991; Raymond et al. 1993; Van de Velde and Kiekens 2004)
	Conductometric titration (Raymond et al. 1993)
	Potentiometric titration (Jiang et al. 2003)
	Differential scanning calorimetry (Guinesi and Cavalheiro 2006)
Average MW and/or MW distribution	Viscometry (Rinaudo et al. 1993; Schipper et al. 1996; Sabnis and Block 2000; Berth and Dautzenberg 2002)
	Gel permeation chromatography (Chen et al. 1996; Kumar 2000; Brugnerotto et al. 2001; Pochanavanich and Suntornsuk 2002)
	Light scattering (Terbojevich et al. 1992; Fee et al. 2003)
	High-performance liquid chromatography (Wu et al. 1976)
Crystallinity	X-ray diffraction (Yen et al. 2009)

their application are poor solubility, low surface area, and porosity. Therefore, the modification of chitin and chitosan became essential owing to their limitations and also to improve their properties. Modification can be done by physical or chemical processes to improve the mechanical and chemical properties. The heteropolymer chitin/chitosan has three types of reactive functional groups: an amino/acetamido group as well as both primary and secondary hydroxyl groups at the C-2, C-3, and C-6 positions, respectively.

Many novel chitin and chitosan derivatives have been obtained by chemical modification using the reactive activities of hydroxyl and amino/acetamido groups (Rinaudo and Reguant 2000). In particular, the amino contents mainly contribute in changing the structure and physicochemical properties and their distribution is random, which makes it easy to generate intra- and intermolecular hydrogen bonds (Hudson and Jenkins 2003). Also, the amino group has nucleophilic property, allowing the easy formation of imine by reaction with aldehyde or corresponding amide derivatives in acylating reagents; in acidic solution, the amino groups showed alkaline properties and receive protons to generate salts, presenting cationic polymer properties. Besides, the amino functional group has also been correlated with chelation, flocculation, and biological functions.

25.5.1 Physical Modifications

The drug delivery process depends on physicochemical properties, such as particle size, shape, porosity, surface energy, and area. Since chitin and chitosan, when used as they are, stick up with limitations, physically modified chitosan is used in different fields of application as powder, flakes, gels, beads, membranes, and fibers (Guibal 2005; Kas 1997; Merrifield et al. 2004; Krajewska 2005). Molecular shape and size are the prime parameters in the diffusion process. Chitosan has been used in preparing films, beads, intragastric floating tablets, microspheres, and nanoparticles in the pharmaceutical field (Berthold et al. 1996; Calvo et al. 1997; Felt et al. 1998; Illum 1998). Chitosan modified into beads and microgranules was used as a drug carrier for disodium diclofenac (Gupta and Kumar 2000) and for 1,3-thiazolidine linker (Liu et al. 2003).

25.5.2 Chemical Modifications

Chitin and chitosan are interesting polysaccharides because of the presence of the amino functionality, which could be suitably modified to impart desired properties and distinctive biological functions, including solubility (Muzzarelli 1977; Roberts 1992; Kurita 2001; Hudson and Smith 1998; Peter 2002; Tharanathan and Kittur 2003; Rinaudo 2006; Campana-Filho 2007). The chemical modification alters the molecular structure and improves the physicochemical properties of chitin and chitosan. It would not change the fundamental skeleton of chitosan and would retain the original physicochemical and biochemical properties while bringing new or improved properties.

The chemical modification is achieved to provide the derivatives of chitin and chitosan that are soluble at neutral and basic pH values, to control the hydrophobic nature and the cationic and anionic properties, as well as to attach various functional groups and ligands. Fortunately, chitosan is amenable to chemical modifications due to having hydroxyl, acetamido, and amine functional groups. Apart from the amino groups, they have two hydroxyl functionalities for effecting appropriate chemical modifications to enhance solubility (Dumitriu 1996; Desai et al. 2004).

Chemically modified chitosan structures results in improved solubility in water and general organic solvents have been reported by some researchers (Suh and Matthew 2000; Kubota et al. 2000; Kumar 2000; Xie et al. 2002; Kim and Rajapakse 2005).

The amine and alcohol functional groups of chitosan are easily modified by many organic reactions such as tosylation (Kurita et al. 1991), alkylation (Kurita et al. 1990), carbocylation (Muzzaralli et al. 1982), sulfonation (Terbojevich et al. 1989), Schiff base reaction (Moore and Roberts 1982), and quaternary salt reaction (Muzzaralli and Tanfani 1985). Trimethylated chitosans (TMCs) were prepared by Kean and Roth (2005) to produce more permanent cationic charges on the chitosan's chains. TMCs were evaluated for the effect of the quaternization on cytotoxicity and transfection efficiency.

Grafting of chitosan allows the formation of functional derivatives by covalent binding of a molecule, the graft, onto the chitosan backbone (Jayakumar et al. 2005). Most recently, Tanodekaew and Prasitsilp (2004) reported the preparation of acrylic grafted chitin for wound-dressing application. Various studies were conducted to make water-soluble derivatives of chitosan by chemical modification techniques, such as polyethylene glycol (PEG) grafting (Ouchi et al. 1998; Sugimoto et al. 1998), sulfonation (Suh and Matthew 2000), partial N-acetylation (Kubota et al. 2000), N-acetylation (Kumar 2000), using chitosan-carrying phosphonic and alkyl groups (Ramos et al. 2003), using hydroxypropyl chitosan (Xie et al. 2002), branching with oligosaccharides (Tommeraas et al. 2002), using chitosan saccharide derivatives (Yang and Chou 2002; Chung et al. 2005), using O-succinyl chitosan (Zhang and Ping 2003), quaternization (Snyman et al. 2002), and carboxymethylation of chitosan (Chen and Park 2003).

PEG-grafted chitosan (chitosan-g-PEG) was synthesized by Jiang et al. (2006) utilizing the reaction between methoxy PEG-nitrophenol carbonate and chitosan. The amine group of chitosan is modified using many chemical methods, including O- and N-carboxymethyl chitosans, chitosan 6-O-sulfate (Terbojevich et al. 1989; Naggi et al. 1986), and more recently as N-sulfated chitosan (Holme and Perlin 1997).

Polymer blending is a well-used technique wherein a macroscopically homogeneous mixture of two or more different species of polymer is obtained. Yuvarani et al. (2012) prepared curcumin-coated chitosan–alginate blend for wound-dressing application. Govindarajan et al. (2011) prepared the blends of nanochitosan with carboxymethyl cellulose to improve the film-forming capacities of nanochitosan. Recently, chitosan was used as a pore former in coated beads for colon-specific drug delivery of 5-aminosalicylic acid (5-ASA) (Omwancha et al. 2013), and the carboxymethyl chitosan hydrogel beads prepared in 30% alcohol–aqueous binary solvent may be a promising delivery system for hydrophobic nutrients or drugs (Yangchao Luo et al. 2013).

25.5.3 ENZYMATIC MODIFICATION

The enzymatic approach to the modification of chitin and chitosan is also better owing to their specificity and environmental advantages compared with chemical modification (Payne et al. 1996; Chen et al. 2000; Wu et al. 2001). The enzymes offer safety and remove the hazards associated with reagents. Payne et al. (1996) reported enzymatic grafting of phenolic compounds onto chitosan to confer water solubility under basic conditions (Shao et al. 1999). Chen et al. (2000) grafted hexyloxyphenol to chitosan by tyrosinase. This enzymatic approach uses tyrosinase for the conversion of phenols into reactive o-quinones, and subsequent reaction of these o-quinones with chitosan to form both Schiff bases and Michael-type adducts (Shao et al. 1999). Partially hydrolyzed chitosan by enzymatic methods seems to have better biochemical significance compared to the chitosan from which they derive, and that are handled very easily (Ilyina et al. 1999; Zhang et al. 1999; Muzzarelli and Muzzarelli 2002). Efforts to improve chitosan solubility continued and this in turn increases its applications, while homogeneously modified chitosan exhibited rheological properties characteristic of associating water-soluble polymers.

25.6 APPLICATIONS OF CHITIN/CHITOSAN FOR DRUG DELIVERY APPLICATIONS

Chitin and chitosan are multipurpose biopolymers, and along with their derivatives have shown a variety of functional properties in various fields, such as medicine, agriculture, textiles, paper, cosmetics, water treatment, food, and pharmaceuticals (Kim and Rajapakse 2005). These functional properties have attracted considerable interest in the pharmaceutical area such as to inhibit fibroplasias in wound healing and to promote tissue growth, differentiation in tissue culture (Le et al. 1997), and drug delivery application (Dodane and Vilivalam 1998).

Being cationic in nature, chitosan entraps lipids and binds with cholesterol in cholesterol-lowering and slimming formulations (Shahidi et al. 1999; Kanauchi et al. 1995; Wuolijoki et al. 1995). Human enzymes, especially lysozyme, metabolize chitosan due to its positive charges at physiological pH. Chitosan is also bioadhesive, which increases retention at the site of application (Berger et al. 2004).

Chitosan has received considerable attention as a possible pharmaceutical excipient in recent decades. Also, it has recently been approved by the authorities, and a monograph relating to chitosan hydrochloride was included in the fourth edition of the *European Pharmacopoeia* (2002). Chitosan has been used extensively for oral and intranasal delivery and for parenteral drug delivery (Gomez and Duncan 1997). It acts as a versatile carrier for biologically active species and drugs due to the presence of free amino groups as well as its low toxicity. Gomez and Duncan (1997) reported that chitosan polymers when used as soluble polymeric carriers for intravenous administration show the potential to induce cellular toxicity.

25.6.1 DRUG DELIVERY SYSTEM AND ANTIMICROBIAL ACTIVITY

Across the world, increasing number of researchers are currently investigating chitosan in micro and nano forms as a controlled delivery device for hormones, vitamins, proteins, and enzymes (Cheng et al. 2005; Grenha et al. 2005; Krishnamchari et al. 2007). Chitosan and its derivatives have been used as a gene carrier owing to their less toxic nature (Lee et al. 1998) and also as a carrier for directly compressed tablets (Kristmundsdottir et al. 1995). They have been used as a blood anticoagulant as well as hypocholesterolemic agents (Agnihotri et al. 2004).

Currently, chitosan-based micro- and nanoparticles play a vital role in sustained and targeted drug delivery (Dev et al. 2010; Sanoj Rejinold et al. 2011; Jayakumar et al. 2011). Controlled release systems showed more advantages when comparing with conventional dosage forms. They reduce the side effects and prolong the efficacy of the drug molecule. This controlled delivery system regulates the release rate of drugs and reduces the regularity of administration of the drug, thus promising better patient compliance. Chitosan–apatite cement delivery system shows promise in the control of infection for skeletal-based surgeries (Takechi et al. 2002). Hassan and Gallo (1993) prepared magnetic chitosan microspheres containing oxantrazole (MCM-OX), an anticancer drug, for the treatment of brain tumors.

25.6.1.1 Antimicrobial Properties

The progress toward the antibacterial activity of chitosan was more and confirmed that chitosan was an effective agent for the inhibition of bacteria. The antimicrobial properties of chitosan mainly depend on the MW of chitosan and the type of bacteria that is chosen for the study. Chitosan shows stronger bactericidal effects for Gram-positive bacteria (*Listeria monocytogenes, Bacillus megaterium, B. cereus, Staphylococcus aureus, Lactobacillus plantarum, L. brevis,* and *L. bulgaris*) than for Gram-negative bacteria (*E. coli, Pseudomonas fluorescens, Salmonella typhimurium,* and *Vibrio parahaemolyticus*) (No et al. 2002). Tsai and Su (1999) reported in an extensive research on the antimicrobial activity of chitosan prepared from shrimp against *E. coli* that higher temperature and acidic pH of foods increased the bactericidal effect of chitosan. They also explained the mechanism of chitosan antibacterial action involving a cross-linkage between polycations of chitosan and the anions on the bacterial surface that changes membrane permeability. No et al. (2002) reported that the chitosan shows higher antibacterial activity character at lower pH ranges that reveals that the addition of chitosan to acidic foods will improve its effectiveness as a natural preservative.

25.6.2 DRUG RELEASE AND RELEASE KINETICS

The release of drug from chitosan-based dosage form depends on the morphology, size, density, and extent of cross-linking of the particulate system, physicochemical properties of the drug, as well as the polymer characteristics such as it is either hydrophilic or hydrophobic, gel formation ability, swelling capacity, mucoadhesive or bioadhesive properties, and also on the presence of other excipient present in the dosage form. The release of drug from particulate systems involves three different mechanisms: (a) erosion, (b) diffusion, (c) and release from the surface of the particle. The release of the drug mostly follows more than one type of mechanism. The burst release of drug can be prevented by the use of cross-linking agents or by washing microparticles with a suitable solvent. Al-Helw et al. (1998) explained a high release of the phenobarbitone in the initial hours and the drug release rate was dependent on the MW of chitosan and particle size of the microspheres. Slow drug release profile was observed when the microspheres were prepared from high-molecular-weight chitosan as compared to those prepared from low-molecular-weight chitosan. Kweon and Kang (1999) prepared the CS-g-poly(vinyl alcohol) copolymer matrix to study the release pattern of prednisolone under various conditions. In this study, drug releases was controlled by the extent of polyvinyl alcohol (PVA) grafting heat treatment or cross-link density and showed that there was a linear relationship between the amount of drug release and square root of time, indicating that release was based on the diffusion mechanism. Recently, the development and *in vitro* evaluation of coated pellets which containing chitosan to potential colonic drug delivery (Ferrari et al. 2013).

25.6.3 ROLE OF CHITIN/CHITOSAN IN VARIOUS DRUG DOSE FORMS

Tozaki et al. (2002) reported that chitosan capsules containing 5-ASA for colon-specific delivery to treat ulcerative colitis. It was observed that chitosan capsules collapsed unambiguously in the large intestine as compared to the control formulation (in the absence of CS), which confirmed the absorption of the drug in the small intestine. El-kamel et al. (2006) prepared chitosan-based vaginal tablets containing metronidazole by directly compressing the natural cationic polymer chitosan, loosely cross-linked with glutaraldehyde. The batch containing 6% chitosan, 24% sodium alginate, 30% sodium carboxymethyl cellulose, and 20% montmorillonite clay (MCC) explained passable release properties in both media and gave lower values of swelling index compared with the other formulations. It also proved to have good adhesion properties with minimum applied weights. Moreover, its release properties (% dissolution efficiency, DE) in buffer pH 4.8, as well as release mechanism (n values), were negligibly affected by aging. Thus, this formula may be considered a good candidate for vaginal mucoadhesive dosage forms.

25.6.4 ROLE OF CHITIN/CHITOSAN TARGETED DRUG DELIVERY

Drug delivery, mainly targeted delivery of drugs, is very important in improving therapeutic efficacy and minimizing side effects. Chitosan has been used as a potential carrier for prolonged delivery of drugs, macromolecules, and targeted drug delivery. The targeted drug delivery system comprises three components: a therapeutic agent, a targeting moiety, and a carrier system. The drug can be incorporated by either passive absorption or chemical conjugation into the carrier system. The pharmacokinetics and pharmacodynamics of the drug mainly depends on the choice of the carrier molecule.

Chitosan is a well-accepted and a promising polymer for drug delivery in colonic part because it can be biodegraded by the microflora present in the human colon. Lorenzo-Lamosa et al. (1998) proposed the design of microencapsulated chitosan microspheres microcores entrapped in enteric acrylic microspheres for colonic drug delivery.

25.6.5 Gene Delivery

Chitosan plays the most prominent role among natural polymer-based gene delivery vectors (Prabaharan and Mano 2005; Mao and Sun 2010). The cationic nature of chitosan under slightly acidic conditions forms a complex with DNA or siRNA through electrostatic interactions (Duceppe et al. 2010; Leong et al. 1998; Mao et al. 2003). The effect of chitosan's MW on gene delivery systems interplays in the complex formation with DNA. For DNA delivery and to ensure high transfection level, chitosan should have an MW of between 10 and 50 kDa (Mao and Sun 2010). Complexation of DNA with low-molecular-weight highly deacetylated chitosan (5 kDa, 99%) has also been reported (Liu et al. 2005). In this type of application, it is necessary to enhance the transfection efficiency of DNAs or siRNAs to target cells. In addition, the biocompatibility and low toxicity of chitosan enable its in vivo use (Han and Mangala 2010). Chew et al. (2003) reported that the chitosan–DNA nanoparticles have also been successfully used to generate and provide immune response to the dust mite allergen (Van Der Lubben et al. 2003).

Zhi-Hong et al. (2006) developed chitosan nanoparticles for effective gene delivery application. The ternary nanoparticles in which pDNA/chitosan particles are coated with hyaluronic acid (HA) is known to specifically bind CD44.

25.6.6 Role of Chitosan in pH Responsive Drug Delivery

pH is an important parameter that changes the physical and chemical properties of polymers, such as swelling and solubility, based on local pH levels (Gupta and Kumar 2001; Qu et al. 2000). Temperature and pH conditions have been comprehensively considered in the biomedical field because these two factors can be easily controlled and are applicable in both *in vitro* and *in vivo* conditions (Matsusaki and Akashi 2005; Lo et al. 2005; Ju et al. 2001). pH-responsive surfaces have been developed and applied to various fields such as drug delivery (Iwata et al. 1998; Schmaljohann 2006). Nanomedicines are more recently included in the pH responsive mechanism of drug release to improve the systemic exposure from greater gastric retention, transepithelial transport, and cellular targeting with surface-functionalized ligands (Colombo et al. 2009; Roger et al. 2010).

pH-dependent hydrophobic-to-hydrophilic transitions may also be used to control polymer dissolution, in which the polymer matrix collapses for drug release. Dias et al. (2008) developed the pH-responsive surfaces for the simultaneous control and analysis of the production of biomimetic apatite and found that the formation of biomimetic apatite was mainly dependent on the conformational changes of chitosan across its critical pH. In the case of bovine blood proteins release, due pH-dependent swelling release was observed very quickly at pH 7.4; in the same case at pH 1.2, it was very slow.

Oral delivery is an attractive drug delivery route for its convenience, patient compliance, and cost-effectiveness. But orally delivered drugs end up with poor systematic exposure. Therefore, it is become a challenging task to achieve sufficient and reliable bioavailability levels for orally administered drugs (Delie and Blanco-Príeto 2005; Morishita and Pappes 2006; Yamanaka and Leong 2008; Sarmento et al. 2007).

25.6.7 Modified Drug Delivery

Chitin and chitosan can be modified in different forms as a new drug delivery system (Table 25.2).

25.6.8 Chitosan for Controlled Drug Delivery

Currently, it is well known that chitosan drug delivery systems offer suitable means of site-specific and/or time-controlled delivery and for that, among various kinds of polymeric systems, chitosans are now widely used as drug containers or release rate controlling barriers. Controlled drug delivery systems have been used to overcome the shortcomings of conventional drug formulations.

TABLE 25.2
Different Types of Chitosan-Based Drug Delivery Systems

Capsules (Capsule Shell)	Film (Solution Casting)	Tablets (Matrix)
Microspheres/microparticles (spray drying, coacervation)	Gel (cross-linking)	Nanoparticle (ionic gelation)

These systems may have the ability to maintain therapeutic concentrations at a target site, reducing the chance for toxicity and collateral damage, and enhancing the drug efficiency (Hoffman 2002). Good film forming and mucoadhesive properties of chitosan make it a suitable coating material in drug delivery applications. Chitosan-coated particulate systems show more advantages such as controlled and prolonged delivery of the drug molecules. Jitendra Kawadkar and Ram (2007) observed the increase of drug loadings as well as bioadhesive properties compared to uncoated particles. Jitendra Kawadkar et al. (2007) reported the preparation of CS-coated microsphere system to treat ulcerative colitis-A, and observed better release of 5-ASA in the colon having ulcerative colitis. Chiou et al. (2001) studied the controlled release of drug from CS-coated polylactic acid (PLLA) microspheres. Chitosan with different MW was used for coating the microspheres to improve the controlled drug release. Shu and Zhu (2000) prepared alginate beads coated with CS to prepare a novel system for sustained drug delivery. Chitosan hydrogels were prepared by grafting chitosan with PEG groups and photo-cross-linkable groups and prepared hydrogels used as an antimicrobial water shield coating agent.

25.6.9 New Drug Delivery Systems Based on Chitin/Chitosan

Although many types of vectors have been developed so far, our focus will only be on chitosan-based delivery systems in the form of chitosan gels and their derivatives (Zarzycki et al. 2008). A targeted and controlled-release system of biologically active substances has gained importance in modern pharmacy and is in the main trend of intensive studies on the preparation of new forms of drugs characterized by higher efficiency and limited side effects (Upadrashta et al. 1992; Knapczyk 1993; Steckel and Mindermann-Nogly 2004; Jess 2007).

Prabaharan and Jayakumar (2009) developed a novel chitosan-g-β-cyclodextrin (chit-g-β-CD) scaffold by a freeze-drying method and used it as a matrix for drug loading and controlled release. This biodegradable scaffold provided a slower release of the entrapped ketoprofen than the chitosan scaffold, which gives a healthy environment and enhances the surrounding tissue regeneration. O-Carboxymethyl chitosan nanocarrier was reported by Anitha et al. (2011) for curcumin delivery. It acts as a promising nanomatrix for drug delivery applications. A nanoformulation of 5-fluorouracil (5-FU) with biodegradable thermoresponsive chitosan-g-poly(N-vinylcaprolactam) biopolymer composite was prepared by ionic cross-linking for its delivery to cancer cells (Sanoj Rejinold et al. 2011). The drug-loaded nanoparticles showed comparatively higher toxicity to cancer cells while being less toxic to normal cells. The results indicated that this could be a promising candidate for cancer drug delivery. Sanoj Rejinold et al. (2011) also reported chitosan for saponin delivery in nano form, for efficient therapeutic agent.

25.7 FUTURE PROSPECTS

There is a considerable promising future for the development and expansion of the innovative chitin and chitosan derivatives for drug delivery applications. Because targeted and sustained delivery is considered as the most required property in the pharmaceutical field, these biopolymers and their derivatives can be increasingly utilized for new drug delivery route for both hydrophilic and hydrophobic drugs.

25.8 CONCLUSION

During the last decade, increasing attention has been paid on chitin and chitosan toward the development of systems to deliver drugs for long periods at controlled rates. The sustainability of these materials in terms of their unique physicochemical characters and bioactivities leads to enable the development of a wide range of fit-for-purpose products and applications. According to the growing number of recently published scientific articles, it can be deduced that chitin and chitosan is a prospective material of the extensive potential for various applications. However, little effort has been given to developing systems for the controlled release of nucleic acids.

REFERENCES

Agnihotri, S.A., N.N. Mallikarjuna, and T.M. Aminabhavi. 2004. Recent advances on chitosan-based micro- and nanoparticles in drug delivery. *Control Release* 100:5–28.

Al Sagheer, F.A., M.A. Al-Sughayer, S. Muslim, and M. Elsabee. 2009. Extraction and characterization of chitin and chitosan from marine sources in Arabian Gulf. *Carbohydrate Polymers* 77:410–419.

Al-Helw, A.A., A.A. Al-Angary, G.M. Mahrous, and M.M. Al-Dardari. 1998. Preparation and evaluation of sustained release cross-linked chitosan microspheres containing phenobarbitone. *Journals of Microencapsulation* 15:373–382.

Anitha, A., S. Maya, N. Deepa, K.P. Chennazhi, S.V. Nair, and R. Jayakumar. 2011. Curcumin-loaded N,O-carboxymethyl chitosan nanoparticles for cancer drug delivery. *Journal of Biomaterial Science*. Polymer Edition [Epub ahead of print].

Berger, J., M. Reist, J.M. Mayer, O. Felt, and R. Gurny. 2004. Structure and interactions in chitosan hydrogels formed by complexation or aggregation for biomedical applications. *European Journal of Pharmaceutics and Biopharmaceutics* 57:35.

Bernkop-Schnürch, A., D. Guggi, and Pinter, Y. 2004. Thiolated chitosans: Development and *in vitro* evaluation of a mucoadhesive, permeation enhancing oral drug delivery system. *Controlled Release* 94:177–186.

Berth, G. and H. Dautzenberg. 2002. The degree of acetylation of chitosans and its effect on the chain conformation in aqueous solution. *Carbohydrate Polymers* 47:39–51.

Berthold, A., K. Cremer, and J. Kreutzer. 1996. Preparation and characterization of chitosan microspheres as drug carrier for prednisolone sodium phosphate as model for anti-inflammatory drugs. *Journal of Controlled Release* 39:17–25.

Brugnerotto, J., J. Desbrieres, G. Roberts, and M. Rinaudo. 2001. Characterization of chitosan by steric exclusion chromatography. *Polymer* 42:09921–09927.

Calvo, P., C. Remunan-Lopez, J.L. Vila-Jato, and M.J. Alonso. 1997. Novel hydrophilic chitosan-polyethylene oxide nanoparticles as protein carriers. *Journal of Applied Polymer Science* 63:125–132.

Campana-filho, S.P. 2007. Extração, estruturas e propriedades de α- e β-quitina. *Quím. Nova* 30:644-650.

Carreno-Gomez, B. and R. Duncan. 1997. Evaluation of the biological properties of soluble chitosan and chitosan microspheres. *International Journal of Pharmaceutics* 148:231–240.

Chen Rong Huei, Horng-Dar Hwa. 1996. Effect of molecular weight of chitosan with the same degree of deacetylation on the thermal, mechanical, and permeability properties of the prepared membrane. *Carbohydrate Polymers* 29(4):353–358.

Chen, T., G. Kumar, M.T. Harris, P.J. Smith, and G.F. Payne. 2000. Graft copolymerized chitosan-present status and applications. *Biotechnology and Bioengineering* 70:564–573.

Chen, X.G. and H.J. Park. 2003. Chemical characteristics of O-carboxymethyl chitosans related to the preparation conditions. *Carbohydrate Polymers* 53(4):355–359.

Chen, Y.L. 2008. Preparation and characterization of water soluble chitosan gel for skin hydration. MPH Thesis, University Sains Malaysia. 1–181.

Cheng, Y.H., A.M. Dyer, and I. Jabbal Gill. 2005. Intranasal delivery of recombinant human growth hormone (somatropin) in sheep using chitosan based powder formulations. *European Journal of Pharmaceutical Sciences* 26:9–15.

Chew, J.L., C.B. Wolfowicz, H.-Q. Mao, K.W. Leong, and K.Y. Chua. 2003. Chitosan nanoparticles containing plasmid DNA encoding house dust mite allergen, Der p 1 for oral vaccination in mice. *Vaccine* 21:2720–2729.

Chiou, S.H., W.T. Wu, Y.Y. Huang, and T.W. Chung. 2001. Effects of the characteristics of chitosan on controlling drug release of chitosan coated PLLA microspheres. *Journals of Microencapsulation* 18:613–625.

Chung, Y.C., C.L. Kuo, and C.C Chen. 2005. Preparation and important functional properties of water-soluble chitosan produced through Maillard reaction. *Bioresource Technology*. 96:1473–1482.

Cohen, E. 1993. Chitin synthase: An enzyme and a selective target for pesticide action. In: Mitasui, T., Matsumara, F., and Yamaguchi, I. (eds.), *Pesticides/Environment: Molecular Biological Approaches*. Pesticide Science Society of Japan, Tokyo, 101–113.

Colombo, P., F. Sonvico, G. Colombo, and R. Bettini. 2009. Novel platforms for oral drug delivery. *Pharmaceutical Research* 26:601–611.

Crini G., E. Guibal., M. Morcellet., G. Torri., and P.M. Badot. 2009. Chitine et chitosane. Préparation, propriétés et principales applications. In: chitine et chitosane. Du biopolymère à l'application, 1st Ed., Presses universitaires de Franche-Comté, France.19–54.

Dash, M., F. Chiellini, R.M. Ottenbrite, and E. Chiellini. 2011. Chitosan-A versatile semi-synthetic polymer in biomedical applications. *Progress in Polymer Science* 36:981–1014.

Delie, F. and M.J. Blanco-Príeto. 2005. Polymeric particulates to improve oral bioavailability of peptide. *Molecules* 10: 65–80.

Desai, T.A., T. West, M. Cohen, T. Boiarski, and A. Rampersaud, 2004. Nanoporous microsystems for islet cell replacement. *Advanced Drug Delivery Reviews* 56:1661–1673.

Dev, A., N.S. Binulal, A. Anitha, S.V. Nair, T. Furuike, H. Tamura, and R. Jayakumar. 2010. Preparation of poly (lactic acid)/chitosan nanoparticles for anti-HIV drug delivery applications. *Carbohydrate Polymers* 80:833–838.

Dias, C. I., J. F. Mano, and N.M. Alves. 2008. pH-Responsive biomineralization onto chitosan grafted biodegradable substrates. *Journal of Materials Chemistry* 18:2493–2499.

Dodane, V. and V.D. Vilivalam. 1998. Pharmaceutical applications of chitosan. *Pharmaceutical Science and Technology Today* 1:246–253.

Duceppe, M.T. 2010. Advances in using chitosan-based nanoparticles for *in vitro* and *in vivo* drug and gene delivery. *Expert Opinion on Drug Delivery* 7:1191–1207.

Dumitriu, S, editor. 1996. *Polysaccharides in Medicinal Application*. Marcel Dekker, New York.

Dutta, P.K., J. Dutta, and V.S. Tripathi. 2004. Chitin and chitosan: Chemistry, properties and application. *Journal of Scientific and Industrial Research* 63:20–31.

El-Kamel, A. H., D. H. Al-Shora, and Y. M. El-Sayed. 2006. Formulation and pharmacodynamic evaluation of captopril sustained release microparticles. *Journal of Microencapsulation* 23(4):389–404.

Fee, M., N. Errington, K. Jumel, L. Illum, A. Smith, and S.E. Harding. 2003. Correlation of SEC/MALLS with ultracentrifuge and viscometric data for chitosans. *European Biophysical Journal* 32:457–464.

Felt, O., P. Buri, and R. Gurny. 1998. Chitosan: A unique polysaccharide for drug delivery. *Journal of Drug Development and Industrial Pharmacy* 24:979–993.

Ferrari, P.C., F.M. Souza, L. Giorgetti, G.F. Oliveira, H.G. Ferraz, M.V. Chaud, and R.C. Evangelista. 2013. Development and in vitro evaluation of coated pellets containing chitosan to potential colonic drug delivery. *Carbohydrate Polymers* 91:244–252.

Govindarajan, C., S. Ramasubramaniam, T. Gomathi, A. Narmadha Devi, and P.N. Sudha. 2011. Sorption studies of Cr (VI) from aqueous solution using nanochitosan-carboxymethyl cellulose blend. *Archives of Applied Science Research* 3:127–138.

Grenha, A., B. Seijo, and C. Remunan-Lopez. 2005. Microencapsulated chitosan nanoparticles for lung protein delivery. *European Journal of Pharmaceutical Sciences* 25:427–437.

Guibal, E., E. Touraud, and J. Roussy. 2005. Chitosan interactions with metal ions and dyes: Dissolved-state vs solid-state application. *World Journal of Microbiology and Biotechnology* 21:913–920.

Guinesi, L. and E. Cavalheiro 2006. The use of DSC curves to determine the acetylation degree of chitin/chitosan samples. *Thermochimica Acta* 444:128–133.

Gupta, K.C., M.N.V.R. Kumar. 2000. Drug release behavior of beads and microgranules of chitosan. *Biomaterials* 21:1115–1119.

Gupta, K.C. and M.N.V.R. Kumar. 2001. pH dependent hydrolysis and drug release behavior of chitosan/poly(ethylene glycol) polymer network microspheres. *Journal of Materials Science: Materials in Medicine* 12:753–759.

Han, H.D. and L.S., Mangala. 2010. Targeted gene silencing using RGD-labeled chitosan nanoparticles. *Clinical Cancer Research* 16:3910–3922.

Hassan, E.E. and J.M. Gallo. 1993. Targeting anticancer drugs to the brain. I: Enhanced brain delivery of oxantrazole following administration in magnetic cationic microspheres. *Journal of Drug Targeting* 1:7–14.

Hoffman, A.S. 2002. Hydrogels for biomedical applications. *Advanced Drug Delivery Reviews* 43:3–17.

Holme, K.R. and A.S. Perlin. 1997. Chitosan N-sulfate. A water-soluble polyelectrolyte. *Carbohydrate Research* 302:7–12.

Hudson, S.M. and D.W. Jenkins. 2003. Chitin and chitosan. In: Mark, H.F. (ed.), *EPST* vol.1 3rd ed. Wiley, New York, 569–580.

Hudson, S.M. and C. Smith. 1998. Polysaccharide: Chitin and chitosan: Chemistry and technology of their use as structural materials. In: Kaplan, D.L., (ed.), *Biopolymers from Renewable Resources*. Springer-Verlag, Newyork. 96–168.

Illum, L.1998. Chitosan and its use as a pharmaceutical excipient. *Pharmaceutical Research* 15:1326–1331.

Ilyina, A.V., N.Y. Tatarinova, and V.P. Varlamov. 1999. The preparation of low-molecular-weight chitosan using chitinolytic complex from Streptomyces kurssanovii. *Process Biochemistry* 34:875–878.

Inmaculada, A., M. Mengíbar, R. Harris, I. Paños, B. Miralles, N. Acosta, G. Galed, and Á. Heras, 2009. Functional characterization of chitin and chitosan. *Current Chemical Biology* 3:203–230.

Iwata, H., I. Hirata, and Y. Ikada. 1998. Atomic force microscopic analysis of a porous membrane with pH-sensitive molecular valves. *Macromolecules* 31:3671.

Jayakumar, R., M. Prabaharan, R.L. Reis, and J.F. Mano. 2005. Graft copolymerized chitosan-present status and applications. *Carbohydrate Polymers* 62(2):142–158.

Jayakumar, R., M. Prabaharan, P.T. Sudheesh Kumar, S.V. Nair, and H. Tamura. 2011. Biomaterials based on chitin and chitosan in wound dressing applications. *Biotechnology Advances* 29:322–337.

Jess, K. and H. Steckel. 2007. The extrusion and spheronization of chitosan. *Pharmaceutical Technology Europe* 19:21.

Jiang, G.B., D. Quan, K. Liao, and Wang, H. 2006. Preparation of polymeric micelles based on chitosan bearing a small amount of highly hydrophobic groups. *Carbohydrate Polymers* 66:514–520.

Jiang, X., L. Chen, and W. Zhong. 2003. A new linear potentiometric titration method for the determination of deacetylation degree of chitosan. *Carbohydrate Polymers* 54:457–63.

Jitendra Kawadkar, J. and A. Ram. 2007. Colon targeted chitosan microsphere compressed matrices for the treatment of ulcerative colitis. *Pharmaceutical Information* 5:4.

Johnson, E.L. and Peniston, Q.P. 1982. Utilization of shell fish waste for chitin, chitosan production. In: Martin, R.E., Flick, G.J., Hebard, C.E., and Ward, D.R., (eds.), *Chemistry and Bio-chemistry of Marine Food Products*. AVI Publishing Co, Westport, CT, pp. 415–428.

Ju Hee Kyung, So Yeon Kim, Young Moo Lee. 2001. pH/temperature-responsive behaviors of semi-IPN and comb-type graft hydrogels composed of alginate and poly (N-isopropylacrylamide). *Polymer* 42:6851–6857.

Kanauchi, O., K. Deuchi, Y. Imasato, M. Shizukuishi, and E. kobayashi. 1995. Mechanism for the inhibition of fat digestion by chitosan and for the synergistic effect of ascorbate. *Journal of Bioscience, Biotechnology, and Biochemistry* 59:786–790.

Kas, H.S. 1997. Chitosan: Properties, preparations and application to microparticulate systems. *Journal of Microencapsulation* 14:689–711.

Kassai, M. 2008. A review of several reported procedures to determine the degree of N-acetylation for chitin and chitosan using infrared spectroscopy. *Carbohydrate Polymers* 71:497–508.

Kast, C.E., W. Frick, U. Losert, and A. Bernkop-Schnürch. 2003. Chitosan-thioglycolic acid conjugate: A new scaffold material for tissue engineering? *International Journal of Pharmaceutics*. 256:183.

Kean, T. and S. Roth. 2005. Trimethylated chitosans as non-viral gene delivery vectors: Cytotoxicity and transfection efficiency. *Control Release* 103:643–653.

Khan, M.M., D.R. Evans, V. Gunna, R.E. Scheffer, V.V. Parikh, and S.P. Mahadik. 2002. Reduced erythrocyte membrane essential fatty acids and increased lipid peroxides in schizophrenia at the never-medicated first-episode of psychosis and after years of treatment with antipsychotics. *Schizophrenia Research* 58:1–10.

Khor, E. and Lim, L.Y. 2003. Implantable applications of chitin and chitosan. *Biomaterials* 24:2339–2349.

Kim, S.K. and N. Rajapakse. 2005. Enzymatic production and biological activities of chitosan oligosaccharides (COS): A review. *Carbohydrate Polymers* 62:357–368.

Knapczyk, J. 1993. Chitosan hydrogel as a base for semisolid drug forms. *International Journal of Pharmaceutics* 93:233.

Knaul, J.Z., S.M. Hudson, and K.A.M. Creber. 1999. Improved mechanical properties of chitosan fibers. *Journal of Applied Polymer Science* 72:1721–1732.

Krajewska, B. 2004. Application of chitin- and chitosan-based materials for enzyme immobilization: A review. *Enzyme and Microbial Technology* 35:126–139.

Krajewska, B. 2005. Membrane-based processes performed with use of chitin/chitosan materials. *Separation and Purification Technology* 41:305–312.

Krishnamchari, Y., M. Parshotam, and L. Senshang. 2007. Development of pH and time-dependent oral microparticles to optimize budesonide delivery to ileum and colon. *International Journal of Pharmacy* 338:238–247.

Kristmundsdottir, T., K. Ingvarsdottir, and G. Saemundsdottir. 1995. Chitosan matrix tablets: The influence of excipients on drug release. *Drug Development and Industrial Pharmacy* 21:1591–1598.

Kubota, N., N. Tastumoto, T. Sano, and K. Toya. 2000. A simple preparation of half N-acetylated chitosan highly soluble in water and aqueous organic solvents. *Carbohydrate Research* 324:268–274.

Kumar, M.N.V.R. 2000. A review of chitin and chitosan applications. *Reaction Function Polymer* 46:1–27.

Kurita, K. 2001. Controlled functionalization of the polysaccharide chitin. *Progress in Polymer Science* 26:1921–1971.

Kurita, K., Y. Koyama, S. Inoue, and S. I. Nishimura.1990. (Diethylamino) ethyl chitins: Preparation and properties of novel animated chitin derivatives. *Macromolecules* 23:2865–2869.

Kurita, K., S. Inoue, and S. I. Nishimura. 1991. Preparation of soluble chitin derivatives as reactive precursors for controlled modifications. Tosyl- and iodo-chitins. *Journal of Polymer Science* 29(6):937–939.

Kweon, D.K. and D.K. Kang, 1999. Drug-release behavior of chitosan-g-poly (vinyl alcohol) copolymer matrix. *Journal of Applied Polymer Science* 74:458–464.

Le, Y., S.C. Anand, and A.R. Horrocks. 1997. Recent developments in fibres and materials for wound management. *Indian Journal of Fibre and Textile Research* 22:337.

Lee, B.J., J.S. Choe, and C.K. Kim. 1998. Preparation and characterization of melatonin-loaded stearyl alcohol microspheres. *Journal of Microencapsulation* 15:775–787.

Leong, K.W., H.Q. Mao, V.L. Truong-Le, K. Roy, S.M. Walsh, and J.T. August.1998. DNA polycation nanospheres as non-viral gene delivery vehicles. *Journal of Controlled Release* 53:183–193.

Liu, W., S. Sun, and Z. Cao. 2005. An investigation on the physicochemical properties of chitosan/DNA polyelectrolyte complexes. *Biomaterials* 26:2705–2711.

Liu, W.G., X. Zhang, S.J. Sun, G.J. Sun, and K.D. Yao. 2003. N-Alkylated chitosan as a potential nonviral vector for gene transfection. *Bioconjugate Chemistry* 74:782–789.

Lo, K.M., G.H. Lin, and Aslue. 2005. pH-Responsive biomineralization onto chitosan grafted biodegradable substrates. *Journal of Controlled Release* 104:477.

Lorenzo-Lamosa, M.L., C. Remunan-Lopez, J.L. Vila-Jato, and M.J. Alonso. 1998. Design of drug delivery. *Journal of Controlled Release* 52:109–118.

Mao, J.S., H.F. Liu, Y.J. Yin, and K.D. Yao. 2003. The properties of chitosan-gelatine membranes and scaffolds modified with hyaluronic acid by different method. *Biomaterials* 24:1621–1629.

Mao, S.R. and W. Sun. 2010. Chitosan-based formulations for delivery of DNA and siRNA. *Advanced Drug Delivery Reviews* 62:12–27.

Matsusaki, M. and M. Akashi. 2005. Novel functional biodegradable polymer IV: pH-sensitive controlled release of fibroblast growth factor-2 from a Poly (γ-glutamic acid)-Sulfonate matrix for tissue engineering. *Biomacromolecules* 6:3351.

Merrifield, J.D., W.G. Davids, J.D. Macrae, and A. Amirbahman. 2004. Uptake of mercury by thiolgrafted chitosan gel beads. *Water Research* 38:3132–3138.

Mincea, M., A. Negrulescu, and V. Ostafe. 2012. Preparation, modification, and applications of chitin nanowhiskers: A review. *Reviews on Advanced Materials Science* 30:225–242.

Moore, G.K. and G.A.F. Roberts.1982. Reactions of chitosan: Preparation of organo soluble derivatives of chitosan. *International Journal of Biological Macromolecules* 4:246–249.

Morishita, M., N.A. Peppas. 2006. Is the oral route possible for peptide and protein drug delivery? *Drug Discovery Today* 11:905–910.

Muzzarelli, R.A.A. and C. Muzzarelli. 2002. *Chitosan in Pharmacy and Chemistry*, Atec Edizioni, Grottammare.

Muzzarelli, R.A.A., F. Tanfani, M. Emanuelli, and S. Mariotti.1982. N-Carboxy-methylidene chitosans and N-carboxymethyl chitosans: Novel chelating polyampholytes obtained from chitosan glyoxylate. *Carbohydrate Research* 107:199–214.

Muzzarelli, R.A.A. 1977. Chitin. Pergamon Press, New York.

Naggi, A.M., G. Torri, T. Compagnoni, and B. Casu. 1986. Synthesis and physico-chemical properties of a polyampholyte chitosan 6-sulfate. In: Muzzarelli R.A.A, Jeuniaux, C., Gooday, G.W., (eds.), *Chitin in Nature and Technology*. Plenum Publishing Corporation, New York, 371–377.

No, H.K., N.Y. Park, S.H. Lee, and S.P. Meyers. 2002. Antibacterial activity of chitosans and chitosan oligomers with different molecular weights. *International Journal of Food Microbiology* 74:65–72.

No, H.K., S.H. Lee, N.Y. Park, and S.P. Meyers. 2003. Comparison of physicochemical, binding, and antibacterial properties of chitosans prepared without and with deproteinization process. *Journal of Agricultural Food Chemistry* 51:7659–7663.

No, H.K. and S.P. Meyers. 1989. Crawfish chitosan as a coagulant in recovery of organic compounds from seafood processing streams. *Journal of Agricultural Food Chemistry* 37:580–583.

Nunthanid, J., S. Puttipipatkhachorn, K. Yamamoto, and G.E. Peck. 2001. Physical properties and molecular behavior of chitosan films. *Drug Development and Industrial Pharmacy* 27:143–147.

Omwancha, W.S., R. Mallipeddi, B.L. Valle, and S.H. Neau. 2013. Chitosan as a poreformer in coated beads for colon specific drug delivery of 5-ASA. *International Journal of Pharmaceutics* 441:343–351.

Ouchi, T., H. Nishizawa, and Y. Ohya. 1998. Aggregation phenomenon of PEG-grafted chitosan in aqueous solution. *Polymer* 39:5171.

Payne, G.F., M.V. Chaubal, and T.A. Barbari. 1996. Enzyme-catalyzed polymer modification: Reaction of phenolic compounds with chitosan films. *Polymer* 37:4643–4648.

Peter, M.G. 2002. Chitin and chitosan from animal sources. In: De Baets, S., Vandamme, E.J., and Steinbuchel, A., (eds.), *Biopolymers, (Polysaccharides II)*. Wiley–VCH, Weinheim, 2, ch 15.

Pochanavanich, P. and W. Suntornsuk. 2002. Fungal chitosan production and its characterization. *Letters in Applied Microbiology* 35(1):17–21.

Prabaharan, M. and J.F. Mano. 2005. In particular, microgels are expected to enrich the pool of drug delivery system. *Drug Delivery* 12:41.

Prabaharan, M. and R. Jayakumar. 2009. Chitosan-graft-β-cyclodextrin scaffolds with controlled drug release capability for tissue engineering applications. *International Journal of Biological Macromolecules* 44:320–325.

Ramos, V.M., N.M. Rodriguez, M.S. Rodriguez, A. Heras, and E. Agullo. 2003. Modified chitosan carrying phosphonic and alkyl groups. *Carbohydrate Polymers* 51:425–429.

Rao, M.S., J. Munoz, and W.F. Stevens. 2000. Critical factors in chitin production by fermentation of shrimp biowaste. *Applied Microbiology and Biotechnology* 54:808–813.

Raymond, L., F.G. Morin, and R.H. Marchessault. 1993. Degree of deacetylation of chitosan using conductometric titration and solid-state NMR. *Carbohydrate Research* 246:331–336.

Rinaudo, M., M. Milas M, and P. Le Dung. 1993. Characterization of chitosan. Influence of ionic strength and degree of acetylation on chain expansion. *International Journal of Biological Macromolecules* 15:281–285.

Rinaudo, M. and J. Reguant. 2000. Natural polymers and agrofibers composites. *Polysaccharide Derivatives* 15-39.

Rinaudo, M. 2006. Chitin and chitosan: Properties and applications. *Progress in Polymer Science* 31:603–632.

Roberts, G.A.F. 1992. Solubility and solution behavior of chitin and chitosan. In: Roberts, G.A.F. (ed.), *Chitin Chemistry*. MacMillan, Houndmills, 274–329.

Roger, E., F. Lagarce, E. Garcion, and J. Benoit. 2010. Biopharmaceutical parameters to consider in order to alter the fate of nanocarriers after oral delivery. *Nanomedicine* 5:287–306.

Sabnis, S. and L.H. Block. 2000. Chitosan as an enabling excipient for drug delivery systems. I. Molecular modifications. *International Journal of Biological Macromolecules* 27:181–186.

Sanoj Rejinold, N., M. Muthunarayanan, K. Muthuchelian, K.P. Chennazhi, Shanti V. Naira, and R. Jayakumar. 2011. Saponin-loaded chitosan nanoparticles and their cytotoxicity to cancer cell lines in vitro. *Carbohydrate Polymers* 84:407–416.

Sarmento, B., A. Ribeiro, F. Veiga, D. Ferreira, and R. Neufeld. 2007. Oral bioavailability of insulin contained in polysaccharide nanoparticles. *Biomacromolecules* 8:3054–3060.

Sato, H., S.I. Mizutani, S. Tsuge, K. Aoi, A. Takasu, M. Okada, S. Kobayashi, T. Kiyosada, and S.I. Shoda. 1998. Determination of the degree of acetylation of chitin/chitosan by pyrolysis-gas chromatography in the presence of oxalic acid. *Analytical Chemistry* 70:7–12.

Schipper, N.G.M., K. Varum, and P. Artursson. 1996. Chitosans as absorption enhancers for poorly absorbable drugs. 1: Influence of molecular weight and degree of acetylation on drug transport across human intestinal epithelial (Caco-2) cells. *Pharmaceutical Research* 13:1686–1692.

Schmaljohann. D. 2006. Thermo- and pH-responsive polymers in drug delivery. *Advanced Drug Delivery Reviews* 58:1655.

Shahidi, F., J.K. Arachchi, and Y.J. Jeon. 1999. Food applications of chitin and chitosans. *Trends in Food Science & Technology* 10:37–51.

Shao, L., G. Kumar, J.L. Lenhart., P.J. Smith, and G.F. Payne. 1999. Enzymatic modification of the synthetic polymer polyhydroxystyrene. *Enzyme and Microbial Technology* 25:660–668.

Shu, X.Z, and K.J. Zhu. 2000. A novel approach to prepare tripolyphosphate/chitosan complex beads for controlled release drug delivery. *International Journal of Pharmaceutics* 201:51–58.

Sinha, V.R., A.K. Singla, W.S. Adhawan, R. Kaushik, R. Kumria, K. Bansal, and S. Dhawan. 2004. Chitosan microspheres as a potential carrier for drugs. *International Journal of Pharmaceutics* 1:274.

Snyman, D., J.H. Hamman, J.S. Kotze, J.E. Rollings, and F. Kotze. 2002. The relationship between the absolute molecular weight and the degree of quartenisation of N-trimethyl chitosan chloride. *Carbohydrate Polymers* 50:145–150.

Sorlier, P., A. Denuzière, C. Viton, and A. Domard. 2001. Relation between the degree of acetylation and the electrostatic properties of chitin and chitosan. *Biomacromolecules* 2:765–772.

Steckel, H. and F. Mindermann-Nogly. 2004. Production of chitosan pellets by extrusion/spheronization. *European Journal of Pharmaceutics and Biopharmaceutics* 57:107.

Sugimoto, M., M. Morimoto, H. Sashiwa, H. Saimoto, and Y. Shigemasa.1998. Preparation and characterization of water soluble chitin and chitosan derivatives. *Carbohydrate Polymers* 36:49–59.

Suh, J.K. and H.W. Matthew. 2000. Application of chitosan- based polysaccharide biomaterials in cartilage tissue engineering: A review. *Biomaterials* 21:2589.

Susana Rodrigues, M. D., C.R. López, and A. Grenha. 2012. Biocompatibility of chitosan carriers with application in drug delivery. *Journal of Functional Biomaterials* 3:615–641.

Takechi, M., Y. Miyamoto, Y. Momota, T. Yuasa, S. Tatehara, M. Nagayama, K. Ishikawa, and K. Suzuki. 2002. The *in vitro* antibiotic release from anti-washout apatite cement using chitosan. *Journal of Materials Science: Materials in Medicine* 13:973–978.

Tanodekaew S. and M. Prasitsilp. 2004. Preparation of acrylic grafted chitin for wound dressing application. *Biomaterials* 25:1453–1460.

Terbojevich, M., C. Carraro, A. Cosani, B. Focher, A. Naggi, and G. Torri. 1989. Solution studies of chitosan 6-O-sulfate. *Molecular Chemistry And Physics* 190:2673–3031.

Terbojevich, M., A. Cosani, B. Focher, A. Naggi, and G. Torri. 1992. Chitosans from Euphausia superba. 1: Solution properties. *Carbohydrate Polymers* 18:35–42.

Terbojevich, M., A. Cosani, M. Scandola, and A. Fomasa.1986. Solution properties and mesophase formation of chitosan. In Zikakis, J.P. (ed.), *Chitin, Chitosan and Related Enzymes*. Academic Press, Inc., Orlendo, 349–351.

Tharanathan, R.N. and F.S. Kittur. 2003. Chitin- the undisputed biomolecule of great potential. *Critical Review in Food Science and Nutrition* 43:61–87.

Tommeraas, K., M. Köping-Höggård, K.M. Varum, B.E. Christensen, P. Artursson, and O. Smidsrod. 2002. Preparation and characterisation of chitosans with oligosaccharide branches. *Carbohydrate Research* 337:2455–2462.

Tozaki, H., T. Odoriba, N. Okada, T. Fujita, A. Terabe, T. Suzuki, S. Okabe, S. Muranishi, and A. Yamamoto. 2002. Chitosan capsules for colon-specific drug delivery: Enhanced localization of 5-aminosalicylic acid in the large intestine accelerates healing of TNBS-induced colitis in rats. *Journal of Controlled Release* 82:51–61.

Tsai, G.J. and W.H. Su. 1999. Antibacterial activity of shrimp chitosan against Escherichia coli. *Journal of Food Protection* 62:239–243.

Upadrashta, S.M., P.R. Katikaeni, and N.O. Nuessle. 1992. Chitosan as tablet binder. *Journal of Drug Development and Industrial Pharmacy* 18:1701.

Van de Velde, K. and P. Kiekens. 2004. Structure analysis and degree of substitution of chitin, chitosan and dibutyrylchitin by FT-IR spectroscopy and solid state 13C NMR. *Carbohydrate Polymers* 58:409–416.

Van Der Lubben, I.M., G. Kersten, M.M. Fretz, C. Beuvery, J.C. Verhoef, and H.E. Junginger. 2003. Chitosan microparticles for mucosal vaccination against diphtheria: Oral and nasal efficacy studies in mice. *Vaccine* 21:1400–1408.

Varum, K.M., M.W. Antohonsen, H. Grasdalen, and O. Smidsrod. 1991. Determination of the degree of N-acetylation and the distribution of Nacetyl groups in partially N-deacetylated chitins (chitosans) by high-field NMR spectroscopy. *Carbohydrate Research* 211:17–23.

Vipin Bansal, P.K.S., N. Sharma, O.P. Pal, and R. Malviya. 2011. Applications of chitosan and chitosan derivatives in drug delivery. *Advances in Biological Research* 5(1):28–37.

Wang, Y.C, M.C. Lin, D.M. Wang, and H. Hsieh. 2003. Fabrication of a novel porous PGA-chitosan hybrid matrix for tissue engineering. *Biomaterials* 24:1047–1057.

Win, N.N., G. Pengju, W.F. Stevens. 2000. Deacetylation of chitin by fungal enzymes. In: Uragant, et al. (eds.), *Advances in Chitin Science*. University of Podstam, Germany, 55–62.

Win, N.N. and W.F. Stevens. 2001. Shrimp chitin as substrate for fungal chitin deacetylase. *Applied Microbiology and Biotechnology* 57:334–341.

Wu, A.C., W.A. Bough, E.C Conrad, and K.E. Alden Jr. 1976. Determination of molecular-weight distribution of chitosan by high-performance liquid chromatography. *Journal of Chromatography* 128:87–99.

Wu, F.C., R.L. Tseng, and R.S. Juang. 2001. Kinetic modeling of liquid phase adsorption of reactive dyes and metal ions on chitosan. *Water Res.* 35:613–618.

Wu, J., X. Wang, J.K. Keum et al. 2007. Water soluble complexes of chitosan-g-MPEG and hyaluronic acid. *Journal of Biomedical Material Research Part A* 80:800–812.

Wu, T. and S. Zivanovic. 2008. Determination of the degree of acetylation (DA) of chitin and chitosan by an improved first derivative UV method. *Carbohydrate Polymers* 73:248–253.

Wuolijoki, E., T. Hirvela, and P. Ylitalo. 1995. Decrease in Serum LDL cholesterol with microcrystalline chitosan. *Methods and Findings in Experimental and Clinical Pharmacology* 21:57–361.

Xie, W.M., P.X. Xu, W. Wang, and Q. Lu. 2002. Preparation and antibacterial activity of water-soluble chitosan derivatives. *Carbohydrate Polymers* 50:35–40.

Yamanaka, Y.J. and K.W. Leong. 2008. Engineering strategies to enhance nanoparticle-mediated oral delivery. *Biomaterials Science, Polymer Edition* 19:1549–1570.

Yang chao Luo, Z.T., X. Wang, and Q. Wang. 2013. Development of carboxymethyl chitosan hydrogel beads in alcohol-aqueous binary solvent for nutrient delivery applications. *Food Hydrocolloids* 31:332–339.

Yang, Z.-H., T. Sato, and Y. Koyama. 2006. Development of chitosan nanoparticles for gene delivery system. *Molecular Therapy* 13:S76–S77.

Yang, T.C. and C.C. Chou. 2002. Preparation, water solubility and rheological property of the N-alkylated mono or disaccharide chitosan derivatives. *Food Research International* 35:707–713.

Yen, M., J. Yang, and J. Mau. 2009. Physicochemical characterization of chitin and chitosan from crab shells. *Carbohydrate Polymers* 75:15–21.

Yuvarani, I., S. Kumar, J. Venkatesan, S.-K. Kim, and P.N Sudha. 2012. Preparation and characterization of curcumin coated chitosan-alginate blend for wound dressing application. *Journal of Biomaterials and Tissue Engineering* 2:54–60.

Zarzycki, R., Z. Modrzejewska, P. Owczarz, and A. Wojtasz-Pająk. 2008. New chitosan structures in the form of the thermosensitive gels. *Derivatives* 13:35–42.

Zhang, C. and Q. Ping. 2003. Preparation of N-alkyl-O-sulfate chitosan derivatives and micellar solubilization of taxol. *Carbohydrate Polymer* 54:137–141.

Zhang, H., Y. Du, X. Yu, M. Mitsutomi, and S-I Aiba.1999. Preparation of chitooligosaccharides from chitosan by a complex enzyme. *Carbohydrate Research* 320:257–260.

Zhang, X.Z., P.J. Lewis, and C. Chu. 2005. Fabrication and characterization of a smart drug delivery system: microsphere in hydrogel. *Biomaterials* 26:3299–3309.

26 Chitosan and Chitosan Derivatives as Potential Adjuvants for Influenza Vaccine

Nguyen Anh Dzung

CONTENTS

26.1 BACKGROUND ON ADJUVANT FOR INFLUENZA VACCINE

26.1.1 ADJUVANTS FOR VACCINE

An adjuvant is an agent that stimulates the immune system and increases the response to a vaccine, without having any specific antigenic effect in it. An immunologic adjuvant is defined as any substance that acts to accelerate, prolong, or enhance antigen-specific immune responses when used in combination with specific vaccine antigens. The presence of an adjuvant in the vaccine can greatly stimulate the immune response to the antigen by activating dendritic cells (DCs), lymphocytes, and macrophage. Advantages of adjuvants are that the enhancement of the immunogenicity of antigen amount for a successful immunization and reduction of the frequency of booster immunization needed.

O'Hagan and Rappuoli (2004) supposed the roles of adjuvants in vaccine development in brief as follows: enhancing the total antibody titer or functional titers; reducing the dose of antigen needed; decreasing the total number of doses of vaccine necessary for complete immunization; overcoming competition in combination vaccines; increasing immune responses in the young or older populations and the speed and duration of the vaccine-specific protective response; inducing potent cell-mediated immunity, mucosal immunity, and broader immune response.

There are many adjuvants used for vaccine such as inorganic adjuvants, organic adjuvants, oil-based adjuvants, virosomes, and some other potential adjuvants such as chitosan and chitosan derivatives, cholera toxin, interleukin 12, and so on in which aluminum phosphate and aluminum hydroxide are common adjuvants used for human vaccines (Amorij et al., 2010). The organic compound squalene (AS03) is commonly used in animal vaccines. Oil based adjuvants are MF59 and cholesterol lipids (Joseph et al., 2006). Other approved adjuvants are virosomes that contain a membrane-bound hemaglutenin and neuraminidase from influenza virus (Gluck et al., 1999).

26.1.2 ADJUVANTS FOR INFLUENZA VACCINE

Influenza is a significant cause of mortality, morbidity, and economic loss. About 5–10% of the world's population is affected by influenza each year. Influenza is a respiratory disease caused by influenza viruses (WHO, 2003). Influenza pandemic alert occurred in 1997 (H5N1 virus), 1999 (H9N2 virus), and 2003 (H5N1 and H7N7 viruses); in these cases, avian influenza viruses caused serious illness and death in humans (WHO, 2005). Pandemic influenza viruses have caused substantial morbidity and mortality worldwide, and threatened public health in which A/H5N1 and A/H1N1 influenza viruses with highly toxic strains was widespread to the public. "Bird flu" caused by A/H5N1 influenza virus has a highly fatal rate (Subbaro and Luck, 2007). WHO report on October 2011 announced a total of 566 confirmed human cases and death of 322 people since 2003.

In March 2009, A/H1N1 influenza pandemic (swine flu) occurred in Mexico and expanded quickly to 74 countries, with 29,696 H1N1 confirmed cases and hundreds of deaths reported until June 12, 2009. WHO had announced that it raised influenza pandemic alert to phase 6 (WHO, 2010).

Vaccination is the best choice for influenza prevention and is the main method of prophylaxis to control pandemic influenza. Using a whole inactivated virus (WIV) vaccine based on hemagglutinin (HA), it is low immunogenic when administered without adjuvant (WHO, 2009). Therefore, influenza antigens are formulated commonly with adjuvants such as interleukin 12 (Arulanandam et al., 1999), cationic cholesterol (Guy et al., 2001), interferon α (Bracci et al., 2006), lectin C (Song et al., 2007), α-galactosynceramide (Ko et al., 2005; Youn et al., 2007), squalene-based oil-in-water emulsions (such as MF59, AS03, and AF03) formulated with H5N1 vaccine (Risi et al., 2011), chitosan, and its derivatives (Illum et al., 2001; Huang et al., 2004; Amidi et al., 2007; Garmise et al., 2007). Some other adjuvants have been investigated such as cationic liposome DNA complexes (CLDC) for H5N1 vaccine (Dong et al., 2012).

Some promising adjuvants have been used for influenza vaccine such as H3N2 and H6N2 influenza antigen loaded on poly hydroxyethyl methacrylate (PHEMA) nanoparticles or amylopectin nanoparticles (Coucke et al., 2009; Shan et al., 2010), interleukin 12 (Arulanandam et al., 1999), cationic cholesterol (Guy et al., 2001), interferon α (Bracci et al., 2006), lectin C (Song et al., 2007), and α-galactosynceramide (Ko et al., 2005; Youn et al., 2007).

Hiep et al. (2008) used chitosan as an adjuvant to loaded A/H5N1 inactivated whole virus (WIV). Risi et al. (2011) studied AS03 adjuvant (water-in-oil emulsion and tocopherol complexes) for A/H5N1 influenza subunit antigen. Dong et al. (2012) used a complex of cation liposome and DNA to induce and stimulate immune response with A/H5N1 influenza vaccine. Galactosylceramide and peptide M2 were also used as adjuvants for A/H5N1 vaccine (Li, 2011). Harada et al. (2011) and Nakayama et al. (2012) studied using aluminum hydroxide gel, a common adjuvant mixed with inactivated whole virion A/H5N1 to formulate the influenza vaccine. Some potential adjuvants for influenza vaccine are listed in Table 26.1.

26.2 ROLES OF CHITOSAN AND CHITOSAN DERIVATIVES FOR STIMULATING IMMUNE SYSTEM

Chitosan is a biopolymer of glucosamine and N-acetyl glucosamine residues, processed from seafood wastes such as shrimp, crab shell, or cell wall of fungi. Chitosan and its derivatives are natural polymers, nontoxic, and have biocompatible, biodegradable, and adsorption properties (Kumar, 2000; Rinaudo, 2006). Chitosan is a nontoxic, biodegradable, biocompatible polymer, and introduces immune-enhancing effect and therefore, chitosan and its derivatives have been applied as potential adjuvant in vaccine.

Chitin and chitosan have been found to be activators for macrophage and NK cells (Nishimura et al., 1984). Chitin (30% DD) and chitosan (70% DD) formulated with incomplete Freund adjuvant (IFA) induced greater delayed type hypersensitivity (DTH) response than IFA alone and increased antigen-specific serum antibody titer in mice and guinea pig by over threefold in comparison to IFA alone.

TABLE 26.1
List of Potential Adjuvants for Influenza Vaccine

Type of Adjuvant	Antigen	Route	Immunity Data Detected	References
Peptide and cationic polysaccharide core	H1N1 and H3N2	Nasal	IgA, IgG, CD4, and CD8 lymphocyte	Von Hoegen (2001)
PHEMA	H6N2 virus	Injection	HI antibody	Shan et al. (2010)
Heat-labile enterotoxin R192G	X47 H3N2 influenza virus	Nasal	HI antibody, IgA, and IgG	Coucke et al. (2009)
Trehalose powder	WIV (H1N1)	Injection	HI antibody, IgA	Huang et al. (2004)
α-Galactosylceramide	Subunit, WIV	Nasal	IgG, sIgA, and interferon γ	Ko et al. (2005),
Peptide and α-galactosylceramide	H5N1 WIV	Intranasal Injection	IgG 1, IgG 2[a]	Youn et al. (2007), Li (2011)
Interleukin-12	Subunit	Intranasal	IgG, sIgA, and interferon γ	Arulanandam et al. (1999)
Cationic cholesterol	Split	Injection	HI, IgG, and sIgA	Guy et al. (2001)
Interferon α	Subunit	Nasal	IgG, IgA	Bracci et al. (2006)
Liposome	Subunit	Nasal	IgG, IgA, and HI	Joseph et al. (2006)
Mistletoe lectin C	WIV	Nasal	IgG, sIgA, and interferon γ	Song et al. (2007)
AS03 adjuvant system	H5N1 subunit Split (H5N1)	Injection	GMT, SPR, and SCR	Risi et al. (2011), Yang et al. (2012)
Cationic liposome–DNA complex	Split (H5N1)	Injection	HI, IgG1, and IgG2a	Dong et al. (2012)
Aluminum adjuvant	WIV (H5N1)	Injection	IFN-α, IL-1β, and IL 6	Nakayama et al. (2012), Harada et al. (2011)
Chitosan oligomer	WIV (H5N1)	Injection	HI	Hiep et al. (2008)
Chitosan	Subunit	Nasal	HI, IgG, sIgA, and interferon γ	Illium et al. (2001), Read et al. (2005)
Cholera toxin	WIV	Intradermal	IgG, sIgA, and interferon γ	Song et al. (2008), Skountzou et al. (2006)
TMC	Ovalbumin	Nasal	IgG, sIgA	Boonyo et al. (2007), Bal et al. (2012)
Matrix™	H5N1 WIV	Injection	HI, MN, and SRH	Cox et al. (2011)
Bacillus subtilis spore	H5N1	Nasal	TNR, NK, and NF	Song et al. (2012)

Shibata et al. (1997) demonstrated that chitin and chitosan were induced to enhance synthesis of γ-interferon and macrophages and supposed that stimulating of immune system depended on molecular weight, degree of deacetylation, and substitute group on chitin and chitosan.

Chitosan emulsion and Zn–chitosan particles used as adjuvant was formulated with β-human chorionic gonadotropin and vaccinated in mice and guinea pig, and the new adjuvant was found to be very effective in sensitizing mice for antigen-specific DTH response and stimulated both B and T lymphocytes (Seferian & Martinez, 2001). Chitosan was able to upregulate expression of CD 69 on B cells and CD4+ T lymphocytes. CD 69 molecule on CD8+ T lymphocytes was expressed only in splenocytes cultured with chitosan (Borges et al., 2007). Chitosan enhanced immune properties of GM-CFS (Zaharoff et al., 2007). Chitosan solution (CS) was also found to enhance both humoral and cell-mediated immune responses to subcutaneous vaccination (Zaharoff et al., 2007). Zaharoff et al. proposed that biodegradability, immunological activity, and high viscosity of chitosan made it an excellent candidate adjuvant for vaccine and chitosan-enhanced antigen-specific antibody titers fivefold and antigen-specific splenic CD+ proliferation over sixfold.

With positive charge and enhancing immune response, chitosan has been used as a novel nasal delivery system for vaccine. Particularly, chitosan nanoparticles were suitable for mucosal drugs and vaccine delivery (Illium et al., 2001). Chitosan nanoparticles can enhance antigen uptake by mucosal lymphoid tissues and induce strong immune responses against the antigens.

Chitosan and trimethyl chitosan chloride (TMC) were used as adjuvants for inducing immune responses to ovalbumin in mice. It was concluded that the molecular weight of chitosan and the degree of quaternization (DQ) of TMC influenced significantly on the level of immune induction. Chitosan with molecular weight of 5×10^5 g/mol could elicit highest IgG response to ovalbumine and TMC with DQ of 40% could induce IgG response higher than ovalbumin in alum (Boonyo et al. 2007).

Chen et al. (2008) also demonstrated that water-soluble chitosan stimulated a specific immunomodulatory effects on dust mite allergen *Dermatophagoides farinae*, monocytes-derived macrophages (MDMs). These effects decreased the production of inflammatory cytokine IL-6 and TNF-α, downregulated CD44 and TLR4 receptor expression, and inhibited T cells proliferation.

A complex of polyoxyethylene–polyoxypropylene copolymer (F127) with chitosan was also used as a novel adjuvant to enhance specific immune response to intranasally administered tetanus toxoid (TT) (Westerink et al. 2002). Mice boosted with TT in F127/chitosan had significant increase in anti-TT antibody, anti-TT IgG, and IgA antibody response compared to the control boosted only with TT in phosphate buffer solution (PBS) and in F127/lysophosphatidyl choline.

Borges et al. (2007) also reported that hepatitis B antigen encapsulated into alginate-coated chitosan nanoparticles showed significant higher values of CD69 expression in CD4+, CD8+, and T lymphocytes in comparison to the control (Borges et al., 2007).

Zaharoff et al. (2007) found that CS enhanced both humoral and cell-mediated immune responses to substaneous vaccination. Chitosan-enhanced antigen-specific antibody titers over five-fold and antigen-specific splenic CD+ proliferation over sixfold and chitosan adjuvant was better than Freund and aluminum hydroxide. Zaharoff supposed that viscous CS created antigen depot and CS remained an antigen in the injection site longer than the saline. In the saline case, less than 9% protein antigen remained in the injection site after 8 h, whereas 60% protein antigen was delivered in chitosan after 7 days. Especially, chitosan induced and expanded lymph nodes up to 67% compared to the control.

As potential characteristics in vaccine application, chitosan nanoparticles and its derivatives have been used as novel adjuvants for some vaccines such as hepatitis B (Borges et al., 2007; Khatri et al., 2008; Borges et al., 2008a,b) and piglet paratyphoid vaccine (Yang et al., 2007).

26.3   CHITOSAN AND CHITOSAN DERIVATIVES AS POTENTIAL ADJUVANTS FOR INFLUENZA VACCINE

As the safe, less or no side effect, and immunity-stimulating activity, chitosan and its derivatives have been demonstrated as promising adjuvant for vaccine and influenza vaccine. Hiep et al. (2008) used chitosan oligomer (MW 3 kDa, 85% DD) to load 30–60 HAU of A/H5N1 influenza WIV compared to aluminum phosphate and Freund adjuvant. The results showed that chitosan oligomer induced early immune response of the mice boosted with the vaccine and higher specific antibody (HI) compared to the aluminum phosphate and Freund adjuvant (see Figure 26.1). Stimulating early immune response is very important and necessary when the threat of an influenza pandemic appears.

Chitosan oligomer adjuvant and inactivated whole virion antigen was also tested safe in mice. The results showed that chitosan oligomer and the vaccine were safe for mice.

Hiep et al. (2008) found that specific antibody in mice boosted with chitosan oligomer (1 mg/mL) and H5N1 antigen was higher than the common aluminum phosphate and the antigen only without adjuvant (see Figure 26.2). It was concluded that chitosan oligomer with MW 3 kDa and 85% DD was a potential adjuvant for H5N1 vaccine.

Chitosan nanoparticles and microparticles having high surface area and high positive zeta potential leads to increase in adsorption of protein antigen on the surface of the particles. Therefore,

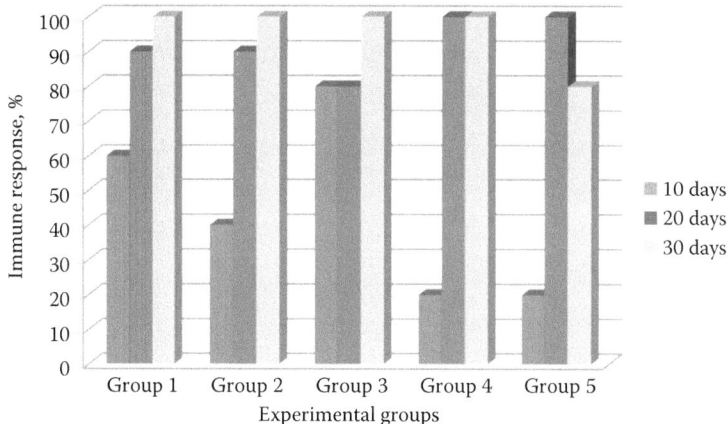

FIGURE 26.1 Effect of various adjuvants on immune response of the mice with A/H5N1 influenza vaccine. Group 1: H5N1 WIV without adjuvant, Group 2: H5N1 WIV with aluminum phosphate, Group 3: H5N1 WIV with chitosan oligomer (1 mg/mL), Group 4: H5N1 WIV with chitosan oligomer (2 mg/mL), Group 5: H5N1 WIV with Freund. (From Hiep, L. V. et al. 2008. *Journal of Chitin and Chitosan,* 13(10), 6–8. With permission.)

chitosan nanoparticles and microparticles have been applied as promising adjuvants for vaccine and influenza vaccine.

In vaccine application, it has been known that efficacy of chitosan microparticles and nanoparticles adjuvant seems to be dependent on molecular weight and degree of deacetylation (van der Lubben et al., 2001; Vila et al., 2004).

Dzung et al. (2011) have used chitosan nanoparticles prepared by gelation with tripolyphosphate (TPP) having different MW 20, 30, and 300 kDa and size from 85 to 100 nm as adjuvant for H1N1 influenza vaccine (see Figure 26.3).

The effect of various conditions such as adjuvant type, molecular weight of chitosan nanoparticles on immune response, and loading efficiency (LE) and loading capacity of H1N1 antigen are shown in Table 26.2.

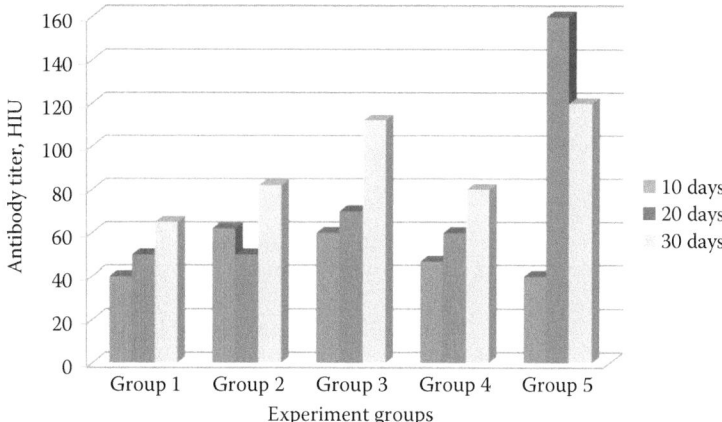

FIGURE 26.2 Effect of various adjuvants on antibody titer (HIU) of mice vaccinated with H5N1 vaccine. Group 1: H5N1 WIV without adjuvant, Group 2: H5N1 WIV with aluminum phosphate, Group 3: H5N1 WIV with chitosan oligomer (1 mg/mL), Group 4: H5N1 WIV with chitosan oligomer (2 mg/mL), Group 5: H5N1 WIV with Freund. (From Hiep, L. V. et al. 2008. *Journal of Chitin and Chitosan,* 13(10), 6–8. With permission.)

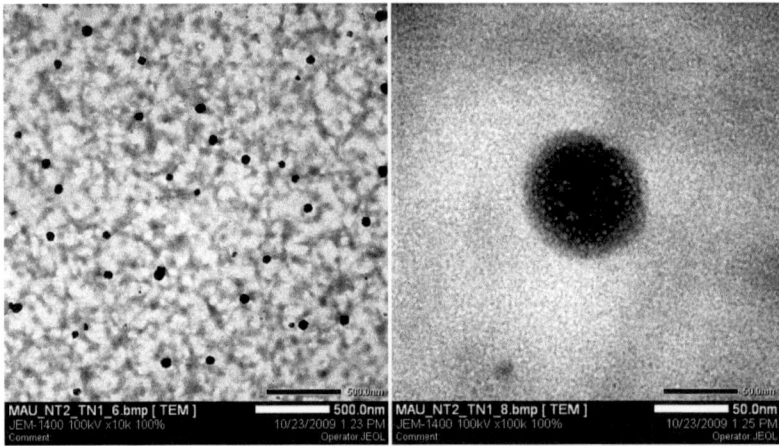

FIGURE 26.3 TEM photograph of chitosan nanoparticles. CS (0.2% w/v) with different molecular weight of 30 kDa was mixed with 0.5% TPP solution (CS:TPP mass ratio was 6:1) at pH = 5.5, room temperature, and magnetic stirring at 1000 rpm. TEM was taken by transmission electro microscope (Jeol, JEM 1400, Japan) at an acceleration voltage of 100 kV. (From Dzung, N. A. et al. 2011. *World Academy of Science, Engineering and Technology*, 1839–1846.)

With nanometer size, high surface area, and high zeta potential (+40 to 45 mV), chitosan nanoparticles loaded over 90% H1N1 antigen are shown in Table 26.2.

The dependence of the immune response on type of adjuvants and molecular weight of chitosan nanoparticles is also investigated. The results shown in Figure 26.4 indicate that the rate of immune response in mice was 60–100% after 10 days of vaccination, particularly the fifth group (CNS 30+H1N1 antigen) and the sixth group (CSN 300+H1N1 antigen) were able to respond to immunity early. After 20 days of vaccination, immune response rate was 100% in all groups except H1N1 antigen. It is clear that chitosan nanoparticles are able to enhance immune response better than aluminum hydroxyte adjuvant and chitosan (CS) solution. Early and stable immune response of the fifth group (CNS 30+H1N1 antigen) and the sixth group (CSN 300+H1N1 antigen) is very necessary when the threat of H1N1 spread to the community.

TABLE 26.2

Effect of Molecular Weight of Chitosan Nanoparticles on LE and Loading Capacity of A/H1N1 Antigen

MW of Chitosan Nanoparticles	Concentration of Chitosan Nanoparticles (mg/mL)	HA Total (HAU)	HA in the Supernatant (HAU)	LE (%)	Loading Capacity LC (HA/mg)
CSN 20	0.5	128	0	100.00	256
CSN 30	0.5	128	4	96.87	248
CSN 300	0.5	128	8	93.75	240
Control	0.0	128	128	0.00	0

Source: Dzung, N. A. et al. 2011. *World Academy of Science, Engineering and Technology*, 1839–1846.

Note: Loading condition: 0.9 mL chitosan nanoparticles (0.5 mg/mL) with different molecular weight (20, 30, and 300 kDa) were mixed with 0.1 mL (3 µg) H1N1 antigen containing 128 HA for 30 min at room temperature. The mixture was centrifuged at 12,000 rpm, 4°C for 10 min.

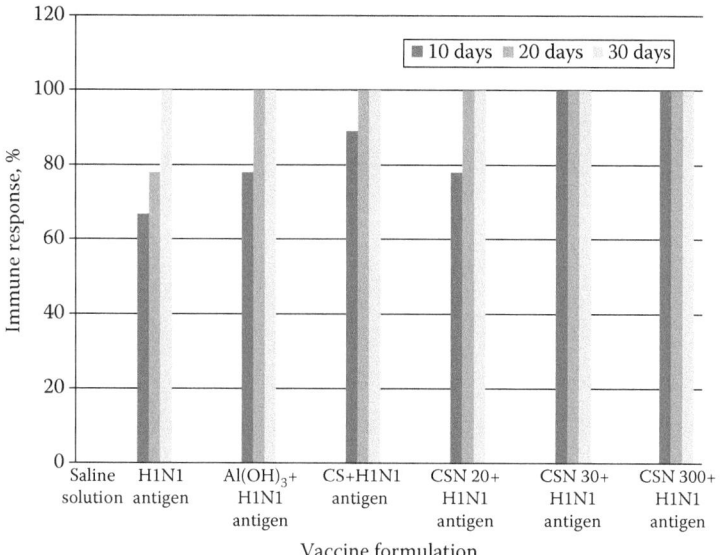

FIGURE 26.4 Effect of vaccine formulation on immune response rate (%) of the mice vaccinated with A/H1N1 influenza vaccine. Ten mice of 4–6 weeks age (14–16 grams (grs)) were injected into their legs with 0.2 mL vaccine containing 0.2 mg chitosan nanoparticles or other adjuvant loaded with 0.375 µg H1N1 antigen. The second vaccination was carried out after 20 days with the same dose as the first. After 10, 20, and 30 days of vaccination, the blood of the mice was collected to prepare sera for HA and HI assay. Control group: saline solution only. First group: H1N1 antigen (Ag). Second group: H1N1 antigen and Al(OH)₃ adjuvant. Third group: H1N1 antigen and CS. Fourth group: H1N1 antigen and 20 kDa chitosan nanoparticles (CSN 20 + H1N1 antigen). Fifth group: H1N1 antigen and 30 kDa chitosan nanoparticles (CSN 30 + H1N1 antigen). Sixth group: H1N1 antigen and 300 kDa chitosan nanoparticles (CSN 300 + H1N1 antigen).

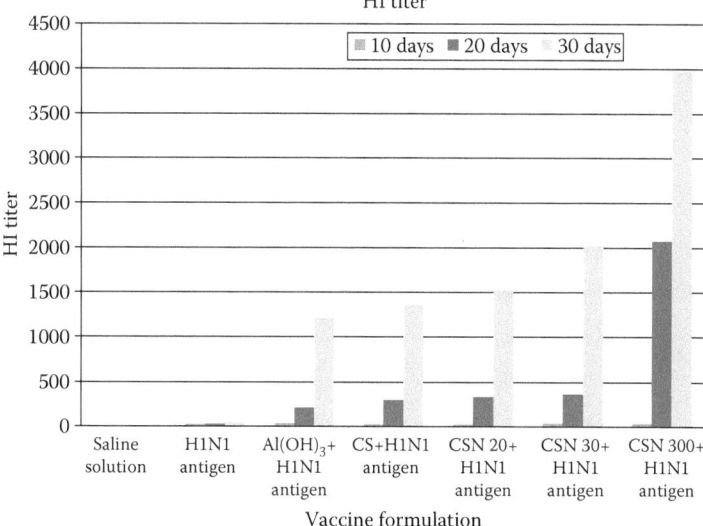

FIGURE 26.5 Effect of formulation of the group on antibody titer of the mice vaccinated with A/H1N1 influenza vaccine. Ten mice of 4–6 weeks age (14–16 grs) were injected into their substaneous legs with 0.2 mL vaccine containing 0.2 mg chitosan nanoparticles or other adjuvant loaded with 0.375 µg H1N1 antigen. The second vaccination was carried out after 20 days with the same dose as the first. After 10, 20, and 30 days of vaccination, the blood of the mice was collected to prepare sera for HA and HI assay.

TABLE 26.3

Application of Chitosan and Its Derivatives in Influenza Vaccine

Adjuvant	Molecular Weight	Material Type	Vaccine Type	Immune Response Detected	Route	References
Chitosan oligomer	3 kDa 85% DD	Solution	H5N1 WIV	HI, % immune response	Subcutaneous injection	Hiep et al. (2008)
Chitosan nanoparticles	20, 30, and 300 kDa, 80–85% DD	Nanoparticles, 80–100 nm size	H1N1 WIV	HI, % immune response	Subcutaneous injection	Dzung et al. (2011)
TMC nanoparticles	28–43 kDa, 90% DD	Nanoparticles, 200–220 nm size	A/PR8/34 (H1N1) WIV	IgG, IgA	Nasal	Hagennars et al. (2009)
TMC nanoparticles	63–94 kDa, 83–96% DD		A/PR8/34 (H1N1) WIV	IgG, IgA	Nasal	Hagennars et al. (2010)
TMC nanoparticles	177 kDa	Nanoparticles, 800 nm size	H3N2 subunit	IgG, IgA	Nasal	Amidi et al. (2007)
Chitosan	—	Solution	A/PR8/34 (H1N1) WIV	IgG, IgG1, IgG2a, and IgA	Nasal	Chang et al. (2004)
Chitosan/polygamma glutamate	LMW	Nanoparticles	H5N1 subunit	IgG, IgG1, IgG2a, and IgA	Nasal	Moon et al. (2012)
Chitosan-coated polycaprolactone	190 kDa	Nanoparticles, 125 nm size	H1N1 subunit	IgG, IgA, and HI	Nasal	Gupta et al. (2011)
N-2-hydroxy-3-trimethylammonium propyl chitosan chloride (HTCC)	780 kDa 95% DD	Hydrogel	H5N1 split	IgG, IgG1, IgG2a, HI, and IFN	Nasal	Wu et al. (2012)
Chitosan/trehalose complex	50–190 kDa	Dry powder complex, 100–200 nm size	A/PR8/34 (H1N1) WIV	IgG, IgG2a, IgA, and HI	Nasal	Huang et al. (2004) Garmise et al. (2007)
Chitosan–plasmid DNA	71.3 kDa, 80% DD	Nanoparticles, 150 nm size	A/H3N2, SIV	HI, IgG	Injection	Zhao et al. (2011)

Dzung et al. (2011) also indicated that there was clear dependence of antibody titer (HI) on MW of chitosan nanoparticles. The result shown in Figure 26.5 indicated that HI titer achieved up to 1270 HIU in the sixth group (CNS 300+H1N1 antigen) after 20 days of vaccination, whereas the fourth group (CNS 20+H1N1 antigen) and the fifth group (CNS 30+H1N1 antigen) was only 325–355 HIU, although the results were higher than the second group (aluminum hydroxyte) and the third group (chitosan only). After 30 days of vaccination, HI titer of all groups increased significantly, 1210 HIU (Al(OH)$_3$), 1350 HIU (chitosan), 1520 HIU (CSN 20+H1N1 antigen), 2000 HIU (CSN 30+H1N1 antigen), and up to 3971 HIU (CSN 300+H1N1 antigen). HI titer of the sixth group was twofold higher than the fifth group and approximately fourfold than the second group. It concluded that chitosan nanoparticles with a high molecular weight (300 kDa) strongly stimulated the rate of immune response and specific HI titer against H1N1 influenza antigen in mice and proposed that it should be studied continuously with higher molecular weight of chitosan nanoparticles.

Recently, chitosan nanoparticles have been used as novel nasal delivery system for drugs, vaccine, and influenza vaccine (MacLaughlin et al. 1998; De Campos et al. 2001; El-Salbouri, 2002; Huang et al., 2004; Kean et al. 2005; Amadi et al., 2007; Zhang et al. 2007; Zheng, 2007; Kumar et al., 2008; Hagenaar et al., 2009, 2010; Wu et al., 2012). Application of chitosan and chitosan nanoparticles in influenza vaccine is listed in Table 26.3 and almost all chitosan nanoparticles and derivatives used as nasal delivery system for influenza vaccine.

Amidi et al. (2007) prepared trimethyl chitosan (TMC) nanoparticles with size of 800 nm and MW of 177 kDa as a novel adjuvant to loaded A/H3N2 influenza vaccine. LE and loading capacity of H3N2 influenza antigen on the nanoparticles were 78% and 13% (w/w). Stimulating immune response in mice with H3N2 influenza vaccine loaded with the TMC nanoparticles was higher than TMC solution only. HI titer and IgG in mice immunized with the H3N2 influenza vaccine loaded on the TMC nanoparticles were 100 times higher than the vaccine loaded on TMC solution only. It was clear that nanosize of the TMC was more effective to stimulate immune response in the mice. DNA of A/H3N2 influenza virus also encapsulated in chitosan nanoparticles. The DNA/chitosan nanoparticles stimulated IgG response and T lymphocyte higher than the naked DNA, blank chitosan nanoparticles, and the control (Zhao et al., 2011).

TMC nanoparticles loaded WIV of A/PR8/34 (H1N1) as nasal vaccine, and IgG, IgG1, and IgG2 antibody in mice vaccinated with TMC nanoparticles were higher than WIV and CS only. This chapter also showed that the WIV and WIV–TMC nanoparticle complex induced minimal local toxicity (Hagennars et al., 2009, 2010). Huang et al. (2004) also formulated WIV of A/PR8/34 (H1N1) with complex of chitosan (MW of 50–190 kDa) and trehalose as influenza vaccine. The particle size around 150–200 μm were formulated with WIV of A/PR8/34 (H1N1) as the vaccine to immunized IM liquid and in IN powder. The results suggested that the IN powder/chitosan complex strongly elicited IgG, IgG1, and IgG2 response that may be very important for protection against influenza (Huang et al., 2004).

Gupta et al. (2011) studied loading of A/H1N1 antigen on chitosan coated with polycaprolactone nanoparticles. The average size of the nanoparticles was 125 nm, loaded with 75% A/H1N1 antigen and used as nasal vaccine. The nanoparticles used as adjuvant strongly stimulated immune system in the mice. The adjuvant enhanced HI titer, IgG titer in the mice immunized with chitosan-coated polycaprolactone nanoparticles compared to antigen only (Gupta et al., 2011).

H5N1 influenza vaccine was also loaded on thermal-sensitive hydrogel prepared by 2-hydroxy-3-trimethylammonium propyl chitosan chloride (HTCC) and glycerophosphate (Wu et al., 2012). It was found that the hydrogel/H5N1 vaccine induced higher antigen-specific immune response and mucosal IgA immunity than MF 59 adjuvant and the naked antigen.

REFERENCES

Amidi, M., Romeijn, S. G., Verhoef, J. C., Junginger, H. E., Bungener, L., Huckriede, A., Crommelin, D. J. A. and Jiskoot, W. 2007. *N*-trimethyl chitosan (TMC) nanoparticles loaded subunit antigen for intranasal vaccination: Biological properties and immunogenicity in a mouse model. *Vaccine*, 25, 144–153.

Amorij, J. P., Hinrich, W. L. J., Frijlink, H. W., Wilschut, J. C. and Huckriede, A. 2010. Needle-free influenza vaccine. *Lancet Infectious Diseases*, 10, 699–711.

Arulanandam, B. P., O' Toole, M. and Metzger, D. W. 1999. Intranasal interleukin-12 is a powerful adjuvant for protective mucosal immunity. *Journal of Infection Disease*, 180, 940–949.

Bal, S. M., Slutter, B., Verheul, R., Bouwstra, J. A., and Jiskoot, W. 2012. Adjuvanted, antigen loaded *N*-trimethyl chitosan nanoparticles for nasal and intradermal vaccination: Adjuvant- and site-dependent immunogenicity in mice. *European Journal of Pharmaceutical Science*, 45, 475–481.

Bracci, I., Caniti, I., and Puzelli, S. 2006. Type I IFN as a vaccine adjuvant for both systemic and mucosal vaccination against influenza virus. *Vaccine*, 24(Suppl.2), 56–57.

Boonyo, W., Junginger, H. E., Waranuch, N., Polnok, A., and Pitaksuteepong, T. 2007. Chitosan and trimethyl chitosan chloride (TMC) as adjuvants for inducing immune responses to ovalbumin in mice following nasal administration. *Journal of Controlled Release*, 168–175.

Borges, O., Tavares, J., de Sousa, A., Borchard, G., Junginger, H. E., and Cordeiro-da-Silva, A. 2007. Evaluation of the immune response following a short oral vaccination schedule with hepatitis B antigen encapsulated into alginate-coated chitosan nanoparticles. *European Journal of Pharmaceutical Science*, 32, 278–290.

Borges, O., Cordeiro-da-Silva, A., Tavares, J., Santarem, N., Sousa, A., Borchard, G., and Junginger, H. E. 2008a. Immune response by nasal delivery of hepatitis B surface antigen and codelivery of a CpG ODN in alginate coated chitosan nanoparticles. *European Journal of Pharmaceutical and Biopharmaceutics*, 69, 405–416.

Borges, O., Borchard, G., de Sousa, A., Junginger, H. E., and Cordeiro-da-Silva, A. 2008b. Induction of lymphocytes activated marker CD69 following exposure to chitosan and alginate biopolymers. *International Journal of Pharmaceutics*, 337, 254–264.

Chang, H. Y., Chen, J. J., Fang, F., and Chen, Z. 2004. Enhancement of antibody response by chitosan, a novel adjuvant of inactivated influenza vaccine. *Chinese Journal of Biology*, 17 (6), 21–24.

Chen, C. L., Wang, Y. M., Liu, C. F., and Wang, J. Y. 2008. The effect of water-soluble chitosan on macrophage activation and the attenuation of mite allergen-induced airway inflammation. *Biomaterials*, 2173–2182.

Coucke, D., Schotsaert, M., Libert, C., Pringels, E., Vervaet, C., Foreman, P., Saelens, X., and Remon, J. P. 2009. Spray-dried powders of starch and crosslinked poly(acrylic acid) as carriers for nasal delivery of inactivated influenza vaccine. *Vaccine*, 27, 1279–1286.

Cox, R. J., Pedersen, G., Madhun, A. S. et al. 2011. Evaluation of a virosomal H5N1 vaccine formulated with Matrix M™ adjuvant in a phase I clinical trial. *Vaccine*, 29, 8049–8059.

De Campos, A. M., Sanchez, A., and Alonso, M. J. 2001. Chitosan nanoparticles: A new vehicle for the improvement of the delivery of drugs to the ocular surface. Application to cyclosporine A. *International Journal of Pharmaceutics*, 224, 159–168.

Dong, L., Liu, F., Fairman, J. et al. 2012. Cationic liposome–DNA complexes (LCDC) adjuvant enhanced immunogenicity and cross protective efficacy of a prepandemic influenza A H5N1 vaccine in mice. *Vaccine*, 30, 254–264.

Dzung, N. A., Ha, N. T. N., Van, D. T. H., Phuong, N. T. L., Quynh, N. T. N., Hiep, D. M., and Hiep, L. V. 2011. Chitosan nanoparticles as a novel delivery system for A/H1N1 influenza vaccine: Safe property and immunogenicity in mice. *World Academy of Science, Engineering and Technology*, 1839–1846.

El-Salbouri, M. H. 2002. Positively charged nanoparticles for improving the oral bioavailability of cyclosporine-A. *International Journal of Pharmaceutics*, 249, 101–108.

Garmise, R. J., Staats, H. F., and Hickey, A. J. 2007. A novel dry powder preparation of whole inactivated influenza virus for nasal vaccination. *AAPS PharmSciTech*, 8, e. Design, characterization and preclinical efficacy of a cationic lipid adjuvant for influenza split vaccine. *Vaccine*, 19, 1794–1805.

Gluck, U., Gebber, J. O., and Gluck, R. 1999. Phase I evaluation of intranasal virosomal influenza vaccine with and without *Escherichia coli* heat labile toxin in adult volunteers. *Journal of Virology*, 73, 7780–7786.

Gupta, N.K, Tomar, P., Sharma, V., and Dixit, V. K. 2011. Development and characterization of chitosan coated polycaprolactone nanoparticles system for effective immunization against influenza. *Vaccine*, 29, 9026–9037.

Guy, B., Pascal, N., and Francon, A. 2001. Design, characterization and preclinical efficacy of a cationic lipid adjuvant for influenza split vaccine. *Vaccine*, 19, 1794–1805.

Hagennars, N., Verheul, R. J., Mooren, I. et al. 2009. Relationship between structure and adjuvanticity of *N, N, N*- trimethyl chitosan (TMC) structural variants in a nasal influenza vaccine. *Journal of Controlled Release,* 140, 126–133.

Hagennars, N., Mania, M., Jong, P. et al. 2010. Role of trimethylated chitosan (TMC) in nasal residence time, local distribution and toxicity of an intranasal influenza vaccine. *Journal of Controlled Release,* 144, 17–24

Harada, Y., Mori, A. N., Takahashi, Y. et al. 2011. Inactivated and adjuvanted whole-virion clade 2.3.4. H5N1 prepandemic influenza vaccine possesses broad protective efficacy against infection by heterologous clades of highly pathogenic H5N1 avian influenza virus in mice. *Vaccine,* 29, 8330–8337.

Hiep, L. V., Thanh, M. T., Van, D. T. H., Khanh, V. T. P., and Dzung, N. A. 2008. Chitosan as a hopeful adjuvant for H5N1 influenza vaccine. *Journal of Chitin and Chitosan,* 13(10), 6–8.

Huang, J., Garmise, R. J., and Crowder, T. M. 2004. A novel dry powder influenza vaccine intranasal delivery technology: Induction of systemic and mucosal immune response in rats. *Vaccine,* 23, 144–153.

Illum, L., Gill, J., Hinchcliffe, M., Fisher, A. N., and Davis, S. S. 2001. Chitosan as a novel nasal delivery system for vaccines. *Advanced Drug Delivery Review,* 51, 81–96.

Joseph, A., Itskovitz-Cooper, N., and Samira, S. 2006. A new intranasal influenza vaccine based on a novel polycationic lipid–ceramide carbamomyl-spermine (CCS) I: Immunogenicity and efficacy studies in mice. *Vaccine,* 24, 3990–4006.

Kean, T., Roth, S., and Thanou, M. 2005. Trimethylated chitosans as non-viral gene delivery vectors: Cytotoxicity and transfection efficiency. *Journal of Controlled Release,* 103, 643–653.

Khatri, K., Goyal, A. K., Gupta, P. N., Mishra, N., and Vyas, S. P. 2008. Plasmid DNA loaded chitosan nanoparticles for nasal mucosal immunization against hepatitis B. *International Journal of Pharmaceutics,* 354, 235–241.

Ko, S. Y., Ko, H. J., Chang, W. S., Park, S. H., Kweon, M. N., and Kang, C. Y. 2005. Alpha-galactosylceramide can act as a nasal vaccine adjuvant inducing protective immune responses against viral infection and tumor. *Journal of Immunology,* 175, 3309–3317.

Kumar, R. M. N. V. 2000. A review of chitin and chitosan applications. *Reactive and Function Polymer,* 46, 1–27.

Kumar, S. R., Ahmed, V. P. I., Parameswaran, V., Sudhakaran, R., Babu, V. S., and Sahul Hameed, A. S. 2008. Potential use of chitosan nanoparticles for oral delivery of DNA vaccine in Asian sea bass (*Lates calcarifer*) to protect from *Vibro (Listonella) anguillarum. Fish and Shellfish Immunology,* 25, 47–56.

Li, K., Luo, J., Wang, C., and He, H. 2011. α-Galactosylceramide potently augments M2e-induced protective immunity against highly pathogenic H5N1 avian influenza virus infection in mice. *Vaccine,* 7711–7717.

MacLaughlin, F. C., Mumper, R. J., Wang, J., Tagliaferri, J. M., Gill, I., Hinchcliffe, M., and Rolland, A. P. 1998. Chitosan and depolymerized chitosan oligomers as condensing carriers for *in vivo* plasmid delivery. *Journal of Controlled Release,* 56, 259–272.

Moon, H. J., Lee, J. S., and Kim, J. C. 2012. Mucosal immunization with recombinant influenza hemagglutinin protein and polygamma glutamate/chitosan nanoparticles induces protection against highly pathogenic influenza A virus. *Veterinary Microbiology,* accepted, DOI: 10.1016/j.vetmic.2012.05.035.

Nakayama, T., Kashiwagi, Y., Kawashima, H., Kumagai, T., Ishii, K. J., and Ihara, T. 2012. Alum-adjuvanted H5N1 whole vireo inactivated vaccine (WIV) enhanced inflammatory cytokine productions. *Vaccine,* 3885–3890.

Nishimura, K., Nishimura, S., Nishi, N. et al. 1984. Immunological activity of chitin and its derivatives. *Vaccine,* 2 (1), 93–97.

O'Hagan, D. T. and Rappuoli, R. 2004. Novel approaches to vaccine delivery. *Pharmacology Research,* 21, 1519–1530.

Read, R. C., Naylor, S. C., Potter, C. W. et al. 2005. Effective nasal influenza vaccine delivery using chitosan. *Vaccine,* 23, 4367–4374.

Rinaudo, M. 2006. Chitin and chitosan: Properties and application. *Progress in Polymer Science,* 31, 603–632.

Risi, G., Frenette, L., Langley, J. M., and Li, P. 2011. Immunological priming induced by two dose series of H5N1 influenza antigen, administered alone or in combination with two different formulation of AS03 adjuvant in adult: Results of a randomized single heterologous booster dose study at 15 month. *Vaccine,* 29, 6408–6418.

Seferian, P. G. and Martinez, M. L. 2001. Immune stimulating activity of two new chitosan containing adjuvant formulation. *Vaccine,* 19, 661–668.

Shan, S., Poinern, E., Ellis, T., Fanwick, S., Le, X., Edward, J., and Jiang, J. T. 2010. Development of a nano vaccine against wild bird H6N2 avian influenza virus. *Procedia in Vaccinology,* 2, 40–43.

Shibata, Y., Foster, L. A., Metzger, W. J., and Myrvic, Q. N. 1997. Alveolar macrophage priming by intravenous administration of chitin particles, polymers of N-acetyl-D-glucosamine, in mice. *Infection and Immunity*, 65(5), 1734–1741.

Skountzou, I., Quan, F. S., Jacob, J., Compans, R. W., and Kang, S. M. 2006. Transcutaneous immunization with inactivated influenza virus induces protective immune responses. *Vaccine*, 24, 6110–6119.

Song, S. K., Moldoveanu, Z., and Nguyen, H. H. 2007. Intranasal immunization with influenza virus and Korean mistletoe lectin C (KML-C) induces heterosubtypic immunity in mice. *Vaccine*, 25, 6359–6366.

Song, J. H., Nguyen, H. H., and Cuburu, N. 2008. Sublingual vaccination with influenza virus protects mice against lethal viral infection. *Proceedings of the National Academy of Science USA*, 105, 1644–1649.

Song, M., Hong, H. A., Huang, J. M. et al. 2012. Killed *Bacillus subtilis* spores as a mucosal adjuvant for an H5N1 vaccine. *Vaccine*, 30, 3266–3277.

Subbaro, K. and Luck, C. 2007. H5N1 viruses and vaccine. *Plos Pathogen*, 3, 1–3.

van der Lubben, I. M., Verhoef, J. C., Borchard. G., and Junginger, H. E. 2001. Chitosan and its derivatives in mucosal drug and vaccine delivery. European Journal of Pharmaceutical Science, 14, 201–207.

Vila, A., Sanchez, A., Jane, K., Behrens, I., Kissel, T., Jato, J. L. V., and Alonso, M. J. 2004. Low molecular weight chitosan nanoparticles as a new carrier for nasal vaccine delivery in mice. *European Journal of Pharmaceutics and Biopharmaceutics*, 57, 123–131.

Von Hoegen, P. 2001. Synthetic biomimetic supra molecular Biovector™(SMBV™) particles for nasal vaccine delivery. *Advanced Drug Delivery Reviews*, 51, 113–125.

Westerink, M. A. J., Smithson, S. L., Srivastava, N., Blonder, J., Coeshott, C., and Rosenthal, G. J. 2002. Projuvant™(Pluronic F127®/chitosan) enhances the immune response to intranasally administered tetanus toxoid. *Vaccine*, 20, 711–723.

WHO. 2003. WHO media influenza factsheet No.211. http:/www.who.int/mediacentre/factsheets/2003/fs211/en/.

WHO. 2005. Weekly epidemio logical record, No.33, Geneva, pp.227–288.

WHO. 2009. Egg-based influenza vaccine manufacturing course manual, Part II.

WHO. 2010. 6th WHO meeting on evaluation of pandemic influenza vaccines in clinical trials, February, 2010.

Wu, Y., Wei, W., Zhou, M., Wang, Y, Wu, J., Ma, G., and Su, Z. 2012. Thermal-sensitive hydrogel as adjuvant-free vaccine delivery system for H5N1 intranasal immunization. *Biomaterials*, 33, 2351–2360.

Yang, Y., Chen, J., Li, H., Wang, Y., Xie, Z., Wu, M., Zang, H. et al. 2007. Porcine interleukin-2 gene encapsulated in chitosan nanoparticles enhances immune response of mice to piglet paratyphoid vaccine. *Comparative Immunology, Microbiology and Infectious Diseases*, 30, 19–32.

Yang, P. C., Yu, C. J., Chang, S. C. et al. 2012. Safety and immunogenicity of a split virion AS03$_A$ adjuvanted A/Indonesia/05/2005 (H5N1) vaccine in Taiwanese adults. *Journal of Formosan Medical Association*, 111, 333–339.

Youn, H. J., Ko, S. Y., and Lee, K. A. 2007. A single intranasal immunization with inactivated influenza virus and alpha-galactosylceramide induces long-term protective immunity without redirecting antigen to the respiratory pathogens. *Vaccine*, 25, 5189–5198.

Zaharoff, D. A., Connie, J. R., Kenneth, W. H., Jeffrey, S., and John, W. G. 2007. Chitosan solution enhances the immunoadjuvant properties of GM-CSF. Vaccine, 25, 8673–8686.

Zhang, J., Chen, X. G., Li, Y. Y., and Liu, C. S. 2007. Self-assembled nanoparticles based on hydrophobically modified chitosan as carriers for doxorubicin. *Nanomedicine: Nanotechnology, Biology, and Medicine*, 3, 258–265.

Zhao, K., Shi, X., Zhao, Y. et al. 2011. Preparation and immunogenicity effectiveness of a swine influenza DNA vaccine encapsulated with chitosan nanoparticles. *Vaccine*, 29, 8549–8556.

Zheng, Y. 2007. Nanoparticles based on the complex of chitosan with polyaspartic acid sodium salt: Preparation, characterization and the use for 5-fluorouracil delivery. *European Journal of Pharmaceutic and Biopharmaceutics*, 67, 621–631.

Index

A